$$A_w = \frac{\pi d_{pw}^2}{4} \qquad \text{for a plug weld} \tag{6.16.11a}$$

$$A_w = (L_{sw} - 0.22d_{sw})d_{sw} \qquad \text{for a slot weld with semicircular ends} \tag{6.16.11b}$$

$$R_{dw} = 0.75\,(0.60F_{EXX})t_e\,L_w = W_d L_w \tag{6.19.4}$$

$$R_d = \min\,[R_{dw}, R_{dBM}] \tag{6.19.8}$$

$$W_d = 0.75\,(0.60F_{EXX})t_e \qquad \text{for a 1-in.-long fillet weld of size } w \tag{6.19.9}$$

$$W_d = 1.392\,D \qquad \text{for a 1-in.-long E70 fillet weld of size } D \tag{6.19.10}$$

$$R_{dw} = 0.75\,(0.6F_{EXX}\,t_e L_w)[1.0 + 0.50\,(\sin\theta)^{1.5}] \tag{6.19.11}$$

$$T_{d1} = \phi_{t1}\,F_y\,A_g = 0.90F_y\,A_g \tag{7.4.5}$$

$$T_{d2} = \phi_{t2}\,F_u\,A_e = 0.75F_u\,UA_n \tag{7.4.6}$$

$$A_h = d_e t \tag{7.6.2}$$

$$A_n = \min\,[A_{n1}, A_{n2}, \dots A_{nk}, \dots A_{nM}] \tag{7.6.5}$$

$$g_{ab} = g_a + g_b - t \tag{7.6.6}$$

$$A_{nk} = A_g - \sum_{i=1}^{n_e} n_i d_e t_i + \sum_{j=1}^{n_d} \frac{s_j^2}{4g_j}\,t_j \tag{7.6.7}$$

$$g_{ab} = (g_a - \tfrac{1}{2}t_w) + (g_b - \tfrac{1}{2}t_f); \qquad t = \tfrac{1}{2}(t_f + t_w) \tag{7.6.8}$$

$$A_n \leq 0.85\,A_g \tag{7.6.9}$$

$$U = \min\left[\left(1 - \frac{\bar{x}_{con}}{L_{con}}\right), 0.9\right] \tag{7.7.2}$$

$$T_{dbs} = \phi F_u A_{nt} + \min\,[\phi(0.6F_y)A_{gv}, \phi(0.6F_u)A_{nv}] \qquad \text{if } \phi F_u A_{nt} \geq \phi(0.6F_u)A_{nv} \tag{7.8.1}$$

$$T_{dbs} = \phi(0.6F_u)A_{nv} + \min\,[\phi F_y A_{gt}, \phi F_u A_{nt}] \qquad \text{if } \phi(0.6F_u)A_{nv} > \phi F_u A_{nt} \tag{7.8.2}$$

$$\frac{L}{r_{min}} \leq 300 \tag{7.9.1}$$

$$A_{g1} = \frac{T_u}{0.9F_y} \tag{7.10.3}$$

$$A_{g2} = \frac{T_u}{0.75F_u U} + \text{estimated loss in area due to bolt holes} \tag{7.10.5}$$

$$A_g \geq \max\,[A_{g1}, A_{g2}] \tag{7.10.6}$$

$$N \geq \frac{T_u}{B_d} \tag{10.8}$$

$$P_y = AF_y \tag{.1.2}$$

$$P_E = P_{cr1} = \frac{\pi^2 EI}{L^2} \tag{4.14}$$

$$P_{cr} = P_e = \frac{\pi^2 EI}{(KL)^2} \tag{8.5.1}$$

$$G_A = \frac{\dfrac{I_c}{L_c} + \dfrac{I_{c1}}{L_{c1}}}{\alpha_{g1}\dfrac{I_{g1}}{L_{g1}} + \alpha_{g2}\dfrac{I_{g2}}{L_{g2}}}, \qquad G_B = \frac{\dfrac{I_c}{L_c} + \dfrac{I_{c2}}{L_{c2}}}{\alpha_{g3}\dfrac{I_{g3}}{L_{g3}} + \alpha_{g4}\dfrac{I_{g4}}{L_{g4}}} \tag{8.5.7}$$

$$F_{ex} = \frac{\pi^2 E}{\left(\dfrac{K_x L_x}{r_x}\right)^2}, \qquad F_{ey} = \frac{\pi^2 E}{\left(\dfrac{K_y L_y}{r_y}\right)^2} \tag{8.5.24}$$

$$F_e = \min\,[F_{ex}, F_{ey}] \tag{8.5.25}$$

$$P_d \equiv \phi_c P_n \geq P_{req} = P_u \tag{8.7.1}$$

STEEL STRUCTURES
Behavior and LRFD

The McGraw-Hill Series in Civil Engineering

STEEL STRUCTURES
Behavior and LRFD

Sriramulu Vinnakota

Marquette University

Boston Burr Ridge, IL Dubuque, IA Madison, WI New York San Francisco St. Louis
Bangkok Bogotá Caracas Kuala Lumpur Lisbon London Madrid Mexico City
Milan Montreal New Delhi Santiago Seoul Singapore Sydney Taipei Toronto

Mc Graw Hill **Higher Education**

STEEL STRUCTURES: BEHAVIOR AND LRFD

Published by McGraw-Hill, a business unit of The McGraw-Hill Companies, Inc., 1221 Avenue of the Americas, New York, NY 10020. Copyright © 2006 by The McGraw-Hill Companies, Inc. All rights reserved. No part of this publication may be reproduced or distributed in any form or by any means, or stored in a database or retrieval system, without the prior written consent of The McGraw-Hill Companies, Inc., including, but not limited to, in any network or other electronic storage or transmission, or broadcast for distance learning.

Some ancillaries, including electronic and print components, may not be available to customers outside the United States.

This book is printed on acid-free paper.

1 2 3 4 5 6 7 8 9 0 DOC/DOC 0 9 8 7 6 5 4

ISBN 0-07-236614-1

Senior Sponsoring Editor: *Suzanne Jeans*
Developmental Editor: *Kate Scheinman*
Senior Marketing Manager: *Mary K. Kittell*
Lead Project Manager: *Peggy J. Selle*
Lead Production Supervisor: *Sandy Ludovissy*
Media Technology Producer: *Eric A. Weber*
Senior Designer: *David W. Hash*
Cover Designer: *Rokusek Design*
(USE) Cover Image: *Miller Park, Milwaukee, WI, ©Jim Brozek*
Lead Photo Research Coordinator: *Carrie K. Burger*
Photo Research: *Karen Pugliano*
Compositor: *International Typesetting and Composition*
Typeface: *10.5/12 Times Roman*
Printer: *R. R. Donnelley Crawfordsville, IN*

Library of Congress Cataloging-in-Publication Data

Vinnakota, Sriramulu.
    Steel structures : behavior and LRFD / Sriramulu Vinnakota. — 1st ed.
        p.   cm.
    ISBN 0-07-236614–1
    1. Building, Iron and steel. 2. Structural design. 3. Steel, Structural. 4. Load factor design.
I. Title.

TA684.V56   2006
    693′.71—dc22

2004061059
CIP

www.mhhe.com

This book is dedicated to my parents Raju A. and Sarada M. Vinnakota for teaching me the importance of hard work, commitment to education and giving back to society, and to my children Rajiv and Jyothi whose achievements and contributions to society have made all of my efforts worthwhile.

ABOUT THE AUTHOR

Sriramulu Vinnakota is Professor of Civil Engineering at Marquette University, Milwaukee, WI. He was born in 1937 in the village of Venuturumilli, Andhra Pradesh, India. He is a graduate of the College of Engineering, Kakinada, Andhra University, where he obtained a degree in Civil Engineering in 1957. After graduation he worked with the Central Water and Power Commission, Ministry of Irrigation and Power, New Delhi, India for 5 years on the design of various gate, hoist and power house structures. In 1962, he received a Swiss Government Scholarship for higher studies in Switzerland. Vinnakota obtained his D.Sc in Structural Engineering in 1967 from The Swiss Federal Institute of Technology (EPF), Lausanne, Switzerland, under the supervision of Prof. M. Cosandey. Subsequently he worked at the EPF-Lausanne where he was named Professor Titulaire in 1978 by the Swiss Federal Council of Ministers. After spending a year at Cornell University, Ithaca, NY and 2 years at the University of Wisconsin, Milwaukee as Visiting Associate Professor, he accepted a faculty position at Marquette University in 1981, where he still remains. He has been involved in teaching, research, design and consulting in structural engineering, particularly in the area of steel structures. He is a registered professional engineer in the state of Wisconsin. Professor Vinnakota has made important contributions in the area of stability of steel members and structures through the publication of numerous papers in the professional journals of international reputation.

Since 1997, Professor Vinnakota has been a member of the Task Committee 4: Member Design of the Committee on Specifications, American Institute of Steel Construction, and a corresponding member of its Committee on Manuals and Textbooks. He is a Fellow of the American Society of Civil Engineers and a member of the International Association of Bridge and Structural Engineers. Also, he is a life member of the Structural Stability Research Council and served as Chairman of its Task Group on Beams from 1982 to 2003. In the 1970s he had the privilege of being the Swiss Delegate to the Commission 5 Plasticity of the European Convention for Constructional Steelworks (ECCS) presided over by the late Professor Charles Massonnet; and he was a member of the Working-Group 8-1, Stability of the ECCS. Dr. Vinnakota contributed to several chapters of the fourth and fifth editions of the SSRC Guide to Stability Design Criteria for Metal Structures (Edited by Prof. T. V. Galambos).

TABLE OF CONTENTS

PREFACE

To design steel structures efficiently, the designer must know steel as a structural material; understand how structures are assembled and braced, and how they sustain and transmit loads; know design philosophies and processes; and learn the proper selection of connectors and connections. The structure of the Load and Resistance Factor Design Specification (LRFDS) developed by the American Institute of Steel Construction (AISC) especially requires that designers have a better understanding of structural behavior since the different limit states of failure must be identified as an integral part of the design process.

The heart and soul of design is the ability for the designer to conceive a structure that will behave as desired and to develop an intuition regarding different framing options. In each chapter of *Steel Structures: Behavior and LRFD,* discussion of theory and the behavior of the member under the various combinations of loads it must resist is followed by a discussion of design applications according to the LRFDS. Practical, fabrication, and erection constraints are indicated where required.

- **Chapter 1** includes a brief description of several steel projects recently built in Milwaukee and elsewhere, so as to develop an interest in students to inspect existing steel structures in their own locality and to visit steel structures under fabrication and erection.
- **Chapter 2** presents steel as a structural material. Topics covered include making structural steels, forming steel shapes, tension tests, residual stresses, corrosion, and painting of structural steels.
- **Chapter 3** is a broad introduction to various types of structures (tier and industrial buildings), structural elements (tension members, compression members, beams, beam-columns, and connections), and structural components (walls, roofs, decking, bracing systems, diaphragms, etc.).
- **Chapter 4** gives the various kinds of loads acting on building structures, as per the ASCE Standards 7 (dead loads, live loads, and wind loads are covered in detail. A brief introduction to snow loads and seismic loads is also given). Probabilistic bases of the LRFD specification are briefly described. Load factors, load combinations, and resistance factors are introduced.
- **Chapter 5** gives simple examples on the calculation of required strengths of typical members of a 4-story, braced multi-story office building and of an unbraced, hinged base portal frame using ASCE Standards 7. These members are designed in later chapters of the book.

- **Chapter 6** covers the behavior and design of connectors used in steel structures (bolts, welds, and pins). Information from the 2000 Specification for Structural Joints Using A325 and A490 Bolts by the Research Council on Structural Connections (RCSC) is included.
- **Chapter 7** treats the behavior and design of tension members.
- **Chapter 8** covers the behavior and design of axially loaded columns.
- **Chapter 9** considers the behavior and design of adequately braced compact beams. Also covered here are the design of bearing plates for beams and base plates for columns, treated as examples of members in flexure.
- **Chapter 10** is concerned with the design of laterally unbraced beams.
- **Chapter 11** treats members under combined forces including beam-columns and biaxially bent beams.
- **Chapter 12** systematically discusses behavior and design of bolted and welded joints.
- **Chapter 13** covers the design of simple-shear connections and moment connections.

How to Use This Book

There is more material in this book than can be covered in a one semester course on steel design to allow the instructor sufficient flexibility in the selection of topics. The book can readily be made to fit courses of different lengths and of different content and objectives. The complete text was prepared with the purpose of offering sufficient material for a one-semester (3-credit) undergraduate, junior/senior level first course in steel design for civil engineering students, and a one-semester (3-credit) undergraduate/graduate level course.

As the title *Steel Structures: Behavior and LRFD* suggests, the book covers not only design but the behavior on which the design specifications are based. Many of the sections that cover behavior are placed on our website (www.mhhe.com/vinnakota), so as not to overwhelm a student taking the first course on steel design. These topics are typically covered in Steel Design 2 (elective undergraduate/graduate) and Advanced Steel Design (graduate level) courses.

Steel Structures: Behavior and LRFD is unique in that it has five introductory chapters (one each on steels, structures, loads, and required strengths, in addition to Chapter 1: Introduction). The coverage of loads in Chapter 4 is added to impress upon students that, in the design process, more errors are committed in the determination of loads and required strengths than in the calculation of design strengths. Chapter 5 on required strengths is added to clearly indicate the complementary nature of analysis and design procedures in the overall iterative process of designing new structures. The required strengths for members of two structures, determined in this chapter utilizing procedures usually learned in analysis courses, are used in later chapters of the book to design these members.

The book is also unique in that it has three large chapters (6, 12, and 13) devoted to connections. Connections are the most important and the least understood components of steel structures. Also, the economy of a steel structure often depends on the proper choice of connections. Further, the choice of the connectors and connections influences the type and magnitude of forces acting on a member. For these reasons, the introductory chapter on connectors, Chapter 6, is placed before the chapters on member design (Chapters 7 to 11).

Each chapter includes a number of example problems which are presented with more complete details than would be required by an experienced designer. Often these examples are selected so as to bring to the attention of the student certain design criteria, or to arrive at certain design tips, usually given as a set of remarks at the end of the solution. In addition, there are several design problems of members and connection elements for which the required strengths are obtained in Chapter 5. These examples help show structural members as components of real-world structures rather than as isolated elements. The author believes that in this way the student will better learn the fundamentals of the design procedure and the sequence of the calculations involved than if they are presented as isolated examples. A large number of problems for assignment are included at the end of each chapter to enable the student to test his or her mastery of the subject. It is absolutely necessary that students have a copy of the third edition *AISC Manual of Steel Construction: Load and Resistance Factor Design* as reference is made throughout the text to various requirements of the LRFD Specification and the tables in the LRFD Manual.

It has been my experience that students who rely on computer programs prepared by others, in a first design course, do not develop a capacity for critical evaluation of the resulting output. Thus, no attempt is made to include any computer software with this text. The present text uses T to represent axial force in a tension member, P to represent axial force in a compression member, B to represent force on a single bolt, and W to represent the force on a unit length weld, etc., with various subscripts added as appropriate. Thus, B_d represents the design strength of a bolt, B_{db} represents the design bearing strength of a bolt, B_{dbe} represents the design bearing strength of an end bolt, and so on. Most of these notations are obvious.

My own experience as an engineer and teacher on three different continents (English units only; metric units only; and U.S. customary units only) indicates that first design courses and textbooks are better if limited to a single system of units only, in order for the student to get a feel for the results to be expected from the design process. Although metrication is inevitable and will likely be the only basis of future AISC Specifications and Manuals, the change has not yet taken place in the steel construction industry. The present text is, therefore, only in U.S. customary units.

Website

The website that accompanies *Steel Structures: Behavior and LRFD* (www.mhhe.com/vinnakota) contains a host of resources to complement the text. Web features include:

- **Extensive "Additional Information" sections.** These sections are integrated throughout the text and can be easily spotted by looking for our marginal website icon. These are **web-only bonus chapters,** covering advanced topics.

- A **downloadable list of important equations.** This is a handy reference guide for students.
- A **comprehensive list of symbols** used throughout the text.
- **Flowcharts** for some of the basic analysis and design procedures. They may be used to guide the student through the steps of a particular analysis or design problem.
- **Tips for instructors** on "Suggested Ways to Use This Book."
- **Historical insight** into the development of the technical specifications governing the structural steel industry.
- A **solutions manual,** available to instructors only, which provides detailed solutions for most of the text problems.

Acknowledgments

This book grew out of lectures which the author gave for a number of years to undergraduate and graduate students of civil engineering at Marquette University. Special thanks are extended to Professors Tom Wenzel, Steve Heinrich, and Chris Foley for their support and encouragement during the preparation of this book. The assistance of Robert Kondrad in the preparation of the numerical examples is gratefully acknowledged. Thanks are also due to Mike Loescher of Computersmith, L.L.C., Milwaukee for his help in the preparation of drawings in Chapters 3, 6, 11, and 12. Special thanks are due to Dr. Shilak Shakya of Pujara Wirth Torke, Inc., Milwaukee for his help in the preparation of the remainder of the drawings.

Former students Shubha Rao and John Peronto contributed to the problem solutions in this book through their homework. I would also like to thank Steve Miller, Bob Schumacher, Jim Hayes, Jason Sorci, Chris Bielefeld, and several other former students who participated in useful discussions about the text during their steel design classes at MU. Nick Hornyak and Carl Schneeman, graduate students; Aleisha Palaniuk and Tom Kennedy of Opus Corporation, Milwaukee; Loei Badreddine and K. Wood of Graef, Anhalt, Schloemer and Associates of Milwaukee; Michael Henke of Construction Supply and Erection, Germantown, Wisconsin; Dr. Surinder Mann (formerly of Ove, Arup and Partners of Los Angeles); Art Johnson of KPFF Consulting Engineers, Portland, Oregon; and Lawrence F. Kruth of Douglas Steel Fabricating Corporation of Lansing, Michigan provided photos and/or information regarding various projects described in this text. My thanks go to all of them. Construction Supply and Erection of Germantown also fabricated and donated the steel connection sculpture at Marquette University (chapter opening photo of Chapter 12).

The cooperation and help of the AISC through Charles J. Carter, Cynthia J. Duncan, and Fromy Rosenberg is greatly appreciated. Special thanks are due to Dr. Nestor Iwankiw and Professor Ted Galambos of SSRC for their encouragement during the preparation of this book.

My special thanks go to my brother Dr. Vara Prasada Rao for his continuous help and encouragement during strenuous times. This book would not have taken this shape had it not been for the constant, constructive criticism of my wife Sreedevi, and I thank her for that.

I would also like to thank the following professors for reviewing the manuscript:

P. K. Basu, *Vanderbilt University*

Charles M. Bowen, *The Oklahoma State University*

Wai-Fah Chen, *University of Hawaii*

Scott A. Civjan, *University of Massachusetts*

John K. Dobbins, *Southern Illinois University Carbondale*

Bruce Ellingwood, *Johns Hopkins University*

Hany J. Farran, *California Polytechnic State University, Pomona*

Larry J. Feeser, *Rensselaer Polytechnic Institute*

Theodore V. Galambos, *University of Minnesota*

Louis F. Geschwindner, *The Pennsylvania State University*

Perry S. Green, *Steel Joint Institute*

Marvin Hallings, *Utah State University*

Robert Hamilton, *Boise State University*

Kenneth G. Kellogg, *Oregon Institute of Technology*

Carl E. Kurt, *University of Kansas*

Roger A. LaBoube, *University of Missouri, Rolla*

Jeffrey Laman, *The Pennsylvania State University*

Le-Wu Lu, *Lehigh University*

Jamshid Mohammadi, *Illinois Institute of Technology*

Husam Najm, *Rutgers University*

Anil K. Patnaik, *South Dakota School of Mines and Technology*

Teoman Pekoz, *Cornell University*

Matthew W. Roberts, *University of Wisconsin, Platteville*

Aziz Saber, *Louisiana Tech University*

Joseph Saliba, *University of Dayton*

Ajay Shanker, *University of Florida*

Avi Singhal, *Arizona State University*

J. Michael Stallings, *Auburn University*

Bozidar Stojadinovic, *University of California, Berkeley*

Habibollah Tabatabai, *University of Wisconsin-Milwaukee*

Christopher Tuan, *University of Nebraska at Omaha*

Chia-Ming Uang, *University of California, San Diego*

Yan Xiao, *University of Southern California*

Finally, I would like to take this opportunity to record my profound appreciation for the McGraw-Hill staff during the planning and production of the book: Suzanne Jeans, Kate Scheinman, Peggy Selle, Carrie Burger, Megan Hoar, Elizabeth Kenyon, and Jill Barrie.

Note The author and the publisher have taken special care to reduce errors to a minimum but cannot hope for perfection in a book including so large a number of design examples and figures; they will welcome notification of errors and suggestions for improvement. The author can be contacted at sriramulu.vinnakota@marquette.edu.

Sriramulu Vinnakota

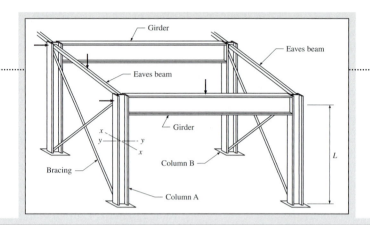

Steel Structures: Behavior and LRFD stresses both the **behavior and design** of steel members and structures under various loading conditions.

The current editions of the **LRFD Specifications** and the **LRFD Manual** are used and extensively referenced throughout the text. Where appropriate, additional design aids are provided.

TABLE 6.8.2

Design Bearing Strength at STD End Bolt Holes, B_{dbe}, for Various End Distances (kips/in. thickness)

	F_{up} = 58 ksi			F_{up} = 65 ksi		
d (in.)	$\frac{3}{4}$	$\frac{7}{8}$	1	$\frac{3}{4}$ 1		$\frac{7}{8}$
$1\frac{1}{2}d$ (in.)	$1\frac{1}{8}$	$1\frac{5}{16}$	$1\frac{1}{2}$	$1\frac{1}{8}$	$1\frac{5}{16}$	$1\frac{1}{2}$
$2.5d + \frac{1}{32}$ (in.)	$1\frac{15}{16}$	$2\frac{1}{4}$	$2\frac{9}{16}$	$1\frac{15}{16}$	$2\frac{1}{4}$	$2\frac{9}{16}$
L_e (in.)↓						
$1\frac{1}{4}$	44.0	40.8	37.5	49.4	45.7	42.0
$1\frac{3}{8}$	50.6	47.3	44.0	56.7	53.0	49.3
$1\frac{1}{2}$	57.1	53.8	50.6	64.0	60.3	56.7
$1\frac{3}{4}$	70.1	66.9	63.6	78.6	75.0	71.3
2	78.3	79.9	76.7	87.7	89.6	85.9
$2\frac{1}{2}$	78.3	91.3	103	87.7	102	115
$\geq 2\frac{3}{4}$	78.3	91.3	104	87.7	102	117

$B_{dbe} = \min [B_{dbo}; B_{dbt}];$ $B_{dbo} = 1.8 F_{up} t;$ $B_{dbt} = 0.9 (L_e - 0.5d_h) F_{up} t$
d = nominal diameter of bolt, in.; d_h = diameter of bolt hole = $d + \frac{1}{16}$ in. for STD punched holes considered
B_{dbe} = design bearing strength at an end bolt hole; L_e = end distance, in.; t = plate thickness = 1 in.
B_{dbo} = strength corresponding to ovalization of bolt hole; B_{dbt} = strength corresponding to shear tear-out of plate
Design strengths controlled by ovalization of bolt hole are shown shaded.

Numerous worked-out **example problems** emphasizing the application of design concepts are included.

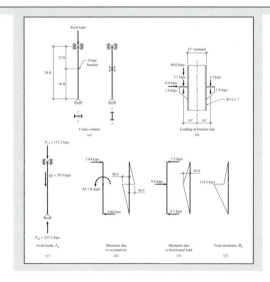

Four hundred and fifty carefully drawn **figures** of structural systems, members, and bolted and welded joints illustrate the text, and **photographs** of real-world construction projects are included.

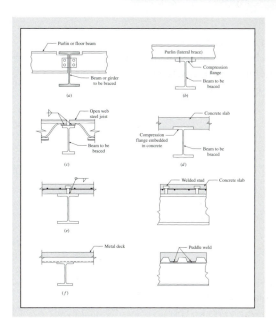

Throughout the text, a web icon references readers to the book's **website** (http://www.mhhe.com/vinnakota), which contains extensive additional coverage of advanced topics.

WWW

Web Chapter 9
Adequately Braced Compact Beams
W9.2: Open Web Steel Joists and Joist Girders

Web Chapter 12
Joints and Connecting Elements
W12.1: Ultimate Strength Method for Bolted Joints in Eccentric Shear

WWW

Our website contains additional **resources for both instructors and students.**

- For instructors, a comprehensive solutions manual as well as tips on how best to use the text for your course.

- For students, a comprehensive list of equations, a detailed list of symbols, and several flowcharts.

A	area of the cross section; area, in.^{2}
A_b	nominal unthreaded body area of bolt or threaded rod, in.^{2}
A_e	effective net area of tension member, in.^{2}
A_f	gross area of one flange, in.^{2}
A_g	gross area, in.^{2}
A_{gt}	gross area subject to tension (limit state of block shear rupture), in.^{2}
A_{gv}	gross area subject to tension (limit state of block shear rupture), in.^{2}
A_n	net area of tension member, in.^{2}
A_{nt}	net area subject to tension (limit state of block shear rupture), in.^{2}
A_{nv}	net area subject to shear (limit state of block shear rupture), in.^{2}
A_w	area of beam web, in.^{2}
B_1, B_2	magnification factors used in determining second-order moment for combined bending and axial compression when first-order analysis is employed
B_d	design strength of a bolt, kips
B_{db}	design bearing strength of a bolt, kips
B_{dbe}	design bearing strength of an end bolt, kips
B_{dbi}	design bearing strength of an interior bolt, kips
B_{dbo}	design bearing strength of connected plate element for the limit state of ovalization of bolt hole, kips
B_{dbte}	design bearing strength of an end bolt for the limit state of shear tear-out of connected plate element, kips
B_{dbti}	design bearing strength of an interior bolt for the limit state of shear tear-out of connected plate element, kips
B_{dv}	design shear strength of a bolt, kips
B_n	nominal strength of a bolt, kips
BF	beam factor
C_b	bending coefficient dependent on moment gradient for lateral-torsional buckling strength of beams
C_d	design strength of connectors in a joint, kips
C_{db}	design bearing strength of connectors in a joint, kips
C_{dv}	design shear strength of connectors in a joint, kips
C_m	coefficient applied to bending term in interaction formula for prismatic members and dependent on column curvature caused by applied moments
C_w	warping torsional constant of a cross section, in.^{6}
D	dead load; fillet weld size in sixteenths of an inch
E	earthquake load
E	modulus of elasticity or Young's modulus ($= 29{,}000$ ksi for steel)
E_s	strain hardening modulus, ksi
E_t	tangent modulus, ksi
F_{BM}	nominal strength of the base material to be welded, ksi
F_{EXX}	classification number of weld metal (minimum specified ultimate tensile stress)
F_L	smaller of $(F_{yf} - f_r)$ or F_{yw}, ksi
F_{cr}	critical stress, ksi
F_e	elastic buckling stress, ksi
F_{ex}	elastic flexural buckling stress about the major axis, ksi
F_{ey}	elastic flexural buckling stress about the minor axis, ksi
F_r	compressive residual stress in flange (10 ksi for rolled shapes; 16.5 ksi for built-up shapes)

F_u	specified minimum ultimate tensile stress of the type of steel being used, ksi
F_{ub}	ultimate tensile stress of bolt material, ksu
F_w	nominal strength of the weld electrode material, ksi
F_y	specified minimum yield stress of the type of steel being used, ksi
F_{uy}	ultimate shear stress, ksi
F_{yv}	shear yield stress, ksi
G	shear modulus of elasticity ($= 11{,}200$ ksi for steel)
G_A, G_B	relative stiffness factors at ends A and B of a column
H	horizontal force, kips
I	impact factor; moment of inertia, in.^{4}
I_c	moment of inertia of a column, in.^{4}
I_g	moment of inertia of a girder, in.^{4}
I_x, I_y	moment of inertia about the x- or y-axis, respectively, in.^{4}
I_p	polar moment of inertia, in.^{4}
J	torsional constant for a section, in.^{4}
K	effective length factor for a column
K_x, K_y	effective length factor for flexural buckling about x axis and y axis
K_z	effective length factor for torsional buckling
L	live load due to occupancy
L	story height or panel spacing; span
L_b	laterally unbraced length; length between points which are either braced against lateral displacement of compression flange or braced against twist of the cross section
L_c	clear distance, in.
L_{ce}	clear distance for an end bolt, in.
L_{ci}	clear distance for an interior bolt, in.
L_{con}	length of connection in the direction of loading, in.
L_e	end distance of a bolt measured in direction of line of force; edge distance, in.
$L_{e,full}$	limiting value of end distance above which limit state of bolt ovalization controls bearing strength, in.
L_{lw}	length of longitudinal weld, in.
L_{tw}	length of transverse weld, in.
L_p	limiting laterally unbraced length for full plastic bending capacity, uniform moment case ($C_b = 1.0$), in.
L_{pd}	limiting laterally unbraced length for plastic analysis, in.
L_r	limiting laterally unbraced length for inelastic lateral-torsional buckling, in.
L_r	roof live load
L_s	side distance of a bolt measured perpendicular to line of force, in.
L_w	length of fillet weld, in.
M	bending moment
M_A	absolute value of moment at quarter point of the unbraced beam segment, in.-kips
M_B	absolute value of moment at centerline of the unbraced beam segment, in.-kips
M_C	absolute value of moment at three-quarter point of the unbraced beam segment, in.-kips
M_{cr}	elastic lateral-torsional buckling moment of a beam or a beam segment
M_{cr}^o	elastic lateral-torsional buckling moment of a beam or beam segment under uniform moment
M_{lt}	maximum first-order factored moment in a beam-column due to lateral frame translation only
M_n	nominal bending strength of a member
M_d	design bending strength of a member
M_{dI}^o	design bending strength of a beam segment under uniform moment, for $L_p < L_b \le L_r$
M_{dE}^o	design bending strength of a beam segment under uniform moment, for $L_b > L_r$
M_{max}	absolute value of the maximum moment within the unbraced length (including the end points)

M_{nt} maximum first-order factored moment in a beam-column assuming there is no lateral translation of the frame

M_p plastic bending moment of a section

M_r bending moment in a section when the extreme fiber stress reaches ($F_y - F_r$)

M_u required bending strength of a member under factored loads

M_u^* required flexural strength of a beam-column under factored loads, including second-order effects

M_{ueq}^o equivalent factored uniform moment capacity for the segment considered

M_y moment corresponding to onset of yielding at the extreme fiber from an elastic stress distribution ($= F_y S$ for homogeneous sections)

M_1 smaller moment at the ends of a laterally unbraced segment of a beam or beam-column

M_2 larger moment at the ends of a laterally unbraced segment of a beam or beam-column

N_s number of shear planes in a joint

P_{cr} critical buckling load

P_d design strength of an axially loaded column, kips

P_E Euler buckling load

P_e elastic buckling load

P_{e1} elastic buckling load used in the determination of magnification factor B_1, kips

P_{e2} elastic buckling load used in the determination of magnification factor B_2, kips

P_n nominal strength of an axially loaded column, kips

P_u axial force under factored loads; required axial strength of a column, kips

P_{ueq} equivalent axial load used in selecting a trial shape for design of a beam column

P_y yield (squash) load of a section, ($= F_y A_g$), kips

Q concentrated transverse load on a member, kips

Q' first moment of area in shear flow formula, in.^{3}

R rain load

R_n nominal strength

S elastic section modulus, in.^{3}; snow load

T_b specified pretension load in high-strength bolt, kips

T_d design strength of a tension member, kips

T_n nominal strength of a tension member, kips

T_u factored tension load, required tensile strength due to factored loads, kips

U reduction factor to account for shear lag

V shear

V_d design shear strength, kips

V_n nominal shear strength, kips

V_u shear force under factored loads, kips

W wind load

W_d design strength of a 1 in. long fillet weld

Z plastic section modulus, in.^{3}

b compression element width perpendicular to load direction, in.

b_f flange width, in.

c distance from neutral axis to extreme fiber where flexural stress is computed

d nominal bolt diameter, in.; d overall depth of member, in.

d_b beam depth, in.

d_c column depth, in.

d_e effective width of bolt hole, in.

d_h diameter of bolt hole, in.

d_i distance between the center of gravity of an element, i of a cross section to the center of gravity of the cross section (for use in parallel axis theorem)

e base of natural logarithm $= 2.71828$; eccentricity of load

f computed compressive stress

f_c' specified 28-day compressive strength of concrete, ksi

g	transverse center-to-center spacing (gage) between bolt gage lines, in.
h	clear distance between flanges less the fillet or corner radius for rolled shapes; and for built-up welded sections, the distance between flanges, in.
k	distance from outer face of flange to web toe of fillet; plate buckling coefficient
p	pitch of bolts, in.
p_{full}	limiting value of pitch above which limit state of bolt ovalization controls bearing strength, in.
q	uniformly distributed transverse load on a member, klf
q_{sv}	shear flow, kli
r	radius of gyration, in.; radial distance
r_x, r_y	radius of gyration about x and y axes, respectively, in.
s	staggered pitch, in.
t	thickness of element, in.
t_f	flange thickness, in.
t_w	web thickness, in.
w	leg size of fillet weld, in.
x	subscript relating symbol to member strong axis
$\bar{x}$	x-coordinate of center of gravity
$\bar{x}_{con}$	connection eccentricity, in.
y	subscript relating symbol to member weak axis
$\bar{y}$	y-coordinate of center of gravity
z	subscript relating symbol to member longitudinal axis
α	shape factor
γ	load factor
δ	deflection
Δ	sway
Δ_{oh}	translation deflection of the story under consideration, in.
ϵ	strain
ϵ_y	yield strain
μ	Poisson's ratio ($= 0.3$ for steel); coefficient of static friction; mean slip coefficient for slip-critical connections
τ	stiffness reduction factor
λ	slenderness parameter
λ_c	column slenderness parameter
λ_p	limiting slenderness parameter for compact element
λ_{pf}	limiting slenderness parameter for the flange of a compact I-shape
λ_{pw}	limiting slenderness parameter for the web of a compact I-shape
λ_r	limiting slenderness parameter for noncompact element
ϕ	resistance factor
ϕ_b	resistance factor for flexure ($= 0.90$)
ϕ_c	resistance factor for compression ($= 0.85$)
ϕ_t	resistance factor for tension
ϕ_v	resistance factor for shear ($= 0.90$)

STEEL STRUCTURES
Behavior and LRFD

Lake Michigan College Fine Arts Center, Benton Harbor, Michigan.

This 1182 ton steel skeleton for the arts center is a complex structure with steep sloping sidewalls, cantilevered, balconies, and jutting canopies.

Photo courtesy Douglass Steel Fabricating Corp., Lansing MI, www.douglassteel.com

CHAPTER 1

Introduction

....

1.1 General

A *structure* may be defined as a system of individual *members* and *connections* arranged so that the entire set remains stable and without appreciable change in form while meeting the prescribed performance criteria. Structures in general, and steel structures in particular, play an important role in our everyday life. We live in buildings, work in office towers and industrial buildings, relax in sports arenas and stadiums, contemplate in and around art museums, study in libraries, and shop in shopping malls, of which a good proportion are built in steel. When we travel by cars on bridges over rivers and valleys, or by planes departing from and arriving at airport concourses, we notice a variety of structures built from structural steel. Tall chimneys, electric transmission towers, erection towers, television masts, and so on are further examples of steel structures. In this text, however, we will be interested primarily in the type of structures that are normally the responsibility of the structural engineer, working either on his own or in collaboration with an architect. Therefore, the emphasis of this text is on steel buildings of all kinds, although much of the material is also applicable to bridges and other structures.

One of the major difficulties the student faces taking a first course on structural steel design is in getting a clear mental picture of the structure with which he or she is entrusted. This is the natural result of a lack of familiarity with the basic structural members, the ways in which they may be arranged to form the skeleton for a building or bridge, and their functions in the completed structure.

The aim of this chapter is to briefly describe several recently built steel structures, so as to develop an interest in students to inspect existing steel structures in their own locality and, more importantly, to visit steel structures under fabrication and erection. To acquire additional knowledge, students should refer to periodicals such as *Modern Steel Construction* (http://www.modernsteel.com) published by the American Institute of Steel Construction (AISC). (Several terms in the descriptions that follow will be explained in later chapters.)

1.2   Miller Park Stadium with Convertible Roof

With its new fan-shaped retractable steel roof, which opens and closes in just 10 minutes, Miller Park Stadium, home of the Milwaukee Brewers baseball team, offers perfect conditions for every game. The stadium is designed to function as an enclosed space to shelter visitors and control temperatures during inclement weather, and to act as an open space to allow air and light in on pleasant days. The structural engineer of record is Ove Arup and Partners of Los Angeles, California. The fan-shaped roof covers about 410,000 square feet and weighs about 12,200 tons. Inside, the roof rises about 240 ft above the playing field, while the height at the peak is 330 ft. The stadium seating bowl alone, with a seating capacity of 42,400, consists of 11,000 tons of exposed steel framing.

The roof system consists of five moving panels and two fixed panels arranged in a fan configuration that conforms to the natural geometry of a baseball field (Fig. 1.2.1). The five moving panels rotate about pivots behind home plate. Moving panel 1 is the highest and most central panel. It is aligned with the stadium centerline when the roof is closed. There are two mirrored versions of each moving panel types 2 and 3 [Hewitt et al., 2002; Chan et al., 2002]. Three of the moving panels open toward left field, and two open toward right field (Photos 1.2.1 and 1.2.2). In the open configuration the moving panels stack over the fixed panels 4. The open and closed configurations are shown in Figs. 1.2.1*a* to 1.2.1*d*. Panels towards the center of the stadium are progressively longer and higher so that they can stack above each other in the open position and provide the maximum clear height at center field in the closed

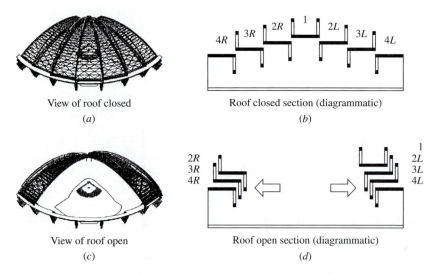

View of roof closed
(*a*)

Roof closed section (diagrammatic)
(*b*)

View of roof open
(*c*)

Roof open section (diagrammatic)
(*d*)

Figure 1.2.1:   Open and closed configurations of the Miller Park Stadium roof [1.5]. *Source: From John Hewitt et al., "Miller Park Stadium Retractable Roof," courtesy Ove Arup & Partners.*

position. The longer edges of each panel are supported on arched trusses that span about 600 ft across the field and seating. At the pivot end, the panels are all supported on a steel frame (Photo 1.2.2). At the outfield end, the panels are supported on curved rails resting on a reinforced concrete track beam, located 170 ft above ground level. The outfield mechanisms include a 60 horsepower electrically powered bogie under each corner of each moving panel. Each panel runs on its own circular track.

Photo 1.2.1: Miller Park Stadium with roof in open position, view from outside.
Photo by S. Vinnakota.

Photo 1.2.2: Miller Park Stadium with roof in open position, view from inside.
Photo by S. Vinnakota.

The main elements of the roof were analyzed and design checks were made using the SAP90 computer program, in accordance with the AISC *Load and Resistance Factor Design (LRFD) Specifications* [Hewitt et al., 2002]. Most of the connections use high-strength bolts in slip-critical joints.

1.3   Milwaukee Art Museum Addition with Movable Sun Screen

The dramatic new Santiago Calatrava–designed addition represents a major expansion of the Milwaukee Art Museum [Badreddine, 2002; Wood, 2001]. The addition consists of three major components: a central building for gallery space; a movable, winglike sun screen in steel—the Burke Brise Soleil; and a cable-stayed pedestrian bridge in steel, which links downtown Milwaukee directly to the museum and the lakefront (Photo 1.3.1). The structural, civil, and environmental engineers of record are Graef, Anhalt, Schloemer and Associates of Milwaukee, Wisconsin.

The reception hall of the new addition, located toward the south end of the gallery space, is provided with a 90-ft-high glass enclosed roof. The east-west axis of this pavilion is aligned with the axis of Wisconsin Avenue and the cable-stayed pedestrian bridge (Photo 1.3.1). The concrete pavilion located over the reception hall can be described as an oval tabletop, with a substantial opening for the glass atrium, and is supported on four legs. The opening is 29 ft in the north-south direction and 136 ft 8 in. in the east-west direction. The pavilion

Photo 1.3.1: Milwaukee Art Museum addition designed by Santiago Calatrava.
Photo by S. Vinnakota.

supports 17 A-frames and the building spine (Photo 1.3.2). Together the spine and the A-frames support the glass roof atop the pavilion and the movable sun screen (*brise soleil*, in French). The building spine consists of a 25-in.-diameter, 0.5-in.-thick steel pipe. Each A-frame takes the shape of an upside-down V with a variable-depth channel cross section. The 17 A-frames lean at 41.64° to the horizon towards the spine. The longest A-frame is 96 ft 3 in., and the shortest is 15 ft 3 in. The steel A-frames are made of custom-built channel sections consisting of 1-in.-thick by 4-in.-deep flanges, and a 1-in.-thick web of variable depth. All webs are 1 ft deep at the apex and 5 ft deep at the bottom.

The most impressive element of the Calatrava addition is the 110-ton movable, steel sun screen, the Burke Brise Soleil, which rests on top of the towering glass atrium above the pavilion (Photos 1.3.2 and 1.3.3). The sun screen has two wings, one extending to the north, and the other to the south. Each wing has 36 ribs (fins) of varying length. The longest rib is 102 ft 2 in., the shortest 22 ft 9 in. The ribs, in steel, have variable-depth tubular cross sections with rounded corners. The ribs have a constant width of 13 in., while the depth varies from 15.75 in. at the tip to 39.37 in. at the connection end. The wall thickness of the tubular sections also varies from 0.118 in. at the tip to 0.472 in. at the connection end. All of the ribs in each wing are connected to a 14-in.-diameter, 2-in.-thick steel rotating spine. Eleven pairs of rotating spine actuator tabs are also welded to the spine. These tabs support the upper pin of the hydraulic rams, and it is by pushing against these tabs that the Brise Soleil moves. From the fully closed position, the fully open position of the sun screen is achieved by rotating the spines through 90° in just over three minutes. When open, the sun screen has a wing span of 217 ft. The original design of the sun screen called for carbon fiber composites for the ribs and steel for the connections and the rotating spines. However, the connections between the composite rib and the steel rotating spine proved difficult, unduly expensive, and required exhaustive testing. Eventually, the composite material was abandoned, and steel was selected for the ribs.

On the same axis of the pavilion spine lies a cable-stayed pedestrian bridge (Photo 1.3.1), which spans 232 ft over three lanes of a major city boulevard. The bridge, consisting of a main span and a back span, links downtown Milwaukee with the museum's new entrance. The cross section of the steel bridge is a five-sided, closed cell, 16 ft 9 in. wide and only 2 ft 4 in. deep. The deck plate is ¾ in. thick, while all other plates of the cross section are in. ⅝ thick. Continuous welds join the plates together. The main span's support is provided by a simple steel frame abutment at the west end and by nine main span cables located along the bridge centerline (Photos 1.3.1 and 1.3.4). Lengths of the main span cables range from 72 ft to a maximum of 352 ft. The cables are connected to a 198-ft-long leaning steel mast. The mast has a circular cross section that varies in diameter from 22 in. at the base to 39 in. at the fifth cable junction and to 12 in. at the top. The mast leans 48.36° against the horizon and therefore is parallel to the building spine. The mast is loaded by the main span cables and supported by 18 (nine pairs of) back span cables that are anchored to a back stay concrete beam.

Photo 1.3.2: Burke Brise Soleil connected to the spine supported by 17 A-frames: open position.
Photo by S. Vinnakota.

Photo 1.3.3: Burke Brise Soleil in closed position with its 36 ribs resting on glass atrium.
Photo by S. Vinnakota.

Photo 1.3.4: Pedestrian bridge connected to a leaning steel mast by nine cables.
Photo by S. Vinnakota.

The back span cables range in length from 68 to 101 ft. The bridge superstructure and above ground portions of the substructure are made of ASTM A572 structural steel plates.

1.4   Al McGuire Center with Cylindrical Roof

The Al McGuire Center is a 128,200 square ft practice facility that is home to Marquette University men's and women's basketball teams as well as the women's volleyball team. The facility consists of a large main arena, a smaller practice gymnasium, and an administration wing. The structural engineer of record is Opus Architects and Engineers of Milwaukee, Wisconsin.

The arena roof is supported by open web steel joists (prefabricated steel trusses) of A36 steel that are 104 in. deep, span 156 ft 6 in. in the north-south direction, and are spaced 9 ft 3 in. on center. The joists have straight and parallel top and bottom chords. To achieve the "cylindrical" shape of the arena roof, at each end of the arena segments of steel beams were bolted to columns of different heights to create a rough radius of the barrel (Photo 1.4.1). The joist seats, whose thicknesses were varied for different joists to achieve the desired radius for the roof, were welded to the beams. The 3 in.-deep acoustical metal decking and roofing material built up the remainder of the structure until the specified radius was attained (Photo 1.4.2). The roof covering of the arena is a standing seam metal product. Because of the unusual length of the steel joists,

Photo 1.4.1: Steel skeleton for Al McGuire Center, view from north.
Photo courtesy of Opus Corporation.

Photo 1.4.2: Steel skeleton for Al McGuire Center, view from west.
Photo courtesy of Opus Corporation.

each joist was shop fabricated and shipped to the job site in two pieces. The two pieces were placed end-to-end on the floor of the arena, and bolted together (Photo 1.4.3). These giant joists were lifted into place by a 350-ton hydraulic crane (Photo 1.4.4). A spreader beam was used for picking so that the top chord of the joist would not buckle during erection.

Photo 1.4.3:   Field assembly of steel joist.
Photo courtesy of Opus Corporation.

Photo 1.4.4:   Erection of steel joist.
Photo courtesy of Opus Corporation.

All of the floors of this structure are made of composite steel-concrete construction. The concrete slab is cast upon 3-in.-deep, 18-gauge cold-formed metal deck, with $4\frac{1}{2}$-in. normal weight concrete topping. The steel beams act compositely with the concrete slab through the use of $\frac{7}{8}$-in.-diameter, 6-in.-long headed shear connectors. The beams supporting the floors were cambered for the weight of the wet concrete.

1.5   Raynor Library, Marquette University

The John P. Raynor, S. J., Library at Marquette University consists of three levels above ground and one below. In addition, a bridge connects the new library to the existing Memorial Library approximately 135 ft away. The structural engineer of record is Opus Architects and Engineers of Milwaukee, Wisconsin. The design of this project conforms to the Wisconsin Administrative Code 1999 Edition, and the AISC LRFD Specification for Structural Steel Buildings [AISC, 1999]. The steel skeleton for this building during erection is shown in Photo 1.5.1.

The wind load for the Raynor Library is 20 psf (pounds per square foot) for the first 50 ft, and 25 psf for the remaining 10 ft. The roof snow load is 30 psf plus snow drifts, as required by code. The roof live load is taken as 20 psf for the library and 25 psf for the bridge. The floor live load is 150 psf in most

Photo 1.5.1: Steel skeleton for Raynor Library Building.
Photo courtesy of Opus Corporation.

Photo 1.5.2: Chevron bracing for Raynor Library Building.
Photo courtesy of Opus Corporation.

areas, a somewhat heavy load due to the weight associated with bookshelves. The live load in stairs, public corridors, and lobbies is 100 psf. The live load in the mechanical equipment room is 125 psf. The dead load has several components: 75 psf for the concrete slab on deck, 5 psf for the structural steel, and 5 psf for miscellaneous materials.

All structural wide flange shapes used for beams and columns in the library are of A992 steel. There are 13 different column types, 8 starting at the basement level and 5 starting at the first (ground) level. The majority of the columns are W12-shapes. The typical interior column starts at the basement level, and is connected (spliced) just above the second level to the upper tier column. Photo 1.5.1 shows a partially erected steel skeleton. At the west corners of the building, two towers rise above roof height. The beams consist predominantly of W18×35's spanning 33 ft and W16×26's spanning 25 ft, while the girders consist predominantly of W24×55's. Braced frames resist the moment induced by wind. The two types of braces used in this design are diagonal braces and Chevron braces (Photo 1.5.2). All of the braces consist of HSS 8×8 with thicknesses ranging from $\frac{1}{4}$ to $\frac{1}{2}$ in.

1.6   Fox Tower, Portland, Oregon

The Fox Tower is a 27-story, high-rise structure located in downtown Portland, Oregon. The structural engineer of record is KPFF Consulting Engineers of Portland. The building consists of five levels of below-grade parking, two levels of retail space on the first and second floors, one level of cinemas on the third floor, 23 floors of office space, and a penthouse [Ambrose, 2001]. The main lateral force resisting system consists of a 30 ft by 80 ft central elevator core shear wall system (Photos 1.6.1 and 1.6.2). In addition, special moment resisting steel frames were provided in the perimeter. The concrete core wall also provided lateral bracing of the structural steel framing during steel erection, which eliminated the need for temporary bracing and further expedited erection. In order to provide the necessary stiffness in the frames to control drift and cut down on weight, built-up column sections were used in combination with standard rolled shapes. The typical composite steel floor framing in the office tower consists of girders spanning 41 ft with beams at 10 ft on center spanning 25 ft between the girders. The typical office floor deck consists of 3-in.-deep, 20-gage composite floor deck with 2.5-in. normal weight concrete topping.

Photo 1.6.1: Steel skeleton for Fox Tower, Portland.
Photo by Hoffman Construction Company; courtesy of KPFF Consulting.

Photo 1.6.2: Construction of steel frame around concrete core, Fox Tower, Portland.
Photo by Hoffman Construction Company; courtesy of KPFF Consulting.

References

1.1 AISC [1999]: *Load and Resistance Factor Design Specification for Structural Steel Buildings,* American Institute of Steel Construction, Chicago, IL, December. (Also referred to as LRFDS in this book.)

1.2 Ambrose, C. [2001]: "Fox Tower Rises over Portland," *Modern Steel Construction,* February, pp. 38–41.

1.3 Badreddine, L. [2002]: "Winged Victory," *Civil Engineering,* ASCE, vol. 72, no. 1, January.

1.4 Chan, J., Gautrey, J., Hewitt, J., Mann, S., Roberts, J. T., and Wells, C. [2002]: "Miller Park," *The Arup Journal,* Ove Arup and Partners, Los Angeles, CA, no. 1.

1.5 Hewitt, J., Roberts, J. T., and Mann, S. [2002]: "Miller Park Stadium Design of the Retractable Roof Structure," Ove Arup and Partners, Los Angeles, CA.

1.6 Wood, K. G. [2001]: "Statement in Steel," *Modern Steel Construction,* AISC, Chicago, March.

Diagram of steel making process.
Photo courtesy American Iron and Steel Institute

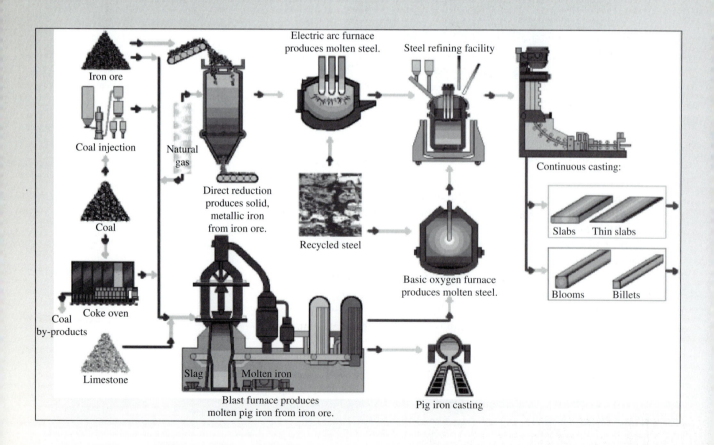

C H A P T E R

Steels

2

2.1 Introduction

The behavior of a structure under load is strongly influenced by the properties of the materials used in the structure's construction. Therefore, in this chapter we will study the properties and behavior of the material (i.e., steel) used in the construction of steel structures. We will consider the production of steels and steel products in Section 2.2, the various available rolled steel shapes in Section 2.3, the heat treatment of steel in Section 2.4, and the classification of structural steels in Section 2.5. Physical tests routinely conducted on steel elements to evaluate their strength, stiffness, and ductility characteristics are described in Section 2.6. The economical advantages of steel as a structural material are given in Section 2.7, while built-up sections are discussed in Section 2.8. Corrosion, which often contributes significantly to the failure of steel members, but which can also be largely precluded by thoughtful design, careful fabrication and erection, and regularly scheduled maintenance, is treated in Section 2.9.

2.2 Production of Steels and Steel Products

The production of steel shapes from raw materials can be divided into several phases: smelting the ore in a blast furnace, making the steel in an electric arc furnace or a basic oxygen furnace (not described in this text), and rolling the shapes in rolling mills.

Blast Furnace

Steel is essentially composed of iron, along with a small amount of carbon and lesser amounts of various other elements. Iron in natural deposits occurs principally in the form of *magnetite* (Fe_3O_4) and *hematite* (Fe_2O_3). These high grade ores contain about 50 percent (or more) pure iron. Earthy impurities and traces of other elements constitute the remaining portion of the ore. The iron is traditionally removed from the ore in a blast furnace during a process called *smelting*. A *blast furnace* is a huge cylindrical, steel-shelled tower. A charge composed of iron ore, coke, and limestone is fed into the furnace, where the iron oxide is reduced to iron by carbon (coke). Limestone, burnt into lime, then acts as the flux which mixes with the earthy matter of the ore and the coke, forming a fusible mixture called *slag.* The slag, being very much lighter than the iron, floats on the surface of the molten iron in the hearth. The slag is tapped off near the top of the melting zone through a cinder hole. The heavier molten iron, called *pig iron,* collects at the bottom of the hearth and is periodically tapped. The largest blast furnaces are capable of producing as much as 8000 to 10,000 tons of pig iron per day. Pig iron contains 4 to 5% C and 0.1 to 0.5% P; both of these values are higher than those permitted by steel specifications. The molten iron may be cast into pigs or transported directly in a molten state by a hot ladle to a continuous casting machine.

Electric Arc Furnace

In the United States, structural steels are produced principally by electric arc furnaces, which provide the most economical means of producing high-grade steels. Electric furnaces used for this purpose have capacities of 20 to 400 tons. Three carbon electrodes, up to 2 ft thick and 24 ft long, extend through the roof and contact the metal charge in the furnace below. Three-phase alternating current flows from one electrode to the metallic charge and then from the charge to the next electrode, providing intense heat and an electric arc. These furnaces operate on currents of up to 80,000 KVA. The charge usually includes carefully sorted scrap metal (such as discarded steel shapes, junked cars, etc.), pig iron from the blast furnace, and small amounts of lime. Desulphurizing, dephosphorizing, and deoxidizing ingredients, in amounts indicated by tests of the metal, are also added. Under the intense heat of the arcs, these materials, together with any impurities in the metal, form a slag which floats on top, leaving pure steel beneath it. The molten steel is poured into a ladle through a tapping spout. The entire furnace is mounted on rockers so that it can be tilted to permit the molten steel to flow through the spout. The time required for refining steel in an electric furnace runs from two to five hours.

At the end of the refining process, the molten steel is tapped into ladles. At this point, alloying elements such as manganese, vanadium, columbium and deoxidizers such as silicon and aluminum are added to the steel. The type and amount of alloying elements determine the type of steel (Section 2.5). Deoxidizers remove oxygen from the steel during the pouring and solidification process. The removal of oxygen from molten metal, to the extent that carbon monoxide does not develop during solidification, is known as *killing.* The amount of deoxidizers that are added may stop, or "kill", the formation of pockets of gas entirely, or simply reduce it to a more acceptable level. Thus, steels are classified by their degree of deoxidzation as: (1) *fully-killed steel* (highest); (2) *semi-killed steel* (intermediate); and (3) *rimmed steel* (steels with little or no oxygen removal). The ladles are hung from traveling cranes which carry them over ingot molds. The molten steel, at about 2800°F, is then poured or **teamed** into these molds. An **ingot mold** is a large steel shell, usually about 7 ft high and 2 ft square, tapered slightly from top to bottom with the big end down to facilitate stripping of the mold. The mold is open at the top for filling, and it has no bottom. These molds are placed on small flat cars, the ladle of molten steel is brought over them, and they are filled through a valve in the bottom of the ladle. When the ingots have cooled sufficiently, the train of filled molds is run under a crane which picks up the molds, leaving the now solidified but red-hot ingots on the cars. A typical ingot weighs about 10 tons. The train is then run to a reheating furnace called a **soaking pit** where the ingots are stored for some hours until they have reached the same temperature throughout (2050 to 2450°F) and are suitable for rolling.

Rolling Mills

The first rolling operations are performed in rolling mills known as **blooming** or **slabbing mills. Rolling** consists of squeezing the steel ingot between two rolls revolving at the same speed but in opposite directions and spaced so that the distance between them is somewhat less than the thickness of the piece of steel entering them. The rolls grip the steel, reduce it in cross-sectional area, and deliver it increased in length. After a few passes through the rolls, a small part at each end, corresponding to the top and bottom of the ingot, is **sheared** or **cropped off** to eliminate the lower grade steel which usually contains segregation, porosities, or pipes. Blooms, billets and slabs are the semifinished products developed in these mills. A **bloom** is mostly square, ranging in cross section from 6×6 to 12×12 in. A **billet** is also mostly square and has a minimum width and thickness of $1\frac{1}{2}$ in. and a cross-sectional area of 2.25 to 36 in.^{2}. A **slab** is rectangular in cross section, has a minimum thickness of $1\frac{1}{2}$ in., a minimum width of twice its thickness, and a cross-sectional area of not less than 16 in.^{2}

The **semifinished** products, after any required reheating are sent to rolling mills known as **finishing mills.** Thus, blooms are passed through structural mills (to form I-shapes, channels, angles, structural tees, and others) and rail mills (to form rails and bulbs). Billets are passed through bar mills (to form round, square, hexagonal, and octagonal bars), wire mills (to form wires and wire ropes), and tube mills (to form tubes and seamless pipes). Slabs are passed through plate mills (to form plates), skelp mills (to form continuous butt-weld pipes), and strip mills (to form hot-rolled sheets and strips). Figure 2.2.1 shows a schematic representation of the rolling process for I-shapes. Each of the commercial forms of steel mentioned earlier, such as an I-shape, rail, round or plate, is made from a bloom, billet, or slab, as appropriate, by repeatedly passing it, while at a red hot temperature, through **roughing rolls** (for rough shaping to the required form) and then through **finishing rolls** (Fig. 2.2.1). This process gradually changes the cross section of the form to that of the desired finished product, while simultaneously increasing the length of the bloom, billet, or slab. These rotating rolls are mounted in a framework comprising a basic unit called a *stand.* Figure 2.2.2 shows some of the successive cross sections an ingot takes as it is rolled into an I-section. Since rolling occurs while the metal is still red hot, the shapes are called *hot-rolled shapes.* When the rolling is done, the member is cut to standard lengths, usually a maximum of 60 to 75 ft.

The casting of steels by the traditional ingot process is inefficient, as part of the ingot top containing major segregations must be cropped (cut) during the early stages of rolling and subsequently remelted. The bulk of structural steel produced in the developed nations now comes from the more efficient continuous casting process, wherein the semifinished products (slabs) are produced directly from liquid steel. The continuous casting process described below, therefore bypasses two steps of the traditional steel making process, namely ingot molds and the primary mills.

Continuous Casting Process

In the continuous casting process (Fig. 2.2.3), liquid steel from furnaces, brought in sequenced ladles, is poured into a tundish and maintained at a pre-determined level. From the tundish the molten metal is fed into a short water-cooled mold at a controlled rate, and as soon as solidification is sufficiently advanced the base of the mold is withdrawn, taking with it the embryo ingot. This embryo is then engaged by a series of support rolls, curving rolls, straightening rolls, and withdrawal rolls which convey the continuous rectangular ribbon of steel to the reduction mill proper. Thus, once the process is started, the

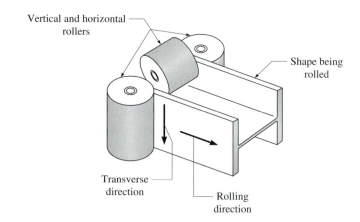

Figure 2.2.1: Schematic representation of rolling process.

Figure 2.2.2: Successive cross sections in rolling an I-beam.

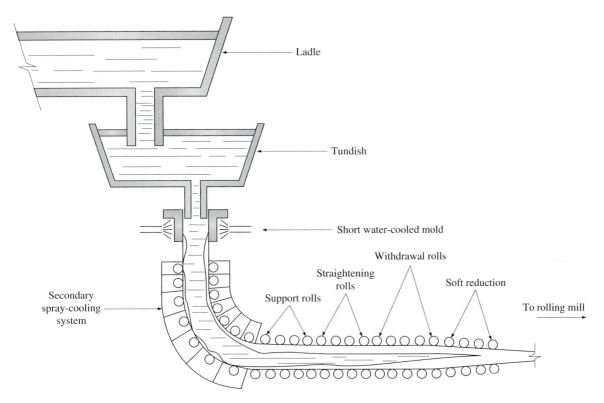

Figure 2.2.3: Continuous casting of steel slab.

outer skin of the steel strand solidifies as it passes through the mold, and this solidification is further assisted by water sprayed on the skin just after the strand exits the mold. The continuous moving ribbon is cut into slabs by a flying saw and held for subsequent rolling in finishing mills. Whatever segregation of impurities that takes place will be uniform and will be further dispersed by the rolling process. Metal loss is extremely small. Once started, casting can proceed virtually nonstop. The whole process is controlled remotely by computers. Also, continuous cast steels are more uniform in metallurgical terms, and there is much less segregation compared to the ingot-based products.

2.3   Rolled Steel Sections

Hot-rolled steel shapes, mostly used in civil engineering structures, are shown in Fig. 2.3.1. Depending on the cross section, they are called I-, C-, L-, or T-shapes and plates, bars, rods and HSS. I-shaped sections may be further classified as wide flange shapes (W-shapes), American standard beams

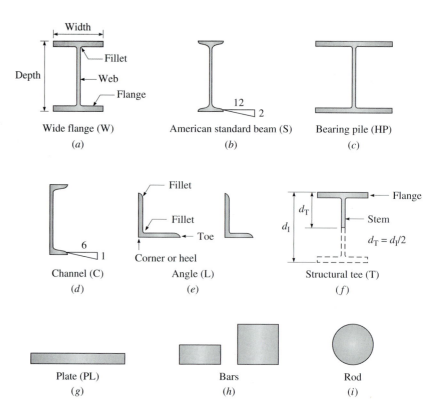

Figure 2.3.1:   Hot-rolled open sections.

(S-shapes), or bearing pile shapes (HP-shapes). They are all described in the sections that follow. The cross sections rolled vary a great deal: the smallest members possess cross-sectional areas less than one square in., while the largest members may have depths exceeding 40 in. Most structural shapes can be obtained in lengths from 60 to 75 ft depending on the producer, while it is sometimes possible to obtain some heavy shapes up to 120 ft in length. The *unit weight of structural steel* is 0.2833 pci (pounds per cubic inch), 490 pcf (pounds per cubic foot), or 3.41 lb/in.^{2}/ft of length, and its specific gravity is 7.85. The ***Manual of Steel Construction: Load and Resistance Factor Design,*** published by the American Institute of Steel Construction (AISC) in 2001, provides detailed dimensions and properties of structural steel shapes [AISC, 2001]. This manual is referred to hereafter as the ***LRFD Manual,*** or just the ***Manual,*** or simply ***LRFDM.*** In Part 1 of the LRFD Manual the dimensions of the shapes are given in decimal numbers and also as fractions. Decimal numbers should be used by designers in calculations. Dimensions in fractions, given to the nearest sixteenth of an inch, are intended for use by steel detailers in fabricating companies that produce shop drawings of structural members and their connections. The cross-sectional nomenclature used in the Manual and in this book is shown in Fig. 2.3.2 for two shapes.

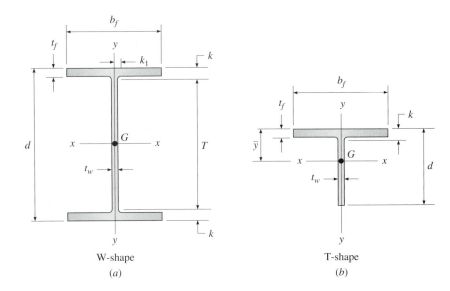

W-shape
(a)

T-shape
(b)

Figure 2.3.2: Cross-sectional nomenclature.

Wide Flange Shapes (W)

A *wide flange shape* has two horizontal rectangular elements, called *flanges,* and a vertical rectangular element called a *web,* connected by *fillets* as shown in Fig. 2.3.1a. The cross section has two axes of symmetry. These shapes have a depth equal to or greater than the flange width, and the flange thickness is generally greater than the web thickness. The inner and outer surfaces of the bottom and top flanges are parallel. The wide flange shape is identified by the alphabetical symbol W, followed by the nominal depth in inches and the weight in pounds per lineal foot of length (plf). Thus, the designation W14×145 represents a wide flange section which is nominally 14 in. deep (actual total depth equals 14.8 in.) and which weighs 145 plf, while W14×808 represents a wide flange section also with a nominal depth of 14 in. (but actual depth now 22.8 in.) and which weighs 808 plf. Thus, the actual depths of wide flange shapes vary within any given nominal depth groupings; however, the T dimension—the distance between web toes of the fillets at the top and bottom of the web shown in Fig. 2.3.2a—remains constant (10 in. for the examples given earlier). This is a result of the rolling process during manufacture: the same size inside rollers are used for all sections of a given nominal web depth. The various sections within a nominal depth grouping result from changing the spacing between the roller sets, thus permitting differing flange and web thicknesses (Fig. 2.3.3a). W-shapes rolled with flange widths approximately equal to their depths are known as *column shapes (W8, W10, W12, and W14),* and are quite efficient as columns. The W12 and W14 series, with weights of 40 to 808 plf, are primarily used for columns in tall buildings. W-shapes having a width of flange smaller than the depth are

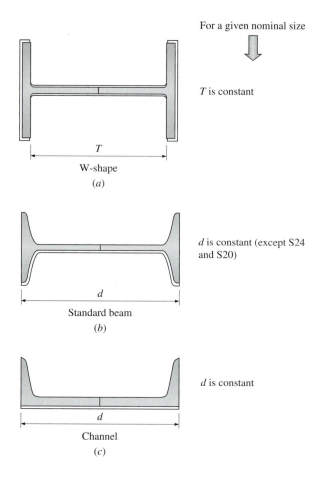

For a given nominal size

T is constant

W-shape

(a)

d is constant (except S24 and S20)

Standard beam

(b)

d is constant

Channel

(c)

Figure 2.3.3: Method of spreading rolls to increase area of sections of same nominal size.

known as **beam shapes.** They range from a depth of 4 in. up to 44 in., and their weights range from 9 to 798 plf. Wide flange shapes are the most commonly used rolled section, and these shapes comprise almost 50 percent of the tonnage of structural steel shapes rolled today.

American Standard Beams (S)

The **American standard beam** has relatively narrower flanges and a thicker web than the wide flange shape. The inner surfaces of the flange have a slope of approximately $16\frac{2}{3}\%$ (2 on 12), as shown in Fig. 2.3.1b. The use of the standard beam is now uncommon because of excessive material in the web and the relatively low lateral stiffness offered by the narrow flange. The standard beam is identified by the alphabetical symbol S. Thus, the designation S12×50 indicates a standard beam of nominal depth 12 in. (actual depth = 12 in.) weighing 50 plf. Unlike wide flange sections, standard beams in a given nominal depth group have the same actual depths. Again, this is a result of the rolling process during manufacture as shapes of greater cross-sectional area are made by spreading one set of rollers only, as indicated in Fig. 2.3.3*b*. Thus the depth remains the same, whereas the width of the flange and the thickness of the web are increased. W-shapes, with wider flanges and thinner webs than the S-shapes, have a larger proportion of their material in the flanges and, therefore, are much more efficient than standard beams. Also, a sufficient variety of S-shapes does not exist to make economical design choices. Consequently, S-shapes are not widely employed in structures today.

Bearing Pile Shapes (HP)

HP-shapes, like wide flange shapes, have two parallel flanges with parallel flange surfaces and a web element. The web and flange thicknesses are equal, and the width of flange and the depth of section are approximately equal to the nominal depth of the section. These shapes are also known as *bearing piles*. The designation HP12×63 indicates a bearing pile shape nominally 12 in. in depth (actual depth =11.94 in.) and 63 plf. The thicker webs, as compared to wide flange shapes, provide better resistance to the impact of pile driving.

Miscellaneous Shapes (M)

The letter M designates I-shapes that cannot be classified as W-, S-, or HP-sections. In this book, we use the generic term *I-shape* to represent W-, S-, M-, or HP-shapes.

Channels (C)

A *channel shape,* like the standard I-beam, has a web and two parallel flanges (Fig. 2.3.1*d*), with the inner surfaces of both flanges having a slope of approximately 16⅔% (2 in. in 12 in.). Also for these shapes, the actual depth equals the nominal depth. A channel shape has only one axis of symmetry. Channel shapes are identified by the letter symbol C. Thus, the designation C12×30 indicates a nominal 12-in.-deep channel (actual depth = 12 in.) having a weight of 30 plf. The letter MC designates channels that cannot be classified as C shapes. Channels are used either individually or in pairs, where a flat side (web) is required for connections to other members. So, channel sections are mostly used as stair stringers, purlins, girts, and diagonal members in trusses.

Angle Shapes (L)

Angle shapes are sections whose cross section is composed of two rectangular elements called *legs,* which are perpendicular to one another (Fig. 2.3.1*e*). The angles having legs of equal lengths are called *equal leg angles,* while those having legs of unequal length are known as *unequal leg angles.* The inner and outer surfaces of each leg are parallel, and the thickness of both the legs is the same no matter whether the angle has equal or unequal legs. Angles are designated by the alphabetical letter symbol L followed by their leg lengths (the longer leg always first) and thickness in inches. Thus, the designation L6×4×½ stands for an unequal leg angle whose long leg is 6 in. long and ½ in. thick, and whose short leg is 4 in. long and $\frac{1}{2}$ in. thick. Similarly, the designation L6×6×$\frac{3}{8}$ stands for an equal leg angle whose legs are 6 in. long and $\frac{3}{8}$ in. thick. The leg length is measured from the corner, or heel, to the toe of the fillet at the other end of the leg. To change the cross-sectional area for an angle of given leg lengths, the thickness of each leg is increased by spreading the rolls. Hence, this method of spreading the rolls changes the leg lengths slightly. Many short span trusses are fabricated with single or double angle members.

Structural Tees (WT, ST, MT)

The cross section of a *structural tee* resembles the letter T (Fig. 2.3.1*f*). The top part of a T-section is called its *flange,* and the lower part (vertical) is called its *stem.* In the United States, T-sections are manufactured at the mill by cutting a W-, S- or M-shape member in half along the center of the web using a rotary shear, resulting in two WT-, ST-, or MT-shapes respectively. Thus WT6×25 designates a structural tee whose depth (tip of stem to outside flange surface) is nominally 6 in. (actual depth is 6.10 in.) and which weighs 25 plf. This tee is therefore cut from a wide flange section W12×50. Structural tees are sometimes used as top and bottom chord members of trusses, and as diagonal bracing members in buildings.

Flat Bars (FLT) and Plates (PL)

Flat bars, plates, strips, and *sheets,* all rectangular in cross section, come in many widths and thicknesses. A flat shape has historically been classified as a bar if its width is less than or equal to 8 in. and as a plate if its width is greater than 8 in. A flat bar is usually designated with the width given before the thickness, for example, a 6 × $\frac{5}{8}$ bar. Conversely, a plate is usually designated with the thickness first, as in a $\frac{1}{2}$ × 9 plate. *Flat bars* are rolled between horizontal and vertical rollers and trimmed to length by shearing or flame cutting on the ends. *Plates* are produced from slabs by squeezing the hot metal between smooth cylindrical rolls adjusted to form the required width and thickness. *Universal mill (UM) plates* are rolled on a mill which, in addition to horizontal rolls, has a set of vertical rolls which provide relatively smooth, straight edges. Such plates are cut only to length. *Sheared plates* are rolled between horizontal rolls and trimmed (sheared or flame cut) on all four edges. The maximum width of UM plates is 60 in. and for sheared plates it is 200 in. Preferably, plate thicknesses are specified in increments of $\frac{1}{16}$ in. for $t \le \frac{3}{8}$ in., $\frac{1}{8}$ in. for $\frac{3}{8} < t \le 1$ in., and $\frac{1}{4}$ in. for $t > 1$ in. Actually, it should be mentioned that very little, if any, structural difference exists between flat bars and plates. Consequently, *plate* is becoming a universally applied term today, and a PL$\frac{3}{8}$×5×1′ 6″, for example, might be fabricated from plate or bar stock.

Hollow Structural Sections (HSS) and Pipes (P)

Hollow structural sections (HSS) and pipes (*P*) may be divided into two categories: *welded* or *seamless.* In the *continuous weld process* of tube forming, coils of sheet steel called *skelp* are welded end-to-end to produce a continuous

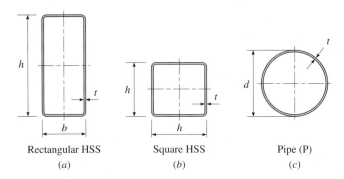

Rectangular HSS   Square HSS   Pipe (P)
(a)   (b)   (c)

Figure 2.3.4: Hot-rolled hollow structural sections.

band of steel, which is passed through a furnace. As the skelp exits the furnace, it is drawn through rolls that bend it into a cylindrical shape, and the tube is closed by a continuous exterior and interior longitudinal weld. The exterior weld is skimmed flush, and then the tube is reheated and passed through a rolling mill where it is stretch-reduced to a circular, square, or rectangular section. The rolling controls both the size of the tube and the wall thickness. In the *seamless process,* a solid round bar of predetermined size is preheated and then pierced by a mandrel while rotating at a high speed. The tube then passes through other rolling operations which bring it to the proper size and thickness. The round tube subsequently passes to a sizing mill where it is formed into a square or rectangular shape. The wall thickness is constant throughout the shape, with the exception of the corner areas (due to local bending effects in the forming process).

A rectangular or square hollow structural section (HSS) is designated by nominal outside dimensions (with the larger dimension first) and then the wall thickness, each in rational numbers, for example, HSS10×4×3/8. The maximum sizes available at present are about 20 by 12 in. rectangular, 16 by 16 in. square, and 20 in. diameter round (Fig. 2.3.4). *Circular tubes* are also known as *pipes.* They are identified by the alphabetical symbol P. Pipes of a given nominal size are produced as standard, extra strong, and double-extra strong by varying the wall thickness and the inside diameter while keeping the outside diameter constant. For example, standard, extra strong, and double-extra strong pipes of 6 in. nominal diameter may be designated as P6, Px6, and Pxx6, respectively.

Comments on Rolled Shapes

1. The flange thickness given in the tables for S-, M-, ST-, MT-, C-, and MC-shapes is the *average* flange thickness.
2. In calculating the theoretical weights, properties, and dimensions of the rolled shapes listed in Part 1 of the LRFD Manual, fillets and roundings have been included for all shapes except angles. The properties of these rolled shapes are based on the *smallest* theoretical size of the fillets

produced; dimensions for detailing are based on the *largest* theoretical size of the fillets produced. These properties and dimensions are either exact or slightly conservative.

3. The moment of inertia of an I-shape is much higher with respect to the *x* axis than with the *y* axis. So the *x* axis is generally known as the **major axis,** and the *y* axis as the **minor axis.**

4. All of the rolled shapes are manufactured to certain tolerances with respect to variations in cross-sectional dimensions, lengths, squareness, flatness, weights, sweep, camber, and so on. The tolerances are contained in ASTM A6 and reproduced in Tables 1-54 to 61 of the LRFDM.

Cold-Formed Steel Shapes

Cold-formed structural steel shapes are produced by passing sheet or strip steel at room temperature through rolls or press brakes and then bending the steel into desired shapes. The thickness of the uncoated sheets in the United States is designated by a nominal gage number. Cold-formed steel production is usually restricted to sheet thicknesses ranging from 30 gage to 4 gage (LRFDM Table 17-10). Cold-formed members may be distinguished from hot-rolled shapes in that their profiles contain rounded corners and slender flat elements, and that all these elements have the same thickness. Once the special set of rolls needed for a new shape is at hand, any conceivable cross-sectional shape can be mass produced in a cold-rolling mill. Thus, a great variety of cold-formed shapes suitable for a particular job or trade have been developed.

Cold-formed shapes can be divided into two types: framing members and surface members. *Framing members* have the general outline of well-established hot-rolled shapes, such as channels, zees, and hat shapes (Fig. 2.3.5*a*). I-shapes are made by spot-welding together two channels or a channel and two angles. The depth of such sections generally ranges from 2 to 12 in. and the thickness from about 18 to 8 gage. These dimensions frequently result in slender plate elements (high flat width-to-thickness ratios) requiring lips or other edge stiffeners along free edges to prevent premature plate local buckling. These light-gage steel members are used in structures subjected to light loads and/or short span lengths. Thus they are extensively used as purlins, girts, wall studs, chord members of open-web steel joists, and so on. For such structures, the use of conventional hot-rolled sections is often uneconomical because the strength (resistance) developed in the smallest available hot-rolled shape may be very high compared to the required strength.

Surface members are load resisting shapes which also provide useful surfaces (Fig. 2.3.5*b*). They are extensively used in roof, floor, wall, and partition construction due to their favorable weight-strength and weight-stiffness ratios. The reduction in dead load possible by using cold-formed steel deck components (compared to concrete, for example) is cumulative in a high-rise structure, and therefore significant material and cost savings are possible. Prefabricating roof, floor, and wall panels using cold-formed steel decking

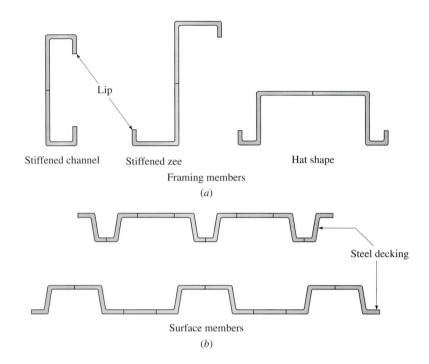

Stiffened channel   Stiffened zee   Hat shape

Framing members

(a)

Steel decking

Surface members

(b)

Figure 2.3.5:   Cold-formed steel sections.

material provides savings in high-cost field labor and usually improves safety in the erection of a building.

Cold-working increases the yield stress of steel above that of the parent metal, but this increase comes at the expense of reduced ductility. Because of the thinness of the cross-sectional elements, problems of instability become a major design concern. The design of these members in the United States is made in accordance with the *Specifications for the Design of Cold-Formed Steel Structural Members* [AISI, 2001] issued by the American Iron and Steel Institute. Properties of many common cold-formed shapes are given in the *Cold-Formed Steel Design Manual* [AISI, 2002].

2.4   Heat Treatment of Steel

As mentioned in Section 2.2, the properties of steel are greatly affected by its chemical composition. But they are also greatly influenced by various treatments to which the steel may be subjected after leaving the furnace in which it was refined and its chemical composition fixed. Many of these treatments involve changes in the temperature of the steel product in the solid state, and the term **heat treatment** is used to cover them. Heat treatments affect the properties of steel through changes which they cause in its crystalline structure (grain size). The heat treatment of articles of steel is usually carried out after they have been given the final form in which they are to be used.

There are three principal heat treatment processes: quenching, tempering, and annealing:

- *Quenching* consists of heating the metal to about 1650°F, and then suddenly cooling it in water, brine, oil, or molten lead. The rapid cooling causes formation of fine grained structures, commonly known as *martensite steel* which is very hard and strong but susceptible to cracking, owing to severe internal stresses (or residual stresses) set up by the rapid cooling. Quenching is therefore followed by tempering. During quenching, heat can be dissipated only through the surface of the article which is in contact with the cooling medium. For thick parts, for which the ratio of surface area to volume is relatively low, the cooling rate for the interior may be so low that the metallurgical changes required for hardening may not be achieved. Only the material near the surface will be hardened and therefore strengthened; the interior of the body will largely be unaffected. This situation is generally referred to as *mass effect* or the *effect of section size.*

- *Tempering* consists of reheating the object to a temperature of at least 1150°F and allowing it to cool in air. This relieves the internal stresses, making the steel more ductile and much tougher, without causing a great reduction in strength or hardness.

- *Annealing,* the third heat treatment, is the opposite of hardening. It is achieved by heating steel above the **transformation range,** and after maintaining this temperature for a sufficient length of time, cooling the steel very slowly either in the furnace itself or in some other medium that ensures a slow cooling rate. This process adds considerably to the ductility of the metal, but reduces its yield stress, tensile strength, and hardness accordingly. The process is also known as ***stress relieving.***

2.5   Classification of Structural Steels

Structural steels available in hot-rolled structural shapes, plates, and bars may be classified as carbon steels, high-strength low-alloy (HSLA) steels, corrosion resistant HSLA steels, quenched and tempered low-alloy steels, and quenched and tempered alloy steels. The American Society for Testing and Materials (ASTM) develops and maintains the relevant material standards for these steels in the United States. They are reissued each year and are contained in the *Annual Book of ASTM Standards* [ASTM, 2003]. The general requirements for delivery of structural steels are covered under ASTM A6 specification [ASTM, 2003]. Structural steels are referred to by ASTM designations which consist of the prefix letter A followed by one, two, or three numerical digits. For example, ASTM A514 refers to the material that is governed by standard number A514 and contained in the *Annual Book of ASTM Standards.*

2.5.1   Carbon Steels

Carbon steels differ from low-alloy and alloy steels in that carbon and manganese are the main strengthening elements. Other alloying elements are not specified. Carbon steels contain less than 1.7% C, 1.65% Mn, 0.60% Si, and 0.60% Cu. Increasing the percentage of carbon raises the yield stress and hardness but reduces ductility and adversely affects weldability. Economical welding without preheat, postheat, or special welding electrodes is possible only when carbon content does not exceed 0.30 percent. Manganese improves strength and notch ductility. Silicon improves strength, but if present in excessive amounts may cause the carbon to occur as graphite flakes. This reduces strength, and so the silicon content is rarely allowed to exceed 0.6 percent. Sulphur and phosphorous have detrimental effects on strength, especially the ductility and weldability of steel. Hence, they are limited to 0.06 percent maximum. Carbon steels have the following deficiencies: (1) low yield strength, (2) poor atmospheric corrosion resistance, and (3) poor notch ductility since they become brittle at temperatures only slightly below room temperature.

Carbon steels can be subdivided into four categories based on carbon content: *low-carbon steels* (less than 0.15% C); *mild-carbon steels* (0.15–0.29% C); *medium-carbon steels* (0.30–0.59% C); and *high-carbon steels* (0.60–1.7% C). Structural carbon steels fall into the mild-carbon category. These steels exhibit a marked yield point.

ASTM A36 or Carbon Structural Steel

Until recently, the basic structural steel commonly used in building and bridge construction has been A36 steel. It has maximum carbon varying from 0.25 to 0.29 percent depending on thickness, and is still the preferred material specification for M-, S-, HP-, C-, MC-, MT-, ST-, and L-shapes and plates (see LRFDM Tables 2-1 and 2-2). A36 steel has a minimum yield stress of 36 ksi, except for plates over 8 in. thick for which the minimum yield stress is 32 ksi. Normally, connection material is specified as A36, regardless of the grade of the primary components themselves. In addition, A36 is the only steel that can be obtained in thicknesses greater than 8 in., although as mentioned above, these plates are only available with a lower specified minimum yield stress of 32 ksi. The ultimate tensile stress of this steel varies from 58 to 80 ksi; for design calculations use the minimum value specified of 58 ksi.

2.5.2   High-Strength Low-Alloy (HSLA) Steels

High-strength low-alloy steels, commonly designated *high-strength steels,* contain moderate amounts of alloying elements other than carbon. Chromium, columbium, copper, manganese, molybdenum, nickel, vanadium, and zirconium are some of the alloying elements. The term *low-alloy steel* is used generically to describe steels for which the total of all the alloying elements does not exceed 5 percent of the total composition of the steel. These steels have been developed as a compromise between the convenient fabrication characteristics and low cost of mild carbon steels, and the high strength of heat treated alloy steels. The alloying elements improve mechanical properties, fabricating characteristics, and other attributes of the steel. Alloy steels are sometimes classified according to the principal alloying element or elements present. Thus, we have nickel steels, chromium steels, chromium vanadium steels, and so on. HSLA steels have yield stresses ranging from 40 to 70 ksi, and some of the high-strength steels offer improved corrosion resistance. HSLA steels, like mild carbon steels, have well-defined yield points. These steels are used in the as-rolled or normalized conditions; that is, no heat treatment is used.

ASTM A572 Steels

ASTM A572 steels are high-strength low-alloy columbium-vanadium steels of structural quality. Specification A572 defines five grades of HSLA steels: 42, 50, 55, 60, and 65. (In ASTM specifications, the term **grade** identifies yield stress level. Thus, Grade 42 represents steel with a yield stress of 42 ksi.) The ultimate tensile stresses of these steels are 60, 65, 70, 75 and 80 ksi, respectively. Increases in minimum yield strengths for A572 grades to 65 ksi are achieved by increasing maximum carbon content from 0.21 percent (Grade 42) to 0.26 percent (Grade 65), plus making other chemistry adjustments within the specifications. The maximum carbon content permitted by the specification depends on both the plate thickness and strength level. These steels are intended for bolted and welded construction of buildings and other structures, excluding bridges.

ASTM A992

The new *ASTM A992* specification covers only W-shapes (rolled wide flange shapes) intended for use in building construction. Note also that for W- and WT-shapes, ASTM A992 is the preferred material specification. It has specified minimum values for F_y and F_u of 50 ksi and 65 ksi, respectively. It also specifies an upper limit on yield stress of 65 ksi, a maximum yield to ultimate tensile stress ratio of 0.85, and a specified maximum carbon equivalent (see Section 6.14.6) of 0.50 percent. In addition, A992 has excellent weldability and ductility characteristics.

2.5.3 Corrosion Resistant HSLA Steels

Corrosion resistant HSLA steels provide enhanced atmospheric corrosion resistance by developing their own dense, hard, tightly adherent oxide film with a pleasing purplish color (instead of normal flaky rust) when exposed to the atmosphere. The tight oxide or *patina,* as it is called, seals the base metal from further oxidation and therefore acts as a coat of paint, protecting the steel from further corrosion. For the patina to form, the steel must be exposed to an alternately dry and wet environment for about two years. The atmospheric corrosion resistance of these weathering steels is approximately two times that of carbon structural steel with copper, or four times that of plain carbon structural steel without copper (0.02% max. Cu). They are easily fabricated and can be welded by all standard welding procedures. These steels are often left unpainted, but if painted the coating life is typically longer than with other steels. Nickel and copper are the principal elements added to HSLA steels to achieve the improved corrosion resistance of weathering steels.

Uncoated weathering steels are not recommended for exposure to concentrated industrial fumes, for exposure in marine locations where salt can be deposited on the steel by either spray or fog, or where the steel is either buried in soil or submerged in water. Also, to achieve the benefits of the enhanced atmospheric corrosion resistance of these bare steels, it is necessary that design, detailing, fabrication, erection, and maintenance practices proper for weathering steels be observed.

ASTM A588 Steels

ASTM A588 is a high-strength, low-alloy weathering steel with a 50 ksi minimum yield point for thicknesses to 4 in. It is also available in greater thicknesses at smaller yield stresses (46 and 42 ksi). This steel is intended primarily for bolted and welded building structures. Its atmospheric corrosion resistance is four times that of A36 steel. Materials of this type were originally known by proprietary names such as Mayari-R and Cor-ten.

2.5.4   Quenched and Tempered Low-Alloy and Alloy Steels

Quenched and tempered steels differ from high-strength low-alloy steels in that they have a higher percentage of alloying elements and in that they rely on heat treatment to develop higher strength levels and other improved mechanical properties. These steels generally have a maximum carbon content of 0.20 percent. Quenching results in a material with a very hard, fine-grained, or martensitic structure. The ductility of these steels is significantly lower than that of the carbon and HSLA steels. Tempering improves ductility. As seen in LRFDM Tables 2-1 and 2-2, quenched and tempered steels are only available in plates. Alloy steels may be quenched and tempered to obtain yield stresses in the range of 90 to 100 ksi, with the alloy providing yield stresses beyond those of carbon. Quenched and tempered steels do not exhibit a well-defined yield point, and yield stress is defined as the stress at 0.2 percent offset strain. Although special welding techniques are usually required, these steels are generally weldable.

ASTM A514 Steels

The *ASTM A514* specification defines several types of quenched and tempered alloy steel plates of structural quality suitable for welding. Specified minimum yield stress is 100 ksi for thicknesses up to and including 2½ in., and 90 ksi for thicknesses over 2½ to 6 in. inclusive. A514 steels can be used for welded high-rise buildings, television towers, welded bridges, water storage tanks, and so forth, where high yield strength-to-weight ratios are required. Although A514 is quite suitable as a structural material, it is not available in wide flange or hot-rolled shapes. Extra precautions must be taken during welding in order to prevent destroying the special heat-induced properties of this steel.

2.6   Physical Properties

Several types of tests are routinely conducted on steel shapes and on specimens taken from shapes in order to determine their physical properties and their suitability for specific uses in structures. Section 3 of the ASTM Specification A370 [ASTM, 2003] covers various tests for steel products.

2.6.1   Test for Strength, Stiffness, and Ductility

Tension Test

A tension test may be used to determine the strength, stiffness, ductility, and toughness of the material. In this test a standardized specimen called a **coupon** is gripped between the jaws of a testing machine and stretched axially (thus subjecting it to tensile forces) until fracture occurs. First, the test specimen is machined so as to have a reduced cross-sectional area in the central part of its length to localize the zone of fracture. Then the specimen is marked, in the parallel length, by two prick points known as **gage points.** The distance between these points (generally 2 in. or 8 in.) is called the **gage length.** Next, the specimen is placed in the testing machine and a tensile force T is applied. The results of such a test are recorded as stress-strain curves. In engineering it is customary to use the **nominal** or **engineering stress** and **nominal** or **engineering strain** defined as follows:

- **Stress** is defined as the load divided by the original cross-sectional area of the specimen at the start of the test, or:

$$f = \frac{T}{A_o} \tag{2.6.1}$$

  where T is the applied axial tensile load in kips (where a kip is 1000 lbs), A_o is the original cross-sectional area of the specimen in in.^{2}, and f is the axial tensile stress in ksi (kips per square inch).

- **Strain** is defined as the elongation of the specimen taken over the gage length. That is,

$$\varepsilon = \frac{e}{L_o} = \frac{L - L_o}{L_o} \tag{2.6.2}$$

  where L_o is the original gage length in inches, L is the distance in inches between gage marks after the load T is applied, e is the elongation of the specimen in inches, and ε is the axial strain in in./in.

Figure 2.6.1 shows a typical engineering stress-strain diagram, while Fig. 2.6.2 shows the initial portion of Fig. 2.6.1 at an enlarged scale and in somewhat

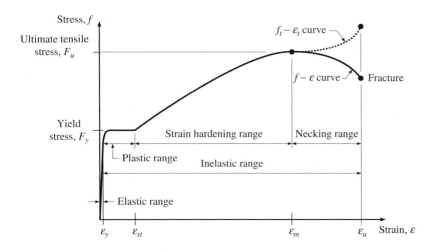

Figure 2.6.1: Typical tensile stress-strain curve for mild carbon steel.

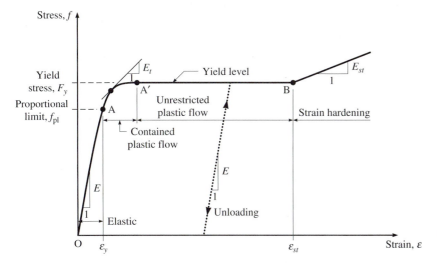

Figure 2.6.2: Initial portion of tensile stress-strain curve for mild carbon steel.

idealized form. In these figures, strain in inches per inch is plotted as the abscissa, and stress in ksi is plotted as the ordinate. Figure 2.6.1 shows the four typical ranges of behavior: the *elastic range,* the *plastic range,* the *strain-hardening range,* and the *necking range* terminating in fracture of the test specimen.

The origin O is a point on the stress-strain curve since for zero stress there is zero strain (Fig. 2.6.2). The curve is linear up to a stress level called the *proportional limit.* Thus, stress is directly proportional to the applied strain below the proportional limit, and the material is said to obey *Hooke's Law.* At a slightly higher load, the material reaches its *yield level.* At this stress level there is a marked increase in strain without any corresponding increase in stress; yielding of the material results from internal slippage at the crystalline level. The yield level can be recognized on the stress-strain diagram by a long, flat plateau called the *yield plateau.* Beyond the yield plateau the stress-strain diagram starts to rise again, indicating that additional stress is again necessary to produce additional strain—this phenomenon is known as *strain hardening.* As the specimen extends and strain hardens, the stress increases, while the cross-sectional area of the specimen simultaneously decreases; the net result is that the rate of increase of load decreases.

In the early stages of the test strain hardening predominates; but this effect decreases with increasing strain, so the stress-strain curve reaches a peak. This maximum stress on the engineering stress-strain diagram is known as the *ultimate tensile stress* (or simply, *tensile strength* or *ultimate strength*) of the material and is obtained by dividing the highest applied load in a tension test by the original cross-sectional area, A_o. After reaching the maximum stress, elongation continues with diminishing load until the specimen breaks or ruptures. The actual magnitude of the stress at the instant of fracture is not especially important as it is a highly variable parameter.

As the tension test proceeds, the actual cross-sectional area of the specimen decreases, and at high stresses this reduction in area becomes appreciable. During the strain hardening range the observer of the test notices that at some point within the body of the specimen, the specimen begins visibly to decrease in diameter (or width and/or thickness) and to noticeably stretch in length. This is called *necking,* and it progresses rapidly until the specimen suddenly breaks at the reduced section. After the rupture, the two halves of the specimen are taken from the grips, the ends are fitted closely together, and the distance L_f between the two gage points is again measured. The strain at which the specimen fails is referred to as the *fracture strain, ε_u.* Note, however, that ε_u is not the strain at which F_u is reached. The strain that occurs before the yield point is known as the *elastic strain, ε_y.* The strain that the specimen undergoes between the attainment of the yield point and the beginning of strain hardening is known as the *plastic strain,* and is typically 6 to 15 times the elastic strain.

Figure 2.6.3 shows the initial portion of typical stress-strain curves for three structural steels, namely a mild carbon steel (A36), a high-strength, low-alloy steel (A572 Grade 50), and a quenched and tempered alloy steel (A514).

The properties of steel of most importance in structural design are described as follows with respect to Figs. 2.6.1 through 2.6.3.

Elastic Limit

At the beginning of loading, when a small force is applied to the specimen, the bonds between the atoms are stretched and the specimen elongates. If such a force is removed, the bonds return to their original length, and the specimen returns to its initial size and shape. With increasing magnitude of the force, this will continue to occur as long as the material behaves elastically.

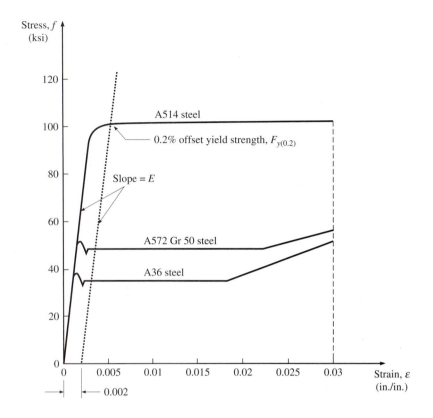

Figure 2.6.3: Initial portion of stress-strain curves for structural steels.

However, eventually a stress will be reached that causes plastic slip to occur, resulting in permanent deformation of the material. Removal of the force allows the recovery of the elastic part of the deformation, but the elongation caused by slip, called **permanent strain,** remains. The corresponding stress level is called the **elastic limit** of the material. In practice, it is difficult to distinguish between the proportional and elastic limit, and they are often considered to occur at the same point on the curve.

Modulus of Elasticity, E
The **modulus of elasticity,** or **Young's modulus,** is the slope of the stress-strain diagram in the elastic region. Thus,

$$E = \frac{f}{\varepsilon} = \frac{\text{stress}}{\text{strain}} \tag{2.6.3}$$

where f is the stress in ksi, ε is the strain in in./in., and E is Young's modulus in ksi. The modulus of elasticity is a measure of the **rigidity** or **stiffness of the material** in the elastic domain, and is related to the attractive force between adjacent atoms in solid material. For atoms of a given material, such as iron, this force has a definite value. As iron accounts for all but one percent or so of the composition of all structural steels, E hardly varies and is practically constant for all structural carbon steels. A value of 29,000 ksi is used in design calculations as the modulus of elasticity for all grades of all structural steels. This stiffness, which is much higher than that of any other common structural material, is an important advantage of steel.

Yield Strength

The stress-strain curve for heat-treated high-strength steels and several other specialty steels is a continuous, fairly smooth curve beyond the initial linear elastic portion and has no well-defined yield point or yield plateau (Fig. 2.6.3). For such materials, the *yield strength* is usually defined as the stress that leaves the material with a specified permanent set (usually 0.002 in./in. or 0.2 percent). The yield strength is established by the *offset method,* in which a line is drawn parallel to the initial tangent of the stress-strain curve, through the point on the abscissa corresponding to the specified permanent set or offset strain as shown in Fig. 2.6.3. The stress corresponding to the intersection of this line with the stress-strain curve gives the yield strength of the material, $F_{y(0.2)}$. Note that, in reporting values of yield strength obtained by this method, the specified value of offset used should always be stated in parenthesis. For these materials, the stress-strain proportionality of Hooke's Law is assumed to be applicable for stresses below yield strength. We also observe that the yield strength so defined, unlike the yield point, does not correspond to a physical property of the material; its value is merely a function of the offset specified.

Yield Stress, F_y

Two definitions of material strength have been presented so far: yield point and yield strength. To reduce the confusion that may arise due to the use of two different definitions, the AISC LRFD specification [1.1] has adopted the term *yield stress* to refer to either one (or both). The symbol F_y is used to designate the yield stress, and it is expressed in ksi. At present, structural steels are readily available with yield stresses from 32 ksi to 150 ksi.

The characteristic stresses of structural steel, such as the proportional limit, yield stress, and ultimate stress for compression tests, have nearly the same values as those for tension.

Tangent Modulus, E_t

The slope of the tangent at a point on the stress-strain curve above the proportional limit is defined as the *tangent modulus* and is designated by the symbol E_t. It represents the stiffness of the material in the inelastic domain.

Strain Hardening Modulus, E_{st}

The slope of the stress-strain curve in the strain hardening range is known as the *strain hardening modulus.* Its highest value occurs at the onset of strain hardening, and this particular value is denoted by E_{st} (Fig. 2.6.2).

The magnitude of the strain hardening modulus E_{st} varies over a much greater range than Young's modulus; the typical values of 600 ksi to 800 ksi have an average value of about 1/50 of Young's modulus for carbon structural steels. The intersection of the yield plateau with the strain hardening portion of the stress-strain curve defines the strain hardening strain, ε_{st}.

Ductility
Ductility is the ability of a material to undergo large deformation without breaking. A measure of ductility is the **percentage elongation** of the gage length of the specimen during a tension test. It is calculated as 100 times the change in gage length divided by the original gage length. Thus,

$$\text{Percentage elongation} = \delta_e = \frac{L_f - L_o}{L_o} \times 100 \qquad (2.6.4)$$

where L_f is the final distance between the gage marks after the specimen breaks, and L_o is the original gage length. A large part of the total strain occurs in the necked down portion of the gage length immediately adjacent to the fracture and within one inch of the fracture. Hence, δ_e is a function of the gage length, with shorter gage lengths resulting in higher values of δ_e. Standard steel material specifications stipulate a minimum percentage of elongation of the specified length of the test specimen (15 to 20 percent in an 8-in. gage length). The value is 18 percent for A992 steel.

As we will see in Chapters 6, 7, 12, and 13 the high ductility of steel enables many structural parts designed using simplified but not necessarily correct assumptions to perform satisfactorily due to the redistribution of stresses made possible by ductility. When ductility is reduced, through poor design details or poor fabrication practices, brittle fracture or fatigue fracture may result. Another advantage of high ductility in a material is that when the structure is overloaded its large deflections give visible evidence of impending failure. For mild carbon steels, the strain at rupture or total strain is between 150 to 200 times the elastic strain. Also, the total strain is approximately 15 times the strain at the beginning of strain hardening.

Poisson's Ratio, μ
Experiments show that if a bar is lengthened by axial tension, there is a simultaneous reduction in the transverse dimensions. For stresses below the proportional limit, the ratio of the strains in the transverse and longitudinal directions is a constant known as **Poisson's ratio.** It is denoted by μ and defined by

$$\mu = -\frac{\varepsilon_x}{\varepsilon_z} = -\frac{\varepsilon_y}{\varepsilon_z} \qquad (2.6.5)$$

where ε_z is the strain due to the applied stress in the z-direction (longitudinal direction), and ε_x and ε_y are the strains induced in the perpendicular directions. The minus sign indicates a decrease in transverse dimensions when ε_z is positive, as in the case of a tension specimen. For steel, Poisson's ratio is approximately 0.3 in the elastic range and 0.5 in the plastic range.

Shear Modulus of Elasticity, G

The *shear modulus of elasticity* is the ratio of shear stress to shear strain within the elastic range and is designated as G. For structural steels, measured values for G vary from 11,500 to 12,000 ksi. A conservative value of 11,200 ksi is used in design calculations as the shear modulus for all structural steels. From the theory of elasticity, we have:

$$G = \frac{E}{2(1 + \mu)} \tag{2.6.6}$$

Effect of Elevated Temperatures

The yield stress, ultimate tensile stress, and modulus of elasticity of all structural steels decrease with increasing temperature, as obtained from short-time elevated temperature tensile tests. Thus, the ratio of elevated temperature yield stress to room temperature yield stress, for carbon and high-strength low-alloy steels, is approximately 0.77 at 800°F, 0.63 at 1000°F, and 0.37 at 1200°F. The modulus of elasticity of structural steel, approximately 29,000 ksi at 70°F, decreases linearly to about 25,000 ksi at 900°F, and then begins to drop at an increasing rate at higher temperatures. The average coefficient of expansion between 70° and 100°F, for all structural steels, is 0.0000065 in./in. for each degree Fahrenheit. For temperatures of 100° to 1200°F, the coefficient is given by the linear formula:

$$\alpha = [6.1 + 0.0019T] \times 10^{-6} \tag{2.6.7}$$

in which

$$\alpha = \text{coefficient of expansion (in./in.)}$$

$$T = \text{temperature in °F}$$

2.6.2   Residual Stresses

Residual or *locked-in stresses* are those that exist in a steel member prior to the application of any external load. They are associated with plastic deformation that occurs during the manufacturing process. For example, these stresses may be due to uneven cooling to room temperature of shapes after hot-rolling or welding; they may be due to operations such as cold straightening by rotarizing or gagging; or they may be due to fabrication operations such as flame-cutting, cold-bending, and so on. *Rotarizing* consists of straightening a crooked member by passing it through a series of rollers at room temperature. Larger shapes are usually straightened by *gagging,* a process which consists of loading a member with hydraulic jacks at certain points along its length. Residual stress patterns due to hot-rolling and welding for several shapes are shown schematically in Fig. 2.6.4.

During the process of cooling from rolling temperatures to room temperature, some regions in the cross sections of rolled shapes cool more rapidly than others. For example, the flange tips of a wide flange shape, exposed to

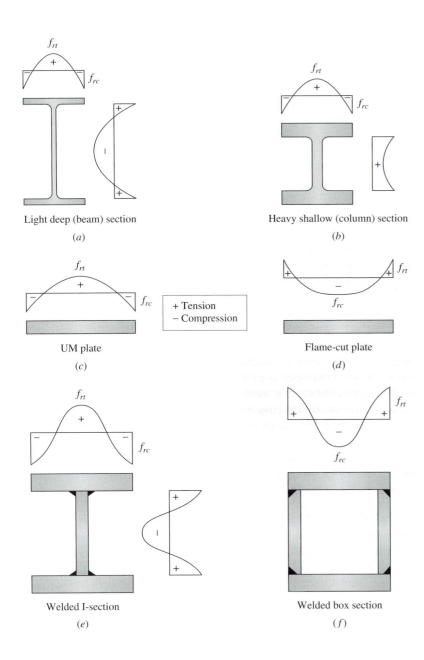

Light deep (beam) section

(a)

Heavy shallow (column) section

(b)

UM plate

(c)

+ Tension
− Compression

Flame-cut plate

(d)

Welded I-section

(e)

Welded box section

(f)

Figure 2.6.4: Residual stress patterns.

the air on three sides, cool more rapidly than the material at the juncture of
the flange and web where the flange metal is exposed to the air only on the
outer face of the flange. The flange tips and the middle of the web contract
freely when they cool since the other regions of the cross section have yet to
cool (and acquire axial stiffness). As the slower cooling portions near the

flange-to-web junction finally begin to cool and attempt to contract, they are not completely free to do so as they are attached to the flange tips and the middle of the web, which have already cooled and acquired axial stiffness. This induces compressive residual stresses on the parts which are already cold. Thus when cooling is complete, the flange tips (and, for thin-webbed shapes, the web centers) have compressive residual stresses, whereas the flange-web junctions are in residual tension (Figs. 2.6.4a and b). Cross-sectional geometry (flange thickness and width, web thickness and depth) influences the cooling rate and hence the magnitude and distribution of residual stresses. The central region of the web of W-shapes with relatively thin webs has residual compression (Fig. 2.6.4a). On the other hand, W-shapes with short thick webs, such as heavy column sections, often have residual tensile stresses throughout the web, with the maximum value occurring at the junction of the flange and web (Fig. 2.6.4b). For rolled I-shapes, the maximum residual compressive stresses that occur at the flange tips are of the order of 10 to 15 ksi, although values as high as 20 ksi have been measured. Residual stresses are essentially independent of the yield stress of the material. For jumbo shapes, the residual stresses vary throughout the thicknesses of the flanges and web.

Residual stresses also develop in welded built-up sections as a result of the localized heat applied during the welding operation and the resultant plastic deformations (Figs. 2.6.4e and f). The stress distribution in a welded cross section is markedly different from that induced in an analogous rolled shape due to cooling. Welded I-shapes generally have high residual compressive stresses at the flange tips, and in the vicinity of the weld there is a tensile residual stress that may approach the yield stress of the material irrespective of the type of steel. On the other hand, a welded box shape develops tensile residual stresses in the corner regions near the welds and compressive stresses in the centerline regions of the plates.

Let us now consider an unloaded member of rolled steel I-section. Let f_r be the residual stress acting on a small elemental area, dA, located at a point (x, y) of the member cross section. As there are no external forces acting on the member, static equilibrium requires that the residual stress distribution should have zero resultant force and zero resultant moment about the principal axes (x and y), at any cross section along the length direction of the member. That is,

$$\int_A f_r \, dA = 0; \quad \int_A f_r \, y \, dA = 0; \quad \int_A f_r \, x \, dA = 0 \qquad (2.6.8)$$

Residual stresses are of particular importance in column design and in the design of plates in compression. Note that the residual stress distribution is significantly different in UM plates versus flame-cut plates (Figs. 2.6.4c and d). Thus, welded built-up wide flange shapes with UM plates are not as strong in compression as those that utilize sheared plates.

2.6.3   Mill Test Reports

Product evaluation occurs continuously in the steel industry. Quality control personnel continually monitor every phase of iron and steel making, aided by computers. One production lot of steel from a steel making unit is known as a *heat (batch) of steel.* Most heats of steel produced in the United States comprise from 50 to 300 tons of steel. A chemical analysis, known as the *heat analysis* (or *ladle analysis*), determines the chemistry of the steel using samples taken from the molten metal. This analysis is normally the controlling analysis for satisfying the specifications. A *check analysis* made using drillings taken from the semifinished or finished product further checks the chemistry of the steel. The check analysis measures certain specified variations from the ladle analysis. After rolling the steel through various stages to shape it to the specified dimensions, a hot saw cuts a section to be used for the physical tests. Coupons taken from this section are tested to determine yield stress, ultimate tensile stress, percent elongation, and other specification requirements.

Steel, as shipped from the mill, is identified by the manufacturer's name or symbol, the heat number, and, when a specified yield point exceeds 36 ksi, the specification number. The manufacturer furnishes to the fabricator a *mill test report* that shows the results of the chemical and mechanical tests and identifies the grade of each piece of steel. The mill test report also lists the customer's order number, number and dimensions of pieces shipped, and the results of any mechanical tests, such as a tension test [see ASTM, 2003; Davis, 1998]. Section A3.1*a* of the LRFD specification states that certified mill test reports made by the fabricator or a testing laboratory in accordance with ASTM A6 or A568, as applicable, shall constitute sufficient evidence of conformity with one of the acceptable ASTM grades of steel.

The mechanical properties reported on the mill test certificate normally exceed the specified properties by a significant amount and merely certify that the steel meets prescribed steel-making specifications. It should be noted, however, that designers should always use the minimum specified properties in the standard steel design specifications, and not the test values reported on the mill test certificates.

2.7   Economics

2.7.1   Advantages of Steel

Steel as a structural material has several desirable qualities such as high strength, high stiffness (resistance to deformation), and high ductility. It is the strongest, most versatile, and economical material available to the construction industry, and its high ductility enables it to withstand large deformations at high stress levels without rupture. Steel is a factory-produced product, manufactured under a stringent factory-based quality control discipline. Unlike other structural materials, it is uniform in strength, dimensionally stable, and its durability is unaffected by alternate freezing and thawing. Unlike concrete, steel has essentially the same compressive and tensile properties. Also, in contrast to concrete, steel properties do not change with time. Steel can be either alloyed or alloyed and heat treated to obtain toughness, ductility and great strength, as the service demands.

Steel is produced in a wide range of shapes, sizes, and grades which provide maximum flexibility in design. Steel members can be fastened together by a variety of rather simple connection devices, such as bolts, pins, and welds. Steel has high strength/unit weight (compared to concrete, wood, masonry, etc.) and provides more strength per dollar than any other structural material. Weights of steel structures will therefore be lower. This fact is important for tall buildings and long span bridges in which dead load is a major contributor to design load, and for structures having poor foundation conditions. Failure or collapse of steel structures is generally preceded by large visible deflections.

Steel offers a number of site benefits too: its comparative low weight can reduce foundation costs; its ease of assembly minimizes the need for on-site supervision. Once erected and braced the steel frame is fully load bearing, giving immediate access to following trades. Steel frames go up rapidly, thus reducing construction financing costs and allowing the building to generate revenue sooner. This is an important factor, particularly during periods of high interest rates.

The design characteristics of structural steelwork include long span capability with minimal column sizes, as well as the flexibility to permit future alterations or extensions to the structure to be carried out. In general, steel structures can be repaired quickly and easily, and can be dismantled and reassembled easily at a different location. Finally, steel has high scrap value and is recyclable.

Despite its advantages, steel is susceptible to corrosion by water and other chemicals. Also, it has greatly reduced strength and stiffness when subjected to the high temperatures of a fire; this property often necessitates fire protection of steel elements. Under certain circumstances steel may fail by brittle fracture rather than in its normal ductile mode. ***Brittle fracture*** takes place with little or no plastic deformation. Because of the absence of plastic deformations, it occurs with little or no prior warning. In addition, the fracture proceeds at very high speed once started, often leading to catastrophic failure of the structure. Brittle fracture is influenced by parameters like low temperature, tensile stress level, and joint restraint in the region around the failure initiation spot [Barsom, 1987]. In general, the strength of steel may be reduced if it is subjected to a large number of stress reversals.

2.7.2   Cost of Steel

When ordered in large quantities, steel can be purchased at the mill at what is known as the ***mill base price.*** Such base prices are published in the *Engineering News Record* (http://www.enr.com/ published by the McGraw-Hill Companies) for various products such as bars, shapes, plates, and sheets. Steel mills often require a minimum order of five tons per section size for rolling. The price of an alloy steel is determined by an ***alloy extra*** that is added to the base price.

When steel is purchased from a steel warehouse, the ***warehouse base price*** for such materials applies. This price is higher than the mill base price because of extra handling charges, transportation, and the warehouse's profit. In addition, various other factors add to the price of steel. For example, the base price applies only to sections of ordinary size; very small or very large sections cost an additional amount known as a ***size extra***. A ***quantity extra*** is added to the price when the amount ordered is less than a specified minimum amount. A ***cutting extra*** is added when stock lengths are not ordered. Other extras, such as machine straightening, annealing, and truck delivery may further increase the per pound price.

2.7.3   Use of High Strength Steels

The strength of structural steels has been continuously rising over the past century with advances in production technology. However, there still is much room for improvement, as the theoretical tensile strength of iron, calculated from the interatomic forces, is found to be about $E/6$, or about 5000 ksi, which is more than 17 times that of the strongest structural steel. Tiny pieces of steel called *whiskers,* with an ultimate tensile stress of 1000 ksi, have been produced in laboratory environments. Flaws at the grain boundaries and fine cracks account for the much lower strength of commercially produced steels.

Steels with 50 ksi yield stress are now widely being used in building construction, replacing A36 steel in many applications. The 50 ksi steels available include the recently introduced A992 steel for W- and WT-shapes, A572 and A913 high-strength low-alloy structural steel, A242 and A588 corrosion resistant high-strength low-alloy structural steels, and A529 high-strength carbon-manganese structural steel. Yield stresses above 50 ksi can be obtained from three grades of A572 steel and three grades of A913 steel as well as A852 and A514 quenched and tempered structural steel plate. With only a slight increase in cost per pound, these higher strength steels may economically reduce the sizes and weights of members. Whenever this weight reduction can be translated into savings in the cost of foundations, supporting structures, or in handling and transportation, these higher strength steels can and should be used. Also, in bridges with relatively long spans, reduction in the weight of the structure (the dead load, which is the dominant design load for such structures) by the use of higher strength steels results in substantial economy of material. Also, whenever the need for built-up sections of carbon steel can be avoided through the use of rolled, higher strength steel sections, a savings in fabricating costs can be realized. A good example is in the heavily loaded, lower tier columns of multistory buildings. However, when sizes of members are controlled by member instability (columns), or plate instability (elements of columns and beams), or by limiting deflections (beams), rather than by stress, no savings can be achieved by the use of higher strength steels. The influence of these parameters (instability and deflections) on material and member selection will be considered in detail in Chapters 8 to 11.

Quenched and tempered steels are generally selected on the basis of special requirements for strength, weight reduction, and superior mechanical properties. Normally, connection material (like gussets, stiffeners, etc.) is specified as A36 steel, unless calculations require material of higher strength. For additional information on the proper selection of structural steels refer to Bjorhovde et al. [2001].

2.7.4   Availability of Shapes and Plates in Various Steels

Hot-rolling of steel changes the grain size from coarse to fine. Also, thin material cools more uniformly than does thick material. For these reasons, a thinner material has a better microstructure and develops higher yield stress and ultimate tensile stress than a thicker material of the same steel. Thus, for some steels (A242 steel, for example), heavier shapes and thicker plates are only available in lower grades (lower yield stress and lower ultimate tensile stress). A36 steel, however, has the same yield stress for all thicknesses up to 8 in. To obtain this result, the chemical composition for A36 steels is varied slightly for shapes and plates and for thin and thick plates. Thicker plates contain more carbon and manganese to raise the yield stress. Such adjustment cannot be done for high-strength steels because of the adverse effect of such increase on ductility and weldability.

As mentioned earlier, the mechanical properties of steel, and in particular F_y and F_u, are a function of the thickness of the material. For this reason ASTM A6 has divided the rolled shapes according to web thickness (leg thickness in the case of angles and stem thickness for tees) into Groups 1 through 5. A listing of shape sizes that fall in each of the five groups is given in *LRFDM* as Table 2-4. Note that the thinnest web sections are in Group 1, and that the heavier the shape, the higher the group number. The heavy W-shapes in Groups 4 and 5 (and WT-shapes cut from them) are often called *jumbo sections.*

Section A3.1a of the AISC LRFDS specification [1.1] lists 18 ASTM specifications for structural steel approved for use in building construction. Twelve of these, namely, carbon steels A36, A53, A500, A501, and A529; HSLA steels A572, A618, A913, and A992; and corrosion resistant HSLA steels A242, A588, and A847 are available in structural shapes. Of these steels five, namely, A36, A529, A572, A242, and A588 are also available in structural plates and bars. Two steels, namely, quenched and tempered low-alloy steel, A852, and quenched and tempered alloy steel, A514, are available only in plates. LRFDM Table 2-1 shows the availability of shapes in different steels, while LRFDM Table 2-2 shows the 10 ranges of thickness of plates and bars available in the various F_y and F_u levels offered by the seven steels.

Material Strength

EXAMPLE 2.7.1

Determine F_y and F_u for a W24×104 of ASTM A242 steel.

Solution

First enter LRFDM Table 2-4, and observe that a W24×104 shape belongs to Group 2. In LRFDM Table 2-1, all the steels available for W-shapes are represented by shaded areas under the W-shape heading.

For A242 steel, the only alloy available for Group 2 shapes is of Grade 50. Also from this table, observe that A242 steel of Grade 50 has a yield stress F_y of 50 ksi and an ultimate tensile stress F_u of 70 ksi.   (Ans.)

2.8   Built-Up Sections

The LRFD Manual tabulates properties for 318 I-shapes, 65 C-shapes, 126 L-shapes, and 302 T-shapes currently available in the market. Rolled steel shapes presently available in the United States can get quite large and heavy, and offer tremendous load-carrying capabilities. For example, the heaviest column shape available, namely, a W14×808 of Grade 50 steel, has an axial compressive strength of about 9330 kips corresponding to an effective length of 13 ft. Similarly, the heaviest beam shape available, a W36×798 of Grade 50 steel has a factored bending strength of 13,400 ft-kips. Thus, in most cases, one of the standard shapes will satisfy the design requirements. If, however, the requirements for strength or stiffness cannot be satisfied by an available section, a ***built-up shape*** or ***compound section*** formed by welding (or, less frequently these days, by bolting) together several plates and/or rolled shapes may be needed to arrive at the desired cross section.

Often, built-up sections are used for beams, when the required bending strength is greater than that of the largest available rolled section. The rolled section may be modified by welding plate material called a ***cover plate*** to the top and/or bottom flanges (Figure 2.8.1a). This increases the bending capacity while maintaining the shear capacity of the original rolled section. This is also an effective way of strengthening an existing structure that is being rehabilitated or modified for a use other than the one it was originally designed for. Built-up sections may also be I-shaped sections, obtained by welding two flange plates to a web plate (Fig. 2.8.1b), or may be box sections, obtained by welding two flange plates and two web plates. Such sections are extensively used in mass produced prefabricated metal buildings.

Although the range of rolled steel shapes available is quite large, for columns in lower stories of buildings in excess of 30 stories or so, and for

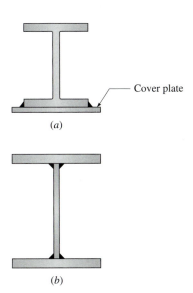

(a)   Cover plate

(b)

Figure 2.8.1:   Built-up sections used as beams.

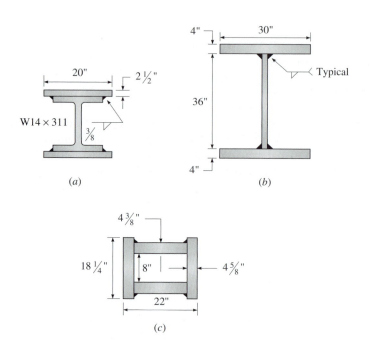

Figure 2.8.2:   Examples of built-up sections used as columns.

compression members in special structures such as convention halls, and stadiums, it becomes necessary to use welded built-up shapes. Figure 2.8.2
gives a brief description of some specially built-up sections used for columns
in large buildings:

- Cover-plated columns are used when the area required is in excess of
  that available in a single rolled section, or when there is an advantage
  in using the same core section in several tiers of a column and
  reinforcing it with plates where added area is necessary (Fig. 2.8.2a).
- The exterior columns for a 54-story office building, One Liberty Plaza
  in New York City, are as shown in Fig. 2.8.2b. Columns of the
  Equitable Life Insurance Building in San Francisco are also similar in
  shape to the section shown in Fig. 2.8.2b. At the base level, the exterior
  columns have two 18×3 in. flange plates welded to a $42 \times 1\frac{1}{2}$ in. web
  plate. Columns of the Detroit Bank and Trust Building in Detroit
  consist of two $24 \times 3\frac{1}{2}$ in. flange plates welded to a 16-in. web plate
  using $\frac{1}{2}$-in. fillet welds.
- For large column sections, four plates can be welded together to form a
  box section. The box section columns for the First Federal Savings and
  Loan Company Building in Detroit were built-up from two $18\frac{1}{4} \times 4\frac{5}{8}$ in.
  flange plates and two $12\frac{3}{4} \times 4\frac{3}{8}$ in. web plates; the latter were recessed
  slightly as shown in Fig. 2.8.2c to permit continuous fillet welds.

The student should realize that fabrication of such special column sections as shown in Fig. 2.8.2 demands low-cost, high production assembly and welding techniques. Also, if it becomes necessary to design a built-up section from rolled steel plates and/or shapes, the cost of fabrication should be included in cost comparisons.

The LRFD Manual lists the section properties for certain often used built-up sections, namely, double angles (LRFDM Table 1-14), double channels (LRFDM Tables 1-15 and 1-16), and I-shapes with cap channels (LRFDM Tables 1-17 and 1-18). For other built-up sections these properties have to be determined. Most readers should be familiar with the method of determining section properties such as moment of inertia, I, section modulus, S, and radius of gyration, r, of a built-up section. However, the necessary procedure is reviewed as follows.

Section Properties
Consider a built-up section of total area A made of n component parts. (A built-up section consisting of two elements, an I-shape, and a channel is shown in Fig. 2.8.3.) The component parts should be selected such that their own section properties can be obtained from tables in the LRFD Manual (I-, C-, T-, L-shapes, and others) or can be easily calculated (rectangles, semicircles, and

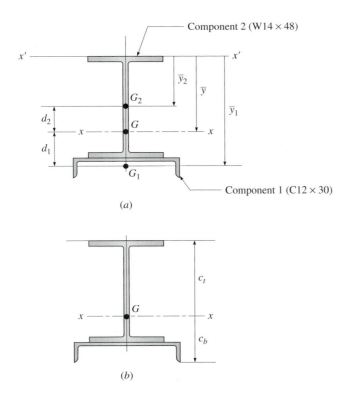

(a)

(b)

Figure 2.8.3:   Determination of section properties for a built-up section.

others). Let G be the center of gravity of the built-up section and x-x an axis through G, about which the moment of inertia is to be calculated. The first step in evaluating I for an area is to locate the centroid of the area, G. If the section has an axis of symmetry, the center of gravity G lies on this axis. If the section has two axes of symmetry, then G coincides with the point of intersection of these two axes. Otherwise, to locate the position of G, we select an arbitrary reference axis x'-x' parallel to the axis x-x and lying in the plane of the section. It is sometimes advantageous to make the axis x'-x' coincide with the face of a component part, if possible, or make it pass through the center of gravity of a component part.

Designate the areas of the component parts as $A_1, A_2, \ldots A_i, \ldots A_n$. Let $\bar{y}_i$ be the distance from the reference axis x'-x' to the center of gravity G_i of the component i. Multiply each component area A_i by the distance $\bar{y}_i$. This is the first moment (or static moment) of the component i. Total the moments of areas of the components and divide this sum by the area of the whole section. The result $\bar{y}$ is the distance of the center of gravity G of the built-up section from the reference axis x'-x'. Thus,

$$A = \sum_{i=1}^{n} A_i \tag{2.8.1}$$

$$\bar{y} = \frac{\sum_{i=1}^{n} A_i \bar{y}_i}{A} \tag{2.8.2}$$

where n = number of component parts
A_i = area of component i
$\bar{y}_i$ = distance of G_i from x'-x'
$\bar{y}$ = distance of G from x'-x'

The required moment of inertia about the centroidal axis x-x is calculated next. For this, determine the distance d_i $(= \bar{y}_i - \bar{y})$ from the axis x-x to the center of gravity G_i of each component. Designate these as $d_1, d_2, \ldots d_i, \ldots d_n$. Determine (or get from tables in the Manual) the moments of inertia of each component about its own center of gravity, *on an axis parallel to the x-x axis*. Designate these as $I_{o1}, I_{o2}, \ldots I_{oi}, \ldots I_{on}$. Using the **parallel axis theorem,** which states that the moment of inertia of an area about any axis is equal to the moment of inertia about a parallel axis passing through the centroid of the area, plus the area multiplied by the square of the distance between these two axes, the moment of inertia of the built-up section about axis x-x is obtained as:

$$I_x = \sum_{i=1}^{n} \left[I_{oi} + A_i d_i^2 \right]_x \tag{2.8.3}$$

where d_i = distance of G_i from x-x = $\bar{y}_i - \bar{y}$
I_{oi} = moment of inertia of component i about its own center of gravity *on an axis parallel to the x-axis*

The radius of gyration r_x is next obtained from the relation

$$r_x = \sqrt{\frac{I_x}{A}} \tag{2.8.4}$$

Also,

$$S_{xt} = \frac{I_x}{c_t}; \quad S_{xb} = \frac{I_x}{c_b} \tag{2.8.5}$$

where S_{xt} and S_{xb} are the section moduli, and c_t and c_b are the distances from G to the extreme top and bottom fibers, respectively (Fig. 2.8.3b).

Note that the moment of inertia of a rectangle of width b and height h, with respect to a centroidal axis parallel to the width is given by:

$$I_o = \frac{1}{12} bh^3 \tag{2.8.6}$$

Section Properties

EXAMPLE 2.8.1

A built-up beam section is formed by welding a C12×30 channel section to the flange of a W14×48 wide flange section to form a singly symmetric section shown in Fig. 2.8.3. Calculate the moments of inertia, radii of gyration, and section moduli. Also, calculate the weight of the beam per ft length.

Solution

The section is composed of two components ($n = 2$). Component 1 is a channel section C12×30 rotated through 90°. Component 2 is a W14×48. We have,

For a C12×30 channel, from Table 1-5 of the LRFDM:

$$A_C = 8.81 \text{ in.}^2; \quad b_{fC} = 3.17 \text{ in.}; \quad \bar{x}_C = 0.674 \text{ in.}$$

$$I_{xC} = 162 \text{ in.}^4; \quad I_{yC} = 5.12 \text{ in.}^4; \quad d_C = 12.0 \text{ in.}$$

For a W14×48, from Table 1-1 of the LRFDM:

$$A_W = 14.1 \text{ in.}^2; \quad d_W = 13.8 \text{ in.}$$

$$I_{xW} = 484 \text{ in.}^4; \quad I_{yW} = 51.4 \text{ in.}^4$$

a. About xx-axis

The centers of gravity G_1, G_2, and G all lie on the vertical axis of symmetry. To locate G, take an arbitrary axis x'-x' coinciding with the top fibers of the top flange of the W-shape as reference axis. With the help of Fig. 2.8.3a,

$$A_1 = 8.81 \text{ in.}^2; \quad \bar{y}_1 = 13.8 + 0.674 = 14.474 \text{ in.}$$

$$A_2 = 14.1 \text{ in.}^2; \quad \bar{y}_2 = \frac{13.8}{2} = 6.90 \text{ in.}$$

[continues on next page]

Example 2.8.1 continues ...

$$A = A_1 + A_2 = 8.81 + 14.1 = 22.9 \text{ in.}^2$$

$$\bar{y} = \frac{A_1\bar{y}_1 + A_2\bar{y}_2}{A_1 + A_2} = \frac{8.81 \times 14.474 + 14.1 \times 6.9}{22.9} = 9.81 \text{ in.}$$

Next, the *d* distances of the components from the centroid of the built-up section are determined as:

$$d_1 = \bar{y}_1 - \bar{y} = 14.474 - 9.81 = 4.66 \text{ in.}$$

$$d_2 = \bar{y}_2 - \bar{y} = 6.9 - 9.81 = -2.91 \text{ in.}$$

The moment of inertia about the *x-x* axis is determined, using the parallel axis theorem and noting that the channel is turned 90°, as:

$$I_x = [I_{yC} + A_1 d_1^2] + [I_{xW} + A_2 d_2^2]$$

$$= [5.12 + 8.81 \times 4.66^2] + [484 + 14.1$$

$$\times (-2.913)^2] = 800 \text{ in.}^4$$

$$r_x = \sqrt{\frac{I_x}{A}} = \sqrt{\frac{800}{22.9}} = 5.91 \text{ in.} \qquad \text{(Ans.)}$$

For bending about the *x-x* axis, the extreme fiber distances are:

$$c_b = 13.8 + 3.17 - 9.81 = 7.16 \text{ in.}; \qquad c_t = 9.81 \text{ in.}$$

The section moduli are:

$$S_{xb} = \frac{I_x}{c_b} = \frac{800}{7.16} = 112 \text{ in.}^3; \quad S_{xt} = \frac{I_x}{c_t} = \frac{800}{9.81} = 81.5 \text{ in.}^3$$

$$\text{(Ans.)}$$

b. About *y-y* axis

As G_1, G_2, and *G* all lie on the same vertical line ($d_1 = d_2 = 0$), the weak axis moment of inertia is simply given by:

$$I_y = [I_{xC} + A_1 d_1^2] + [I_{yW} + A_2 d_2^2] = 162 + 51.4 = 213 \text{ in.}^4$$

$$\text{(Ans.)}$$

$$r_y = \sqrt{\frac{213}{22.9}} = 3.05 \text{ in.} \qquad \text{(Ans.)}$$

For bending about the *y-y*-axis:

$$\text{Extreme fiber distance} = 6.0 \text{ in.}$$

$$\text{Section modulus, } S_y = \frac{I_y}{c} = \frac{213}{6.0} = 35.5 \text{ in.}^3 \qquad \text{(Ans.)}$$

c. Weight

Weight of the built-up section = 48 plf of W-shape + 30 plf of C-shape = 78 plf.

Alternatively, the weight of the built-up section can be calculated from its cross-sectional area. From LRFDM Table 1-20, a $1'' \times 1''$ square bar weighs 3.4 plf. So the weight of the built-up section, with $A = 22.91$ in.^{2}, equals $22.91 \times 3.4 = 78.0$ plf.   (Ans.)

2.9   Corrosion, Painting, and Galvanizing

2.9.1   Corrosion

Corrosion may be defined as the destruction of a metal by chemical or electrochemical reaction with its environment. The amount of corrosion depends on the chemical composition of the metal and the environment. When sections are rolled and stocked, the iron on the steel surface may react with oxygen in the atmosphere, forming an oxide skin called *mill scale.* Mill scale is very hard but also brittle and liable to crack. In normal moist atmospheric conditions, a chemical reaction takes place in which the iron ion Fe^{+++} and the hydroxyl group $3(OH)^-$ combine to form *hydrated ferric oxide* $Fe(OH)_3$ or *rust.* It will only form if both oxygen and moisture are present. However, in practice the amount of moisture required may be rather small. The rusting takes place under the mill scale and, as fully corroded steel occupies seven times the volume of uncorroded steel, causes the mill scale to flake off. The process continues, causing pitting of the metal's surface and a resulting loss of cross-sectional area, strength, and stiffness. Exposure to corrosive environments such as industrial applications, deicing salts, proximity to ocean, among others, may require the use of special corrosion protection systems. It should be noted that weathering steels are not recommended for applications in such corrosive environments.

There are many ways to prevent or control corrosion of steel structures. Increased corrosion resistance may be obtained by the addition of alloying elements to the steel. Thus, we have copper steels (with corrosion resistance twice that of carbon steels), high-strength low-alloy steels (with corrosion resistance four times that of carbon steels), and stainless steels. In general, the superior corrosion resisting properties are accompanied by the added advantage of greater strength with only a moderate cost premium.

Painting and galvanizing are the two methods uilized for protecting steel against corrosion. Although such protection systems are expensive, their extra cost may be insignificant compared with the eventual expense of the frequent maintenance necessary with an initially cheaper but less durable system.

2.9.2   Painting

The function of *painting* is to interpose essentially neutral layers of substances between the steel surface and the corroding media. Its main limitations include the degradation (cracking and wear) of the protective film and the need for repainting at intervals. Before painting, the steelwork must be thoroughly cleaned of all loose mill scale, loose rust and other foreign matter. The cleaning may be done by hand or power-driven wire brushes; by flame descaling; by sand, shot, or grit blasting; or by pickling. Shop painting the elements of a steel structure prior to erection is usually much cheaper than field painting the assembled structure, especially when access towers and scaffolding become necessary. Moreover, in the field the steelwork may eventually be hidden by finishes or be otherwise inaccessible for maintenance.

The major ingredients of wet paint include binders, pigments, and solvents. The ***binders*** make up the liquid portion of the surface coating that bonds the paint elements to each other and to the surface. Binders may be made from naturally occurring materials such as linseed oil or from air-drying resins such as alkyds, vinyls, epoxies, or acrylics. Some binders such as the hard resins contribute hardness to the paint, whereas others such as soft drying oils contribute flexibility, durability and adhesion. Often several binders are present in a paint formulation. ***Pigments*** are finely divided particles of solid material added to the binder to make it opaque and give color, cover, consistency, water resistance, and durability. The materials used include metals (zinc and aluminum), metallic oxides (zinc, iron, chromium, and titanium), chromates (lead and zinc), and other materials such as carbon black. The zinc pigments additionally provide cathodic protection at breaks or pin holes where the steel is exposed. As many binders are highly viscous, ***solvents,*** also called ***thinners,*** are added to liquify and dilute the paint and make it more workable. The mix is designed so that as the paint dries, the solvent evaporates freely, leaving a dry film of binder and pigment. Principal solvents consist of hydrocarbons, such as aromatics, napthenes, and paraffins, or oxygenated solvents, such as alcohol and acetone.

Paints are normally applied in multiple-coat systems starting with primers, followed by under coat(s), and then by a finishing coat. Each coat performs a different function within the system. The ***primer,*** as the name indicates, is the first coat to be applied to the bare metal. It is important that the prime coat thoroughly covers the whole metal surface quickly and easily. Primer is applied in the shop and is intended to protect the steel for only a short period of exposure till the field coats are applied. The ***finishing coat*** is the first line of defense against corrosion. It is fully exposed to the deleterious effects of ultraviolet radiation, rain, frost, and atmospheric pollution. This coat must provide a tough, water repellent surface. ***Intermediate coats,*** if necessary, are applied to build the coating thickness and increase resistance to moisture penetration. Each coat is applied to a controlled wet film thickness (t_w) which becomes the dry film thickness (t_d) when the coat dries.

In building structures, steel need not be primed or painted if it will be enclosed by a building finish, coated with a contact type fireproofing, or in contact with concrete. When enclosed, the steel is trapped in a controlled environment, and the products required for corrosion are quickly exhausted. In effect, LRFD Specification Section M3.1 states that "shop paint is not required unless specified by the contract documents."

2.9.3   Galvanizing

Galvanizing is the coating of steelwork with molten zinc. The zinc forms a metallurgical bond with the bare metal, and the resulting protection system can give a long life before maintenance is required. The surface preparation for galvanizing requires *pickling* the steel in dilute hydrochloric or sulfuric acid and then applying a flux of ammonium chloride which absorbs any remaining impurities and keeps the metal clean until it is immersed into the bath of virtually pure molten zinc. The zinc reacts with the steel to form a hard zinc iron alloy layer and, as the steel is being withdrawn from the bath, a layer of soft, pure zinc skin forms. The alloy layer provides abrasion resistance, while the pure zinc layer acts as a cushion against impact damage. The thickness of the coating depends on the surface roughness, the chemical composition of the steel, the temperature of the molten zinc, and the time of immersion. Usual specified zinc coatings vary from 1 to 2 oz per square ft.

The period over which protection is afforded by zinc coatings is a linear function of the thickness of the coating. It is independent of the method by which the zinc is applied provided the coating is uniform. In a nonindustrial atmosphere the corrosion rate of zinc is about 1/15th to 1/20th that of steel, and a life of 20 to 25 years may be expected from coatings of 75 to 100 microns in thickness. In industrial environments the corrosion rate generally varies between 1/5th to 1/10th that of steel depending on the sulphur content of the atmosphere, and, except in extreme conditions, lives of five to six years may be expected.

References

2.1   AISC [2001]: *Manual of Steel Construction: Load and Resistance Factor Design,* American Institute of Steel Construction, 3rd edition. (Also referred to as LRFDM in this book.)

2.2   AISI [2001]: *Specification for the Design of Cold-Formed Steel Structural Members,* American Iron and Steel Institute, Washington, DC.

2.3   AISI [2002]: *Cold-Formed Steel Design Manual,* American Iron and Steel Institute, Washington, DC.

2.4   ASTM [2003]: *2003 Annual Book of ASTM Standards,* American Society for Testing and Materials, Philadelphia, PA.

2.5   Barsom, J. M. [1987]: "Material Considerations in Structural Steel Design," *AISC Engineering Journal,* Chicago, vol. 24, no. 3, pp. 127–139.

2.6   Bjorhovde, R., Engstrom, M. F., Griffis, L. G., Kloiber, L. A., and Malley, J. O. [2001]: *Structural Steel Selection Considerations: A Guide for Students, Educators, Designers, and Builders,* ASCE, Reston, VA.

2.7   Davis, J. R., Editor [1998]: *Metals Handbook, Desk Edition,* American Society for Metals, OH.

PROBLEMS

P2.1. A W24×192 of A242 steel is to be used as a beam in a building structure. What are the values of the yield stress, F_y, ultimate tensile stress, F_u, and modulus of elasticity, E to be used in the design of this beam?

P2.2. A PL3×8 of A514 steel is to be used as a tension member in a truss. What are the values of the yield stress, F_y, and ultimate tensile stress, F_u, to be used in the design of this member?

P2.3. A chandelier weighing 2 kips hangs from the dome of a theater. The rod from which it hangs is 20 ft long and has a diameter of $\frac{1}{2}$ in. Calculate the stress and strain in the rod and its elongation. Neglect the weight of the rod itself in the calculations.

P2.4. The rod in Problem P2.3 is of A36 steel having a stress-strain diagram (see Figs. 2.6.1 and 2.6.2) with $\varepsilon_{st} = 0.012$ and $\varepsilon_u = 0.18$. Determine the load that causes the rod to yield and the load that causes the rod to fracture. Also, determine the elongation of the bar corresponding to strains of ε_y, ε_{st}, and ε_u. Comment on your results.

P2.5 to P2.7. Locate the principal axes for the beam sections given in Figs. P2.5 through P2.7. Also, calculate the cross-sectional area, weight per lineal foot, moment of inertia, and section moduli about the x and y axes.

P2.8 to P2.11. Locate the principal axes for the column sections given in Figs P2.8 through P2.11. Also, calculate the cross-sectional area, weight per lineal foot, moment of inertia, and radius of gyration about the x and y axes.

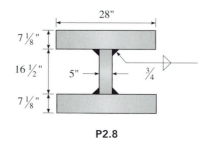

P2.8

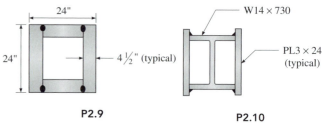

P2.9

P2.10

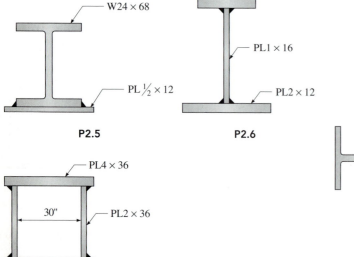

P2.5

P2.6

P2.7

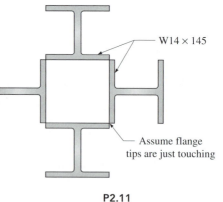

P2.11

Ford research facility, Dearborn, MI

This Six-story, 232,000 sq. ft. research facility used 1576 tons of structural steel. Moment resistant frames with fully restrained beam-to-column connections were used.

Photo courtesy Douglass Steel Fabricating Corp., Lansing MI, www.douglassteel.com

CHAPTER

Structures

3

3.1 Introduction

Familiarity with basic structural members, the ways in which they may be arranged to form the skeleton for a structure, and their functions in the completed structure is a prerequisite to good design. The aim of this chapter is to describe several fundamental structural forms and the ways in which they are often put in place in common types of building structures by means of sketches. Thus, behavior of these structural forms will be briefly discussed in Section 3.2 to provide information that will assist the novice engineer in the preliminary selection, analysis, and design of steel members and structures. The designer must also have a clear concept of the way in which the transfer of vertical and horizontal loads acting on the structure, from the points of their applications to the ground, takes place. This information is provided in Section 3.3 for framed multistory buildings and in Section 3.4 for industrial buildings. The soundness of this structural concept will, in effect, determine the success or failure of the design. The design of any structure involves provision for strength, stability, and stiffness. Several arrangements of structural elements and connections to provide for them will also be given in the Sections 3.3 and 3.4.

To idealize the structure of a building for analysis and design, and to estimate the loading on the components, students should have a knowledge of various forms of construction used for roofs, floors, and walls; thus, brief details with sketches are given for some components of building structures such as steel decking (Section 3.5.1), steel joists (Section 3.5.2), roofing (Section 3.5.3), flooring (Section 3.5.4), ceiling (Section 3.5.5), wall systems (Section 3.5.6), composite beams (Section 3.5.7), and composite columns (Section 3.5.8). Many proprietary building components and systems are now available. The designer should consult appropriate manufacturers' literature for detailed information on these systems. The fabrication processes used and the method of erection adopted may influence the design of a structure. So, fabrication and erection of steel structures will be briefly described in Section 3.6.

3.2   Classification of Structures and Members

Based on their usage, steel structures may be broadly classified as:

Building structures: residential, commercial, assembly, institutional, storage, industrial.

Bridges: pedestrian, highway, railway.

Towers: radio and T.V., transmission, lighting.

Storage structures: water towers, bins, silos, pressure vessels.

Special structures: radio telescopes, satellite tracking dishes, minehead frames, penstocks, offshore drilling structures, etc.

Based on the load carrying system used, steel structures can be classified as frames, trusses, tension structures, arches, and surface structures. In what follows, a brief description of some of these structural systems, pertinent to building structures, is given.

Frames

A *frame* may be defined as a structure composed of two or more members that are joined together by connections, some or all of which are moment resisting, to form a rigid configuration. Some frequently used steel framed structures are shown in Fig. 3.2.1. They include a single-span, single-story rectangular frame, also known as a *portal frame* (Fig. 3.2.1*a*); a multispan, single-story frame (Fig. 3.2.1*b*); a multispan, multistory rectangular frame (Fig. 3.2.1*c*); and a gable frame (Fig. 3.2.1*d*). The members of the frames may be loosely classified as columns, beams, and rafters. The members of framed structures are primarily flexural members subjected mainly to bending moments and shear forces. In addition, the vertical and inclined members are also subjected to axial forces. The most common use of framed structures is in the construction of office buildings, hotels, apartments, houses, warehouses, school buildings, and similar structures.

The connections of steel frames considered in this book are detailed using welds and/or high-strength bolts (Fig. 3.2.2). (Their behavior and design will be considered in Chapters 6, 12, and 13.) The connection elements may be substantially rigid so that the ends of all members connected not only translate but also rotate essentially by identical amounts. Such a connection, known as a *rigid-jointed connection* or *moment resisting connection,* is capable of transmitting both flexural moment and shear force from one member to the other through the connection (Fig. 3.2.2*a*). Sometimes lighter connections are provided that are only semi-rigid. In *semi-rigid connections,* there is a relative rotation between the end of the beam and the joint to which it is connected, and the moment transmitted by the connection is a known and dependable function of this connection rotation (Fig. 3.2.2*c*). Compared to rigid-jointed connections, semi-rigid connections are easier to fabricate and erect, and are more cost effective. However, semi-rigid connections may be used only where the connection moment-rotation properties have been included explicitly in the

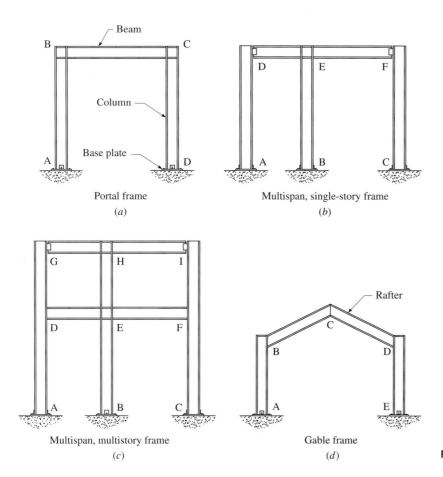

Portal frame
(a)

Multispan, single-story frame
(b)

Multispan, multistory frame
(c)

Gable frame
(d)

Figure 3.2.1: Examples of frames.

analysis and design. Also, their ductility generally limits their use to low-rise buildings. Sometimes the connection elements are so flexible that they transmit only negligible moment through the connection. They are known as *simple shear connections* (Fig. 3.2.2b). At the supports, the columns rest on a rectangular steel plate, called a *base plate,* which is connected to a concrete footing by two or more pairs of *anchor rods.* Column bases may be either *hinged-base* or *fixed-base.* If columns are considered fixed at their bases, the anchorage, including anchor rods and their connections, should be strong enough to resist bending moments at the base of columns.

Figure 3.2.3 shows the idealization of the frames shown in Fig. 3.2.1 for analysis purposes, wherein the members are represented by their longitudinal centroidal axes. This idealization is possible because the frames shown in Fig. 3.2.1 (similar to the majority of real frames) have a vertical plane of symmetry, and we consider the behavior of the frame centerline within this plane. When frames are represented by line diagrams, as in this figure,

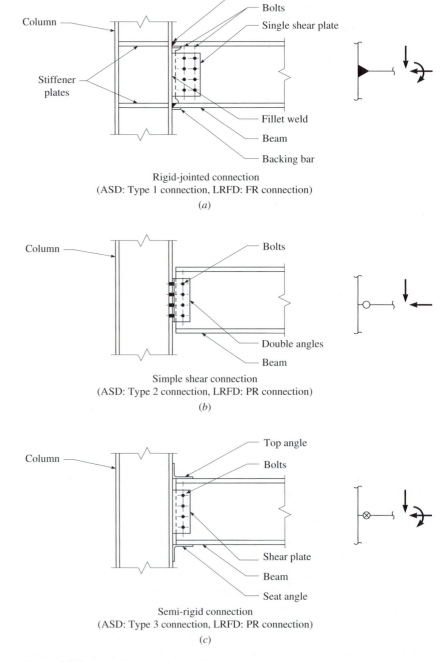

Figure 3.2.2: Classification of connections.

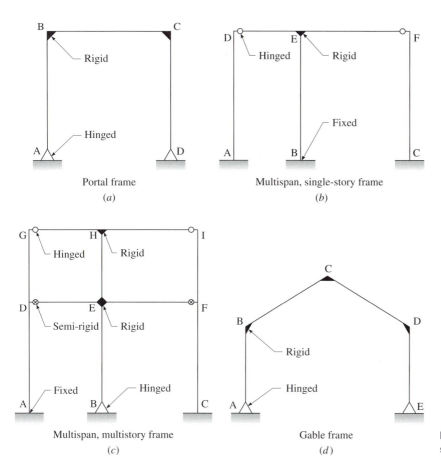

Portal frame
(a)

Multispan, single-story frame
(b)

Multispan, multistory frame
(c)

Gable frame
(d)

Figure 3.2.3: Schematic representation of frames shown in Figure 3.2.1.

fully restrained connections are indicated by showing little triangular fillets between the member-end and the joint to which it is connected. Similarly, the *partially restrained connections* are shown by the symbol ⊗, and *simple shear connections* are shown as small empty circles ○.

When analyzing and designing rectangular frames, it is useful to classify them into two broad categories: braced and unbraced. Figure 3.2.4 shows examples of braced and unbraced rigid-jointed frames subjected to vertical and lateral loads. Also shown in the figure are the deformed shapes of the frames (shown by the dotted lines, drawn to an exaggerated scale). In *braced frames* subjected to horizontal and vertical loads, the joints rotate but remain essentially in place. Thus, there is no relative translation of the ends of a column when a braced frame is subjected to lateral loads. The *maximum chord deflection of a column* is indicated by δ (Fig. 3.2.4a). In *unbraced frames* subjected to horizontal and vertical loads, the joints are free to translate horizontally in addition to rotation. The relative translation of the top end of a column with respect to its bottom end is known as *drift, side sway, or sway*

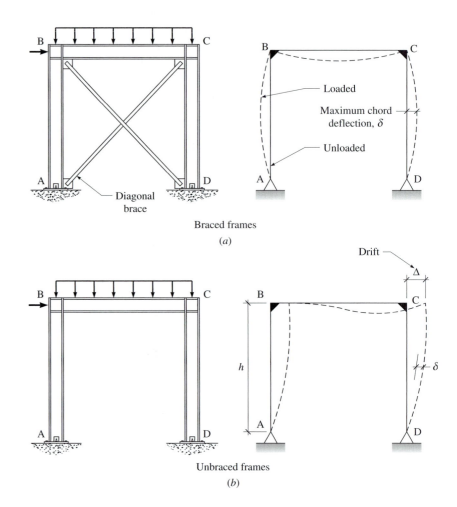

Figure 3.2.4: Braced and unbraced frames.

and is denoted by Δ (Fig. 3.2.4*b*). Thus, unbraced frames are also known as *sway permitted frames,* and braced frames as *sway prevented frames.* The ratio Δ/h, where *h* is the height of the column, is known as *drift index.* The drift Δ plays an important role in the design of members in unbraced frames as will be seen in Chapters 5, 8, and 11.

Because of their widespread use and great importance in steel construction, the design and behavior of conventional beam and column systems will be discussed in some detail in Sections 3.3, 3.4, and 3.5.

Trusses

A *truss* is a set of linear elements arranged in the form of a triangle, or combination of triangles, to result in a rigid planar structure. Two examples of truss configurations that are in common use in building structures, namely, the Pratt and Warren types, are shown in Fig. 3.2.5. The member at the top

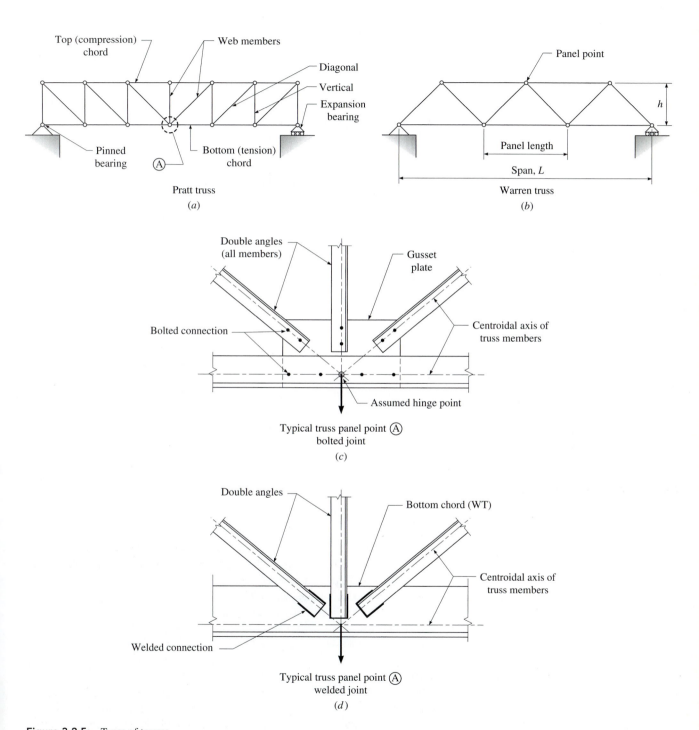

Figure 3.2.5: Types of trusses.

of a truss is known as the ***top chord,*** and the member at the bottom of the truss is known as the ***bottom chord.*** The vertical and the diagonally oriented members that form the triangular pattern between the top and bottom chord are known as the ***web members.*** Steel trusses are fabricated in a manner such that the centroidal axes of all members coming together at any joint have a common point of intersection. Further, the members and supports are customarily arranged such that all loads and reactions occur only at the joints, and the line of action of the applied load at a joint also passes through that common point of intersection (Figs. 3.2.5*c* and *d*). A point where a web member meets a chord is known as a ***panel point.*** The distance between two adjacent panel points on a chord is called the ***panel length.***

The usual form of a statically determinate support system for a truss is to use a pinned bearing at one end (which permits the rotation of the joint as the truss deforms under load) and a roller bearing at the other end (which enables the truss to contract or expand, as well as permitting the joint at this end to rotate), as shown in Figs. 3.2.5*a* and *b*. Under loads, a truss behaves like an I-beam in that the truss top and bottom chord (similar to the flanges of an I-beam) resist the bending moment, while the web members resist the shearing forces. In the analysis of trusses, the individual members are typically assumed to be joined at their ends with frictionless, pinned connections (indicated by small empty circles o in Fig. 3.2.5). Under these assumptions, the members of a truss are subjected only to axial tension or axial compression. For simply supported trusses subjected to vertical, downward acting loads, compressive forces

Warren truss supporting a restaurant over highway.
Photo by Jyothi Vinnakota Robertson.

develop in the upper chord members and tensile forces in the lower chord members. Either type of force may develop in the web members, depending upon their location and orientation, and upon the location of the loads.

If a flat roof or nearly flat roof is to be provided, the Pratt or Warren type trusses are often selected. The **Pratt truss** (Fig. 3.2.5*a*) has the advantage that the longest web members, the diagonals, are almost always in tension, while the shorter verticals are in compression. This arrangement saves some weight and cost, as lower design stresses must be used in compression as compared with tension, and these compressive stresses decrease with the length of the member. The **Warren truss** (Fig. 3.2.5*b*) consists of a series of equilateral triangles and has the practical advantage that all the web members are of the same length. The ratio of depth of truss, h, to span, L, is generally between 1/5 and 1/10. Also, for economy, the diagonals should slope at approximately 30° to 45° to the horizontal.

In reality, joints in trusses are rarely pin connected; instead they are either bolted or, more commonly these days, welded. A panel point of the Pratt truss of Fig. 3.2.5*a* is shown in Fig. 3.2.5*c*. In that detail, all of the members at the panel point are bolted separately to a common plate called a **gusset plate.** If welding is used instead of bolting, it may be possible to join all of the members at a panel point directly without the use of a gusset plate. For example, in welded roof trusses, structural tees are often used for the top-chord and bottom-chord members, and the web members (channels or angles) are welded directly to the stem of these tees. Figure 3.2.5*d* shows a welded alternative to the bolted connection shown in Fig. 3.2.5*c*. Note that the choice of connectors (bolts or welds) influences the choice of sections for the members (for example, a WT instead of double angles for the bottom chord of the Pratt truss considered in Fig. 3.2.5*a*).

Based on the number of gusset plates at each joint, trusses may also be classified as single-plane or double-plane. In **single-plane trusses,** the connections between the various members are made with a single gusset plate at each joint, the gusset plate coinciding with the central plane of the truss (Fig. 3.2.5*c*). **Double-plane trusses** require two gusset plates at each joint, one on each side of the joint and parallel to the central plane of the truss. In both cases, the thickness of the gusset plate is generally kept constant for the entire truss. Single-plane trusses are commonly used in lightly loaded trusses in industrial buildings, while double-plane trusses are required for heavily loaded trusses, like the ones used in stadium or convention center roofs.

At present, the cost of machined pinned joints is warranted only in special cases, such as for the supports of very large bridge structures. Pins are also used for special purposes, where some angular change between the members is anticipated.

In buildings, trusses can be used as roof supporting structures to transfer gravity loads; as vertical bracing structures to transfer wind loads to the foundation; and as story-deep trusses in high-rise structures where they are known as *belt, hat,* and *outrigger trusses* and used to reduce drift. Trusses are

Belt trusses and hat truss, US Bank Building, Milwaukee.
Photo by Jyothi Vinnakota Robertson.

also widely used in the form of prefabricated trusses called *open-web steel joists,* and in other roof systems in which there is a need to span a large distance without interior supports such as for gymnasiums and factories. The panel length of chord members should not exceed 30 to 40 ft. Feasible and economical spans of steel building trusses are in the range of 30 to 300 ft. The individual members of the trusses may be single or double angles, tees, channels, tubes, or W-shapes. Fabrication and erection costs are usually high with truss systems as compared to framed systems, except for open-web steel joists which are mass produced (see Section 3.5.2). Also, maintenance costs of trusses are higher than that of framed structures.

Tension Structures

In *tension structures,* cables, ties, or hangers are the primary load carrying members. *Suspended multistory building systems,* a subset of tension structures used in buildings, consist of a cantilever beam or truss system at the top of one or more of the building cores, from which the outer ends of floor beams are suspended by means of a series of steel hangers along the perimeter of the building (Fig. 3.2.6). Plates, bars, or shapes are used as hangers. The core, consisting of concrete walls or vertical steel trusses, generally houses the services and carries all of the loads on the structure, both vertical and horizontal. The suspended system uses material efficiently by using hangers (tension members) instead of columns to carry the floor loads. This type of construction is

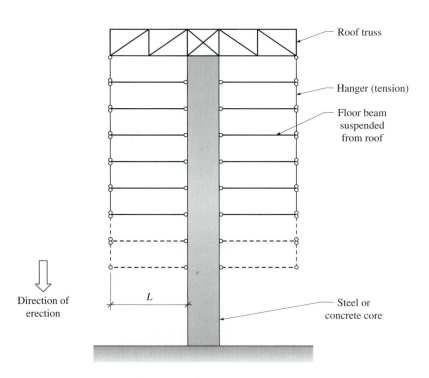

Figure 3.2.6: Suspended multistory building structure.

used where there are restrictions on the location of foundations due to obstructions such as roads, railways, and tunnels, or where the architect wants to leave an open space all around the core at ground level. The configuration of the structure, therefore, resembles that of a tree. This is helpful when the building is located in the city center and occupies a small site. This system has also been utilized for urban renewal projects where a minimum dislocation of existing facilities is permitted. The structure is erected working from the top down using hoisting equipment mounted on the cantilevered girder or truss system. One disadvantage of the suspended system is that all floor loads are carried up by the hangers to the roof level first, before they are transferred down to the foundations through the core. Also, to account for any difference in the elongation of the steel hangers and the shortening (including creep) of the core concrete, installation of ancillary elements—such as enclosure, partitioning, and mechanical work—may have to be delayed until satisfactory vertical dimensional stability is attained.

Beam-to-column connections.
Photo by S. Vinnakota.

Member Classification for Structural Design

Even though most engineering structures are three-dimensional, the arrangement of members is generally such that they may reasonably be broken into a series of component planar structures in two orthogonal directions. Such a structure is commonly analyzed in one plane for loads within that plane, and then again in a perpendicular plane for loads in that second plane, both analyses using only simple planar statics.

Often a given member must be considered as part of more than one component planar structure. For example, the column D2 of the three-dimensional building frame of Fig. 3.2.7 is part of the lateral, planar frame 2-2 and also part of the longitudinal, planar frame D-D. The analysis and design of column D2

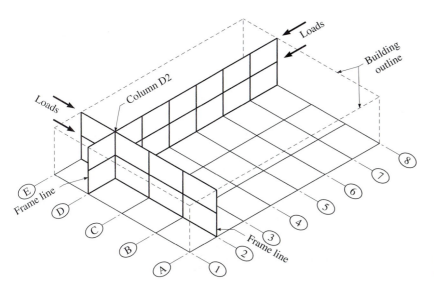

Figure 3.2.7: Three-dimensional building structure composed of two sets of planar frames.

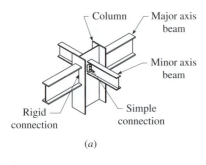

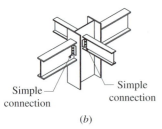

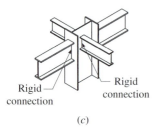

Figure 3.2.8: Some connection types in tier buildings.

would therefore have to include loadings along both frame lines 2 and D for their effects on the column. The analysis of each frame would remain a two-dimensional problem, but the combination of these two two-dimensional analyses would account for the total three-dimensional loads acting on column D2. Similarly, the beams within each frame may be analyzed in two dimensions, even though the roof or floor loads tributary to these beams come from the third direction.

The beams and columns of a building may be (1) rigidly connected in one direction and simply connected in the other direction (Fig. 3.2.8a), (2) simply connected in both directions (Fig. 3.2.8b), or (3) rigidly connected in both directions (Fig. 3.2.8c). In addition, the building structure may be unbraced in one direction and braced in the other direction, braced in both directions, or unbraced in both directions.

In some three-dimensional structures like transmission towers, guyed masts, and derricks, among others, the behavior of a member in one plane is influenced substantially by the behavior of members lying in a different plane, so that the analysis of such a three dimensional structure cannot be broken down into the study of component planar structures. In such cases, a three-dimensional analysis of the structure in its entirety is necessary.

From what we have seen in this section, it is evident that in most structures the members have as their primary loading axial tension, axial compression, transverse forces, or torsional moment, and probably most of them have secondary loading of some other type. It is common practice to name a member according to its major primary loading, thus ignoring the fact that there may be a combination of two or more of these basic types. An exception, however, has to be made for the case where a member is subjected to

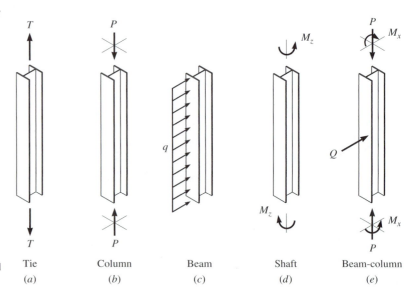

Figure 3.2.9: Member classification for structural design.

Tie	Column	Beam	Shaft	Beam-column
(a)	(b)	(c)	(d)	(e)

axial compressive loads in addition to flexural loads. In effect, the influence of these two systems of loads acting simultaneously is greater than the sum of the influences of the two systems acting separately, as will be seen in Chapters 5, 8, 9, 10, and 11. Thus, there are five basic types of load-carrying members: ***ties*** subjected to axial tension (Fig. 3.2.9*a*) studied in Chapter 7; ***columns*** subjected to axial compression (Fig. 3.2.9*b*) studied in Chapter 8; ***beams*** carrying transverse loads (Fig. 3.2.9*c*) studied in Chapters 9 and 10; and ***shafts*** subjected to torsion (Fig. 3.2.9*d*) and ***beam-columns*** subjected to axial compressive loads and transverse loads and/or moments (Fig. 3.2.9*e*) studied in Chapter 11. No matter how complicated a steel structure may appear to be, it must consist of some combination of these basic elements.

3.3   Framed Multistory Buildings

3.3.1   Introduction

For office buildings, schools, hotels, apartment houses, and similar multistoried buildings, beam and column framed steel construction is often used. In a typical framed building structure, horizontal members called ***beams*** carry gravity loads from a concrete slab or steel decking and transfer them to horizontal members called ***girders*** (Fig. 3.3.1). The girders transfer these loads to vertical members called ***columns.*** The columns in turn are supported on concrete footings resting on the earth, which ultimately supports all building structures. In a tier building, columns are generally arranged at regular intervals on a square or rectangular grid called the ***column grid,*** and the column grid is kept the same for all floors. This allows standardization of roof, floor, ceiling, and wall elements. Columns generally extend the full height

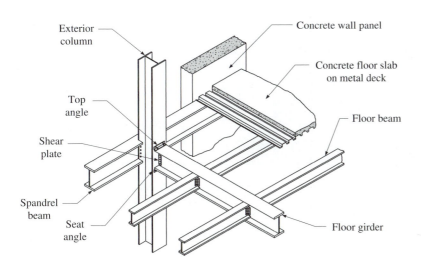

Figure 3.3.1: Typical steel framing in a tier building.

of the building. Thus, in a multistory or tier building, loads accumulated at the roof and at each floor are carried progressively down the columns into the foundations. Also in tier buildings, the horizontal members occur at fairly regular intervals (equal to the story height) throughout the height of the building.

The completed building will have external cladding, usually of precast concrete panels, brickwork, or steel curtain walls, and internal partition walls of masonry units or lightweight (stud) construction. If these wall elements are also used to support the floors and roof, and to brace the structure, they are referred to as *load-bearing walls.* However, in the majority of present-day buildings the functional and structural requirements of walls are served through different components, thus making it possible to use different materials to serve the different requirements. The so-called *curtain walls* of the tier buildings consist of thin, vertical metal struts called *mullions,* which encase the large glass, prestressed concrete, aluminum, or granite panels constituting the wall surface. The curtain walls, built for lighting and thermal conditioning purposes, do not contribute to the strength or stability of the building but serve only as enclosure elements, and hence are referred to as *non-load-bearing walls.* Mullions are normally attached to the floor slab or to the external beams on the outside perimeter of the building called *spandrel beams* or *wall beams.* Spandrel beams support the exterior curtain walls of each story, besides their tributary floor load. Spandrel beams are sometimes connected to the outside face of the wall columns, and not to their webs (see Fig. 3.3.1).

Semi-rigid or rigid connections (capable of transferring girder-end shear force, moment, and axial force) or simple shear connections (capable of transmitting only girder-end shear force) may be used to connect girders to columns. The choice is generally based on the lateral load transfer mechanism chosen for the structure, in the direction parallel to the girder. The type

Inverted V bracing. Note the gusset plate welded to the bottom flange of a girder, double channels bolted to gusset used for braces, and stiffening of the girder web at the connection.
Photo by S. Vinnakota.

Connection of a rod bracing to the face of a column. Note the threaded end of the rod, clevis, pin and the pin plate.
Photo by S. Vinnakota.

of connection chosen controls the nature of the loading applied to the members. Thus, for example, use of simple shear connections between floor girders and columns ensures that each column carries a primarily axial load and the floor girders are not loaded by wind bending moments.

Columns for multistory buildings are normally fabricated in sections several stories high (usually, two or three). Structural calculations may indicate that it is possible to reduce the column size at each story up the building, but then additional connections (splice material) are required at each floor level and may add significantly to the overall cost.

3.3.2 Floor Framing Layouts

The entry level of a multistoried building often includes particular architectural features such as assembly areas and shopping areas, which are not repeated on the floors above. So the first-floor layout may require special structural details. Motors for lifts, cooling towers for the conditioning plant, and other mechanical equipment required to service the building are normally housed in plant rooms on the roof. Hence, the roof layout may also require special structural details. In tall symmetrical buildings, the floors in between the entry level and the roof are normally structurally similar so that one floor framing layout and one slab and deck drawing may serve several floors.

The *floor framing layout,* or the arrangement of floor beams and girders, is usually dictated by the column spacing, which in turn is determined more or less rigidly by the requirements for a certain area of uninterrupted floor space by the prospective tenants of the building. Also, the foundation conditions may control the layout and spacing of the columns and should be considered at an early stage of the design process. The arrangement and spacing of the floor beams are determined by the type of flooring (e.g., steel deck and concrete slab).

Connection of a channel bracing to the web of a column. Note the gusset plate welded to the web of the column and the web stiffener, and double-channels used as brace bolted to the gusset.
Photo by S. Vinnakota.

Connection of an angle bracing to the flange of a column. Note the gusset plate welded to the flange face, and double angles used as brace bolted to the gusset.
Photo by S. Vinnakota.

Each type of flooring has an economical span suited to that particular form of construction. The most widely used floor framing, having rectangular bays of size $L_b \times L_g$ is shown in Fig. 3.3.2a. It uses a concrete slab on steel decking continuous over floor beams spaced at a uniform interval, S_b. All floor beams of span L_b, including AB, would carry the same floor load, while AC and BD would be heavy girders of span L_g. Note that the spans are measured to the center lines of the supporting girder or column, as the case may be. A rectangular

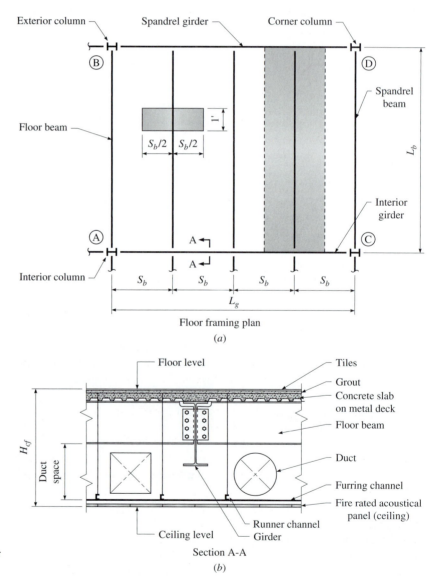

Figure 3.3.2: Typical floor framing in a tier building.

bay with a length-to-width ratio of approximately 1.15 to 1.50 is generally found to be most efficient.

Building service system elements that run horizontally are generally arranged beneath the primary structural system and enclosed with a dropped ceiling (Fig. 3.3.2*b*). Though highly efficient from a constructional standpoint, total building height is increased when this system is used. When service elements to be accommodated are few in number and/or small in dimension, the floor beams and/or girders can be fabricated with minor penetrations in webs. Another alternative is to use trusses or open-web steel joists that can easily accommodate horizontal service elements, since they can simply pass between web members.

Vertical service and functional elements (such as HVAC ducts and electrical, telephone and T.V. cables) are grouped into clusters and located in core units that fit into individual structural bays. This part of the structure is treated separately by designing special local framing in this bay. The advantage of this option is that the general structural arrangement in other bays is undisturbed by vertical penetrations and hence could be designed more efficiently.

3.3.3   Horizontal Diaphragms and Shear Bents

In modern multistory buildings of light construction, the walls and partitions are thin and not generally an integral part of the structural system; thus, the steel framing must provide all resistance to the lateral forces. When wind forces act laterally on the side of a multistory building, these forces are transmitted by the cladding to mullions (described in Section 3.3.1), and then to the horizontal slab elements. The floor slabs, stiffened by the horizontal members of the floors, namely floor beams and floor girders, form *horizontal diaphragms* uniting the columns at each floor level. These horizontal diaphragms act as *rigid plates in their plane,* and deliver the wind load increments that arise in each story to a set of vertical elements, called *shear bents,* which in turn transfer these horizontal loads to the foundation by cantilever action (Fig. 3.3.3). A *shear bent* is any vertical, planar structural system (such as a wall, a moment frame, etc.) that is capable of transmitting in-plane shear forces from one level to a lower level. A shear bent has negligible (zero) resistance to horizontal forces applied normal to its plane (i.e., in the direction of its thickness).

The multistory structure, composed of horizontal roof and floor diaphragms and vertical shear bents, must be organized such that sufficient lateral load carrying mechanisms exist to render the resultant assembly stable under any conceivable type of lateral loading conditions. For a floor diaphragm to be stable under all directions of in-plane loading, at least three shear bents must be provided (Fig. 3.3.4*a*). As a shear bent is only effective in resisting loads in its own plane, the three shear bents must not all be parallel. In effect, if the shear bents were all parallel (Fig. 3.3.4*b*), there would be no mechanism to resist (transfer) a load applied normal to the direction of

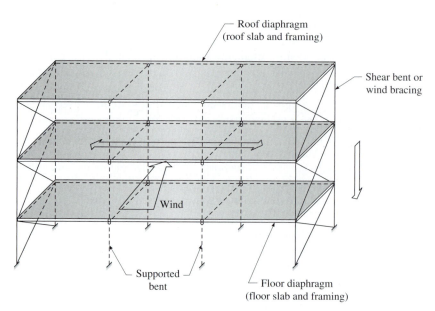

Figure 3.3.3: Transfer of lateral loads to the foundation.

Note: Wind bracing in the longitudinal direction is not shown.

the shear bents. In addition, the lines of action of the three shear bents must not meet at a point, since any in-plane load applied other than through their intersection point O will cause rotation of the floor diaphragm because none of the shear bents can provide a balancing moment about O (Fig. 3.3.4c). Even if the lines of action nearly intersect at a point, the resulting forces on the bents to produce rotational equilibrium of the floor diaphragm will be

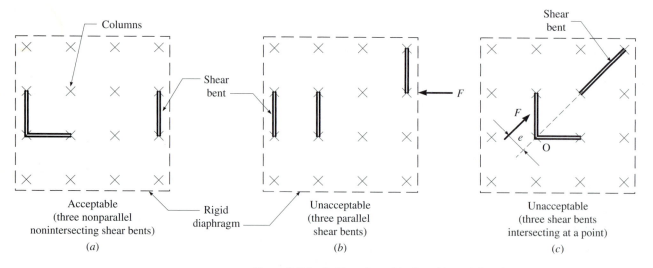

Figure 3.3.4 Stable and unstable dispositions of shear bents.

considerable. A good solution is to have two of the shear bents parallel to one another, a reasonable distance apart, and the third bent perpendicular to the others. The minimum requirements for the in-plane stability equilibrium of the diaphragm plate under all directions of in-plane loading may therefore be summarized as three reaction components, which are not all parallel, nor have lines of action which meet at a point (Fig. 3.3.4a). If more than three such shear bents are provided, there is a further increase in structural stiffness and a certain redundancy.

In a multistory building, lateral wind loading will be acting at each floor level as shown in Fig. 3.3.5. In every story there must be sufficient shear bents to transfer the cumulative shear force from the floor diaphragm above to the floor diaphragm below. Thus each story is similar to the simple, single-story model discussed earlier, indicating that three shear bents will be required at each story. The rigid horizontal diaphragm maintains the shape of the building at each level as it transmits horizontal loads to the shear bents below. Moving down the building, the cumulative shear to be transferred increases. It is not necessary for the position of the three shear bents to be the same in each story (Fig. 3.3.5a), since the transfer of shear through any one floor may be treated as an isolated problem (as part of the floor diaphragm design). It is, however, frequently more advantageous to adopt a repetitive arrangement over the height (Fig. 3.3.5b).

A *shear bent* in a multistory steel building generally consists of two adjacent tiers of columns, the connecting horizontal girders, and a certain number of inclined members connecting them, called *bracing.* A shear bent that is provided with any form of bracing is called a *braced bent.* Ordinarily, simple beam-to-column connections, capable of transferring shear only and having little bending resistance, are generally employed when full panel bracing is used, as the requirement of joint rigidity is not crucial.

During construction, unless the beam-to-column connection restraint is sufficient to provide the shear transfer requirement, it is essential that *temporary bracing* be provided until the permanent shear bents and shear diaphragms are erected and able to resist lateral loads; otherwise the wind forces on the unclad frame could be high enough to cause a collapse of the structure.

As mentioned earlier, the roof and floor diaphragms must possess the necessary strength and rigidity under loads in their own plane to deliver the wind loads to the shear bents in the floor below. If the roof or floor is incapable of serving as such a horizontal distribution unit, explicit horizontal bracing in the plane of the diaphragm should be provided between the shear bents. Most of the customary systems of floor construction used in fire protected buildings possess the requisite strength for this purpose.

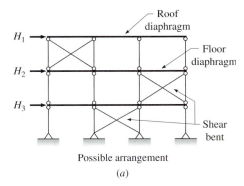

Possible arrangement

(a)

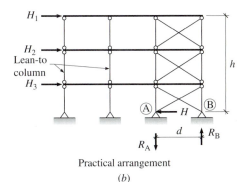

Practical arrangement

(b)

Note: There are at least two other shear bents at each level which are not shown here.

Figure 3.3.5: Disposition of shear bents in elevation.

3.3.4   Types of Shear Bents

Shear bents can be divided into four types: concentrically braced shear bents, eccentrically braced shear bents, shear walls, and moment resistant shear bents (Fig. 3.3.6). ***Concentrically braced shear bents (CBSB)*** are vertical cantilever trusses (Fig. 3.3.6*a*) with members loaded axially when resisting lateral loads; they consist of vertical members (columns), horizontal members (beams), and inclined members (braces). ***Eccentrically braced shear bents (EBSB)*** have at least one end of a brace engaging a beam at a location other than at the intersection of the beams and columns (Fig. 3.3.6*b*). ***Shear walls (SW)*** are vertical, planar, thin-plate elements which transmit in-plane shear loads acting at the top edge to the parallel, bottom edge (Fig. 3.3.6*c*). ***Moment resistant shear bents (MRSB),*** also known as ***unbraced shear bents (UBSB)*** or ***moment resistant frames (MRF),*** are composed of beams and columns assembled by rigid or semi-rigid connections (Fig. 3.3.6*d*)

In a modern multistory building the elevators, stairs, water pipes, heating, HVAC ducts, electrical, and other services take up much space. Under code regulations most of these must be enclosed in fire proof walls. Thus, grouping these utilitarian services together into one or more relatively compact vertical shafts called ***cores*** results in overall economy. In most tier buildings these cores

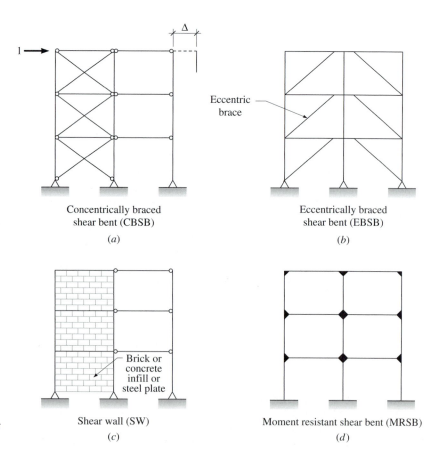

FIGURE 3.3.6: Types of shear bents in tier buildings.

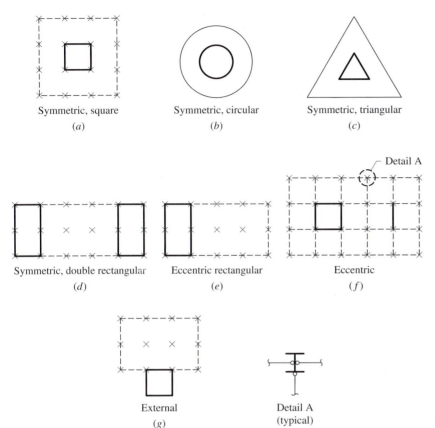

Symmetric, square
(a)

Symmetric, circular
(b)

Symmetric, triangular
(c)

Symmetric, double rectangular
(d)

Eccentric rectangular
(e)

Eccentric
(f)

External
(g)

Detail A
(typical)

Detail A

Figure 3.3.7: Disposition of shear cores in plan.

cover 20 to 25 percent of floor area, and they occur as a continuous vertical shaft for the full height of the building. The core therefore offers an excellent location for providing shear bents. Some arrangements of core systems are shown in Fig. 3.3.7. An attempt is always made to place the core or cores symmetrically in the plan, to provide the building with reasonably uniform stiffness. Core shafts generally have a rectangular or square cross section. But in the case of a round building, they may be circular, or they may be triangular for the case of a building with a triangular foot print. Shafts, located centrally (Figs. 3.3.7a, b, and c), or symmetrically (Fig. 3.3.7d) in the structure, are an ideal solution for shear cores. However, if they are positioned towards one end of a building, another shear bent should preferably be provided towards the other end to help relieve the heavy torsional moments that would otherwise act on the shaft (Figs. 3.3.7e and f).

End or side walls and service cores provide a convenient location for shear bents. If the floors effectively act as rigid diaphragms in their plane, there is no maximum limit for spacing of shear bents. No restriction needs to be placed on carrying all the lateral force to the end or sidewalls, if the vertical bracing there is sufficient to provide for the entire wind force. Shear bents

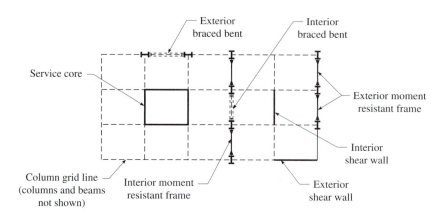

Figure 3.3.8: Nomenclature of lateral load resisting systems of tier buildings.

may be external (i.e., located in the plane of the exterior walls, facade, or perimeter of the building), internal (i.e., located in planes within the building), or incorporated into core walls, as shown in Fig. 3.3.8.

Concentrically Braced Shear Bents (CBSB)

Concentrically braced shear bents, or simply ***braced bents,*** are vertical cantilever trusses designed to transmit lateral loads acting on a building to the foundations. They derive their lateral stiffness from the axial stiffness of the individual members including the diagonal bracing members. Their stiffness does not involve the flexural deformations of members. Concentrically braced bents result in very stiff structures. They are the most economical systems for resisting lateral loads in low and medium-rise buildings. Steel diagonal members are generally slender and may be readily incorporated into partitions. By proper choice of the system, the bracing may be arranged to fit around door and window openings. However, as the height of the building increases, and with it the size of the inclined members, special problems may arise in the fitting of partitions in and around these inclined members. Also, diagonal members, when used in the facade, may necessitate windows having unusual locations, sizes, or shapes. The primary use of concentrically braced bents is, therefore, in and around cores, where they can be placed in unseen and nonarchitectural spaces. Sometimes, instead of being concealed, diagonal bracing is provided in the external walls and is made an architectural feature of the finished building. Concentrically braced shear bents have a great amount of stiffness but low ductility; thus, they are used in areas of low seismic activity, where high ductility is not essential.

Various configurations of concentrically braced shear bents are shown in Fig. 3.3.9. Full-length ***double diagonals bracing,*** or ***X-bracing*** (Fig. 3.3.9*a*), produces the most effective type of bracing. Double diagonal bracing usually cannot be placed in exterior walls as it is likely to interfere with windows. For this reason, shear bents with a single diagonal member capable of resisting

John Hancock building, Chicago. Note the exterior diagonal bracing.
Photo by Rajiv Vinnakota.

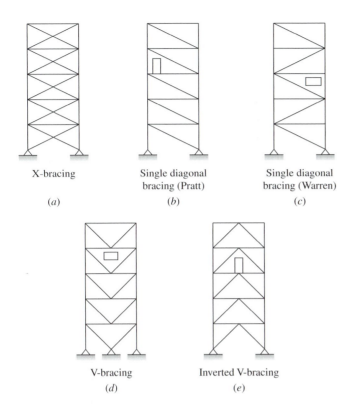

X-bracing

(a)

Single diagonal
bracing (Pratt)

(b)

Single diagonal
bracing (Warren)

(c)

V-bracing

(d)

Inverted V-bracing

(e)

Figure 3.3.9: Configurations of concentrically braced shear bents.

both tensile and compressive forces (wind from either direction) are sometimes used (Figs. 3.3.9*b* and *c*). In the arrangement shown in Fig. 3.3.9*b*, the horizontal member (beam or girder) at any level has to transfer the horizontal shear in the panel above from one end to its other end. Hence, the horizontal member has to be designed as a beam-column. The V- or inverted V-shaped ***chevron bracing system*** (Figs. 3.3.9*d* and *e*), in which the horizontal member (beam or girder) is supported at midspan by inclined members, is more efficient. The chevron system allows more freedom on the use of aisle space, since it is possible to fit doors or windows beneath its apex. The V- and inverted V-type of bracing provide intermediate vertical support to the horizontal members, resulting in a reduction in their bending moments both from dead load and live load. This reduction in the size of the girders results in economy, which will partially compensate for the cost of the wind bracing.

   Rod bracing is very economical for X-bracing in low industrial buildings where the bracing forces are small. Single angles, channels, or tees may be used when somewhat greater rigidity is required. They allow easy back-to-back connections where members cross in X-bracing. In high-rise buildings, wide flange shapes may be required to accommodate the large lateral loads. The shape selected for the diagonal is also influenced by connection considerations. Rod bracing may be connected to a column through a clevis, pin, and a gusset plate welded to the column (Fig. 3.3.10*a*). Angle, channel, or tee

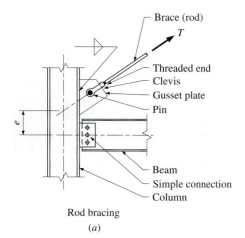

Rod bracing
(a)

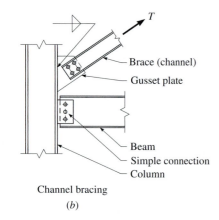

Channel bracing
(b)

Figure 3.3.10: Connections of diagonal bracing.

braces may be bolted to a gusset plate that is welded to the face of a column (Fig. 3.3.10b). As shown in the figure, the beam-to-column connections in concentrically braced shear bents are typically simple shear connections and thus provide no moment transfer.

The braced bent tends to resist the horizontal forces in the manner of a vertical cantilever truss, anchored to the foundation, with the columns forming the chords, and the bracing and the girders forming the web (Fig. 3.3.3). The columns on the windward side of the axis of the bent become tension chords and those on the leeward side, compression chords.

In buildings of usual proportions, the height-to-width ratio of shear bents is such that the gravity loads transferred to the bent are generally sufficient to resist bodily overturning. The effect of lateral wind loads is therefore to relieve the compression in the windward columns of the bent and to increase the compression in the leeward columns, if the building is braced diagonally from top to bottom. In the case of exceptionally tall buildings (high height-to-width ratios of braced bents), the factored wind loads may create tension in the windward columns exceeding the compression due to the factored gravity loads, and so produce an actual uplift on the foundations of the windward columns (see Load Combination LC-6 in Section 4.10). It therefore becomes necessary to anchor these columns down to the foundations (rock, caissons, etc.).

Eccentrically Braced Shear Bents (EBSB)

As mentioned earlier, *eccentrically braced shear bents* have at least one end of a brace connecting a beam at a location other than at the intersection of the beams and columns (Fig. 3.3.6b). The eccentric brace is not permitted to intersect a column at any location except at the intersection with a beam. The segment of the beam between the end of the brace and a column, or between the ends of two braces intersecting near the midpoint of the beam is termed a *link beam.* The beams and columns of the eccentrically braced bents are flexurally deformed, which lowers the stiffness but increases the ductility, compared to a CBSB.

The eccentrically braced bents possess a favorable combination of the characteristics, namely, the stiffness of the centrally braced bents and the ductility and energy dissipation capacity of the unbraced bents. So, EBSBs are used extensively in areas of high seismic activity.

Shear Walls (SW)

A *shear wall* may be a wall connecting two or more columns, or a panel-wall filling the openings of an unbraced bent (Fig. 3.3.6c). Steel plates have been used as shear walls in steel frames. The steel plate may be either welded or bolted to the surrounding frame. The plate must be stiffened horizontally and vertically at regular intervals by steel bars, angles, or tees to prevent buckling. Steel plate shear walls are useful for buildings in seismic zones. Concrete walls, either cast-in-place or precast, have great shear strength and are extensively used as shear walls. Simple connections are generally used to connect adjacent steel members to the shear walls and core walls to allow the columns and core to move vertically, relative to one another, without inducing stresses.

Moment Resistant Shear Bents (MRSB)

Where the full usable space is required in shear bents, moment resistant frames are the only solution. A *moment resistant frame* derives its lateral stiffness essentially from the flexural stiffness of its individual members and the flexural rigidity of the connections between them. Unbraced bents may be internal, external, or in core walls. The beam-to-column connections may be fully rigid or semi-rigid (Fig. 3.2.2).

Wide flange shapes are extensively used as beams and columns of moment resistant frames. Sometimes tubes and built-up shapes are also used. The columns, particularly in the lower stories, must resist heavy moments due to lateral loads, and so will require a heavier section than in structures where the wind loading is resisted by braced bents or shear walls. Moment resistant frames are less efficient than shear walls or braced bents as a lateral load-transfer mechanism. However, they can be located in the exterior walls where unobstructed bays for glazing are required. Moment resistant frames are commonly used in low- to medium-rise buildings below 20 stories. *For further information on lateral load resisting systems in framed multistory buildings, refer to Section W3.1 on the website http://www.mhhe.com/Vinnakota.*

HSS brace field welded to gusset plate. Note the erection bolt used to hold the brace in place during erection and field welding.
Photo courtesy of OPUS Corporation.

3.3.5   Orientation of Column Cross Section

Sometimes one type of lateral load-carrying mechanism is used in one direction of the building, and another is used in the other direction. In rectangular buildings, the greatest problem with lateral load transfer is in the short direction of the building. In effect, the relatively smaller magnitudes of the lateral forces acting on the end faces of the building could be transferred entirely by the more efficient diagonally braced bents arranged in the longitudinal walls. However, the larger magnitudes of lateral forces acting on the longitudinal face of the building may not be able to be transferred by the diagonally braced bents in end walls alone. Thus, lateral load transfer responsibility in this direction has to be shared by frame action of some or all transverse bents. Under these conditions, wide flange members used as columns are oriented such that, about their strong axis, they function as part of the moment resistant frame in the short direction, and about their weak axis they function as part of the diagonally braced structural system in the long direction. In the short direction, lateral loads are transferred by frame action inducing high bending in these members (see Fig. 3.2.8a).

3.4   Industrial Buildings

Introduction

Though highly prestigious, high-rise buildings account for a very small portion of the total number of structures being built everyday in every city and country in the world. The overwhelming majority of steel structures being built at any time are low-rise, laterally braced, statically determinate structures. Industrial buildings, a subset of low-rise buildings, are generally of one story only and normally used for warehouses, assembly plants, factories, and steel plants. In these buildings large, clear working floor areas are required; therefore, interior columns, walls, and partitions are eliminated or kept to a minimum. These buildings generally require adequate headroom for use of an overhead traveling crane.

Single-story industrial structures have parallel portal frames, multispan single-story frames, gable frames, or trusses as their primary load carrying elements. (A truss is shown in Fig. 3.4.1.) The portion of the structure consisting of a roof truss and its supporting columns is called a **bent.** The distance between two successive bents of an industrial building is called a **bay.** Thus, the length of the building is divided into bays. The space between two rows of columns is called an **aisle** or **span.** An industrial building may be single-span or multispan in type. The two chords of the roof truss may be connected to the columns of the bent, thus providing it with lateral stiffness. Trusses carrying light roofs are usually spaced from 20 to 40 ft apart.

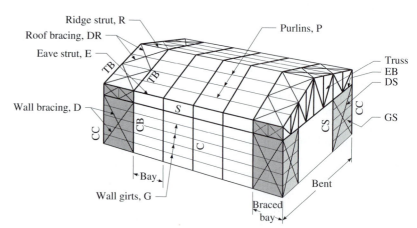

Truss on column system with diaphragms and trussed shear bents

(a)

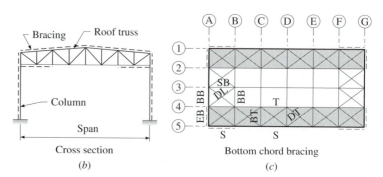

Cross section

(b)

Bottom chord bracing

(c)

FIGURE 3.4.1:   Framing for an industrial building.

Flat trusses of the Warren type generally have an end depth of $L/10$ to $L/8$ and a top chord slope of about 1 in. per ft. Figure 3.4.1 shows framing for a typical single-span, multibay industrial building with a truss-on-column system, using the two end bays as braced bays.

The external surfaces of industrial buildings are normally non-load bearing. Light metal sheeting, ribbed for flexural strength and backed by insulation material, is the most widely used form of covering. Such cladding is rarely strong enough to span between the main frames of the building, and secondary structural members called *purlins,* or *joists,* in the roof and *girts* in the walls are required to support the covering. The purlins or joists are supported on roof beams, rafters, or top chords of roof trusses. Ceilings are generally omitted in industrial buildings, and structural members and ducts are left exposed. The different subsystems and elements will be discussed in Section 3.5.

A summary of notations used in Figs. 3.4.1*a* and *c*, and elsewhere is given below and will be elaborated in sections to follow:

R: ridge strut
E: eave strut
P: purlin
CC: corner column
C: column in longitudinal wall
G: girt in longitudinal wall
CB: column in braced bay
CS: column in side wall
GS: girt in side wall
S: strut in longitudinal wall
D: diagonal in longitudinal wall
DT: diagonal in transverse bracing at truss bottom chord
DL: diagonal in longitudinal bracing at truss bottom chord
SB: strut in longitudinal bracing at truss bottom chord
TB: truss top chord in a braced bay
BB: truss bottom chord in a braced bay
EB: truss bottom chord in side wall
T: tie in bottom chord bracing
DS: diagonal in side wall
DR: diagonal in roof
BT: bottom chord in transverse bracing

Purlins and Sag Rods

Purlins are flexural members supporting the roof decks of industrial buildings and spanning from one bent to the next (Fig. 3.4.2*a*). Hot-rolled or cold-formed sections such as channels, I-shapes, or zees are used as purlins. Channels are often selected as purlins due to the ease with which they can be erected. Short clip angles are previously bolted or welded to the top chord of the truss above which the purlin may rest while it is being connected (Figs. 3.4.2*b* and *c*).

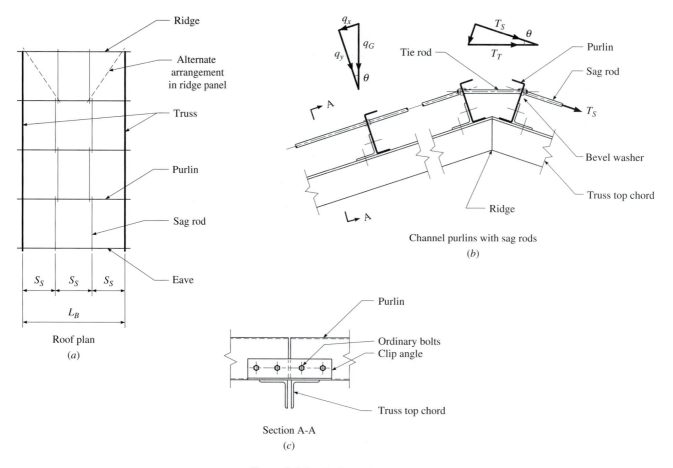

Roof plan

(a)

Channel purlins with sag rods

(b)

Section A-A

(c)

Figure 3.4.2: Purlins and sag rods.

The depth of the purlins is generally not less than 1/30 of their span. To give rigidity to the roof, purlins are usually made in two or three span lengths with alternate purlins spliced at alternate trusses.

The spacing of purlins depends on the type of roof covering, and the spacing in turn determines, to some extent, the design of the roof trusses. The loads from the roof are delivered by the purlins to the trusses. Usually the purlin spacing equals the distance between panel points of the top chord. Sometimes, the purlin spacing is made less than the panel length, and the truss top chord in this case also acts as a beam to transmit purlin loads to the truss panel points. The top chord in this case, must therefore be designed for both axial force and bending moments.

Roof surfaces are often sloped so as to shed rain water easily. On a sloping roof, gravity loads (q_G) have a component (q_x) parallel to the roof surface, and this component must also be supported by the purlins (Fig. 3.4.2b).

However, the purlins are generally weak in the direction parallel to the roof surface, and if the slope is more than just a few degrees, *sag rods* are used to shorten the span and reduce the bending moment in this direction. Sag rods are usually $\frac{1}{2}$ to 1 in. diameter plain steel rods with threaded ends. They are inserted through holes in the webs of the purlins and are held in place with common nuts. In the weak direction then, each purlin becomes a continuous beam with each span, S_S, equal to one-half or one-third of the distance between the bents, L_B, depending on whether one or two lines of sag rods are used in each bay. The sag rod itself is a tension member and must be proportioned to take the accumulated tangential components of all purlins below the ridge of the truss and on one side of the roof. The sag rods are not continuous across the building, as each rod extends only from one purlin to the next as shown in Fig. 3.4.2*a*. At the ridge, the downward pull of the rod on one side of the roof is balanced by the downward pull in the opposite direction on the other side.

When properly designed, connected, and anchored, purlins act as lateral bracing to the compression chord of the roof truss or compression flange of the rafter beam to which they are connected, with consequent increase in the design strength of the compression chord or rafter beam, as the case may be. For further details see Sections 7.12.4 and 9.5.2.

Girts

Wall panels in industrial buildings are rarely strong enough to span between the main frames of the building. Thus, they are often supported on horizontal members called *girts,* which span from one bent to the next (Fig. 3.4.3). The wall panel, therefore, acts as a beam continuous over several spans and transfers the lateral loads due to wind to the girts. Where girts are necessary to support wind loads normal to the wall surface, they are usually also required to support the dead weight of the wall. However, the girts are weak in the vertical direction, and thus sag rods similar to those in the roof are used (Fig. 3.4.3). The girts, therefore, behave like continuous beams with regard to vertical loads, supported at the columns and at the sag rods. The load in the sag rods is carried upward to the eave strut. As in the roof, the sag rods are plain rods with threaded ends which pass through the webs of the girts and are held with common nuts.

It is through girts that the wind loads are transmitted to the structure, and they should be positioned in such a way that the loads are introduced to the columns at the most convenient points. More importantly, when properly designed, connected, and anchored, girts can be used to reduce the so-called effective length of columns to which they are connected, and consequently increase the design strength of the columns, as will be seen in Section 8.5.3.

Lateral Load Resisting Systems in Industrial Buildings

To transfer the wind loads in the lateral and longitudinal directions of the building to the foundations, a system of diagonal bracing (or shear diaphragms) is generally included in the upper and lower chord planes of the roof trusses, in the side walls, and in the longitudinal walls. *For further information on such*

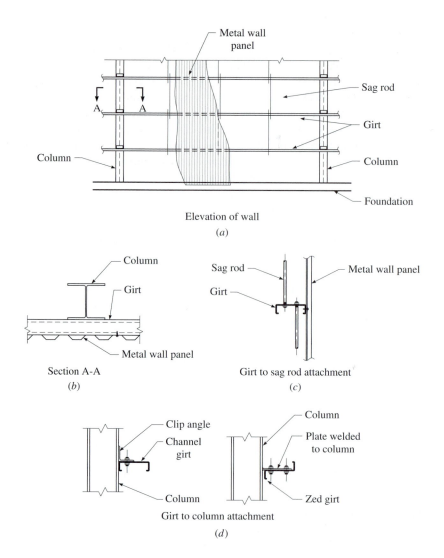

Elevation of wall

(a)

Section A-A

(b)

Girt to sag rod attachment

(c)

Girt to column attachment

(d)

Figure 3.4.3:   Girts and sag rods.

   lateral load resisting systems in industrial buildings, refer to Section W3.2 on the website http://www.mhhe.com/Vinnakota.

3.5   Components of Buildings

3.5.1   Steel Decks

Light-gage, cold-formed, ribbed steel decking, the most popular roof and floor deck material, is used for about 70 percent of new industrial buildings and many other building types including high-rise buildings. Steel decks are manufactured by cold drawing flat steel strips through dies to produce the required section. ASTM A446 sheet steel is used to make painted decks, while A445 steel is used for galvanized decks. Steel decks function as substrata for built-up roofs *(roof deck)*, as stay-in-place concrete forms *(form deck),* or as composite form that furnishes the slab reinforcement for positive bending *(composite deck).* Typical steel decks selected as roof decks are shown in Fig. 3.5.1. Steel deck units are anchored to supporting flexural members by puddle welds, powder-activated and pneumatically driven fasteners, and self-drilling screws. Puddle welds are made

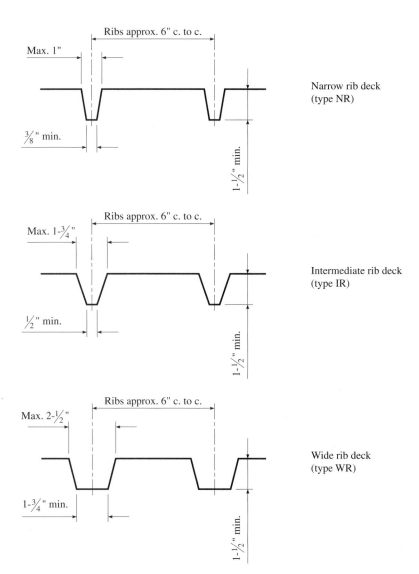

Figure 3.5.1: Types of roof deck.

Steel decking, shear studs, and pour stop in place for a floor slab.
Photo courtesy of OPUS Corporation.

from the top side of the deck with the welder immediately following the deck placement crew.

Steel decks come in different thicknesses, depths, rib spacing, widths, and lengths. They are available with or without stiffening elements, with or without acoustical material, and in cellular and noncellular forms. Some usual values are:

Thickness: 16, 18, 20, and 22 gage down to 28 gage.
Depth: $1\frac{1}{2}$, 2, 3, $4\frac{1}{2}$, 6, and $7\frac{1}{2}$ in.
Ribs spacing: 6, $7\frac{1}{2}$, 8, 9, or 12 in.
Width: 18, 24, 30, or 36 in.
Lengths: 20 ft or more.

Steel decks can be erected in most weather conditions, eliminating the costly delays that can occur with other types of roof and floor systems. The steel deck can be erected as soon as beams and girders have been erected. When suitably designed and installed, the steel deck also acts as a safe working and storage platform for the various trades that follow and provides protection for workmen on lower levels. Form decks and composite decks eliminate the need for erection and removal of temporary forms for the cast-in-place slab. Cellular decks can be used to provide electrical, telephone, and cable wiring and to serve as ducts for air distribution, without significant increase in floor thickness. With properly specified factory and field coatings, steel decking is easy to maintain, durable, and aesthetically pleasing. Installation of stud connectors used to develop composite beam action is simplified by the use of steel floor decking.

When properly anchored to supporting flexural members, steel decks provide lateral stability to the top flange of the supporting structural members and resist the uplift forces due to wind in the construction stage. The steel deck acts as a one-way beam and, depending upon the purlin or joist spacing and the available lengths of decking, may be considered as simply supported or continuous over several supports.

When designed as a composite deck, the steel deck carries construction loads and the weight of the wet concrete, acting as a beam. Once the concrete hardens the deck acts as the tensile reinforcement for the concrete slab. The composite section has to carry only the live loads and the dead loads applied after the concrete hardens (flooring, ceiling, and partitions). Hence, its loading condition changes from the construction stage to the service stage. Some reinforcement is always needed to resist the negative moments above the supports, to prevent formation of large temperature and shrinkage cracks, and for fire safety reasons.

The load carrying capacity of the deck is influenced by the depth of the cross section, the thickness of the metal (gage), the span length, and the continuity. The Steel Deck Institute's *Design Manual* [SDI, 2001] provides standard load tables for steel decking having a yield strength $F_y = 33$ ksi, subjected to uniformly distributed load.

3.5.2   Steel Joists

Open-web steel joists are standardized, predesigned, prefabricated, welded steel trusses used as simply supported beams (Fig. 3.5.2). The lighter joists consist of longitudinal rods, top and bottom, connected by rod trussing welded to them. Other joists may be made from hot-rolled shapes, plates, or bars, or from cold-rolled shapes, depending on the manufacturer's preference. All connections and splices use welding. Open-web joists, usually employing a Warren-type web configuration, may be obtained in various spans and depths. They are generally hung from the supporting girders or walls; they should be tack welded or bolted to the flanges of the supporting members to prevent movement during construction. To prevent buckling of the steel joist during construction, lateral members called *bridging* are provided at intervals. The prefabrication of joists ensures accuracy of shop fabrication, reduces the number of pieces to be handled, and speeds erection, which requires only a minimum of site labor needing few specialized skills.

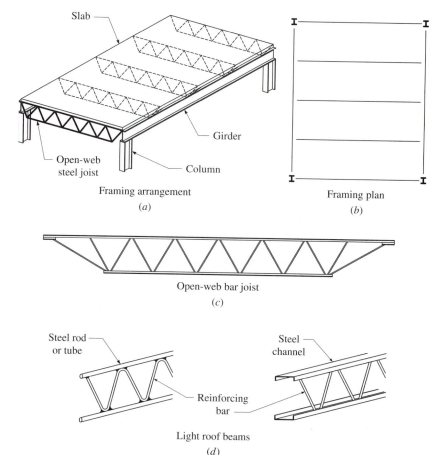

Slab

Girder

Open-web
steel joist

Column

Framing arrangement
(a)

Framing plan
(b)

Open-web bar joist
(c)

Steel rod
or tube

Steel
channel

Reinforcing
bar

Light roof beams
(d)

Figure 3.5.2: Open-web steel joists.

Open-web steel joists are very economical for supporting lightweight floors of long span. They possess an exceptionally high strength-to-weight ratio in comparison with other building systems. Solid-web beams may be less costly than joists for certain spans and loads, but openings required for pipes and ducts may increase the ceiling-to-floor height H_{cf}, as defined in Fig. 3.3.2, and hence favor selection of joists. In effect, the open webs of the joist trusses permit the ready passage and concealment of pipes, ducts, and electric and telephone conduits within the depth of the joists. In high-rise buildings this can result in a reduced overall building height, which leads to considerable cost savings. Open-web steel joists have been used in composite action with flat soffit concrete slabs and metal deck slabs supporting concrete fill. The design of these systems is primarily based on manufacturer test data.

Ordinarily, the project engineer does not design steel joists, but simply selects from standard tables, joists appropriate for the given span and load. The standard specifications [SJI, 2002] developed by the Steel Joist Institute (SJI) give details for three types of open-web steel joists (standard K-series, longspan LH-series, and deep longspan DLH series) and for open-web joist girders (G-series). For additional information on designing with steel joists and joist girders, refer to Fisher et al. [1991]. Selection of steel joists and joist girders is considered in Section 9.8.

3.5.3   Roofing

A modern ***built-up roof system*** consists of three basic components: structural deck, thermal insulation, and membrane [NRCA, 2003]. The ***structural deck*** transmits gravity, wind, and earthquake forces to the roof framing. ***Thermal insulation*** cuts heating and cooling costs, increases comfort, and prevents condensation on interior building surfaces. The ***membrane*** comprises the waterproofing component of the roof system. A fourth component, a ***vapor retarder,*** is sometimes required for roofs over humid interiors in the northern climates. ***Flashing,*** although not a basic component of the built-up roof system, is an independent accessory.

Structural roof decking may be made of steel, concrete, or gypsum. Light-gage, cold-formed steel roof decking is the most popular structural deck material. In cross section, steel decking is ribbed, with ribs generally spaced at 6 in. on centers and $1\frac{1}{2}$ in. or 2 in. deep. The sloped-side ribs measure 1 in. across at the top for narrow rib decking, $1\frac{3}{4}$ in. for intermediate decking, and $2\frac{1}{2}$ in. for wide rib decking (Fig. 3.5.1). Wide rib decking is by far the most popular, and most insulation boards 1 in. or thicker can be used in the wide rib deck. Thinner insulation boards may require a narrower deck rib opening. The wide rib deck has higher section properties (per inch thickness) than the other patterns, and therefore it can span greater distances.

Concrete decks come in two forms: cast-in-place and precast. A ***cast-in-place structural deck*** is continuous, except where interrupted by an expansion joint or another building component. ***Precast concrete decks*** come in a variety of cross sections such as: single T, double T, and inverted channels. Single or double T's vary in width from about 2 to 8 ft, with spans up to 60 ft or so. Inverted channels come in depths of 3 to 12 in. with spans ranging up to 20 ft or so. Gypsum decks, like concrete, are either cast-in-place or precast. A cast-in-place gypsum deck is poured on gypsum form boards, spanning flanges of closely spaced steel tees. Like a poured concrete deck, a cast-in-place gypsum deck presents a large, seamless expanse of roof surface. Precast gypsum planks come in 2-in.-thick panels, with metal tongue-and-groove edges. Because gypsum decks are nailable, insulation can be either nailed or bonded to them with hot bitumen.

The Asphalt Roofing Manufacturer's Association (ARMA) Built-up Roofing Committee recommends a minimum $\frac{1}{4}$ in./ft slope for roofs. The most dependable and often most economical way to provide slope on a new building's roof is to slope the structural framing and the deck (especially for cast-in-place concrete applications). Compared with sloped framing and constant insulation thickness, tapered insulation necessitates excessive insulation thickness to provide for adequate slope. The roof deck should always be checked for snow drifting, additional dead load from ballasted roof system, and maintenance loads.

3.5.4   Flooring

Concrete slabs, usually of constant thickness, may be cast in place on the upper flanges of joists, beams, or girders by means of removable formwork. Cast in place concrete slabs require on-site labor for formwork, laying of reinforcement, casting of concrete, and formwork stripping—a wet process which requires long periods of setting and hardening of the concrete. The slab may be designed for composite action with the steel beams. For roof slabs the finish is usually grout with asphalt topping and gravel. For floor slabs grout is used to obtain a level, semifinished floor. This can be covered with commercial floor materials such as tiles, and so on, to provide a finished surface.

The thickness of the concrete slab will be determined by the span and loading conditions, or alternatively, by the minimum thickness required for fire resistance. In effect, all multistory buildings must have a standard fire resistance period, which varies with the size, location, and occupancy of the building. The slab acts as a rectangular plate of dimensions $S_b \times L_b$ where S_b is the spacing of floor beams and L_b, the span of floor beams (Fig. 3.3.2a). Since the length of the long side of this rectangle is generally two or more times that of the short side, the slab can be assumed to act as a series of parallel continuous beam strips spanning in the short direction, and it is reinforced accordingly. Excessively thick slabs are to be avoided to reduce the dead weight of the floor. This is achieved by keeping the spans short (normally 6 to 10 ft and rarely greater than 12 ft) and by using lightweight concrete. Lightweight concrete weighs 30 to 50 pcf less than normal weight concrete. The thickness of the slab generally varies from about $\frac{1}{30}$ to $\frac{1}{15}$ of the span. For office buildings, a lightweight concrete slab of 4 in. thick is often used for clear spans up to 8 ft. Where heavier loads or longer spans occur, the slab is thickened to 5 or 6 in. The resulting reduction of dead load using lightweight concrete can produce significant savings in the cost of steel framing and foundations. However, normal weight concrete instead of lighter weight concrete may be required in locations where vibration transmission is to be prevented (e.g. rooms with office equipment). Nominal reinforcement is provided in the long direction (parallel to the floor beam) to prevent formation of large temperature and shrinkage cracks. The spans using composite metal deck are generally in the range of 8 to 15 ft. Some major high-rise projects have used $2\frac{1}{2}$-in.-thick concrete on 3-in.-deep metal deck spanning as much as 15 ft.

3.5.5   Ceiling

The function of a *ceiling* is to present an aesthetically pleasant finish to the underside of a structural floor. In addition, a *suspended ceiling,* by far the most common type of ceiling used in high-rise buildings, conceals the underside of the floor structure and any services that may be located under that floor (Fig. 3.3.2*b*). It also acts as acoustic absorption surface and a fire protection barrier. Structural steel floor framing with metal decking topped with concrete, which normally would have to be spray-coated to achieve the necessary fire ratings, can have the spray eliminated if an appropriate fire-rated ceiling is installed.

Every community has zoning regulations and codes that may include height limitations for buildings. If a maximum number of floors are needed to increase the rentable area, the floor-to-floor heights must be kept to a minimum. The space between the bottom level of ceiling and top level of the flooring above, H_{cf} in Fig. 3.3.2*b*, is wasted (from the owner's point of view) and increases the overall height of the building. Consequently, not only is the structural framing more costly but so are the other building components whose costs vary with building height, such as exterior walls, vertical ducts, service core walls, and additional lengths of column steel. Note that six inches saved in height H_{cf}, in each story, will reduce the required height of a 24-story building by one entire story.

3.5.6   Wall Systems

Tier Buildings
An important nonstructural or non-load bearing architectural element of a tier building is the *exterior wall system,* also known as the *facing* or *cladding system.* Common materials used for cladding are plastic or PVC coated steel, aluminum, bronze, stone, marble, or granite (cut into thin slabs and backed with insulating material), precast concrete, and glass. A metal *curtain wall* is a building's exterior wall which is made of metal or glass and other surfacing materials and is supported by or within a metal framework. The curtain wall carries no superimposed vertical loads but must withstand wind forces (wind pressure and suction), and must absorb all movements of the building structure. The window wall element, or glass curtain wall, must protect the internal environment of the

building from rapidly changing external conditions. It should control the transfer of heat and the passage of light, and prevent the penetration of moisture, rain, dirt, and sounds. It must also provide fire protection and be capable of easy cleaning and maintenance. In one type of curtain wall system, the full floor height vertical mullion members are installed first. Then the framed units, pre-assembled in the factory and of modular width, are placed between the mullions. The framed sections may be either full floor height or divided into vision panel and spandrel unit. All connections of the curtain walls should be corrosion-proof for the life of the building.

Where no primary or secondary framing is employed in the system, that is, where wall panels span from floor to floor, the cladding units must integrally accommodate the loadings. In framed systems, all cladding and glazed units must transfer their loads effectively to the primary framing. The vertical elements, or mullions, are assumed to be simply supported members. Allowable deflections for framing members are generally in the range of $L/240$ to $L/360$.

Industrial Buildings

The walls of industrial buildings may be constructed of masonry, sheet metal, or other surfacing material. Precast concrete panels, large enough so that one may cover a whole bay, are sometimes used. At the present time three-layer sandwich panels with light-gage cold-formed metal faces and foam-in-place cores are being used extensively in commercial and industrial buildings in the United States as building enclosures. The rapid rise in their use is due to their structural efficiency, thermal and sound insulation capabilities, mass productivity, transportability, durability, and reusability. The wall panels are usually 2 in. thick, but they have high insulation values and thermal properties exceeding those of traditional construction methods, which may require double or even four times as much thickness. They are available in 24, 30, and 36 in. widths. The wall system offers fast erecting one-piece construction. Its exterior and interior steel skins encapsulate a foamed-in-place isocyanurate insulation core. The interior and exterior steel skins are cold-formed steel conforming to ASTM A446, Grade A, with a minimum yield stress of 33 ksi.

3.5.7   Composite Beams

In *composite beam* construction, the concrete deck is attached to the top flanges of the steel beams and girders by the use of suitable mechanical shear transfer devices called *shear connectors* or *shear studs* (Fig. 3.5.3). The concrete deck then acts as part of the compression flange of the resulting composite beam. As a result, the neutral axis of the section shifts upward, making the bottom flange of the steel beam more effective in tension. The composite action of the floor beam results in an increase in strength and stiffness for a given steel beam size. The concrete deck may be a flat soffit reinforced concrete slab directly resting on the top flange (Fig. 3.5.3a) or a concrete slab on metal decking (Figs. 3.5.3b and c). The beam may be a wide flange shape,

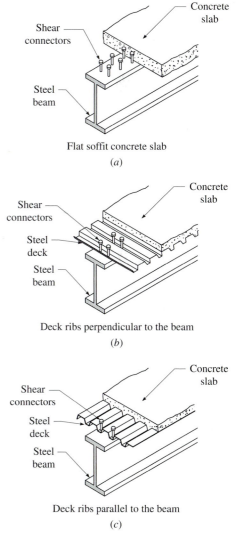

Flat soffit concrete slab
(a)

Deck ribs perpendicular to the beam
(b)

Deck ribs parallel to the beam
(c)

Figure 3.5.3: Composite beams.

open-web steel joist, or a truss. Since the slab already serves as part of the floor system, the only additional cost to achieve composite action will be that of the shear connectors. In multistory buildings, composite action is used mainly for simply supported beams resisting gravity loads. A composite system is most economical for long spans and heavy loads. The metal decking may be oriented perpendicular (Fig. 3.5.3*b*) or parallel to the supporting beam (Fig. 3.5.3*c*). The metal deck slab itself may be either composite or noncomposite (form deck). The floor beams are normally spaced such that the metal decking, acting as the concrete form, spans between the beams without requiring any additional shoring.

The **shear connector** is a short length of round steel bar welded to the steel beam at one end and having an anchorage provided in the form of a round upset head at the other end. The most common diameters, in building structures, are ⅝, ¾, and ⅞ in. The upset head thickness of the stud is usually ⅜ or ½ in., and the diameter is ½ in. larger than the stud diameter. The length of the stud is dependent on the rib depth and should extend at least 1½ in. above the top of the decking. The studs are welded to the beam with an automatic stud welding gun and, when properly executed, the welds are stronger than the steel studs. The shear connectors resist horizontal shearing forces between the slab and beam, provide anchorage for the slab, and prevent any tendency for it to flex independently of the beam. They also help in the collection and distribution of diaphragm shear from the slab to the beams and girders.

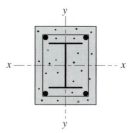

W-shape encased in reinforced concrete

(*a*)

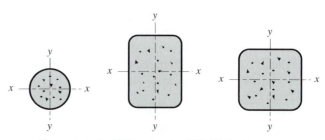

Pipe, rectangular HSS, and square HSS filled with concrete

Figure 3.5.4:   Composite columns.

(*b*)

3.5.8   Composite Columns

Steel concrete composite columns are being used more and more in building structures. They may be broadly classified as those with rolled or built-up steel shapes encased in concrete or those with concrete placed in steel pipes or HSS. Concrete encased columns are either rectangular or square in cross section. A composite column with a W-shape embedded in concrete and provided with regular reinforcement at the corners is shown in Fig. 3.5.4*a*. In addition, the column is provided with lateral ties at intervals. A steel pipe, a rectangular HSS, and a square HSS filled with concrete are shown in Fig. 3.5.4b.

3.6   Fabrication and Erection of Steel Buildings

Structural steel work is normally prepared by specialist contractors, called **steel fabricators,** who fashion the individual members of the structure, such as beams, columns, braces, and connection elements in their workshops known as **fabricating shops.**

To begin, the fabricator prepares a bill of material for the main material (preferably from the shop drawings but usually from engineer drawings) and sends it to the mill. The fabricator designs nonstandard connections in accordance with the loads (reactions, member forces, and transfer forces between members) shown on the structural drawings. This includes general configuration, size of plates and angles, number and size of bolts, and length and location of welds. These details are submitted to the structural engineer for approval. At this time, structural details will be investigated for possible combinations of pieces, reducing the total number of pieces to be erected, duplication of fabrication, and feasibility of erection. Erection procedures affecting the structural stability are also investigated at this stage. In effect, on large size projects, equipment necessary for the erection process may require heavier connections and/or members than those shown on the engineer drawings. An erection scheme would determine the location and weight of cranes and other equipment. An erection plan, a two-dimensional line drawing showing the framing at each floor, is prepared next. It indicates the location of each piece of steel to be placed, and each piece is marked with a letter or number or a combination of both. The same marks will be painted on the pieces at the fabrication plant.

3.6.1   Fabrication

Each fabricating shop has its own characteristics for handling materials. The flow of the material from the **stockyard** at one end of the plant to the **dispatch bay** at the other end of the plant is generally through the **material preparation bay,** to the **material assembly bay,** and then to the **painting bay.**

The material is shipped from the mill to the fabricating shop in ordered dimensions, shapes, grades and quantities and is stored in the stockyard until required for fabrication. The pieces then are transferred through the shop by roller conveyors or, less frequently these days, by overhead cranes through a series of saws, punches, and drills. The sequence of operations on a piece should minimize the movement of material from one operation to the other to reduce the material handling expenses and simplify production control.

The first operation in the material preparation bay is to cut the exact lengths of the main members from multiple length material. For beams and columns of rolled shapes this is done by an automatic sawing machine. **Cold saws** utilize a blade traveling at a slow rate that actually cuts away the material. A **friction saw** is a high-speed saw in which the teeth actually melt away the material. Plates or flat bars under a certain thickness are generally cut on a guillotine-type machine called a **shear.** Automatic **cropping machines** capable of shearing both legs of an angle in one stroke and punching holes, all in a single operation, are used extensively. When plates are ordered in multiple widths, the cutting of the narrow widths is done by **flame cutting** or **oxygen cutting** with a plate stripper. A traveling gantry carries several gas cutting torches set at desired widths. The plates may be cut to a required camber with this equipment. Shape cutters also make use of multiple cutting torches to cut gusset plates or other shapes, to cut beam copes, and to cut circular and rectangular holes in beam webs. The torch can be tilted to provide a beveled edge. Four axis structural coping machines incorporate three or four oxy-fuel-cutting torches that cut top and bottom flanges and the web in one pass. Using this machine, an operator positions the leading end of a section against the inside roller-feed measuring disk and then initiates the automatic cutting cycle. The operator enters parameters for section size, cut shape, and cut dimensions into the computer-numerically-controlled (CNC) program for coping centers. The software assigns preheat times, cutting speed, and torch path geometry. The CNC program activates vertical and horizontal clamps to secure the part. Clamps adjust automatically to accommodate sections to 50 in. wide. The CNC program adjusts roller clamp height and directs beam positioning. Each torch assembly carries an electronic noncontacting probing system. As commanded by the program, probes find the leading and trailing edges of the beam and the top and bottom edges of the flanges to establish the zero point for directing preheat and for gaging the cope cuts. Flange torches automatically move to the correct position based on section depth. Web torches probe the web surface to set cutting tip-to-web distance. With the part probed, flange torches move in to cut. Cutting torches preheat the surface to be cut, and then the CNC program signals the flow of cutting oxygen. The software sets a programmable pressure valve to adjust oxygen pressure for cutting of a range of material thicknesses. Contact pads on the torches maintain tip-to-mark distance. Torch assemblies are guided. Each flange-torch assembly rotates up to 45° to make bevel cuts. The slender construction of the web torch permits it to move to within 5/16 in. of the flanges. Using a cursor on the screen, the operator can notch, cope, block, cutoff, cutout, miter, bevel, make slots, and trim flanges. The operator can also command longitudinal cuts for beam splitting, castellation, and cutting of haunches.

Swinging of a floor beam into place in a tier building. *Photo courtesy of OPUS Corporation.*

A roof beam held in position by a crane, being connected to the roof girders.
Photo courtesy of OPUS Corporation.

Beams and columns are next passed through automatic, computer-controlled punching and/or drilling machines to cut holes for bolts. Punching is the most commonly used method of making bolt holes in steel. Normally, mild carbon steel material up to a thickness of $\frac{1}{8}$ in. greater than the bolt diameter can be punched. Holes are drilled otherwise. High-strength steels are somewhat harder, and punching may be limited to thinner material. Several holes can be punched simultaneously in shapes, plates, or large angles using an automatic computer-controlled multiple punch that has a number of punches arranged in a transverse row. Fully automatic drilling machines use a computer to direct multiple drill heads. In this, three axis movements of each drill can be independently programmed. Such machines are used to drill several holes in the web and/or flanges, in one operation. Also, the computer directs the piece being drilled to move a specified amount forward when a group of holes are drilled. This operation is automatically repeated until the total length of the piece being drilled is finished. Built-up sections for beams and columns (such as plate girders) are fabricated in a special bay, where component plates and/or shapes are assembled and welded together manually or automatically. When there is sufficient duplication of large assemblies, *jigs* are used. These jigs may be stationary and rotatable. When it is necessary to access both sides of an assembly for bolting or welding, the jig may be mechanically revolved. The use of jigs also assures accuracy in fabrication.

The work in the assembly bay consists of collecting the main and detail pieces of an assembly when fabrication is completed and connecting them to form a single assembled piece. Here, connection material like base plates and beam seats are attached, and splice material is placed. Also, holes are matched and reamed to full size, if necessary, to permit insertion of bolts. The assemblies are bolted or welded together next. The ends of columns are machined, if necessary, to smooth square ends so that close contact between upper and lower lengths can be obtained, on milling machines, and planers. A *milling machine* has a movable head fitted with one or more high-speed, carbide tipped rotary cutters. The head moves over a bed which securely holds the piece in proper alignment during the finishing operation.

As much fabrication as possible should take place in the workshop, where good working conditions exist, rather than on the site where conditions are generally not ideal to achieve good workmanship. However, the size of the fabricated pieces may be limited by the capacity of the fabrication shop in regard to space, machine capacity, or maximum loading of available lifting equipment. The size of fabricated pieces may also be limited by the size, capacity, and clearance restrictions of the transportation mode used (truck, rail, barge, etc.). Next, all work is checked for accuracy of overall dimensions, locations and dimensions of connections, proper assembly of fittings, and proper installation of bolts. In addition to *visual inspection,* the soundness of welds may be verified by *magnetic particle tests, dye penetration tests, ultrasonic tests,* or *radiographic tests,* if stipulated in the job specification. Cleaning and painting or galvanizing are generally done in a separate bay or building. Normally, cleaning is done to remove loose mill scale, dirt, grease and other foreign material from the steel, prior to painting. Blast cleaning may be manually or automatically applied using shot or grit. Certain automatic facilities can blast clean and apply a rust inhibitor. The fabricated pieces are next stored in the dispatch bay or shipping yard until they are shipped to the erection site in trucks, railroad cars, or barges.

3.6.2   Erection

It is essential that a close liaison is maintained between the fabricating shop and the erection team. The fabricator should ensure that the steel work is delivered to the site in the correct sequence in accordance with the erection program as storage area on site is generally limited.

The first step in the erection of a steel building is to set the base plates to line and grade. Next, a few basement columns are set up. The sling for each column is put around the member at convenient distance above the center of gravity so that the column will be almost vertical when it is picked up by the boom of the crane. The operator can bring the column exactly over the base plate on which it is to be set. The column is then fastened tightly to the anchor rods, which will hold it so that it will stand in the air without further support. This operation is repeated for several columns. The columns are spliced just above the second floor, and only the lifts below the splices are raised first.

When several columns are standing, the next step is to tie them together with various beams. For this, the sling is placed on a beam around its center of gravity so that the member will hang horizontally when it is lifted off the ground. It is customary to tie a hemp rope, known as a *hand line* or *tag line,* to one end of the beam that is being hoisted, for use in guiding the beam into place. As many of the first and second floor beams as possible are set in place

Miller Park Stadium roof structure during erection.
Photo by S. Vinnakota.

so that the erected columns will be well tied together. At the time of erection, each connection is bolted up to about 50 percent; that is, bolts are inserted in only half the number of holes available for final bolting.

With the lower lengths of several columns erected and connected to one another by beams, the next step in erection is to set the upper length of each column that was previously put up. After all the upper columns are erected, the beams that frame between those columns are put in their proper places. After the steelwork for the entire structure is erected and partially bolted up, the frame is plumbed to make the columns vertical and the beams horizontal. For this, a wire rope containing a turnbuckle is run from a point near the top of a column at one end of the line of columns to a point near the foot of a column at the other end of the line, and the rope is tightened by operating the turnbuckle. In practice, two such ropes are run in opposite directions between the same two columns so as to form an X-brace, allowing both columns to be plumbed in one direction by adjusting the length of those ropes. Since all the columns in any line are tied together by the horizontal beams, it may usually be assumed that when the columns at the ends of the line are vertical, the intermediate columns in the line are also satisfactory. However, any interior columns that must be absolutely vertical, say columns next to elevators, can be adjusted separately.

The plumbing operation is performed on each parallel row of columns in succession with all the braces being left in place. Generally, the structure is plumbed starting at the center bay and working outward. After the columns are made vertical in one direction, the process is repeated in the perpendicular direction. Thus, the entire structure is kept truly plumb, temporarily, by the system of X-bracing.

If the connections are bolted, it is necessary to place and tighten the bolts in all the connections throughout the building while the X-bracing is still in position. The plumbness of a column can usually be checked by use of a transit. If the holes for the bolts can be aligned exactly, all bolts can be inserted in the holes without difficulty. Sometimes holes are misaligned by a small amount. However, a bolt hole may be expanded about 1/64 in. to accommodate the bolt by driving a ***drift pin*** through the hole. A drift pin is a cigar-shaped piece of steel 6 to 8 in. long.

For bolting up or tightening bolts to a snug tight condition, the most convenient tool is a spud wrench. The jaws on a spud wrench are made of tough steel and are accurately machined to fit the nut of a bolt. The plain end of a spud wrench is tapered from a comparatively sharp point to a diameter equal to that of the bolt for which the jaws are designed. By inserting this end in a set of holes for a bolt, a person can easily line up the holes in the connection and hold the pieces in a proper position while a bolt is being inserted in an adjacent set of holes.

A small erection group would consist of a foreman in charge of the steel group, a crane operator, two connectors, and two hookers-on. The connectors work in the air and bolt the beams to the various columns. The hookers-on are responsible for putting slings on various pieces of steel to be hoisted, for

Lifting of a roof truss element, Miller Park Stadium. *Photo by S. Vinnakota.*

hooking the slings to the hook block on the crane, and for handling the hand line used in guiding the pieces to the connector's hand.

Buildings in the low- to medium-rise range can be erected with independent tower cranes. However, for tall buildings erection must be carried out using the building itself. Guy derricks are often used for erecting tall building frames tier by tier. They are of the high-lift type. Each tier represents a column height of usually two or three building floors. The first step in erection of tier buildings is to set up the derrick(s) in the basement of the building and to raise, by means of (each) derrick, all of the steel within its reach. The steel farthest from the derrick is set first, and the steelwork gradually closes in around the derrick so that it is finally trapped. When this happens, the next step is to raise the derrick, or jump it, to a new level so that it is supported on the part of the frame previously erected. The operations involved in raising steel are then repeated from this position of the derrick, and the steel frame is carried up to a higher level. The erection process continues by alternatively jumping the derrick and raising more steel.

The structural members in the floor of the building and columns affected by the guy derrick must be capable of providing support for the hoist loads of the derrick in addition to the dead load and live load. The structure must be analyzed for the vertical and horizontal components of guy forces. Care must be taken in analyzing composite beams regarding the horizontal distribution of guy forces. Uplift as well as gravity loads on the building structure must be checked. From the description given above, it is clear that erection engineering must include investigation of the entire structure under erection conditions. Special bracing may have to be installed and removed. Sometimes it is also necessary to leave plumbing guys in place to provide stability.

Where composite floor beams and girders are used, the steel beams alone have less load carrying capacity and resistance to lateral buckling than would beams not designed for composite construction. Hence, erectors should be furnished with the actual capacity of the beams without the concrete slab. In effect, several beam failures have occurred where, before the concrete slab was cast, beams were overloaded by steel erectors who did not appreciate the reduced capacity of a steel beam alone.

References

3.1   Fisher, J. M., West, M. A. and Van de Pas, J. P. [1991]: *Designing with Steel Joists, Joist Girders, Steel Deck*, Vulcraft.

3.2   NRCA [2003]: *Roofing and Waterproofing Manual*, The National Roofing Contractors Association, Chicago, IL.

3.3   SDI [2001]: *Design Manual for Composite Decks, Form Decks*, and *Roof Decks*, Steel Deck Institute, Fox River Grove, IL.

3.4   SJI [2002]: *Standard Specifications and Load Tables for Steel Joists and Joist Girders*, Steel Joist Institute, Myrtle Beach, SC 41st edition, Aug.

PROBLEMS

P3.1.   What is the difference between braced and unbraced frames? Illustrate with the help of a single-span, two-story frame.

P3.2.   In a tier building, the W16×36 floor beams are 30 ft long and spaced at 6 ft intervals, while the W21×62 girders are 24 ft long. The girders are connected to the flanges of W12×65 columns. Show the floor framing layout for an interior bay.

P3.3.   Illustrate any two types of shear bents in tier buildings.

P3.4.   Sketch any three types of concentrically braced shear bents, for a four-story building.

P3.5.   What are purlins and sag rods?

Snow loading on roofs can be quite large in certain localities, as shown here.

A major lake effect snowstorm dumped nearly seven feet of snow on the Buffalo, NY area in a five day period, in December 2001.

4

Design Loads and Design Philosophy

4.1   Introduction

Structural design is a process by which an optimum solution for a structural form and the sizes of its elements is obtained. This solution must satisfy the functional, economic, aesthetic, sociological, and other requirements, using one or more criteria such as minimum overall cost, minimum weight, minimum construction time, and so on. The goal of the structural engineer of a project is, therefore, to select an efficient assembly of structural elements so as to:

*   Bring to practical reality the concept and form desired by the architect and the owner.
*   Achieve this reality at minimum capital cost to the client.
*   Ensure satisfactory service life of the building by ensuring that it conforms to accepted standards of strength, deflection, vibration, and maintenance characteristics.

Steel structures, like all structures, must be designed to resist a variety of forces to which they are likely to be subjected during their lifetime. They include forces due to gravity, environmental loads (wind, snow, rain, earthquake), and loads due to use (live, traffic, crane) and misuse (blast). The intensity of loads used for the design of building structures may be found in documents called *standards, specifications, model codes,* and *codes.*

4.2   Specifications

Specifications refer to rules and guidelines written by an architect or engineer pertaining to one particular building under construction. They are legal documents. More frequently, *specifications* refer to documents developed by various engineering organizations. Specifications present the best opinion of a group of experts in a specific field of study (such as steel, concrete, timber, steel decks, and steel joists) as to what represents good engineering practice in that field. These specifications, however, have no legal standing unless they are made a part of a particular contract by reference or embodied in the local building code. Specifications are necessarily being updated continuously because of newly developed knowledge brought about by tests, theoretical developments, and experiences gained. Specifications, unlike the codes to be discussed in Section 4.3, contain detailed procedures and specific guidance for the design of structural members and connections. Designers, especially student designers, must understand the behavior for which a particular specification rule applies and its limitations before using that rule.

Specifications are written as guides to the best current practice among engineers who are recognized as experts in their particular field. However, most specifications are in the writing stage for a long time (5 to 10 years is not unusual), so it is not feasible to keep them completely up to date at all times during their currency. Hence, they do become somewhat overtaken by changes in practice with the passage of time. Specifications are written for the benefit of those designing small-to medium-size structures of a simple nature, leaving the design of large and complex structures to the experienced engineer who would often tend to work beyond the confines of a code by obtaining permissions from the approving authorities. Some of the specification writing organizations in the United States are ASCE, AISC, AASHTO, and AREA.

The American Society of Civil Engineers (ASCE) is a society comprised of individuals in the field of civil engineering. In addition to publishing several technical journals, they have developed standards including the *Minimum Design Loads for Buildings and Other Structures*. This standard is referred to hereafter as the **ASCE Standard 7** or simply **ASCES** [ASCE, 2000]. A more recent version of the standard is now available [ASCE, 2003].

The American Institute of Steel Construction (AISC), founded in 1921, is a nonprofit organization of structural steel fabricators, steel rolling companies, and individuals interested in steel design and research. It is engaged in research and dissemination of information pertaining to the use of steel products. Two specifications, known as the Allowable Stress Design Specification **(ASDS)** and the Load and Resistance Factor Design Specification **(LRFDS)** [**AISC,** 1989a, 1999], and two manuals, the *Manual of Steel Construction: Allowable Stress Design,* and the *Manual of Steel Construction: Load and Resistance Factor Design (LRFDM)* [AISC, 1989b, 2001] are among the many publications developed and maintained by the AISC. Because the emphasis of this book is on the design of structural steel buildings, the AISC Specifications and Manuals will be considered in detail in later sections.

The American Association of State Highway and Transportation Officials (AASHTO) is a national association of state highway officials. *Standard Specifications for Highway Bridges* [AASHTO, 2002] developed by AASHTO governs the design and construction of steel, reinforced concrete, and timber highway bridges. The AASHTO specifications are considered a legal set of documents since they have been adopted by state highway departments. Another national organization, the American Railway Engineering Association (AREA), is an association of individuals and companies interested in the field of railroad engineering. The design, fabrication, and erection of steel railway bridges in the United States are performed in accordance with the Specifications for Steel Railway Bridges [AREA, 1992] adopted by the AREA.

4.3   Building Codes

A ***building code*** is a legal document that, when adopted by a governmental body of a city, county, state, or country attains the force of law. A building code lays down a framework of rules within which each professional engineer must work. Thus, federal, state, county, and city governments, having legal responsibility for public safety, have developed building codes by which they control the construction of various structures under their jurisdiction. They are broadly based documents that give rules relating to safety on aspects such as structural design, fire protection, heating and air conditioning, plumbing and sanitation, lighting and transportation (elevators and stairs), and access to the physically impaired, as well as other specific design considerations. More importantly, building codes prescribe the minimum loads for which structures are to be designed.

Most building codes in small-to medium-sized communities in the United States are typically adoptions of one of the nationally recognized documents called ***model building codes.*** Large cities like New York and Chicago quite often have their own building code, but even the provisions of these codes are increasingly coming into harmony with the model codes. Until recently, there were three different building code organizations in the United States preparing three national model codes: the National Building Code (NBC), the Standard Building Code (SBC), and the Uniform Building Code (UBC). The three model building code organizations coordinated their efforts over years to develop a single code called the *International Building Code* [IBC, 2000] that now serves as the national model building code. The three model code organizations will continue to exist, but there will be no further editions of the NBC, SBC, or UBC. More recently, a new organization, the National Fire Protection Association (NFPA) developed the *Building Construction and Safety Code* [NFPA, 2002].

The IBC 2000 and NFPA 5000 adopted various reference standards with minimum modifications or exceptions to them, thus making it less complicated for users familiar with the adopted reference standards to use the code. For example, the structural section of IBC 2000 consists of nine chapters, of which Chapter 16 (Structural Design) was coordinated with the ASCE7-98, and Chapter 22 (Steel) adopted the latest AISC LRFD Specification without modification. These two chapters are the only sections of the IBC 2000 that are relevant to the subject of this text book.

The following is a brief summary of differences between the IBC and NFPA [see Miller, 2004].

"Structural Design" (Chapter 16, IBC) vs. "Structural Design" (Chapter 35, NFPA)

The IBC provides its own values for basic load combinations in both LRFD and ASD, but it recommends review of additional requirements in the ASCE 7 for design criteria. Both codes use "nature of occupancy" for determination of building categories, but the NFPA provides additional definitions in the occupancy categories. Regarding the wind loads, the IBC and the NFPA both refer to Section 6 of the ASCE 7. However, there are several modifications to the seismic requirements in the IBC compared to what is given in the NFPA and ASCE 7.

"Steel" (Chapter 22, IBC) vs. "Steel" (Chapter 44, NFPA)

Both codes have adopted the AISC-ASD and AISC-LRFD specifications without modifications.[*]

As IBC, NFPA, and LRFDS refer to ASCE 7-98, the present text refers to ASCE 7-98 as well and uses it in determining the required strengths of several structural elements in Chapters 4 and 5. Note that for these problems use of ASCE 7-2002 instead of ASCE 7-98 will not change the results.

4.4   Loads

The loads and forces for which a building structure has to be designed can be divided broadly into three classes: dead loads, use or occupancy-related loads, and environment-related loads. *Dead loads* represent the weight of permanent (fixed) materials of construction, including the self-weight of the structural elements, and other accessory parts of the building necessary for giving it the desired utility. *Occupancy-related* or *live loads* represent loads related to construction, occupancy, use, or maintenance of the structure. Some examples are loads from people, furniture, stored materials, movable partitions, cranes, movable equipment, impact, and blast. *Environment-related loads* are loads imposed on the structure due to the environment. Examples include wind, snow, ice, rain, and earthquake loads, as well as forces developed by hydrostatic pressure, soil pressure, and temperature effects. Since environmental forces vary with geographic location, a structure must be specifically designed to sustain prescribed environmental loads within the locality where it is to be built. Not all of these loads will necessarily apply to any given building structure, nor will they all be additive. The loads a particular building must be designed for are usually stipulated by an applicable building code. Where such a code is nonexistent, use of the ASCE Standard 7, *Minimum Design Loads for Buildings and Other Structures,* is recommended. *For ease of reference Tables 1-1, 4-1, 4-2, 6-1, 6-5, 6-6, 6-7, 7-2, 7-3, 7-4, C3-1, C3-2; and Figures 6-1, 6-3, 7-1, 7-2, and 7-5 from the ASCE Standard 7-98 are made available as Section W4.1 on the website http://www.mhhe.com/Vinnakota.*

---

[*]From: *The Structural Engineer and the Changing Building Codes,* by Ramond T. Miller, Structure Magazine, March, 2004. Reprinted by permission of Structure.

Classification of Buildings and Other Structures

In the United States, the life of a structure is expected to be 50 years; hence, a 50-year reference period forms the basis for many of the environmental load models that have been developed. The *basic environmental loads* given in the ASCES are loads with a 2 percent annual probability of being exceeded (50-year mean recurrence interval) and are used for the design of all structures of normal importance (i.e., most permanent structures). More stringent loading criteria are generally required for structures in which the consequence of failure may be severe. Thus, for the purpose of determining environmental loads, structures are classified into four categories as illustrated in Table 1-1 of the ASCES. Structures having normal occupancies and functions are classified as Category II and are designed for basic environmental loads. These basic environmental loads are given in the ASCES, and some of them will be discussed further in this chapter. Higher-risk (Categories III and IV) and lower-risk (Category I) structures are required to be designed by using 100-year and 25-year recurrence intervals, 1 percent and 4 percent of annual probability of being exceeded, respectively. These higher-and lower-risk situations, however, could be accounted for more conveniently by using the basic environmental loads given in the ASCES multiplied by the importance factors, *I*, given there. The specific importance factors differ according to the statistical characteristics of the environmental load under consideration and the manner in which the structure responds to that load.

4.5   Dead Loads, *D*

Dead load is the vertical load due to the weight of the various structural members and all nonstructural components that are permanently attached to the structure. Dead load does not vary with time, with regards to position and weight. Dead load in a building may consist of the weight of the roof, purlins, floor decks, floor slabs, beams, girders, columns, exterior walls and cladding, floor finishes and the underfill, suspended ceilings and their supports, permanent partitions, plumbing, and interior walls. Services in floors, such as weights for cables, HVAC ducts, water pipes, and so on, are generally included by an appropriate allowance in the floor dead loads. The dead loads calculated must also include fire protection, if required.

ASCES, LRFDM, and many building codes list the weights of building components in psf (pounds per square foot) and of building materials in pcf (pounds per cubic foot). For example, Tables C3-1 and C3-2 of the ASCES provide such summaries. There have been cases reported where the actual weight of members and construction materials in a structure has exceeded the dead load values used in its design, by as much as 20 percent. Therefore, tabular values should be used with caution. Also, allowance should be made for such factors as an increase in floor-slab thickness due to formwork and floor-framing deflections when pouring level floors, and for the weight of future wearing surfaces where there is a good possibility that they may be applied.

4.6 Live Loads

Live loads are those loads imposed on a structure by the use and occupancy of that structure and are the result of human actions. This sets them apart from natural forces, such as wind and snow loads, that may be assumed to conform to physical laws and to be predictable within specifiable limits on the basis of past experience. Human actions will never be predictable in the same way. Live loads vary with time in position and/or magnitude. They may be classified as either movable loads or moving loads. *Movable loads* are loads that may be transported from one location to another on a structure without any dynamic effect, for example, human occupants, furniture, books in a library, equipment, goods stored on a warehouse floor, and movable partitions. *Moving loads,* on the other hand, are loads that move continuously over the structure, for example, cranes on the crane runways of a building and trucks and trains on bridges. Loads incidental to construction, maintenance, and repair should also be treated as live loads.

4.6.1 Basic or Unreduced Live Loads, L_0

Because buildings, or zones of buildings (in tall buildings), are usually designed for a particular kind of usage, it is traditional to base recommended floor loadings on the type of occupancy. Such classifications could be:

1. Residential (houses, apartments, hotels).
2. Offices (offices, banks).
3. Educational (schools, colleges).
4. Public assembly (theaters, auditoriums, halls, restaurants, churches).
5. Institutional (hospitals, prisons).
6. Retail (department stores, shops, sales rooms).
7. Storage (warehouses, libraries).
8. Industrial (workshops, factories, manufacturing, fabricating and assembly plants).
9. Parking.

ASCES Tables 4-1 and C4-1 provide *minimum uniformly distributed live loads,* L_o, also known as *basic live loads* or *unreduced live loads,* which are based on the occupancy or use of the building. These tables should be used when no other code or specification governs. When selecting live loads, consideration should be given to present uses and to probable increases in the future that might result from changes in the use and occupancy of the building. The live load for which a building is designed should, wherever possible, include a reasonable allowance for such possible changes. Note that the public areas of a building, such as corridors and halls, which at times may be crowded with people, must be designed for larger live loads than private rooms. Furthermore, ASCES Section 4.2.2 requires that a provision for partition weight (usually 10 to 20 psf) shall be made in office buildings where partitions might be subject to erection and rearrangement (whether or not partitions are shown on the plans), unless the specified live load exceeds 80 psf. Also, note that heavier live loads must be used on floors where filing and storage space is provided.

4.6.2   Nominal Live Load, *L*

The live load on a building floor is not uniform, as assumed for design, but consists of different areas having different loading intensities. Most codes allow a live load reduction to reflect the fact that members that support large floor areas have a reduced probability that they will be subjected to full live loading over the entire area supported. The area used in a design to compute the total load that must be supported by a structural member is known as the *tributary area A_T*. The *influence area A_I*, on the other hand, is the actual area over which any applied load would have its effect felt by the member under consideration. No portion of a load applied outside of the influence area would be carried by the member under consideration. In multistory frames, influence areas for columns and foundations supporting more than one floor are summed. The ratio of the influence area of a member to its tributary area is designated as the *live load element factor K_{LL}* of that member. ASCES Fig. C4-1 illustrates typical influence areas and tributary areas for a structure with regular bay spacings. For members having an influence area of more than 400 sq. ft, Eq. 4-1 of the ASCES allows a reduced live load, also known as *nominal live load, L*, given by:

$$L = r_L L_o \tag{4.6.1}$$

with

$$r_L = \left[0.25 + \frac{15}{\sqrt{K_{LL} A_T}} \right] \quad \text{and} \quad r_{L\,\min} \le r_L \le 1.0 \tag{4.6.2}$$

where L = nominal live load or reduced design live load per sq. ft of area supported by the member, psf

L_o = basic live load or unreduced design live load assigned to the type of occupancy, psf

K_{LL} = live load element factor

A_T = tributary area, sq. ft

A_I = influence area, sq. ft ($= K_{LL} A_T$)

r_L = live load reduction multiplier

$r_{L\,\min}$ = minimum value to be used for the live load reduction factor, r_L

= 0.5   for members with a contributory load from just one floor such as floor beams and floor girders

= 0.4   for members supporting two or more floors such as columns and foundations

Table 4.6. 1 adapted from Table 4-2 of the ASCES gives the values of K_{LL} for a variety of structural members.

▓ TABLE 4.6.1
▓ Live Load Element Factor, K_{LL}

K_{LL}	Element
4	Interior columns; exterior columns without cantilever slabs
3	Edge columns with cantilever slabs
2	Interior beams; edge beams without cantilever slabs; corner columns with cantilever slabs
1	Edge beams with cantilever slabs; cantilever beams; two-way slabs
1	All other members not identified above

From ASCE Standards, ASCE 7-98. "Minimum Design Loads for Buildings and Other Structures." Copyright © 2000 ASCE. Reproduced by permission of the publisher, ASCE. www.pubs.asce.org.

From Eq. 4.6.2 it can be verified that the maximum reductions in live loads correspond to $r_L = 0.5$ and 0.4, and apply to A_I values greater than 3600 and 10,000 sq. ft, respectively. In heavy storage and commercial occupancies, for which live loads exceed 100 psf, the floors are more likely to be fully loaded. So the ASCES does not permit any live load reduction for members supporting one such floor ($r_L = 1.0$), and permits a reduction of 20 percent only for those members supporting more than one such floor ($r_L = 0.8$). For analogous reasons, in garages for passenger cars only, basic live loads on members supporting more than one floor may be reduced only 20 percent. No reduction is permitted by ASCES for areas of public assembly with live loads of 100 psf or less ($r_L = 1.0$).

4.6.3   Impact Loads, *I*

A live load that is suddenly applied often produces a dynamic effect, which is taken into account approximately, by specifying what is known as impact load. **Impact loads** are equal to the difference between the magnitude of the effect actually caused by a dynamically applied load and the magnitude of the effect had the load been statically applied. In buildings, impact loads are to be considered in the design of hangers supporting floors and balconies, and in the design of supports for elevators, monorails, and cranes. Such impact loading may result from the sudden stopping of a downward moving elevator, from the wheels of a traveling overhead crane as they pass over irregularities such as rail joints, and so on. Grandstands, stadiums, and similar public assembly structures may be subjected to loads caused by crowds swaying in unison, jumping to their feet, or stomping, and the possibility of such dynamic loads should be considered in design. ASCES Section 4.7 stipulates that in the design of supports for elevators, the elevator loads shall be increased by 100 percent for impact. The impact for hangers supporting floors or balconies is 33 percent.

4.6.4   Partial Loading

A partial live load pattern may create the most critical results (forces, moments, deflection, etc.) in the element or section under consideration. Section 4.6 of the ASCES stipulates that *the full intensity of the appropriately reduced live load applied only to a portion of a structure or member shall be considered if it produces a more unfavorable effect than the same intensity L applied over the full*

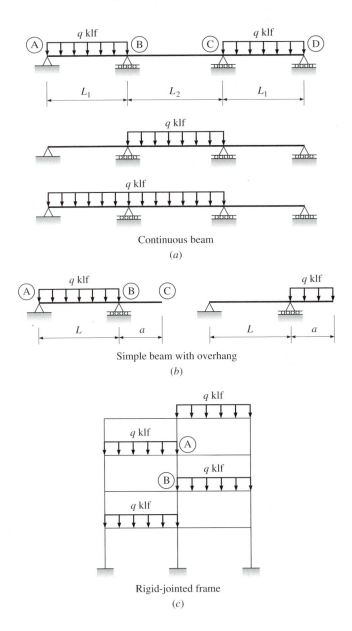

Continuous beam

(*a*)

Simple beam with overhang

(*b*)

Rigid-jointed frame

(*c*)

Figure 4.6.1: Partial live load patterns for critical loading.

structure or member [ASCE, 2000]. Thus, in many cases, a given structural member must be investigated for various positions of the live load so not to overlook a potential failure mode. For example, partial-length loads on a simple beam or truss will produce higher shear on a portion of the span than a full-length load. Several other examples are illustrated in Fig. 4.6.1. Thus, to get the maximum positive moment in the end span of a continuous beam of three spans, live loads should be put on the end spans only. For positive moment in the middle span, only that span would be loaded, whereas for maximum negative moment at the first interior support, only the middle and end span adjoining this support would be loaded with the live loads (Fig. 4.6.1*a*). For a simple beam with overhang on one side (Fig. 4.6.1*b*), maximum positive bending moment is obtained by placing the live load on the main span only; to evaluate the maximum deflection of the cantilever tip, the live load is placed only on the overhang portion of the beam. Checkerboard loading on multistoried, multispan rigid-jointed frames will produce higher positive moments (Fig. 4.6.1*c*) than full loads. A more exact determination of the positioning of the live load that will produce maximum effects can be obtained by constructing influence lines.

EXAMPLE 4.6.1   Live Load Reduction

A four-story office building has interior columns spaced 25 ft apart in one direction and 24 ft apart in the perpendicular direction. Determine the reduced live load supported by a typical interior column located at ground level.

Solution

From ASCES Table 4-1, the basic uniformly distributed live load is $L_o = 50$ psf for office areas.
Tributary area of column/floor = 25(24) = 600 ft^2.
The column considered gets live loads from three floors.
Tributary area of column, $A_T = 3(600) = 1800$ ft^2.
For interior columns, live load element factor $K_{LL} = 4$.
From Eq. 4.6.2 live load reduction multiplier,

$$r_L = 0.25 + \frac{15}{\sqrt{K_{LL} A_T}} = 0.25 + \frac{15}{\sqrt{4.0(1800)}} = 0.427$$

For members supporting two or more floors, $r_{L\,min} = 0.4$.
As $r_{L\,min} < r_L < 1.0$, use $r_L = 0.427$.
Design live load, $L = r_L L_o = 0.427(50) = 21.4$ psf.
Reduced live load in the interior column at ground level,

$$P_L = L A_T = \frac{21.4(1800)}{1000} = 38.5 \text{ kips} \qquad \text{(Ans.)}$$

4.7   Wind Loads

4.7.1   Introduction

Like any moving fluid, wind exerts pressure (force per unit area) on the surface of any body with which it comes in contact. ASCES defines a building or structure whose fundamental frequency is greater than or equal to 1 Hz as a *rigid building* or *structure.* Since most buildings are very rigid, with high natural frequencies (greater than 10 Hz), and require enormous amounts of energy from loading mechanisms to produce a dynamic response of any magnitude, it is satisfactory to treat wind as a static load in the majority of structures. Slender structures with low natural frequencies, however, are susceptible to developing a dynamic mode of response to wind. Thus, for television, radio, and transmission towers, tall chimneys, tall narrow buildings, and buildings with flexible roofs, wind loadings are of major importance. Also, wind loading may be the controlling factor in the design of the wall cladding, roof covering, girts, and purlins. Connections fastening these elements to the frame of the building must be strong enough to resist either inward or outward forces produced by the wind.

The magnitudes of wind loads on buildings vary with geographical location, height above ground, type of terrain surrounding the building including the size and type of other nearby structures [ASCE, 2000].

4.7.2   Basic Wind Speed, V

The *basic wind speed, V*, used to determine the design wind loads on buildings and other structures is defined as a three-second duration gust wind speed in miles per hour at 33 ft (10 m) above ground level in Exposure C category, as defined in Section 6.5.6.1 of the ASCES, and is associated with an annual probability of 0.02 of being equaled or exceeded for a 50-year mean recurrence interval. A map of the United States with basic wind speed contours or isolines for 50-year mean recurrence intervals is given in Fig. 6-1 of the ASCES. Note that the basic wind speed used for design in the United States ranges from 90 to 150 mph.

4.7.3   Importance Factor for Wind Loads, I_w

The wind speed map of the ASCES is based on an annual probability of 0.02 that the wind speed is exceeded (50-year mean recurrence interval). It is recommended that basic wind speeds associated with a 100-year mean recurrence interval be used for the design of structures where a high degree of hazard to life and property exists and when these structures are considered to be essential facilities. Also, structures that represent a low hazard to life and property may be designed using basic wind speeds associated with a 25-year mean recurrence interval. The importance factor for wind, I_w, given in Table 6-1 of the ASCES, adjusts the design wind speed to annual probabilities of being exceeded for values other than an annual probability of 0.02.

4.7.4   Velocity Pressure, q_z

The **velocity pressure,** q_z, defining the design wind climate, irrespective of the structure, is given by ASCES Eq. 6-13, namely:

$$q_z = 0.00256\, K_z\, K_{zt}\, K_d\, I_w\, V^2 \qquad (4.7.1)$$

where V = basic design wind speed, mph
   I_w = importance factor that accounts for different mean recurrence intervals of design wind speeds, dimensionless
   K_z = velocity pressure exposure coefficient that accounts for terrain and height above ground, dimensionless
   K_{zt} = topographic factor (= 1.0 for flat terrain), dimensionless
   K_d = wind directionality factor (= 0.85 for buildings), dimensionless
   z = height above ground level where the velocity pressure is being evaluated, ft

Values of K_z are given in Table 6-5 of the ASCES to 500 ft above ground, for Exposure categories A, B, C, and D. Note that, below the height of 15 ft, the value of K_z is taken as a constant (K_z determined at 15 ft) because of increased turbulence near the ground. For Exposure C at 33 ft above ground, the value of K_z is 1.0 because the basic wind speed of the standard is associated with this exposure and height above ground. The value of K_z varies from a minimum value of 0.32 (for 0–15 ft, Exposure A) to a maximum value of 1.89 (for 500 ft height, Exposure D). *For further information on velocity pressure, refer to Section W4.2 on the website http://www.mhhe.com/Vinnakota.*

The **wind directionality factor,** K_d, accounts for the reduced probability of maximum winds coming from any direction and the reduced probability of the maximum pressure coefficient occurring for any given wind direction (Table 6-6 of the ASCES). It equals 0.85 for buildings. The wind speed-up effect at isolated hills, ridges, and escarpments constituting abrupt changes in the general topography is included in the design by introducing the **topographic factor,** K_{zt} (ASCES Section 6.5.7). It equals one for flat terrain.

4.7.5   Design Wind Pressure, p

On a gable framed building of rectangular plan, wind exerts pressure on the windward wall, suction on the leeward and sidewalls, suction on the leeward slope of low gable roofs, and either suction or pressure on the windward slopes of the roof (Fig. 4.7.1). Also, as buildings are not completely airtight, the interior may also be subjected to pressure or suction. ASCES gives several equations for **design wind pressure,** p, which is the equivalent static pressure used in the determination of wind loads for buildings. The choice of equation depends on the height and flexibility of the structure, and whether the design is for the main wind-force resisting system or for the building's components and cladding. For example, the design wind pressure on rigid

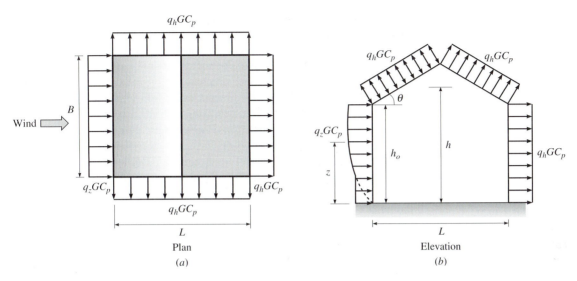

Figure 4.7.1:   Wind pressure distribution on external surfaces of a building for design of main wind-force resisting system.

buildings of all heights, for the design of main wind-force resisting systems, is given by a two-termed equation (ASCES Eq. 6-15) composed of both external and internal pressures which may be expressed as:

$$p = p_e - p_i \qquad (4.7.2)$$

or, more explicitly and conservatively as,

$$
\begin{aligned}
p &= q_z\,GC_p - q_h\,(GC_{pi}) \quad \text{for windward wall} && (4.7.3a)\\
&= q_h\,GC_p - q_h\,(GC_{pi}) \quad \text{for leeward wall, side walls,}\\
&&& \text{and roof} && (4.7.3b)
\end{aligned}
$$

where p = design wind pressure, psf

  p_e = external wind pressure, psf

  p_i = internal wind pressure, psf

  z = height above ground level

  q_z = velocity pressure evaluated at height z above ground (from ASCES Table 6-5, Case 2), psf

  h = mean roof height of a building (eave height must be used for roof angle θ of less than or equal to 10°)

  q_h = velocity pressure evaluated at height h (from ASCES Table 6-5, Case 2), psf

  C_p = external pressure coefficient from Fig. 6-3 of the ASCES

  G = gust effect factor from Section 6.5.8 of the ASCES

  GC_{pi} = internal pressure coefficient from Table 6-7 of the ASCES

The design pressure must be applied simultaneously on windward and leeward walls and on roof surfaces, normal to the surface.

External Pressure Coefficients, C_p

Design wind pressures on structures are determined by multiplying the applicable velocity pressures by appropriate pressure coefficients. The *external pressure coefficients* define the pressure or suction acting normal to the surface at local portions on the external surface of a building or structure. A negative pressure coefficient indicates suction (directed away from the surface) as distinct from a positive pressure (directed towards the surface). Both positive and negative coefficients are specified for C_p, and the worst combination must be used in the design. Pressure coefficients have been assembled from boundary layer wind tunnel and full scale tests, and from the previously available literature.

Pressure coefficients given in Fig. 6-3 of the ASCES are used for the determination of wind pressure on main wind-force resisting systems of all enclosed buildings. The values are given for windward, leeward, and side walls and windward and leeward roofs. Since the resulting pressures are for the main wind-force resisting system, the wind pressures on side walls are generally not critical. Note that the pressure coefficients for the windward wall are associated with velocity pressure (q_z), evaluated at height (z) above ground, which results in design wind pressures varying with height. For the leeward wall, however, the coefficients are used with velocity pressure (q_h) evaluated at mean roof height h, resulting in design wind pressures that are uniform over the entire height of the leeward wall. The eave height, h_o, must be substituted for the mean roof height, h, if the roof slope (θ) from the horizontal is less than or equal to 10 degrees (Fig. 4.7.1). The pressure coefficient values are provided for wind parallel to two major axes of the building. The main wind force resisting system (such as frames, bracing system, and diaphragms) should be designed to resist winds from these directions. The negative pressures at eaves and corners are higher than the average pressure over the side or roof of the building. This produces potential points of initial failure, where the roofing or even the roof or wall could separate from the main frame of the building. These locations must be designed for higher pressures than those used to design the main frames.

Gust Effect Factor, G

For rigid structures, the value of the gust effect factor G may be taken as 0.85 or calculated by methods given in Section 6.5.8 of the ASCES.

Internal Pressure Coefficients, GC_{pi}

The internal pressures are caused by openings in walls, and their magnitude and sense depend on the area of these openings. They could be permanent openings or openings caused by windstorms or air leakage. Operable windows and doors should be considered as openings since they may be inadvertently left open during high winds. Also, if windows and doors are likely to break during a storm by the action of wind borne debris, they should be considered as openings. For the purpose of calculating internal wind pressures, buildings are classified in ASCES Section 6.2 as open, partially enclosed, or enclosed.

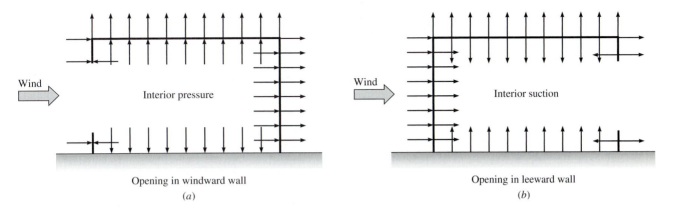

Interior pressure

Interior suction

Opening in windward wall

Opening in leeward wall

(a)

(b)

Figure 4.7.2: Influence of opening in one wall on wind pressure.

Windward wall openings cause an increase in pressure within the building (Fig. 4.7.2a). The increase in internal pressure combines with the outward pressure already acting across the leeward wall, the roof, and the sidewalls to intensify the net, outward-acting pressures across these surfaces. Conversely, an opening in a side wall or leeward wall causes a decrease in pressure inside the building because the air is drawn out (Fig. 4.7.2b). The decrease in internal pressure combines with the inward-acting pressure across the windward wall to produce a larger inward-acting pressure across this surface. The decrease in internal pressure tends to relieve outward-acting pressures across the roof, sidewalls, and leeward walls.

The internal wind pressures for main wind-force resisting systems may conservatively be calculated using velocity pressure q_h evaluated at height h. Values of GC_{pi} are given in ASCES Table 6-7. They are 0.0 for open buildings, ± 0.55 for partially enclosed buildings, and ± 0.18 for enclosed buildings. G and C_{pi} should not be separated since the combined values are measured from wind tunnel tests.

Remarks

ASCES 7-98 also defines zones of high-suction pressure along wall corners from ground to roof and along roof lines, and gives the pressure coefficients used in these regions for the design of cladding and components such as girts, purlins, and so on. (For details see ASCES Figs. 6-5 to 6-8.) These pressure coefficients may result in design pressures on some building components that are up to five times the pressures used for building frame design.

4.7.6 Design Wind Loads

The process of evaluating wind loads on a building begins with establishing a basic wind speed in miles per hour, based on the geographical location of the structure. This velocity, after making allowance for exposure category, importance of the structure, variation in the velocity with height, wind directionality, and topography is then converted to velocity pressure in psf. Next, a variety of modifying coefficients are applied to arrive at a design wind pressure (in psf) that is imposed on the whole building or specific parts of it.

The design wind pressure distribution up the side of the building is converted to wind loads acting on the structural framework, taking into account the way the cladding is supported. In multistory buildings, the wind loads are almost always applied joint loads. Half of the wind load acting on the rectangular wall panel bounded by two adjacent column lines and two adjacent floor levels is applied to the spandrel beam at the top of this wall panel, and the other half of this load is applied to the spandrel beam at the bottom of this wall panel. The spandrel beam spans between the columns and transfers the wind loads as concentrated loads on the columns at the floor levels (at the joints of the framework). Similar reasoning indicates that, in the case of single-story industrial buildings, wind loads are transferred to the columns as concentrated loads at the girt levels and at the eaves level (see Figs. 3.4.1 and 3.4.3).

EXAMPLE 4.7.1

Wind Pressure Distribution

Determine the wind pressure distribution on the windward and leeward faces and the roof of a four-story office building located on level ground in a suburban area of Milwaukee, Wisconsin. The building is 60 ft wide, 100 ft long, and 48 ft high. The wind direction is normal to the long face. There are no areas where more than 300 people can congregate.

Solution

Height of building, $h = 48$ ft
Plan dimensions of the building:

$$\text{Parallel to wind direction, } L = 60 \text{ ft}$$

$$\text{Normal to wind direction, } B = 100 \text{ ft}$$

a. Velocity pressures, q
From Fig. 6-1 of the ASCES, the basic wind speed, V, for Milwaukee is 90 mph. The building is located in a suburban area. Therefore, use Exposure B (See ASCES Section 6.5.6).

The building function is office space. It is not considered an essential facility or likely to be occupied by 300 people in a single area at one time. Therefore, the building is of Category II, from Table 1-1 of the ASCES.

Note: For a copy of ASCES tables and figures referred to in this Example go to Web Section W4.1.

TABLE X4.7.1

z (ft)	K_z	q_z (psf)	p_{ez} (psf)
12	0.57	10.0	6.8
24	0.65	11.4	7.8
36	0.74	13.0	8.8
48	0.80	14.1	9.6

The importance factor I_w equals 1.0 for Category II structures (see ASCES Table 6-1). The topographic factor K_t is taken to be 1.0 because the building is on a level ground. The wind directionality factor K_d equals 0.85 for buildings, from Table 6-6 of the ASCES. Velocity pressure at height z above ground level (Eq. 4.7.1):

$$q_z = 0.00256\, K_{zt}\, K_d\, V^2\, I_w\, K_z = (0.00256)(1.0)(0.85)(90^2)(1.0\, K_z)$$

$$= 17.6\, K_z$$

Velocity pressure exposure coefficients K_z for the main wind-force resisting system (Case 2) and Exposure B may be obtained from ASCES Table 6-5. Values of K_z and the resulting velocity pressures q_z at floor and roof levels are given here in Table X4.7.1.

The velocity pressure at mean roof height, q_h, is 14.1 psf.

b. External wind pressures, p_e, for the main wind-force resisting system are calculated using Eq. 4.7.3 as

$$p_e = q_z\, GC_p \quad \text{for windward wall}$$

$$= q_h\, GC_p \quad \text{for leeward wall and roof}$$

The least width of the building is $L_{min} = 60$ ft, thus:

$$\frac{h}{L_{min}} = \frac{48}{60} = 0.8 < 4.0, \text{ indicating that it is a rigid structure.}$$

Therefore, from ASCES Section 6.5.8.1, the gust factor is $G = 0.85$.

The values for the external pressure coefficients C_p are obtained from Fig. 6-3 of the ASCES:

- The windward wall pressure coefficient is 0.8. So the external wind pressures, $p_{ez} = q_z\, G\, C_p = 0.85(0.8)q_z$, for the windward wall are as given in Table X4.7.1.   (Ans.)
- The leeward wall pressure coefficient is a function of the L/B ratio. For wind normal to the 100 ft face, $\frac{L}{B} = \frac{60}{100} = 0.6$. Therefore, the leeward wall pressure coefficient is -0.5.

The leeward wall pressure $= q_h\, GC_p = (14.1)(0.85)(-0.5)$

$$= -6.0 \text{ psf} \qquad \text{(Ans.)}$$

[*continues on next page*]

Example 4.7.1 continues ...

- For the given structure $\theta = 0$ and $\frac{h}{L} = 0.8$. Use, conservatively, roof pressure coefficients corresponding to $\theta = 0$ and $h/L = 1.0$ for which two roof zones are specified for roof wind pressures:

$$\text{For } 0 \text{ to } \frac{h}{2}, \text{ that is from 0 to 24 ft,} \quad C_p = -1.3$$

$$\text{For } \frac{h}{2} \text{ to } L, \text{ that is from 24 to 60 ft,} \quad C_p = -0.7$$

The $C_p = -1.3$ may be reduced with the area over which it is applicable.

$$\text{Area} = 24(100) = 2400 \text{ ft}^2 > 1000 \text{ ft}^2$$

From Fig. 6-3 of the ASCES, reduction factor = 0.8.

$$\text{Reduced } C_p \text{ for 0 to 24 ft} = -1.3 \times 0.8 = -1.04.$$

Roof wind pressure:

$$0 - 24 \text{ ft} = (14.1)(0.85)(-1.04) = -12.5 \text{ psf} \qquad \text{(Ans.)}$$
$$24 - 60 \text{ ft} = (14.1)(0.85)(-0.7) = -8.4 \text{ psf} \qquad \text{(Ans.)}$$

c. **Internal wind pressures** p_i **for the main wind-force resisting system**
The building is classified as a partially enclosed building. From Table 6-7 of the ASCES, $GC_{pi} = \pm 0.55$, for partially enclosed buildings.

$$\text{Internal wind pressures, } p_i = q_h (GC_{pi}) = \pm 14.1(0.55)$$

$$= \pm 7.8 \text{ psf} \qquad \text{(Ans.)}$$

4.8   Snow Loads, *S*, and Rain Loads, *R*

In the cold northern, eastern, and Rocky mountain states of the United States, roof design is generally controlled by snow loads. Freshly fallen dry snow may weigh as little as 6 pcf, whereas wet snow can weigh 12 pcf, and compact snow might approach 50 pcf. In general, the ground snow load varies from 30 to 40 psf in the northern and eastern states and 20 psf or less in the southern states. It should be noted, however, that lake-effect storms may produce ground snow loads in excess of 75 psf along portions of the Great Lakes, and that in some areas of the Rocky Mountains, ground snow loads may exceed 200 psf.

The proper design snow load to be used for a structure will depend on a number of factors, geographic location being the most important. Some other factors include the occupancy and function of the structure, geometry of the roof, roof exposure, roof thermal condition, and character of the roof surface. Snow may drift against parapets or build up in valleys. Snow may slide off a higher roof onto a lower one. Also, wind may blow snow off one side of a sloping roof, resulting in unbalanced snow loads on a symmetrical gable frame. Many failures of roofs during winters in northern areas of the United States may be attributed to snow loads and snow drifts.

The snow loads are specified as so many pounds on the horizontal projection (plan area) of the roof surface. In many areas, the local building code authority establishes the design snow load. In its absence, ASCES Figure 7-1 could be used to obtain ground snow loads, p_g, for the United States. For calculation of design flat-roof snow loads, p_f, and design sloped-roof snow loads, p_s, refer to Section 7 of the ASCES.

For flat roofs, particularly those in warmer climates, rain loads may be more critical than snow loads. Rain loads, R, are considered in Section 8 of the ASCES. If water on a flat roof accumulates faster than it is drained, the result is called **ponding.** The increased load due to the retained rain water causes the roof to deflect into a dish shape that can accumulate more water, which causes even greater deflections. This process may continue until equilibrium is reached, or until collapse occurs, if the roof framing does not possess adequate stiffness to prevent progressive deflection. The best method of preventing this **ponding instability** or **ponding failure** is to have an appreciable slope of the roof ($\frac{1}{4}$ in./ft or more) together with a good draining system.

Snow Loads

EXAMPLE 4.8.1

Determine the balanced and unbalanced snow loads for the gabled frame of an industrial building. The frame has a span of 96 ft and a rise of 24 ft. The building length is 300 ft. It is located in Milwaukee, Wisconsin in a suburban area with minimal obstruction from surrounding buildings and the terrain.

Solution

a. Span, $L = 96$ ft
Rise, $h = 24$ ft

$$\text{Roof slope, } \theta = \tan^{-1}\left(\frac{h}{L/2}\right) = \tan^{-1}\left(\frac{24}{48}\right) = 26.57°$$

From Fig. 7-1 of the ASCES the ground snow load p_g for Milwaukee is 30 psf. The exposure factor C_e can be taken equal to 0.9 from Table 7-2 of the ASCES because there are minimal obstructions (fully exposed conditions) and Exposure B criteria apply for a suburban area (see ASCES Section 6.5.6.1). Thermal factor C_t is 1.0 from Table 7-3 of the ASCES because the building has to be a heated structure.

....

Note: For a copy of ASCES tables and figures referred to in this Example go to Web Section W4.1.

[continues on next page]

Example 4.8.1 continues ...

The building function is industrial space. It is not considered an essential facility or likely to be occupied by 300 persons in a single area at one time. Therefore, building Category II is appropriate. Thus, importance factor I equals 1.0 from Table 7-4 of the ASCES.

From Eq. 7-1 of the ASCES, the flat-roof snow load is obtained as

$$p_f = 0.7\, C_e\, C_t\, I\, p_g = 0.7(0.9)(1.0)(1.0)(30) = 18.9 \text{ psf}$$

From Fig. 7-2 of the ASCES, the roof slope factor C_s is 1.0 for warm roofs with roof slope $\theta = 26.57° < 30°$.

From Eq. 7-2 of the ASCES, the sloped-roof snow load is obtained as

$$p_s = C_s\, p_f = 1.0(18.9) = 18.9 \text{ psf} \tag{Ans.}$$

This load is called the *balanced snow load* and is applied to the entire roof of the structure.

b. Unbalanced snow load

Roof length parallel to the ridge line $L = 300$ ft.
Horizontal distance from eave to ridge $W = 48$ ft.

$$\frac{L}{W} = \frac{300}{48} = 6.25 > 4$$

So, from Eq. 7-3 of the ASCES, the gable roof drift parameter β equals 1.0
From Eq. 7-4 of the ASCES, snow density,

$$\gamma = 0.13\, p_g + 14 = 0.13\,(30) + 14 = 17.9 \text{ pcf}$$

$$\text{Minimum slope, } \theta_{min} = \frac{275\ \beta p_f}{\gamma W} = \frac{275(1.0)(18.9)}{17.9(48)} = 6.1°$$

As $W > 20$ ft and $\theta > \theta_{min}$, the structure must be designed to resist an unbalanced uniform snow load on the leeward side (ASCES Section 7.6.1 and ASCES Fig. 7-5).
Intensity of unbalanced snow load on the windward side is:

$$p_{uw} = 0.3\, p_s = 0.3(18.9) = 5.7 \text{ psf} \tag{Ans.}$$

Intensity of unbalanced snow load on the leeward side is:

$$p_{ul} = 1.2\left(1 + \frac{\beta}{2}\right)\frac{p_s}{C_e} = 1.2\left(1 + \frac{1.0}{2}\right)\frac{18.9}{0.9}$$
$$= 37.8 \text{ psf} \tag{Ans.}$$

4.9   Earthquake Loads, *E*

For a brief introduction to earthquake loads on steel structures, refer to Section W4.3 on the website http://www.mhhe.com/Vinnakota.

www

4.10 Design Philosophies

4.10.1 Limit States

A *limit state* is a condition that represents a boundary of structural usefulness beyond which the structure ceases to fulfill its intended function. Limit states may represent the actual collapse of whole or part(s) of a structure due to fracture or instability. They may also be conceptual, such as formation of a plastic hinge or a plastic mechanism; they may be dictated by functional requirements, such as maximum deflections or drift; or they may be arbitrary, such as maximum levels of stress beyond which the actual stresses should not rise. Limit states may be divided into two types: *limit states of strength* (or *ultimate limit states*) and *limit states of serviceability.*

Limit states of strength relate to safety against the extreme loads during the intended life of the structure and are concerned with strength of the element, member, or structure. The following limit states of strength, many of which will be defined in later chapters, are the most common:

>Fracture of a tension member
>Buckling of a column
>Onset of yielding
>Formation of a plastic hinge
>Formation of a plastic mechanism
>Lateral buckling of a beam
>Flexural-torsional buckling of a beam-column
>Local buckling of plates
>Overturning as a rigid body
>Frame instability
>Torsional buckling of a column
>Rupture of connection element
>Ponding

Limit states of strength may vary from member to member, and several limit states may apply to a given member. The controlling limit state is the one that results in the lowest design strength. For example, some of the possible strength limit states for a rolled steel beam are flexural strength, shear strength, lateral buckling strength, and local buckling of flange plate or web plate.

Limit states of serviceability are concerned with the functional requirements of the structure under normal service conditions. They are formulated to insure that malfunctions during everyday use of the structure are rare. Although malfunctions do not result in structural failure, they can cause discomfort to occupants, distress to nonstructural elements, and they can reduce or even eliminate any economic gain. Serviceability limit states include deflection limitations of a beam, drift limitation of a column, rotation limitations for a connection, and vibrations of a floor beam, among others. Many serviceability criteria are common sense or practice tested rules relating to limiting dimensions such as length-to-depth ratio restrictions for beams. In general, the consequence of occasionally exceeding these serviceability criteria is not catastrophic and thus can be tolerated. It is therefore accepted that in designing against wind-induced building deflections, for example, the design wind has a relatively short return period of 10 years.

4.10.2 LRFD Design Criteria

In the Load and Resistance Factor Design Specifications (LRFDS) the design strength of each structural component or assemblage must equal or exceed the required strength based on the factored nominal loads [Galambos and Ravindra, 1976; Ellingwood et al., 1982]. That is,

$$\phi R_n \geq \sum \gamma_i Q_i \qquad (4.10.1)$$

$$\text{Factored resistance} \quad \geq \quad \text{Factored load effects}$$

or

$$R_d \geq R_{\text{req}} \qquad (4.10.2)$$

Design strength provided $\geq$ **Required strength**

where $\sum$ = summation
   i = type of load (dead load, live load, wind, etc.)
   Q_i = nominal load effect
   γ_i = load factor corresponding to Q_i
   R_n = nominal strength
   ϕ = resistance factor corresponding to R_n
   $R_d = \phi R_n$ = design strength
   $R_{\text{req}} = \sum \gamma_i Q_i$ = required strength

The right side of Eq. 4.10.1 is the ***required strength*** of the element under consideration (beam, column, connection, etc.) computed by an analysis of the structure under factored loads in the combination considered. The summation accounts for loads from different sources and allows for a different load factor to be assigned to each load. The right side is essentially material independent. The load factors γ reflect the fact that the actual load effects may deviate from the nominal values of Q_i computed from the specified nominal loads. These load factors account for uncertainties in the determination of loads and inaccuracies in the theory. They provide a margin of reliability to provide for unexpected loads. However, they do not account for gross error or negligence.

The left side of Eq. 4.10.1 represents a limiting structural capacity or resistance, also known as ***design strength,*** provided by the selected element. The resistance factor ϕ is always less than or at most equal to 1.0 because the actual resistance may be less than the nominal value, R_n, computed by the equations given in the LRFDS. The design criteria, expressed by Eq. 4.10.1, ensures that a limit state is violated only with an acceptable small probability by using the specified load and resistance factors and nominal load and resistance values.

The LRFD specification, like all other structural specifications, treats almost exclusively the limit states of strength because of the overriding considerations of public safety for people and property. This does not signify that the limit states of serviceability are not important, and the designer must equally ensure functional requirements. However, these latter considerations permit more exercise of judgement on the part of the designers. Minimum considerations of public safety, on the other hand, are not matters of individual judgement, and thus codes and specifications dwell more on the limit states of strength than on the limit states of serviceability.

For further information on probabilistic bases for LRFD, refer to Section W4.4 on the website http://www.mhhe.com/Vinnakota.

4.10.3   Load Combinations (LC-1 to LC-7)

ASCES 7 stipulates that the following loads are to be considered in the design of a steel structure:

D = dead load
L = live load due to occupancy
L_r = roof live load
S = snow load
R = load due to initial rain water or ice exclusive of the ponding contribution
W = wind load
E = earthquake load
F = load due to fluids
H = load due to lateral earth pressure, ground water pressure, etc.
T = self-straining force

Each of these load quantities represents a ***mean-maximum lifetime value.*** They are also known as ***nominal loads or service loads.*** For design purposes, the loads stipulated by the applicable code under which the structure is designed (or dictated by the conditions involved) shall be taken as ***nominal loads.*** In the absence of a code, the nominal loads shall be those given in the ASCES.

The product of a nominal load, Q_i, and the appropriate load factor, γ_i, is called a ***factored load.*** Factored loads must be used when checking ultimate limit states. The load factors γ provide for:

* Load variations with time.
* Uncertainties about load location on structure.
* Design idealizations such as assumption of uniform load distribution.
* Uncertainties in structural analysis.

According to the LRFD Specification Section A4, structures, components, and foundations shall be designed so that their design strength equals or exceeds the effects of the factored loads in the seven load combinations defined in ASCES Section 2.3.2. Underlying the list of factored load combinations used in the strength design or LRFD method is the idea that it is very unlikely that two types of load effects will simultaneously reach their lifetime maximum values. Thus each design load combination includes the dead load, which is always present, one of the other loads at its **maximum lifetime value,** and the rest of the loads at their **arbitrary point-in-time value**—the value which can be expected to be acting on the structure at any instant of time. A condensed version of the seven load combinations given in Section 2.3.2 of the ASCES, with fluid loads, load due to lateral earth pressure, and self-straining forces removed, are given below and referred to in this text as **load combinations LC-1 to LC-7:**

$$1.4D \tag{LC-1}$$
$$1.2D + 1.6L + 0.5(L_r \text{ or } S \text{ or } R) \tag{LC-2}$$
$$1.2D + 1.6(L_r \text{ or } S \text{ or } R) + (0.5L \text{ or } 0.8W) \tag{LC-3}$$
$$1.2D + 1.6W + 0.5L + 0.5(L_r \text{ or } S \text{ or } R) \tag{LC-4}$$
$$1.2D + 1.0E + 0.5L + 0.2S \tag{LC-5}$$
$$0.9D + 1.6W \tag{LC-6}$$
$$0.9D + 1.0E \tag{LC-7}$$

(4.10.3)

For garages, areas occupied as places of public assembly, and all areas where the live load is greater than 100 psf, the load factor on L in combinations LC-3, LC-4, and LC-5 shall equal 1.0 instead of 0.5. The notation (L_r or S or R) indicates that the worst effect from roof live load, snow load, or rain load shall be considered where appropriate, but the three loads need not be assumed to act simultaneously. Where applicable, L shall be determined from the reduced design live load specified for the given member, by using the appropriate live load reduction factor based on the influence area. For structures carrying live loads that induce impact (see Section 4.6.3), the nominal live load used shall be increased to provide for this impact in combinations LC-2 and LC-3 only.

The notations D, L, L_r, S, R, W, and E in the load combinations represent either these loads themselves or the load effects such as the forces and moments caused by these loads acting on the structure. The results will be the same, provided the principle of superposition is valid. This is usually true when deflections are small and the stress-strain behavior is linear elastic; consequently, second-order effects (defined in Chapter 5) can be neglected. The linear elastic assumption is usually valid under normal in-service loads and, although not true at the strength limit states, is permissible as a design assumption.

Each load combination in Eq. 4.10.3 represents the total design loading condition when a different type of load is at its maximum lifetime (critical) value. Thus, we have:

Combination	Critical Load
LC-1	Dead load, D (during construction)
LC-2	Live load, L
LC-3	Roof load, L_r or S or R
LC-4	Wind load, W (acting in direction of dead load, D)

LC-5	Earthquake load, E (acting in direction of dead load, D)
LC-6	Overturning (W opposing dead load, D)
LC-7	Overturning (E opposing dead load, D)

The maximum lifetime value of a variable is higher than its mean maximum value by an amount depending on its variability. The magnitude of load factors also vary depending on the type and combination of loads. The load factor value of 1.4 used for dead loads is smaller than the value of 1.6 used for live loads because of the smaller variability of dead loads. The load factors specified are based on a survey of reliabilities inherent in existing design practice.

The nominal loads are substantially higher than the arbitrary point-in-time values. To avoid having to specify both a maximum lifetime and an arbitrary point-in-time value for each load type, only nominal values are used in the load combinations; thus some of the specified load factors are less than unity in combinations LC-2 through LC-7. For example, the mean value of arbitrary point-in-time live load, L_a, is of the order of 0.24 to 0.4 times the mean maximum lifetime live load, L, for many occupancies. The load factor 0.5 assigned to L in equation LC-3 reflects the statistical properties of L_a, and the ratio L_a/L introduced so that L instead of L_a could be used in the third, fourth, and fifth load combinations too. The arbitrary point-in-time wind load, W_a, acting in conjunction with the maximum lifetime live load, L, is the maximum daily wind. It turns out that $\gamma_{Wa} W_a$ is a negligible quantity. Consequently, combination LC-2 does not include the term W.

In some circumstances, a lesser load, rather than a higher load, may cause worst-case effects in terms of stresses, deflections, instability. For example, the tensile forces in anchor rods under lateral wind loads will be a maximum when the sustained vertical load is reduced. Load combination LC-6 takes this into consideration by using a load factor of 0.9 on dead load. Load combinations LC-6 and LC-7 are included to study the limit states of uplift and overturning of structures and to evaluate the tensile forces in anchorages and column splices.

Remarks

- Different load combinations may control the design of different elements of a given structure.
- For a given member, different limit states may be controlled by different load combinations.
- The most unfavorable effect may occur when one or more of the contributing loads are not acting.
- Special load factors and load combinations can be stipulated for unusual structures or circumstances, by the engineer in charge or the owner.
- Often, a structure has to be designed for several load combinations, rather than for just one combination.

When considering members subjected to gravity loads only, such as floor-framing members, the number of applicable load combinations reduces to the following two:

$$1.4D \quad \text{and} \quad 1.2D + 1.6L$$

We observe that the first load combination can only control when the dead load, D, exceeds eight times the live load, L. Designs with such loadings are infrequent; one possible case is an unshored composite beam during construction, when the steel beam alone has to resist all the construction (essentially dead) loads. Thus, in routine gravity-load design of floor framing members, only the load combination $1.2D + 1.6L$ need be considered to determine the required strength. Similarly, in routine gravity load design of roof-framing members, only the load combination $1.2D + 1.6 L_{rSR}$ need be considered to determine the required strength.

EXAMPLE 4.10.1

Load Combinations

The axial forces in a diagonal member of a roof truss have been calculated as 200 kips from dead load, 50 kips from roof live load, 180 kips from snow load, 100 kips from rain load, ± 160 kips from wind, and ± 60 kips from earthquake. For this case, tensile forces are considered positive. Determine the required strength of the tension member as per LRFDS.

Solution
We have:

Dead load, D	=	200 kips
Roof live load, L_r	=	50 kips
Snow load, S	=	180 kips
Rain load, R	=	100 kips
Wind load, W	=	± 160 kips
Earthquake load, E	=	± 60 kips
Live load, L	=	0

Also, let:

$$L_{rSR} = \max (L_r, S, R) = \max (50, 180, 100) = 180 \text{ kips}$$

The different load combinations to be considered are shown in the table. The critical load combination is that given by load combination LC-3 (maximum snow load) and

$$\text{Required axial tensile strength, } T_{\text{req}} = 656 \text{ kips}$$

The negative result of load combination LC-6 indicates that there is stress reversal in the member; therefore the required axial compressive strength of the member is $P_{\text{req}} = 76$ kips.

Combination	Factored Axial Force (kips)	
LC-1	$1.4D = 1.4(200)$	$= 280$
LC-2	$1.2D + 1.6L + 0.5\,L_{rSR} = 1.2(200) + 0.5(180)$	$= 330$
LC-3	$1.2D + 1.6L_{rSR} + 0.8W = 1.2(200) + 1.6(180)$	
	$+ 0.8\,(160)$	$= \mathbf{656}$
LC-4	$1.2D + 1.6W + 0.5L + 0.5L_{rSR} = 1.2(200) + 1.6(160)$	
	$+ 0.5\,(180)$	$= 586$
LC-5	$1.2D + 1.0E + 0.5L + 0.2S = 1.2(200) + 1.0(60)$	
	$+ 0.2\,(180)$	$= 336$
LC-6	$0.9D + 1.6W = 0.9(200) - 1.6(160)$	$= \mathbf{-76}$
LC-7	$0.9D + 1.0E = 0.9(200) - 1.0(60)$	$= 120$

Note: Load combinations LC-3 and LC-6, shown in bold face, represent the critical load combinations.

Note: Several additional examples on load combinations are worked out in Chapter 5.

4.10.4   Resistance Factors, ϕ

Resistance factor, ϕ, accounts for the uncertainties in predicting the resistance of a structural element. The randomness in the resistance, R, is due to the variability inherent in the mechanical properties of the materials, and variations in dimensions due to rolling. The randomness in resistance includes the influence of the variations in geometrical properties introduced by fabrication tolerances and welding tolerances, and variations in erection. The randomness in the resistance, R, also reflects the uncertainties of the assumptions used in determining the resistance from design models. Among others, these uncertainties can be the result of using approximations for theoretically exact formulas, making assumptions such as ideal elasticity, ideal plasticity, and/or using beam theory instead of the theory of elasticity. The resistance factor ϕ is less than or at most equal to 1.0 because there is always a chance for the actual resistance to be less than the nominal value R_n computed by the equations in the LRFDS. For uniform reliability, the greater the scatter in the test data for a given nominal resistance, the lower its resistance factor will be.

4.10.5   Allowable Stress Design (ASD)

If, in Eq. 4.10.1, all loads in a load combination are assumed to have the same average variability, one may substitute $\gamma = \gamma_i$ for all i, and rewrite the expression as:

$$\sum Q_i \le \frac{R_n}{(\gamma/\phi)} \quad \longrightarrow \quad \sum Q_i \le \frac{R_n}{\Omega} = R_a \qquad (4.10.4)$$

where $\Omega = \gamma/\phi$ is known as the factor of safety for the combination considered. If, in addition, the load effects ΣQ_i are linearly related to the resistance, and both are expressed as stress, then the above equation can also be written as:

$$\sum f_i \leq \frac{F_n}{\Omega} = F_a \qquad (4.10.5)$$

where Σf_i = resultant stress for load combination i
Ω = factor of safety
F_a = allowable stress for the limit state considered

Eqs. 4.10.4 and 4.10.5 are the basis of design formats for the allowable stress design method.

The allowable stress design philosophy has been the principal design approach used by AISC during the past 70 years. In the **allowable stress design (ASD) method,** the limits of structural usefulness are allowable stresses that must not be exceeded when the forces in the structure under service loads (working loads) are evaluated by an elastic analysis. Thus,

$$\text{Load Effect} \quad \leq \quad \text{Resistance}$$
$$f \quad \leq \quad F_a = \frac{F_{\lim}}{\Omega} \qquad (4.10.6)$$

where f is the actual stress; F_a is the allowable stress; $F_{\lim}$ is a stress that denotes a limit of usefulness such as the minimum specified yield stress F_y of the material, the ultimate tensile stress F_u at which the material fractures, a critical stress F_{cr} at which a compressed element or member buckles, or the stress range F_{sr} in fatigue. The ratio of a limit state stress to an allowable stress is called the **factor of safety.** The factors of safety, Ω, are numbers always greater than one. A different factor of safety is stipulated for each conceivable type of failure. Thus, for the ductile failure of a tension member the Allowable Stress Design Specification of the AISC (ASDS) requires a Ω of 1.67, for the fracture of a tension member a Ω of 2.0, while for the buckling failure of columns a Ω of 1.67 to 1.92 is stipulated. However, the ASDS does not explicitly state the factor of safety. Instead, the allowable stress is given as a multiple of the appropriate limiting stress, $F_{\lim}$.

In the allowable stress design philosophy, the entire variability of the loads and the resistance is placed on the strength side of the equation in the form of a factor of safety. The factors of safety given in ASDS were not determined by statistical analysis. Instead they evolved as a result of experience, experiments, and judgement over a number of years. In its application the structure is loaded with the stipulated service loads or working loads. Elastic theory is used in the analysis for the various load cases, and these are combined by superposition to give the worst design case. Sections are sized using elastic theory to ensure that Eq. 4.10.6 is satisfied everywhere.

4.10.6   LRFD Specification

The *Load and Resistance Factor Design Specification for Structural Steel Buildings* [AISC, 1999] gives the nominal stresses and resistance factors for different limit states for different types of members. Calculated or actual stresses under appropriate load combinations are expressed by lower case letters, f. The nominal stresses given are designated by capital letters, usually F, with an appropriate subscript. Nominal stresses are given for different cases of loading:

F_t, for tension;
F_b, for bending;
F_v, for shear;
F_p, for bearing;
F_{cr}, for compression;
and so on.

The LRFD Specification is divided into 14 chapters, namely Chapter A to Chapter N (for example, Chapter F. Beams and Other Flexural Members). Within each chapter, major headings are labeled with the chapter designation followed by a number (for example, F1. Design for Flexure). Further subdivisions are numerically labeled (for example, F1.2. Lateral-Torsional Buckling). Several appendices and numerical tables follow the body of the Specification. They are considered to be official parts of the Specification and carry the same authority as material in the body. The tables are followed by a commentary (referred to as **LRFDC** in this book), which is nonmandatory and gives background and elaboration on many of the provisions of the Specification. The commentary uses the same numbering scheme as the Specification, so material applicable to a particular section can be easily located.

4.10.7   LRFD Manual

The third edition of the *LRFD Manual of Steel Construction* contains design tables, design information for structural members, and information on connections. The Manual has been divided into 17 parts. Preceding each section or table in each of the first 15 parts is a short text explaining the use of the information that follows. Part 16 contains several specifications and codes. In particular, Part 16.1 consists of the LRFD Specification and commentary.

4.11   Design Process

Conceptually the design of a building consists of a series of iterative and often overlapping steps:

1. One or more conceptual designs of the structure are developed by the architect. These functional designs will give the general shape, form, and size of the building based on the needs of the client, code requirements,

physical and monetary constraints (rental space, floor areas, clearances, etc.), and aesthetics. Structural form and materials are identified, and some preliminary sizing of members is done. Provision is made for adequate transportation facilities (elevators, stairways, crane runways); lighting, heating, ventilation, and air conditioning; and fire protection.

2. The forces and loads that are likely to act on the structure are determined by the structural design team. In some cases, such as live loads on floors or snow loads on roofs, they are specified by regulatory authorities. In other cases, loads such as the wind load on a projected tall building may depend on the shape and size of the proposed structure in relation to its existing built environment and may best be determined by wind tunnel tests on models. The structural engineer should also be aware of the needs of the mechanical engineer (layout of duct work for heating, ventilating, and airconditioning), electrical engineer (conduits for electrification and cables), fire proofing engineer (sprinklers and water pipes), as well as others, and include the loads assessed by these professionals on the structure.

3. Knowing the loads acting on the structure, the engineer can analyze the structure under different load combinations to evaluate the behavior of the structure and to determine the load effects (bending moments, axial forces, shears, and torsional moments) on the various components. The analysis made in this step can have varied degrees of sophistication in different iterations. For preliminary designs it may be relatively crude, such as $qL^2/8$, while for final design, first- or second-order analyses using matrix methods and computers are likely to be used.

4. Finally, the principal elements such as slabs, steel decking, cladding, steel joists, beams, columns, and foundations, as well as bracing can be designed. A given member may have to be designed for different load combinations and for different limit states. The structural response of the resulting structure (deflection, drift, etc.) must be checked to satisfy the code requirements. If not, the designer selects new proportions and/or members and repeats the procedure in steps 3 and 4, until satisfaction is obtained. For large structures several design alternatives using different methods of framing, different bracing systems, and/or different grades of steels may have to be studied before arriving at an appropriate arrangement for detailed design. Basic information needed by the fabricator to detail the structure is provided by the structural engineer in the form of ***structural scheme drawings.*** These typically include separate details for the framework which show all member sizes and line drawings of the frame on which the forces and moments at each connection are given. The option to bolt, weld, or to use a combination of both should preferably be the decision and choice of the fabricator and/or erector of the steel, subject to the approval of the structural engineer in charge.

4.11.1   Detail Drawings

The general contractor of a building normally sublets the fabrication and erection of the steel frame. Several fabricators may be invited to submit tenders for the work. The connections between members in the frame constitute a major part of steelwork costs. The practice by which the fabricator is responsible for the choice of connections allows each fabricator to propose connection methods with which he or she is accustomed to working and for which he or she has the necessary fabrication and erection equipment.

Once the subcontract has been awarded, the fabricator receives the design drawings and specifications. The fabricator may suggest changes in detail in order to simplify fabrication or erection, and when all information is complete *shop drawings* are created. These drawings include details such as the required cut lengths of the different members, location and sizes of holes, welds, stiffeners, and painting and galvanizing information. Separate drawings are normally prepared for each different group of elements—beams, columns, braces, and so forth. In addition, the fabricator prepares a set of erection plans which include plan grid dimensions, levels of the erected frame, as well as other data pertinent to the construction. Each separate piece or subassembly of pieces has an assigned shipping or erection mark to identify and place it in its correct position in the framework. The erection plans also include an anchor rod plan. The erection drawings should also show the positions on the structure where temporary bracing or restraints are necessary until walls, floors, and diaphragms are in position. Computerized drafting, detailing, and preparation of input data for computer-controlled automatic cutting, punching, drilling, and welding machines are used more and more by fabricators today.

The shop drawings are submitted by the fabricator to the structural engineer for approval before fabrication operations begin. The structural engineer must ensure that the details are such that the structure acts in the way he or she has idealized it for design. Acceptance of the shop drawings by the engineer indicates approval of the details, design, and adequacy of the connections.

The person who signs the structural drawings (usually the structural engineer in charge) remains professionally responsible and personally liable for the proper design of that structure. In the event of a structural failure, the designer cannot seek immunity from legal action on grounds that the design meets code requirements. There is no assumption of liability by the city, if the code's minimum requirements prove inadequate for any one condition. The structural engineer remains personally responsible to determine whether code minimum standards are in fact adequate for the project under consideration, or whether additional measures must be included. If a structure has many components, each with its own designer, overall stability must be the particular responsibility of one of those designers. There must be no doubt where this responsibility lies.

References

4.1   AASHTO [2002]: *Standard Specifications for Highway Bridges,* 17th ed., American Association of State Highway and Transportation Officials, Washington, D.C.

4.2   AISC [1989*a*]: *Specification for Structural Steel Buildings: Allowable Stress Design and Plastic Design,* American Institute of Steel Construction, Chicago, IL. (Also referred to as **ASDS** in this book.)

4.3   AISC [1989*b*]: *Manual of Steel Construction: Allowable Stress Design,* 9th edition, American Institute of Steel Construction, Chicago, IL.

4.4   AISC [1999]: *Load and Resistance Factor Design Specifications for Structural Steel Buildings,* American Institute of Steel Construction, Chicago, IL. (Also referred to as **LRFDS** in this book.)

4.5   AISC [2001]: *Manual of Steel Construction: Load and Resistance Factor Design,* 3rd ed., American Institute of Steel Construction, Chicago, IL. (Also referred to as **LRFDM** in this book.)

4.6   AREA [1992]: *Specifications for Steel Railway Bridges,* American Railway Engineering Association, Chicago, IL.

4.7   ASCE [2000]: *Minimum Design Loads for Buildings and Other Structures,* ASCE 7-98, American Society of Civil Engineers. (Also referred to as **ASCES** in this book.)

4.8   ASCE [2003]: *Minimum Design Loads for Buildings and Other Structures,* ASCE 7-02, American Society of Civil Engineers.

4.9   Ellingwood, B. E., McGregor, J. G., Galambos, T. V., and Cornell, C. A. [1982]: "Probability-Based Load Criteria: Load Factors and Load Combinations," *Journal of the Structural Division,* ASCE, vol. 108, no. 5, pp. 978–997.

4.10 Galambos, T. V., and Ravindra, M. K. [1976]: "Proposed Criterion for Load and Resistance Factor Design of Steel Building Structures," Research Report no. 45, Civil Engineering Department, Washington University, St. Louis, MO, May.

4.11 IBC [2000]: *International Building Code 2000,* International Conference of Building Officials, Whittier, CA.

4.12 Miller, R. T. [2004]: "The Structural Engineer and the Changing Building Codes," *Structure Magazine,* National Council of Structural Engineers Associations, vol.11, no. 3, March, Reedsburg, WI.

4.13 NFPA [2002]: *NFPA 5000—Building Construction and Safety Code,* National Fire Protection Association, Quincy, MA.

P R O B L E M S

P4.1. Determine the wind pressures on the windward wall, leeward wall, and roof of a 40 ft by 80 ft by 20 ft building with a flat roof. The building is located on level ground in a suburban area of Chicago. Steel roof deck connected to open-web steel joists and girders acts as roof diaphragms.

P4.2. Determine the external pressure acting over the windward wall, leeward wall, and roof of a gabled building, located on open flat terrain in Boston, for the wind load perpendicular to the ridge of the building. The fully enclosed agricultural building has plan dimensions of 60 by 120 ft, height to eave of 20 ft, and a roof slope of 35°. Also, determine the internal pressure in the building.

P4.3. Determine the design snow load acting on the building given in Problem P4.1. The building is in an industrial park with no trees or other structures offering shelter.

P4.4. Determine the balanced and unbalanced design snow loads for the roof of the gabled frame of Problem P4.2.

P4.5. The axial force on a building column from the code-specified loads have been determined as 180 kips dead load, 30 kips from the roof snow load, 140 kips (reduced) floor live load, 120 kips from wind, and 40 kips from earthquake. Determine the required design strength of the column. If the resistance factor ϕ is 0.85, what is the required nominal strength?

P4.6. Repeat Problem P4.5 for a garage column.

P4.7. Loads acting on a roof deck include a dead load of 30 psf, a roof live load of 20 psf, a snow load of 35 psf, and a wind pressure of 18 psf (upward or downward). Determine the governing loading on the decking.

P4.8. The axial forces in a diagonal member of a roof truss are as follows: 50 kips dead load (T), 40 kips snow load (T), 20 kips roof live load (T), 15 kips rain load (T), and 18 kips wind load (C). Determine the required design strength of the diagonal member.

P4.9. A beam is part of the framing system for the floor of a residential building. The end moments caused by the service loads are 160 ft-kips (clockwise) from dead load, 230 ft-kips (clockwise) from live load, and 150 ft-kips (clockwise or anticlockwise) from wind load. Determine the maximum factored bending moment. What is the controlling ASCES load combination? Also, if the resistance factor ϕ is 0.9, what is the required nominal bending strength in ft-kips?

P4.10. The maximum moments caused by the service loads on the roof beam of an office building are as follows: dead load, 58 ft-kips; snow load, 75 ft-kips; roof live load, 50 ft-kips; rain load, 25 ft-kips; and wind load −30 ft-kips. All these moments occur at the same location on the beam (center) and can therefore be combined. Determine the required bending strength of the roof beam.

P4.11. A beam-column is subjected to the following forces by the service loads indicated. Axial compression, P: dead load, 120 kips; live load, 220 kips. Bending moments, M: dead load 180 ft-kips, live load, 240 ft-kips. Determine the required axial compressive and bending strengths.

Kansas City Hyatt Regency walkways collapse, July 17, 1981.

Two 120-ft-long walkways within the atrium area of the hotel collapsed leaving 113 people dead and 186 injured. The second floor walkway was suspended from the fourth floor walkway. In turn, this fourth floor walkway was suspended from the atrium roof framing by a set of six $1\frac{1}{4}$-in.-dia. hanger rods. A change in hanger rod arrangement, from a continuous rod (as detailed in the contract drawings) to interrupted rods (as erected), essentially doubled the load to be transferred by the fourth floor box beam-to-hanger connections. The connection, consisting only of a nut and a washer, pulled through the flanges of the box beam, built-up from two MC8×8.5 channels.

5

Structural Analysis and Computation of Required Strengths

5.1 Introduction

Load and resistance factor design is a method of proportioning structural elements (members, connectors, and connecting elements) and assemblages such that no applicable limit state is exceeded when the structure is subjected to all appropriate load combinations. That is, the design strength of each structural element or assemblage shall equal or exceed the required strength based on the factored loads. The **design strength, R_d ($=\phi R_n$)**, for each applicable limit state is calculated as the product of the nominal strength, R_n, multiplied by a resistance factor ϕ. Nominal strengths R_n and resistance factors ϕ are given in LRFDS Chapters D through K for various structural elements, and will be discussed in detail in Chapters 6 through 13 of this text. The **required strength, R_{req}**, of each structural element or assemblage is determined for each applicable load combination (such as load combinations LC-1 to LC-7 discussed in Section 4.10) by structural analysis.

Structural analysis is the process of determining forces in each element of a structure when the configuration of elements and the loads acting on that structure are already known, while structural design is the process of configuring elements to resist forces whose values are already known. Thus, analysis and design are complementary procedures in the overall process of designing new structures. After assuming a preliminary configuration of the structure and the loads acting on it, the designer performs a preliminary analysis to determine the forces in each element. Once the forces in the elements are known, the elements can be designed based on the procedures described in detail in Chapters 6 through 13. After designing all of the elements, the designer performs, if necessary, a more accurate analysis of the structure using the configuration chosen. The process iterates between analysis and design until convergence is achieved.

Though methods of structural analysis are typically treated in textbooks on structural analysis, this chapter summarizes a few analysis procedures to clearly indicate the complementary nature of analysis and design. Thus, simplified analysis procedures, useful for preliminary analysis, will be briefly described in Section 5.2 for braced building structures with simple beam-to-column connections, and in Sections 5.3 for unbraced portal frames. These analysis procedures will be illustrated with the help of two numerical examples:

- A **braced building structure** with simple beam-to-column connections shown in Fig. 5.1.1, and
- An **unbraced portal frame** shown in Fig. 5.1.2.

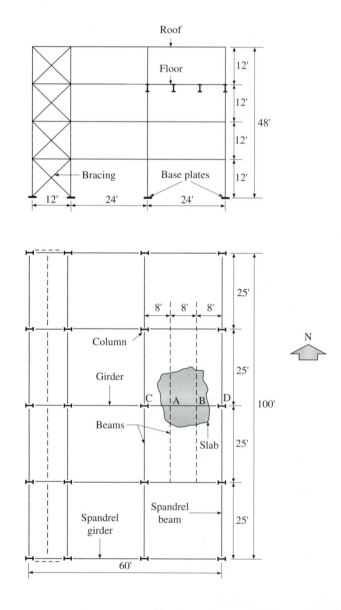

Figure 5.1.1: Example of a braced building with simple beam-to-column connections.

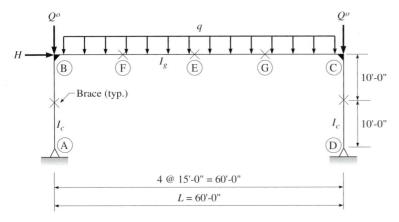

Note: ✕ indicates brace location

Figure 5.1.2: Example of an unbraced portal frame.

Several beams, columns, beam-columns, tension members, base plates, simple connections, and moment connections selected from these two structures will be designed in later chapters, using the required strengths calculated in this chapter.

5.2   Braced Frames with Simple Beam-to-Column Connections

Low-rise steel building structures generally consist of a rectangular grid of horizontal beams and girders, and vertical columns. Simple connections, capable of transferring shear only, are used to join the beams to girders and girders to columns, while the columns are made continuous. Vertical truss bracing is provided in the transverse and longitudinal planes within certain bays of such buildings to transfer wind loads to the foundation and to provide stability. Analysis of such structures under gravity and wind loading will be considered in this section.

5.2.1   Beams and Girders in Braced Frames

Figures 3.3.1 and 3.3.2 show a concrete floor slab which is carried on beams. The beams are supported by girders, and the girders in turn are supported by columns. The loads which are applied to the slab, beams, and girders may be classified as distributed and concentrated. A ***distributed load*** is one which extends over a considerable length of the beam or girder. It may extend throughout the length of the member, or over only a part of the length. A distributed load may be uniform, as would be the case for a beam supporting a slab of constant thickness, or nonuniform, as would be the case if the slab thickness varied over the beam length. Similarly, where a beam has the same

cross section throughout its length, the weight of the beam itself is a uniformly distributed load extending along the length of that member (for
example, 50 plf for a W16×50 beam).

A **concentrated load** is a load applied over such a small portion of the length
of a member that, without appreciable error in the calculated effects of the load,
it may be assumed to act at a point. The loads actually applied to the girders by
the beams are usually applied through the bolts in the beam-to-girder connections (Figs. 5.2.1a and b). That is, the reaction of each beam is applied to the girder in two vertical lines, separated by the distance between the gage lines of bolts
in the two connection angles at the end of the beam as detailed in Fig. 5.2.1b. The
standard practice, however, is to treat the beam reaction on the girder as if it
were a single concentrated load applied to the girder at its intersection with the
vertical axis of the beam as shown in Fig. 5.2.1a.

Another way in which concentrated loads are applied to beams is shown in
Fig. 5.2.2 which shows an axially loaded W8×28 column supported on the top
flange of a 20 ft long W14×90 beam. Actually the load, Q, is distributed over

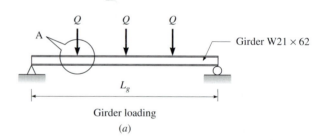

Girder loading

(a)

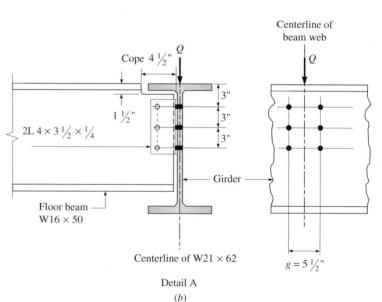

Detail A

(b)

Figure 5.2.1: Load transfer from floor beam to girder.

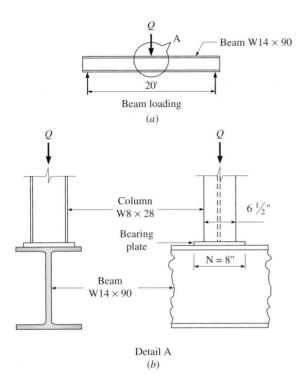

Beam W14 × 90

20'

Beam loading

(a)

Column
W8 × 28

6 ½"

Bearing
plate

N = 8"

Beam
W14 × 90

Detail A

(b)

Figure 5.2.2: Column supported on the top flange of a beam.

a 8 in. length of beam (equal to the length, N, of the bearing plate shown in Fig. 5.2.2*b*), but no appreciable error results in the design of the beam from assuming that the load is applied at a single point, namely, at the intersection of the axis of the column with the axis of the beam.

Let us consider a multiple bay building having a typical bay size of L_b by L_g, with each bay consisting of n beams at a uniform spacing S_b (Fig. 3.3.2*a*). Here the concrete slab on metal deck is supported by floor beams located at a spacing, S_b, and these in turn are supported by the girders at the ends. All of the connections are assumed to be simple connections (Fig. 3.2.2*b*). LRFDS Section B8 stipulates that beams, girders, and trusses designed on the basis of simple spans shall have an effective length equal to the distance between the centers of gravity of the members to which they deliver their end reactions. Thus, for design purposes, the span of the beams (L_b) equals the center-to-center distance of the supporting girders or columns, while the span of the girders (L_g) is the center-to-center distance of the supporting columns. In most situations the difference resulting from considering a member length to be the distance between centers of gravity of supporting members rather than its actual length, measured as center-to-center of end connections, is small. In some cases, however, there may be sufficient difference to merit computing the actual length. Regardless of the length used for design of the

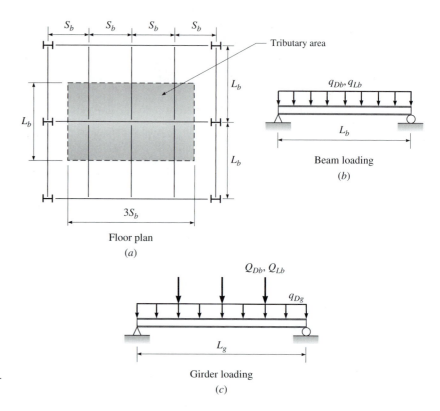

Figure 5.2.3: Tributary area and loads on a typical interior girder.

supported member, the actual connection detail may cause an eccentric load, or moment, to act on the supporting member. Such effects must be taken into account.

The tributary width for a typical floor beam is $\frac{1}{2} S_b + \frac{1}{2} S_b = S_b$. The tributary area per unit length of a typical beam equals S_b by 1, while the tributary area of that beam equals $S_b L_b$ (shown cross-hatched in Fig. 3.3.2a). The load carried by a floor beam is usually taken to be uniform and composed of (1) its own weight, (2) the weight of the floor (slab, floor finish, ceiling, etc.) for a width S_b, and (3) the live load for the same tributary width, including an allowance for partitions if necessary. The end reactions on this beam would then be applied as concentrated loads to the girders.

A girder must carry a uniform load caused by its own weight. It also must carry a set of load concentrations from the reactions of the floor beams. The tributary area of a typical interior girder supporting $(n - 1)$ beams is $(n - 1) S_b L_b$, shown shaded in Fig. 5.2.3. As this area is considerably larger than the tributary area for a typical interior beam, a reduced live load intensity, p_{Lg}, may be appropriate. The reactions from the floor beams used in the girder design should reflect this reduction.

The following notations are used in the analysis of floor beams:

L_o = basic (unreduced) uniformly distributed live load (see Section 4.6.1), psf

S_b = spacing of beams, ft

L_b = span of beams, ft

A_{Tb} = tributary area of an interior beam = $S_b L_b$

A_{Ib} = influence area of an interior beam = $K_{LLb} A_{Tb} = 2A_{Tb}$

K_{LLb} = live load element factor for an interior beam (see Section 4.6.2)

p_{Db} = dead load per unit area on the tributary area of an interior beam, psf

q_b = self-weight per unit length of an interior beam, plf

p_{Lb} = design live load intensity on the tributary area of an interior beam ($\leq L_o$), psf

q_{Db} = dead load per unit length of an interior beam, plf

q_{Lb} = live load per unit length of an interior beam, plf

q_{ub} = factored uniformly distributed load on a beam, plf

As the beam gets its live load from one level only, we have from Section 4.6.2:

$$p_{Lb} = \max\left\{ \left[0.25 + \frac{15}{\sqrt{A_{Ib}}} \right] L_o, 0.5L_o \right\} \tag{5.2.1}$$

Also,

$$q_{Db} = (q_b + p_{Db}S_b) \tag{5.2.2}$$

$$q_{Lb} = (p_{Lb}S_b) \tag{5.2.3}$$

$$q_{ub} = (1.2q_{Db} + 1.6q_{Lb}) \tag{5.2.4}$$

In addition, the following notations are used in the analysis of floor girders:

L_g = span of girders, ft

A_{Tg} = tributary area of an interior girder = $(n - 1) S_b L_b$

A_{Ig} = influence area of an interior girder = $K_{LLg} A_{Tg} = 2A_{Tg}$

K_{LLg} = live load element factor for an interior girder (see Section 4.6.2)

p_{Lg} = design live load intensity on the tributary area of an interior girder ($\leq p_{Lb}$), psf

q_g = self-weight per unit length of an interior girder, plf

Q_{Db} = dead load transmitted from beams to an interior girder, lbs

Q_{Lb} = live load transmitted from beams to an interior girder, lbs

q_{ug} = factored uniformly distributed load on a girder, plf

Q_{ug} = factored concentrated load transmitted from beams to an interior girder, lbs

As the girder gets its live load from one level only, we have from Section 4.6.2:

$$p_{Lg} = \max\left\{\left[0.25 + \frac{15}{\sqrt{A_{Ig}}}\right]L_o, 0.5L_o\right\} \tag{5.2.5}$$

Also,

$$Q_{Db} = (q_b + p_{Db}S_b)L_b \tag{5.2.6}$$

$$Q_{Lb} = (p_{Lg}S_b)L_b \tag{5.2.7}$$

$$Q_{ug} = (1.2Q_{Db} + 1.6Q_{Lb}) \tag{5.2.8}$$

$$q_{ug} = 1.2q_g \tag{5.2.9}$$

EXAMPLE 5.2.1

Required Strengths for an Interior Roof Beam

In the office building shown in Fig. 5.1.1, roof beams are spaced at 8 ft intervals and are connected to girder webs by simple connections (Fig. X5.2.1). The spacing of the girders is 25 ft. The roofing consists of $4\frac{1}{2}$-in.-thick lightweight concrete slab, five-ply felt and gravel, and suspended metal lath and gypsum plaster ceiling. The live load on the roof is 20 psf; the snow load, 40 psf; and the rain load, 25 psf. Self-weight of the roof beam is estimated to be 30 plf. Making allowance for mechanical ducts and miscellaneous (4 psf), determine the required shear and bending strengths for which an interior roof beam is to be designed. Use the LRFD Specification and the ASCE standard 7.

Solution

Spacing of the beams, $S_b = 8$ ft
As the end connections are simple, the beam is designed as a simply supported beam.
Span, L_b of the roof beam = spacing of girders = 25 ft

Tributary width per foot length of the beam $= \left(\dfrac{8}{2} + \dfrac{8}{2}\right) = 8$ ft

Loads.   From the ASCES Table C3-1: Minimum Design Dead Loads:

Five-ply felt and gravel	= 6 psf
$4\frac{1}{2}$-in.-thick lightweight concrete slab $4\frac{1}{2}(8)$	= 36 psf
Suspended metal lath and gypsum plaster ceiling	= 10 psf
Mechanical duct allowance	= 4 psf
Miscellaneous (given)	= 4 psf

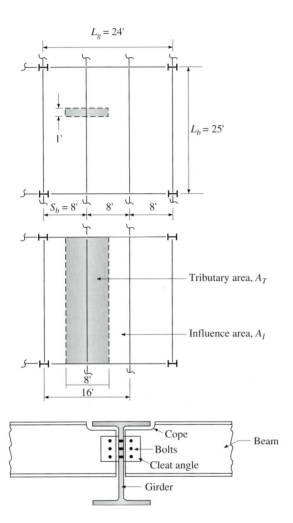

Figure X5.2.1

	Total =	60 psf
Dead load intensity of the roofing, p_{Db}	=	60 psf
Self-weight of the roof beam, q_b (given)	=	30 plf

Dead load on the beam, $q_{Db} = \dfrac{1}{1000}[30 + 60(8)] = 0.510$ klf

Roof live load, L_r	=	20 psf
Snow load, S	=	40 psf
Rain load, R	=	25 psf

Controlling live load, $p_{Lb} = \max(L_r, S, R)$
$$= \max(20, 40, 25) \quad = 40 \text{ psf}$$

Live load on the roof beam, $q_{Lb} = \dfrac{40(8)}{1000} \qquad = 0.320$ klf   [*continues on next page*]

Example 5.2.1 continues ...

Load Combinations.   From Eq. 4.10.3, the factored loads for different load combinations are:

$$\text{LC-1} \rightarrow 1.4q_{Db} = 1.4(0.510) = 0.714 \text{ klf}$$

$$\text{LC-3} \rightarrow 1.2q_{Db} + 1.6q_{Lb} = 1.2(0.510) + 1.6(0.320) = 1.12 \text{ klf}$$

Design load on the roof beam, $q_{ub} = \max [0.714, 1.12] = 1.12$ klf
From Case 1 in LRFDM Table 5-17: Shears, Moments, and Deflections, observe that:

$$\text{Maximum shear, } V_{\max} = \frac{q_{ub}L_b}{2} = \frac{1.12(25)}{2} = 14.0 \text{ kips}$$

$$\text{Maximum bending moment, } M_{\max} = \frac{q_{ub}L_b^2}{8} = \frac{1.12(25^2)}{8} = 87.5 \text{ ft-kips}$$

Required shear strength, $V_{\text{req}} = V_{\max} = 14.0$ kips   (Ans.)
Required flexural strength, $M_{\text{req}} = M_{\max} = 87.5$ ft-kips   (Ans.)

E X A M P L E   5 . 2 . 2

Required Strengths for an Interior Floor Beam

In the office building shown in Fig. 5.1.1, floor beams are spaced at 8 ft intervals and are connected to girder webs by simple connections. The spacing of the girders is 25 ft. The flooring consists of a 5-in.-thick stone concrete slab resting on the beams and a suspended metal lath and gypsum plaster ceiling. Make allowance for mechanical ducts, finishing, and carpet (6 psf) and partitions (10 psf). Self weight of the floor beam is estimated to be 40 plf. Determine the required shear and bending strengths for which an interior floor beam is to be designed. Use the LRFD Specification and the ASCE standard 7.

Solution

Spacing of the beams, S_b = 8 ft

Tributary width per foot length of the beam = $\left(\dfrac{8}{2} + \dfrac{8}{2}\right)$ = 8 ft

Span of the floor beam, L_b = spacing of girders = 25 ft

Dead Load.   From the ASCES Table C3-1: Minimum Design Dead Loads:

5-in.-thick stone concrete slab = 5(12)	= 60 psf
Suspended metal lath and gypsum plaster ceiling	= 10 psf
Mechanical duct allowance	= 4 psf
Finishing + carpet (given)	= 6 psf
Partitions (given)	= 10 psf
Total	= 90 psf

Dead load intensity from floor, p_{Db} = 90 psf

Self-weight of the floor beam, q_b (given)   = 40 plf

Dead load on the floor beam, $q_{Db} = \dfrac{1}{1000}[40 + 90(8)] = 0.760$ klf

Live Load.   From the ASCES Table 4-1, for an office building, basic live load, L_o = 50 psf

For the floor beam, tributary area, $A_T = 8(25) = 200$ ft^2

Influence area, $A_I = K_{LL}A_T = 2(200) = 400$ ft^2

From Section 4.8 of the ASCES, there is no live load reduction permitted for influence areas $\leq$ 400 sq. ft. So,

Design live load, $L = L_o = 50$ psf

Live load on the floor beam, $q_{Lb} = \dfrac{50(8)}{1000} = 0.40$ klf

Load Combinations.   From Eq. 4.10.3 the factored loads for different load combinations are:

LC-1 $\rightarrow$ $1.4q_{Db} = 1.4(0.760) = 1.06$ klf

LC-2 $\rightarrow$ $1.2q_{Db} + 1.6q_{Lb} = 1.2(0.760) + 1.6(0.400) = 1.55$ klf

The second load combination, which gives the maximum factored load, governs.

Design load on the floor beam, $q_{ub} = \max[1.06, 1.55] = 1.55$ klf

Required Strengths.   The beam is to be designed for a uniformly distributed load of 1.55 klf. As the floor beam is simply supported:

$$V_{max} = \frac{q_{ub}L_b}{2} = \frac{1.55(25)}{2} = 19.4 \approx 20.0 \text{ kips}$$

$$M_{max} = \frac{q_{ub}L_b^2}{8} = \frac{1.55(25^2)}{8} = 121 \text{ ft-kips}$$

Thus, the required strengths of the floor beam are:

Shear strength, $V_{req} = V_{max} = 20.0$ kips   (Ans.)

Flexural strength, $M_{req} = M_{max} = 121$ ft-kips   (Ans.)

Note: Design of this floor beam will be considered in Chapter 9 (Example 9.7.5).

Required Strengths for an Interior Floor Girder   E X A M P L E   5 . 2 . 3

Determine the required strengths for an interior floor girder supporting floor beams considered in Example 5.2.2. Self-weight of the girder is estimated to be 60 plf. The girders are connected to column flanges by simple shear connections. Use the LRFD Specification and the ASCE standard 7.   *[continues on next page]*

Example 5.2.3 continues ...

Solution

As the end connections are simple, the girder is designed as a simple beam.
Span, L_g, of the floor girder = spacing center-to-center of columns = 24 ft
Span of the floor beams, L_b = 25 ft
The girder is subjected to concentrated reactions from floor beams A1,
A2 at point A and from floor beams B1, B2 at point B (Fig. X5.2.3). In

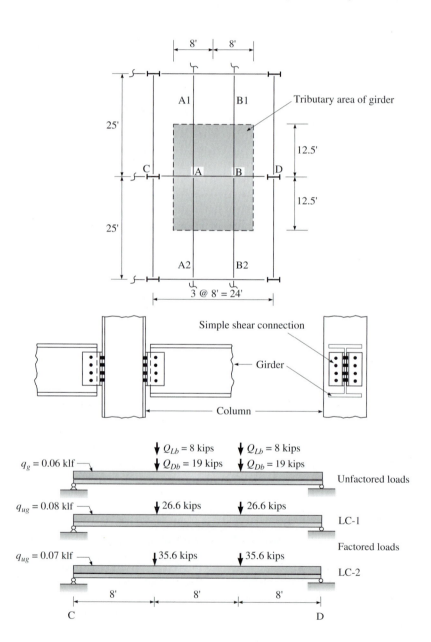

Figure X5.2.3

addition, it is subjected to a uniformly distributed load from self-weight of the girder, over its entire length, CD. The girder is typically designed as a simply supported beam, with a span equal to the center-to-center distance of the columns (LRFDS Section B8).

Loads.   From Example 5.2.2, dead load on floor beams, $q_{Db} = 0.760$ klf. So, contribution to concentrated loads at points A and B from dead load of floor beams is:

$$Q_{Db} = [0.760(12.5)](2) = 19.0 \text{ kips}$$

The tributary area of the interior girder is shown shaded in Fig. X5.2.3.

Tributary area, $A_T = [8(12.5)] [2(2)] = 400 \text{ ft}^2$
Influence area, $A_I = K_{LL}A_T = 2A_T = 2(400) = 800 \text{ ft}^2$
Basic live load, $L_o = 50 \text{ psf}$

$$\text{Design live load, } L = \max\left\{\left[0.25 + \frac{15}{\sqrt{A_I}}\right]L_o, 0.5L_o\right\}$$

$$= \max\left\{\left[0.25 + \frac{15}{\sqrt{800}}\right]50, 0.5(50)\right\}$$

$$= \max\{0.780(50), 25\} = 39.0 \text{ psf}$$

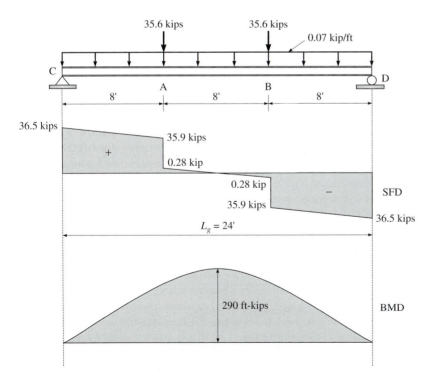

Figure X5.2.3: Continued

[continues on next page]

Example 5.2.3 continues ...

Live load from floor beams at points A and B,

$$Q_{Lb} = \frac{39.0(8)(12.5)(2)}{1000} = 7.80 \approx 8.00 \text{ kips}$$

Self-weight of girder, $q_g = 60$ plf (given) $= 0.06$ klf
As the only uniformly distributed load acting on the girder is the self-weight of the girder, we have, $q_{Dg} = q_g = 0.06$, and $q_{Lg} = 0.0$.
The unfactored loads acting on the beam are shown in Fig. X5.2.3.

Load Combinations.   The two possible load combinations are summarized in the table below:

Load Combination	Q_{ug} (kips)	q_{ug} (klf)
LC-1 → 1.4D	1.4(19.0) = 26.6	0.084
LC-2 → 1.2D + 1.6L	1.2(19.0) + 1.6(8.00) = 35.6	0.072

These factored loads are also shown in Fig. X5.2.3. It is evident that the combination LC-2 will control the design of the girder.

Required Strengths.   The shear and bending moment diagrams for the girder CD under factored loads are shown in Fig. X5.2.3. From symmetry of the structure and the loading,

$$R_C = R_D = 35.6 + \frac{24}{2}(0.072) = 36.5 \text{ kips}$$

The maximum bending moment occurs at the center of the girder and is given by:

$$M_{max} = 36.5(12) - 35.6(4) - 0.072(12)(6)$$
$$= 290 \text{ ft-kips}$$

Maximum shear, $V_{max} = R_C = 36.5 \approx 37.0$ kips.
So, the required strengths for the girder are:

$$M_{req} = 290 \text{ ft-kips and } V_{req} = 37.0 \text{ kips} \tag{Ans.}$$

Note: Design of this floor girder will also be considered in Chapter 9 (Example 9.7.5).

5.2.2   Columns in Braced Frames

In the preliminary design of a multistory building with simple beam-to-column connections and with bracing provided in both directions, the design procedure for columns starts at the roof level and proceeds downwards to the foundation. The steel columns are generally fabricated and erected in tiers of two-story

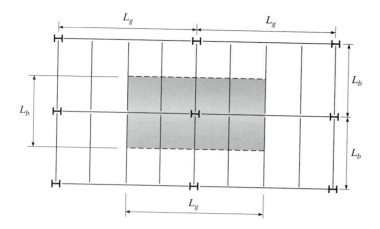

Figure 5.2.4: Area contributed by one floor to the tributary area of an interior column.

lengths. Thus, only the lower story segment of each tier, which is generally the more critically loaded segment, is designed. The axial force in a column at a given level is obtained by summing the girder end-shears and beam end-shears at the roof level and at all floor levels for which the column is the vertical supporting member. Figure 5.2.4 shows the area contributed by one floor level to the tributary area of an interior column. The dead load calculated should include an estimated value for the self-weight of the column. Where permitted, the live load used in these calculations may be the reduced live load based on the influence area for the column at the level considered. Using this procedure, we obtain the required axial strength for each column, in each tier. The columns are first designed for "gravity-critical" load combinations only (load combinations LC-1, LC-2, and LC-3). Later, for those columns that are part of the bracing system in one or both directions, the required strengths are calculated and the design revised, if necessary, for "wind-critical" load combinations (load combinations LC-4 and LC-6), as discussed further in Section 5.2.3.

In designing exterior columns, the weight of the outside wall has to be included in the design loads. This weight will depend on the material of which the wall is constructed and on the number and size of window openings, if any. At the roof level, there will generally be a parapet wall and a cornice, and the weight of these should be included in the calculation of the dead load on the exterior column.

Let

m = number of floors supported by the column, at the level considered

A_{Tc} = tributary area of an interior column = $mL_b L_g$

A_{Ic} = influence area of an interior column = $K_{LLc} A_{Tc} = 4A_{Tc}$

p_{Lc} = design live load intensity on the tributary area of an interior column ($\leq p_{Lg}$), psf

From Section 4.6.2, for a column that receives live loads from more than one floor level:

$$p_{Lc} = \max\left\{ \left[0.25 + \frac{15}{\sqrt{A_{Ic}}} \right] L_o, 0.4L_o \right\} \qquad (5.2.10)$$

Figure 5.2.5 shows the disposition of beams and girders at interior, corner, and wall columns. The simple connections generally used to connect girders and beams to columns in braced frames in low-rise structures (see Fig. X5.2.3)

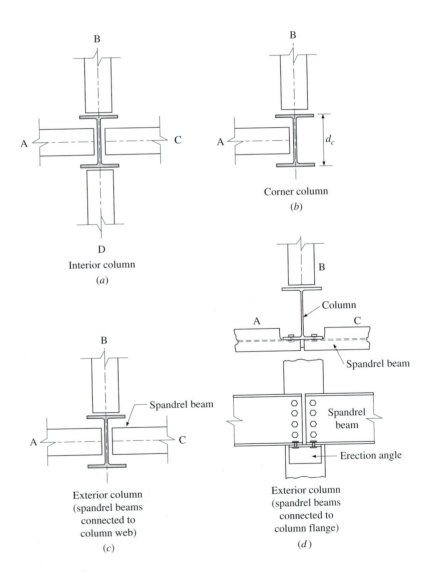

Figure 5.2.5:   Eccentric loads on columns.

will transmit little or no bending moment from the beam or girder to the column. Consequently, it is generally assumed that the reaction from a beam framing into the flange of a column (i.e., a "flange-beam"), say beam B of Fig. 5.2.5b, is delivered to the column at its face; in other words, the force is applied with an eccentricity $e = d_c/2$, where d_c is the depth of the column section. The eccentricity of the reaction of a beam framing into the web of a column (i.e., a "web-beam"), say beam A of Fig. 5.2.5b, is practically zero. Thus, when beams having equal reactions frame into the column opposite each other, as would be the case of an interior column of a structure with regular spacing of beams, girders, and columns, the column loads may be assumed to be concentric, that is, applied along the axis of the column. Thus, for the interior column shown in Fig. 5.2.5a, if the reactions of the flange-beams B and D are the same, their moments neutralize each other. If, however, flange-beam D were omitted as shown in Fig. 5.2.5c, or if the reaction of D was considerably less than that of B in Fig. 5.2.5a, it is evident that the loads on the column would no longer be symmetrical and that the column would be subjected to a bending moment about the major axis, in addition to axial load. This eccentric loading condition occurs frequently in exterior columns of buildings, where a roof beam or floor beam is supported on the interior flange without a corresponding load on the exterior flange. Figure 5.2.5d indicates one way of framing sometimes used to lessen this eccentricity, or even to balance the moments, if the total reaction of the two spandrel beams is nearly the same as that of the floor beam.

   Usually columns support beams which have similar or identical connection eccentricities at every floor level. Since the column shaft is made continuous by its splices, a column in a braced frame using simple beam connections acts as a (vertical) continuous beam on simple supports and, hence, any external moment applied to the column by beam reactions at any floor level will affect the internal moments in all stories of the column. A common practice in the design of columns for braced frames with simple beam-to-column connections is to distribute the moments, M^o, arising from eccentric beam reactions, into the columns above and below that floor, each in proportion to its stiffness, I/L. This assumption is slightly on the nonconservative side. Another common practice is to design each story of a column for a bending moment M^o. This is evidently on the conservative side and will compensate for any moment transmitted from the beam by the partial rigidity of actual simple connections.

Required Axial Strength for an Interior Column

EXAMPLE 5.2.4

Determine the required axial strength of the interior column, C, at ground floor level of the building shown in Fig. 5.1.1. Assume an average value (over the two tiers) for the self-weight of the column as 40 plf. Use the LRFD Specification and the ASCE standard 7.

[*continues on next page*]

Example 5.2.4 continues ...

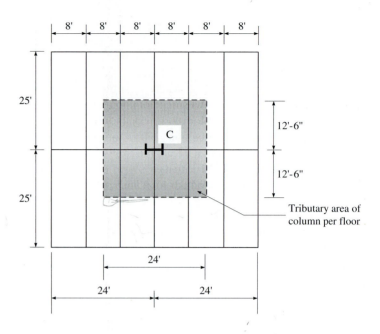

Figure X5.2.4

Solution

The column segment considered receives loads from the roof and the three floors below:

$$\text{Tributary area contributed at each level} = \left(\frac{24}{2} + \frac{24}{2}\right)\left(\frac{25}{2} + \frac{25}{2}\right) = 600 \text{ ft}^2$$

For an interior column, live load element factor, $K_{LL} = 4$

Influence area contributed at each level, $= 4(600) = 2400 \text{ ft}^2$

For the column segment considered, tributary area, $A_T = 600(3) = 1800 \text{ ft}^2$

For the column segment considered, influence area, $A_I = 2400(3) = 7200 \text{ ft}^2$

Also, we observe from Fig. 5.2.4 that the column is loaded by six beams and two girders at each level. From Examples 5.2.1, 5.2.2, and 5.2.3 we obtain roof dead load (60 psf), floor dead load (90 psf), self-weight of roof beams (30 plf), floor beams (40 plf), and floor girders (60 plf). We will assume a lineal weight of 50 plf for the roof girders.

Loads.

Dead load from roof	$\dfrac{60(600)}{1000}$	$= 36.0 \text{ kips}$
Weight of roof beams	$\dfrac{30(12.5)(2)(3)}{1000}$	$= 2.25 \text{ kips}$

Weight of roof girders	$\dfrac{50(12)(2)}{1000}$	= 1.20 kips
	Total dead load from roof	= 39.5 kips
Dead load from floor	$\dfrac{90(600)}{1000}$	= 54.0 kips
Weight of floor beams	$\dfrac{40(12.5)(2)(3)}{1000}$	= 3.00 kips
Weight of floor girders	$\dfrac{60(12)(2)}{1000}$	= 1.44 kips
	Total dead load from one floor	= 58.4 kips
Self-weight of column	$\dfrac{40(12)(4)}{1000}$	= 1.92 kips
Axial force in the column due to dead loads, P_D		= [39.5 + 3(58.4) + 1.92] = 217 kips

As the column is subjected to live load from more than one floor level,

$$\text{Design live load, } L = \max\left\{\left[0.25 + \frac{15}{\sqrt{A_{Ic}}}\right]L_o, \, 0.4\,L_o\right\}$$

$$= \max\left\{\left[0.25 + \frac{15}{\sqrt{7200}}\right]50, \, 0.4(50)\right\}$$

$$= \max\{0.427(50), \, 20\} = 21.4 \text{ psf}$$

Axial force in the column due to live load, $P_L = \dfrac{21.4(1800)}{1000} = 38.5$ kips

Axial force in the column due to snow load, $P_S = \dfrac{40(600)}{1000} = 24.0$ kips

Load Combinations.   From Eq. 4.10.3, the factored loads for the different load combinations are:

LC-1 → $1.4D = 1.4(217) = 304$ kips

LC-2 → $1.2D + 1.6L + 0.5S = 1.2(217) + 1.6(38.5) + 0.5(24.0)$
$$= 334 \text{ kips}$$

LC-3 → $1.2D + 1.6S + 0.5L = 1.2(217) + 1.6(24.0) + 0.5(65.0)$
$$= 331 \text{ kips}$$

Thus, the required axial strength of the column is:

$$P_u = \max[304, 334, 331] = 334 \text{ kips} \approx 335 \text{ kips}\qquad\text{(Ans.)}$$

Note: This column will be designed in Chapter 8 (Example 8.10.2). Also, the column base plate will be designed in Chapter 9 (Example 9.10.1).

5.2.3   Members of Vertical Truss-Bracing Systems

Figure 5.1.1 shows schematically a multistory, multibay frame using simple beam-to-column connections. Braced bents in the two end walls provide lateral stiffness to the system in the east-west direction (Note: the bracing arrangement in the north-south direction and the framing for stairs and elevators are not shown in Fig. 5.1.1 for simplicity). As per LRFDS Section C2.1, the columns, beams, girders, and diagonal members, when used as the vertical bracing system of a braced frame, may be considered to form a vertically cantilevered pin-connected truss in the lateral load analyses. Such articulation of joints symbolizes that the model does not consider the contribution to frame lateral stiffness provided by the bending stiffness of the primary members, that is, by the continuous columns and any rigidity provided by the beam-to-column connections. Only lateral stiffness due to the internal bracing system is considered. Hence, the axial forces in beams, columns, and bracing members of a braced bent obtained on the basis of this model will be on the conservative side.

Two examples of vertical bracing systems used for low- and medium-rise buildings are shown in Figs. 5.2.6 and 5.2.7. In each case, the four-story braced bent consists of a vertical planar cantilever truss-form made up of vertical elements (columns) and horizontal elements (beams) that are braced against the horizontal loads by the addition of the diagonal elements. Circular rods, bars, single angles, tees, or channels are often used as diagonals. Trussed bracing for braced frames could be provided by a single diagonal in each of the stories of the frame as shown in Fig. 5.2.6, or by providing double diagonals as shown in Fig. 5.2.7. In these figures, only wind loads (either wind from the left or wind from the right) are considered.

The senses of the reactions and internal forces (C represents compression, and T indicates tension) for the single diagonal braced bent, for wind from the left and right, are as shown in Figs. 5.2.6a and b, respectively. Note the reversal of direction of the reactions and internal forces that occurs with the change in the direction of the wind. Thus, in this case, the single diagonal in any story must be designed both as a compression member and a tension member in order to brace the structure for wind either from the left or from the right. Because of its great length and the associated specification requirements for minimum stiffness against buckling, and the lower design strengths for compression members as compared to tension members, this element is likely to be quite heavy in proportion to the actual load it must carry. Thus, although the use of a single diagonal is possible, it often results in an uneconomical design due to the necessity to provide a very long compression element. This is a major reason for the use of double diagonals.

Theoretically, the use of two diagonals in each story makes the truss structure statically indeterminate, since the extra diagonals are redundant for simple static equilibrium of the truss. However, note that while the wind from a single direction will tend to induce tension in one of the diagonals of

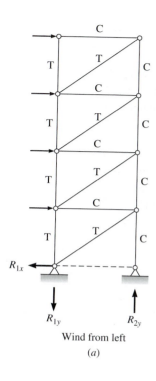

Wind from left

(a)

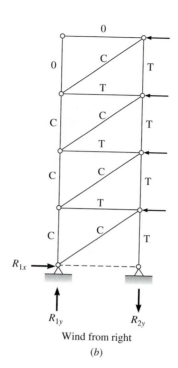

Wind from right

(b)

Figure 5.2.6: Behavior of single diagonal trussed bracing.

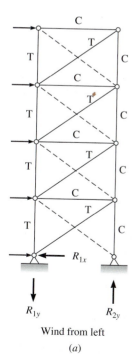

Wind from left

(a)

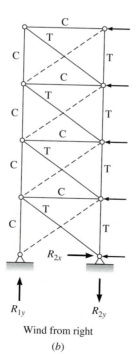

Wind from right

(b)

Figure 5.2.7: Behavior of double diagonal trussed bracing.

a story and compression in the other diagonal, the compression diagonal is assumed to buckle under the compression force, leaving the tension diagonal to resist the entire story shear. With this assumption the wind load from the left, as shown in Fig. 5.2.7a, is assumed to be taken by one diagonal in each story, while the opposing diagonal buckles (as indicated by the dashed lines in the figure). With wind load from the right, the roles of the diagonals reverse (Fig. 5.2.7b). The individual diagonals are designed as tension members. Note that with the assumption that only the tension diagonals are effective, the entire horizontal shear will be transmitted to the column support on the windward side. Since the wind may come from either side, this necessarily results in some over design for the column supports, as each must be capable of resisting the entire horizontal load applied.

Figure 5.2.8 shows the factored gravity and wind loads acting on a braced bent under a "wind-critical" load combination (load combinations LC-4 and LC-6). It should be noted that in addition to the axial force due to gravity loads, the leeward columns receive additional axial compressive force due to the vertical component of the tension in the diagonal member. These additional forces

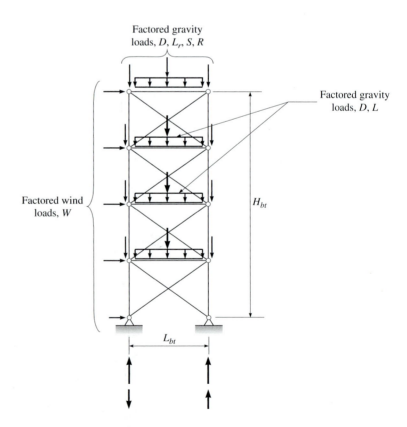

Figure 5.2.8:   Loads acting on a braced bent.

in the column are a function of the aspect ratio of the braced bent, namely, H_{bt}/L_{bt}. Here, H_{bt} is the height of the truss and L_{bt} is the width of the truss, as shown in Fig. 5.2.8. For low-rise buildings this contribution to column axial force due to horizontal loads may not be of great consequence. However, for multistory buildings with cantilevered, trussed bents with high aspect ratios, the additional axial force in the columns due to factored wind load may be quite significant (keeping in mind that it cumulatively increases with building height), to the extent that wind load combination LC- 4 may control the column design.

Horizontal members (beams) that are part of the trussed bents (Fig. 5.2.8) must resist the axial force developed in the member due to the factored lateral loads on the braced bent, in addition to the moments and shears produced by the concurrent factored gravity load component of the load combination acting on the member itself (LRFDS Section C2-1). That is, for certain load combinations, these horizontal members have to be designed as beam-columns (Chapter 11), not simply as beams (Chapters 9 and 10).

Note further that gravity loads produce only compressive forces in the columns, whereas horizontal (wind) loads may produce either compressive (leeward side) or tensile (windward side) forces in these members. When the wind direction reverses, the sense of the wind force in these members also reverses.

When a structure is subjected to a combination of gravity and wind loads, it is usually necessary to consider the effects on the structure of different load combinations:

1. Factored gravity loads (dead, live, snow, etc.). These loadings are likely to be critical for columns in low- to mid-rise structures (load combinations LC-1, LC-2 and LC-3).
2. Factored gravity loads plus factored wind from left (load combination LC-4).
3. Factored gravity loads plus factored wind from right, only if the structure is not symmetrical (load combination LC-4).
4. Reduced dead load and factored wind from left (load combination LC-6).
5. Reduced dead load and factored wind from right, only if the structure is not symmetrical (load combination LC-6).

Consideration of these various combinations produces the critical design forces for the elements and supports of the structure. In most cases it is necessary to consider two different net results for each element of the bracing system: the first result is the maximum design force; the second result is the so-called minimum design force. Of course each element must be designed for the maximum force. However, when another loading combination produces a force of opposite sense, it may also be critical. Thus, the bottom segment of a column in a braced bent of high aspect ratio may experience a sense reversal of force as the wind reverses. However, if this segment is designed as a compression member under load combination LC-4, it is not likely to be critical as a tension member developed under load combination LC-6. The more critical concern would be for the connection elements to this segment such as column splices, base plates, and anchor rods, which must be designed for a vertical uplift as well as the compression.

Required Strength of Stability Bracing

Provisions in LRFDS Section C3-2 give detailed means by which to access the strength and stiffness to be provided by the bracing. LRFDM Part 2, in the section "Design of Beams," states:

> **As an alternative to the more rigorous and generally applicable provisions for stability bracing in LRFDS Section C3, the historically approximate and conservative procedure of designing the bracing element to resist a force equal to 2 percent of the factored compressive force in the braced member will normally suffice.**

This approximate approach to designing stability bracing will be considered in this text. Thus, the diagonals in the trussed bent of Fig. 5.1.1 will be designed (in Example 5.2.5) for the shear due to the factored wind (as discussed in the previous section) and a shear equal to 2 percent of the gravity loads acting simultaneously on the structure in the load combination considered.

E X A M P L E   5 . 2 . 5

Required Axial Strength for a Diagonal Bracing

The building considered in Fig. 5.1.1 is provided with diagonal bracing in the two end walls to resist wind acting normal to the length of the building. Assume a net wind pressure, representing the combined pressure on the windward face and suction on the leeward face, of 30 psf. Determine the required tensile strength of the diagonals. Use the LRFD specification.

Solution

Assume that only the tension diagonal is effective in resisting the story shear. Thus, if V is the story shear and T is the tension in the diagonal brace, we have from Fig. X5.2.5c:

$$T \cos \theta = V \rightarrow T \cos 45° = V \rightarrow T = 1.4142\ V$$

Any dead load, live load, or snow load on the structure does not produce forces in the diagonal members of the braced bent. However, these gravity loads generate destabilizing shear forces that the diagonal bracing has to resist. So, for design of stability bracing, the load combination LC-4 of Eq. 4.10.3, for shear in a story may be rewritten as:

$$\text{LC-4: } 1.6W + 0.02\ (1.2G_D + 0.5G_L + 0.5G_S) = V_W + V_{P\Delta}$$

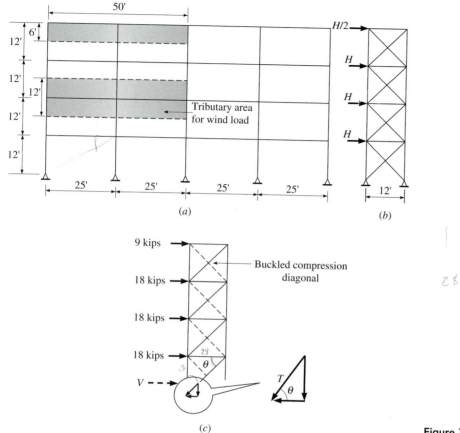

(a)

(b)

(c)

Figure X5.2.5

Here, G_D, G_L, and G_S are the gravity loads on the structure due to dead, live, and snow loads, W is the wind load on the braced bent, V_W is the story shear under factored wind loads, and $V_{P\Delta}$ is the additional story shear due to the destabilizing effects of gravity loads.

a. Tension due to wind load

Story height = 12 ft

Width of the building normal to wind = 100 ft

Each braced bent receives wind acting on one-half the length of the building.

Tributary area for joint at a typical floor level $= 50\left(\dfrac{12}{2} + \dfrac{12}{2}\right) = 600 \text{ ft}^2$

Wind load at a typical floor level, $H = \dfrac{600(30)}{1000} = 18.0 \text{ kips}$

Tributary area for joint at roof level $= \dfrac{600}{2} = 300 \text{ ft}^2$

[*continues on next page*]

Example 5.2.5 continues ...

Wind load at roof level $= \dfrac{H}{2} = 9.0$ kips

Total shear in the braced bent, at ground floor level, $V_W = 3(18.0) + 9.0 = 63.0$ kips

The tension in the brace due to factored wind load is:

$$T_W = 1.4142(1.6V_W) = 1.4142(1.6)63.0 = 143 \text{ kips}$$

b. Tension due to gravity loads

From the symmetry of the structure and the bracing system, each trussed bent has to resist destabilizing forces generated by gravity loads on one-half of the structure.

Tributary area for each level $= 60(50) = 3000 \text{ ft}^2$

Dead Load

Dead load from roof (from Example 5.2.1)	$= 60$ psf
Provision for weight of roof beams, girders, columns, and connections (assumed)	$\doteq 8$ psf
Total dead load from roof	$= \dfrac{(60 + 8)(3000)}{1000} = 204$ kips
Dead load from floor (from Example 5.2.2)	$= 90$ psf
Provision for weight of floor beams, girders, columns, and connections (assumed)	$= 12$ psf
Dead load from floor	$= \dfrac{(90 + 12)(3000)}{1000}$
	$= 306$ kips
Weight of wall (assuming 50 psf)	$= \dfrac{50(50 + 60 + 50)(12)}{1000}$
	$= 96.0$ kips
Total dead load from one floor level	$= 306 + 96.0 = 402$ kips
Total dead load at ground level, G_D	$= 204 + 3(402) = 1410$ kips
Total snow load from roof, G_S	$= \dfrac{40(3000)}{1000} = 120$ kips

Live Load

Tributary area, A_T	$= 3(3000) = 9,000 \text{ ft}^2$
Design live load, $L =$	

$$\max\left\{ \left[0.25 + \frac{15}{\sqrt{18{,}000(1)}} \right](50),\ 0.4(50) \right\}$$
$$= \max\{0.362(50), 20\} = 20.0 \text{ psf}$$

Total live load, $G_L = \dfrac{20(3000)(3)}{1000} = 180$ kips

Total factored gravity load in load combination LC-4:

$$= 1.2G_D + 0.5G_L + 0.5G_S = 1.2(1410) + 0.5(180)$$

$$+ 0.5(120) = 1840 \text{ kips}$$

Additional horizontal shear due to destabilizing effects of gravity loads is:

$V_{P\Delta} = 2\%$ of factored gravity loads $= (0.02)(1840) = 36.8$ kips

So, the additional tensile force in the diagonal brace, induced by the destabilizing effects of gravity loads is:

$$T_{P\Delta} = \frac{36.8}{\cos 45°} = 52.1 \text{ kips}$$

c. Required strength
So, the required tensile strength of the diagonal brace is:

$$T_{\text{req}} = T_W + T_{P\Delta} = 143 + 52.1 = 195 \text{ kips} \qquad \text{(Ans.)}$$

Note: This diagonal member will be designed for a tensile force of 200 kips, in Chapter 7 (Example 7.10.1).

Stairs require special structural framing around the stairwell. The load from stair stringers can be estimated and the supporting members designed. Their strength requirements are often relatively small, and a 8- or 10-in. channel usually proves adequate. Framing around other openings, such as elevator shafts, present similar problems. Landings may be supported by hangers from above, by columns from below, or by cantilever beams at their own levels. Such design involves complicated details but does not constitute a serious structural problem.

5.3   Unbraced Portal Frames

A *portal frame* is a single-bay, single-story, rigid jointed rectangular frame. It consists of a horizontal girder rigidly connected to two vertical columns of equal length and size. The column bases may be either hinged (Fig. 5.3.1) or fixed (Fig. 5.3.3). The lengths of the members are taken as the distance between the intersections of their neutral axes. Let h be the height of the columns, L the span of the girder, I_c the moment of inertia of the column, and I_g the moment of inertia of the girder. Further, let us define the *member relative stiffness factor* as:

$$\beta = \frac{I_c/h}{I_g/L} \qquad (5.3.1)$$

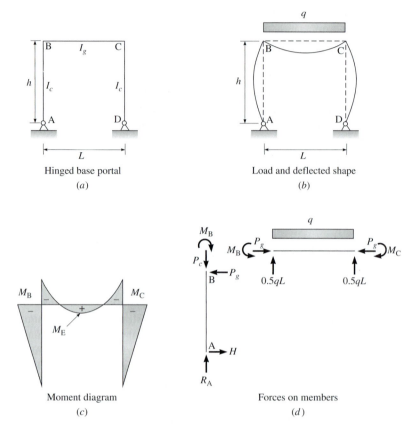

Figure 5.3.1:   Hinged base portal frame under uniformly distributed gravity load, q.

The change in axial length of a member is assumed to be negligible. Thus, when the frame and the gravity load acting on the girder are both symmetrical about midspan, the tops of the columns do not move laterally.

5.3.1   Hinged Base Portal Frames

First, let us consider the hinged base portal frame ABCD shown in Fig. 5.3.1a. Since four unknown reaction components exist at the supports but only three equilibrium equations are available for solution, the hinged base portal frame is statically indeterminate to the first degree. The bending moment diagram in the hinged base frame under a uniformly distributed gravity load, q, may be determined by using, for example, the slope-deflection method [Leet and Uang, 2002]. The bending moments at end B, and mid-point E of span BC may be shown to be given by:

$$M_B = -\frac{qL^2}{12}\frac{3\beta}{3\beta + 2} \tag{5.3.2}$$

$$M_E = +\frac{qL^2}{24}\frac{3\beta + 6}{3\beta + 2} \tag{5.3.3}$$

It is observed from Eqs. 5.3.2 and 5.3.3 that as the relative stiffness factor β approaches infinity, that is, as the rigidity of the columns increases, the moments M_B and M_E approach the values of those in a fixed beam; while as β approaches zero, that is, as the columns decrease in rigidity, M_B approaches zero and M_E approaches the simply supported moment $qL^2/8$.

At its end B, column AB (Fig. 5.3.1d) is subjected to a moment M_B, an axial force P_c equal in magnitude to the girder reaction, $qL/2$, and a shear force equal to the axial force in the girder, P_g. For vertical equilibrium, the hinge at A reacts with an upward reaction $R_A = qL/2$. Moment equilibrium about A of all the forces acting on the column, results in:

$$P_g = \frac{M_B}{h} = \frac{qL}{12}\left(\frac{3\beta}{3\beta + 2}\right)\left(\frac{L}{h}\right) \tag{5.3.4}$$

Fig. 5.3.1d shows how both the columns and the girder are under bending, compression, and shear. Also, Eq. 5.3.4 indicates that the thrust in the girder is usually a small fraction of the transverse load qL on the frame.

The maximum vertical deflection which occurs at the center of the girder is given by:

$$\delta_E = \frac{1}{384}\left(\frac{3\beta + 10}{3\beta + 2}\right)\left(\frac{qL^4}{EI_g}\right) \tag{5.3.5}$$

Under a horizontal load H applied at joint B, the hinged base portal frame deflects antisymmetrically as shown in Fig. 5.3.2a with horizontal reactions $H/2$ at the column bases, so that:

$$M_B = -M_C = \frac{Hh}{2} \tag{5.3.6}$$

The bending moment diagram is shown in Fig. 5.3.2b. The side sway of the frame may be shown to be given by:

$$\Delta = \frac{2\beta + 1}{12\beta}\frac{Hh^3}{EI_c} \tag{5.3.7}$$

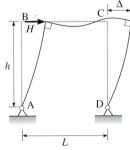

Load and deflected shape

(a)

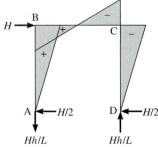

Moment diagram

(b)

Figure 5.3.2: Hinged base portal frame under lateral load, H.

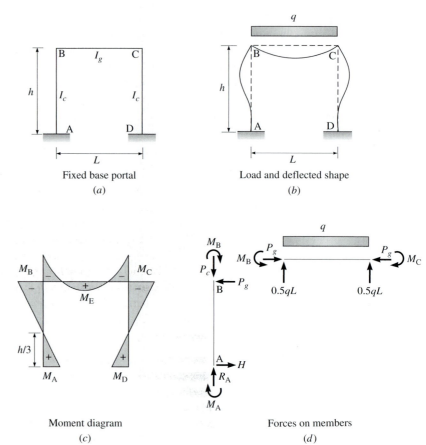

Figure 5.3.3: Fixed base portal frame under uniformly distributed gravity load, q.

Moment diagram
(c)

Forces on members
(d)

5.3.2 Fixed Base Portal Frames

Figure 5.3.3 shows a fixed base portal frame under a uniformly distributed gravity load, q. The frame is indeterminate to the third degree since there is a total of six unknown support reactions but only three equilibrium equations with which to solve them. The deflected shape of the frame is shown in Fig. 5.3.3b, while the bending moment diagram is shown in Fig. 5.3.3c. We have:

$$M_B = -\frac{qL^2}{12}\left(\frac{2\beta}{2\beta + 1}\right); M_A = -\frac{1}{2}M_B \tag{5.3.8}$$

$$M_E = \frac{qL^2}{24}\left(\frac{2\beta + 3}{2\beta + 1}\right) \tag{5.3.9}$$

Since $M_A = -\frac{1}{2}M_B$, the column has an inflection point at $h/3$. Summing moments of all the forces acting on column AB about A, and noting that M_B

and M_A are both clockwise, the thrust P_g in the girder is given by:

$$P_g = \frac{qL}{8} \frac{2\beta}{2\beta + 1} \left(\frac{L}{h}\right) \tag{5.3.10}$$

It is observed from Eqs. 5.3.8 and 5.3.10 that the thrust in the fixed frame is larger than in the hinged base frame. Since the inflection point in column AB of the fixed frame is at $h/3$ from A and is at A in the hinged frame, it follows that for any other condition of partial fixity of the column base, the inflection point for the uniformly distributed gravity load falls within the lower third of the column.

The maximum vertical deflection that occurs at the center of the girder may be shown to be given by:

$$\delta_E = \frac{1}{384} \left(\frac{2\beta + 5}{2\beta + 1}\right) \frac{qL^4}{EI_g} \tag{5.3.11}$$

Under a horizontal load H applied at joint B, the fixed base frame also deflects antisymmetrically, as shown in Fig. 5.3.4a. The moments may be

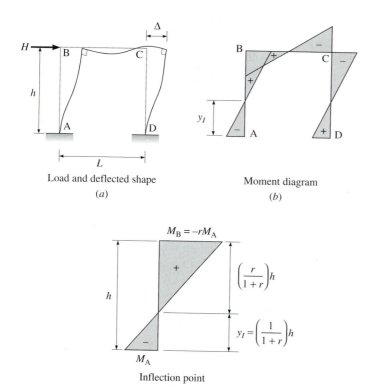

Load and deflected shape
(a)

Moment diagram
(b)

Inflection point
(c)

Figure 5.3.4:   Fixed base portal frame under lateral load, H.

shown to be given by:

$$M_B = -M_C = \frac{Hh}{2}\left(\frac{3}{\beta + 6}\right) \tag{5.3.12}$$

$$M_A = -M_D = -\frac{Hh}{2}\left(\frac{\beta + 3}{\beta + 6}\right) \tag{5.3.13}$$

The ratio r of the absolute value of M_B to that of M_A is:

$$r = \left|\frac{M_B}{M_A}\right| = \frac{3}{\beta + 3} \tag{5.3.14}$$

and the height of the inflection point from the base of the column is:

$$y_I = \left(\frac{1}{1 + r}\right)h \longrightarrow \frac{y_I}{h} = \frac{\beta + 3}{\beta + 6} \tag{5.3.15}$$

It is observed from this relation that the inflection point is always in the upper half of the column and approaches the column midpoint as the girder stiffness becomes large in comparison with the column stiffness. Since $\beta < 1$ for most frames, and since for $\beta < 1$ the inflection point in the column is close to $0.5L$, fixed frames (including multistory rigid jointed frames) subjected to a horizontal force H at their tops behave, for all practical purposes, as frames with hinges (inflection points) at $h/2$ in the columns and at $L/2$ in the girder (Fig. 5.3.4a). In this case, the fixed frame is statically determinate, and $M_B = -M_C = Hh/4$.

Finally, the horizontal deflection of the fixed base portal frame under a horizontal load H may be shown to be given by:

$$\Delta = \frac{2\beta + 3}{12\beta + 72}\left(\frac{Hh^3}{EI_c}\right) \tag{5.3.16}$$

Remarks

1. In any symmetrical frame subjected to lateral loads or asymmetrical gravity loads, or in any loaded asymmetrical frame, the joints will not remain fixed in position; that is, the joints translate. The resulting transverse displacement, Δ, of the top end of a column relative to its bottom end is known as the **drift**, or **sidesway** of the column, and the ratio of the drift to the height of the column is known as the **drift index**.
2. Note that for the portal frame considered in this section, the internal forces on each element are functions of a single parameter, β, defined in Eq. 5.3.1. So, by assuming an appropriate value for β, based on the geometry and the loading on the frame, and intuition, the design process can be initiated. This will be illustrated with the help of Examples 5.3.1 and 11.15.11.

Required Strengths for Members of a Portal Frame

Determine the internal forces in the members of the hinged-base portal frame ABCD shown in Fig. 5.1.2. The girder BC is subjected to a uniformly distributed service load of 2.5 klf [(1 klf) D + (1.5 klf) S]. In addition, the frame is subjected to a concentrated dead load of 30 kips at the column tops B and C. The wind load on the frame consists of a 15 kip horizontal force acting at the joint B. The dead loads given include provision for self-weight of members. The loads given are nominal loads. Use the LRFD Specification. Assume the moment of inertia of girder BC to be twice that of the columns, and determine the required strengths of the members. Also, calculate the maximum vertical deflection, and drift under unit vertical and horizontal loads, as a function of the moments of inertia I_g and I_c.

Solution

$$\text{Span, } L = 60 \text{ ft;} \qquad \text{Height, } h = 20 \text{ ft;} \qquad I_g = 2I_c$$

a. Member forces and deflections under unit loads on the frame
   From Eq. 5.3.1, member relative stiffness factor for the frame is

$$\beta = \frac{I_c/h}{I_g/L} = \frac{1/20}{2/60} = 1.5$$

For unit uniformly distributed load on girder BC of the hinged base portal frame ABCD (Fig. 5.3.1), we obtain from Eqs. 5.3.2 and 5.3.3:

$$M_{C(1)} = -\frac{qL^2}{12}\frac{3\beta}{3\beta + 2} = \frac{-1(60^2)}{12}\left[\frac{3(1.5)}{3(1.5) + 2}\right]$$
$$= -207.7 \text{ ft-kips}$$

$$M_{E(1)} = +\frac{qL^2}{24}\frac{(3\beta + 6)}{3\beta + 2} = +\frac{1(60^2)}{24}\left[\frac{[3(1.5) + 6]}{[3(1.5) + 2]}\right]$$
$$= 242.3 \text{ ft-kips}$$

$$P_{CD(1)} = \frac{qL}{2} = \frac{1(60)}{2} = 30.0 \text{ kips}$$

The maximum vertical deflection under the uniformly distributed gravity load occurs at the midspan, E, of the girder. For a unit distributed load of 1 klf this value may be calculated, with the help of Eq. 5.3.5, as:

$$\delta_{E(1)} = \frac{1}{384}\left(\frac{3\beta + 10}{3\beta + 2}\right)\left(\frac{qL^4}{EI_g}\right)$$

$$= \frac{1}{384}\left[\frac{3(1.5) + 10}{3(1.5) + 2}\right]\left[\frac{1(60^4)(12^3)}{29,000I_g}\right]$$

$$= \frac{4486}{I_g} \text{ in.}$$

[continues on next page]

Example 5.3.1 continues ...

For a unit horizontal load at joint B of the hinged base portal frame ABCD (Fig. 5.3.2), we obtain from Eq. 5.3.6:

$$M_{C(1)} = -\frac{Hh}{2} = -\frac{1(20)}{2} = -10 \text{ ft-kips}$$

$$M_{E(1)} = 0$$

$$P_{CD(1)} = \frac{Hh}{L} = \frac{1}{3} \text{ kip}$$

The sidesway of the frame under the unit horizontal load, may be calculated from Eq. 5.3.7 as:

$$\Delta_{(1)} = \frac{2\beta + 1}{12\beta} \left(\frac{Hh^3}{EI_c}\right) = \frac{2(1.5) + 1}{12(1.5)} \left[\frac{1(20^3)(12^3)}{29,000 \, I_c}\right] = \frac{105.9}{I_c} \text{ in.}$$

b. Factored loads on the frame

Service dead load on the girder, q_D = 1.0 klf
Nominal snow load on the girder, q_S = 1.5 klf
Service dead load at the column top, Q_D = 30 kips
Nominal wind load at joint B, Q_W = 15 kips

With only dead, snow, and wind loads acting on the structure the load combinations considered in Section 4.10 reduce to:

$$\text{LC-1} \longrightarrow \{1.4 \, D\}$$
$$\text{LC-3} \longrightarrow \{1.2D + 1.6S\} + 0.8W$$
$$\text{LC-4} \longrightarrow \{1.2D + 0.5S\} + 1.6W$$

Note that, in each load combination, the term within curly brackets represents the vertical component of the loading, while the remainder represents the horizontal component of the loading. The symmetric frame, under the symmetric vertical component of the loading given, deforms with no joint translation. Hence, the moments produced by these vertical loads are moments with no lateral translation of the frame, or the so called M_{nt} moments. On the other hand, under the horizontal loads, the joints translate or drift. Hence, the moments produced by the horizontal component of the loading are the moments due to a lateral translation of the frame, or the so called M_{lt} moments (see Section 11.9, for details).

Substituting the values of the service loads in the load combinations, we obtain the vertical and horizontal components of the factored loads for the three load combinations considered. They are given in Table X5.3.l*a*.

c. Member forces under factored loads on the frame

Using the vertical and horizontal components of factored loads in the three load combinations given in Table X5.3.1*a* and the internal forces that develop in the members under unit loads calculated in part *b*, we can determine the internal forces in the members under the factored loads of the three load combinations. The results for column CD are summarized in Table X5.3.1*b*.

TABLE X5.3.1*a*

Components of the Factored Loads

	Vertical Component		Horizontal Component
Load Combination	q_{uV} (klf)	Q_{uV} (kips)	Q_{uH} (kips)
LC-1	1.4	42	0
LC-3	3.6	36	12
LC-4	2.0	36	24

TABLE X5.3.1*b*

Internal Forces in the Frame Under Factored Loads

	Vertical Component			Horizontal Component		Total	
	q_{uV}		Q_{uV}	Q_{uH}			
	M_C	P_{CD}	P_{CD}	M_C	P_{CD}	M_C	P_{CD}
LC-1	−291	42	42	0	0	−291	84
LC-3	**−748**	**108**	**36**	**−120**	**4**	**−868**	**148**
LC-4	−415	59	36	−240	8	−655	103

Note: Moments are in ft-kips, and axial load is in kips.

Note: The girder BC and the column CD will be designed in Chapter 11 (Example 11.15.11), and the girder-to-column connection will be designed in Chapter 13 (Example 13.10.1).

5.4   Unbraced Frames

Approximate Methods of Analysis

The usual procedure in the design of any structure is to calculate, from the loads acting on the structure, the internal moments, shears, and axial forces in the members and then to proportion them accordingly. From the single-bay single-story frame discussed in Section 5.3, it is evident that in any indeterminate frame, the distribution of internal forces depends upon the relative sizes of the members of the frame, and consequently an accurate analysis of the internal forces cannot be made until the member sizes have been approximately determined.

Approximate analysis methods that reduce the given indeterminate rigid frame to a statically determinate frame, by introducing a sufficient number of hinges, are often used in practice. The hinges represent the points of inflection, that is, points of zero internal moment, in the deformed frame, and their locations are worked out by reasonable assumptions based on experience, results from more exact analyses, and knowledge of the true behavior of structures. Since the way a rigid jointed frame behaves is quite different for vertical loads versus horizontal loads, different sets of assumptions are made for the study of frames under these two different sets of loads. Hence, separate approximate methods are used for gravity loads and lateral loads.

A student engineer may feel that in this age of electronic computation there is no need to learn about approximate methods. However, all computer programs used to analyze indeterminate frames require some initial input with regard to the stiffness of columns and girders. The approximate methods do therefore serve a useful purpose in that they can provide preliminary member sizes upon which to initiate the computer analysis process, and the reader is referred to standard structural analysis texts for these methods [Leet and Uang, 2002; Schueller, 1977; Timoshenko and Young, 1945].

For final design, the designer makes a first- or second-order analysis of the entire structure, using matrix methods and commercially available computer programs, to determine the internal forces on the members [for details see for example, McGuire et al., 2000].

First-Order Elastic Analysis

First-order elastic analysis is the traditional approach used in the analysis and design of framed structures. Here, the material is assumed to be linearly elastic. Equilibrium is formulated on the initial, undeformed geometry of the frame. This type of analysis provides a simple estimate of the distribution of internal forces in the structural system; it does not give any information on the stability of the frame.

Second-Order Elastic Analysis

In a *second-order elastic analysis,* the material is still assumed to be linearly elastic, but the equilibrium equations are formulated in the deformed state for the members and the structure. The moments in the members are magnified by the product of the axial force and the deflection. This type of analysis includes both $P\delta$ (or member curvature) effects, and $P\Delta$ (or member chord rotation) effects. For an introduction to second-order moments in structures refer to Appendix A5.1.

References

5.1   Leet, M. K., and Uang, C-M. [2002]: *Fundamentals of Structural Analysis,* McGraw-Hill Co., NY, 2002.

5.2   McGuire, W., Gallagher, R. H., and Ziemian, R. D. [2000]: *Matrix Structural Analysis,* 2nd ed., John Wiley & Sons, NY, 2000.

5.3   Schueller, W. [1977]: *High-Rise Building Structures,* John Wiley & Sons, NY, 1977.

5.4   Timoshenko and Young [1945]: *Theory of Structures,* McGraw-Hill Book Company, Inc., NY, 1945.

PROBLEMS

P5.1. Figure P5.1a shows a plan view of an office building in a suburb of Chicago. All beam-to-column connections are simple, and the lateral loads are resisted by the braced bents only, which are located in the end walls. Two suggested bracing configurations are shown in Figs. P5.1b and P5.1c. The roof and floor slabs consist of a $3\frac{1}{2}$ in. normal weight concrete (144 lb/ft³) over a 3-in.-deep metal deck resting on steel beams and girders. Use ASCE 7-98 and determine the required strengths of typical beams, girders, columns, and braces.

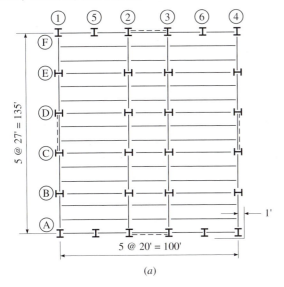

5 @ 27' = 135'

5 @ 20' = 100'

(a)

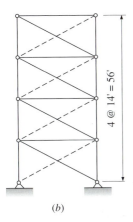

(b)

4 @ 14' = 56'

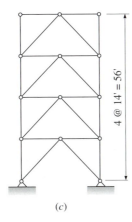

(c)

4 @ 14' = 56'

P5.1

P5.2. A typical interior truss, for a parking garage in Milwaukee, Wisconsin, is shown in Fig. P5.2. Trusses are spaced at 24 ft apart. Simply supported steel beams span between the trusses at the upper chord panel points. The 6-in. concrete deck is supported by the steel beams. Cars can be parked on the roof in any arrangement, and design live load is to be taken as 100 psf. Determine the required strengths of the beams, and typical truss members. Use the ASCE 7-98 Standard for loads.

P5.3. to P5.5. Use the equations for member forces and deformations of portal frames given in Section 5.3, to solve these problems.

P5.3. Repeat Example 5.3.1, assuming $I_g = I_c$.

P5.4. Repeat Example 5.3.1, assuming bases are fixed and $I_g = 2I_c$.

P5.5. Repeat Problem 5.3.1, assuming bases are fixed and $I_g = I_c$.

P5.6. Figure P5.6 shows an L-shaped, rigid-jointed frame ABC proposed for an annex building, wherein AB = 12 ft, and BC = 24 ft. The spacing of the frames is 24 ft. Assume $I_g = 2I_c$, elastic behavior, and determine the end moments, and axial forces in both members, for the following load conditions.

    (a) A uniformly distributed load of 1 klf on the girder BC only, with horizontal displacement of joint C prevented. Also calculate the maximum deflection of the roof girder BC.

    (b) A horizontal load of 1kip at joint B. Also, calculate the horizontal drift of the joint C.

    (c) The roof is subjected to a dead load of 50 psf, a roof live load of 33 psf, and the wall is subjected to a wind load of ±20 psf. The dead load given includes provision for the self weight of the girder. Note that the wind load given may act in both directions. Based on the results obtained in parts (a) and (b), tabulate the values of the moments and axial forces to be used in the design of the column and the girder, for the ASCE 7-98 load combination LC-3.

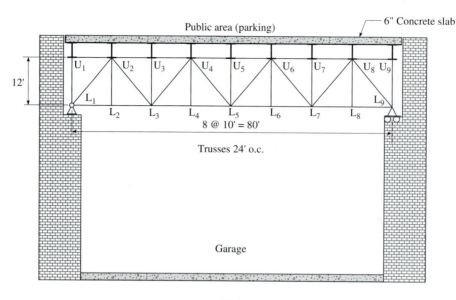

P5.2

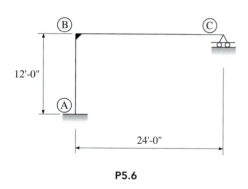

P5.6

P5.7. A symmetric, single-story, two-bay, fixed base, unbraced, rigid-jointed steel frame is shown in Fig. P5.7. We have: AB = CD = EF = 16 ft, and BD = DF = 32 ft. W14×90 shapes are used for exterior columns, W14×68 shape for interior columns, and W24×76 shape for roof girders. Assume elastic behavior. Determine the end moments and axial forces in all members for the following load conditions.

(*a*) A uniformly distributed load of 1 klf on the roof girders BD and DF. Also calculate the maximum deflection of the roof girder.

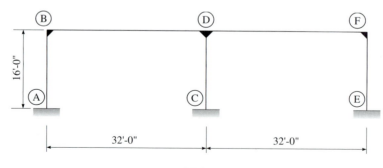

P5.7

(b)   A uniformly distributed load of 1 klf on the roof girder BD only, with the translations of roof joints prevented by a fictitious support at F. Also calculate the maximum deflection of the roof girder BD, and the horizontal reaction at the fictitious support at F.

(c)   A horizontal load of 10 kip at joint B. Also, calculate the horizontal drift of the joint F.

(d)   Roof girders are subjected to a dead load $q_D = 2$ klf and a snow load, $q_S = 3$ klf. In addition, concentrated dead loads of 100, 200, and 100 kips act at joints B, D, and F respectively. The wind load on the frame consists of a horizontal force of 6 kips acting at joint B. The loads given are service loads. Based on the results obtained in parts (a), (b), and (c), tabulate the values of the moments and axial forces to be used in the design of the columns and the girders, for the ASCE 7-98 load combination LC-3. For columns, calculate separately the effects of the gravity-load-component and the lateral-load-component of the load combinations.

CNC structural processing systems like the Peddinghaus BDL 1250/9 drill/saw tandem system shown here, *on right,* are used for detail work in steel fabricating shops.

(*Below left*) Three drill spindles enhance productivity as no time is lost to change drill sizes.

(*Below right*) Some drilled beams that were processed on the BDL1250 drill system.

Photos courtesy Peddinghaus Corporation, Bradley, IL, www.peddinghaus.com

C H A P T E R

Connectors

6

In steel structures, connection of members to each other is usually accomplished by the use of bolts, pins, or welds collectively known as **connectors** or **fasteners.** A **bolt** is a metal rod of circular cross section with a head formed at one end and the body or shank threaded at the other end, in order to receive a nut. Bolts are used to join elements of steel: the bolts are inserted through holes in these elements, and the nuts are tightened at the threaded ends. Structural bolts may be classified, based on material and strength, as ordinary structural bolts and high-strength bolts. At present, high-strength bolts are the most commonly used fasteners for structural steel. Unlike a bolted joint, a pin connection has only one connector at a joint and hence no redundancy if it fails. Pins are used for hinged joints or connections at which it is desired to have zero moment. Structurally, a **pin** is a short, cylindrically shaped steel beam connecting several members converging at a joint of the structure. The pin fits snugly in its hole drilled in each of the component members. **Welding** is a process of connecting pieces of steel together by molten metal produced by the application of intense heat. Often, the welding heat is obtained by passing an electric arc between the pieces to be welded and a steel wire or rod called an **electrode**. Rivets, one time the primary means of connecting members, are obsolete now and will not be considered in this text.

As shown in Fig. 6.1.1, from the viewpoint of load transfer, connectors may also be classified as **point** connectors (e.g., bolts) where the load transfer between connected elements occurs at discrete points; **line** connectors (e.g., fillet welds) where the load transfer occurs along a line or number of lines (along the axis of the weld); and **surface** connectors (e.g., slot welds) where the load transfer occurs over an area.

The behavior of the common types of connectors—bolts, welds, and pins—will be covered thoroughly in this chapter. The behavior of groups of connectors in a plane, called **joints,** will be studied under various loads in Chapter 12. Armed with a knowledge of the behavior of connectors and joints, the structural designer should be able to cope with common connection design. Some typical bolted connections are shown in Fig. 6.1.2, while some

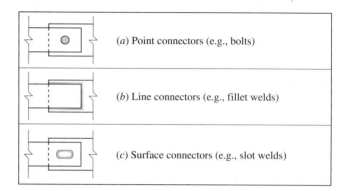

Figure 6.1.1: Basic types of steel connectors.

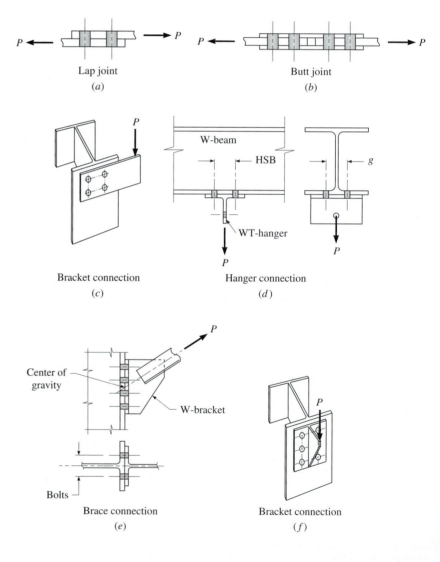

Figure 6.1.2: Typical bolted connections.

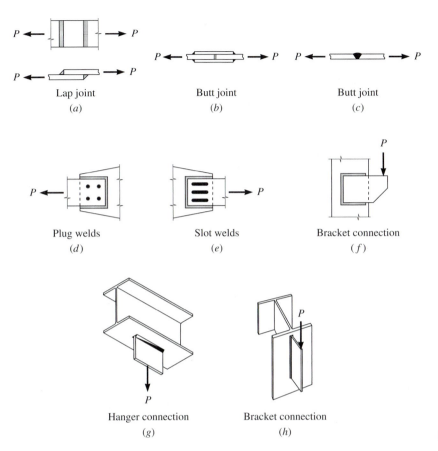

Figure 6.1.3: Typical welded connections.

typical welded connections are given in Fig. 6.1.3. Detailed consideration of these connections will be postponed to Chapter 13. The LRFD Specification deals with connectors (welds, bolts, pins, and rivets), joints, and connections in Chapter J.

6.2   High-Strength Bolts

The first specification governing the use of high-strength bolts for structural connections was issued in 1951 by the Research Council on Riveted and Bolted Structural Joints (RCRBSJ) of the Engineering Foundation. Research developments under the sponsorship of RCRBSJ, now renamed the Research Council on Structural Connections (RCSC), led to several editions of the bolt specifications, the latest one being the 2000 edition [RCSC, 2000], included as Part 16.4 of the LRFDM. The design requirements for bolted joints, in Chapter J of the LRFDS, are highly influenced by the 1994 edition of this

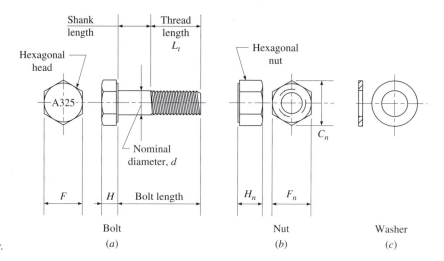

Figure 6.2.1: High-strength bolt, nut, and washer.

reference [RCSC, 1994]. An excellent review of the research developments in the field of riveted and bolted connections was published by Kulak, Fisher, and Struik [1987].

Bolts

A simple bolt assembly normally consists of a bolt, a nut, and if required, a washer (Figs. 6.2.1 and 6.2.2). A washer, when required by the specification, is put under the element which is to be turned (usually the nut). ***The nominal size of a bolt*** is the diameter, d, of the bolt shank in the unthreaded part (see Fig. 6.2.1). High-strength bolts have ***heavy-hexagonal heads,*** or simply,

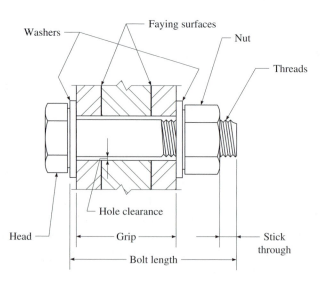

Figure 6.2.2: High-strength bolt assembly.

heavy-hex heads. They range in diameter from $\frac{1}{2}$ in. to $1\frac{1}{2}$ in. Bolt, nut, and washer dimensions for high-strength bolts are listed in Table 7.1 of the LRFDM.

At present, high-strength bolts supplied for structural joints meet the requirements for two principal strength grades:

- The *ASTM A325-97 Specification:* Structural Bolts, Steel, Heat-Treated, 120/105 ksi Minimum Tensile Strength [2.4]; or
- The *ASTM A490-97 Specification:* Heat-Treated Steel Structural Bolts, 150 ksi Minimum Tensile Strength, for higher load applications [2.4].

A325 bolts are heat-treated medium carbon steel. A490 bolts are also heat-treated but are of alloy steel. They were developed for use with high-strength steel members. ASTM Specifications permit galvanizing of A325 bolts, but not A490 bolts. The A325 bolt is by far the most commonly used structural bolt. The most common bolts used in building structures are $\frac{3}{4}$-in. and $\frac{7}{8}$-in. dia. A325 bolts, whereas the most common bolts in bridge structures are $\frac{7}{8}$-in. and 1-in. dia. A490 bolts. Alternate design fasteners meeting the requirements of the *ASTM F1852-98 Specification:* Twist-Off-Type Tension-Control Structural Bolt/Nut/Washer Assemblies, Steel, Heat Treated, 120/105 ksi Minimum Tensile Strength (equivalent to ASTM A325) are also available [2.4].

Nuts

High-strength bolts use *heavy-hexagonal nuts (heavy-hex nuts)* of the same nominal size as the bolt head so that the erector may use a single size wrench or socket on both the head and the nut. The nut is an important part of the bolt assembly. The ASTM A563 Grade C nut is the most widely used heavy-hex nut for structural bolting, and it is the recommended nut for use with A325 high-strength structural bolts. The ASTM A563 Grade DH nut is a heavy-hex nut recommended for use with A490 high-strength structural bolts.

Washers

The primary function of a *washer* is to provide a hardened nongalling surface under the turned element (i.e., the nut, or less frequently, the bolt head) in pretensioning, particularly for those installation procedures which depend upon torque for control or inspection [Section 6, RCSC 2000]. Thus, specifications require that hardened washers be used under the turned element for torque-based pretensioning methods such as the calibrated wrench pretensioning method and the twist-off-type tension-control bolt pretensioning method. Hardened washers are required under both the bolt head and nut when pretensioned A490 bolts are used to connect material with a yield stress of less than 40 ksi. When the outer face of a bolted joint has a slope that is greater than 1 in 20 (such as the tapered flanges of I- and C-shapes), a beveled washer is used against the inclined flange surface to furnish a nor-mal seating for the nut or bolt. Flat circular washers and square or rectan-gular beveled washers must conform to the ASTM F436 Specification [2.4].

Marking

High-strength bolts, nuts, and washers are required by ASTM specifications to be distinctively marked. These identifying marks are indicated in RCSCS Fig. C-2.1 (see page 16.4–8 of the LRFDM). Certain markings are mandatory. For example, all A325 bolt heads are marked "A325" as depicted in Fig. 6.2.1. All A490 bolts are marked "A490".

Bolt Length

The *grip* of a bolt is equal to the sum of the thicknesses of the plate elements joined as shown in Fig. 6.2.2. The *length of a bolt,* L, is calculated as the grip plus allowances for washers and L' (given in RCSCS Table C-2.2) and then rounded to the next longer $1/4$-in.-length increment ($1/2$-in.-length increment for lengths exceeding 5 in.). Values of L for different bolt diameters and grip lengths are given in Table 7-2 of the LRFDM. The value, L, provides for *full thread engagement* with an installed heavy-hex nut, defined as having the end of the bolt flush with the face of the nut. Bolts longer than 5 in. are gen-erally available only in $1/2$-in. increments. A length of 8 in. is generally the maximum stock length available. High-strength bolts are generally cold-headed up to about 9 in. in length and hot-headed when over that length.

Mechanical Properties of Bolts

Bolt performance is strongly influenced by the presence of threads, and since their influence is not proportional either to the diameter or to the area of the shank, it is more appropriate to use the total force rather than stresses when discussing bolt behavior. Mechanical properties of A325 and A490 high-strength bolts are determined by applying axial tension loading to a full-size bolt. To determine the load-elongation curve of a bolt, a bolt is placed in a suitable holding device and a pure tensile load—between the head of the bolt and the nut—is applied. The resulting curve consists of an initial linear elas-tic region ending in the proportional limit, where the elongation is no longer linear, followed closely by the elastic limit. Loads that cause the bolt to exceed the elastic limit produce some permanent set when the bolt is unloaded. The maximum tension reached, through additional loading, is called the *ultimate strength* or *tensile strength* of the bolt. Because bolt material exhibits a load-deformation behavior that has no well defined yield point, a so-called *proof load* of the bolt is established by using a 0.2% offset strain. This value is con-sidered to be equivalent to the yield strength of the bolt.

In the 2003 ASTM specifications, the ultimate tensile stress for A325 bolts is prescribed to be 120 ksi for $1/2$ to 1-in.-dia. bolts, and 105 ksi for $1\frac{1}{8}$- to $1\frac{1}{2}$-in.-dia. bolts. The yield stress, measured at 0.2% offset, is required to be 92 ksi minimum for $1/2$- to 1-in.-dia. bolts, and 81 ksi for $1\frac{1}{8}$- to $1\frac{1}{2}$-in.-dia. bolts. For A490 bolts, the ultimate tensile stress is prescribed to be

150 ksi minimum, and the yield stress, obtained by 0.2% offset method, is required to be 130 ksi minimum, for all diameters. It should be noted that the ratio of proof load to ultimate tensile stress for the A325 bolts is about 0.75, whereas for A490 bolts it is 0.85. We have, from LRFDM Table 2-3:

$$
\begin{aligned}
F_{ub} &= 120 \text{ ksi} \quad \text{for } \tfrac{1}{2}\text{- to 1-in.-dia. A325 bolts} \qquad (6.2.1) \\
&= 105 \text{ ksi} \quad \text{for } 1\tfrac{1}{8}\text{- to } 1\tfrac{1}{2}\text{-in.-dia. A325 bolts} \\
&= 150 \text{ ksi} \quad \text{for A490 bolts of all diameters}
\end{aligned}
$$

where F_{ub} = minimum specified ultimate tensile stress of bolt material

Bolt Holes

Standard holes for bolts are circular and are made $\tfrac{1}{16}$ in. bigger in diameter than the nominal size of the bolt body. This provides a certain amount of play in the holes, which compensates for small misalignments in hole location or assembly, and helps in the shop and field installation of bolts. The standard holes also provide some latitude for adjustment in plumbing up a frame during erection. However, certain conditions encountered in field erection of particularly large joints may require greater adjustment than the standard clearance can provide. Consequently, the LRFDS allows the limited use of three kinds of larger holes, namely, oversized, short-slotted, and long-slotted holes (Fig. 6.2.3). An *oversized hole* provides the same extra clearance in all

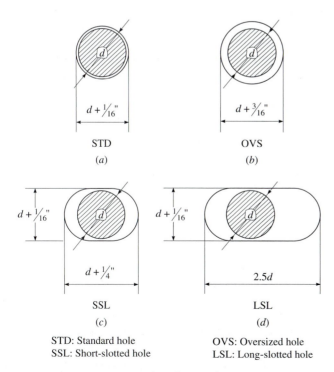

STD: Standard hole   OVS: Oversized hole
SSL: Short-slotted hole   LSL: Long-slotted hole

Note: Sizes given are for $d = \tfrac{5}{8}$ ", $\tfrac{3}{4}$ ", and $\tfrac{7}{8}$ "

Figure 6.2.3: Types and sizes of bolt holes.

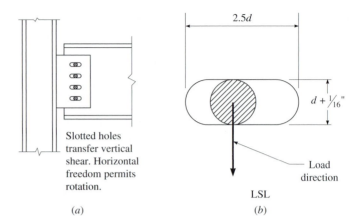

Figure 6.2.4: Use of slotted holes.

directions, as compared to a standard hole, to meet tolerances during erection. If, however, an adjustment is needed in a particular direction, **short-slotted holes** or **long-slotted holes** can be used. Depending on the direction of the slots with respect to the direction of the applied load, slotted holes are identified by their parallel or transverse alignment. LRFDS Table J3.3 gives the maximum size of bolt holes for **standard** (STD), **oversized** (OVS), **short-slotted** (SSL), and **long-slotted** (LSL) conditions. Note that the width of the SSL and LSL holes is the same as the diameter of the STD holes for the same bolt diameter. The application of slotted holes in a simple beam-to-column connection is illustrated in Fig. 6.2.4. The use of these four different hole types is controlled by LRFDS Section J3.2. In the absence of approval by the *Engineer of Record* for use of other hole types, standard holes must be used in all high-strength bolted joints.

Holes for bolts are either punched, subpunched and reamed, or drilled (LRFDS Section M2.5). The common practice is to use punched holes for buildings and ordinary highway bridge work. Holes may be punched when the thickness of the material is not greater than the nominal diameter of the bolt plus $\frac{1}{8}$ in. The standard holes are punched $\frac{1}{16}$ in. larger than the nominal diameter, d, of the bolt. Punching causes the metal piece to stretch, and the amount of stretch depends upon the thickness of the metal and the number of holes. As shown in Fig. 6.2.5, punching damages some material around the hole.

When the material thickness is greater than the nominal diameter of the bolt plus $\frac{1}{8}$ in. reaming is required. This consists of subpunching holes to $\frac{1}{16}$ in. undersize, and then reaming the holes to the bolt diameter plus $\frac{1}{16}$ in. after the pieces being joined are assembled. The object of reaming is to remove the material surrounding the hole, which is more or less injured in punching and to ensure a better fit and matching of holes. Where the steel used is very thick or of very high strength, holes are usually drilled. Thus, holes in A514 steel plates over $\frac{1}{2}$-in. thick must be drilled. Some specifications require all holes

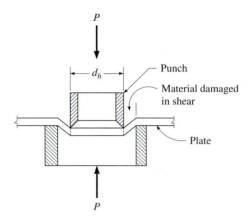

Figure 6.2.5: Punching of holes.

to be drilled in certain situations. Drilling is the most expensive of the three techniques.

Bolt Spacing

A *row of bolts* generally refers to a line of bolts, placed parallel to the line of stress (generally parallel to the longitudinal axis) of the member. Figure 6.2.6 shows a plate member in tension with four rows of bolts, namely A, B, C, and D. The holes on rows A and B are aligned, while the holes on rows C and D are staggered with respect to those on lines A and B. *Bolt gage, g,* is the distance

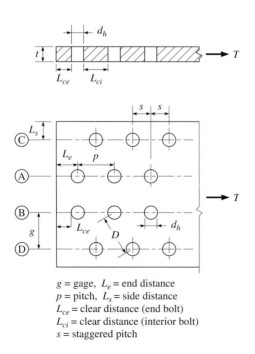

g = gage, L_e = end distance
p = pitch, L_s = side distance
L_{ce} = clear distance (end bolt)
L_{ci} = clear distance (interior bolt)
s = staggered pitch

Figure 6.2.6: Spacing of bolts with standard holes.

measured perpendicular to the longitudinal axis of the member between two adjacent rows of bolts. ***End distance, L_e***, is the distance from the center of a bolt hole to the plate edge, measured parallel to the line of stress. ***Side distance, L_s***, is the distance from the center of a bolt hole to the plate edge, measured perpendicular to the line of stress. Quite often end distance and side distance are known by the common name ***edge distance. Pitch, p,*** is the distance center-to-center of two adjacent bolt holes on a given gage line. ***Staggered pitch*** or simply ***stagger, s,*** is the longitudinal center-to-center spacing of any two consecutive bolt holes on adjacent gage lines. Note that the distances to bolts and the distances between them, shown in design and shop drawings, are always drawn to the centers of the holes. The diameter of the bolt hole is d_h. The ***clear distance for an end bolt, L_{ce}***, is the distance in the direction of load, between the edge of an end hole and the edge of the material. The ***clear distance for an interior bolt, L_{ci}***, is the distance in the direction of load, between the edge of a bolt hole and the edge of the adjacent hole. Thus, for standard holes:

$$L_{ce} = L_e - 0.5\, d_h \qquad (6.2.2)$$

$$L_{ci} = p - d_h \qquad (6.2.3)$$

Bolts may be placed in a variety of patterns depending upon the shape of the members to be connected, available connection area, magnitude of forces to be transferred, and so on. Generally speaking, weight will be saved and appearance improved if joints are made as short and compact as possible. This means that the bolts should be placed as closely as possible. Thus, LRFDS Section J3.3 specifies that the ***minimum bolt spacing,*** namely, the distance, D, between the centers of two adjacent bolt holes in any direction be not be less than $2\frac{2}{3}$ times the nominal diameter of the bolt, but preferably not less than three diameters. This minimum spacing applies to the distance between any two bolts, in line or not. Clearances required by wrenches to enter and tighten the nuts determine the minimum spacing that must be maintained between bolts and other parts projecting from the plate being bolted. For examples, see Fig. 6.2.7. The specifics of such entering and tightening clearances for high-strength bolts are provided in LRFDM Tables 7-3*a* and 7-3*b*.

It is possible that bolt holes might be punched so closely to each other that the metal between them would be injured to such an extent that its strength would be reduced to unacceptable levels. Also, edge failure may result from placing a bolt hole so close to the edge of the member that a local bulge is formed when the bolt hole is punched. So, LRFDS Sect. J3.4 specifies that the ***minimum edge distance*** from the center of a standard hole to an end or to a side of a connected part be not less than the applicable value from LRFDS Table J3.4. The minimum edge distances stipulated are dependent on the edge condition of the material. Rolled, machine flame-cut, sawed, or planed edges are relatively uniform. Sheared or hand flame-cut edges are

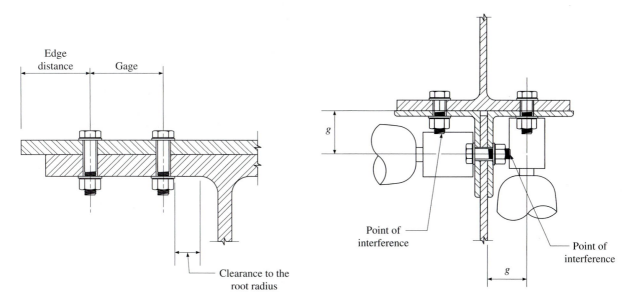

Figure 6.2.7: Bolt clearances.

slightly rough and, therefore, extra allowance has to be made to ensure that the basic minimum distance is achieved.

At the other extreme, when bolts are placed too far apart, the elements joined might not be in close contact between bolts, leaving a space for water and dust to accumulate. As a result, rust might develop. As fully corroded steel occupies seven times the volume of uncorroded steel, high local stresses could develop and/or the plates could buckle. Thus, LRFDS Section J3.5 stipulates that the ***maximum edge distance*** from the center of any bolt hole to the nearest edge of parts in contact shall be twelve times the thickness of the connected part under consideration, but shall not exceed 6 in. (Fig. 6.2.8). Further, for built-up members, the longitudinal spacing of connectors between a plate and a shape or two plates in continuous contact shall not exceed:

*   Twenty-four times the thickness of the thinner plate or 12 in. for painted members or unpainted members not subject to corrosion.
*   Fourteen times the thickness of the thinner plate or 7 in. for unpainted members of weathering steel subject to atmospheric corrosion.

Thus:

$$L_e \leq \min\,[12t_p,\ 6\ \text{in.}] \tag{6.2.4a}$$

$$p \leq \min\,[24t_{p1},\ 12\ \text{in.}]\ \ \text{for painted members} \tag{6.2.4b}$$

$$\leq \min\,[14t_{p1},\ 7\ \text{in.}]\ \ \ \text{for members of weathering steel} \tag{6.2.4c}$$

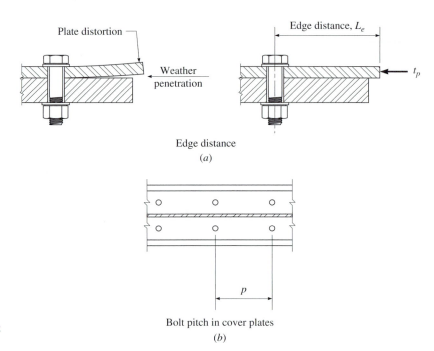

Figure 6.2.8: Maximum edge distance and bolt spacing.

where t_{p1} is the thickness of the thinner plate, L_e is the edge distance and p is the pitch.

Bolting in flanges of I-, T-, and C-shapes, and in angle legs is usually done along lines called **gage lines. Workable gages** in flanges of W-, C-, and WT-shapes that provide for entering and tightening clearances and for edge distance and spacing requirements are given in LRFDM Tables 1-1, 1-5, and 1-8, respectively. The workable gages used in bolting angles are given in Table 6.2.1, adapted from the table on page 10-10 of the LRFDM. These gages are dependent on the length of the leg in which the holes are located

TABLE 6.2.1								
Workable Gages in Angle Legs (in.)								
Leg	8	7	6	5	4	3½	3	2½	2
g_1	4½	4	3½	3	2½	2	1¾	1⅜	1⅛
g_2	3	2½	2¼	2	—	—	—	—	—
g_3	3	3	2½	1¾	—	—	—	—	—

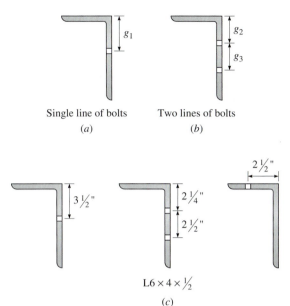

Single line of bolts   Two lines of bolts
(a)   (b)

L6 × 4 × ½
(c)

Figure 6.2.9: Workable gages in angle legs.

and on the number of lines of holes in that leg. For leg sizes greater than 4 in., two lines of bolts could be used in that leg. Gage distance g_1 applies when only one line of bolts is used in the leg under consideration, while gage distances g_2 and g_3 are used when two lines of bolts are used in a leg (Figs. 6.2.9a and b). Thus, for a 6 by 4 by ½ in. angle, gage distance g_1 = 3½ in., corresponding to a leg size of 6 in., is used with a single line of bolts in the long leg, while gage distances g_2 = 2¼ in. and g_3 = 2½ in., corresponding to a leg size of 6 in., are used with two lines of bolts in the long leg (Fig. 6.2.9c). In either case, if bolts are provided in the short leg also, gage distance g_1 = 2½ in., corresponding to a leg size of 4 in., is to be used for that leg. Unless special situations dictate, the gages given in Table 6.2.1 are adhered to, as any deviation from these standards will result in higher fabrication costs.

6.3   Bolt Loading Cases

There are three stresses acting at any point on a surface—the normal stress and the two orthogonal shear stresses (Fig. 6.3.1a). A stress resultant is defined as the integral of a stress over the area of a cross section or the integral of the moment caused by stresses on elemental areas, about a chosen axis, over the area of a cross section. It is usual to consider stress resultants acting on cross sections normal to the axis of the bolt. There are six such possible stress resultants as shown in Fig. 6.3.1b. With the type of bolted connections

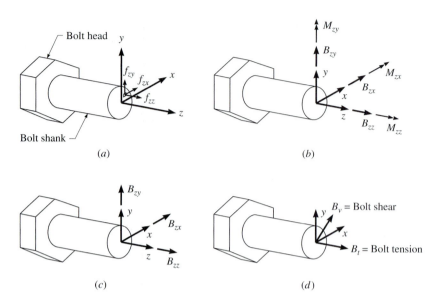

Figure 6.3.1: Basics of bolt loading.

used in steel structures, in which all connected plates are in contact with each other and bolt length-to-diameter ratio is small, torsional and bending moments on bolt cross sections are either zero or negligible. Thus, the stress resultants on a bolt section reduce to three forces (Fig. 6.3.1c): a normal force, B_{zz}, and two shear forces, B_{zx} and B_{zy}. Due to the circular shape of the bolt section, B_{zx} and B_{zy} could be replaced by their resultant, $B_v = \sqrt{B_{zx}^2 + B_{zy}^2}$. So, the stress resultants on a structural bolt reduce to a tensile force $B_t\,(\equiv B_{zz})$ acting along the axis of the bolt and/or a shear force B_v acting at right angles to the bolt axis (Fig. 6.3.1d). Depending on the resultant force(s) acting on a bolt due to external loads, it could be classified as a bolt in shear only, a bolt in tension only, or a bolt in combined shear and tension.

6.4 Behavior of Shear Loaded, Bolted Joints

There are two types of shear loaded, bolted joints—lap joints and butt joints. In a **lap joint,** the two plates to be connected are lapped over one another and fastened together by one or more rows of bolts, with one or more bolts in each row, as shown in Fig. 6.4.1a. In a **butt joint,** the two plates to be connected, called **main plates,** are butted together and joined by two **cover plates** bolted to each of the main plates, as shown in Fig. 6.4.1b. Again, one or more rows of bolts, with one or more bolts in each row, can be used. For both lap and butt joints, the number of bolts in each row identifies the joint as single-bolted, double-bolted, and so on. Examples of double-bolted joints are

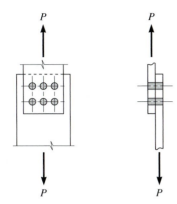

Double-bolted lap joint, no stagger

(*a*)

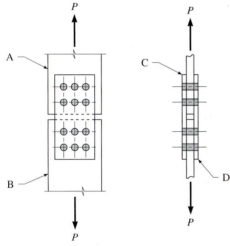

Double-bolted butt joint, no stagger

(*b*)

Figure 6.4.1:   Lap and butt joints.

shown in Fig. 6.4.1. For the butt joint (Fig. 6.4.1*b*) when a tensile force is applied to the main plate A, it is transmitted from A to the bolts which pass through A, then to the cover plates C and D, then to the bolts which pass through plate B, and then to the main plate B.

Consider the elementary lap joint with a single bolt in a standard hole ($\frac{1}{16}$ in. larger than bolt diameter) shown in Fig. 6.4.2*a*. The bolt is assumed to be placed concentrically in the hole initially. When the bolt is tightened, usually by turning the nut or, less often, by turning the head of the bolt, the bolt elongates by a small amount. This elongation introduces tension, B_o, in the bolt which in turn creates a clamping (compressive) force, C_o, on the joint.

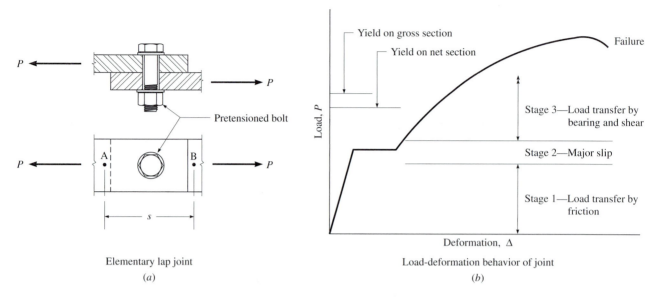

Figure 6.4.2: Schematic behavior of a shear loaded bolted joint.

The initial tension in the bolt is called the **bolt preload.** The free-body diagram in Fig. 6.4.2c shows that the total compressive force, C_o, acting on the connected part is numerically equal to the preload, B_o, in the bolt. These normal forces press the plates of the joint tightly together. The plane of contact between two plies of a joint is known as a *faying surface.*

The joint is next subjected to an in-plane loading, P, through the centroid of the bolt group, and the load is increased gradually. Let Δ represent the elongation of the joint as measured between sections A and B separated by a distance, s, as shown in Fig. 6.4.2a. Three characteristic loading stages exist as shown on the load-deformation curve in Fig. 6.4.2b.

In Stage 1, owing to the inter-surface compression (normal forces) between the plates, there will be frictional resistance to any motion of the plates past one another. So, when external loads P are applied to the plates, the transfer of load between the plates occurs through frictional resistance, F, on the faying surface. The amount of friction between two mechanical parts is proportional to the normal compressive force, C_o, clamping the two parts together and the coefficient of friction at the interface, μ. The coefficient of static friction of steel on steel may vary widely over the range 0.1 to 0.6, depending on the roughness of the faying surfaces. Also, galvanizing, painting, and surface grinding generally reduce the friction. In a bolted joint, the normal force is produced by the preload in one or more bolts (one shown in Fig. 6.4.2a). The total friction force developed in the joint is called the *slip resistance* of the joint. The slip resistance is also a function of the number of slip surfaces, N_s, involved in the joint ($N_s = 1$ for the lap joint under study).

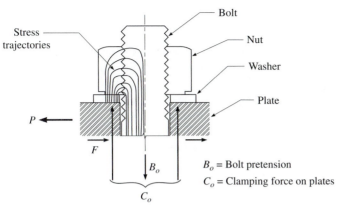

B_o = Bolt pretension
C_o = Clamping force on plates

Pretension and precompression

(c)

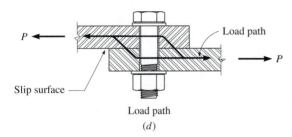

(d)

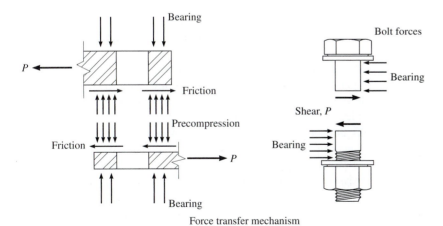

Force transfer mechanism

(e)

Figure 6.4.2 (*Continued*)

Thus, we have:

$$P = F \leq S = \mu\, C_o\, N_s = \mu\, B_o\, N_s \tag{6.4.1}$$

where P = applied shear load on the joint
      = applied axial force in the plates
   F = frictional force developed
   S = slip resistance of the joint
   μ = static coefficient of friction
   N_s = number of slip surfaces (faying surfaces)
   B_o = preload in a bolt

In Stage 1, the axial load is transferred from one plate to the other plate as if the joint were cut from a solid block, as shown in Fig. 6.4.2d (neglecting, for the moment, the local stress concentrations caused by the holes or by the clamping forces produced by individual bolts). Observe that in this stage, the bolt shank does not bear on the connection plates. Hence, the bolt itself is not subjected to any shear under the action of load P. As long as Eq. 6.4.1 is satisfied, friction is capable of transferring the entire load P. When an axial load P is applied to the main plate A of a butt joint (Fig 6.4.1b), connected to cover plates C and D by a pretensioned bolt, the force is transmitted by friction between the surfaces of plates A and C and plates A and D. In other words, for the same clamping force, two friction surfaces are developed for the symmetric butt joint shown in Fig. 6.4.1b, and only one in the unsymmetrical lap joint shown in Fig. 6.4.1a. Since twice as much frictional area is developed in the butt joint with two faying surfaces, twice as much force could be transferred by friction. Actual connections consist of more than one bolt, generally of the same size, material, and pretension. This means, as long as friction alone can transfer the load, all bolts participate equally in transmitting the load P.

In Stage 2, the load equals and then slightly exceeds the slip resistance of the joint; the plates move over each other or *slip* until prevented from further motion by the sides of the bolts. Thus, slip brings the connected plates to bear against the sides of the bolts in Stage 2.

In Stage 3, the applied load is transmitted by bearing of the plates on the bolts and by shear in the bolts. Note that bearing stress differs from, say, the compressive stress in a column in that the latter is the internal stress caused by a compressive force, whereas bearing stress is a contact pressure between separate bodies, such as bolt and plate. Returning to the lap joint, the bolt exerts bearing pressure on the plate. This pressure is evidently equal and opposite to the pressure exerted on the bolt by the plate. As depicted in the free-body diagrams of the bolt in Fig. 6.4.2e, the transfer of the load P between the plates, in Stage 3, is actually made via shear on the bolt at the faying surface. In addition to shear and bearing, the bolt is subjected to slight rotation (bending). However, in ordinary bolted joints where the plates are actually in contact with one another, the effect of bending on bolt stresses is negligible and neglected in design.

In Stage 3, the bolts and plates initially deform elastically, and consequently the load-deformation relationship remains linear. However, as the load is increased, the plates, bolts, or both yield, and the stiffness of the joint decreases. The load-deformation response becomes nonlinear and reaches a maximum before failure occurs as a result of bolt shearing, plate fracture, or any one of the many failure modes to be described in Section 6.7.

Overlapping effects may make the distinctions between stages less clear cut than indicated by Fig. 6.4.2b. However, in many tests on bolted joints, these stages were recognized clearly [Bahia and Martin, 1981]. Tests also indicate that the load at which the joint finally fails is independent of the initial pretension in the bolts, or of the coefficient of friction between faying surfaces. The total movement range (slip) that is theoretically possible in a joint fitted with bolts in STD holes is twice the specified clearance. This would normally be $2 \times (1/16) = \frac{1}{8}$ in. of movement. However, this theoretical slip typically cannot occur in practice, since the relative mismatch within any group of holes causes some bolts to go into bearing prior to the remaining bolts. The total slip possible in practical joints is actually on the order of the hole clearance.

6.5 Slip-Critical Joints and Bearing-Type Joints

The performance of a bolted joint depends on the manner in which the bolts are tightened and on the faying surface conditions.

Bolts in the vast majority of joints in building structures need only be tightened to what is known as the *snug-tightened condition,* where the nuts are tightened sufficiently to prevent play in the connected members and loosening of the nut. LRFDS defines *snug-tightened condition* as the tightness that is attained with a few impacts of an impact wrench or the full effort of an iron worker using an ordinary spud wrench to bring the connected plies in a joint into firm contact. The initial tension developed may or may not be substantial and is often negligible. In other joints, called *slip-critical joints* and *pretensioned joints,* bolts must be tightened beyond the snug-tightened condition to attain the minimum pretension by one of four methods: turn-of-nut pretensioning, direct-tension-indicator pretensioning, twist-off-type tension-control bolt pretensioning, and calibrated wrench pretensioning. The *specified minimum bolt pretension* for high-strength bolts is equal to 70 percent of the specified minimum tensile strength of the bolt, rounded to the nearest kip. Thus, we have:

$$T_b = 0.7 F_{ub} A_b \tag{6.5.1}$$

where T_b = specified minimum bolt pretension, kips

A_b = area of bolt shank, in.^{2}

F_{ub} = minimum specified ultimate tensile stress of the bolt material (Eq. 6.2.1), ksi

Numerical values of the specified minimum bolt pretension loads, T_b, are given in LRFDS Table J3.1 for various bolt diameters, for both A325 and A490 bolts (reproduced partially in Table 6.7.1). For example, the pretension load specified for a $\frac{3}{4}$-in.-dia. A325 bolt is 28 kips. Note that T_b nearly equals the proof load for A325 bolts, and is about 85 to 90 percent of the proof load for A490 bolts.

Based on the connection faying surface condition two broad categories of bolted joints are provided for in the LRFD Specifications, namely, *slip-critical joints* and *bearing-type joints.* Bearing-type joints may be further subdivided into *fully-pretensioned bearing-type joints* (or, simply, *pretensioned joints*), and *snug-tightened bearing-type joints* (or, simply, *snug-tightened joints*). *Slip-critical joints* are those joints that have specified faying surface conditions that, in the presence of clamping provided by pretensioned bolts, resist a design load in the plane of the joint solely by friction and without slip at the faying surfaces. Slip-critical joints, therefore, have a low probability of slip at any-time during the life of the structure. In a *bearing-type joint,* slip is acceptable, and shear and bearing actually occur. Under the provisions of LRFDS, certain bearing-type joints are required to be pretensioned, but are not required to be slip-critical. Such joints are known as *pretensioned joints.* The specified minimum bolt pretension, T_b, for bolts in pretensioned joints is the same as that required for bolts in slip-critical joints (as given by Eq. 6.5.1 and tabulated in LRFDS Table J3.1). All other joints are *snug-tightened joints.*

Thus, when pretension is required to prevent slip, a slip-critical joint should be specified. When pretension is required for reasons other than to prevent slip, a pretensioned joint should be specified. In all other joints, the bolts could be ordinary bolts (A307) or high-strength bolts (A325, F1852, or A490), and the bolts need only be snug-tightened. Snug-tightening the bolts induces only small clamping forces in the bolts (less than $0.3\,T_b$). Also, these forces can vary considerably because elongations are still within the elastic range. So, no frictional resistance on the faying surfaces is assumed, and for design purposes slip in bearing-type joints is assumed to occur as soon as external loads are applied.

In a slip-critical joint, the fully-pretensioned bolt creates resistance to slip through the friction on the faying surface between two connected parts. This slip resistance, as seen from Eq. 6.4.1, is a function of the slip coefficient μ of the faying surface. The RCSC Specification [RCSC, 2000] defines three classes of surface preparation: unpainted clean mill scale steel faying surfaces (or surfaces with Class A coatings on blast-cleaned steel) as Class A surfaces with $\mu = 0.33$; unpainted blast-cleaned faying surfaces (or surfaces with Class B coating on blast-cleaned steel) as Class B surfaces with $\mu = 0.50$; and hot-dip galvanized and roughened surfaces as Class C surfaces with $\mu = 0.35$. It is important to remember that the surface requirements for slip-critical joints apply only to the faying surfaces, and do not include the surfaces under the bolt, washer, or nut.

For information on bolt installation and pretensioning methods, refer to Section W6.1 on the website http://www. mhhe.com/Vinnakota.

WWW

For additional information on the use of slip-critical and bearing-type joints, refer to Section W6.2 on the website http://www.mhhe.com/Vinnakota.

6.6   Failure Modes and Limit States

There are two broad categories of bolted joint failure: failure of the connected parts and failure of the bolt. Some of the possible limit states or failure modes that may control the strength of a bolted joint are:

- Tensile fracture of the connected plate elements.
- Tensile yielding of the connected plate elements.
- Shear failure of the bolt.
- Bearing failure of the connected plate elements.
  - Ovalization of bolt hole.
  - Shear tear-out of the connected plate elements.
- Bearing failure of bolt.
- Stripping of bolt thread or nut thread.
- Tensile failure of the bolt.
- Bending failure of the bolt.
- Slip of connected plate elements.

Some elemental types of failures are illustrated in Fig. 6.6.1 by considering a single bolted lap joint (shown in Fig. 6.6.1*a*). Figure 6.6.1*b* represents a failure caused by the tension fracture of one of the plates. This failure occurs on a section through the bolt hole, called the **net section,** which obviously has minimum fracture resistance. Figure 6.6.1*c* represents a limit state resulting in excessive elongations due to member yielding. These two failure modes of the plate are related to the design of the member itself and as such will be considered extensively in Chapter 7.

In the lap joint, a **shear failure of bolt** necessitates that the bolt be sheared through once, at the section where the faces of the two plates are in contact with one another as shown in Fig. 6.6.1*d*. The bolt in the lap joint is said to be in **single shear.** In a butt joint, a shear failure of bolts involves pulling the main plate out from between the cover plates; this cannot occur without shearing of the bolts at the two sections where the faces of the cover plates are in contact with the main plate. The bolt in a butt joint is said to be in **double shear.** Occasionally, we come across joints in heavy structures in which more than three plate elements are bolted together, putting the bolts in multiple shear. Thus, the number of elements being clamped together and the nature of the joint determine whether the bolt will be subjected to single shear, double shear, or multiple shear. Shear failure of bolts will be considered in detail in Section 6.7.

Bearing failure of the connected plate elements at bolt holes is often due to the piling up of the plate material behind the bolt as shown in Fig. 6.6.1*e* and due to the resulting **ovalization (enlargement) of the bolt hole** and the

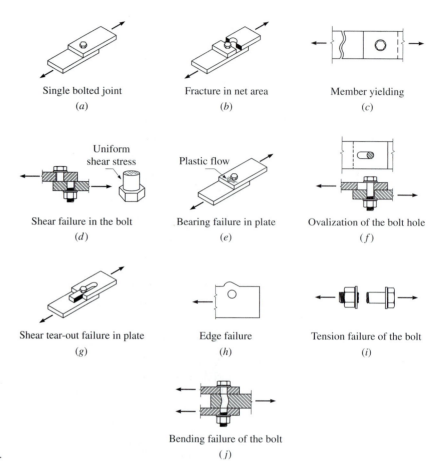

Figure 6.6.1: Limit states of failure of bolted joints.

consequent relative movement between the plates, as shown in Fig. 6.6.1*f*. For an end bolt with small clear distance, L_{ce} (see Eq. 6.2.2), there is a tendency for a block of metal included between the hole and the edge of the plate to tear out along the two horizontal lines tangent to the sides of the bolt hole (Fig. 6.6.1*g*). This limit state is known as ***shear tear-out of the connected plate element.*** This is likely to occur if the hole is located too close to the end of the member. By making the end distance large enough, this failure can be avoided. The limit state then is that of hole distortion or ovalization. Bearing failure is likely to control the design when the plates are either thin compared to the bolt diameter or fabricated from a material of low tensile strength, or when the bolt is in double or multiple shear. Bearing failure of the connected parts will be studied in detail in Section 6.8.

Edge failure may occur from placing a bolt so close to the edge of the member that a bulge is formed when the bolt hole is punched (Fig. 6.6.1*h*). The bolt thread and the nut thread should sustain the tensile force in the bolt without ***stripping of the thread material,*** a form of shearing. In practice, bolt failure, which is ductile, is preferable to nut failure. So, the stripping strength of the bolt-nut assembly is deliberately made greater than the tensile strength of the bolt used (to ensure that thread stripping or other types of nut failures do not occur). Such a pair is said to be a ***matched bolt-nut assembly.*** A ***regular hexagonal nut*** has a thread length equal to $7/8$ times the nominal diameter of bolt, while a ***heavy-hex nut*** has a length approximately equal to the nominal diameter of the bolt (see dimension H_n shown in Fig. 6.2.1). Fracture limit state of a matching bolt-nut assembly in tension is shown in Fig. 6.6.1*i*.

In ordinary bolted joints in building structures, where the plates are actually in contact with one another, the effect of bending is negligible (Fig. 6.6.1*j*) and thus neglected. More specifically, we assume that the bolts do not bend but remain essentially straight. This is possible only if there are equal deformations of the connected plates between adjacent rows of bolts. This is not generally true in the elastic domain. But since the plates are generally specified to be ductile, equal plastic deformations can occur as the stresses approach the yield stress. In a lap joint, there is a bending moment on the bolt which results from the eccentricity between the two plate forces. Again, the effect of this eccentricity on bolt design is neglected.

In a slip-critical joint, the potential failure mode is slipping of the plates, which will bring the bolts into bearing. In a bearing-type joint, slipping of plates is not considered a failure limit state. Several of the failure modes described above will be considered in detail, in the sections that follow.

6.7   Design Shear Strength of a Bolt (Bearing-Type Joint)

In a bearing-type joint, the applied load is transmitted by the bearing of the plates on the bolts and by the shear in the bolts. The shear resistance of a bolt is directly proportional to the available bolt area in the shearing plane (plane of joint). So, the greatest shear strength is obtained when the full shank of the bolt is in the shear plane. Further, the available shear area in the threaded part of a bolt is equal to the root area, which for most commonly used bolts, is about 70 percent of the shank area of the bolt. So, when a shear plane passes through the threaded portion, the shear capacity of the bolt may theoretically be reduced to as little as 70 percent of the full shank strength. In order to reflect this difference in strengths, LRFD Specification uses designations ***A325-X*** and ***A490-X*** to indicate the situation where threads are e**X**cluded from shear planes and designations ***A325-N*** and ***A490-N*** where threads are i**N**cluded in the shear planes (Fig. 6.7.1). Tests by Yura, Frank, and Polyois [1987] produced an average value for the ratio of nominal shear strength for bolts with threads in the shear plane to the nominal shear strength for bolts with threads excluded from the shear plane to be 0.833 with a standard deviation of 0.03. The LRFDS uses a value of 0.80, as a factor to account for the

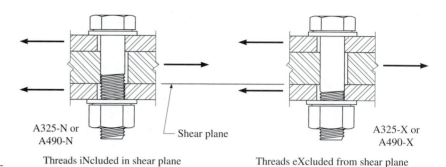

A325-N or
A490-N

Shear plane

A325-X or
A490-X

Figure 6.7.1: Bolt classification based on the relative location of the shear plane.

Threads iNcluded in shear plane

(a)

Threads eXcluded from shear plane

(b)

reduced shear strength of a bolt with threads in the shear plane. That is,

$$A_v = A_b \qquad \text{for \quad X-type bolts} \qquad (6.7.1)$$
$$= 0.80\, A_b \quad \text{for \quad N-type bolts}$$

where A_v = net shearing area per shear plane
    A_b = cross-sectional area based upon the nominal diameter of bolt
    $= \pi\, d^2 / 4$

From tests, the shear strength of the bolt material [Wallaert and Fisher, 1965] was found to be approximately 62 percent of the tensile strength of the bolt material. Also, this percentage was found to be independent of the bolt grade. However, in shear connections with more than two bolts in the line of force, deformation of the connected material resulted in a nonuniform bolt shear force distribution. As a consequence, the strength of the connection, in terms of the average bolt strength, goes down as the joint length increases (see Section 6.13.2). Rather than provide a function that reflects this decrease in average bolt strength with joint length, a single reduction factor of 0.80 was applied to the 0.62 multiplier. The result will accommodate bolts in all joints up to about 50 in. in length without seriously affecting the economy of very short joints. Thus, the shear strength of the bolt material is given by:

$$F_{uvb} = 0.8(0.62\, F_{ub}) \approx 0.50\, F_{ub} \qquad (6.7.2)$$

where F_{uvb} = averaged value for the ultimate shear stress of the bolt material for joint lengths up to 50 in.
    F_{ub} = ultimate tensile stress of the bolt material

Because of the short thread length on high-strength structural bolts, threads cross shear planes only when the outer ply of gripped material at the nut end is less than $\frac{3}{8}$- or $\frac{1}{2}$-in. thick, depending on the bolt diameter (see Section 6.2). Hence, in practice, threads will almost always be excluded from the shear plane. For convenience, both X- and N-types of bolts are designed on the basis of a shear stress based on the nominal area (shank area) of the bolt; the reduced capacity for N-type of bolts is thus obtained by using a reduced value for the nominal shear stress, F_{nv}. The design shear strength of a bolt in a bearing-type joint, with the help of Eqs. 6.7.1 and 6.7.2, may be written as:

$$B_{dv} = \phi B_{nv} = \phi (F_{uvb} A_v) N_s = \phi (F_{nv} A_b) N_s \qquad (6.7.3)$$
$$= 0.75(0.50 F_{ub}) A_b N_s \quad \text{for X-type bolts}$$
$$= 0.75(0.40 F_{ub}) A_b N_s \quad \text{for N-type bolts}$$

where B_{dv} = design shear strength of a bolt in a bearing-type joint
   ϕ = resistance factor (= 0.75)
   B_{nv} = nominal shear strength of a bolt
   A_b = cross-sectional area based upon the nominal diameter of bolt
   N_s = number of shear planes (= 1 for a bolt in single shear)
      (= 2 for a bolt in double shear)
   F_{ub} = ultimate tensile stress of the bolt material
   F_{uvb} = averaged value for the ultimate shear stress of the bolt material
      = $0.50 F_{ub}$ (from Eq. 6.7.2)
   F_{nv} = nominal shear strength per unit area of the bolt shank
      = F_{uvb} for X-type bolts
      = $0.80 F_{uvb}$ for N-type bolts

Remarks
- Both A325 and A490 bolts have shorter thread lengths than comparable ordinary bolts to reduce the likelihood of threads crossing the shearing plane (faying surface). LRFDM Table 7-2 lists the minimum thickness of plate, closest to the nut, required to exclude threads from shear plane for different bolt diameters and grip lengths for assemblies with and without washers. As an example, consider the outside plate element of a joint, for a connection having a grip of $1\frac{3}{8}$ in. and utilizing $\frac{7}{8}$-in.-diameter high-strength bolts. Unless the plate is less than $\frac{3}{8}$-in. thick, the bolt threads will not extend across the shear plane.
- When the bolt length is equal to or shorter than four times the nominal diameter, ASTM A325 now permits specifying that the bolt be threaded for the full length of the shank. These bolts are required to be marked with the symbol A325T. This exception to the A325 Specification is permitted to increase economy through simplified ordering and inventory control in the fabrication and erection of structures using relatively thin material. In these joints the threads are always in the shear plane (N-type).
- Also, as mentioned earlier, Section 5.1 of the RCSCS stipulates that the design shear strength of a bolt shall not be reduced by the installed bolt pretension.
- Table 6.7.1 gives the design shear strengths of $\frac{5}{8}$, $\frac{3}{4}$, $\frac{7}{8}$ and 1 in.-dia.-bolts.

TABLE 6.7.1

Design Strengths of High-Strength Bolts (kips)

	Description of Load and Bolt		d (in.) A_b (in.^{2})	5/8 0.307	3/4 0.442	7/8 0.601	1 0.785
1.	Pretension	T_b	$0.70\,F_{ub}A_b$				
	A325			19.0	28.0	39.0	51.0
	A490			24.0	35.0	49.0	64.0
2.	Tensile strength	B_{dt}	$0.75(0.75)\,F_{ub}A_b$				
	A325		$67.5\,A_b$	20.7	29.8	40.6	53.0
	A490		$84.8\,A_b$	26.0	37.4	51.0	66.6
3.	Shear strength: Bearing-type joints. Standard holes						
	N-type	B_{dv}	$0.75(0.40\,F_{ub})\,A_b\,N_s$				
	A325-N: S		$36.0\,A_b$	11.0	15.9	21.6	28.3
	D		$72.0\,A_b$	22.1	31.8	43.3	56.5
	A490-N: S		$45.0\,A_b$	13.8	19.9	27.1	35.3
	D		$90.0\,A_b$	27.6	39.8	54.1	70.7
	X-type	B_{dv}	$0.75(0.5\,F_{ub})\,A_b\,N_s$				
	A325-X: S		$45.0\,A_b$	13.8	19.9	27.1	35.3
	D		$90.0\,A_b$	27.6	39.8	54.1	70.7
	A490-X: S		$56.3\,A_b$	17.3	24.9	33.8	44.2
	D		$112.6\,A_b$	34.5	49.7	67.6	88.4
4.	Shear strength: Slip-critical joints. Class A faying surface, standard holes						
	Service loads	B_{dss}	$1.0\,F_v\,A_b\,N_s$				
	A325-SC: S		$17.0\,A_b$	5.22	7.51	10.2	13.4
	D		$34.0\,A_b$	10.4	15.0	20.4	26.7
	A490-SC: S		$21.0\,A_b$	6.44	9.28	12.6	16.5
	D		$42.0\,A_b$	12.9	18.6	25.3	33.0
	Factored loads	B_{dsf}	$1.13(0.33)\,T_b\,N_s$				
	A325-SC: S			7.09	10.4	14.5	19.0
	D			14.2	20.9	29.1	38.0
	A490-SC: S			8.95	13.1	18.3	23.9
	D			17.9	26.1	36.5	47.7

For additional information on behavior of a bolt in shear in a bearing-type joint, refer to Section W6.3 on the website http://www.mhhe.com/Vinnakota.

WWW

6.8   Design Bearing Strength at Bolt Holes

The plate material in bearing is confined and cannot fracture. Therefore, bearing failure can only be defined rather arbitrarily on the basis of judgment as to when deformation due to crushing becomes excessive. The actual failure mode in bearing depends on the clear distance, L_c. The ***clear distance, L_c,*** is the distance from the edge of the hole to the unstressed end of the connected plate element or to the edge of the adjacent hole, measured along the line of pressure from the bolt (Fig. 6.2.6 and Eqs. 6.2.2 and 6.2.3). For large values of L_c, the hole will elongate due to excessive deformations developed in the plate material in front of the hole (Figs. 6.6.1*e* and *f*). However, if the clear distance is inadequate, the bolt tears through the end of the connected plate element (Fig. 6.6.1*g*). Observe that, in both failure modes, the bearing strength of the bolt is actually the bearing strength of the connected plate element, as the connected material is always critical, compared to the bolt.

6.8.1   Limit State of Ovalization of Bolt Hole

Although the actual bearing stress distribution is not known, the assumption is made in bolt design that the bearing stress is uniformly distributed over a projected area of the bolt shank on the connected plate element. A rectangle, of dimensions of the thickness of the connected plate element and the diameter of the bolt, constitutes this area. Thus, the nominal bearing strength of a bolt in a bearing-type joint is:

$$B_{nb} = F_{nb}\, dt \qquad (6.8.1)$$

where F_{nb} = nominal bearing stress, assumed to be uniform over the projected area of the bolt on the connected plate element, ksi
   d = nominal diameter of the bolt, in.
   t = thickness of the connected plate element, in.
   B_{nb} = nominal bearing strength of the connected plate element, kips

The actual bearing stress on the plate is not uniformly distributed, nor is the area of contact equal to the assumed area. However, joints designed on this basis and the specified nominal bearing stresses are safe.

Tests by Frank and Yura [1981] have shown that hole elongations less than 0.25 in. will develop for values of the nominal bearing stress, F_{nb}, equal to $2.4F_{up}$ for bolts in standard, oversized, or short-slotted holes. Here, F_{up} is the ultimate tensile stress of the connected plate material. Tests also showed that the total elongation of a standard hole, loaded to obtain nominal bearing stresses equal to $3F_{up}$, was on the order of the diameter of the bolt. This apparent hole elongation results largely from bearing deformation of the material that is immediately adjacent to the bolt. So, if deformation around the bolt hole is not a design consideration and if adequate end distance or spacing is provided, the nominal bearing stress may be increased to $3F_{up}$. When long-slotted holes are oriented with the long dimension perpendicular to the direction of the load, the bending component of the deformation in the material between adjacent holes or between the hole and the edge of the plate is increased. The nominal bearing stress in this case is limited to $2F_{up}$, which again provides a bearing strength limit state that is attainable at reasonable deformation. So, whenever $L_c \geq 2.0d$, the design bearing strength of the connected plate element corresponding to the **limit state of ovalization of bolt hole** is given by:

$$B_{dbo} = \phi\, F_{nb}\, dt \qquad\qquad (6.8.2)$$

$$= 0.75(2.4\, F_{up})\, dt \quad \text{for bolts in STD, OVS, and SSL holes independent of the direction of loading when deformation of the bolt hole at service load is a design consideration}$$

$$= 0.75(3.0\, F_{up})\, dt \quad \text{when deformation of the bolt hole at service load is not a design consideration}$$

$$= 0.75(2.0\, F_{up})\, dt \quad \text{for bolts in LSL holes with the slot perpendicular to the load}$$

with

B_{dbo} = design bearing strength of connected plate element corresponding to the limit state of ovalization of the bolt hole, kips

ϕ = resistance factor (= 0.75)

F_{nb} = nominal bearing strength of the connected plate element, ksi

F_{up} = ultimate tensile stress of the connected plate element, ksi

d = nominal diameter of bolt, in.

t = thickness of the connected plate element, in.

We will normally assume that the deformations around the bolt holes are to be restricted. Thus, unless specifically stated otherwise, the value of $2.4F_{up}$ will be used for bearing calculations rather than the value of $3.0F_{up}$. It should be realized that these seemingly high bearing stresses (several times the material strength) are allowed because experiments indicate that material so confined has a very high resistance to crushing.

6.8.2   Limit State of Shear Tear-Out of the Connected Plate Element

As mentioned earlier, if the clear distance is relatively small, the bolt tears through the end of the connected plate as shown in Fig. 6.6.1g [Kim and Yura, 1996; Lewis and Zwernmann, 1996]. For small clear distances, the

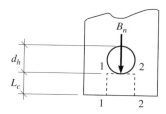

shear tear-out of the plate is assumed to occur by shear fracture along the lines 1-1 and 2-2, and tension yield along line 1-2, as shown in Fig. 6.8.1. As a lower bound for shear tear-out strength, the tensile resistance along line 1-2 could be taken as zero. The nominal shear tear-out strength of the connected plate element is then given as:

$$B_{nbt} = 2L_c t \tau_{up} \tag{6.8.3}$$

where τ_{up}, is the shear fracture stress of the connected material; and B_{nbt} is the nominal bearing strength of the connected plate element, corresponding to the limit state of shear tear-out of the plate. For most commonly used steels, the shear fracture stress is about 60 percent of ultimate tensile stress F_{up}, and the nominal bearing strength becomes:

$$B_{nbt} = 2L_c t (0.60F_{up}) \tag{6.8.4}$$

The **design bearing strength of an end bolt** corresponding to the **limit state of shear tear-out of the connected plate element** is given by:

$$B_{dbte} = 1.2\phi F_{up}L_c t = 0.9F_{up}(L_e - 0.5d_h)t \tag{6.8.5}$$

where B_{dbte} = design bearing strength of an end bolt corresponding to the limit state of shear tear-out of the connected plate element, kips

ϕ = resistance factor (= 0.75)

F_{up} = ultimate tensile stress of the plate material, ksi

L_c = clear distance, in.

L_e = end distance, in.

t = thickness of the connected plate element, in.

d_h = diameter of the bolt hole (STD or OVS holes), in.

Let L_c be the distance measured in the line of force from the edge of a bolt hole to the nearest edge of an adjacent bolt hole, and p the distance between the centers of these two STD bolt holes (i.e., pitch). By analogy to Eq. 6.8.5, the **design bearing strength of an interior bolt** corresponding to the **limit state of shear tear-out** can be written as:

$$B_{dbti} = 1.2\phi F_{up}L_c t = 0.90F_{up}(p - d_h)t \tag{6.8.6}$$

where B_{dbti} = design bearing strength of an interior bolt corresponding to the limit state of shear tear-out of the connected plate element, kips

p = pitch of bolts, in.

and all other terms are as defined for Eq. 6.8.5.

We observe from Eqs. 6.8.2 and 6.8.5 that the failure mode for an end bolt changes from a shear tear-out type to one in which material deformation occurs and results in ovalization of the hole, for values of L_c greater than or equal to 2.0d (or equivalently, for values of L_e greater than or equal to a value $L_{e,full} = 2d + 0.5d_h$, for STD and OVS holes). Similarly, we observe from Eqs. 6.8.2 and 6.8.6 that the failure mode for an interior bolt changes from a shear tear-out type failure to one in which material deformation occurs, resulting in ovalization of the hole for values of L_c greater than or equal to 2.0d (or equivalently, for values of p greater than or equal to a value $p_{full} = 2d + d_h$ for STD and OVS holes). In particular, for standard punched holes we have:

$$L_{e,full} = 2.5d + \frac{1}{32}\text{ in.;}\quad p_{full} = 3d + \frac{1}{16}\text{ in.} \qquad (6.8.7)$$

where $L_{e,full}$ = limiting value of the end distance, beyond which the design bearing strength is controlled by the limit state of material deformation that results in ovalization of the hole

p_{full} = limiting value of the bolt pitch, beyond which the design bearing strength is controlled by the limit state of material deformation that results in ovalization of the hole

Values of $L_{e,full}$ and p_{full} are given in the LRFDM Tables 7-13 and 7-12, respectively (where they are indicated as $L_{e,full}$ and s_{full}, respectively).

Although the bolt itself is subject to the same magnitude of compressive forces as those acting on the side of the hole, tests have shown that the bolt is not critical. Thus, the same bearing value applies to joints assembled by all high-strength bolts, regardless of fastener shear strength or the presence or absence of threads in the bearing area. Also, the nominal bearing strength per unit projected area at a bolt hole is the same for double shear bearing and single shear bearing. While there is a difference in the stress distribution in the two cases, particularly at low load levels, tests have shown that this difference is reduced by plastic redistribution to the extent where it has no apparent effect on the ultimate behavior.

To summarize: If the plate is thick and the clear distances are large, the bearing resistance of the plate will be high and failure will occur by shearing of the bolt. If the plate is thin, low bearing resistance will result. The failure will then occur either by excessive hole elongation and plate material piling up behind the bolt if the clear distances are large, or by the shear tear-out of the plate material behind the bolt when these distances are small.

Design Bearing Strength at STD Interior Bolt Holes, B_{dbi}, for Various Bolt Spacings (kips/in. thickness)

		F_{up} = 58 ksi			F_{up} = 65 ksi	
d (in.)	3/4	7/8	1	3/4	7/8	1
$2\frac{2}{3}d$ (in.)	2	2⅓	2⅔	2	2⅓	2⅔
$3d$ (in.)	2¼	2⅝	3	2¼	2⅝	3
$3d + \frac{1}{16}$ (in.)	2 5/16	2 11/16	3 1/16	2 5/16	2 11/16	3 1/16
p (in.) ↓						
2	62.0	—	—	69.5	—	—
2¼	75.0	—	—	84.1	—	—
2½	78.3	81.6	—	87.7	91.4	—
2¾	78.3	91.3	88.1	87.7	102	98.7
3	78.3	91.3	101	87.7	102	113
≥ 3¼	78.3	91.3	104	87.7	102	117

$B_{dbi} = \min [B_{dbo}; B_{dbti}]$; $B_{dbo} = 1.8 F_{up}t$; $B_{dbti} = 0.9 (p - d_h)F_{up}t$
d = nominal diameter of bolt, in.; d_h = diameter of bolt hole = d + 1/16 in. for STD punched holes considered
B_{dbi} = design bearing strength at an interior bolt hole; p = pitch, in.; t = plate thickness = 1 in.
B_{dbo} = strength corresponding to ovalization of bolt hole; B_{dbti} = strength corresponding to shear tear-out of plate
Design strengths controlled by ovalization of bolt hole are shown shaded.
— indicates spacing less than minimum spacing required per LRFD Specification Section J3.3.

Design Bearing Strength at STD End Bolt Holes, B_{dbe}, for Various End Distances (kips/in. thickness)

		F_{up} = 58 ksi			F_{up} = 65 ksi	
d (in.)	3/4	7/8	1	3/4	7/8	1
$1\frac{1}{2}d$ (in.)	1⅛	1 5/16	1½	1⅛	1 5/16	1½
$2.5d + \frac{1}{32}$ (in.)	1 15/16	2¼	2 9/16	1 15/16	2¼	2 9/16
L_e (in.) ↓						
1¼	44.0	40.8	37.5	49.4	45.7	42.0
1⅜	50.6	47.3	44.0	56.7	53.0	49.3
1½	57.1	53.8	50.6	64.0	60.3	56.7
1¾	70.1	66.9	63.6	78.6	75.0	71.3
2	78.3	79.9	76.7	87.7	89.6	85.9
2½	78.3	91.3	103	87.7	102	115
≥ 2¾	78.3	91.3	104	87.7	102	117

$B_{dbe} = \min [B_{dbo}; B_{dbte}]$; $B_{dbo} = 1.8 F_{up}t$; $B_{dbte} = 0.9 (L_e - 0.5d_h)F_{up}t$
d = nominal diameter of bolt, in.; d_h = diameter of bolt hole = d + 1/16 in. for STD punched holes considered
B_{dbe} = design bearing strength at an end bolt hole; L_e = end distance, in.; t = plate thickness = 1 in.
B_{dbo} = strength corresponding to ovalization of bolt hole; B_{dbte} = strength corresponding to shear tear-out of plate
Design strengths controlled by ovalization of bolt hole are shown shaded.

Thus, we have:

$$B_{dbe} = \min [B_{dbo}, B_{dbte}] \quad \text{for an end bolt} \tag{6.8.8a}$$

$$B_{dbi} = \min [B_{dbo}, B_{dbti}] \quad \text{for an interior bolt} \tag{6.8.8b}$$

where B_{dbe} = design bearing strength of an end bolt
 B_{dbi} = design bearing strength of an interior bolt

It is evident that spacing and/or end distance may be increased to provide for a required bearing strength, or bearing force may be reduced to satisfy a spacing and/or end distance limitation. Table 6.8.1 gives the design bearing strength at bolt holes for various bolt spacings, p, while Table 6.8.2 gives the design bearing strength at bolt holes for various edge distances, L_e.

··

6.9   Design Strength of a Bolt in Tension

The weakest section of any bolt in tension is the threaded portion. Consequently, the net tensile area that remains after threading is known as the **stress area.** For threads used in the USA the stress area is given by:

$$A_s = \frac{\pi}{4}\left[d - \frac{0.97432}{n_t}\right]^2 \tag{6.9.1}$$

where d is the nominal bolt diameter (in.), n_t equals the number of threads per in., and A_s is the stress area of the bolt (in.^{2}).

The tensile capacity of a bolt is equal to the product of stress area A_s and the ultimate tensile stress F_{ub} of the bolt material. The ratio of stress area to nominal bolt area, A_b ($= \pi d^2/4$) varies from 0.75 for $\frac{3}{4}$-in.-diameter bolts to 0.79 for $1\frac{1}{8}$-in.-diameter bolts [Kulak et al., 1987]. Accordingly, to simplify calculations, a lower bound reduction of 0.75 is incorporated in the LRFDS. Also, for design purposes, it is more convenient to specify a nominal tensile stress on the basis of the nominal area of the bolt, A_b, rather than the stress area, A_s. Thus, the nominal strength of a bolt in tension is:

$$B_{nt} = F_{ub} A_s = F_{ub} (0.75 A_b) = 0.75 F_{ub} A_b = F_{nt} A_b \tag{6.9.2}$$

where B_{nt} is the nominal tensile strength of a bolt, A_s is the stress area of the bolt, A_b is the cross-sectional area based upon the nominal diameter of the bolt, F_{ub} is the ultimate tensile stress of the bolt material, and F_{nt} is the nominal tensile strength per unit area of the bolt shank.

The design tensile strength of a high-strength bolt (LRFDS Section J3.6) is:

$$B_{dt} = \phi B_{nt} = \phi F_{nt} A_b = 0.75(0.75 F_{ub}) A_b \tag{6.9.3}$$

where B_{dt} = design tensile strength of a bolt, kips
  B_{nt} = nominal tensile strength of the bolt, kips
  ϕ = resistance factor (= 0.75)
  F_{nt} = nominal tensile strength per unit area (LRFDS Table J3.2), ksi
  = $0.75\, F_{ub}$
  = 90 ksi   for A325 bolts
  = 113 ksi   for A490 bolts

6.10   Design Strength of a Bolt in Combined Shear and Tension

Bolts in wind bracing connections (Fig. 6.1.2*e*) are often subjected to both shear and tension under applied loads. The possible reduction in bolt strength due to such simultaneous loading conditions has to be considered in the design. Specifically, the strength interaction relation for combined shear and applied tension, for high-strength bolts in bearing-type joints, is assessed by applying the elliptical relationship [Eq. 5.2 of the RCSC Specification, RCSC, 2000]:

$$\left(\frac{B_{tu}}{B_{dt}}\right)^2 + \left(\frac{B_{vu}}{B_{dv}}\right)^2 \le 1 \qquad (6.10.1)$$

where B_{tu} = tensile component of applied factored load for combined shear and tension loading
  B_{vu} = shear component of applied factored load for combined shear and tension loading
  B_{dt} = design strength of the bolt in tension, when the bolt is subjected to tension only
  B_{dv} = design strength of the bolt in shear, when the bolt is subjected to shear only

The elliptic relation, shown in Fig. 6.10.1, can be replaced with only minor modifications by three linear relations, namely:

$$\frac{B_{tu}}{B_{dt}} + \frac{B_{vu}}{B_{dv}} \le C; \quad \frac{B_{tu}}{B_{dt}} \le 1.0; \quad \frac{B_{vu}}{B_{dv}} \le 1.0 \qquad (6.10.2)$$

where C is a constant, taken as 1.30 by the LRFDS. The trilinear representation of Eq. 6.10.1, in stress format, becomes:

$$f_{tu} \le F'_{dt} = \phi F'_{nt} \qquad (6.10.3)$$

Here, F'_{nt} is used instead of F_{nt} to signify that the nominal tensile strength per unit area accounts for the presence of a simultaneously acting shear

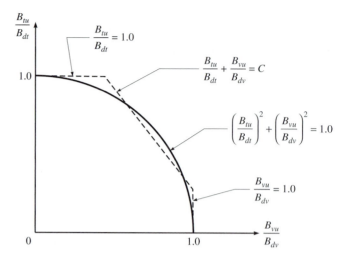

Figure 6.10.1: Linear representation of bolt strength in combined shear and tension.

stress f_{vu} computed at factored-load level. As mentioned earlier, values of F_{nv} and F_{nt} are given in LRFDS Table J3.2. The linear interaction formulae for combined shear and tension, in stress format, are given in the LRFDS Table J3.5 as:

$$F'_{nt} = 117 - 2.5 f_{vu} \leq 90 \quad \text{for } f_{vu} \leq 36 \text{ ksi, for A325-N type} \quad (6.10.4a)$$

$$= 117 - 2.0 f_{vu} \leq 90 \quad \text{for } f_{vu} \leq 45 \text{ ksi, for A325-X type} \quad (6.10.4b)$$

$$= 147 - 2.5 f_{vu} \leq 113 \quad \text{for } f_{vu} \leq 45 \text{ ksi, for A490-N type} \quad (6.10.4c)$$

$$= 147 - 2.0 f_{vu} \leq 113 \quad \text{for } f_{vu} \leq 56 \text{ ksi, for A490-X type} \quad (6.10.4d)$$

Here f_{vu} is the shearing stress in the bolt resulting from the factored loads ($B_{vu} /(N_s A_b)$). A graphical representation of Eqs. 6.10.4 is shown in Fig. 6.10.2. By multiplying both sides of Eqs. 6.10.4a to d, by A_b, the nominal strength of a high-strength bolt in tension, for combined shear and tension loading, is obtained as:

$$B'_{nt} = C_1 - C_2 B_{vu} \leq B_{nt} \quad \text{for } B_{vu} \leq B_{dv} \quad (6.10.5)$$

where C_1 and C_2 are constants for a given type of bolt and joint.

For additional information on design strength of a bolt in combined shear and tension, refer to Section W6.4 on the website http://www.mhhe.com/ Vinnakota.

www

Note that the LRFD Specification requires in Section J3.1 that A490 high-strength bolts in such connections, subjected to tension under external loads shall be pretensioned to a bolt tension T_b specified in LRFDS Table J3.1.

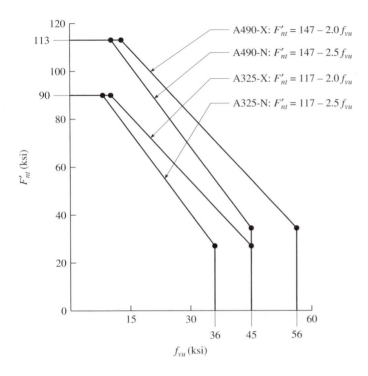

A490-X: $F'_{nt} = 147 - 2.0 f_{vu}$

A490-N: $F'_{nt} = 147 - 2.5 f_{vu}$

A325-X: $F'_{nt} = 117 - 2.0 f_{vu}$

A325-N: $F'_{nt} = 117 - 2.5 f_{vu}$

Figure 6.10.2: Combined tension and shear of bolts in bearing-type joints.

6.11 Behavior and Design of Bolts in Slip-Critical Joints

For information on the slip resistance of a bolt in a slip-critical joint, refer to Section W6.5 on the website http://www.mhhe.com/Vinnakota.

For information on the behavior of a precompressed bolt-plate assembly in direct tension, refer to Section W6.6 on the website http://www.mhhe. com/Vinnakota.

For information on the behavior of a bolt in a slip-critical joint subject to shear and tension, refer to Section W6.7 on the website http:// www.mhhe.com/Vinnakota.

6.12 Ordinary Bolts

Ordinary bolts, also called **unfinished** or **common** bolts, are forged from rolled steel round rods. They are made of A307 mild steel with stress-strain characteristics very similar to those of A36 steel. They are available in diameters from $\frac{5}{8}$ in. to $1\frac{1}{2}$ in. in $\frac{1}{8}$-in. increments. They have either square or hexagonal heads, and are available in regular and heavy sizes. The nuts are also either square or hexagonal and are available in regular and heavy sizes. Square heads and nuts are slightly cheaper, but hexagonal heads are easier to hold and turn with a wrench, require less turning space, and are more attractive. The usual practice is to use bolts with hexagonal heads and square or hexagonal nuts.

Since the tension developed in tightening ordinary bolts is small and uncertain, no frictional resistance on the faying surfaces is assumed, and slip may occur at low shearing loads. So, joints using ordinary bolts are always designed as bearing-type joints. As the ordinary bolts have relatively large tolerances in shank and thread dimensions, their design strengths are proportionally lower than those for high-strength bolts. Because of the uncertainty as to whether or not the threaded portion of unfinished bolts extend into the shear plane, their strength is always based on the assumption that threads cross the shear plane, as in the N-type configuration. They are primarily used in light structures subjected to static loads only, and for secondary members such as purlins, girts, and connections in small trusses. Ordinary bolts are acceptable in tension-type connections (such as hangers) under static loads only. Heavy size nuts may be needed for bolts carrying tension loads. The nominal stresses in shear and tension for A307 bolts are tabulated in LRFDS Tables J3.2 and J3.5. The design strengths of A307 bolts in shear, bearing, and tension are included in Tables 7-10, 7-12, and 7-14 of the LRFDM, respectively, for bolt diameters of $\frac{5}{8}$ in. to $1\frac{1}{2}$ in.

The discussion relating to bearing-type joints using high-strength bolts, presented in earlier sections, can easily be adapted to the other fastener types, particularly ordinary bolts and rivets. The design procedures are identical for all of these fasteners, design stresses being the only significant variation.

6.13   Bolt Design

Bolt diameters in building structures should be limited to a 1-in. maximum when possible. In effect, the industry standard $\frac{3}{4}$-in., $\frac{7}{8}$-in., and 1-in. bolt diameters provide adequate design strength for the vast majority of connections in steel structures. Accordingly, commonly available bolt installation equipment has been designed with a capacity to fully tension 1 in. diameter A490 bolts, when required; larger bolts will usually require special equipment and/or effort. In addition, bolt diameters larger than 1 in. require larger clearances, edge distances, and spacings than are standard. Therefore, bolt diameters larger than 1 in. should be avoided, when possible, to avoid potential bolt tensioning difficulties. The usual (industry standard) pitch and gage are 3 in. and the usual edge distances are $1\frac{1}{2}$ in. and 2 in.

A325 bolts with full length threading (ASTM A325T bolts) may be specified for A325 bolts of length less than or equal to four times the bolt diameter only. Note that if A325T bolts are specified, it is impossible to exclude the threads from the shear plane, and the design must be based upon the strengths corresponding to N-type bolts. Whenever X-type bolts are used, the designer must specify and make provision to assure that the threads are, in fact, excluded from the shear plane as assumed. When the connected plates are relatively thin, it is usually prudent to design the connection for the worst case of threads included in the shear plane (N-type). Note that for a bolt in double- or multiple-shear, if threads occur in one shear plane, the conservative assumption is made in the LRFDS that threads are in all shear planes.

For joints with bolts that are loaded in shear or combined shear and tension, the *Engineer of Record* must specify the joint type as snug-tightened, pretensioned, or slip-critical. For slip-critical joints, the required class of slip resistance (A, B, or C), and whether resistance to shear at service loads is verified using factored loads or service loads, must also be specified. For joints that are loaded in tension only, the EOR must specify the joint type as snug-tightened or pretensioned.

6.13.1   Design Tables and Design Aids

LRFDS Section J3.4 specifies that the distance from the center of a standard hole to an end (L_e) or to a side (L_s) of a connected part must be no less than the applicable value from LRFDS Table J3.4. The designer should avoid using the minimum edge distance and minimum spacing for bolts in joints that may be critical in bearing, as the erector may be required to ream the holes in the field thus decreasing the bolt strength below the design values. The usual, workable, or industry standard gages used in bolting angles are given in Table 6.2.1, adapted from Figure 10-6 of the LRFDM, while tables in Part 1 of the LRFDM give workable gages in flanges of W-, C-, and T-shapes. Tables 7-3*a* and 7-3*b* of the LRFDM give entering and tightening clearances for high-strength bolts. Several tables in Part 7 of the LRFDM can be used to determine bolt areas.

The nominal shear strength per unit area of bolts, for bolts in bearing-type joints (that is, values of F_{nv}), are given in Table J3.2 of the LRFDS. These values are 48 ksi for A325-N bolts, 60 ksi for A325-X and A490-N bolts, and 75 ksi for A490-X bolts. It should be noted that a value of $F_{ub} = 120$ ksi is used for all A325 bolts. That is, for A325 bolts, no distinction is made between small and large diameters, even though the minimum tensile strength, F_{ub} is lower (105 ksi) for bolts with diameters larger than 1 in. It was felt that such a refinement of design was not justified, particularly in view of the low value set for the resistance factor ϕ and other compensating factors. Design shear strength values for bolts in bearing-type joints, B_{dv}, can be calculated using Eq. 6.7.3. LRFDM Table 7-10 provides the values of B_{dv} (in kips) for A325 and A490 bolts of nominal diameters $d = \frac{5}{8}$ to $1\frac{1}{2}$ in. Single and double shear values are given for bearing-type joints, for N-type and X-type of bolts. Note that the design shear strength of a A490 bolt is about 25 percent greater than that of a A325 bolt. Also note that the design shear strength of a A325-X type bolt is same as that of a A490-N type bolt of the same diameter. Also, when the length between extreme bolts in a bearing-type joint, measured parallel to the line of force exceeds 50 in., the bolt shear strength has to be reduced by 20 percent (see LRFDS Table J3.2, footnote e).

Bearing strength of connected plate elements at bolt holes—as a function of the end distance, L_e, and pitch of bolts, p—was considered in Section 6.8. As indicated earlier, along a line of transmitted force, the distance from the center of a hole to the end of the connected part should be no less than $L_{e,full}$ ($= 2\frac{1}{2} d + \frac{1}{32}$ in. for a STD hole, for example) to attain the maximum bearing strength for an end bolt. Also, the distance measured in the line of force, from the center of any bolt hole to the center of an adjacent hole, should be no less than p_{full} ($= 3d + \frac{1}{16}$ in. for STD holes, for example) to ensure maximum design bearing strength for an interior bolt. Design bearing strength values, B_{dbo}, corresponding to the limit state of ovalization of bolt holes (when $L_e \geq L_{e,full}$ for an end bolt, or $p \geq p_{full}$ for an interior bolt) can be calculated using Eq. 6.8.2. Also, design bearing strength values, B_{dbt}, corresponding to the limit state of shear tear-out of plate material (when, $L_e < L_{e,full}$ for an end bolt, or $p < p_{full}$ for an interior bolt) can be calculated using Eqs. 6.8.5 and 6.8.6, respectively. Design bearing strength of a bolt in a bearing-type joint, B_{db}, is the smaller of B_{dbo}, and B_{dbt}, as given by Eq. 6.8.8. Values of B_{db} may also be obtained from LRFDM Tables 7-12 and 7-13 for nominal bolt diameters of $\frac{5}{8}$ to $1\frac{1}{2}$ in. for a material thicknesses of 1 in. Values are given for F_{up} of the connected materials equal to 58 ksi and 65 ksi. LRFDM Table 7-12 gives values of B_{db} for $p = 2\frac{2}{3}d$ and 3 in., and values of B_{dbo}; also tabulated here are the values of p_{full} and $2\frac{2}{3}d$ for different bolt diameters, d. LRFDM Table 7-13 gives values of B_{db} for $L_e = 1\frac{1}{4}$ and 2 in., and values of B_{dbo}; also tabulated here are the values of $L_{e,full}$. In both tables, values are given, with appropriate modifications, for STD, OVS, SSL, LSLP, and LSLT holes. For material thicknesses other than 1 in., the bearing strength can be calculated by multiplying the value tabulated for 1-in.-thick material by the actual thickness of the connected material.

Table 6.7.1 provides a partial summary of LRFDM Tables 7-10, 7-14, 7-15 and 7-16 for $\frac{5}{8}$-,$\frac{3}{4}$-,$\frac{7}{8}$- and 1-in.-diameter high-strength bolts. Also, Table 6.8.1 gives the design bearing strength at bolt holes for various bolt spacings, p, while Table 6.8.2 gives the design bearing strength at bolt holes for various edge distances, L_e.

6.13.2   Simple Joints

Joints where the line of action of the resultant load lies in the plane of the joint and passes through the center of gravity of the connectors are known as *simple joints* (see Figs. 6.4.1a and 6.4.1b). Only such simple joints will be considered in this section. Study of joints subjected to all other types of loading will be postponed until Chapter 12.

For information on the load distribution in such axially loaded bolted joints, refer to Section W6.8 on the website http://www.mhhe.com/Vinnakota.

www

The design strength of a simple joint can be found by adding the design strengths of individual connectors in the joint. Thus, for a symmetric and symmetrically loaded joint consisting of connectors of equal size and type, the design shear strength can be found by multiplying the capacity of a single connector by the total number of connectors in the joint. Also, the design bearing strength may be taken as the sum of the bearing strengths of the end bolts and the bearing strengths of the interior bolts, if any. Thus, we have:

$$C_{dv} = NB_{dv} \qquad (6.13.1)$$

$$C_{db} = n_e B_{dbe} + n_i B_{dbi} \qquad (6.13.2)$$

$$C_d = \min [C_{dv}, C_{db}] \qquad (6.13.3)$$

where

C_d = design strength of connectors in a joint
C_{dv} = design shear strength of connectors in the joint
C_{db} = design bearing strength of connectors in the joint
n_e = number of end bolts in the joint
n_i = number of interior bolts in the joint
N = total number of bolts in the joint ($= n_e + n_i$)
B_{dbe} = design bearing strength of an end bolt (Eq. 6.8.8a)
B_{dbi} = design bearing strength of an interior bolt (Eq. 6.8.8b)
B_{dv} = design shear strength of a bolt (Eq. 6.7.3)

If a symmetric and symmetrically loaded bolt group consists of bolts of different strengths, the group connector strength is taken as the sum of the individual bolt strengths. Several numerical examples are worked out below to illustrate the use of the design equations and the use of available design tables.

Strength of Connectors

EXAMPLE 6.13.1

Determine the factored axial tensile load P that the bolts in the bearing-type connection shown in Fig. X6.13.1 can transfer from plate A to plate B. The two $\frac{3}{8}$-in.-thick A572 Grade 50 steel plates of the lap joint are connected by six $\frac{7}{8}$-in.-dia. A325 bolts. The holes are of standard size, and threads are eXcluded from the shear plane. Assume gas cut edges and:

a. $L_e = 2\frac{1}{2}$ in. and $p = 3$ in.

b. $L_e = 1\frac{1}{4}$ in. and $p = 2\frac{1}{2}$ in.

Solution

Plate material: A572 Grade 50
From Table 2-2 of the LRFDM, we obtain:

$$F_{yp} = 50 \text{ ksi}, \quad F_{up} = 65 \text{ ksi}$$

[continues on next page]

Example 6.13.1 continues ...

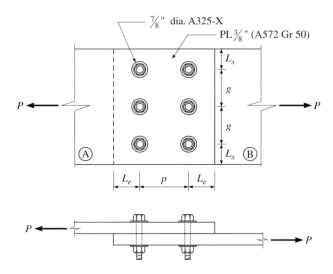

Figure X6.13.1

Bolt material: A325
From Table 2-3 of the LRFDM, for A325 steel: $F_{ub} = 120$ ksi
Bolt diameter, $d = 7/8$ in.; $2d = 1.75$ in.; $3d = 2.63$ in.
There are 6 bolts consisting of 3 end bolts and 3 interior bolts. That is:

$$n_e = 3; \quad n_i = 3; \quad N = 6$$

For gas cut edges, minimum edge distance for $7/8$ in. dia. bolt from LRFDS Table J3.4, $L_{e, \min} = 1.13$ in.
Edge distance provided, $L_e = 2.50$ in. > 1.25 in.   O.K.
Minimum spacing of bolts (from LRFDS Section J3.3):

$$2\frac{2}{3}d = \frac{8}{3}\left(\frac{7}{8}\right) = 2.33 \text{ in.}$$

Preferred spacing of bolts (from LRFDS Section J3.3):

$$3d = 3\left(\frac{7}{8}\right) = 2.63 \text{ in.}$$

Pitch provided, $p = 3.0$ in. > 2.63 in.   O.K.

Bolt area, $A_b = \pi\dfrac{d^2}{4} = \dfrac{\pi}{4}\left(\dfrac{7}{8}\right)^2 = 0.601$ in.^{2}

Bolts in a lap joint are in single shear. So, $N_s = 1$

For standard holes, diameter of hole, $d_h = d + \dfrac{1}{16} = \dfrac{7}{8} + \dfrac{1}{16} = \dfrac{15}{16}$ in.

As threads are excluded from the shear plane, the nominal shear strength per unit area of bolt material (from LRFDS Table J3.2):

$$F_{nv} = 0.5F_{ub} = 0.5(120) = 60.0 \text{ ksi}$$

Design shear strength of a single bolt, from Eq. 6.7.3:

$$B_{dv} = \phi F_{nv} N_s A_b = 0.75(60.0)(1)(0.601) = 27.1 \text{ kips}$$

Alternatively, the value of B_{dv} may be obtained from Table 6.7.1 (or from LRFDM Table 7-10). Corresponding to a $\frac{7}{8}$-in.-dia. A325-X bolt in single shear the value is found to be 27.1 kips.
The design shear strength of the connectors in the joint (from Eq. 6.13.1) is:

$$C_{dv} = NB_{dv} = 6(27.1) = 163 \text{ kips}$$

The design bearing strength of a connected plate element corresponding to the limit state of ovalization of the bolt hole is given by Eq. 6.8.2. For a $\frac{7}{8}$-in.-dia. bolt in a $\frac{3}{8}$-in.-thick A572 Grade 50 plate material, we have:

$$B_{dbo} = 2.4\phi F_{up} \, dt$$

$$= 2.4(0.75)(65.0)\left(\frac{7}{8}\right)\left(\frac{3}{8}\right) = 102.4 \text{ kips/in.} \left(\frac{3}{8} \text{ in.}\right)$$

$$= 38.4 \text{ kips}$$

a. Joint with $L_e = 2\frac{1}{2}$ in. and $p = 3$ in.
Clear distance for an end bolt,

$$L_{ce} = L_e - 0.5d_h = 2\frac{1}{2} - \frac{15}{32} = 2.03 \text{ in.} > 2d$$

So, shear tear-out of the plate will not control the bearing strength of the end bolts.
Clear distance for an interior bolt,

$$L_{ci} = p - d_h = 3 - \frac{15}{16} = 2.06 \text{ in.} > 2d$$

So, shear tear-out of the plate will not control the bearing strength of the interior bolts either. The design bearing strength of the end and interior bolts is, therefore, given by the limit state of ovalization of the bolt hole. That is,

$$B_{db} = B_{dbo} = 38.4 \text{ kips}$$

Alternatively, the design bearing strengths for $\frac{7}{8}$-in.-dia. bolts at standard bolt holes in plate material with $F_u = 65$ ksi may be obtained from Tables 6.8.1 and 6.8.2. This value is 102 kips/in. thickness from Table 6.8.1 for an interior bolt with a pitch $p = 3$ in., and 102 kips/in. thickness from Table 6.8.2 for an end bolt with end distance $L_e = 2\frac{1}{2}$ in. Multiplying

[*continues on next page*]

Example 6.13.1 continues ...

these values by the thickness of the plate, $\frac{3}{8}$ in., the design bearing strength of the interior and end bolts is again obtained as 38.4 kips.

From Eq. 6.13.2, design bearing strength of connectors in the joint is:

$$C_{db} = 3(38.4) + 3(38.4) = 230 \text{ kips}$$

From Eq. 6.13.3, design strength of connectors in the joint with $L_e = 2\frac{1}{2}$ in. and $p = 3$ in. is:

$$C_d = \min [C_{dv}, C_{db}] = \min [163, 230] = 163 \text{ kips} \qquad \text{(Ans.)}$$

b. Joint with $L_e = 1\frac{1}{4}$ in. and $p = 2\frac{1}{2}$ in.

Clear distance, $L_{ce} = L_e - 0.5d_h = 1.25 - (^{15}\!/_{32}) = 0.781$ in. $< 2d$

So, the design bearing strength of the end bolts is determined by the limit state of shear tear-out of the plate. Using Eq. 6.8.5, we obtain:

$$B_{dbe} = B_{dbte} = 1.2 \, \phi \, F_{up} \, L_{ce} \, t$$

$$= 1.2(0.75)(65.0)(0.781)\left(\frac{3}{8}\right) = 45.7 \text{ kips/in.} \left(\frac{3}{8}\right) \text{ in.}$$

$$= 17.1 \text{ kips}$$

Clear distance, $L_{ci} = p - d_h = 2.5 - (^{15}\!/_{16}) = 1.563$ in. $< 2d$

So, the design bearing strength of the interior bolts is also determined by the limit state of shear tear-out of the plate. Using Eq. 6.8.6, we obtain:

$$B_{dbi} = B_{dbti} = 1.2 \, \phi \, F_{up} \, L_{ci} \, t$$

$$= 1.2 \, (0.75)(65)(1.56)\left(\frac{3}{8}\right) = 91.4 \text{ kips/in.} \left(\frac{3}{8}\right)\text{in.} = 34.3 \text{ kips}$$

From Eq. 6.13.2, the design bearing strength of connectors in the joint consisting of three end bolts and three interior bolts is:

$$C_{db} = n_e \, B_{dbe} + n_i \, B_{dbi} = 3(17.1) + 3(34.3) = 154 \text{ kips}$$

From Eq. 6.13.3, the design strength of connectors in the joint with $L_e = 1\frac{1}{4}$ in. and $p = 2\frac{1}{2}$ in. is:

$$C_d = \min[C_{dv}, C_{db}] = \min[163, 154] = 154 \text{ kips} \qquad \text{(Ans.)}$$

EXAMPLE 6.13.2

Strength of Connectors

Determine the maximum axial tensile load P (30 percent dead load and 70 percent live load) that can be transmitted by the bolts in the butt splice shown in Fig. X6.13.2. The main plates are $\frac{1}{2}$-in. thick, and the cover plates are $\frac{3}{8}$-in. thick. Assume 1-in.-dia. A490 bolts in standard holes with threads

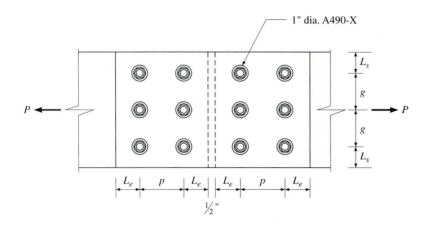

1" dia. A490-X

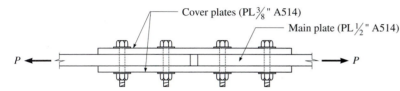

Cover plates (PL⅜" A514)

Main plate (PL½" A514)

Figure X6.13.2

e**X**cluded from the shear planes. The plates are of A514 Gr 100 steel. $L_e =$ 1¾ in. and $p = 3½$ in. Consider the joint as a bearing-type joint.

Solution

Plate material: A514 Gr 100
From Table 2-2 of the LRFDM, we obtain:

$$F_{yp} = 100 \text{ ksi}, \ F_{up} = 110 \text{ ksi}$$

Bolt material : A490
From Table 2-3 of the LRFDM: $F_{ub} = 150$ ksi
Bolt diameter, $d = 1.0$ in.; $2d = 2.0$ in.; $3d = 3.0$ in.
There are 6 bolts in the butt joint, consisting of 3 end bolts and 3 interior bolts. So

$$n_e = 3; \ n_i = 3; \ N = 6$$

Bolt area, $A_b = \dfrac{\pi}{4}(1^2) = 0.785$ in.^{2}

Diameter of the hole, $d_h = d + \dfrac{1}{16} = 1.06$ in.

Bolts in a butt joint are in double shear. So, $N_s = 2$
Assuming sheared edges, the minimum edge distance, $L_{e,min}$, for a 1-in.-dia. bolt may be obtained from LRFDS Table J3.4 as 1.75 in.
Edge distance provided, $L_e = 1.75$ in.   O.K.   *[continues on next page]*

Example 6.13.2 continues ...

Recommended minimum spacing of bolts (from LRFDS Section J3.3) = $3d = 3.0$ in.

Spacing provided, $p = 3.5$ in.   O.K.

Bolt type: A490-X

As threads are excluded from the shear plane, the nominal shear strength per unit area of bolt (from LRFDS Table J3.2):

$$F_{nv} = 0.5F_{ub} = 0.5(150) = 75.0 \text{ ksi}$$

Design shear strength of a 1-in.-dia. A490-X type bolt in double shear, from Eq. 6.7.3, is:

$$B_{dv} = \phi F_{nv} N_s A_b = 0.75 (75.0)(2)(0.785) = 88.4 \text{ kips}$$

Alternatively, the value of B_{dv} may be obtained from Table 6.7.1 (or, from LRFDM Table 7-10). Corresponding to a 1-in.-dia. A490-X bolt in double shear the value is found to be 88.4 kips.

Design shear strength of the connectors in the joint, from Eq. 6.13.1, is:

$$C_{dv} = NB_{dv} = 6(88.4) = 530 \text{ kips}$$

As the main plate and cover plates are of the same material and as the combined thickness of the cover plates ($\frac{3}{8} + \frac{3}{8} = \frac{3}{4}$ in.) is greater than the thickness of the main plate ($\frac{1}{2}$ in.), the bearing strength of the bolts will be controlled by the thickness of the main plate.

The design bearing strength of the connected plate material, corresponding to the limit state of ovalization of the bolt hole, is given by Eq. 6.8.2. For a 1-in.-dia. bolt in a $\frac{1}{2}$-in.-thick plate with $F_{up} = 110$ ksi, we obtain:

$$B_{dbo} = 2.4\phi F_{up}\, dt = 2.4(0.75)(110)(1.0)(\tfrac{1}{2}) = 99.0 \text{ kips}$$

Bolt pitch, $p = 3\frac{1}{2}$ in.

Clear distance, $L_{ci} = p - d_h = 3.5 - 1.06 = 2.44 > 2d$,

indicating that shear tear-out of the plate will not control the design strength of interior bolts. Hence, the design bearing strength of an interior bolt is:

$$B_{dbi} = B_{dbo} = 99.0 \text{ kips}$$

End distance, $L_e = 1\frac{3}{4}$ in.

Clear distance, $L_{ce} = L_e - 0.5d_h = 1.75 - 0.5(1.06) = 1.22$ in. $< 2d$

indicating that the bearing strength of the end bolts is limited by the shear tear-out of the plate. From Eq. 6.8.5, we obtain:

$$B_{dbe} = B_{dbte} = 1.2\phi F_{up} L_{ce}\, t = 1.2(0.75)(110)(1.22)(0.5)$$
$$= 60.4 \text{ kips}$$

Design bearing strength of the connectors in the joint (from Eq. 6.13.2) is:

$$C_{db} = n_e\, B_{dbe} + n_i\, B_{dbi} = 3(60.4) + 3(99.0) = 478 \text{ kips}$$

Design strength of the connectors in the joint (from Eq. 6.13.3) is:

$$C_d = \min[C_{dv},\, C_{db}] = \min[530,\, 478] = 478 \text{ kips}$$

If P_s is the service load the connectors are capable of transferring, we have:

$$1.2(0.30\, P_s) + 1.6(0.70\, P_s) \le C_d = 478$$

$$\rightarrow P_s \le 323 \text{ kips}$$

Hence,

$$P_{s,\,max} = 323 \text{ kips} \qquad\qquad \text{(Ans.)}$$

Design of Connectors

E X A M P L E   6 . 1 3 . 3

A lap joint connecting two $\tfrac{1}{2}$-in. plates transmits axial service tensile loads $P_D = 60$ kips and $P_L = 60$ kips using 1-in.-dia. A325 high-strength bolts in standard holes with threads iNcluded in the shear plane. Assume A572 Gr 50 steel and $L_c > 2d$ for all bolts. Determine the number of bolts required for a bearing-type joint.

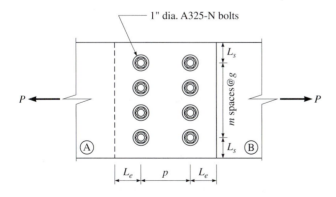

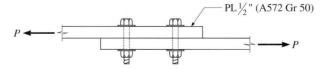

Figure X6.13.3

[continues on next page]

Example 6.13.3 continues ...

Solution

Factored load, $P_u = 1.2P_D + 1.6P_L = 1.2(60) + 1.6(60)$
$$= 168 \text{ kips}$$

Diameter of bolt, $d = 1.0$ in.; $2d = 2.0$ in.; $3d = 3.0$ in.

Diameter of bolt hole, $d_h = 1.06$ in.

Cross sectional area, $A_b = 0.785$ in.^{2}

Type of bolt: A325-N. So, $F_{ub} = 120$ ksi

In a lap joint the bolts are in single shear. So, $N_s = 1$

Plate material, A572 Grade 50 steel. From Table 2-2 of the LRFDS, $F_{up} = 65$ ksi

Plate thickness, $t = \frac{1}{2}$ in.

Design shear strength of an N-type bolt (from Eq. 6.7.3):

$$B_{dv} = \phi(0.4\,F_{ub})\,N_s\,A_b = 0.75(0.4)(120)(1)(0.785) = 28.3 \text{ kips}$$

As $L_c > 2d$ for the interior and end bolts, the design bearing strength of the connected plates at interior and end bolts is given by the limit state of ovalization of bolt hole. That is:

$$B_{db} = B_{dbo} = 2.4\,\phi F_{up}\,dt$$

$$= 2.4(0.75)(65.0)(1.0)\left(\frac{1}{2}\right) = 58.5 \text{ kips}$$

Design strength of a single bolt is,

$$B_d = \min[B_{dv}, B_{db}] = \min[28.3, 58.5] = 28.3 \text{ kips}$$

Number of bolts required $= \dfrac{168}{28.3} = 5.94$

Provide 6 bolts. That is 3 bolts in each vertical row ($m = 2$ in Fig. X6.13.3).

(Ans.)

6.14   Welding and Welding Processes

6.14.1   Introduction

In structural steel fabrication, ***arc welding***—the fusing together of metals utilizing the heat generated by an electric arc—is used extensively. Figure 6.14.1a shows schematically the electrical circuit basic to almost all arc welding processes. An electrical power source is connected to the work-piece to be welded by the ***ground cable.*** A second cable from the power source, called the ***electrode cable,*** is connected to the electrode holder and then to the ***electrode*** (welding wire). The ***arc*** is initiated at the end of the electrode when the electrode is touched to the work-piece, thus closing the circuit, and then

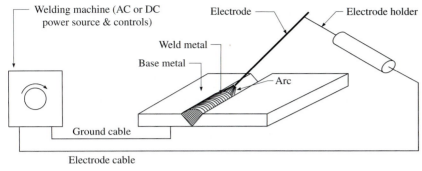

Arc welding circuit

(a)

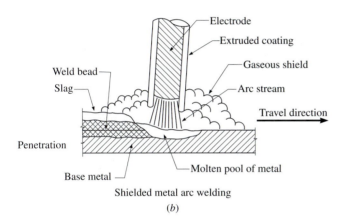

Shielded metal arc welding

(b)

Figure 6.14.1: Shielded metal arc welding (SMAW) process.

raised slightly above it. An intense heat is generated by the arc, sufficient to reduce steel to the molten state. Typically, temperatures in elements being welded are over 3000°F, while within the arc the temperature may be as high as 10,000°F. The electromagnetic field generated carries the molten globules of metal from the electrode into the molten base material, and on cooling, they are united with the elements to be joined. Thus, the **weld** is really composed of a mixture of the base material and the electrode metal. The electrode is moved along the path of the weld at the proper speed either manually by the welder or automatically by a welding machine. Arc welding processes require a continuous supply of electric power sufficient in amperage and voltage to maintain the arc. The power source may supply either alternating current (AC) or direct current (DC). High voltages generally provide high welding rates.

Structural welds are usually made either by the shielded metal arc welding process or by the submerged arc welding process, which are described in Sections 6.14.2 and 6.14.3, respectively [AWS, 2001; Blodget, 1966].

6.14.2 Shielded Metal Arc Welding (SMAW) Process

In the *shielded metal arc welding process* (Fig. 6.14.1*b*), a specially coated metal electrode (stick), which is consumed in the process, is used. The *coating* consists of a clay-like mixture of silicate binders and powdered materials, such as carbonates, fluorides, oxides, metal alloys, and cellulose. The resistance of the air or gases within the gap, to the passage of current, transforms the electrical energy into heat. The coating melts in the arc, thereby releasing inert gases. The electric arc in this process is shielded by gases and thus its name. The arc melts electrode metal and base material, liquefying them in a common pool of molten metal called the *crater.* As the arc is moved along its path, the pool solidifies behind it to form a homogeneous weld, which is fused with and becomes an integral part of both of the components being connected together. Metal globules transfer from the electrode to the pieces being welded by molecular attraction and surface tension, rather than by gravity, so the shielded metal arc welding process can be used in overhead welding as well. The electrode diameters range from $\frac{5}{32}$ to $\frac{1}{2}$ in.

The coating is converted partially into shielding gases, partially into slag, with the remaining material absorbed by the weld metal. The coating serves several important purposes:

1. The coating is consumed at a slower rate than the metal core, thereby forming a protruding sheath which serves to direct and channel the arc stream.
2. Protective gases are produced by the combustion and decomposition of coating constituents. Gas formation is due to the presence of limestone and cellulose in the coating. In general, the gases surrounding the arc consist of carbon monoxide, carbon dioxide, hydrogen, and water vapor. All of these gasses are considered protective because they shield the weld from oxygen and nitrogen in the air, thereby preventing the formation of undesirable oxides and nitrides which cause loss of ductility (weld embrittlement), low strength, and poor corrosion resistance.
3. Other constituents in the coating, such as alumina, magnesia, manganese oxide, and silica, form slag in the molten metal. The molten slag in contact with the molten weld metal attracts and removes impurities which may have formed in the weld metal. The slag, which is lighter than the molten base material, rises to the surface and protects the weld from the air while the weld cools.
4. Electrode coatings normally contain deoxidants such as ferro-manganese, ferro-silicon, and ferro-titanium. Deoxidants are substances that chemically react with oxygen. These deoxidants provide added protection to the weld metal against contamination due to possible breakdown of the shielding gas and slag systems. They also refine the grain structure of the weld metal.
5. The electrode coating also serves as a means of transferring desirable alloying elements into the weld pool during welding so that any specific weld metal compositions can be obtained.

The flux is self-cleansing; that is, as the weld cools, the differential in cooling rates between the weld and the slag is sufficient to cause the slag to free itself from the weld on its own accord. The slag must be removed by brushing or peening before painting or depositing additional beads of weld.

The electric power used with the SMAW process may utilize either direct or alternating current. With direct current, either straight or reverse polarity may be used. For *straight polarity,* the base material is the positive pole, and the electrode is the negative pole of the welding arc. For *reverse polarity,* the base material is the negative pole, and the electrode is the positive pole. Electrical equipment with a welding current rating of 400 to 600 amps is usually used for structural steel fabrication.

Shielded metal arc welding, sometimes referred to as *manual, hand, or stick welding,* is an important welding process in both shop fabrication and field assembly of steel structures. The equipment needed is less costly and more portable, and the electrodes may be purchased off the shelf in most locations. Also, there are more people qualified to perform this type of welding and thus it is easier to find welders who have the experience. The SMAW process is used when the work involves repetitive starting and stopping, as in the jobs where intermittent fillet welds are used.

6.14.3   Submerged Arc Welding (SAW) Process

The *submerged arc welding process* (Fig. 6.14.2) uses a continuous *bare wire electrode* and granular material called *flux* spread over the seam, instead of a coated *stick electrode.* The bare electrode wire is fed automatically from a reel, through the welding head at a rate sufficient to maintain a constant arc length.

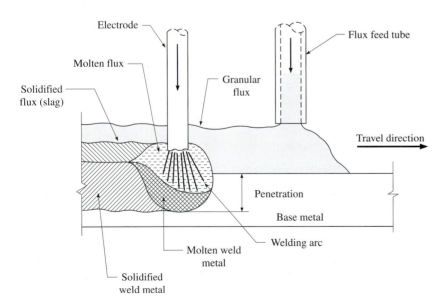

Figure 6.14.2: Submerged arc welding (SAW) process.

Flux is laid automatically from a hopper onto the work area by gravity, ahead of the advancing electrode in an amount sufficient to submerge the arc completely (Fig. 6.14.2). The constituents of submerged arc fluxes are similar to shielded metal arc electrode coatings, and therefore have similar electrical, metallurgical, and physical properties. The electric arc in this process is submerged under a heap of granular flux and hence its name. Electrical power is transmitted to the electrode wire through copper contact tips close to the arcing point. The electrode, flux hopper, and power source are attached to a frame which is on rollers and which advances the weld at a predetermined rate. Some of the flux melts to form a covering of slag over the weld, stabilizes the arc, and excludes air from the work-piece. After welding, the slag is removed by burning or peening, and the unmelted flux is recovered for reuse.

The submerged arc welding process is always applied in the flat or horizontal position because the granular flux and the typically large molten weld and slag pool would be uncontrollable in other positions. This process is extensively used to shop fabricate structural steel, using automatic or semiautomatic equipment. SAW is particularly well suited to long welding runs (30 ft or more). It can be used on thin or thick sections of metal and is capable of producing high-quality fillet, partial penetration, or full penetration groove welds.

Generally, higher electric currents are used in the SAW process as compared to the SMAW process. This results in significantly more melting of the base material and deeper penetration of the weld. Also, for SAW the joint opening need not be as great as that necessary for SMAW. Thus less filler metal is required, and a faster deposition rate of weld is possible with SAW than with SMAW.

For even greater welding speed at reduced cost, multiple electrode welding—using two to ten electrodes of small diameter wires—is used where larger weld sizes and longer weld lengths justify such equipment. Progress made in automatic manipulators enables the welding head to be properly aligned with the joint in a matter of seconds.

Descriptions of other welding processes, such as the gas metal arc welding (GMAW) process and the flux cored arc welding (FCAW) process, are available in the *Welding Handbook* [AWS, 2001]. Gas welding is welding performed by using an oxyacetylene flame to heat the base material and melt the welding rod.

6.14.4   Resistance Welding

Resistance welding is a heat and squeeze process, wherein the parts to be welded are heated to the temperature of fusion by the electrical resistance generated from the passage of very high current (up to 100,000 A). Once the welding temperature has been reached, pressure is applied mechanically to bring about the union. Resistance welding comprises several processes, the most important of which are spot welding and seam welding. The *spot weld* is made by overlapping the parts and clamping the pieces between two opposing electrodes, through which current is passed and pressure applied,

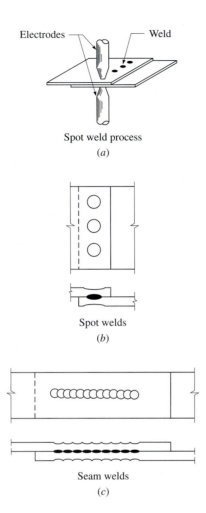

Spot weld process
(*a*)

Spot welds
(*b*)

Seam welds
(*c*)

Figure 6.14.3:   Resistance welding.

to make the weld in a single spot (Figs. 6.14.3*a* and *b*). The size of the final weld is a function of the electrode size. The ***seam weld*** is similar to the spot weld except that circular, rolling electrodes are used to produce the effect of a continuous seam (Fig. 6.14.3*c*). Cold-formed structural steel members are usually shop fabricated by resistance welding. Resistance welding is also often used on-site to connect cold-formed members, such as metal decking, to other cold-formed members, such as chords of open web steel joists, or to hot-rolled framing members.

6.14.5   Welding Electrodes and Fluxes

Welding electrodes are the rods or wires that are used in making welds. The type of electrode used affects weld properties such as strength, ductility, and corrosion resistance. Welding electrodes must conform to American Welding Society (AWS) specifications [AWS, 2000].

The electrodes used in shielded metal arc welding are specified in AWS A5.1 [AWS, 1991] and A5.5 [AWS, 1996]. These electrodes are classified as E60XX, E70XX, E80XX, E90XX, E100XX, and E110XX. The "E" denotes electrode. The first two digits (or the first three digits if the first digit is a "1") following the "E" indicate the ultimate tensile stress of the weld metal in ksi. Thus, the ultimate tensile stress of the weld electrode material, F_{uw}, ranges from 60 to 110 ksi. The "X"s represent numbers specifying the usage of the electrode. The first "X" indicates the position code; that is, the welding position in which the electrode can be used (see Section 6.15.2). Many electrodes are specified for use in a particular position and should be used in that position to obtain the best results. Only some electrodes are suitable for use in all positions. If the first "X" is a "1", this means that the electrode may be used for all welding positions. The number "2" indicates that the electrode may be used in the flat and horizontal positions only. A "3" means that the electrode is only to be used in the flat position. The second "X" denotes the type of coating, the type of current (AC or DC), and the polarity (straight or reversed). The size of the electrode (core wire diameter) depends primarily on the joint detail, welding position, and welding equipment available. Electrode sizes of $\frac{1}{8}$, $\frac{5}{32}$, $\frac{3}{16}$, $\frac{7}{32}$, $\frac{1}{4}$, and $\frac{5}{16}$ in. are commonly used. Small size electrodes are 14-in. long, and the larger sizes are 18-in. long.

Fluxes are fused or agglomerated (finely powdered) constituents glued together with silicates. They are classified in AWS specifications according to the weld metal properties produced in the standard specification weld tests. For submerged arc welding (SAW) the combinations of fluxes and electrodes, which also serve as filler material, are specified separately. They are specified in AWS A5.17 [AWS, 1998] and A5.23 [AWS, 1997], and are designated F6X-EXXX, F7X-EXXX, F8X-EXXX, F9X-EXXX, F10X-EXXX, and F11X-EXXX. The letter "F" designates a granular flux. The first digit (or the first two digits if the first digit is a "1") following the "F" indicates the ultimate tensile stress of the weld metal. For example, "6" means $F_{uw} = 60$ ksi, and "10" means $F_{uw} = 100$ ksi. The next digit gives the Charpy V-notch impact strength of the test weld. This is followed by a set of letters and numbers denoting the type of bare wire electrode used with this flux.

The welding rod (for SMAW) or flux-electrode combination (for SAW) chosen is determined by the characteristics of the base material and the properties desired in the weld itself. If the properties of the electrode material are comparable to the properties of the base material, the electrode is said to be a *matching electrode.* AWS D1.1 Table 4.1.1 [AWS, 2000] lists matching electrodes for various ASTM structural steels and is referenced in LRFDS Table J2.5. Table 6.14.1 indicates which electrodes should be matched with each particular structural steel. *Matching* as used here is the assignment of certain electrode materials to base metals in accordance with AWS code and based on the ultimate tensile stress of the base metal and weld metal. Use of electrodes

TABLE 6.14.1

Matching Filler Metal Requirements

Group	Base Material (ASTM Specification)	Welding Process	
		SMAW	SAW
I	A36	E60XX or E70X	F6X or F7X
II	A242 A572 Grades 42 and 50 A588 A992	E70XX	F7X
III	A572 Grades 60 and 65	E80XX	F8X
IV	A514 (over 2 ½ in. thick)	E100XX	F10X
V	A514 (2 ½ in. and under)	E110XX	F11X

Note: Adapted from AWS Table 4.1.1 [AWS, 2000]

one strength-level higher than "matching" is permitted. Typical structural steel grades with F_y equal to 36 ksi and 50 ksi are normally welded with electrode material of 70 ksi nominal strength, indicated as E70XX for SMAW or F7X-EXXX for SAW.

Because strength is usually the most important parameter to the structural designers, we usually specify electrodes simply as E70, E80, F7, F8, and so on.

6.14.6   Weldability of Structural Steels

The *weldability* of a steel is a measure of the relative ease of producing a satisfactory, crack-free, structurally sound joint possessing adequate strength and ductility. Various elements in the chemical composition of steels influence their mechanical properties such as hardness. As carbon has the greatest effect on hardness and as hardness is related to weldability and susceptibility to cracking, the influence of the chemical elements in steel on its weldability is expressed through a *carbon equivalent (CE)* formula. Many different carbon equivalent formulas are used as a guide for preheat requirements and welding procedures. The AWS D1.1:2000 Appendix X1 Formula for carbon equivalent is used for structural steels, namely:

$$CE = C + \frac{(Mn + Si)}{6} + \frac{(Cu + Ni)}{15} + \frac{(Cr + Mo + V)}{5} \qquad (6.14.1)$$

The variables on the right side of this equation represent the percentages of the respective elements in the steel at hand: C = Carbon; Mn = Manganese; Si = Silicon; Ni = Nickel; Cu = Copper; Cr = Chromium; Mo = Molybdenum; V = Vanadium. The symbol *CE* on the left side of the equal sign stands for carbon equivalent in percent.

Good weldability is virtually assured if the relation $CE \leq 0.50$ is satisfied. Steels with a carbon content less than or equal to 0.30 percent are well suited to high-speed welding. Steels with a carbon content greater than 0.35 percent require special care during welding. Most of the ASTM specified structural steels can be welded without special procedures or precautions. However the need for special procedures, such as preheating, increases with plate thickness and with the amount of alloying elements in its composition. *Preheating* of the elements being welded reduces the temperature gradient between the molten metal of the weld and the cooler material of the parent metal. The more uniform temperature gradient that results ensures cooling without the possibility of cracking. The temperature used for preheating is a function of the thickness and chemical composition of the steel [AWS, 2000]. The typical welding preheat ranges from 75° to 200°F, depending on the size of the pieces that are welded.

6.14.7   Advantages and Disadvantages of Welding and Bolting

Welding offers many advantages over bolting:

1. With welding, connecting elements, such as gusset plates, cover plates, and splice plates, are reduced or eliminated since they often are not required. This results in considerable savings in weight and in fewer pieces to be fabricated, handled, and erected.
2. Welded connections of tension members result in weight savings for these members because no deductions need be made for bolt holes (the net section is the gross section, as we will see in Section 7.6).
3. With the use of welding, fabrication costs and time are reduced since operations such as punching, reaming, and drilling are eliminated.
4. The typical welded joint produces a smooth, uncluttered connection that can be left exposed without detracting from the architectural appearance of the structure.
5. The fused joints obtained by welding result in a more rigid structure as compared to the more flexible structure made with bolted joints. Rigid connections, in turn, often result in reduced beam depth and weight.
6. Welding is the only plate-joining procedure that results in joints that are intrinsically airtight and watertight. Hence, it is ideal for fabricating pressure vessels, water tanks, penstocks, and so on.
7. Welded structures can be erected in relative silence, a great advantage when building near hospitals, schools, and office buildings, or when making additions to existing and occupied buildings.
8. The taller the building grows, the greater the role of welding. This applies to the shop fabrication of columns, girders, and other structural members, and also to the field welding associated with erection.

9. Welding makes connection of members to a curved or sloping surface economically feasible. Examples are structural connections to tube and pipe columns.

10. Welding simplifies the rehabilitation and strengthening of existing bolted or welded structures.

Disadvantages of welding:

1. Welding requires skilled workers.

2. Inspection of the finished weld requires considerable expertise and experience.

3. Fabrication tolerances, generally, are more stringent than those of bolted connections.

Advantages of bolting:

1. High-strength bolts require fewer and less-skilled workers, thereby reducing labor costs.

2. Bolting requires less equipment.

3. Installation techniques are simple, and a worker can be trained in hours.

4. No fire hazard exists with bolted construction.

The equipment available in the fabricator's shop and the training of workers will generally have an important bearing on the bid price and the choice of either bolted or welded work.

6.15   Weld Classifications

6.15.1   Types of Welds

Welds used for structural steel are classified according to the shape of their cross section as fillet, groove, plug, and slot (Fig. 6.15.1). For welded structural steel connections, fillet welds are used approximately 80 percent of the time, groove welds, 15 percent, and slot and plug welds, 5 percent.

Fillet Welds
Fillet welds are theoretically triangular in cross section and join two surfaces approximately at right angles, formed by lapping or intersecting parts of structural members. Thus, they may be found in lap, tee, and corner joints (discussed in Section 6.15.3). Fillet welds are the most commonly used weld, especially for light loads, and the most economical, as little preparation of plate material is needed. Also, fillet welds do not require the same level of operator skill as groove welds.

Groove Welds
Groove welds are welds deposited in a groove or gap between adjacent ends, edges, or surfaces of two parts to be joined. They are generally used

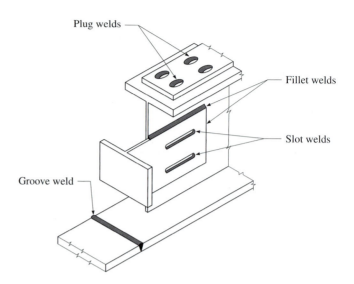

Figure 6.15.1:   Fillet, groove, plug, and slot welds.

to connect two plates lying in the same plane (butt joint), but also used in tee or corner joints. Groove welds generally require special *edge preparation,* that is, the machining or flame cutting of the mating parts into the proper shapes to facilitate welding. Based on the shape given to the edges to be welded, groove welds are classified into *square, bevel, Vee, J,* and *U* (Fig. 6.15.2). With the exception of the square, these groove welds are further subdivided into *single groove* and *double groove.* The edges of one or both plates to be groove welded are usually prepared by flame cutting, edge cutting, edge planing or arc-air gouging. The strength of a groove weld is not dependent upon the type of edge preparation, as long as the required preparation is properly executed. Groove welds are commonly used to fabricate plate girders, to make flange and column splices, to connect beam flanges to columns in FR connections (Fig. 3.2.2a), and to form built-up box beams. Groove welds must extend the full width of the plates joined, and intermittent groove welds are not permitted. Groove welds require less weld metal than fillet welds of equal strength. Also, they frequently eliminate the need for extra metal in the form of connecting plates, angles, or other structural shapes (compare Figs. 6.1.3b and c). Groove welds are generally more expensive than fillet welds because of the cost of edge preparation. Also, the butting together of sections to provide groove welds requires the cutting of members to more or less exacting lengths, while the use of lap joints with fillet welds permits larger tolerances. So, many structural connections are made by fillet welding. However, groove welds are the most economical for heavy loads, as the full strength of the base material can be easily achieved. Distortion due to welding of heavy fabricated members can be reduced by choosing butt

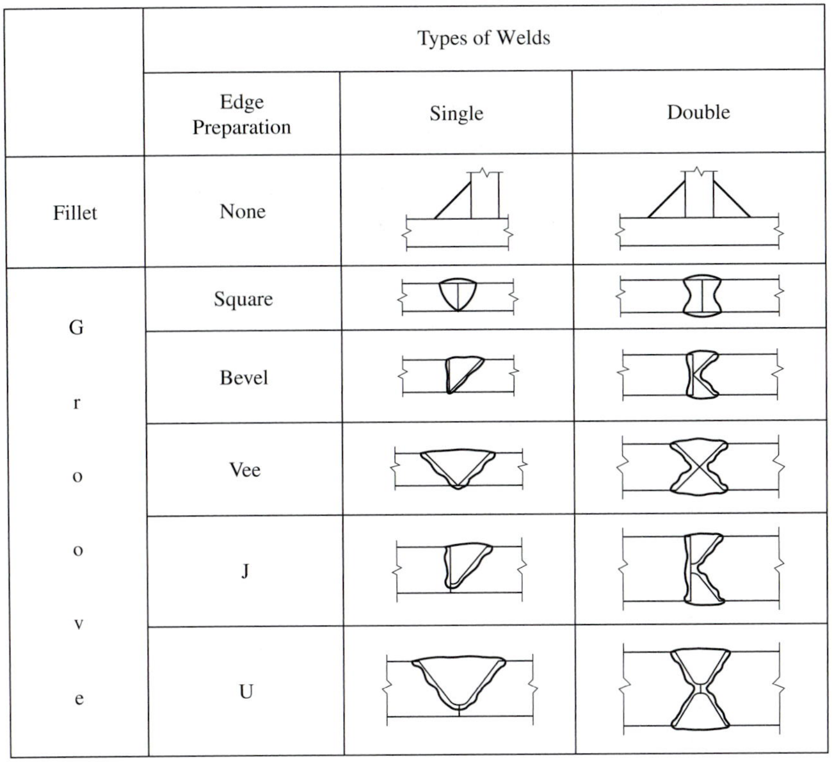

	Types of Welds		
	Edge Preparation	Single	Double
Fillet	None		
G r o o v e	Square		
	Bevel		
	Vee		
	J		
	U		

Figure 6.15.2: Types of fillet and groove welds.

welds rather than fillet welds. Groove welds are preferable to fillet welds in joints subjected to dynamic loads.

Plug and Slot Welds

Plug and *slot welds* are made by depositing weld metal in circular or slotted openings formed in one of two members being joined (Fig. 6.15.1). The openings may be partly or completely filled depending on the plate thickness. Plug and slot welds are used to transmit shear in lap joints, to prevent buckling of wide overlapping plates in compression members, and as stitch welds to join components of built-up members. Plug and slot welds are rarely used as primary welds. Instead, they are generally used to gain additional strength when there is insufficient room to place the needed length of fillet weld (Fig. 6.1.3*d* and *e*). Plug and slot welds are not permitted on A514 steel.

6.15.2   Welding Positions

The position of the electrode relative to the joint during welding affects the ease of making the weld, the size of the electrode chosen, the current required, and the thickness of each weld layer deposited in multipass welds.

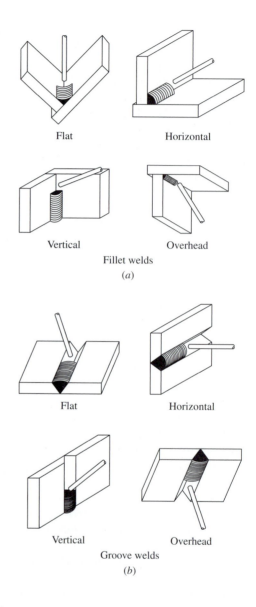

The following basic weld positions, shown in Fig. 6.15.3, are defined as follows:

1. For *flat welds,* the weld face is nearly horizontal, and welding is performed from above the joint.
2. For *horizontal welds,* the axis of the weld is horizontal. For groove welds, the face of the weld is approximately vertical; for fillet welds, the face is usually at 45° to the horizontal and vertical surfaces.
3. For *vertical welds,* the longitudinal axis of the weld is nearly vertical and welds are made by moving the electrode in the upward direction.

4. For **overhead welds,** the longitudinal axis of the weld is horizontal, the electrode is nearly vertical, and welding is performed from the underside of the joint.

In the *flat* position, the base metal provides support for the molten pool of weld metal. Therefore, this position provides for the fastest deposition rate and the most economical weld. Welding in the *horizontal* position is similar, but slightly less efficient. Welding in the *vertical* or *overhead* position requires slower deposition rates to maintain the integrity of the molten pool against the effects of gravity. As mentioned earlier, gravity is not necessary for depositing weld in place, but it does speed up the process. The globules of the molten electrodes can be forced against gravity into the overhead weld, and good welds will generally result. The vertical and overhead welding processes are slow, however, and thus the cost increases accordingly. Also, electrode diameters above $\frac{5}{32}$ in. produce weld pools with surface tension and arc forces that are unable to overcome the pull of gravity, causing the weld metal to run. Thus, it is desirable to avoid overhead welds wherever possible. The flat position is the most preferred welding position because weld metal can be deposited faster and more easily. A $\frac{5}{16}$-in. manual fillet weld, for example, may require 50 percent more time to deposit in the horizontal position than in the flat position. Vertical and overhead welds may take four times as long as the same weld made in the flat position. In modern structural fabricating shops, **jigs** and **fixtures** are used to support and position the component parts to be welded so that joints may be welded in a flat or horizontal position.

The maximum size of a fillet weld made in one pass is $\frac{3}{8}$ in. in the flat position, $\frac{5}{16}$ in. in the horizontal or overhead positions, and $\frac{1}{2}$ in. in the vertical position. For a fillet weld, the electrode ordinarily should bisect the angle between the two legs of the weld. Also, it must lean about 20° in the direction of travel. Field welding seldom permits positioning, and vertical and overhead welds often cannot be avoided. However, careful planning in the drafting room can minimize the need for such welds by arranging field welded joints for flat or horizontal welding whenever possible.

6.15.3   Types of Joints

In welded connections, the **joint** is that portion of a surface common to the two elements to be connected. There are five basic types of welded joints, based on the relative position(s) of the plates being joined: lap, butt, tee, corner, and edge joints (Fig. 6.15.4).

Lap Joints

In a **lap joint,** the plates to be connected are lapped over one another and welded together. Edges of pieces being lapped do not require any special edge preparation. They are usually flame cut or sheared. The plates can be shifted slightly to accommodate minor errors in fabrication. Plates of different

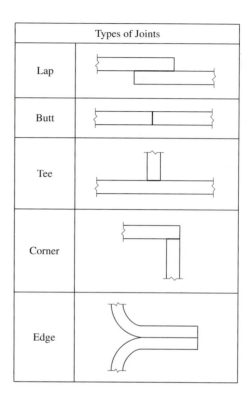

Types of Joints	
Lap	
Butt	
Tee	
Corner	
Edge	

Figure 6.15.4: Types of welded joints based on disposition of elements connected.

thicknesses can be joined easily. Only fillet, plug, and slot welds can be used in lap joints.

Butt Joints

Butt joints are used to connect structural members that are aligned in the same plane and butted together. Edges to be connected must usually be specially prepared and very carefully aligned prior to welding. Only groove welds can be used on butt joints. Edges to be joined should be of the same or nearly same thickness. Butt joints are usually made in the fabricating shop.

Tee Joints

Tee joints are used to join the end of one plate to the surface of another plate. They are used to fabricate built-up sections such as plate girders and tees from plates and to connect members such as welded hangar plates (Fig. 6.1.3g), bracket plates (Fig. 6.1.3h), and stiffeners, to main members. Either fillet welds or groove welds can be used.

Corner Joints

Corner joints are used to fabricate built-up rectangular box sections from plates. Groove welds are generally used.

Note that neither the geometry of the weld itself nor the method of edge preparation has any influence on the basic definition of the joint. For example, the tee joint could be either fillet welded or groove welded. A single bevel groove weld could be used in a butt, tee, or corner joint.

6.16   Welding Definitions and Geometry

6.16.1   Fillet Welds

The cross section of a typical fillet weld is a right triangle with equal legs, and the *size of a fillet weld* is defined as the leg size, *w* (Fig. 6.16.1). In the case of a concave or convex fillet weld, the leg size is measured by the largest right triangle which can be inscribed within the weld. This triangle is called the *diagrammatic fillet weld*. The most commonly used fillet welds increase in size by sixteenths of an in. from $\frac{1}{8}$ to $\frac{1}{2}$ in. and by eighths of an in. for sizes greater than $\frac{1}{2}$ in. The smallest practical weld size is about $\frac{1}{8}$ in. and the most economical size is probably about $\frac{5}{16}$ in. The $\frac{5}{16}$-in. weld is about the largest size that can be deposited in a single pass with the SMAW process, and the $\frac{1}{2}$ in. with the SAW process. Larger welds must be laid in multiple passes by depositing one bead of weld on top of another. Each pass must cool, and the slag must be removed, before the next pass is made. Thus, multiple pass fillet welds require appreciably more time and

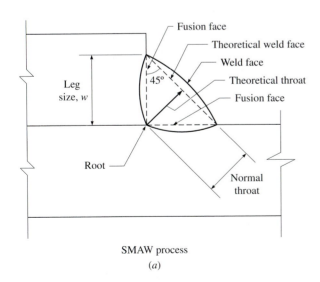

SMAW process

(*a*)

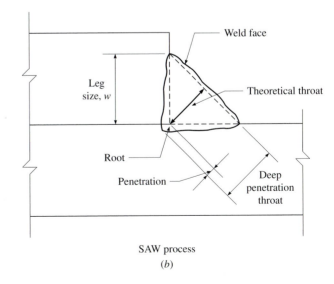

SAW process

(*b*)

Figure 6.16.1:   Fillet weld terminology.

labor, at increased cost, than a weld that can be placed with a single pass of the electrode.

6.16.2   Minimum Size of Fillet Welds

Solidified but still-hot filler metal contracts significantly as it cools to room temperature. The restraint thick-material provides to such weld metal shrinkage may result in weld cracking. Furthermore, if the weld is small, the total amount of heat at the weld is small, and if one or both parts being joined is thick, this heat is conducted away so quickly that the weld may be chilled and become very brittle. This quench effect of thick material on small welds may thus result in a loss of ductility of the weld. The weld may actually crack because of the combination of these two effects. To prevent formation of cracks and minimize distortion, specifications provide for a minimum size of weld, w_{min}. The minimum sizes of fillet welds are given in LRFDS Table J2.4 as a function of the thickness of the thicker of the two parts joined. The recommended sizes are based on experience and also provide some margin of safety for uncalculated stresses encountered during fabrication, transportation, and erection. As an illustration, if a $\frac{1}{2}$-in. plate is welded to a $\frac{7}{8}$-in. plate, it is seen from LRFDS Table J2-4 that the minimum permissible fillet-weld size is $\frac{5}{16}$-in., even if a $\frac{1}{4}$-in. weld might provide adequate strength.

Because a $\frac{5}{16}$-in. fillet weld is the largest that can be deposited in a single pass by the SMAW process, the LRFDS prescribes the $\frac{5}{16}$-in. minimum size for all plates $\frac{3}{4}$ in. or greater in thickness, but with added requirements for minimum preheat and interpass temperature. Also, while the minimum size of weld need not exceed the thickness of the thinner part, care must be taken to provide sufficient preheat for soundness of the weld. Most high-strength steel elements to be welded also require preheat.

6.16.3   Maximum Size of Fillet Welds

Even though there is no specific limitation as to the maximum size of fillet welds, the disposition of the joint material sometimes limits the maximum size of the weld that can be properly deposited and measured along the edge of a plate element. It is necessary that the weld inspector be able to identify the edge of the plate to position the weld gage. To assure this, the weld is required to be back at least $\frac{1}{16}$ in. from the corner. Thus, in such joints, the ***maximum size of a fillet weld,*** w_{max}, is determined by the edge thickness of the member along which the weld is deposited (LRFDS Section J2.2b). Along the edge of material less than $\frac{1}{4}$-in. thick, the maximum leg size of a fillet weld shall be equal to the plate thickness. Along the edge of material $\frac{1}{4}$-in. thick or more, the maximum size of weld shall be equal to the plate thickness, t_p, less $\frac{1}{16}$ in. (Fig. 6.16.2), unless noted on

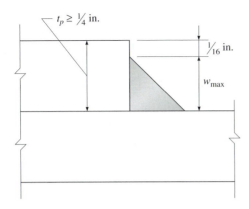

Figure 6.16.2:   Maximum size of fillet welds.

the drawing that the weld is to be built out to obtain full-throat thickness. That is:

$$w_{max} = t_p \qquad \text{for } t_p < \tfrac{1}{4} \text{ in.}$$

$$\leq t_p - \frac{1}{16} \qquad \text{for } t_p \geq \tfrac{1}{4} \text{ in.} \qquad (6.16.1)$$

The size of fillet weld actually used must be within the range of minimum and maximum weld sizes given in LRFDS Section J2.2b, and discussed in Sections 6.16.2 and 6.16.3.

6.16.4   Throat Size of Fillet Welds

The **root** of a fillet weld is the point at which the faces of the original metal pieces intersect (Fig. 6.16.1a). The part of a weld which is assumed to be effective in transferring stress is called the **throat.** For fillet welds, a line perpendicular to the theoretical weld face and passing through the root locates the weld throat. The length of this line, from the root to the theoretical weld face, is the **normal** or **theoretical throat** (Fig. 6.16.1a). It is the shortest distance from the root of the joint to the face of the diagrammatic weld. Fillet welds made by the SAW process have greater penetration into the base material compared to welds made by the SMAW process, as shown schematically in Fig. 6.16.1b. Thus for SAW, the throat extends beyond the root, into the base material, and defines the **deep penetration throat.**

Tests on fillet welds using matching electrodes have shown that the weld will fail through its effective throat before the material will fail along the weld leg. For an equal-legged fillet weld of size, w, made by the shielded metal arc welding process, the effective throat thickness, as per LRFDS

Section J2.2a, equals the normal throat. That is, for fillet welds made by the SMAW process:

$$t_e = w \sin 45° = 0.707w \qquad (6.16.2)$$

where w = leg size of a fillet weld, in.
t_e = effective throat thickness, in.
= normal throat for a fillet weld made by the SMAW process

For fillet welds made by the submerged arc process, the effective throat thickness t_e equals the deep penetration throat and may be taken as equal to the leg size, w, for $\frac{3}{8}$-in. and smaller welds. Since the added penetration is primarily due to the first pass, automatic submerged arc welds made by more than one pass can count only on the initial pass penetration for added size. Hence, for weld sizes larger than $\frac{3}{8}$ in., the effective throat thickness is equal to the theoretical throat plus 0.11 in. Thus, for fillet welds made by the SAW process:

$$t_e = w \qquad\qquad \text{for } w \le \tfrac{3}{8} \text{ in.} \qquad (6.16.3a)$$

$$= 0.707\, w + 0.11 \qquad \text{for } w > \tfrac{3}{8} \text{ in.} \qquad (6.16.3b)$$

where w = weld size, in.
t_e = effective throat thickness, in.
= deep penetration throat for a fillet weld made by SAW process

The strength of a fillet weld is in direct proportion to its throat size, and hence, its leg size. However, the volume of deposited metal, and thus the cost of the weld, increases as the square of the weld size. A $\frac{1}{2}$-in. fillet weld contains four times the volume of metal required for a $\frac{1}{4}$-in. weld of equal length, yet it is only twice as strong. So, it is often preferable to specify a long narrow weld, rather than a more costly, thick short weld, when possible. Also, the heat input is a function of the volume of weld metal deposited, so use of thin long welds to transmit a given force reduce the possibility of warping and distortion. Weld sizes of $\frac{3}{16}$, $\frac{1}{4}$, and $\frac{5}{16}$ in. are favored in practice because they can be made by a single pass of the electrode. Very large fillet welds may be so costly that it might be less expensive (even after accounting for the extra cost of edge preparation) to specify a groove weld, which generally uses a much smaller amount of weld metal than the large fillet.

6.16.5 Effective Area of Fillet Welds

When a fillet weld is started and terminated, small sections near the ends are not fully effective due to craters and stress concentrations (Fig. 6.16.3a). The gross length of a fillet weld is the distance from the crater at the starting end to the crater at the terminating end. The **effective length** of a fillet weld, L_w,

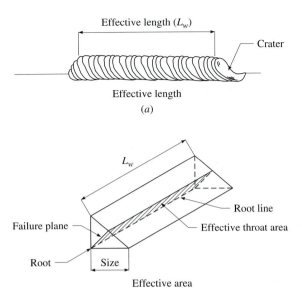

Effective length (L_w)

Crater

Effective length

(a)

L_w

Failure plane

Root line

Effective throat area

Root

Size

Effective area

(b)

Figure 6.16.3:   Effective length and effective area of a fillet weld.

is the distance from end-to-end of the full-sized fillet, measured parallel to its root line. Thus, the effective length of a fillet weld is generally taken as the gross length minus twice the nominal weld size to allow for craters. However, the designer makes no allowance for craters, as by usage, weld lengths shown on the drawings are effective weld lengths, and the welder provides the necessary additional length to account for craters.

$$L_w = L_g - 2w \qquad (6.16.4)$$

where L_g = gross length of a fillet weld
 w = leg size of the fillet weld
 L_w = effective length of the fillet weld

For curved fillet welds, the effective length is equal to the throat length, measured along the line bisecting the throat area. The effective length of a fillet weld designed to transfer a force may not be less than four times the nominal size. Otherwise, the effective leg size, w_e, of the fillet weld shall be considered equal to ¼ of the actual length, L_w. That is:

$$L_w \geq L_{w,\,min} = 4w \quad \text{or else} \quad w_e = L_w/4 \qquad (6.16.5)$$

Thus, for example, the shortest length of ¼-in. fillet weld which is permitted to be considered to transmit load is 1 in.

A plane passed through the throat and the root lines contains the **_effective throat area._** The effective area of a fillet weld, A_w, is the product of the

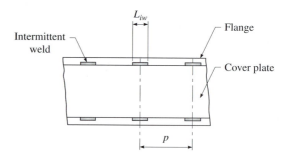

Figure 6.16.4: Intermittent fillet welds.

effective length of the weld times the effective throat thickness of the fillet weld (Fig. 6.16.3*b*). Thus:

$$A_w = L_w \, t_e \qquad (6.16.6)$$

where L_w = effective length of fillet weld, in.
t_e = effective throat thickness of the fillet weld, in.
A_w = effective area of the fillet weld, in.^{2}

6.16.6   Intermittent Fillet Welds

A required fillet weld may be provided as a continuous weld or as an intermittent weld (LRFDS Section J2.2*b*). An ***intermittent weld*** is one in which relatively short uniform lengths of fillet weld are separated by regular spaces (Fig. 6.16.4). Such welds are sometimes used when a continuous fillet weld of the smallest permitted size would provide considerably more strength than required, or where it is required to stitch together component parts of a built-up member. Intermittent welding is done by the manual (SMAW) process only, since the continual starting and stopping of the arc does not lend itself to automatic welding. Intermittent welding is not recommended for situations involving exposed structures where corrosion could be a problem, as the gaps between adjacent intermittent welds can trap moisture and initiate corrosion. Intermittent groove welds are not permitted. The effective length, L_{iw}, of any segment of intermittent fillet weld designed to transfer a force shall not be less than four times its leg size, with a minimum of $1\frac{1}{2}$ in. That is:

$$L_{iw} \geq \; \max (4w; \; 1\tfrac{1}{2} \text{ in.}) \qquad (6.16.7)$$

6.16.7   Longitudinal, Transverse, and Inclined Welds

Based on the angle between the axis of the weld and the line of the applied force, fillet welds can be classified as longitudinal, transverse, or inclined (Fig. 6.16.5). ***Longitudinal*** or ***parallel welds*** have forces applied parallel to their axis. ***Transverse welds*** have forces applied transversely or at right angles to their axis. All welds other than longitudinal and transverse welds are called ***inclined*** or ***oblique welds.***

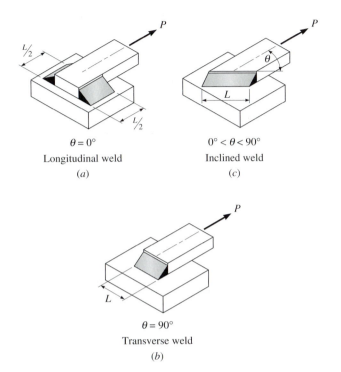

$\theta = 0°$

Longitudinal weld

(a)

$0° < \theta < 90°$

Inclined weld

(c)

$\theta = 90°$

Transverse weld

(b)

Figure 6.16.5: Longitudinal, transverse, and inclined welds.

In the case of longitudinal fillet welds, the throat is stressed only in shear. For an equal-legged longitudinal fillet weld, the maximum shear stress occurs on the 45° throat. In the case of transverse fillet welds, the throat is stressed both in shear and in tension (or compression). For an equal-legged transverse fillet weld, it could be shown that the maximum shear stress occurs on the 67.5° throat, and the maximum normal stress occurs on the 22.5° throat. Tests have shown that a transverse fillet weld is much stronger than a longitudinal fillet weld of the same size. The reason probably is due to the fact that the force on the throat of the transverse weld has both a shearing and a normal component as mentioned earlier. Therefore, it is reasonable to expect the strength of the transverse weld to be intermediate between the shearing and the tensile strengths of the weld metal, which is, in fact, actually the case. While this increased strength is ignored in LRFD Specification Section J2.4, the provisions of LRFD Specification Appendix J2.4 take into account this higher strength (see Equation 6.19.11).

6.16.8   Additional Considerations for Fillet Welds

For information on additional considerations for fillet welds, refer to Section W6.9 on the website http://www.mhhe.com/Vinnakota.

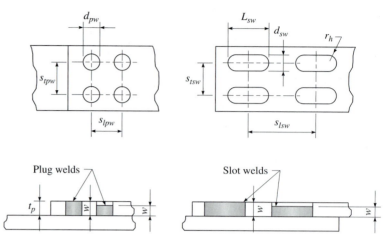

Figure 6.16.6:   Size requirements for plug and slot welds.

6.16.9 Plug and Slot Welds

The use of plug and slot welds is limited to transferring shear loads in joint planes parallel to the faying surfaces. Plug and slot welds should not be subjected to tensile stresses or to stress reversal. Size requirements for plug and slot welds are given in LRFDS Section J2.3 and summarized in Fig. 6.16.6.

The thickness of plug or slot welds in material up to $\frac{5}{8}$-in. thick must equal the plate thickness. In material over $\frac{5}{8}$-in. thick, the weld thickness shall be at least one-half the thickness of the material but not less than $\frac{5}{8}$ in. Thus:

$$w = t_p \qquad\qquad\qquad \text{for } t_p \leq \tfrac{5}{8} \text{ in.} \qquad (6.16.8a)$$

$$w \geq \max(t_p/2, \tfrac{5}{8} \text{ in.}) \qquad \text{for } t_p > \tfrac{5}{8} \text{ in.} \qquad (6.16.8b)$$

where t_p = thickness of the plate material, in.
         w = thickness of plug or slot weld, in.

The diameter of the holes for plug welds shall be not less than the thickness of the part containing it plus $\frac{5}{16}$ in. The hole diameter shall not be greater than $2\frac{1}{4}$ times the thickness of the weld metal, w, nor greater than the minimum diameter plus $\frac{1}{8}$ in. The diameter used is rounded to an odd $\frac{1}{16}$ in. to permit use of standard structural punches in fabricating shops. The minimum center-to-center spacing of plug welds shall be four times the diameter, d_{pw}, of the hole. Thus:

$$d_{pw,\,min} = t_p + \tfrac{5}{16}; \qquad d_{pw,\,max} = \min[d_{pw,\,min} + \tfrac{1}{8} \text{ in.}; \, 2\tfrac{1}{4}w] \quad (6.16.9)$$

$$d_{pw,\,min} \leq d_{pw} \leq d_{pw,\,max}; \qquad s_{lpw} \geq 4d_{pw}; \quad s_{tpw} \geq 4d_{pw}$$

where d_{pw} = diameter of plug weld, in.
$\ s_{tpw}$ = transverse spacing of plug welds, in.
$\ s_{lpw}$ = longitudinal spacing of plug welds, in.

The width of a slot for a slot weld may not be less than the thickness of the part containing it plus $\frac{5}{16}$ in. Also, the width must be less than or equal to $2\frac{1}{4}$ times the weld thickness, w. The value selected is made an odd multiple of $\frac{1}{16}$ in. The maximum length permitted for a slot weld is 10 times the weld thickness. The ends of the slot shall be semicircular or shall have the corners rounded to a radius not less than the thickness of the part containing it, except those ends which extend to the edge of the part. The minimum center-to-center spacing of lines of slot welds in a direction transverse to their length shall be four times the width of the slot. The minimum center-to-center spacing of slots in a longitudinal line shall be two times the length of the slot. Thus:

$$d_{sw,min} = t_p + \tfrac{5}{16}; \quad d_{sw,max} = \min[d_{sw,min} + \tfrac{1}{16} \text{ in.}; 2\tfrac{1}{4}w]$$

$$d_{sw,min} \leq d_{sw} \leq d_{sw,max} \tag{6.16.10}$$

$$L_{sw} \leq 10w; \quad r_h \geq t_p$$

$$s_{tsw} \geq 4d_{sw}; \quad s_{lsw} \geq 2L_{sw}$$

where d_{sw} = width of slot for a slot weld, in.
$\ L_{sw}$ = length of slot for a slot weld, in.
$\ s_{tsw}$ = transverse (center-to-center) spacing of slot welds, in.
$\ s_{lsw}$ = longitudinal (center-to-center) spacing of slot welds, in.

The ***effective shearing area of a plug or slot weld***, A_w, is the nominal cross-sectional area of the hole or slot in the plane of the faying surface. Thus:

$$A_w = \frac{\pi d_{pw}^2}{4} \qquad \text{for a plug weld} \tag{6.16.11a}$$

$$A_w = (L_{sw} - 0.22d_{sw})d_{sw} \quad \text{for a slot weld with semicircular ends} \tag{6.16.11b}$$

The limitations for the maximum size of plug and slot welds are imposed, to limit the detrimental shrinkage that occurs around these types of welds when they exceed certain sizes. For this reason, when the holes are larger than the limits specified above, fillet welds around the periphery of the hole or slot are specified, rather than plug or slot welds. Fillet welds placed around the inside of a hole or slot should not to be considered as plug or slot welds. In the case of such fillet welds, the effective shearing area is the product of the effective throat thickness and the effective length—the length of the center line of the weld measured along the center of the plane through the throat.

6.17 Welding Symbols

The American Welding Society has standardized a system of welding symbols [AWS, 1993]. Use of this system in design and shop drawings will assure that the correct welding instructions are transmitted from the designer to the fabricator and welder without ambiguity and in concise form.

The basic elements of the system are symbols graphically depicting the different types of welds (Fig. 6.17.1). Any combination of these symbols can be built up to represent any given situation found in a welded joint. Another basic symbol is a horizontal line, called the *reference line,* with information on the type, size, length, and pitch of the weld placed on or below it. At one end of the reference line is an inclined arrow pointing to the joint or member to be welded. The side of the joint to which the arrow points is called the *near side,* and the other side of the joint is called the *far side.* The arrow for a bevel or J groove weld points with a definite break toward the element to be grooved.

If the symbol for the weld is below the reference line, the weld is on the arrow side of the joint, that is, the part of the joint that the arrow is touching. If the symbol is above the reference line, the weld is on the other side of the joint, which may or may not be hidden from view in the drawing. If both sides of the joint are to be welded, the symbols should be shown on both sides of the reference line. As shown in Fig. 6.17.1, a right triangle with the vertical on the left side is used to indicate a fillet weld. A number to the left of the triangle indicates the size of the weld in inches. A number to the right of the triangle indicates the length of the weld; the absence of a number to the right specifies a weld that is to be the full length of the member. When two numbers such as 4-12 are used to the right of the triangle, this calls for intermittent fillets 4-in. long spaced 12 in. center-to-center. "Both side" welds of the same type are the same size unless otherwise shown. Note that welding symbols apply between abrupt changes in geometry of the joint, unless otherwise designated. All welds are continuous unless otherwise shown.

The specification controlling the type of electrode and other welding specifications or instructions to the welder may be placed in the tail of the arrow as shown in Fig. 6.17.1. The arrow tail may be omitted when a reference is not needed. Dimensions of weld sizes, increment lengths, and spacings are in inches. A circle at the bend in the reference line is an instruction to weld continuously all around the periphery of the joint. A *flag* placed at the bend in the reference line indicates that the weld is to be made in the field. Whenever a particular connection is used many times throughout a structure, it may only be necessary to show the detail once and label it as *typical (typ.).*

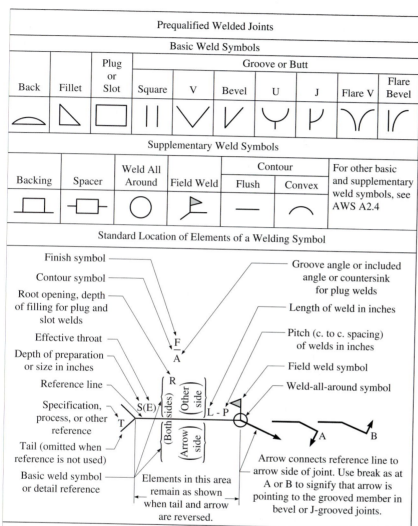

Prequalified Welded Joints									
Basic Weld Symbols									
		Plug or Slot	Groove or Butt						
Back	Fillet		Square	V	Bevel	U	J	Flare V	Flare Bevel

Supplementary Weld Symbols

Backing	Spacer	Weld All Around	Field Weld	Contour		For other basic and supplementary weld symbols, see AWS A2.4
				Flush	Convex	

Standard Location of Elements of a Welding Symbol

Finish symbol

Contour symbol

Root opening, depth of filling for plug and slot welds

Effective throat

Depth of preparation or size in inches

Reference line

Specification, process, or other reference

Tail (omitted when reference is not used)

Basic weld symbol or detail reference

Groove angle or included angle or countersink for plug welds

Length of weld in inches

Pitch (c. to c. spacing) of welds in inches

Field weld symbol

Weld-all-around symbol

F
A
R
S(E)
T
L - P
(Both sides)
(Other side)
(Arrow side)

Elements in this area remain as shown when tail and arrow are reversed.

Arrow connects reference line to arrow side of joint. Use break as at A or B to signify that arrow is pointing to the grooved member in bevel or J-grooved joints.

A   B

Note:

Size, weld symbol, length of weld, and spacing must read in that order, from left to right, along the reference line. Neither orientation of reference nor location of the arrow alters this rule.

The perpendicular leg of ⊿, V, ⊬, ⏀, weld symbols must be at left.

Arrow and other side welds are of the same size unless otherwise shown. Dimensions of fillet welds must be shown on both the arrow side and the other side symbol.

The point of the field weld symbol must point toward the tail.

Symbols apply between abrupt changes in direction of welding unless governed by the "all around" symbol or otherwise dimensioned.

These symbols do not explicitly provide for the case that frequently occurs in structural work, where duplicate material (such as stiffeners) occurs on the far side of a web or gusset plate. The fabricating industry has adopted this convention: that when the billing of the detail material discloses the existence of a member on the far side as well as on the near side, the welding shown for the near side shall be duplicated on the far side.

Figure 6.17.1: Basic weld symbols [LRFDM, 2001].

Source: From LRFD Manual. Copyright ©

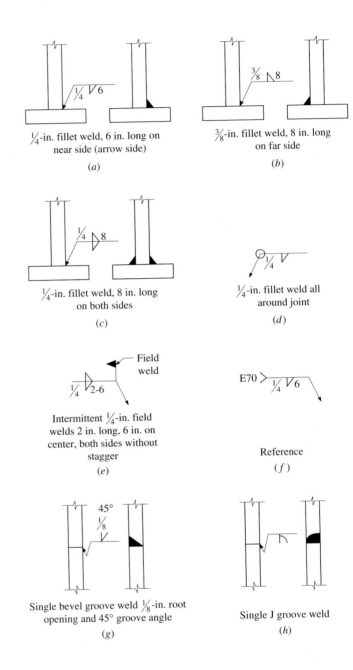

Figure 6.17.2:   Sample examples of weld symbols.

The student designer should devote some time to becoming familiar with the welding symbols shown in Fig. 6.17.1. To further clarify the use of these symbols, several examples of welds are shown in Fig. 6.17.2, together with an explanation of each. For additional examples and clarification see [AISC, 2002; Blodgett, 1966].

6.18   Behavior of Fillet Welds

For information on the behavior and load-deformation response of fillet weld elements, refer to Section W6.10 on the website http://www.mhhe.com/ Vinnakota.

6.19   Design Strength of Welds

The design strength of welds is determined in accordance with LRFDS Sections J2, J4, and J5. Two limit states, namely, the limit state of weld-metal strength and the limit state of base-metal strength, must be checked following LRFDS Table J2.5. From LRFDS Section J2.4, the weld metal design strength is:

$$R_{dw} = \phi F_w A_w \qquad (6.19.1)$$

where R_{dw} = design strength of the weld corresponding to the limit state of the weld metal failure
$\quad\quad A_w$ = effective cross-sectional area of the weld
$\quad\quad \phi$ = resistance factor
$\quad\quad F_w$ = nominal strength of the weld electrode material

The design strength of the base material is:

$$R_{dBM} = \phi F_{BM} A_{BM} \qquad (6.19.2)$$

where R_{dBM} = design strength of the weld corresponding to the limit state of the base material failure
$\quad\quad \phi$ = resistance factor
$\quad\quad A_{BM}$ = cross-sectional area of the base material
$\quad\quad F_{BM}$ = nominal strength of the base material

Design strength of the weld is given by:

$$R_d = \min[R_{dw}, R_{dBM}] \qquad (6.19.3)$$

More explicit relations for R_{dw} and R_{dBM} are given in the following sections for fillet welds and for slot and plug welds.

6.19.1   Design Strength of Fillet Welds

It is the usual practice to always consider the force on a fillet weld as a shear on the throat, irrespective of the direction of the applied load relative to the throat. Therefore, the limit state of weld metal for a fillet weld always corresponds to shear fracture through the throat of the fillet with the resistance factor taken as 0.75. The design strength of weld metal for a fillet weld is therefore given by:

$$R_{dw} = 0.75(0.60F_{EXX})t_e L_w \qquad (6.19.4)$$

When the load is in the same direction as the weld axis, the base material must also be investigated for shear capacity. The design shear rupture strength of the adjacent base material is given by LRFDS Section J4.1 as:

$$R_{dBM1} = 0.75(0.60 F_{uBM}) t_p L_w = 0.45 F_{uBM} t_p L_w \qquad (6.19.5a)$$

where t_p = thickness of base material along which weld is placed
    F_{uBM} = ultimate tensile stress of base material

The design shear yielding strength of the adjacent base material is given by LRFDS Section J5.3 as:

$$R_{dBM2} = 0.90 (0.60 F_{yBM}) t_p L_w = 0.54 F_{yBM} t_p L_w \qquad (6.19.5b)$$

where F_{yBM} = yield stress of the base material

The base material design shear strength may be written as:

$$R_{dBM} = \min [R_{dBM1}, R_{dBM2}] \qquad (6.19.6)$$

The design shear strength of a fillet weld is therefore given by:

$$R_d = \min [R_{dw}, R_{dBM}] \qquad (6.19.7)$$

Note that, for all structural steels other than the A514 quenched and tempered alloy steels, Eq. 6.19.5b results in a lower design strength for the base material.

Design Shear Strength of Unit-Length Fillet Weld

In most fillet welded joint problems, it is advantageous to work with the strength per unit length of weld. We have, with the help of Eq. 6.19.4:

$$R_{dw} = [0.75(0.6 F_{EXX}) t_e] L_w = W_d L_w \qquad (6.19.8)$$

with

$$W_d = 0.75(0.60 F_{EXX}) t_e = 0.45 F_{EXX} t_e \qquad (6.19.9)$$

where W_d = design shear strength of a unit-length fillet weld, kli
    t_e = effective throat thickness of the fillet weld, in. (Eqs. 6.16.2 and 6.16.3)

For example, the design shear strength of a unit-length, $\frac{1}{16}$-in. fillet weld produced by the shielded metal arc welding process using E70 electrodes, with the help of Eqs. 6.19.9 and 6.16.2, is:

$$W_{d(1/16)} = 0.45(70) t_e = 31.5 t_e = 31.5 (0.707w) = 22.27w$$

$$= 22.27(1/16) = 1.392 \text{ kli}$$

As the strength of a SMAW fillet weld is proportional to the leg size w, we have for a E70 fillet weld:

$$W_d = 1.392D \qquad (6.19.10)$$

where W_d = design shear strength of a unit length E70 fillet weld of size w, kli

D = number of sixteenths of an inch in the fillet weld size ($w = \dfrac{D}{16}$)

For example, the design shear strength of a ⅜-in. E70 fillet weld can be conveniently determined as $1.392(6) = 8.35$ kli. Values of design shear strengths, W_d, of unit-length fillet welds for various electrode strengths are given in Table 6.19.1 for SMAW welds, and in Table 6.19.2 for SAW welds.

Alternative Design Strength

LRFDS Appendix J2.4 provides an alternative approach to determine the design strength of fillet welds as a function of angle of loading, θ, in lieu of the constant design strength given in LRFDS Table J2.5 (see also Section 6.18).

TABLE 6.19.1

Design Shear Strength of Unit-Length SMAW Fillet Weld, W_d (kli)

w	t_e			F_{EXX}	(ksi)			L_w^*
(in.)	(in.)	60	70	80	90	100	110	(in.)
$\dfrac{1}{8}$	0.088	2.38	2.77	3.17	3.56	3.96	4.36	12.5
$\dfrac{3}{16}$	0.133	3.58	4.18	4.77	5.37	5.97	6.56	18.8
$\dfrac{1}{4}$	0.177	4.77	5.57	6.36	7.16	7.95	8.75	25.0
$\dfrac{5}{16}$	0.221	5.97	6.96	7.95	8.95	9.94	10.9	31.3
$\dfrac{3}{8}$	0.265	7.16	8.35	9.54	10.7	11.9	13.1	37.5
$\dfrac{7}{16}$	0.309	8.35	9.74	11.1	12.5	13.9	15.3	43.8
$\dfrac{1}{2}$	0.354	9.54	11.1	12.7	14.3	15.9	17.5	50.0

F_{EXX} = minimum tensile strength of weld material, ksi; w = nominal size of weld, in.
W_d = $0.75t_e\,(0.60F_{EXX})$; t_e = effective throat = $0.707w$
   = design shear strength of equal leg fillet weld per in. length, kli
L_w^* = maximum length of end-loaded fillet weld below which the effective length equals the actual length (LRFDS Section J2.2b), in.

TABLE 6.19.2

....

Design Shear Strength of Unit-Length SAW Fillet Weld, W_d (kli)

w	t_e			F_{EXX} (ksi)				L_w^*
(in.)	(in.)	60	70	80	90	100	110	(in.)
$\frac{1}{8}$	0.125	3.38	3.94	4.50	5.06	5.63	6.19	12.5
$\frac{3}{16}$	0.188	5.06	5.91	6.75	7.59	8.44	9.28	18.8
$\frac{1}{4}$	0.250	6.75	7.88	9.00	10.1	11.3	12.4	25.0
$\frac{5}{16}$	0.313	8.44	9.84	11.3	12.7	14.1	15.5	31.3
$\frac{3}{8}$	0.375	10.1	11.8	13.5	15.2	16.9	18.6	37.5
$\frac{7}{16}$	0.419	11.3	13.2	15.1	17.0	18.9	20.8	43.8
$\frac{1}{2}$	0.463	12.5	14.6	16.7	18.8	20.9	22.9	50.0

....

t_e = effective throat, in.; w = nominal size of weld, in.
  = w for $w \leq 3/8$ in.; F_{EXX} = minimum tensile strength of weld
  = $0.707 w + 0.11$ for $w > 3/8$ in.   material, ksi
W_d = $0.75t_e (0.6 F_{EXX})$
  = design shear strength of equal leg fillet weld per in. length, kli
L_w^* = maximum length of end-loaded fillet weld below which the effective length equals the actual length (LRFDS Section J2.2b), in.

Thus, for a linear fillet weld loaded in-plane through its center of gravity, the design strength is:

$$R_{dw} = 0.75(0.6\, F_{EXX}\, t_e\, L_w)\,[1.0 + 0.50\,(\sin\theta)^{1.5}] \quad (6.19.11)$$

where R_{dw} = design strength of a fillet weld, as a function of load inclination
       t_e = effective throat thickness, in.
       L_w = effective length of weld element, in.
       θ = angle of loading measured from the weld longitudinal axis

Maximum Effective Fillet Weld Size, w_{BM}^*

For fillet welds loaded in longitudinal shear only, a **maximum effective fillet weld size,** w_{BM}^*, above which weld strength is limited by the strength of the base metal rather than the weld metal, can be defined. That is, no useful purpose will be served by using a bigger size weld than w_{BM}^*.

For information on maximum effective fillet weld size, refer to Section W6.11 on the website http://www.mhhe.com/Vinnakota.

6.19.2   Design Strength of Slot and Plug Welds

As mentioned earlier, the use of plug and slot welds is restricted to the transfer of shear forces in joint planes parallel to the faying surfaces. The design shear strength of these welds is given by:

$$R_{dw} = \phi \, (0.6 F_{\text{EXX}}) \, A_w \qquad (6.19.12)$$

where R_{dw} = design shear strength of a plug or slot weld, kips
ϕ = resistance factor (= 0.75)
F_{EXX} = minimum specified tensile strength of the weld metal, ksi
A_w = effective shearing area of plug or slot weld, in.^{2} (Eq. 6.16.11)

6.19.3   Design Strength of Groove Welds

For information on groove weld definitions and their design strength, refer to Section W6.12 on the website http://www.mhhe.com/Vinnakota.

6.20   Design Aids for Welding

Matching filler metal requirements are given in AWS [2000] (see Table 6.14.1). Note that for the commonly used grades of steel and SMAW process, only two electrodes need be considered: E70 electrodes with steels that have a yield stress less than 60 ksi and E80 electrodes with steels that have a yield stress of 60 ksi or 65 ksi. Minimum sizes of fillet welds to be used are given in LRFDS Table J2.4, as a function of the thickness of the thicker of the two parts joined. The maximum weld size along the edge of material is given by Eq. 6.16.1. Weld sizes of $\frac{3}{16}$, $\frac{1}{4}$, and $\frac{5}{16}$ in. are favored in practice because they can be made by a single pass of the electrode. Pound per pound, weld metal is the most expensive material in a steel structure. So, the designer should aim to minimize the volume (size by length) of the required weld deposit. The effective throat thickness of equal legged fillet welds is given by Eqs. 6.16.2 and 3. The design strength of fillet welds is given by Eqs. 6.19.4, 6.19.5, and 6.19.7. The design strength of 1-in.-length, equal-leg, SMAW fillet weld of $\frac{1}{16}$-in. leg size is given by Eq. 6.9.10. As mentioned earlier, transverse welds are 50 percent stronger than longitudinal welds. LRFDS Appendix J2.4 gives an alternative design approach that allows this additional strength to be taken advantage of (see Eq. 6.19.11).

There are three types of lapped welded joints used to transmit tensile forces from one plate to another. The first type uses two longitudinal fillet welds only, the second uses two transverse fillet welds only, and the third uses two longitudinal welds and a transverse weld.

LRFDS Section J1.7 requires that connections providing design strength be designed to support a factored load not less than 10 kips, except for lacing, sag rods, or girts.

EXAMPLE 6.20.1

Effective Throat Thickness

Determine the effective throat thickness of a:

a. $\frac{5}{16}$-in. fillet weld produced by SMAW process.

b. $\frac{5}{16}$-in. fillet weld produced by SAW process.

c. $\frac{7}{16}$-in. fillet weld produced by SAW process.

Solution

a. SMAW process. Weld size, $w = \frac{5}{16}$ in.
From Eq. 6.16.2, effective throat thickness:

$$t_e = 0.707w = 0.707\left(\frac{5}{16}\right) = 0.221 \text{ in.} \qquad \text{(Ans.)}$$

b. SAW process. Weld size, $w = \frac{5}{16}$ in. $< \frac{3}{8}$ in.
From Eq. 6.16.3a, effective throat thickness:

$$t_e = w = \frac{5}{16} = 0.313 \text{ in.} \qquad \text{(Ans.)}$$

c. SAW process. Weld size, $w = \frac{7}{16}$ in. $> \frac{3}{8}$ in.
From Eq. 6.16.3b, effective throat thickness:

$$t_e = 0.707w + 0.11 = 0.707\left(\frac{7}{16}\right) + 0.11 = 0.419 \text{ in.} \qquad \text{(Ans.)}$$

EXAMPLE 6.20.2

Strength of Fillet Welds

Determine the design shear strength of a 4-in.-long, $\frac{5}{16}$ in. fillet weld. Assume SMAW process and E70 electrodes. Assume that the applied load passes through the center of gravity of the weld. The weld is: (*a*) a longitudinal weld, (*b*) a transverse weld, (*c*) an oblique weld, with the load inclined at 30° with the axis of the weld. Use: (1) LRFDS Table J2.5; (2) LRFDS Appendix J2.4.

Solution

Weld size, $w = \frac{5}{16}$ in.
Effective length, $L_w = 4.0$ in.

SMAW process. From Eq. 6.16.2 effective throat thickness:

$$t_e = 0.707w = 0.707\left(\frac{5}{16}\right) = 0.221 \text{ in.}$$

E70 electrodes. So, $F_{EXX} = 70.0$ ksi

As no details are given, assume that the base material does not control the design of weld.

1. Strength based on LRFDS Table J2.5

   In this approach, the design strength of the weld is independent of the orientation of the applied load. From Eq. 6.19.4, the design strength of the fillet weld is

   $$R_{dw} = 0.45F_{EXX}\, t_e\, L_w = 0.45(70.0)(0.221)(4.0)$$
   $$= 27.9 \text{ kips} \qquad\qquad \text{(Ans.)}$$

2. Strength based on LRFDS Appendix J2.4

   The design strength of a linear weld loaded in-plane through the center of gravity using Eq. 6.19.11 (or from LRFDS Appendix J2.4) is:

   $$R_{d\theta} = 0.45F_{EXX}\, t_e\, L_w[1.0 + 0.50 \sin^{1.5}\theta]$$
   $$= 0.45(70)(0.221)(4.0)[1.0 + 0.50 \sin^{1.5}\theta]$$
   $$= 27.85\,[1.0 + 0.50 \sin^{1.5}\theta]$$

   where θ is the inclination of the load measured from the weld axis.

   a. Longitudinal weld

      For longitudinal weld, $\theta = 0 \rightarrow \sin\theta = 0.0$

      $$R_{d(\theta=0)} = 27.85(1.0 + 0.0) = 27.9 \text{ kips} \qquad \text{(Ans.)}$$

   b. Transverse weld

      For transverse weld, $\theta = 90.0 \rightarrow \sin\theta = 1.0$

      $$R_{d(\theta=90°)} = 27.85[1.0 + 0.50(1.0^{1.5})] = 41.8 \text{ kips} \qquad \text{(Ans.)}$$

   c. Oblique weld

      For an oblique weld with $\theta = 30° \rightarrow \sin\theta = 0.50$

      $$R_{d(\theta=30°)} = 27.85[1.0 + 0.50(0.50^{1.5})] = 32.8 \text{ kips} \qquad \text{(Ans.)}$$

   Observe that the transverse weld is 50 percent stronger than the longitudinal weld and the oblique weld 17.7 percent. Also note that the design method (traditionally used) given in LRFDS Table J2.5 ignores this additional strength.

EXAMPLE 6.20.3

Plug Weld

Determine the design strength of a plug weld connecting a $\frac{1}{2}$-in. plate by SMAW process using E70 electrodes.

Solution

Thickness of the plate, $t_p = \frac{1}{2}$ in. $< \frac{5}{8}$ in.
So, from Eq. 6.16.8, thickness of plug weld, $w = t_p = \frac{1}{2}$ in.
If d_{pw} is the diameter of the hole, from Eqs. 6.16.9:

$$d_{pw,\,min} = t_p + \frac{5}{16} = \frac{1}{2} + \frac{5}{16} = \frac{13}{16} \text{ in.}$$

$$d_{pw,\,max} = \min\left[d_{pw,min} + \frac{1}{8}, 2\frac{1}{4}w\right]$$

$$= \min\left[\frac{13}{16} + \frac{1}{8}, \frac{9}{4}\left(\frac{1}{2}\right)\right] = \min\left[\frac{15}{16}, \frac{18}{16}\right] = \frac{15}{16} \text{ in.}$$

Select the smallest size, namely, $d_{pw} = \frac{13}{16}$ in.

Shear area, $A_w = \frac{\pi}{4} d_{pw}^2 = \frac{\pi}{4}\left(\frac{13}{16}\right)^2 = 0.519 \text{ in.}^2$

From Eq. 6.19.12, design strength of the plug weld is:

$$R_{dw} = 0.45 F_{EXX}\, A_w$$

$$= 0.45\,(70.0)(0.519) = 16.3 \text{ kips} \qquad\qquad \text{(Ans.)}$$

6.21   Pins

As mentioned earlier, a *pin* is a short, cylindrically shaped beam connecting two or more plates. Pin connections offer no resistance to rotation and can transfer forces in any direction perpendicular to the axis of the pin. Pin connections are presently used in the design of two-hinged and three-hinged arches in buildings and bridges. Pins are also often used in bridge bearing supports to permit end rotation. Also, tension rods used as diagonal bracing in building structures are often provided with a clevis at each end (Fig. 3.3.10a), connected to a gusset plate by means of a small size pin (usually a turned bolt). Dimensions and design strengths of clevises are given in LRFDM Table 15-3, while compatibility of clevises with various rods and pins is given in LRFDM Table 15-4. The design strength of available clevises varies from 8.75 kips for a Number 2 clevis to 338 kips for a Number

8 clevis. Unlike a bolted joint, a pin connection has only one connector and hence no redundancy. So, a low resistance factor, $\phi = 0.5$, is used in the determination of the design strengths tabulated. Pins range from 2 to 10 in. in diameter. The loading on the pin consists of the forces in the connected plates and therefore may be applied from various directions defined by the center lines of the members connected by the pin. It is customary to resolve all loads into the vertical and horizontal planes and to design the pin for their resultant values. Bending moments in the pin may be conservatively evaluated on the assumption that forces from the connected plates are concentrated at the centers of their thicknesses. The design of the pin and the design of the connected members are interdependent, and hence their design is by trial and error.

References

6.1 AISC[2002]: *Detailing for Steel Construction,* AISC, Chicago.

6.2 AWS[2000]: *Structural Welding Code—Steel (AWS D1.1:2000),* 17th ed., American Welding Society, Miami, FL.

6.3 AWS[2001]: *Welding Handbook,* 9th ed., vol. 2 and 3, American Welding Society, Miami, FL.

6.4 AWS[1991]: "Specification for Carbon Steel Electrodes for Shielded Metal Arc Welding," AWS A5.1-91.

6.5 AWS[1993]: "Symbols for Welding, Brazing and Nondestructive Examination (A2.4-93)," American Welding Society, Miami, FL.

6.6 AWS[1996]: "Specification for Low-Alloy Steel Covered Arc Welding Electrodes," AWS A5.5-96.

6.7 AWS[1997]: "Specification for Low-Alloy Steel Electrodes and Fluxes for Submerged Arc Welding," AWS A5.23-97.

6.8 AWS[1998]: "Specification for Carbon Steel Electrodes and Fluxes for Submerged Arc Welding," AWS A5.17-98.

6.9 Bahia, C. S. and Martin, L. H. [1981]: "Experiments on Stressed and Unstressed Bolt Groups Subject to Torsion and Shear," *Proceedings Joints in Structural Steelwork,* Teeside Polytechnic, April, John Wiley & Sons, NY.

6.10 Blodgett, O. W. [1966]: *Design of Welded Structures,* James F. Lincoln Welding Foundation, Cleveland, Ohio.

6.11 Frank, K. H. and Yura, J. A. [1981]: "An Experimental Study of Bolted Shear Connections," FHWA/RD-81/148, December.

6.12 Kim, H. J., and Yura, J. A. [1996]: "The Effect of End Distance on the Bearing Strength of Bolted Connections," PMFSEL Report No. 96-1, University of Texas at Austin.

6.13 Kulak, G. L., Fisher, J. W., and Struik, J. H. A. [1987]: *Guide to Design Criteria for Bolted and Riveted Joints,* 2nd ed., John Wiley & Sons, New York, NY.

6.14 Lewis, B. E., and Zwernemann, F. J. [1996]: *Edge Distance, Spacing, and Bearing in Bolted Connections,* Oklahoma State University, July.

6.15 RCSC[1994]: *Load and Resistance Factor Design Specification for Structural Joints Using ASTM A325 or A490 Bolts,* AISC, Chicago, IL, June.

6.16 RCSC[2000]: *Specification for Structural Joints Using ASTM A325 or A490 Bolts,* AISC, Chicago, IL, June (also Part 16.4 of the LRFD Manual 2001).

6.17 Wallaert, J. J. and Fisher, J. W. [1965]: "Shear Strength of High-Strength Bolts," *Journal of the Structural Division,* ASCE, vol. 91, ST3, June, pp. 99–125.

6.18 Yura, J. A., Frank, K. H., and Polyois, D. [1987]: "High-Strength Bolts for Bridges," PMFSEL Report No. 87-3, University of Texas at Austin, May.

P R O B L E M S

P6.1. A 4 by $\frac{5}{8}$ in. tension member is connected to a $\frac{5}{8}$-in.-thick gusset plate, by three bolts as shown in Figure P6.1. The bolts are $\frac{3}{4}$-in.-diameter A307 bolts, and A36 steel is used for both the member and the gusset. Assume gas cut edges.

(*a*) Check the spacing and edge distances for compliance with the LRFDS.

(*b*) Calculate the design shear strength of the connectors.

(*c*) Calculate the design bearing strength of the connectors.

(*d*) Calculate the design strength of the connectors.

P6.2. Repeat Problem P6.1 if 1-in.-diameter A307 bolts are used instead.

P6.3. Determine the design strength of the connection shown in Figure P6.3, based on shear and bearing of the bolts. The bolts are $\frac{7}{8}$-in.-diameter A325-N type, and A572 Grade 42 steel is used for both the member and the gusset.

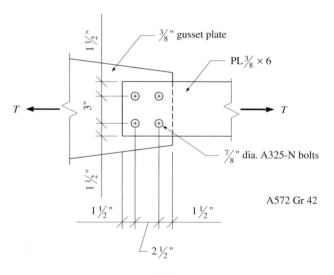

P6.1

P6.3

P6.4. Repeat Problem P6.3 if the bolts are of A490-X type.

P6.5. A 1-in.-thick plate tension member is spliced using two, $\frac{5}{8}$-in.-thick splice plates as shown in Fig. 6.5. The connection is made with eight 1-in.-diameter A490-N type bolts. The main plate and splice plates are of A514 steel. Determine the design strength of the connectors based on the limit states of bearing and bolt shear. Assume that the edges of the plates are sheared edges.

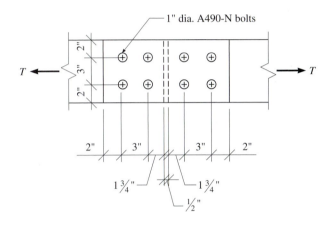

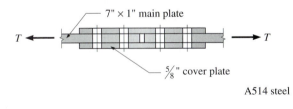

P6.5

P6.6. Repeat Problem P6.5 if all of the steel is A572 Grade 65.

P6.7. A $\frac{5}{8}$-in.-thick plate tension member is connected to a $\frac{5}{8}$-in.-thick gusset plate using a single line of $\frac{3}{4}$-in.-diameter bolts. The plates are of A242 Grade 50 steel. The connection must resist a service dead load of 48 kips and a service live load of 80 kips. Assume that adequate bolt spacing and end distances are provided and that bearing strength is controlled by the limit state of the ovalization of bolt hole. Consider the limit states of bolt shear and bolt bearing only, and determine how many bolts are required, using:

(a) A307 bolts.
(b) A325-X type bolts.
(c) A490-X type bolts.

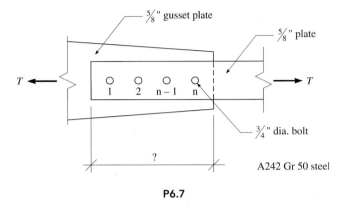

P6.7

P6.8. A tension member consisting of two angles 5″ by 3″ by $\frac{7}{16}$″ with long legs back-to-back is connected to a $\frac{5}{8}$-in.-thick gusset plate, as shown in Fig. P6.8. The connection must resist a dead load of 48 kips, a roof live load of 20 kips, and a live load of 80 kips. Determine the number of bolts required. Assume $\frac{7}{8}$-in.-diameter A325-N type bolts.

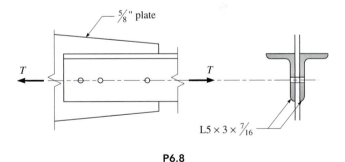

P6.8

P6.9. Repeat Problem P6.8 if the architect requires the shortest size for connection using $\frac{7}{8}$-in.-diameter high-strength bolts.

P6.10. Determine the service load tension capacity of the connection of Fig. P6.10, for the case assigned by the instructor. The diameter of the bolt, d, the type of bolt, and the type of steel used for the main and cover plates are given on the following page.

	Steel		d	Type	% of Load	
	Main	Cover			D	L
1	A36	A36	$\frac{3}{4}$	A325-X	20	80
2	A572 Gr 65	A572 Gr 65	$\frac{7}{8}$	A490-X	40	60
3	A572 Gr 50	A572 Gr 50	$\frac{7}{8}$	A325-N	30	70
4	A572 Gr 50	A36	$\frac{7}{8}$	A325-SC	30	70
5	A36	A36	$\frac{3}{4}$	A325-SC	20	80
6	A572 Gr 50	A36	$\frac{7}{8}$	A490-SC	40	60

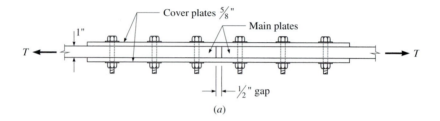

(a)

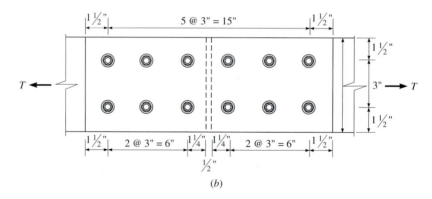

(b)

P6.10

P6.11. Determine the design strength of the welded connection shown in Fig. P6.11. A36 steel is used for both the tension member and the gusset plate. The welds are $\frac{1}{4}$-in.-fillet welds using E70 electrodes.

P6.12. Determine the maximum factored load that can be applied to the welded joint shown in Fig. P6.12. The 8 by $\frac{1}{2}$ in. plate member and the $\frac{1}{2}$-in. gusset are of A514 steel. Assume $\frac{5}{16}$-in. welds and E110 electrodes. Use:

    (a)   LRFDS Section J2.4.

    (b)   LRFDS Appendix J2.4.

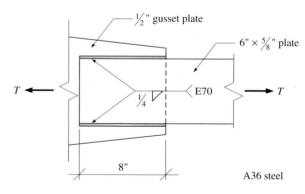

P6.11

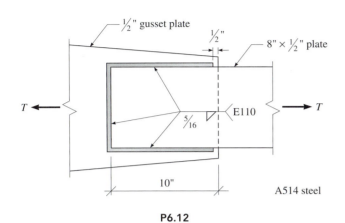

P6.12

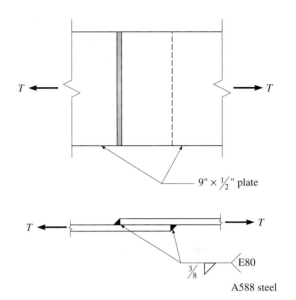

P6.14

P6.13. Repeat Problem P6.12, if the SAW process using F11 flux is used to prepare the welds.

P6.14. Determine the design strength of the lapped joint shown in Fig. P6.14. Each component is a plate 9 by $\frac{1}{2}$ in. of A588 steel. The weld is a $\frac{3}{8}$-in. fillet weld using E80 electrodes. Use:

    (*a*)   LRFDS Section J2.4.

    (*b*)   LRFDS Appendix J2.4.

P6.15. The end of a tension member consisting of a $\frac{5}{8}$ by 10 in. A572 Grade 42 plate is lap welded to the bottom chord member of a truss (stem of a WT), which allows a lap of 8 in. The load is 30 kips dead load and 80 kips live load. Use a transverse weld, and two equal-length longitudinal welds, using E70 electrodes. Determine the required size of the welds.

P6.16. Design the welded end connection required to transmit a dead load of 70 kips and a snow load of 210 kips through two C10×20 to a $\frac{7}{8}$-in. gusset plate as shown in Fig. P6.16. All material is A572 Grade 50. Welds are to be deposited using E70 electrodes.

P6.17. Determine the service load, T, permitted on the welded connection in Fig. P6.17 if the load is 75 percent live load and 25 percent dead load. The weld consists of two 8-in.-long $\frac{5}{16}$-in. longitudinal fillet welds and two $1\frac{1}{2}$-in.-diameter plug welds. The welds are made of the SAW process and F7XX electrode.

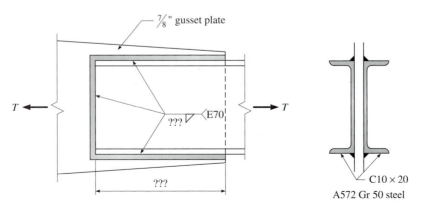

P6.16

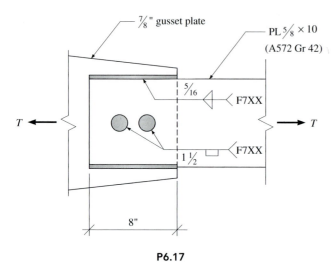

P6.17

P6.18.   Determine the adequacy of the welds connecting a C8×13.75 to the web of a WT6×39.5 as shown in Fig. P6.18. The weld consists of a 8-in.-long transverse weld, two 4-in.-long longitudinal welds, and a 2 by ¾ in. slot weld with semicircular ends. The fillet welds are of ³⁄₁₆ in. Assume A572 Grade 50 steel, and welding by the SMAW process using E70 electrodes. The service load consists of 40 kip dead load and 50 kip live load.

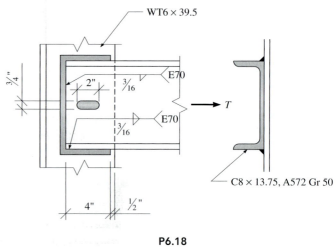

P6.18

An iron worker guiding a Pratt truss in place for the Safeco Baseball Stadium, Seattle, WA.

Tension Members

7

7.1 Introduction

Tension members are structural elements that are subjected to direct axial forces that tend to elongate the member. An axially loaded tension member is subjected to uniform, normal tensile stresses at all cross sections along its length. Many building, bridge, and tower structures contain members that are loaded primarily in tension. Examples of tension members include the diagonal and bottom chord members of typical trusses (Fig. 3.2.5), bracing members in building structures (Fig. 3.2.4*a*), suspenders in suspended multistory buildings (Fig. 3.2.6), and sag rods in industrial buildings (Figs. 3.4.2 and 3.4.3).

Any available steel shape may be used as a tension member. The choice of section used is governed to a large extent by the type of end connection used to connect the member to the rest of the structure. Ideally, the end connection is designed such that the tension force is concentrically applied to the member. In welded structures, connections to adjacent members can usually be made directly by lapping or butting. However, in bolted structures each connection usually requires a *gusset plate.* A single gusset plate at each joint is sufficient for light roof trusses *(single-plane trusses),* but two parallel gusset plates may be required in heavier long-span roof trusses *(double-plane trusses).*

Design of tension members is covered in Chapter D of the LRFD Specification. Design requirements that are common with those of other types of members are covered in Chapter B of the LRFDS.

7.2 Types of Tension Members

There are four types of tension members: single structural shapes and built-up members, rods, eyebars and pin-plates, and cables. Figures 7.2.1, 7.2.2, and 7.2.3 show some of the various types of sections used as tension members. These will be described in detail in the following sections.

7.2.1   Single Shapes and Built-Up Members

Single rolled sections are usually more economical than built-up sections and are ordinarily used whenever they provide adequate strength, rigidity, and ease of connection. Flat bars, tees, channels, angles, W- and S-shapes are the most commonly used shapes for tension members (Fig. 7.2.1).

Structural tees are extensively used as tension chords in light welded trusses, as web members of the truss can readily be bolted or welded to the tee stems. Single channels are often employed as web members in trusses because a channel has a smaller eccentricity than a single angle of comparable cross-sectional area. Channels can be conveniently bolted or welded to chord members. While slightly less efficient, single angles are used extensively as tension members in roof trusses when loads are light and member lengths are not excessive. They may be bolted to a gusset plate at each end, or they may be welded directly to the stems or flanges of tee or wide flange chord members, respectively. Large tension loads in heavy trusses are often carried by W- or S-shapes.

Built-up members consist of two or more structural shapes or a combination of rolled shapes and plates connected together at intervals so that they behave as a single unit. Some of the more common built-up shapes are shown in Figure 7.2.2. As depicted by the dotted lines of the figure, spacer plates, and tie plates are used to hold members in their built-up configurations. While these plates are not considered to add load-carrying capacity in and of themselves, they do provide rigidity and more evenly distribute the load among the main elements.

A built-up member may be necessary:

- When the required area cannot be provided by a single rolled shape.
- When a greater moment of inertia can be obtained with a built-up section than what can be provided with a single rolled shape having the same cross-sectional area.

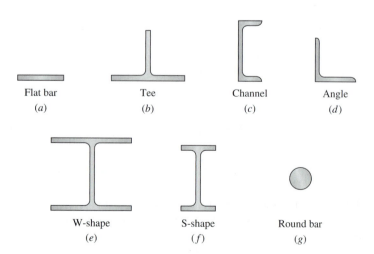

Flat bar	Tee	Channel	Angle
(a)	(b)	(c)	(d)

W-shape	S-shape	Round bar
(e)	(f)	(g)

Figure 7.2.1:   Typical rolled shapes used as tension members.

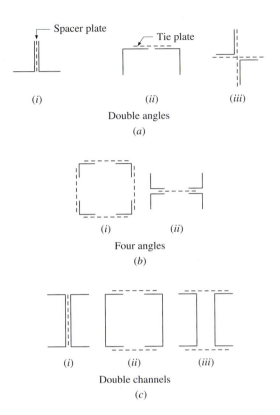

Double angles
(*a*)

Four angles
(*b*)

Double channels
(*c*)

Figure 7.2.2: Typical built-up sections used as tension members.

- When the width or depth of a member, necessary for a proper connection, cannot be obtained with a standard rolled section.

Another advantage of built-up members is that they can be made sufficiently stiff to carry compression as well as tension, making them desirable when stress reversals might occur.

Double angles or double channels are often used in bolted trusses. They are placed back-to-back with appropriate space between them to permit the insertion of gusset plates for connection purposes.

7.2.2 Rods

The simplest tension member is the rod or round bar (Fig. 7.2.1*g*). Rods are usually employed as members subjected to small design loads. Threaded tension rods are often used as:

- Hangers for supporting balconies, walkways, sign posts, and so on.
- Diagonal wind bracing in walls, roofs, and water towers (see Fig. 3.4.1).
- Tie rods for resisting the thrust of arches and gable frames.
- Sag rods for providing intermediate support to purlins in industrial buildings (see Fig. 3.4.2).

Four 4-in.-dia. steel cables collectively serve as a tension member in the Miller Park Stadium roof trussed-arch.
Photo by S. Vinnakota.

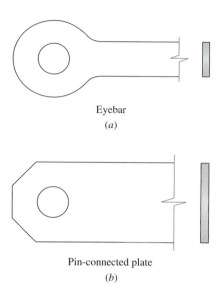

Eyebar
(a)

Pin-connected plate
(b)

Figure 7.2.3:   Eyebar and pin-connected plate.

- Sag rods for providing intermediate vertical support to girts in walls of industrial buildings (see Fig. 3.4.3).

The main disadvantage of rods is their low stiffness, which sometimes results in their noticeable sag under their own weight or that of workers during erection. Also, their compressive strength is negligible.

7.2.3   Eyebars and Pin-Connected Plates

Eyebars are plates of uniform thickness that have a widened circular head at each end (Fig. 7.2.3a). The periphery of each head is concentric with a pin hole and is connected by a transition curve to the body of the eyebar. Eyebars are not provided with any additional reinforcement at the ends. *Pin-connected plates* are tension members consisting of a plate of constant width (Fig. 7.2.3b). Eyebars and pin-connected plates are usually formed by flame cutting the edges and boring the pin holes.

7.3   Behavior of Tension Members

7.3.1   Load-Elongation Response

Consider an annealed steel member of length L and uniform cross-sectional area A subjected to an axial tensile force T applied at each end (Fig. 7.3.1a). The idealized material stress-strain plot used in this discussion is shown in Fig. 7.3.1b. When loaded, the member elongates by an amount Δ. Figure 7.3.1c shows the load-elongation response of the member as T is

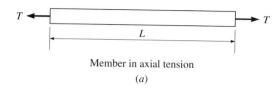

Member in axial tension

(a)

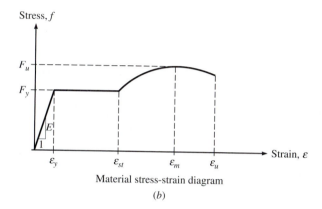

Material stress-strain diagram

(b)

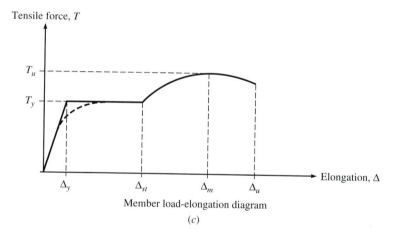

Member load-elongation diagram

(c)

Figure 7.3.1: Response of a member in axial tension.

increased gradually. As expected, the T-Δ diagram for the member is similar to the stress-strain diagram of the material. Thus, the initial portion of the curve shows a linear elastic response typical of a ductile material such as steel. That is, unloading of the member anywhere in that region causes the member to return to its original undeformed state. The force T, the strain ε, and the elongation Δ are related in the elastic region by the expressions:

$$\Delta = \varepsilon L = \frac{f}{E} L = \frac{TL}{AE} \qquad (7.3.1)$$

where E is the Young's modulus of elasticity of the material. Linear behavior continues until the stress reaches the yield stress of the material, F_y. The yield load of the member in tension is given by:

$$T_y = AF_y \qquad (7.3.2)$$

The maximum elastic elongation occurs just before the attainment of the yield load and is given by:

$$\Delta_y = \varepsilon_y L = \frac{F_y}{E} L \qquad (7.3.3)$$

When the applied load reaches the yield load, elongation will increase suddenly without any increase in load (corresponding to the yield plateau of the stress-strain diagram) until the fibers begin to strain harden. The corresponding elongation of the member is:

$$\Delta_{st} = \varepsilon_{st} L \qquad (7.3.4)$$

where ε_{st} is the strain at the onset of the strain hardening range. After strain hardening commences, the load can be increased slowly until it reaches the ultimate strength of the member in tension:

$$T_u = AF_u \qquad (7.3.5)$$

Here, F_u is the ultimate tensile stress of the material. The corresponding elongation of the member is given by:

$$\Delta_m = \varepsilon_m L \qquad (7.3.6)$$

where ε_u is the strain corresponding to F_u. Beyond T_u a local cross section of the member necks down and the load capacity decreases. Finally, fracture occurs, corresponding to an elongation Δ_u given by:

$$\Delta_u = \varepsilon_u L \qquad (7.3.7)$$

If the member contains residual stresses due to rolling or welding processes, local yielding starts before the yield load T_y is reached, as shown by the dotted curve in Fig. 7.3.1c. Thus, the range over which the load-elongation behavior is linear decreases.

Let us consider a 20-ft-long, A36 steel ($F_y = 36$ ksi, $E = 29,000$ ksi) tension member. The maximum elastic elongation of the member based on Eq. 7.3.3 is:

$$\Delta_y = \frac{36}{29,000}(20)(12) = 0.3 \text{ in.}$$

Using an approximate value for ε_m of 0.12 in Eq. 7.3.6, the elongation corresponding to the attainment of ultimate tensile stress of the material is obtained as:

$$\Delta_m = 0.12(20)(12) = 28.8 \text{ in.}$$

Such a large elongation of the member can precipitate failure of the structural system of which it is a part (adjacent compression members and appurtenances, for example). So, as we will see in Section 7.4, the design of tension members is based on the yield strength of the member rather than its ultimate strength.

Next, let us consider a 2-ft-long, A36 steel tension member. The maximum elastic elongation of this member is 0.03 in., and the elongation corresponding to the attainment of ultimate tensile stress of the material is 2.88 in. The short length of this member produces only tolerably small elongations till the stresses reach the ultimate tensile stress of the material. The 2-ft-long member could represent two 1-ft-long connections at the end of a typical tension member. These observations will be used in explaining the design criteria for tension members in Section 7.4.

In the discussion above, we tacitly assumed that the end connection for load application is made to all elements of which the cross section is composed. If the end connection to such an axially loaded member is by bolts, the stress distribution in a cross section through bolt holes is found to be highly nonuniform at low levels of loading, due to the phenomenon of stress concentrations. However for ductile materials like steel, the stress distribution in the cross section through bolt holes is again found to be essentially uniform at failure.

For further information on distribution of stresses in plate sections with bolt holes refer to Section W7.1 on the website http://www.mhhe.com/ Vinnakota.

7.3.2   Load Distribution in Axially Loaded Bolted Joints

Tests indicate that, in short connections having few bolts per line, almost complete equalization of bolt loads occurs due to the ductility of the bolt and the plate material. However, each bolt may deform inelastically by a different amount before the ultimate load is attained. Furthermore, tests indicate that in longer connections the end bolts can reach a critical shear deformation and fail before the full strength of each bolt is attained. This premature, sequential failure of bolts progresses inward from the ends of the joint and is called **unbuttoning.** Earlier studies indicate that it is desirable to arrange a joint compactly in order to equalize the loads on the bolts as much as possible. Thus, conventional design of axially loaded bolted joints assumes all bolts in such a joint to be equally loaded.

7.3.3   Load Transfer at End Connections

Consider a plate tension member connected at its end to a gusset plate by five bolts of equal size and type arranged as shown in Fig. 7.3.2*a*. The member is subjected to an axial tensile force, *T*. One of the basic assumptions made in the analysis of bolted joints is that each of the bolts in a group of equal-sized bolts transfers an equal share of the load whenever the bolts are arranged symmetrically about the centroidal axis of the tension member. For the connection under consideration, the load resisted by each bolt equals *T*/5. Free body diagrams of the member and the gusset are shown in Figs. 7.3.2*b* and *c*, respectively. Also,

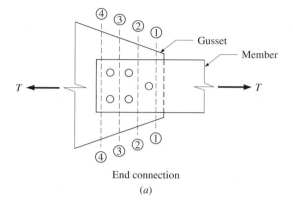

End connection

(a)

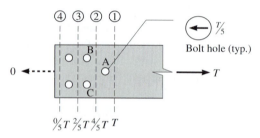

Free body diagram of the member

(b)

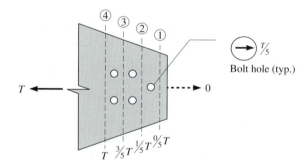

Figure 7.3.2: Load transfer at end connection of a tension member.

Free body diagram of the gusset

(c)

shown here are four sections indicated 1-1, 2-2, 3-3, and 4-4, where the internal forces in the member and the gusset are calculated below.

Considering the free body diagram of the member shown in Fig. 7.3.2b, it is observed that the full tensile force T in the member acts on section 1-1. Examination of the section 2-2 of the member indicates that a capacity of $4T/5$ is needed at section 2-2, as the bolt A located to the right of this section has already transferred its share of the load $T/5$ from the member to the gusset.

Similarly, examination of section 3-3 of Fig. 7.3.2b indicates that a capacity of only 2T/5 is needed at that section as the bolts A, B, C located to the right of this section have already transferred their share of the load 3T/5 from the member to the gusset. Finally, there must be zero tensile force acting on the member at section 4-4, as the force T must have been entirely transferred to the gusset by the five bolts that are all to the right of the section.

Thus, it is observed that as one goes from the sections at the front end of the connection (such as section 1-1) to the sections at the back end of the connection (such as section 4-4), the required strength at the section decreases. Thus, failure of the member in the connection is more likely to occur at sections located at the front end of the connection than at sections located at the back end.

Next, considering the free body diagram of the gusset shown in Fig. 7.3.2c, the resistance to be provided by the gusset at sections 1-1, 2-2, 3-3 and 4-4 is calculated as 0T/5, 1T/5, 3T/5, and 5T/5, respectively.

7.3.4   Shear Lag

The stress distribution in an axially loaded tension member is uniform at sections away from the connections, that is, in the body of the member. At failure, the stress distribution in the net area of the connection zone is also found to be essentially uniform when the end connection is made to all of the plate elements of which the cross section is composed. However, for tension members other than plates, the end connection is often made to only some elements of the section. In such cases, tests have shown that when failure by fracture of the net area occurs, the experimental failure load divided by the net area is generally found to be less than the ultimate tensile stress of the steel. In effect, as shown in Fig. 7.3.3a, when some elements are connected and others are not, stress must flow out of the plate elements that are not connected and into the ones that are connected, so as to reach the bolts and subsequently into the gusset(s) and adjoining members. The resulting crowding of the stress trajectories towards the connected element develops higher stresses in those parts of the section and reduces the efficiency of the connection.

Fig. 7.3.3b shows the web of the I-section of Fig. 7.3.3a in the unloaded state, and Fig. 7.3.3c shows that same web in the loaded state. The four forces shown in Fig. 7.3.3c are the resultants of the bolt shears in the connections. Since the ends of the web are free, the distortion will be as shown. The introduction of shear loads along the edges tends to cause the plate to warp slightly. Therefore, an element A in the unloaded web will be deformed as shown in Fig. 7.3.3c when the member is loaded. This is a shear deformation, and the stress in the web is said to lag because of it. This phenomenon of nonuniform straining of the unconnected elements and the concentration of shear stress in the vicinity of the connection (Fig. 7.3.3d), is often referred to as **shear lag.**

Shear lag reduces the effectiveness of member components that are not directly connected to a gusset plate or other anchorage and, hence, reduces the design strength of the member. Shear lag is applicable to both bolted

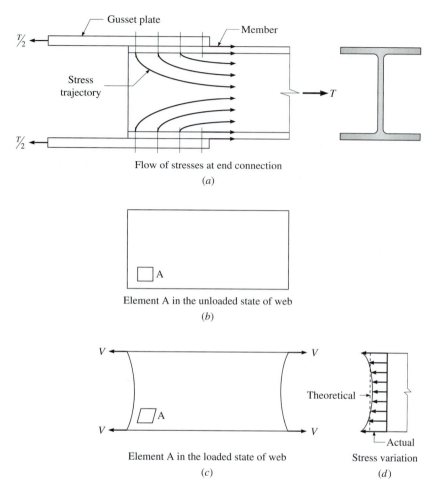

Figure 7.3.3: Shear lag effect at end connection to the flanges of an I-shape.

and welded tension members and will be considered in detail in Section 7.7.

7.4   Strength Limit States of a Tension Member

Typically, tension members are connected at their ends to gusset plates by bolts or welds. When bolted connections are used, the cross-sectional area of the member is reduced due to the presence of bolt holes. Further, as discussed in Section 7.3, end connections frequently only connect a portion of the actively stressed area to the gussets. For example, Fig. 7.4.1a shows a C-shape used as the diagonal bracing in the braced bent of a building structure. These diagonals are designed as tension members. Only the web of the C-shape is bolted to the

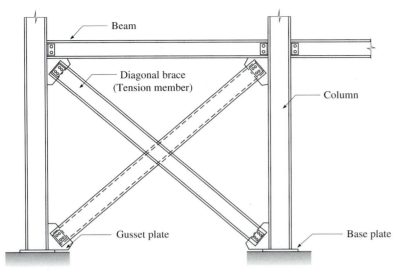

Tension member used as a brace

(*a*)

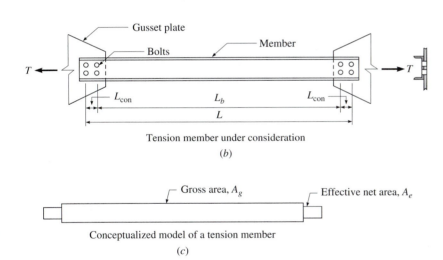

Tension member under consideration

(*b*)

Conceptualized model of a tension member

(*c*)

Figure 7.4.1: Tension member.

gusset plate, while the flanges are unconnected. Because of shear lag, the efficiency of the member to resist tension is less than 100 percent. Therefore, a tension member connected at its ends to plates by bolts or welds may be conceptually visualized as being composed of two segments:

1. The body of the member of length L_b between the end connections having a cross-sectional area equal to the gross area of the member, A_g (Figs. 7.4.1*b* and *c*).

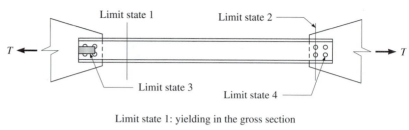

Limit state 1: yielding in the gross section

Limit state 2: fracture in the net section

Limit state 3: block shear rupture

Limit state 4: failure of connectors

Figure 7.4.2:   Limit states for tension members.

2. The connected length (two small portions at each end of the member of length L_{con}) over which the cross-sectional area is less than the gross area. The reduced cross-sectional area is referred to as the effective net area, A_e. The reduction accounts for the presence of bolt holes, stress concentrations, and any connection inefficiency that may result if only some of the member's cross-sectional elements are actually attached to the gusset (Figs. 7.4.1*b* and *c*).

There are five limit states or failure modes to be considered in the design of tension members. Four of them are shown in Fig. 7.4.2.

Limit State 1

As observed in Section 7.3.1, a ductile steel member loaded in axial tension can resist a force greater than the yield load without fracturing. However, the resulting large elongations due to uncontrolled yielding of the member can cause failures in adjacent members and may precipitate the failure of the structural system of which it is a part. Hence, as shown in Fig. 7.4.3, *yielding in the gross section* constitutes one failure limit state and is intended to limit excessive elongation of members. It represents a limit state for which the failure is gradual.

Limit State 2

Because the length over which the effective net area of a member applies is negligible relative to the total length of the member, yielding of the net section does not constitute a failure mode. The short length produces only tolerably small elongations till the stresses reach the ultimate tensile stress of the material, as seen in Section 7.3.1. Also, before yielding occurs in the body of the member, the connection region of the member end may experience strain hardening, and fracture can possibly occur in this region. Hence, *fracture in the net section* constitutes a second failure limit state, and is detailed in Fig. 7.4.4. This fracture occurs with little deformation.

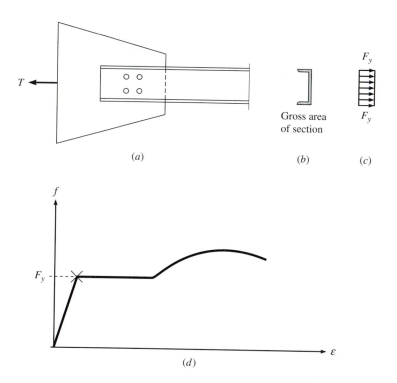

Figure 7.4.3: Limit state of yielding in the gross section.

Limit State 3

When the connection of a member to a gusset is made by a small number of closely spaced, large-diameter, high-strength bolts, a rectangular block of material in the connected part may tear out from the rest of the member, as shown in Fig. 7.4.2. This mode of failure is called *block shear failure*.

Limit State 4

The connecting bolts or welds may fail in one of the several modes described in Chapter 6. Such failures will be considered to constitute limit state 4.

Limit State 5

Finally, connecting elements such as gusset plates and splice plates may fail prior to the tension member, thus precluding the other limit states. This will be indicated as limit state 5.

The nominal strengths corresponding to these five limit states are:

T_{n1} = yielding in the gross section in the body of a member
T_{n2} = fracture in the net section within a connected part

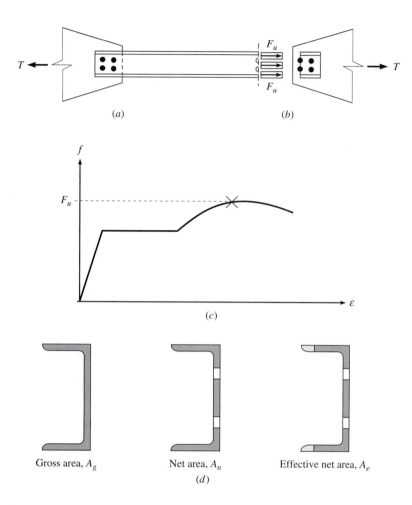

Figure 7.4.4: Limit state of fracture in the net section.

T_{n3} = block shear rupture strength
T_{n4} = strength of fasteners (bolts or welds) in a connection
T_{n5} = strength of a connecting gusset plate

The design strengths corresponding to these nominal strengths are obtained with the general relation:

$$T_{di} = \phi_{ti} \, T_{ni} \qquad (7.4.1)$$

where ϕ_{ti} is the appropriate resistance factor. The design strength of the tension member is then given by:

$$T_d = \min [T_{d1}, T_{d2}, T_{d3}, T_{d4}, T_{d5}] \qquad (7.4.2)$$

Usually, gussets are designed such that failure occurs in the member rather than in the gusset. Evaluation of gusset strength will be discussed in Chapter 12.

So, for the purposes of the present discussion, Eq. 7.4.2 reduces to:

$$T_d = \min\ [T_{d1}, T_{d2}, T_{d3}, T_{d4}] \qquad (7.4.3)$$

Through proper selection of the type, number, and arrangement of connectors, it is usually possible (and always desirable) to assure that block shear rupture strength and connector strength do not limit the carrying capacity of the member. This discussion will therefore initially concentrate on the first two limit states. Equation 7.4.3 therefore reduces to:

$$T_d = \min\ [T_{d1}, T_{d2},] \qquad (7.4.4)$$

As a result of the preceding information, Section D1 of the LRFD Specification states that the design strength of a tension member, T_d, shall be the lower value obtained according to the limit states of yielding in its gross section and fracture at its weakest effective net section.

$$T_{d1} = \phi_{t1} F_y A_g = 0.90 F_y A_g \qquad (7.4.5)$$

$$T_{d2} = \phi_{t2} F_u A_e = 0.75 F_u A_e \qquad (7.4.6)$$

where A_e = effective net area of the member
  A_g = gross area of the member
  F_y = yield stress of the material
  F_u = ultimate tensile stress of the material
  ϕ_{t1} = resistance factor for the limit state of tension yielding = 0.90
  ϕ_{t2} = resistance factor for the limit state of tension fracture = 0.75

In general, it is preferable that the controlling limit state is yielding of the body, a ductile failure, rather than fracture of the connection, a brittle failure.

The difference in resistance factors, namely 0.90 and 0.75 adopted by the LRFDS for ϕ_{t1} and ϕ_{t2}, is due to the difference in type of failure, namely, ductile versus brittle. Values of F_y and F_u for various structural steels are given in Table 2 of the LRFDS. When a range of values is given (such as 100–130 ksi for A514 Grade 90 steel, for example), the value to be used in design is the minimum value (100 ksi, in this case). Values of F_y, F_u are also given in Table 7.4.1 for several steels.

7.5   Gross Area, A_g

If a cut is made perpendicular to the longitudinal axis of a member, the area of that cross section represents the **gross area** of the member, A_g. The cross-sectional areas for rolled sections tabulated in Part 1 of the LRFD Manual are gross areas. These values include the areas of any web-to-flange fillets. Alternately, such fillets may be ignored, and the gross area of a member may be approximated as the sum of the products of the thickness and the width of each rectangular element comprising the cross section.

TABLE 7.4.1

Design Stresses as a Function of F_y and F_u for Various Structural Steels

Material		Tension Member		Block Shear Rupture			Rods
		$0.9\,F_y$	$0.75\,F_u$	$0.75(0.6)\,F_y$	$0.75(0.6)\,F_u$	$0.75\,F_y$	$0.75(0.75)\,F_u$
F_y (ksi)	F_u (ksi)	(ksi)	(ksi)	(ksi)	(ksi)	(ksi)	(ksi)
A36 36	58	32.4	43.5	16.2	26.1	27.0	32.6
A992 50	65	45.0	48.8	22.5	29.3	37.5	36.6
⎡A242⎤ 42	63	37.8	47.3	18.9	28.4	31.5	35.4
⎢ ⎥ 46	67	41.4	50.3	20.7	30.2	34.5	37.7
⎣A588⎦ 50	70	45.0	52.5	22.5	31.5	37.5	39.4
42	60	37.8	45.0	18.9	27.0	31.5	33.8
50	65	45.0	48.8	22.5	29.3	37.5	36.6
A572 55	70	49.5	52.5	24.8	31.5	41.3	39.4
60	75	54.0	56.3	27.0	33.8	45.0	42.2
65	80	58.5	60.0	29.3	36.0	48.8	45.0
90	100	81.0	75.0	40.5	47.0	67.5	56.3
A514 100	110	90.0	82.5	47.0	49.5	75.0	61.9

7.6 Net Area, A_n

If the end connection of a tension member is to be made by bolting, then material must be removed from the cross section to form bolt holes. Only a portion of the gross area of the member remains to carry load. This remaining area is termed the **net area of the member.** Thus:

$$A_n = A_g - \text{Area lost due to bolt holes} \qquad (7.6.1)$$

Holes for bolts are permitted to be punched whenever the thickness of the material is not greater than the nominal bolt diameter, d, plus $\frac{1}{8}$ in. (see LRFDS Section M2.5). When the material thickness is greater than this, holes are either drilled or subpunched and reamed. Subpunched holes are made with dies at least $\frac{1}{16}$ in. smaller than the nominal diameter of the bolt. Holes are then reamed until they attain the desired diameter. For standard holes, the hole diameter d_h is made $\frac{1}{16}$ in. greater than the nominal diameter of the bolt d, to provide for erection tolerance (LRFDS Table J3.3). The punching process distorts and damages the metal immediately around the edges of the hole (Fig. 6.2.5). Thus, an additional $\frac{1}{16}$ in. is added to the nominal hole diameter to account for the damaged material that is assumed to be noneffective in transmitting load (In the case of slotted holes, $\frac{1}{16}$ in. should be added to the actual width of the hole). As a result, in computing the net area for tension, the effective width of a standard punched hole is taken as $\frac{1}{8}$ in. greater than the nominal bolt diameter. The area of metal effectively removed from the cross section by a bolt hole is then given as:

$$A_h = d_e t \qquad (7.6.2)$$

where A_h = area of cross section lost due to a bolt hole, in.^{2}
d_e = effective width of a bolt hole, in.
  = $d + \frac{1}{8}$ in. for standard punched holes
  = $d + \frac{1}{16}$ in. for drilled or subpunched and reamed STD holes
d = nominal diameter of the bolt, in.
t = thickness of the plate material, in.

For example, a standard punched hole for a $\frac{7}{8}$-in.-diameter bolt in $\frac{1}{2}$-in.-thick material reduces the cross-sectional area by $\frac{1}{2}$ $(\frac{7}{8} + \frac{1}{8}) = 0.5$ in.^{2}

7.6.1 A_n for Bolt Arrangements without Stagger

The simplest fastener pattern for an end connection of a tension member is one in which the holes are not staggered in the longitudinal direction. Thus, each transverse section through the holes perpendicular to the longitudinal axis of the member contains the same number and arrangement of holes. When such a member is subjected to tension, it is intuitively expected that the critical path along which the fracture of the member occurs is the first transverse section through the bolt holes. Thus, in the case of the rectangular plate with three lines of bolts shown in Fig. 7.6.1a, the critical section is section 1-1. The net area at that section equals the gross area minus the area of cross section lost due to the three bolt holes. More generally, the net area of a tension member with a repetitive arrangement of bolt holes is obtained by deducting from the gross area of the member the rectangular areas of cross section lost due to bolt holes. Hence:

$$A_n = A_g - \sum_{i=1}^{n_e} n_i \, d_e \, t_i \qquad (7.6.3)$$

where A_g = gross area of the member, in.^{2}
A_n = net area at the critical section, in.^{2}
n_e = number of rectangular elements in the cross section
i = element number
n_i = number of bolt holes in element i
t_i = thickness of element i, in.

For example, the net area of the W-shape shown in Fig. 7.6.1b is obtained by subtracting from the gross area four bolt holes in the flanges $(4d_e t_f)$ and three bolt holes in the web $(3d_e t_w)$.

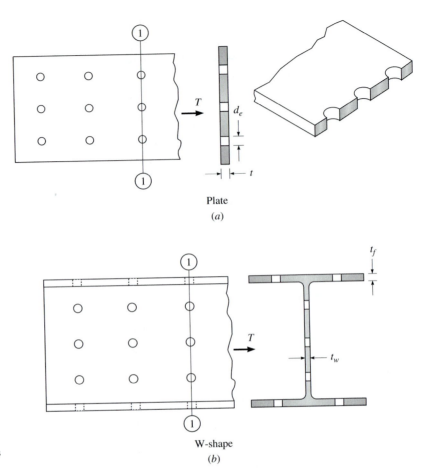

Plate

(a)

W-shape

(b)

Figure 7.6.1:   Failure paths for bolt arrangements without stagger.

7.6.2   A$_n$ for Bolt Arrangements with Stagger

Plates

When there are two or more rows of bolt holes in a member, as is the case in the plate shown in Fig. 7.6.1a, it is possible to achieve a larger net area by using a staggered arrangement of bolts as shown in Fig. 7.6.2a. When bolt holes are staggered, the section along which fracture of the member occurs, the critical path, is no longer obvious, and several possible failure paths must be considered. Inspection of the figure indicates that failure may occur in one of the two ways. One possible failure path is a-b-c-d. In this case, the net area is equal to the gross area minus the area effectively lost due to two bolt holes. Another possible failure path is the zig-zag line a-b-e-c-d. In this case, the stress along the inclined or diagonal segments b-e and e-c contains both tension and shear rather than simple tension. Compared with the path a-b-c-d with two holes, the net area of the zig-zag path a-b-e-

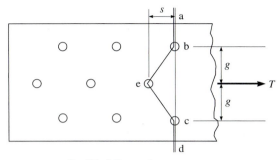

Possible failure paths

(a)

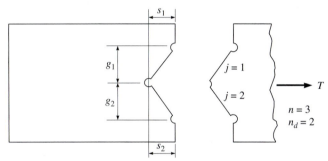

Fracture along a path with diagonal segments

(b)

Figure 7.6.2: Failure paths for bolt arrangement with stagger.

c-d has less area due to the presence of a third hole but also increased area due to the sloped segments b-e and e-c. The strength of the member along path a-b-e-c-d is therefore somewhere between the strength obtained by using a net area computed by subtracting three holes from the transverse cross-sectional area, along path 1-1 of Fig. 7.6.1a, and the value obtained by subtracting two holes from the transverse cross section corresponding to path a-b-c-d of Fig. 7.6.2a.

A rigorous calculation of the net area for staggered bolt arrangements is quite complex, as it must include appropriate failure theories for combined stress states [McGuire, 1968]. A simple empirical method of computing the net width of plates with staggered holes was presented by Cochrane [1922]. It agrees with research findings and is accepted by several specifications, including the LRFD Specification. In this method, the net area of a plate is calculated by taking the gross area of the plate and subtracting the rectangular areas of cross section lost due to all bolt holes in the path being considered. A correction is then made by adding the quantity given by the expression $(\frac{s^2}{4g}t)$ for each gage space with a ***diagonal segment*** in the failure path (Fig. 7.6.2b).

In this expression, s is the longitudinal spacing of any two consecutive holes in the failure path or **staggered pitch,** and g is the transverse spacing of the same holes or **gage.** The distances s and g are measured between hole centers. Thus:

$$A_{nk} = A_g - nd_e t + \sum_{j=1}^{n_d} \frac{s_j^2}{4g_j} t \qquad (7.6.4)$$

where A_{nk} = net area along possible failure path k in.^{2}
       n = number of holes in the failure path considered
      n_d = number of gage spaces with a diagonal segment in the failure path
      s_j = staggered pitch for the jth diagonal segment, in.
      g_j = gage for the jth diagonal segment, in.

For a given bolted joint there may be several possible failure paths, any one of which may be critical. All these possible paths must be investigated, and the one having the smallest net area defines the critical path and the net area of the member. Thus:

$$A_n = \min [A_{n1}, A_{n2}, \dots A_{nk}, \dots A_{nM}] \qquad (7.6.5)$$

where A_n = net area of the member, in.^{2}
      M = number of possible failure paths
      A_{nk} = net area along possible failure path k, given by Eq. 7.6.4 $(1 \leq k \leq M)$

Angles

As discussed in Section 6.2, bolt holes in steel angles are normally provided along specific lines called **gage lines.** These locations or gages are dependent on the length of the leg in which the holes are located and on the number of lines of holes in that leg. For leg sizes greater than 4 in., two lines of bolts can be used. Table 6.2.1 shows workable gages for angles as given by Fig. 10-6 of the LRFDM. Gage distance g_1 applies when only one line of bolts is used in the leg under consideration, while gage distances g_2 and g_3 are used when two lines of bolts are used (Figs. 6.2.9a and b).

In some applications an angle may be connected by bolts in both legs. If the bolt holes in the two legs are staggered (as shown in Fig. 7.6.3a), the usual procedure is to transform (bend) the cross section into an equivalent flat plate by revolving the plates about the center lines of the component parts (Fig. 7.6.3b). The critical net section can then be determined by the procedure described in the previous section. For a diagonal segment joining staggered holes on two legs of an angle using gage distances g_a and g_b, the gage length g_{ab} for use in the $(\frac{s^2}{4g})$ expression is the distance between the two lines of bolts measured along the centerline of the angle thickness.

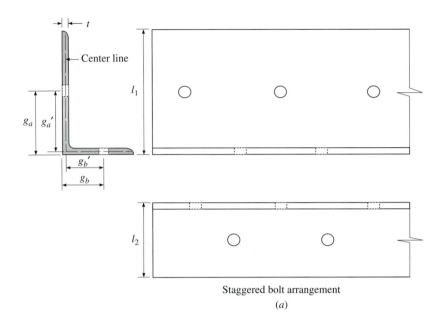

Staggered bolt arrangement

(a)

Angle developed into a plate

(b)

Figure 7.6.3: Staggered holes on two legs of an angle.

That is:

$$g_{ab} = g_a' + g_b' = \left(g_a - \frac{t}{2} \right) + \left(g_b - \frac{t}{2} \right) \qquad (7.6.6)$$

$$= g_a + g_b - t$$

where g_a' and g_b' are defined as shown in Fig. 7.6.3. Thus, the gage distance g_{ab} is the sum of the gages measured from the back (heel) of the angle in each leg minus the thickness of the angle.

I-, T-, and C-Shapes

Part 1 of the LRFD Manual provides the workable gages for flanges of I-, T-, and C-shapes. For these shapes, the web and flange thicknesses are not the same. So, if a diagonal segment goes from a flange hole to a web hole, the thickness changes at the junction of the flange and the web. For such diagonal segments the $(\frac{s^2}{4g})$ term is multiplied by the average of the flange and web thicknesses in the approximate calculation of the net area. Thus:

$$A_{nk} = A_g - \sum_{i=1}^{n_e} n_i d_e t_i + \sum_{j=1}^{n_d} \frac{s_j^2}{4g_j} t_j \qquad (7.6.7)$$

where A_g = gross area of the member, in.^{2}
  A_{nk} = net area along possible failure path k, in.^{2}
  d_e = effective width of a bolt hole, in.
  n_e = number of elements in the cross section
  i = element number
  n_i = number of bolt holes in element i along path k
  t_i = thickness of element i, in.
  n_d = number of gage spaces with a diagonal segment along path k
  s_j = staggered pitch for the jth diagonal segment, in.
  g_j = gage for the jth diagonal segment, measured along the center-line of each element, in.
  t_j = average element thickness corresponding to the jth inclined segment, in.

For example, Fig. 7.6.4a shows a channel with bolt holes in its flanges staggered with those in its web. The channel is assumed to be flattened out into a single plate as shown in Fig. 7.6.4b. We have:

$$g_{ab} = g'_a + g'_b = \left(g_a - \tfrac{1}{2} t_w\right) + \left(g_b - \tfrac{1}{2} t_f\right); \; t = \tfrac{1}{2}(t_f + t_w)$$

$$(7.6.8)$$

The net area of the channel along the zig-zag path 1-1 is then obtained by first subtracting from the gross area the areas lost due to the flange holes and web holes, and then adding the $(\frac{s^2}{4g_{ab}} t)$ terms for each of the two diagonal segments.

7.6.3 A_n for Connection Elements

Connection elements to members loaded in tension—such as splice plates, gusset plates, and cover plates—are themselves short tension members. Being short, these members have poor stress flow. So based on test results, LRFD Specification Section J5.2b places additional restrictions on their design by limiting the net area of a connection element to be no greater than 85 percent of its gross area [Fisher and Struik, 1974]. That is:

$$A_n \leq 0.85 A_g \qquad (7.6.9)$$

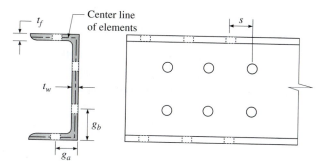

Staggered bolt arrangement

(a)

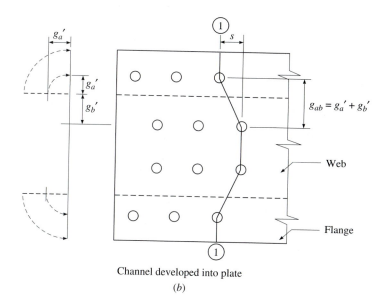

Channel developed into plate

(b)

Figure 7.6.4: Staggered holes in the web and flanges of a channel shape.

7.6.4 A_n for Welded Connections

If the end connection of a tension member is made by welding, there generally is no resulting reduction in cross section, and the net area of the member is equal to the gross area. That is:

$$A_n = A_g \qquad (7.6.10)$$

However, even in welded construction, if the erector puts holes in tension members to facilitate erection, the loss in cross section must be considered as for members with bolted connections. Also, in determining the net area across plug or slot welds, the weld metal shall not be considered to replace to the net area the area lost due to the slot (LRFDS Section B2).

7.7   Effective Net Area, A_e

At sections located away from the end connections of an axially loaded tension member (i.e., at sections within the body of the member), the stresses are uniform across the entire section. A connection disturbs the uniformity of this stress field, however, due to such effects as shear lag and stress concentrations at bolt holes and welds (see Web Section W7.1), and almost always weakens the tension member. The influence of this weakening is accounted for by using a *reduction coefficient,* U, and defining an effective net area of the member. The *effective net area* A_e is the portion of the net area of a tension member that is participating effectively in the transfer of the force. Thus:

$$A_e = UA_n \qquad (7.7.1)$$

where A_e = effective net area of a tension member, in.^{2}
         U = reduction coefficient
         A_n = net area, in.^{2}

7.7.1   Reduction Coefficient, *U*, for Bolted Members

The value of the reduction coefficient is influenced by the length of the connection, the shape of the cross section, and the geometry of the cross-sectional elements that are not directly connected to the gusset(s). This is illustrated with the help of a single angle tension member shown in Fig. 7.7.1. One leg of the angle is connected to a gusset by two lines of bolts without stagger as indicated in Figs. 7.7.1*a* and *b*. Figure 7.7.1*c* schematically shows the flow of stresses in the unconnected leg of the angle when three bolts are used in each line to connect the member, while Fig. 7.7.1*d* shows the flow of stresses when the number of bolts per line is increased to five. Also shown shaded in these figures is the material that is ineffective at the critical section in each case (section 1-1 in Figs. 7.7.1*c* and *d*). As indicated, the member with three bolts, having a shorter length of connection, has less effective net area than the longer connection. It follows that an increase in the *length of the connection,* L_{con}, results in more uniform stress on the net area at the section where the transfer of force from the member to the gusset plate first occurs (front end of the connection).

The distance from the contact plane between the gusset plate and the connected element (fastener plane) to the center of gravity of the tributary connected area of the member, shown as $\bar{x}_{con}$ in Fig. 7.7.1*b*, represents the *connection eccentricity.* It is a measure of the ratio of the unconnected area to the total cross-sectional area of the member. When, for example, an angle is connected to a gusset by one leg, the tributary area for that connection is the entire area of the angle (Fig. 7.7.1*b*).

The effect of these two parameters, L_{con} and $\bar{x}_{con}$, on members with bolted end connections can be expressed empirically by a reduction coefficient, U,

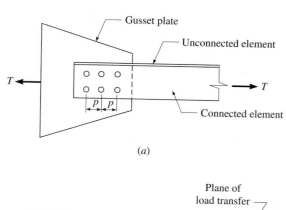

(a)

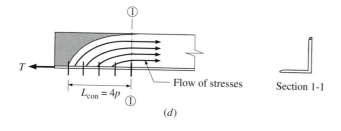

(b)

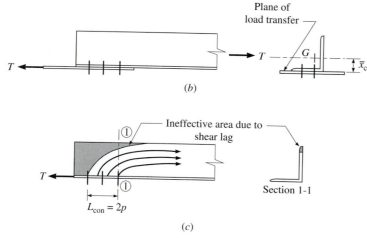

(c)

(d)

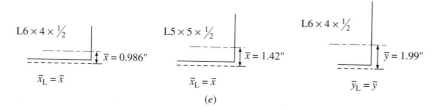

(e)

Figure 7.7.1:   Parameters influencing reduction coefficient, U.

given by [Chesson and Munse 1957; Munse and Chesson, 1963]:

$$U = \min\left[\left(1 - \frac{\bar{x}_{con}}{L_{con}}\right), 0.9\right] \qquad (7.7.2)$$

where L_{con} = length of the connection in the direction of the loading, in.
$\bar{x}_{con}$ = connection eccentricity, in.

Determination of Connection Eccentricity, $\bar{x}_{con}$

For a single angle connected to a gusset (Fig. 7.7.2a), the value of $\bar{x}_{con}$ to be used in Eq. 7.7.2 will be the dimension $\bar{x}_L$. Here, $\bar{x}_L$ is the distance from the

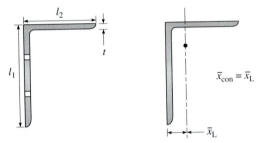

Connection to the long leg of an angle

(a)

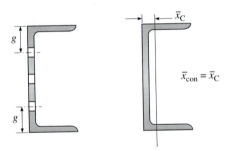

Connection to web of a C-shape

(b)

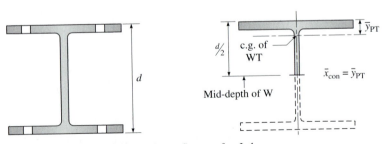

Connection to flanges of an I-shape

(c)

Figure 7.7.2:   Determination of $\bar{x}_{con}$ for reduction coefficient U.

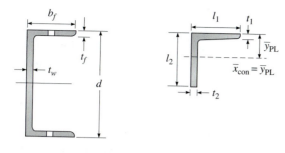

Connection to flanges of a C-shape

(d)

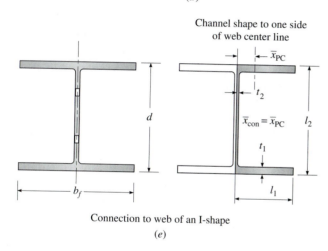

Connection to web of an I-shape

(e)

Figure 7.7.2: Continued.

gusset-angle interface to the center of gravity of the given angle. This value is given by the dimension $\bar{x}$ tabulated for angle shapes in Table 1-7 of the LRFDM if the connection is to the long leg, or the dimension $\bar{y}$ in Table 1-7 of the LRFDM if the connection is to the short leg. Figure 7.7.1e shows three angles having essentially the same cross-sectional area but different proportions of unconnected material. We have:

- For the L6×4×½ with its long leg connected to gusset,
  $\bar{x}_{con} = \bar{x}_L = 0.986$ in.
- For the L5×5×½ with one of its legs connected to gusset,
  $\bar{x}_{con} = \bar{x}_L = 1.42$ in.
- For the L6×4×½ with its short leg connected to gusset,
  $\bar{x}_{con} = \bar{y}_L = 1.99$ in.

If, in addition, all three angle members considered have the same connection length (L_{con}), of 10 in., the reduction coefficient U from Eq. 7.7.2 is 0.90, 0.86, and 0.80, respectively. This example clearly shows the advantage of connecting more material directly to the gusset.

For a channel shape connected by a gusset to its web only (Fig. 7.7.2b), the value of $\bar{x}_{con}$ to be used in Eq. 7.7.2 is the dimension $\bar{x}_C$, the distance from the outside edge of the web to the center of gravity of the channel shape. This value is represented by the dimension $\bar{x}$ tabulated for channel shapes in Tables 1-5 and 1-6 of the LRFDM. For example, for a C15×40 tension member with its web connected to a gusset, we have:

$$\bar{x}_{con} = \bar{x}_{C15 \times 40} = 0.778 \text{ in.}$$

For a double plane member such as an I- or C-shape connected by gussets to both flanges, the area tributary for each fastener plane is the area of half the cross section, and the coordinate $\bar{x}_{con}$ is measured from each fastener plane to the centroid of each half cross section.

Thus, for an I-shape connected by gussets to its flanges only (Fig. 7.7.2c), it is assumed that the section is split into two structural tees of equal size. The value of $\bar{x}_{con}$ to be used in Eq. 7.7.2 will be the distance $\bar{y}_{PT}$ measured from the outside edge of each flange to the center of gravity of each pseudo-tee. This length is given by the dimension $\bar{y}$ tabulated for structural tees in Tables 1-8, 1-9, and 1-10 of the LRFDM. For example, for a W10×60 tension member with its flanges connected to gussets, we have:

$$\bar{x}_{con} = \bar{y}_{PT} = \bar{y}_{WT5 \times 30} = 0.884 \text{ in.}$$

For a channel shape connected by gussets to its flanges only (Fig. 7.7.2d), it is assumed that the section is split into two equal angles. Thus, if b_f is the flange width of the channel; t_f, the flange thickness; d, the depth; and t_w, the web thickness, the tributary area for each gusset is a pseudo-angle of leg sizes ($l_1 = b_f$, $t_1 = t_f$) and ($l_2 = \frac{1}{2} d$, $t_2 = t_w$). The value of $\bar{x}_{con}$ to be used in Eq. 7.7.2 will be the distance $\bar{y}_{PL}$ measured from the outside edge of each flange to the center of gravity of the pseudo-angle, also shown in Fig. 7.7.2d. For this type of connection, $\bar{y}_{PL}$ is not tabulated in the LRFDM and must be calculated from basic theory.

Finally, for an I-shape connected by gussets to its web (Fig. 7.7.2e), the value of $\bar{x}_{con}$ to be used in Eq. 7.7.2 is the dimension $\bar{x}_{PC}$. The dimension $\bar{x}_{PC}$ is the distance from the web centerline to the center of gravity of the pseudo-channel-shaped area created by splitting the I-shape through the centerline of the web. This pseudo-channel has flange dimensions ($l_1 = \frac{1}{2} b_f$; $t_1 = t_f$) and web dimensions ($l_2 = d$; $t_2 = \frac{1}{2} t_w$). Here, b_f, t_f, d, and t_w are the dimensions of the given I-shape. For this type of connection, $\bar{x}_{PC}$ is not tabulated in the LRFDM and must be calculated from basic theory.

Connection Length, L_{con}

The connection length, L_{con}, is the distance, parallel to the line of force, between the first and last bolts, and is measured along the line with the

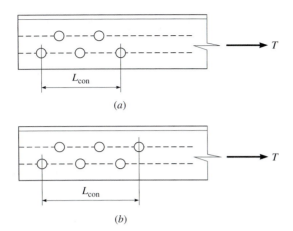

Figure 7.7.3:   Connection length, L_{con}, for reduction coefficient U.

maximum number of bolts (Fig. 7.7.3*a*). For staggered arrangement of bolts, L_{con} is the out-to-out dimension between the centers of the extreme bolts (Fig. 7.7.3*b*).

For any given profile and connecting elements, the connection eccentricity is a fixed geometric property. Connection length, however, is dependent upon the number of bolts required to develop the given factored tensile load, and that in turn is dependent upon the mechanical properties of the member and the capacity of the bolts used. From Eq. 7.7.2, it is seen that as the connection length increases, the shear lag decreases, and the connection efficiency increases. While increasing the length of the connection increases the efficiency of the net section of the member, it decreases the efficiency of the bolts due to unbuttoning, as mentioned in Section 7.3.2. The designer must consider both of these opposing effects in the design of a tension member and its connections.

Lower Bound Values for Reduction Coefficient, *U*

For a selected bolt pitch, the number of fasteners per line, N_l, is a measure of the length of the connection, L_{con}. So, Section B3 of the LRFDC (page 16.1-177 of the LRFDM) provides a set of numerical values for U as a function of N_l. The values provide reasonable ***lower bound values for U*** for the profile types and connection means described and can be used in preliminary designs. Thus, for bolted connections:

1. When the load is transmitted through all of the elements of the cross section, $U = 1$.
2. For all members having N_l equal to 2, $U = 0.75$.
3. For an I-shape connected only at the member's flanges, having a flange width not less than two-thirds the depth of the section, and with no fewer

than three bolts per line in the direction of stress (i.e., if $b_f \geq \frac{2}{3}\,d$, $N_l \geq 3$), then $U = 0.9$.

4. For a T-shape connected only at the member's flange, having a flange width not less than four-third the depth of the stem, and with no fewer than three bolts per line in the direction of stress (i.e., if $b_f \geq \frac{4}{3}\,d$, $N_l \geq 3$), then $U = 0.9$.

5. For I- and T-shapes not meeting the above conditions, and for all other shapes, including built-up cross sections, $U = 0.85$.

Figure 7.7.4 shows several examples of bolted end connections. Also indicated in this figure are the lower bound values for U as per the recommendations given above. For design problems, the use of lower bound values for U suggested here are more convenient since $\bar{x}_{con}$ and L_{con} are unknown before the section has been chosen and the connector group has been designed. However, for analysis problems use of Eq. 7.7.2 is to be preferred.

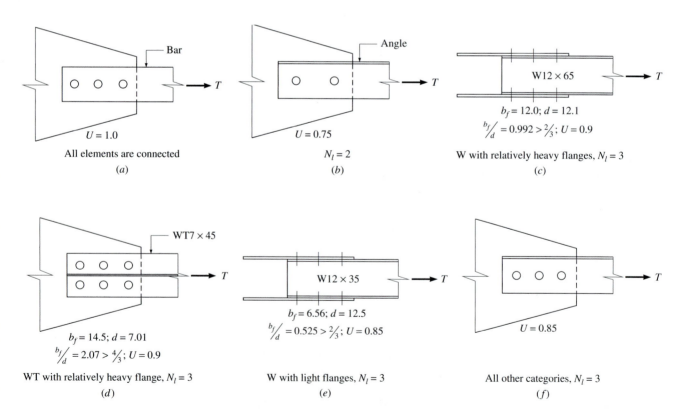

Figure 7.7.4: Examples of lower bound values for U.

7.7.2   Reduction Coefficient, *U,* for Welded Members

Shear lag is a factor in determining the effective area of welded connections whenever welds directly connect some, but not all, of the elements of a tension member. The following is a summary of *U* factors for welded connections as per LRFD Specifications.

1. When tensile load is transmitted by welds to all elements of the cross section of the member, $U = 1$, except when a plate is connected by longitudinal welds only.
2. When a plate is connected by longitudinal fillet welds alone and no transverse weld (as shown in Fig. 7.7.5*a*), the following values of *U* are to be used:

$$
\begin{array}{llll}
U & = 0.75 & \text{when} & 1.0 \leq (L_{lw}/W_{pl}) < 1.5 \\
& = 0.87 & \text{when} & 1.5 \leq (L_{lw}/W_{pl}) < 2.0 \\
& = 1.00 & \text{when} & (L_{lw}/W_{pl}) \geq 2.0
\end{array}
\tag{7.7.3}
$$

where L_{lw} = length of longitudinal weld, in.
$\quad\quad W_{pl}$ = plate width, in.

Note: The length of each longitudinal fillet weld may not be less than the plate width, as per LRFDS Section J2.2*b*.

3. When tensile load is transmitted by transverse welds only (as shown in Fig. 7.7.5*b* for an angle shape), the effective net area A_e shall be taken to be the area of the directly connected element(s). That is, in Eq. 7.7.1:

$$
A_n = A_{ce}; U = 1.0
\tag{7.7.4}
$$

where A_{ce} = area of the directly connected elements using transverse welds, in.^{2}

4. When tensile load is transmitted only by longitudinal welds to a member other than a plate, or by longitudinal welds in combination with transverse welds (as shown in Fig. 7.7.5*c* for an angle shape):

$$
A_n = A_g; U = \min\left[\left(1 - \frac{\bar{x}_{con}}{L_{con}}\right), 0.9\right]
\tag{7.7.5}
$$

where A_g = gross area of the member, in.^{2}
$\quad\quad L_{con}$ = connection length, taken as the length of the longer longitudinal weld, in.
$\quad\quad\quad$ = max $[L_{lw1}, L_{lw2}]$

For combinations of longitudinal and transverse welds, L_{con} is still determined by the length of the longitudinal weld, because the transverse weld has little or no effect on the shear lag problem. That is, a transverse weld does little to transmit load into the unattached elements of the member.

Figure 7.7.5: Members with welded end-connections.

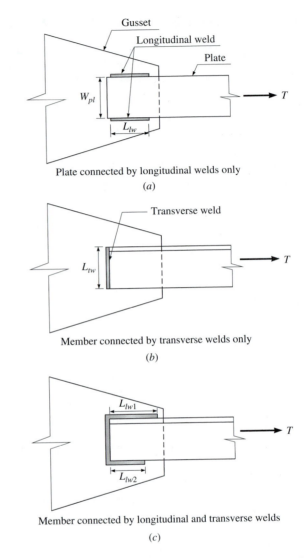

Plate connected by longitudinal welds only

(a)

Member connected by transverse welds only

(b)

Member connected by longitudinal and transverse welds

(c)

Plates welded only longitudinally can fail prematurely due to shear lag if the distance between the welds is too great. Thus, a minimum weld length equal to the plate width, W_{pl}, is stipulated in specifications. Of course, that minimum required length does not apply if the plate is also welded transversely.

EXAMPLE 7.7.1

Plate Member Welded to Gusset

A tension member consists of a 9 by $\frac{1}{2}$ in. A242 Gr 50 steel plate connected to a gusset plate at each end by transverse and longitudinal welds. Determine the design strength of the member.

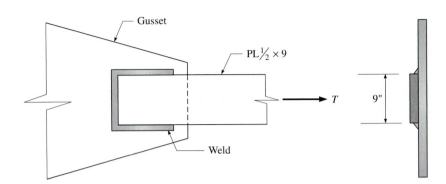

Figure X7.7.1

Solution

Gross area, $A_g = \frac{1}{2}(9) = 4.50$ in.^{2}
As there are no bolt holes, the net area is $A_n = A_g = 4.50$ in.^{2}
As there is only one element in the cross section, and since both longitudinal and transverse welds have been used for this element, the reduction coefficient U equals 1.0.
Effective net area, $A_e = UA_n = 1.0(4.50) = 4.50$ in.^{2}
From LRFDM Table 2-2, for A242 Grade 50 steel we have: $F_y = 50$ ksi and $F_u = 70$ ksi
The design strength of the member corresponding to the limit state of:

tension yielding, $T_{d1} = 0.9F_y A_g = 0.9(50)(4.50) = 203$ kips

tension fracture, $T_{d2} = 0.75F_u A_e = 0.75(70)(4.50) = 236$ kips

So, the design tensile strength of the member is:

$$T_d = \phi T_n = \min[T_{d1}, T_{d2}] = \min[203, 236]$$
$$= 203 \text{ kips} \qquad\qquad \text{(Ans.)}$$

Plate Member Bolted without Stagger

EXAMPLE 7.7.2

A tension member consists of a 9 by $\frac{1}{2}$ in. steel plate connected to a gusset plate at each end by $\frac{7}{8}$-in.-dia. bolts through standard punched holes as shown in Figs. X7.7.2 *a* and *b*. Assuming A242 Gr 50 steel, determine the design tensile strength of the member.

Solution

Gross area, $A_g = 9(\frac{1}{2}) = 4.50$ in.^{2}
Diameter of bolt, $d = \frac{7}{8}$ in.

[continues on next page]

Example 7.7.2 continues ...

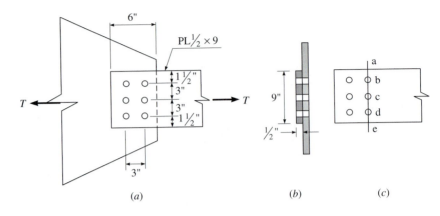

Figure X7.7.2

For standard punched holes, effective width of hole, $d_e = d + \frac{1}{8} = \frac{7}{8} + \frac{1}{8} = 1$ in.

As there is no stagger in the bolt arrangement, there is only one critical path, a-b-c-d-e as shown in Fig. X7.7.2c.

Net area, $A_n = A_g - nd_e t = 4.50 - 3(1)(\frac{1}{2}) = 3.00$ in.^{2}

As there is only one element in the cross section, and it is connected to the gusset plate by bolts, the reduction coefficient U equals 1.0.

Effective net area, $A_e = UA_n = 3.00$ in.^{2}

The design strength of the member corresponding to the limit state of:

$$\text{tension yielding, } T_{d1} = 0.9 F_y A_g = 0.9(50)(4.50) = 203 \text{ kips}$$

$$\text{tension fracture, } T_{d2} = 0.75 F_u A_e = 0.75(70)(3.00) = 158 \text{ kips}$$

So, the design tensile strength of the member is:

$$T_d = \min [T_{d1}, T_{d2}] = \min [203, 158]$$

$$= 158 \text{ kips} \qquad\qquad \text{(Ans.)}$$

EXAMPLE 7.7.3

Plate Member Bolted to Gusset, Staggered Arrangement

A tension member consists of a 9 by $\frac{1}{2}$ in. steel plate connected to a gusset plate at each end by $\frac{7}{8}$-in.-dia. bolts through standard punched holes. The bolts are staggered as shown in Fig. X7.7.3. Assuming A242 Gr 50 steel, determine the design tensile strength of the member.

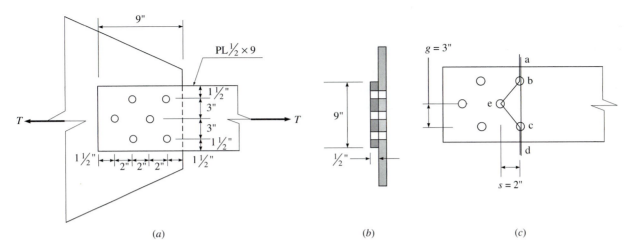

Figure X7.7.3

Solution

Bolts: $d = \frac{7}{8}$ in., $d_e = 1$ in. for STD punched holes
For A242 Gr 50 steel: $F_y = 50$ ksi, $F_u = 70$ ksi
Gross area of plate, $A_g = 9(\frac{1}{2}) = 4.50$ in.^{2}
Two possible critical paths are identified in Fig. X7.7.3c, which provides for two possible net areas.

path a-b-c-d: $A_{n1} = A_g - nd_e t = 4.50 - 2(1)(\frac{1}{2}) = 3.50$ in.^{2}

path a-b-e-c-d: $A_{n2} = A_g - nd_e t + \sum \dfrac{s_j^2}{4g_j} t$

$$= 4.50 - 3(1)(\tfrac{1}{2}) + \left[\frac{2^2}{4(3.0)}\right](\tfrac{1}{2})(2) = 3.33 \text{ in.}^2$$

So, the net area is $A_n = \min [A_{n1}, A_{n2}] = \min [3.50, 3.33] = 3.33$ in.^{2}
Reduction coefficient, $U = 1.0$
Effective net area, $A_e = UA_n = 1.0(3.33) = 3.33$ in.^{2}
The design strength of the member corresponding to the limit state of:

tension yielding, $T_{d1} = 0.9F_y A_g = 0.9(50)(4.50) = 203$ kips

tension fracture, $T_{d2} = 0.75F_u A_e = 0.75(70)(3.33) = 175$ kips

So, the design tensile strength of member is:

$$T_d = \min [203, 175]$$

$$= 175 \text{ kips} \qquad\qquad \text{(Ans.)}$$

EXAMPLE 7.7.4

Plate Member Bolted to Gusset, Staggered Arrangement

Determine the tensile capacity of a $\frac{1}{2}$-in.-thick A514 Grade 100 steel plate connected to a gusset plate as shown in Fig. X7.7.4a. Use $\frac{7}{8}$-in.-dia. bolts and the LRFD Specification.

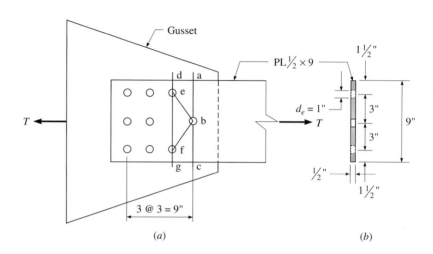

Figure X7.7.4   (a)   (b)

Solution

Bolts: $d = \frac{7}{8}$ in., $d_e = 1$ in. assuming STD punched holes.
(Note: LRFDS Section M2.5 requires drilled holes for A514 steel plates with thicknesses greater than $\frac{1}{2}$ in.)
From Table 2-2 of the LRFDM, for A514 Grade 100 steel,

$$F_y = 100 \text{ ksi}; \; F_u = 110 \text{ ksi}$$

Gross area, $A_g = 9(\frac{1}{2}) = 4.50$ in.²

For net area calculations, three possible failure paths are identified in Fig. X7.7.4a and the relevant data is tabulated below:

Path	A_n	A_n	Force	A_n for 100% T
a-b-c	$4.50 - 1(1)(\frac{1}{2})$	4.00	T	4.00
d-e-b-f-g	$4.50 - 3(1)(\frac{1}{2}) + 2\left[\dfrac{3^2}{4(3)}\right](\frac{1}{2})$	3.75	T	3.75 critical
d-e-f-g	$4.50 - 2(1)(\frac{1}{2})$	3.50	$8T/9$	3.93

Note that on path d-e-f-g only a force $8T/9$ is presumed to act, as the connector b located to the right of the path has already transferred its share $(T/9)$ of the load. The 3.5 sq. in. net area for $0.89T$ acting on this path is equivalent to $3.5/0.89 = 3.93$ sq. in. for 100 percent of T. A comparison of the net areas 4.00, 3.75 and 3.93 for 100 percent of T, in the last column, shows that path d-e-b-f-g governs. Therefore, $A_n = 3.75$ in.^{2}.

As the only element of the section is connected, $U = 1$, and $A_e = UA_n = 3.75$ in.^{2}.

The design strength of the member for the limit state for yielding and fracture are:

tension yielding, $T_{d1} = 0.9F_y A_g = 0.9(100)(4.50) = 405$ kips

tension fracture, $T_{d2} = 0.75F_u A_e = 0.75(110)(3.75) = 309$ kips

Thus, the design tensile strength of the member, T_d, equals 309 kips.

(Ans.)

Angle Member Bolted to Gusset

EXAMPLE 7.7.5

An $8 \times 4 \times \frac{1}{2}$ single angle tension member has two bolt lines in its long leg and one in its short leg. Bolts having $\frac{3}{4}$ in. dia. and standard punched holes are used. They are arranged with a pitch of 3 in. and a stagger of $1\frac{1}{2}$ in. on standard gage lines as shown in Fig. X7.7.5a. The force T is transmitted to the gusset by bolts in both legs, using a lug angle not shown in the figure. Assuming the member is made of A588 Gr 50 steel, determine the design tensile strength of this member.

Solution

Section: L8×4×½
Gross area, $A_g = 5.80$ in.^{2}
From Table 6.2.1, the gage distances are:

for 4-in. leg: $g_1 = 2\frac{1}{2}$ in.

for 8-in. leg with two lines of bolts: $g_2 = 3$ in., $g_3 = 3$ in.

The angle is flattened out as shown in Fig. X7.7.5b.
The gage distance between the bolt lines closest to the heel measured along the centerline of the plate thickness is given by:

$$g_{12} = g_1 + g_2 - t = 2\frac{1}{2} + 3 - \frac{1}{2} = 5 \text{ in.}$$

[continues on next page]

Example 7.7.5 continues ...

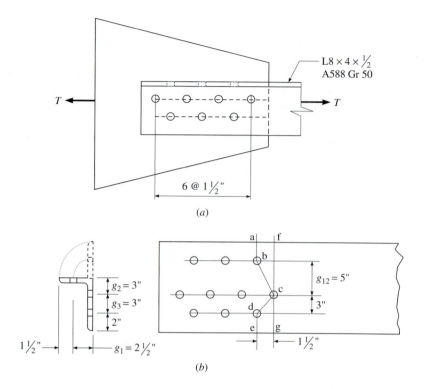

Figure X7.7.5

Two possible failure paths are shown in Fig. X7.7.5, for which the net areas are:

$$\text{path a-b-c-d-e } A_{n1} = 5.80 - 3(\tfrac{7}{8})(\tfrac{1}{2}) + \left[\frac{1.5^2}{4(5)}\right]\left(\frac{1}{2}\right) + \left[\frac{1.5^2}{4(3)}\right]\frac{1}{2}$$

$$= 4.64 \text{ in.}^2$$

$$\text{path f-c-g } A_{n2} = 5.80 - 1(\tfrac{7}{8})(\tfrac{1}{2}) = 5.36 \text{ in.}^2$$

Thus the net area is $A_n = \min [4.64, 5.36] = 4.64$ in.^{2}
As both legs of the member are connected to the gusset, the reduction coefficient U equals 1.0.
Effective net area, $A_e = UA_n = 1.0(4.64) = 4.64$ in.^{2}
The design strength of the member for the limit states for yielding and fracture are:

$$\text{tension yielding, } T_{d1} = 0.9(50)(5.80) = 261 \text{ kips}$$

$$\text{tension fracture, } T_{d2} = 0.75(70)(4.64) = 244 \text{ kips}$$

So, design tensile strength of the member,

$$T_d = \min [261, 244] = 244 \text{ kips} \qquad \text{(Ans.)}$$

W-Shape Bolted to Gussets

EXAMPLE 7.7.6

Determine the design tensile strength of a W10×60 member of A572 Grade 65 steel. As shown in Fig. X7.7.6, the member end connection has two lines of ⅞-in.-dia. A490-X high-strength bolts in each flange, six in each line. Neglect block shear strength. Use (*a*) the lower bound value for *U* suggested in the LRFDC and (*b*) the Eq. 7.7.2 for *U*.

Solution

Shape: W10×60
Steel: A572 Grade 65
From LRFDM Table 2-4, a W10×60 belongs to Group 2 shapes. Also, LRFDM Table 2-1 indicates that this shape is available in A572 Grade 65 steel, though this is not the preferred material specification. In addition, $F_y = 65$ ksi, and $F_u = 80$ ksi, from this table.
From Table 1-1 of the LRFDM for a W10×60 section:

$$A = 17.6 \text{ in.}^2, \qquad t_f = 0.680 \text{ in.}$$

$$b_f = 10.1 \text{ in.}, \qquad d = 10.2 \text{ in.}$$

Gross area, $A_g = 17.6$ in.^{2}
Design strength, $T_{d1} = 0.9F_y A_g = 0.9(65)(17.6) = 1030$ kips
Diameter of bolt, $d = ⅞$ in., therefore $d_e = 1$ in. assuming STD punched holes.

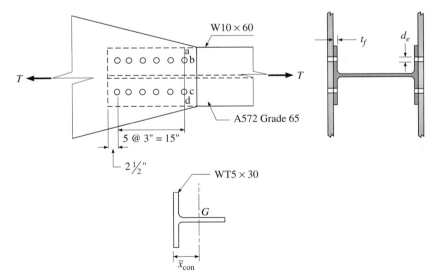

Figure X7.7.6

[continues on next page]

Example 7.7.6 continues ...

From Fig. X7.7.6 it is observed that the critical section passes through four bolt holes (two in each flange). Hence:

$$A_n = A_g - n\, d_e\, t_f = 17.6 - 2(2)(1.00)(0.680) = 14.9 \text{ in.}^2$$

a. Reduction coefficient U from LRFDC

As the number of bolts per line, $N_l = 6 > 3$, and the connection is to the flanges of a W-shape with the ratio $\frac{b_f}{d} = \frac{10.1}{10.2} = 0.99 > \frac{2}{3}$, the reduction coefficient, $U = 0.9$.

Effective net area, $A_e = UA_n = 0.9 \times 14.9 = 13.4 \text{ in.}^2$

Design strength, $T_{d2} = 0.75\, F_u\, A_e = 0.75(80)(13.4) = 803$ kips

From LRFDM Table 7-10 for a $\frac{7}{8}$-in.-diameter A490-X bolt in single shear, the design shear strength, B_{dv} equals 33.8 kips. Also, for a $\frac{7}{8}$-in.-diameter bolt through standard holes and plate material with $F_u = 65$ ksi, the design bearing strength is given as 102 kips/in. thickness, for an interior bolt with $p = 3$ in. (from Table 6.8.1), and also for an end bolt with $L_e = 2\frac{1}{2}$ in. (from Table 6.8.2). Hence, for 0.68-in.-thick flange with $F_u = 80$ ksi,

Design bearing strength, $B_{db} = 102(\frac{80}{65})(0.680) = 85.4$ kips

Design strength of bolt, $B_d = \min[B_{dv}, B_{db}] = \min[33.8, 85.4] = 33.8$ kips

Design strength of bolts, $T_{d4} = NB_d = 24\,(33.8) = 811$ kips

Design strength of the member, $T_d = \min[T_{d1}, T_{d2}, T_{d4}]$

$$T_d = \min[1030, 803, 811] = 803 \text{ kips} \qquad \text{(Ans.)}$$

b. Reduction coefficient U from Eq. 7.7.2

As the W10×60 shape is connected by two gussets, the tributary area corresponding to each gusset is a WT5×30. From LRFDM Table 1-8, for a WT5×30, the distance $\bar{y}$ of the center of gravity of the T-shape from the flange face is 0.884 in. Thus:

$$\bar{x}_\text{con} = \bar{y}_\text{PT} = \bar{y}_{\text{WT5} \times 30} = 0.884 \text{ in.}$$

Length of connection, $L_\text{con} = 5p = 5(3.0) = 15.0$ in.

From Eq. 7.7.2:

$$U = \min\left[\left(1 - \frac{0.884}{15.0}\right), 0.9\right] = \min[0.941;\, 0.9] = 0.9$$

As the U value is same, there is no change in the design strength.

(Ans.)

EXAMPLE 7.7.7

Strength of a Splice Plate

Determine the design tensile strength of the 12 by $\frac{3}{8}$ in. splice plates shown in Fig. X7.7.7a. Holes are punched for $\frac{7}{8}$-in.-dia. bolts. Use 572 Grade 50 steel.

Solution

Gross area of splice plates, $A_g = 2(12)(\frac{3}{8}) = 9.00$ in.^{2}

Diameter of bolt, $d = \frac{7}{8}$ in.

Effective width of hole using standard punched holes, $d_e = \frac{7}{8} + \frac{1}{8} = 1$ in.

Note that for the splice plates shown, the internal forces increase from zero at the ends to a maximum of T at the center.

Three possible failure paths are identified in Fig. X7.7.7. Note that the path a-b-c-d is located at the front end of the splice plates. The possible net widths are:

Figure X7.7.7

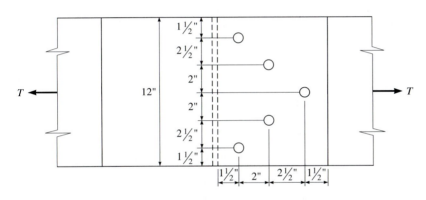

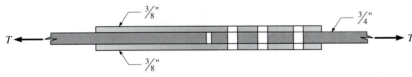

Splice plate assembly

(a)

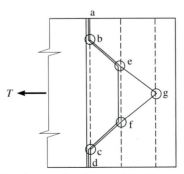

Splice plate showing three possible failure paths

(b)

[continues on next page]

Example 7.7.7 continues ...

$$\text{path a-b-c-d} \rightarrow w_{n1} = 12 - 2(1.0) = 10.0 \text{ in.}$$

$$\text{path a-b-e-f-c-d} \rightarrow w_{n2} = 12 - 4(1.0) + 2\left[\frac{2^2}{4(2.5)}\right] = 8.80 \text{ in.}$$

$$\text{path a-b-e-g-f-c-d} \rightarrow w_{n3} = 12 - 5(1.0) + \frac{2(2^2)}{4(2.5)} + \frac{2(2.5^2)}{4(2)} = 9.36 \text{ in.}$$

Net width, $w_n = \min [w_{n1}, w_{n2}, w_{n3}] = \min [10.0, 8.80, 9.36] = 8.80$ in.
Net area of the splice plates, $A_n = 8.80(\frac{3}{4}) = 6.60$ in.^{2}
According to LRFDS Section J5.2, for fracture of connecting elements, such as the splice plates considered in this example, the net area should not exceed $0.85A_g = 0.85(9.00) = 7.65$ in.^{2}
So, the net area $A_n = 6.60$ in.^{2}
Effective net area, $A_e = UA_n = 1.0(6.60) = 6.60$ in.^{2}
For A572 Grade 50 steel: $F_y = 50$ ksi, $F_u = 65$ ksi
Design strength corresponding to the applicable limit states are:

$$\text{tension yielding, } T_{d1} = 0.9F_y A_g = 0.9(50)(9.00) = 405 \text{ kips}$$

$$\text{tension fracture, } T_{d2} = 0.75F_u A_e = 0.75(65)(6.60) = 322 \text{ kips}$$

Design tensile strength of the splice plates is:

$$T_d = \min [T_{d1}, T_{d2}] = \min [405, 322] = 322 \text{ kips} \qquad \text{(Ans.)}$$

7.8   Block Shear Rupture Strength

Block shear rupture is a failure mode in which one or more blocks of plate material tear out from the end of a tension member or a gusset plate. Such blocks are generally rectangular in shape when bolts are arranged without stagger, and are bounded by the centerlines of bolt holes. In the case of welded connections, the blocks are bound by the centerlines of fillet welds. Block shear resistance may be defined as the sum of the shear resistance provided by the side(s) of the block parallel to the tensile force, and the tensile resistance provided by the side(s) of the block perpendicular to the load. Design criteria for block shear strength are given in LRFDS Section J4.3. The specification was based on the following research work: Birkmoe and Gilmor [1978], Ricles and Yura [1983], Hardash and Bjorhovde [1985], Epstein [1992], and Easterling and Gonzales [1993]. The case of a single failure block in the web of a channel section member is shown in Figure 7.8.1. Here the shaded rectangular block abcd would tend to tear out by shear along the longitudinal surfaces ab and cd, and by tension on the transverse surface, bc. Block shear failure occurs when high tensile forces are transmitted through relatively thin material and a short connection length. Thus, connections detailed using relatively few large diameter bolts and minimum

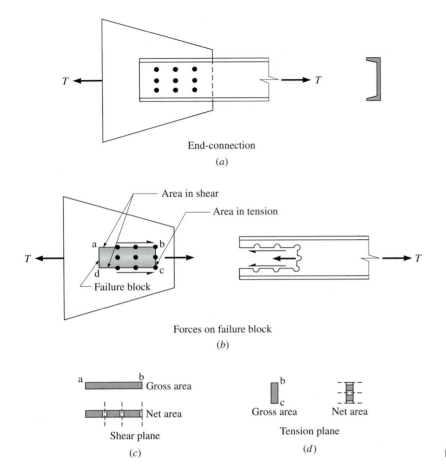

Figure 7.8.1:   Block shear rupture in tension.

values specified for bolt pitch, p, end distance, L_e, and gage, g, should be checked for adequate block shear strength.

When the bolts are staggered, the fracture path may consist of a series of segments, some loaded primarily in tension, including inclined segments, others primarily in shear. The net area of segments normal to the tensile force and any diagonal segments inclined to the tensile force are determined as in Section 7.6.2. For the latter, the net area is increased by $(s^2 t/4g)$ for each diagonal segment.

Based on test results, the LRFD Specification has adopted a model for predicting the block shear strength wherein shear strength along one plane is added to the tension strength of a perpendicular plane. Block shear is essentially a fracture phenomenon, not a yield phenomenon. Thus, block shear failure involves a combination of fracture of the member in one plane, along with a simultaneous yielding or fracture in a perpendicular plane. Therefore, two possible block shear strengths T_{bs1} and T_{bs2} can be calculated. T_{bs1} is the fracture strength on the net tensile area combined with the smaller of yielding or fracture strength on the shear plane. T_{bs2} is the fracture strength on the net shear area combined with the smaller of yielding or fracture strength on the tensile area. The controlling equation is the one that contains the larger first term. Gross areas are used for the limit states of yielding, while net areas are used for the limit states of fracture.

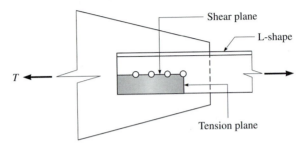

Block shear rupture with one failure block

(a)

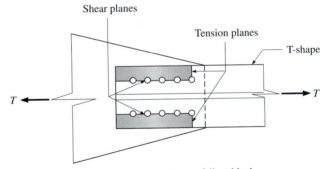

Block shear rupture with two failure blocks

(b)

Block shear rupture in a welded connection

(c)

Figure 7.8.2:   Examples of block shear rupture.

Thus, these formulas are consistent with the philosophy contained in LRFDS Chapter D for tension members. Based on the energy-of-distortion theory, the shear yield stress is taken to be $0.6F_y$, and the shear fracture stress is taken to be $0.6F_u$. Thus:

$$T_{dbs} = T_{bs1} = T_{fnt} + \min[T_{ygv}, T_{fnv}] \quad \text{if } T_{fnt} \geq T_{fnv} \qquad (7.8.1)$$

$$T_{dbs} = T_{bs2} = T_{fnv} + \min[T_{ygt}, T_{fnt}] \quad \text{if } T_{fnv} > T_{fnt} \qquad (7.8.2)$$

where
$$\begin{aligned}
T_{ygt} &= \phi F_y A_{gt} & (7.8.3a)\\
T_{fnt} &= \phi F_u A_{nt} & (7.8.3b)\\
T_{ygv} &= \phi(0.6F_y)A_{gv} & (7.8.3c)\\
T_{fnv} &= \phi(0.6F_u)A_{nv} & (7.8.3d)
\end{aligned}$$

wherein T_{dbs} = design block shear rupture strength, kips
T_{fnt} = tension rupture component, kips
T_{fnv} = shear rupture component, kips
T_{ygt} = tension yielding component, kips
T_{ygv} = shear yielding component, kips
ϕ = resistance factor = 0.75
A_{gv} = gross area subjected to shear, in.^{2}
A_{gt} = gross area subjected to tension, in.^{2}
A_{nv} = net area subjected to shear, in.^{2}
A_{nt} = net area subjected to tension, in.^{2}

Block shear failure should also be checked around the periphery of welded connections. However, as there are no bolt holes involved in such connections, net areas equal gross areas. That is, for welded connections:

$$A_{nt} = A_{gt}; A_{nv} = A_{gv} \qquad (7.8.4)$$

Some additional examples of block shear mode of failure in bolted and welded tension members are shown in Fig. 7.8.2.

Block Shear Rupture Strength, Bolted Connection

EXAMPLE 7.8.1

Determine the design strength of a single C15×50 fastened to a ¾-in. gusset plate. Use A36 steel. Assume A490-X high-strength bolts of ⅞-in.-dia. and standard punched holes. Include block shear strength.

Solution

Bolts: $d = ⅞$ in., $d_e = 1$ in.
A36 Steel: $F_y = 36$ ksi, $F_u = 58$ ksi

[continues on next page]

Example 7.8.1 continues ...

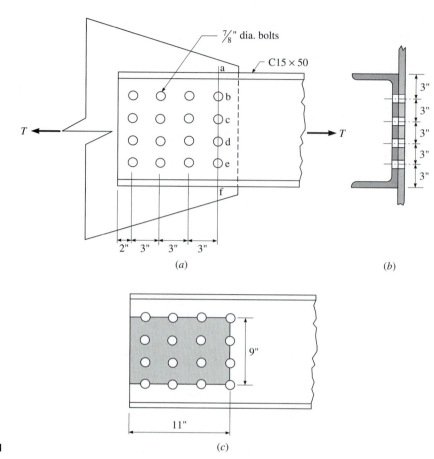

Figure X7.8.1

C15×50 Section: $A = 14.7$ in.^{2}; $t_w = 0.716$ in.; $\bar{x} = 0.799$ in.
Gross area, $A_g = 14.7$ in.^{2}
Design strength corresponding to tension yielding,

$$T_{d1} = 0.9F_y A_g = 0.9(36)(14.7) = 476 \text{ kips}$$

The critical path is path a-b-c-d-e-f shown in Fig. X7.8.1a.
Net area, $A_n = 14.7 - 4(1)(0.716) = 11.8$ in.^{2}
Length of the connection, $L_{con} = 3p = 3(3) = 9.00$ in.
Reduction coefficient,

$$U = \min\left[\left(1 - \frac{\bar{x}_{con}}{L_{con}}\right), 0.9\right] = \min\left[\left(1 - \frac{0.799}{9.00}\right), 0.9\right] = 0.9$$

Effective net area, $A_e = UA_n = 0.9(11.8) = 10.6$ in.^{2}
Design strength corresponding to tension fracture,

$$T_{d2} = 0.75F_u A_e = 0.75(58)(10.6) = 461 \text{ kips}$$

Block shear strength:

Gross area in tension, $A_{gt} = 9.00(0.716) = 6.44$ in.^{2}

Net area in tension, $A_{nt} = 6.44 - (\frac{1}{2} + 1 + 1 + \frac{1}{2})(1.0)(0.716)$
$$= 4.29 \text{ in.}^2$$

Gross area in shear, $A_{gv} = 2(11.0)(0.716) = 15.8$ in.^{2}

Net area in shear, $A_{nv} = 15.8 - 2(1 + 1 + 1 + \frac{1}{2})(1)(0.716) = 10.8$ in.^{2}

$\phi F_u A_{nt} = 0.75 (58)(4.29) = 187$ kips

$\phi(0.6F_u)(A_{nv}) = 0.75(0.6)(58)(10.8) = 282$ kips ← controls

$\phi F_y A_{gt} = 0.75(36)(6.44) = 174$ kips

$\phi(0.6F_y) A_{gv} = 0.75(0.6)(36)(15.8) = 256$ kips

So,

$$T_{dbs} = \phi(0.6F_u) A_{nv} + \min [\phi F_y A_{gt}, \phi F_u A_{nt}]$$
$$= 282 + \min [174, 187] = 456 \text{ kips}$$

The design strength of the member corresponding to block shear failure is therefore:

$$T_{d3} = T_{dbs} = 456 \text{ kips}$$

Design strength of a single bolt (from LRFDM Table 7-10), $B_d = B_{dv} = 33.8$ kips

Design strength of bolts in the connection, $T_{d4} = NB_d = 16(33.8) = 541$ kips

Design strength of the member,

$$T_d = \min [T_{d1}, T_{d2}, T_{d3}, T_{d4}] = \min [476, 461, 456, 541]$$

$$= 456 \text{ kips (controlled by block shear strength)} \qquad \text{(Ans.)}$$

Block Shear Rupture Strength, Welded Connection

EXAMPLE 7.8.2

A L6 × 3$\frac{1}{2}$ × $\frac{1}{2}$ is used as the web member of a roof truss. As shown in Fig. X7.8.2, $\frac{7}{16}$ in. longitudinal and transverse fillet welds are used to connect the long leg of the angle to the stem of a WT10.5×50.5 used as the truss top chord. Determine the design tensile strength of the angle member. Assume A572 Gr 50 steel and E70 electrodes.

Solution

From Table 1-7 of the LRFDM for a L6×3$\frac{1}{2}$×$\frac{1}{2}$:

$A = 4.48$ in.^{2}; $\bar{x} = 0.833$ in.

Design tensile strength of the member corresponding to the limit state of yielding in the gross section,

$$T_{d1} = 0.9F_y A_g = 0.9(50)(4.48) = 202 \text{ kips}$$

[continues on next page]

Example 7.8.2 continues ...

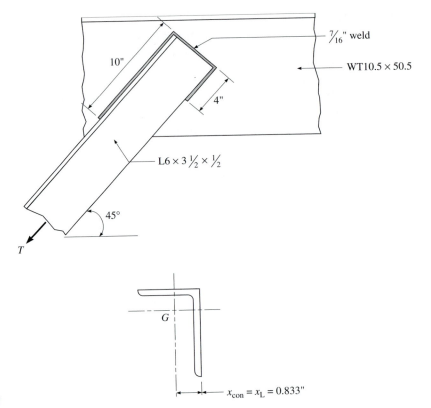

Figure X7.8.2

Length of connection, L_{con} is conservatively taken as the average length of the two longitudinal welds. Thus: $L_{con} = \frac{1}{2}(10 + 4) = 7.00$ in.
Also, for a L6×3½×½ with its long leg welded to the gusset: $\bar{x}_{con} = \bar{x} = 0.833$ in.

Reduction coefficient, $U = 1 - \dfrac{\bar{x}_{con}}{L_{con}} = 1 - \dfrac{0.833}{7.00} = 0.881 < 0.9$

Effective net area, $A_e = UA_g = 0.881(4.48) = 3.95$ in.²
Design tensile strength of the member corresponding to the limit state of fracture in the net section is:

$$T_{d2} = 0.75F_u A_e = 0.75(65)(3.95) = 193 \text{ kips}$$

From Table 6.19.1, the design strength of a unit length $\frac{7}{16}$-in. E70 longitudinal fillet weld is 9.74 kips. Noting that the design strength of a transverse weld is 1.5 times that value:

$$T_{d4} = (10.0 + 4.00)(9.74) + 6.00(9.74)(1.5) = 224 \text{ kips}$$

Block shear strength noting that the thickness of stem of a WT10.5×50.5 is ½ in., is:

Gross area in shear, $A_{gv} = (10.5 + 4.50)(0.5) = 7.50$ in.^{2}
Gross area in tension, $A_{gt} = 6.00 (0.5) = 3.00$ in.^{2}
Net area in shear, $A_{nv} = A_{gv} = 7.50$ in.^{2}
Net area in tension, $A_{nt} = A_{gt} = 3.00$ in.^{2}
$\phi F_u A_{nt} = 0.75(65)(3.00) = 146$ kips
$\phi(0.6F_u) A_{nv} = 0.75(0.6)(65)(7.50) = 219$ kips $\leftarrow$ controls
$\phi F_y A_{gt} = 0.75(50)(3.00) = 113$ kips
$\phi(0.6F_y) A_{gv} = 0.75(0.6)(50)(7.50) = 169$ kips

So,

$$T_{bs} = \phi(0.6F_u) A_{nv} + \min [\phi F_y A_{gt}, \phi F_u A_{nt}]$$

$$= 219 + \min [113, 146] = 332 \text{ kips}$$

The design strength of the member corresponding to block shear failure is thus:

$$T_{d3} = T_{dbs} = 332 \text{ kips}$$

Design tensile strength of the member,

$$T_d = \min [T_{d1}, T_{d2}, T_{d3}, T_{d4}] = \min [202, 193, 332, 224]$$
$$= 193 \text{ kips}$$

(Ans.)

7.9   Slenderness Limitations

Tension members that are too slender may be damaged during shipping and erection, and may deflect excessively under their own weight. Such members used in open trusses and towers exposed to wind, and in structures supporting mechanical equipment, may vibrate undesirably if they are too slender. To prevent excessive sag (deflection) and flutter (lateral vibration) and to provide adequate rigidity, Section B7 of the LRFD Specification suggests that slenderness ratios for tension members be limited to a maximum value of 300. It should be noted that this is a suggested value, not a mandatory requirement. The slenderness ratio is defined as the ratio of the unbraced length of the member, L, to the least radius of gyration, r_{min}. Thus:

$$\frac{L}{r_{min}} \le 300 \qquad (7.9.1)$$

The unbraced length is considered to be the distance between adjacent points of bracing or other lateral support, measured along the longitudinal axis of the member.

For common rolled shapes, values of r_x and r_y about the principal axes x and y are tabulated in Part 1 of the LRFDM, and need not be computed. For doubly symmetric shapes such as I-shapes and rectangular HSS, $r_{min} = r_y$

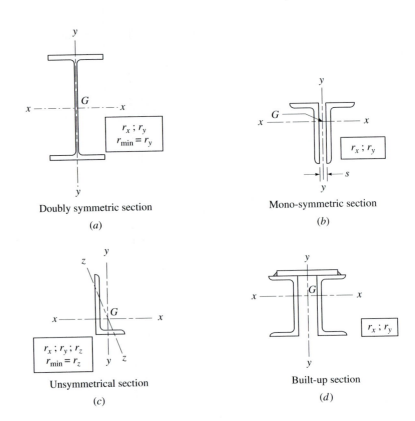

Figure 7.9.1: Minimum radius of gyration of sections.

(Fig. 7.9.1a). For mono-symmetric shapes such as T-, and C-shapes, $r_{min} = \min[r_x, r_y]$. For single angles, the principal axes are inclined with respect to the legs of the angle (axis z-z of Fig. 7.9.1c), and $r_{min} = r_z$. In the case of double angles having a back-to-back spacing of s (Fig. 7.9.1b), r_y is often r_{min} and is a function of separation, s. The spacing, s, represents the thickness of the gusset plate used in the connection. Values of r_y for double angles, for three commonly used spacings (0, $\tfrac{3}{8}$, and $\tfrac{3}{4}$ in.) are tabulated in Table 1-14 of the LRFDM. When a member is built up from two or more plates or shapes (Fig. 7.9.1d), for which the section properties are not tabulated in the Manual, the minimum moment of inertia of the compound section, I_{min}, must be found before r_{min} can be determined (see Section 2.8).

EXAMPLE 7.9.1

Maximum Recommended Member Lengths

Determine the maximum length recommended by the LRFD Specification for a tension member having a cross section consisting of (a) W10×60, (b) WT5×16.5, (c) 5×3½ × ⅝ single angle, (d) two 6×4× ½ angles with long legs connected to a $\frac{3}{8}$-in. gusset, and (e) two 6×4× $\frac{1}{2}$ angles with long legs connected to a $\frac{5}{8}$-in. gusset.

Solution

For tension members, $L_{\text{max}} = 300\ r_{\text{min}}$

a. W10×60

From LRFDM Table 1-1, $r_x = 4.39$ in., $r_y = 2.57$ in. $\rightarrow r_{\text{min}} = r_y = 2.57$ in.

$$L_{\text{max}} = 300(2.57) = 771 \text{ in.} = 64.3 \text{ ft} \qquad \text{(Ans.)}$$

b. WT5×16.5

From LRFDM Table 1-8, $r_x = 1.26$ in., $r_y = 1.94$ in. $\rightarrow r_{\text{min}} = r_x = 1.26$ in.

$$L_{\text{max}} = 300(1.26) = 378 \text{ in.} = 31.5 \text{ ft} \qquad \text{(Ans.)}$$

c. L5×3$\frac{1}{2}$×$\frac{5}{8}$

From LRFDM Table 1-7, $r_x = 1.56$ in., $r_y = 0.987$ in., $r_z = 0.746$ in.

$$r_{\text{min}} = r_z = 0.746 \text{ in.}$$

$$L_{\text{max}} = 300(0.746) = 223.8 \text{ in.} = 18.7 \text{ ft} \qquad \text{(Ans.)}$$

d. 2L6×4×$\frac{1}{2}$×$\frac{3}{8}$ LLBB

From LRFDM Table 1-7, $r_x = 1.92$ in., and from LRFDM Table 1-14, for a separation, $s = \frac{3}{8}$ in., $r_y = 1.64$ in. $\rightarrow r_{\text{min}} = r_y = 1.64$ in.

$$L_{\text{max}} = 300(1.64) = 492 \text{ in.} = 41.0 \text{ ft} \qquad \text{(Ans.)}$$

e. 2L6×4×$\frac{1}{2}$×$\frac{5}{8}$ LLBB

From LRFDM Table 1-7, $r_x = 1.92$, while r_y may be interpolated from the values given in LRFDM Table 1-14 for $s = \frac{3}{8}$ in. and $s = \frac{3}{4}$ in., namely 1.64 and 1.78. Thus: $r_y = 1.64 + (1.78 - 1.64)\ [\frac{2/8}{3/8}] = 1.73$ in. Or, it could be calculated more exactly as follows:

For a single $6 \times 4 \times \frac{1}{2}$ angle from Table 1-7 of the LRFDM:

$$I_y = I_o = 6.21 \text{ in.}^4, \bar{x} = 0.986 \text{ in.}, A = 4.72 \text{ in.}^2$$

So, for the double angles with $s = \frac{5}{8}$ in.,

$$d_1 = \frac{s}{2} + \bar{x} = \frac{1}{2}\left(\frac{5}{8}\right) + 0.986 = 1.30$$

$$I_y = \sum(I_o + Ad^2)_i = 2[6.21 + 4.72\ (1.30)^2] = 28.4 \text{ in.}^4$$

$$r_y = \sqrt{\frac{28.4}{2(4.72)}} = 1.73 \text{ in.}$$

[*continues on next page*]

Example 7.9.1 continues ...

So,

$$L_{max} = 300(1.73) = 519 \text{ in.} = 43.3 \text{ ft} \qquad \text{(Ans.)}$$

7.10   Design of Tension Members

In design problems, the required tensile strength of a member, T_u, is known. It is obtained from an analysis of the entire structure subjected to factored loads as discussed in Chapters 3, 4 and 5. The design task then consists of selecting a section and end connections such that the design tensile strength, T_d, is greater than or equal to the required strength, T_u. The dimensions of the section and end connections should be such that they fit the overall architectural dimensions of the structure. Thus, for design:

$$T_d = \min \ [T_{d1}, T_{d2}, T_{d3}, T_{d4}] \geq T_u \qquad (7.10.1)$$

or

$$T_{d1} \geq T_u; \quad T_{d2} \geq T_u; \quad T_{d3} \geq T_u; \quad T_{d4} \geq T_u \qquad (7.10.2)$$

Here,

T_u = required tensile strength of the member (see Chapter 5)
T_d = design strength of the member
T_{d1} = design strength corresponding to the limit state of yielding in gross section (Eq. 7.4.5)
T_{d2} = design strength corresponding to the limit state of fracture in the net section (Eq. 7.4.6)
T_{d3} = design strength corresponding to the limit state of block shear rupture (Eqs. 7.8.1 and 2)
T_{d4} = design strength of the connectors (see Chapter 6)

If necessary, it is often possible to increase the block shear rupture strength, T_{d3}, by making changes to the connection layout (e.g., by increasing the end distance, L_e, of the bolts). Similarly, by changing the type, number, and/or arrangement of bolts, it is usually possible to increase T_{d4} so that it will not limit the strength of the member. Thus, member selection is usually dictated by the calculated values of T_{d1} and T_{d2}. For design purposes, it is convenient to rewrite the first two relations of Eq. 7.10.2. To satisfy the limit state of yielding in the gross section, the gross area must satisfy the relation:

$$A_{g1} \geq \frac{T_u}{0.9F_y} \qquad (7.10.3)$$

while to satisfy the limit state of fracture in the net section, the net area must satisfy the relation:

$$A_n \geq \frac{T_u}{0.75F_u U} \qquad (7.10.4)$$

The gross area required to satisfy the limit state of tension fracture must be greater than (or at least equal to) the sum of the net cross-sectional area and the area associated with the bolt holes. This may be expressed as:

$$A_{g2} \geq \frac{T_u}{0.75F_uU} + \text{estimated loss in area due to bolt holes} \quad (7.10.5)$$

Equation 7.10.5 is preferable since the required size of the member, that is, gross area required to resist the factored load, T_u is sought. Given the factored load, the designer must first select a trial cross section that satisfies Eq. 7.10.3. Then a trial connection must be designed so that the reduction coefficient, U, can be obtained from Eq. 7.7.2 or from the table of permissible lower bound values discussed in Section 7.7.1. Inserting the appropriate value for U and an estimated loss in area due to bolt holes into Eq. 7.10.5, the designer must then verify that the gross area of the trial section is indeed greater than or equal to the right-hand side of the equation. Several iterations of this process may be necessary in order to obtain a section that satisfies A_{g2}. Since the actual number of bolts that will ultimately be used in the final design is not known at the start of the process, the following rules of thumb are often found beneficial:

1. For single angle members, deduct one hole if the size of the connected leg is estimated to be 4 in. or less, and two holes for larger sizes.
2. For double angle members, deduct twice the area suggested in (1).
3. For plates, deduct one hole for every 3 in. of width.
4. For channels connected at the flanges, deduct one hole for each flange.
5. For T-sections connected at the flange, deduct two holes for the flange.
6. For I-shapes connected at the flanges, deduct two holes for each flange (a total of four holes).
7. For a connection to the stem of a T, the web of a C, or the web of a W-shape, deduct one hole for every 3 in. of depth.

Exceptions are bound to occur, and redesigns are often needed. As previously mentioned, the type of cross section selected often depends on the type of end connection required. That is, a particular shape may be chosen simply because it facilitates less expensive and/or easier connections. Connections should be arranged such that the eccentricity of the connection is minimized. That is, the distance between the center of gravity of the connection and the center of gravity of the member should be minimized so as to reduce the influence of end moments upon the member. Recall, the member has been idealized as being subjected to simple uniaxial tension only. Also, to reduce shear lag and increase the efficiency of the member, as much of the section as possible should be connected to gusset plates. Thus, for a single angle member used as a tension member, the lightest section will consist of an unequal leg angle, with the long leg connected to the gusset (see Fig. 7.7.1e).

Using the above rules of thumb, the designer may use the larger value of the gross area required from Eqs. 7.10.3 and 7.10.5 as an initial size estimate.

That is:

$$A_g \geq \max \ [A_{g1}, A_{g2}] \tag{7.10.6}$$

The maximum recommended slenderness ratio, L/r_{min}, for tension members is 300, as per the LRFDS. So, only sections that satisfy the following relation are retained for further consideration in design.

$$r_{min} \geq \frac{L}{300} \tag{7.10.7}$$

The relation in Eq. 7.10.2 expressing the limit state of strength of bolts in the end connection can be rewritten, to isolate the required number of bolts, as follows:

$$N \geq \frac{T_u}{B_d} \tag{7.10.8}$$

where B_d is the design strength of a single bolt. The plates bearing on the bolts are usually thick enough and have clear distances large enough that the bolt bearing on the member and on the gusset do not limit the capacity of the connection. The design strength B_d is therefore the design shear strength of a single bolt, B_{dv}. The N bolts should be arranged to give the largest possible net area for the tension member. Common pitches for bolts are $2\frac{1}{2}$, 3, and $3\frac{1}{2}$ in. for $\frac{3}{4}$, $\frac{7}{8}$, and 1 in. diameter bolts, respectively.

In general, it is preferable that yielding in the gross section occurs before fracture in the net section, as yielding provides a visible signal of structural distress (large deformations). Yielding has the additional advantage that a steel member at yield can still carry some additional load prior to total failure due to the increase in strength that occurs as the material strain hardens. However, this preference is more difficult to meet with high-strength steels having ultimate tensile stresses that are not much greater than their yield stresses (e.g., A514 Grade 100 steel with $F_y = 100$ ksi and $F_u = 110$ ksi).

7.10.1   Design Aids

Values of F_y and F_u for all presently available structural steels are given in Table 2 of the LRFD Specification. LRFDM Table 9-1 gives values of area reduction for different plate thicknesses and bolt diameters, for standard, oversized, short-slotted, and long-slotted holes. Design stresses for tension members as a function of F_y and F_u are given in Table 7.4.1 for various structural steels.

E X A M P L E   7 . 1 0 . 1

Design of a WT Member Bolted to Gusset

Select the lightest WT5 needed for the diagonal brace studied in Example 5.2.5 to carry a factored tensile load, T_u, of 200 kips. Use $\frac{3}{4}$-in.-dia. A325-X bolts and A992 steel.

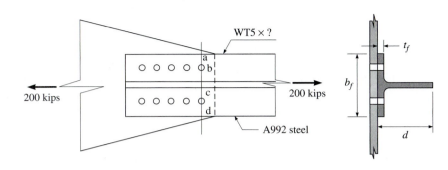

WT5 × ?

a
b
c
d

200 kips

200 kips

A992 steel

t_f

b_f

d

Figure X7.10.1

Solution

A992 steel: $F_y = 50$ ksi, $F_u = 65$ ksi

For the limit state of tension yielding, the gross area required,

$$A_{g1} \geq \frac{T_u}{0.9F_y} = \frac{200}{0.9(50)} = 4.44 \text{ in.}^2$$

Diameter of the bolts, $d = \frac{3}{4}$ in.

Assuming standard punched holes, $d_e = \frac{3}{4} + \frac{1}{8} = \frac{7}{8}$ in.

From Table 1-8 of the LRFDM for WT5 sections with A_g slightly greater than the required gross area of 4.44 sq in., the ratio b_f/d is approximately $8.0/5.0 = 1.60$. As the connection is to the flange of the WT; $b_f/d = 1.60 > \frac{4}{3}$; and the assumed number of fasteners per line, $N_l \geq 3$, the reduction coefficient U from LRFDC Sec. B3 is 0.90. Hence, the net area required for the limit state of tension fracture is:

$$A_n \geq \frac{T_u}{0.75F_u U} = \frac{200}{0.75(65)(0.90)} = 4.56 \text{ in.}^2$$

From Fig. X7.10.1, it is seen that the critical net section corresponds to the transverse section a-b-c-d through two bolt holes in the flange. Hence,

$$A_n = A_g - 2(\tfrac{7}{8})t_f \geq 4.56 \quad \text{and} \quad A_{g2} \geq 4.56 + 1.75t_f$$

where t_f is the thickness of the flange.

Thus, a WT5 must be selected such that:

$$A_g \geq \max [A_{g1}, A_{g2}] = \max [4.44, 4.56 + 1.75t_f] = 4.56 + 1.75t_f$$

Iteration begins with a WT5×16.5, the lightest section that provides an area greater than 4.56 in.². If necessary, one must proceed to heavier WT5 sections until the second requirement for A_g is also satisfied. The following tabular procedure may be found useful in making the selection:

[continues on next page]

Example 7.10.1 continues ...

Section	$A_g = A_{WT}$ (in.²)	t_f (in.)	$A_{g\,req} =$ (4.56 + 1.75 t_f) (in.²)	Remarks
WT5×16.5	4.85	0.435	5.32	N.G.
WT5×19.5	5.73	0.530	5.49	O.K.

Thus, a WT5×19.5 is the lightest WT5 that satisfies both requirements. Check $\frac{b_f}{d} = \frac{7.99}{4.96} = 1.61 > \frac{4}{3}$ as assumed in the determination of the coefficient U.

From LRFDM Table 7-10, the design shear strength for a $\frac{3}{4}$-in.-dia. A325-X bolt in single shear is 19.9 kips. Assuming that bearing of the bolt on the flange will not be critical, the number of bolts required is:

$$N = \frac{T_u}{B_d} = \frac{200}{19.9} \approx 10.0$$

Provide ten $\frac{3}{4}$-in.-dia. A325-X type bolts (5 in each row). Select typical pitch of 3 in. and end distance of 2 in. The design bearing strength of a $\frac{3}{4}$-in.-dia. interior bolt with a pitch of 3 in. on a plate with $F_u = 65$ ksi is obtained from Table 6.8.1 as 87.7 kips/in. thickness. Similarly, the design bearing strength of a $\frac{3}{4}$-in.-dia. end bolt with an end distance of 2 in. on a plate with $F_u = 65$ ksi is obtained form Table 6.8.2 as 87.7 kips/in. thickness. The design bearing strength of these bolts on a 0.53 in. thick A992 steel flange plate is:

$$B_{db} = 0.53 \times 87.7 = 46.9 \text{ kips} > 19.9 \text{ kips}$$

indicating that bearing does not control the bolt design as assumed. Length of the connection, $L_{con} = 4(3.0) = 12.0$ in.

Also, $\bar{x}_{con} = \bar{y}_{WT5 \times 19.5} = 0.876$ in.

Reduction coefficient, $U = \min\left[\left(1 - \frac{0.876}{12.0}\right), 0.9\right] = 0.90$

The assumption is valid, so use a WT5×19.5 of A992 steel.   (Ans.)

EXAMPLE 7.10.2

Design of a W-Member Bolted to Gussets

A member of a truss is required to transmit a factored tensile load of 415 kips. The member is 30 ft long. A588 Grade 50 steel shapes are readily available. Use $\frac{7}{8}$-in.-dia. bolts. Assume that the end connections are to the flanges only and that there are two lines of bolts in each flange. Neglect block shear strength.

a. Assume that there are at least three bolts in each line and select the lightest W16-shape.

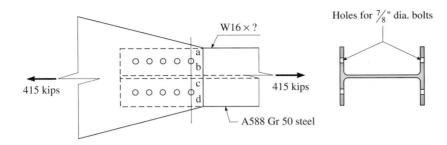

Holes for $\frac{7}{8}$" dia. bolts

Figure X7.10.2

b. For the shape selected, determine the minimum length of connection to be provided to achieve the assumed U.

Solution

a. Member selection
A588 Grade 50 steel: $F_y = 50$ ksi, $F_u = 70$ ksi
Required member strength, $T_u = 415$ kips
Design strength corresponding to tension yielding is $T_{d1} = 0.9F_y A_g \geq T_{req}$.
Therefore,

$$A_{g1} \geq \frac{T_u}{0.9F_y} = \frac{415}{0.9(50)} = 9.22 \text{ in.}^2$$

As the member length is 30 ft, $r_{min} \geq \dfrac{L}{300} = \dfrac{30.0(12)}{300} = 1.20$ in.

From Table 1-1 of the LRFD Manual, it is seen that the lightest W16 section satisfying these two requirements is a W16×36 section with:

$$A_g = 10.6 \text{ in.}^2 > 9.22 \text{ in.}^2 \qquad\qquad \text{O.K.}$$
$$r_y = 1.52 \text{ in.} > 1.20 \text{ in.} \qquad\qquad \text{O.K.}$$
$$b_f = 6.99 \text{ in.}, \ d = 15.9 \text{ in.}, \ \frac{b_f}{d} = 0.439$$
$$t_f = 0.430 \text{ in.}, \ r_x = 6.51 \text{ in.}$$

As $N_l \geq 3$ and $b_f/d < \frac{2}{3}$, the reduction coefficient for preliminary design may be assumed to be 0.85. As there are four lines of bolts connecting the two flanges to the gussets, the net area provided is:

$$A_n = A_g - 4d_e t_f = 10.6 - 4(1)(0.430) = 8.88 \text{ in.}^2$$

Design strength corresponding to the limit state of tension fracture:

$$T_{d2} = 0.75F_u U A_n = 0.75(70)(0.85)(8.88)$$
$$= 396 \text{ kips} < 415 \text{ kips. N.G.}$$

Select the next heavier section, a W16×40, which provides (details not shown here) a design strength $T_d = 428$ kips > 415 kips.   O.K.
(Ans.)

[*continues on next page*]

Example 7.10.2 continues ...

b. Minimum length of connection for a W16×40

The tributary area for each gusset connection corresponds to a WT8×20. From LRFDM Table 1-8, $\bar{y} = 1.81$ in. for a WT8×20. The length of connection required to achieve a U value of 0.85 can be determined from the relation:

$$U = 1 - \frac{\bar{x}_{con}}{L_{con}} \longrightarrow 0.85 = 1 - \frac{1.81}{L_{con}} \longrightarrow L_{con} \approx 12.0 \text{ in.}$$

So, select a W16×40 section of A588 Grade 50 steel, and provide a connection length of at least 12 in.

(Ans.)

EXAMPLE 7.10.3

Design of a Bolted Double-Angle Member

A tension member is composed of two unequal leg angles back to back and is connected to ¾-in.-thick gusset plates by ¾-in.-dia. bolts through standard holes. The eight bolts required at each end-connection are arranged in two rows without stagger in the 6 in. legs. If the tensile force in the member under factored loads is 390 kips, and the member length is 36 ft, select the lightest section. Use A572 Grade 42 steel. Select suitable bolts.

Solution

A572 Grade 42 steel: $F_y = 42$ ksi, $F_u = 60$ ksi from LRFDM Table 2-2
Bolts: $d = \frac{3}{4}$ in., $d_e = \frac{3}{4} + \frac{1}{8} = \frac{7}{8}$ in.
Required tensile capacity of member, $T_u = 390$ kips

Limit state of tension yielding: $A_{gl} \geq \dfrac{T_u}{0.9F_y} = \dfrac{390}{0.9(42)} = 10.3 \text{ in.}^2$

As only one leg of the angle is connected to the gusset, and $N_l = 4 > 3$, the reduction coefficient $U = 0.85$.

Limit state of tension fracture: $A_n \geq \dfrac{T_u}{0.75F_uU} = \dfrac{390}{0.75(60)(0.85)}$

$= 10.2 \text{ in.}^2$

From Fig. X7.10.3, it is seen that the critical net path is the transverse section a-b-c-d through two bolt holes in each connected leg.

$$A_n = A_g - 2(2)(\tfrac{7}{8})\,t \geq 10.2 \text{ in.}^2 \longrightarrow A_{g2} \geq 10.2 + 3.5t$$

where t is the thickness of the angle. Also, the minimum r required to satisfy LRFDS Section B7 for tension members is found from Eq. 7.10.7:

$$r_{min} \geq \frac{L}{300} = \frac{36(12)}{300} = 1.44 \text{ in.}$$

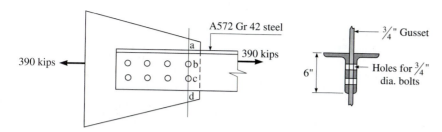

Figure X7.10.3

Thus, the designer must select a pair of unequal leg angles having 6-in. legs separated by a distance of ¾ in. that satisfies the following conditions:

$$A_g \geq A_{g\,req} = \max\,[A_{g1}, A_{g2}] = \max[10.3 \text{ in.}^2,\ 10.2 + 3.5t \text{ in.}^2]$$

$$r_{min} \geq 1.44 \text{ in.}$$

The following tabular procedure may be found useful in making the selection, using the data for single and double angles given in the LRFDM Tables 1-7 and 1-14, respectively.

Angle	A_L (in.²)	$A_g = A_{2L}$ (in.²)	t (in.)	$A_{g\,req}$ (in.²)	r_x (in.)	r_y (in.)	Remarks
L6×4×⅝	5.83	11.7	⅝	12.4	1.90	1.81	N.G.
L6×4×¾	6.90	13.8	¾	12.8	1.88	1.83	O.K.

Tentatively assume a connection with $p = 3$ in. and $L_e = 2$ in.
Length of the connection, $L_{con} = 3p = 3(3.0) = 9.00$ in.
For a L6×4×¾ with the long leg bolted to the gusset, $\bar{x}_{con} = \bar{x}_L = 1.08$ in. From Eq. 7.7.2, reduction coefficient,

$$U = \min\left[\left(1 - \frac{1.08}{9.00}\right),\ 0.9\right] = 0.880 > 0.85 \text{ assumed. O.K.}$$

Provide 2L6×4×¾ LLBB of A572 Grade 42 with long legs connected to a ¾-in.-thick gusset.   (Ans.)
Required strength of a bolt, $B_{req} = 390/8 = 48.8$ kips
The ¾-in.-dia. bolts are in double shear.
So, from LRFDM Table 7-10, tentatively select A490-X bolts providing a design shear strength, $B_{dv} = 49.7$ kips > 48.8 kips O.K.
Design bearing strength,

$$B_{db} = 0.75(2.4)(60)(¾)(¾) = 60.8 \text{ kips} > 48.8 \text{ kips O.K.}$$

So, select eight A490-X bolts. Use $p = 3$ in. and $L_e = 2$ in.   (Ans.)

EXAMPLE 7.10.4

Design of a Plate Member Bolted to a Gusset

Design the lap joint shown in Fig. X7.10.4a. The tension member is subjected to a dead load of 20 kips and a live load of 60 kips. Use A36 steel plates and $\frac{7}{8}$-in.-dia. A307 bolts. Each bolt center line shown in Fig. X7.10.4a represents a column of bolts across the transverse plate width.

Solution

Service loads: $D = 20$ kips; $L = 60$ kips
Factored load, $T_u = 1.2D + 1.6L = 1.2(20) + 1.6(60) = 120$ kips
Bolts: $d = \frac{7}{8}$ in.; $d_h = \frac{15}{16}$ in.; $d_e = 1$ in.
First, calculate the number of bolts required based on shear capacity. Then select a suitable plate width based on the minimum gage distance and side distance required. Next determine a suitable plate thickness to provide adequate strength for the limit states of tension yielding and tension fracture. Then calculate the length of overlap and check block shear.

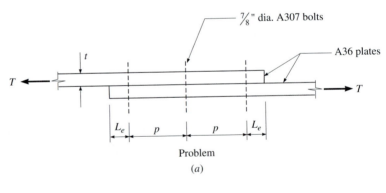

Problem
(a)

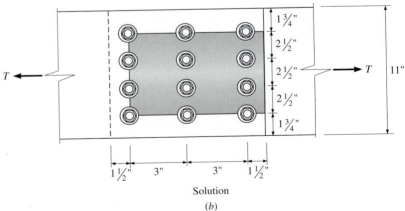

Solution
(b)

Figure X7.10.4

a. Bolt selection
The strength of a $\frac{7}{8}$-in.-diameter A307 bolt in single shear is:

$$B_{dv} = \phi F_v N_s A_b = 0.75(24)(1)(0.601) = 10.8 \text{ kips}$$

Number of bolts required for shear: $N \geq \dfrac{T_u}{B_{dv}} = \dfrac{120}{10.8} = 11.1$

Use 12 bolts, 4 in each transverse line.

b. Plate selection
Minimum bolt spacing (LRFDS Section J3.3) $= 2\frac{2}{3}d = 2.33$ in.
Preferred minimum bolt spacing $= 3d = 3.0(0.875) = 2.63$ in.
Minimum edge distance for sheared edges from LRFDS Section J3.4 and Table J3.4:

$$L_{s\ min} = 1\frac{1}{2} \text{ in.}$$

Use a gage dsistance, $g = 2\frac{1}{2}$ in. and side distance, $L_s = 1\frac{3}{4}$ in., resulting in a plate width, $w = 3(2\frac{1}{2}) + 2(1\frac{3}{4}) = 11$ in.

Assuming punched holes, the net width of the plate is:

$$w_n = w - nd_e = 11.0 - 4(1.0) = 7.00 \text{ in.}$$

If t is the thickness of the plate, the limit state of tension yielding results in:

$$T_{d1} = 0.9F_y wt \geq T_u \longrightarrow t \geq \frac{120}{0.9(36)(11.0)} = 0.337 \text{ in.}$$

The limit state of tension fracture results in:

$$T_{d2} = 0.75F_u U w_n t \geq T_u \longrightarrow t \geq \frac{120}{0.75(58)(1.0)(7.00)} = 0.394 \text{ in.}$$

Select a $\frac{7}{16}$-in. plate $(0.438 \geq 0.394)$.

c. Check bearing
Use a pitch, p, of 3 in. and end distance, L_e, of $1\frac{1}{2}$ in. Thus:

$$p = 3.0 \geq 3d = 2.63 \text{ in.} \qquad\qquad \text{O.K}$$
$$L_e = 1.5 \geq 1\frac{1}{2}d = 1.5(\tfrac{7}{8}) = 1.31 \text{ in.} \qquad\qquad \text{O.K.}$$

The clear distance for an interior bolt is as follows:

$$L_{ci} = p - d_h = 3.0 - 0.938 = 2.06 > 2d = 1.75 \text{ in.}$$

So, the design bearing strength of an interior bolt corresponds to the limit state of ovalization of the bolt hole:

$$B_{dbi} = \phi(2.4F_u)\,dt = 0.75(2.4)(58)(\tfrac{7}{8})\left[\frac{7}{16}\right] = 40.0 \text{ kips}$$

[continues on next page]

Example 7.10.4 continues ...

The clear distance for an end bolt is as follows:

$$L_{ce} = L_e - 0.5d_h = 1.5 - 0.5(0.938) = 1.03 < 2d = 1.75 \text{ in.}$$

So, the design bearing strength of an end bolt corresponds to the limit state of shear tear-out of the plate.

$$B_{dbe} = 0.75(1.2L_{ce}F_ut) = 0.75(1.2)(1.03)(58)\left[\frac{7}{16}\right] = 23.5 \text{ kips}$$

Since $B_{db} \geq B_{dv}$, for both the interior and end bolts, bearing strength of the connected material does not control the design strength of the bolt. Overlap of plates $= 2p + 2L_e = 2(3.0) + 2(1.5) = 9.00$ in. The plates selected are shown in Fig. X7.10.4b.

d. Check block shear

$$A_{gv} = 2 [2(3) + 1.5]\left[\frac{7}{16}\right] = 6.56 \text{ in.}^2; A_{gt} = 3(2.5)\left[\frac{7}{16}\right] = 3.28 \text{ in.}^2$$

$$A_{nv} = 2 [7.5 - 2^1/_2(10)]\left[\frac{7}{16}\right] = 4.38 \text{ in.}^2$$

$$A_{nt} = [7.5 - (2 + {}^1/_2 + {}^1/_2)(1.0)]\left[\frac{7}{16}\right] = 1.97 \text{ in.}^2$$

$\phi F_u A_{nt} = 0.75(58)(1.97) = 85.7$ kips
$\phi F_y A_{gt} = 0.75(36)(3.28) = 88.6$ kips
$\phi(0.6)F_u A_{nv} = 0.75(0.6)(58)(4.38) = 114$ kips
$\phi(0.6)F_y A_{gv} = 0.75(0.6)(36)(6.56) = 106$ kips
Since $\phi(0.6)F_u A_{nv} > \phi F_u A_{nt}$, use Eq. 7.8.2
$T_{dbs} = \phi(0.6)F_u A_{nv} + \min [\phi F_y A_{gt}, \phi F_u A_{nt}]$
$\qquad = 114 + \min [88.6; 85.7]$
$\qquad = 200 \text{ kips} > 120 \text{ kips}$
Thus, block shear does not control.
So, select $^7/_{16}$ in. by 11 in. A36 plates with twelve $^7/_8$-in.-dia. A307 bolts through standard holes with g $= 2^1/_2$ in. and an overlap of 9 in. Use:

$$L_s = 1^3/_4 \text{ in.; } p = 3.0 \text{ in.; } L_e = 1^1/_2 \text{ in.} \qquad \text{(Ans.)}$$

7.11   Built-Up Tension Members

As mentioned in Section 7.2.1, built-up sections are used as tension members when the designer is unable to provide sufficient area or rigidity via a single rolled shape. One of the more common built-up configurations for tension members is the double-angle section. Because the use of this configuration is quite widespread, tables of properties of various combinations of angles are included in LRFDM Tables 1-14 and 3-7.

The elements of a built-up tension member may be in continuous contact, closely spaced (Fig. 7.11.1a), or widely spaced (Fig. 7.11.1b).

The longitudinal spacing of bolts or intermittent fillet welds connecting two or more rolled shapes in continuous contact shall not exceed 24 in. The maximum longitudinal spacing of bolts or intermittent welds between elements in continuous contact, consisting of a plate and a shape or two plates, shall not exceed the following provisions of (LRFDS Section J3.5):

- 24 times the thickness of the thinner plate or 12 in. for painted members or unpainted members not subject to corrosion.
- 14 times the thickness of the thinner plate or 7 in. for unpainted members or members of weathering steel subject to atmospheric corrosion.

Components of built-up members with elements not in continuous contact must be adequately tied together at their ends and also at regular intervals between the ends. This is to ensure that:

- All elements maintain the correct geometrical relationship with respect to one another.
- The force distribution between components does not vary along the length of the member.
- All elements act together as a single unit.

Bolts used between member ends for the above purposes to connect closely spaced elements are known as **stitch bolts.** These bolts pass through ring fills or washers of the same thickness as the gusset plate. Alternatively, stitching may be achieved by welding **spacer plates** between the two components at appropriate intervals (Fig. 7.11.1a). The spacing of stitch bolts (or spacer plates) shall be such that the slenderness ratio of any component in the length between the bolts should preferably not exceed 300.

Open sides of built-up tension members should be tied together with **tie plates.** Tie plates do not contribute to the cross-sectional area of the section, and they do not theoretically carry any part of the load. The LRFD Specification requires that tie plates of tension members have a longitudinal length, L_{tp}, not less than $\frac{2}{3}$ the distance G between the lines of bolts or welds connecting them to the components of the member (Fig. 7.11.1b). The thickness of a tie plate, t_{tp}, should be at least $\frac{1}{50}$ of the distance G. Longitudinal spacing, s, of bolts or intermittent welds to tie plates shall not exceed 6 in. The minimum spacing and the minimum and maximum edge distances of bolts shall conform to Sections J3-3, 4, and 5 of the LRFD Specification. Spacing of tie plates, S_{tp}, shall be such that slenderness ratio of any component in the length between tie plates should preferably not

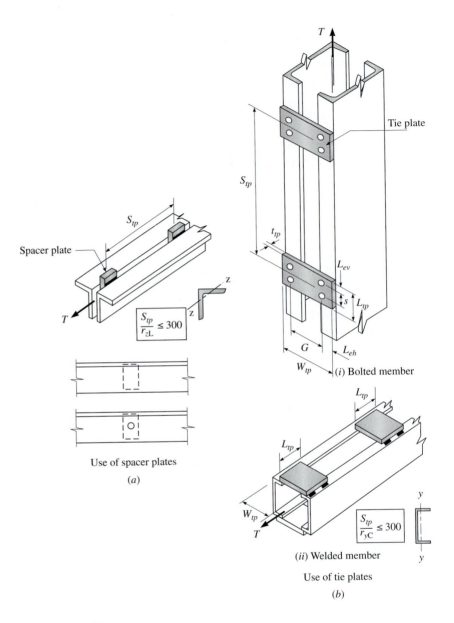

Figure 7.11.1: Built-up tension members.

exceed 300. Thus:

$$L_{tp} \geq \frac{2}{3}\, G; \quad t_{tp} \geq \frac{G}{50}; \quad s \leq 6 \text{ in.}$$

$$S_{tp} \leq 300 r_{\min}; \quad W_{tp} = G + 2L_{eh} \tag{7.11.1}$$

where W_{tp} is the width of the tie plate.

Note that, in this text, lengths of tie plates are always measured parallel to the length of the member, and the widths are always measured normal to the length of the member. Minimum values for edge distances L_{ev} and L_{eh} are as given in LRFDS Table J3.4. For example, for a $\frac{7}{8}$-in.-diameter bolt, these values are $1\frac{1}{2}$ in. at sheared edges and $1\frac{1}{8}$ in. at rolled or gas cut edges.

Built-Up Section with Tie Plates

EXAMPLE 7.11.1

A tension member is composed of two C12×25 A36 steel channels, 8 in. back-to-back, held together by tie plates and connected at the ends by double gusset plates (not shown), one gusset bolted to the web of each channel. The web-to-gusset connection shall use three lines of $\frac{7}{8}$-in.-dia. A325-N high-strength bolts without stagger. Use a pitch of 3 in. and a gage of 3 in. There are four bolts in each line. The tie plates are connected to each flange by three bolts staggered with respect to the web bolts by $1\frac{1}{2}$ in. at the ends. Determine the design strength of the member. Dimension the tie plates to satisfy the LRFD requirements, given that the length of the member is 36 ft.

Solution

a. Section properties

For the C12×25, from LRFDM Table 1-5:

$$A_C = 7.34 \text{ in.}^2, \qquad d_C = 12 \text{ in.}, \qquad t_{wC} = 0.387 \text{ in.},$$
$$b_{fC} = 3.05 \text{ in.}, \qquad t_{fC} = 0.501 \text{ in.},$$
$$r_{xC} = 4.43 \text{ in.}, \qquad r_{yC} = 0.779 \text{ in.} = r_{min\,C}$$
$$\bar{x}_C = 0.674 \text{ in.}, \qquad I_{yC} = 4.45 \text{ in.}^4$$

Also from LRFDM Table 1-5, the workable gage, g, for the flange of a C12×25 is 1.75 in.

For the built-up section:

$$A = 2(7.34) = 14.7 \text{ in.}^2$$
$$I_y = \sum (I_o + Ad^2)_y$$
$$= 2[4.45 + 7.34(4.0 + 0.674)^2] = 330 \text{ in.}^4$$
$$r_y = \sqrt{\frac{330}{14.7}} = 4.74 \text{ in.}$$
$$r_x = r_{xC} = 4.43 \text{ in.}$$
$$r_{min} = r_x = 4.43 \text{ in.}$$

b. Design strength of the member

Gross area of the section, $A_g = A = 14.7 \text{ in.}^2$

[continues on next page]

Example 7.11.1 continues ...

The gage distance between bolts f and b (Fig. X7.11.1a), measured along the middle surface of the elements is:

$$g_{12} = g_1 + g_2 - \left(\frac{t_f}{2} + \frac{t_w}{2}\right) = 1.75 + 3.0 - (0.501 + 0.387)/2$$

$$= 4.31 \text{ in.}$$

The average thickness, $t = 0.5(t_f + t_w) = 0.5(0.501 + 0.387) = 0.444$ in. Net area along:

path a-b-c-d-e: $A_{n1} = 14.7 - 2(3)(1)(0.387) = 12.4$ in.^{2}

path f-b-c-d-h: $A_{n2} = 14.7 - 2(3)(1)(0.387) - 2(2)(1)(0.501)$

$$+ \left[\frac{2(2)(1.5^2)}{4(4.31)}\right]0.444 = 10.6 \text{ in.}^2$$

Note that for the diagonal segments fb and hd, which run from holes in the flanges to holes in the web, the $(s^2/4g)$ term has been multiplied by the average of the flange and web thicknesses.

So, the net area, $A_n = \min [12.4, 10.6] = 10.6$ in.^{2}.

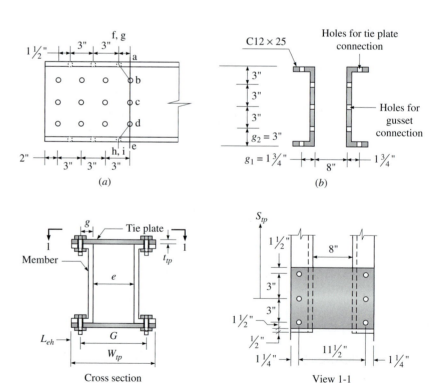

Figure X7.11.1

Reduction coefficient U may be calculated from Eq. 7.7.2. The length of the connection L_{con} equals 9 in. and $\bar{x}_{con} = \bar{x}_C = 0.674$ in. So:

$$U = 1 - \frac{\bar{x}_{con}}{L_{con}} = 1 - \frac{0.674}{9.00} = 0.925 > 0.90$$

Use $U = 0.90$
Effective net area, $A_e = 0.90(10.6) = 9.54$ in.^{2}.
Design strength corresponding to the limit state of:

$$\text{tension yielding, } T_{d1} = 0.9(36)(14.7) = 476 \text{ kips}$$

$$\text{tension fracture, } T_{d2} = 0.75(58)(9.54) = 415 \text{ kips}$$

c. Check block shear rupture
Gross area in tension, Agt $= 2(6.0)(0.387) = 4.64$ in.2
Net area in tension, $A_{nt} = 4.64 - 2[2(1.0)(0.387)] = 3.09$ in.^{2}
Gross area in shear, $A_{gv} = 2(2)(11.0)(0.387) = 17.0$ in.^{2}
Net area in shear, $A_{nv} = 17.0 - 2[2(3.5)(1.0)(0.387)] = 11.6$ in.^{2}

$$\phi F_u A_{nt} = 0.75(58)(3.09) = 134 \text{ kips}$$

$$\phi F_{uv} A_{nv} = 0.75(34.8)(11.6) = 303 \text{ kips} \leftarrow \text{controls}$$

$$\phi F_y A_{gt} = 0.75(36)(4.64) = 125 \text{ kips}$$

$$\phi F_{yv} A_{gv} = 0.75(21.6)(17.0) = 275 \text{ kips}$$

So:

$$T_{dbs} = \phi F_{uv} A_{nv} + \min [\phi F_y A_{gt}, \phi F_u A_{nt}]$$

$$= 303 + \min [125, 134] = 428 \text{ kips}$$

The design strength of the member corresponding to block shear failure is thus:

$$T_{d3} = T_{dbs} = 428 \text{ kips}$$

d. Bolt strength
Design strength of a single bolt is 21.6 kips (LRFDM Table 7-10)
Design strength of bolts, $T_{d4} = NB_d = 2(12)(21.6) = 518$ kips
Design strength of member, $T_d = \min [T_{d1}, T_{d2}, T_{d3}, T_{d4}]$
$= \min [476, 415, 428, 518]$
$= 415$ kips   (Ans.)
Maximum recommended length of the member,

$$L_{max} = 300 r_{min} = \frac{300(4.43)}{12} = 111 \text{ ft} > L = 36 \text{ ft} \text{O.K.}$$

e. Tie plates
Distance between bolt lines, $G = 8 + 2(1.75) = 11.5$ in.

Length of tie plate, $L_{tp} \geq \frac{2}{3} G = \frac{2}{3}(11.5) = 7.67$ in.   *[continues on next page]*

Example 7.11.1 continues ...

In order to provide three bolts through each end of each tie plate with a spacing, s of 3 in. and an edge distance of $1\frac{1}{2}$ in., the tie plates must have a length of 2 by 3 plus 2 by $1\frac{1}{2}$ or 9 in. $\geq$ 7.67 in.

Thickness of the tie plate, $t_{tp} \geq \dfrac{G}{50} = \dfrac{11.5}{50} = 0.23$ in. Use $\frac{1}{4}$ in.

Width of tie plate, $W_{tp} = G + 2(1\frac{1}{4}) = 11.5 + 2.5 = 14.0$ in.

Tie plates are $14 \times 9 \times \frac{1}{4}$

Spacing of tie plates, $S_{tp} \leq 300\, r_{yC} = 300(0.79) = 237$ in. $= 19.8$ ft

Provide three sets of tie plates, one at each end of the member and the third at midlength of the member. Spacing provided, $S_{tp} = 17'6''$ < 19.8 ft O.K.

(Ans.)

7.12   Tension Rods

7.12.1   Introduction

Tension rods are bars which are circular in cross section with threads at one or both ends. The minimum diameter for rods used in building structures is $\frac{5}{8}$ in., as smaller rods are often damaged during transportation and erection. Also, the diameter of a rod is preferably not less than $\frac{1}{500}$ of its length in order to ensure some rigidity even when strength calculations may permit much smaller sizes.

LRFD Specification Section A3.4 permits the use of unheaded rod material from the following ASTM Specification as threaded rods: A36, A354, A572, and A588. Two types of round rods are used: *rods of constant diameter* and *rods with upset ends* (Fig. 7.12.1). In the first type, threads are cut from a rod of constant diameter. Rods are said to be *upset* when the threaded ends are made larger than the main body of the rod, and the threads are cut in the enlarged part. The diameter of the enlarged end is such that the cross-sectional area at the root of the threads is at least equal to gross area of the rod in the body of the member. Upset rods are costly because of additional labor costs, and are likely to be economical only if a large quantity of long rods are required so that the savings in material exceeds costs of labor. Good practice requires that the upset ends have an area at the root of the threads about 20 percent greater than that of the body of the rod. Hence, the controlling net area of an upset rod is the gross area of the central portion.

Usually rods of constant diameter are simply cut to length from stock readily available at the mill or warehouse. The threaded ends are often held in place by nuts. Information about standard sizes and details of nuts may be found in Tables 7-1 and 7-6 of the LRFDM.

Pin-ended connections for rods can be made by using a clevis. A *clevis* is a mechanical fastener, one end of which is threaded and screwed onto the threads of a rod. The other end consists of two identical parallel prongs, each of which contains a hole of diameter p (see Fig. 7.12.2). The two prongs straddle a gusset plate that also contains a hole of diameter p. A bolt or pin

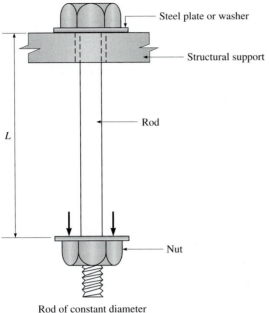

Rod of constant diameter
(a)

Rod with upset end
(b)

Figure 7.12.1:   Tension rods.

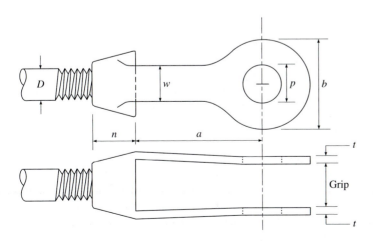

Figure 7.12.2:   Clevis.

is passed through the prongs and the gusset to complete the connection (Fig. 3.3.10*a*). Clevises vary in length from 6 to 18 in. depending on the diameter of the rod. Standard clevises are available to fit typical threaded rods and vary in design strengths from 8 to 330 kips. The appropriate clevis is selected by comparing the maximum rod diameter that each clevis can accommodate to the rod chosen. Dimensions, weights, and design strengths of clevises may be found in LRFDM Table 15-3, and the compatibility of clevises with rods and pins of various diameters is given in LRFDM Table 15-4.

When tension rods are used as wind bracing, it is good practice to subject them to initial tension. Such prestressed diagonals are taut when erected, and thereby tighten up the structure and reduce or eliminate potential flutter of the rods. One way to obtain initial tension is to intentionally fabricate the member slightly shorter than the required length. Typically, members are shortened $\frac{1}{16}$ in. for every 20 ft of member length. These members are then stretched into place by the use of tapered pins called *drift pins*. This process is referred to as **drifting.** The typical prestress in such members can then be calculated as follows:

$$f = \varepsilon E = \frac{1}{16}\left[\frac{1}{20(12)}\right](29,000) \approx 8 \text{ ksi}$$

Another way to introduce initial tension into a rod is to fit it with clevises at both ends. In such cases, one end of the rod is fitted with left-handed threads, and the other end of the rod is fitted with right-handed threads. A pipe wrench can then be used to twist the rod and thereby vary the tension.

When tension rods are used as hangers supporting floors or balconies, ASCES7-98 Section 4.7.2 requires that a $33\frac{1}{3}$-percent allowance for impact be considered in the design.

7.12.2   Effective Area of Threaded Parts

The strength of a threaded tension rod is controlled by the threads. The thread size is specified by giving the number of threads per in., n. A large number of tensile tests have shown that a threaded rod has approximately the same tensile strength as an unthreaded rod having a a cross-sectional area equal to the **net tensile area** (A_e) of the threaded rod. Values of A_e are listed in the LRFDM Table 7-4 for different rod diameters. For the basic Unified National Coarse (UNC) thread series used in the United States, A_e is closely given by:

$$A_e = \frac{\pi}{4}\left[d_R - \frac{0.9743}{n}\right]^2 \qquad (7.12.1)$$

where d_R is the nominal diameter of the rod, and n is the number of threads per in. Further, it can be verified from LRFDM Table 7-4, that for coarse threads from $\frac{3}{4}$ to $2\frac{1}{2}$ in. diameter, A_e is about 75 to 80 percent of the gross area A_g. Accordingly, to avoid the necessity of looking up the net area at the base of the screw thread, LRFDS uses a lower bound reduction of 0.75.

Equation 7.12.1 may now be rewritten as:

$$A_e \approx 0.75 \, \frac{\pi d_R^2}{4} = 0.75 A_R \qquad (7.12.2)$$

where A_R is the cross-sectional area based on the nominal diameter.

7.12.3   Strength Limit States of a Rod

The tensile force T in a tension rod is transmitted from the body of the member to the threaded portions of the rod (Figs. 7.12.3a and b). From there it

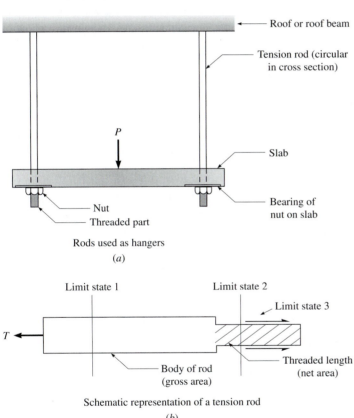

Roof or roof beam

Tension rod (circular in cross section)

P

Slab

Nut

Threaded part

Bearing of nut on slab

Rods used as hangers

(a)

Limit state 1

Limit state 2

Limit state 3

T

Body of rod (gross area)

Threaded length (net area)

Schematic representation of a tension rod

(b)

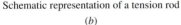

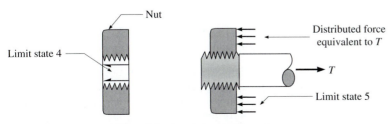

Nut

Limit state 4

Distributed force equivalent to T

T

Limit state 5

Nut and the threaded part of the rod

(c)

Figure 7.12.3: Limit states of strength for tension rods.

passes to the threads of the rod and then to the threads of the nut via thread bearing (Fig. 7.12.3c). Finally, the load is transferred from the nut to the rest of the structure by bearing of the nut (Fig. 7.12.3a). As shown in Fig. 7.12.3, five different limit states of strength may be identified:

T_{d1} = design strength corresponding to the limit state of yielding in the gross section in the unthreaded part (body) of the rod (= $\phi_1 T_{n1}$)

T_{d2} = design strength corresponding to the limit state of fracture in the net section in the threaded part of the rod (= $\phi_2 T_{n2}$)

T_{d3} = design strength corresponding to the limit state of the nut stripping the threaded portion of the rod (= $\phi_3 T_{n3}$)

T_{d4} = design strength corresponding to the limit state of the rod stripping the threaded portion of the nut (= $\phi_4 T_{n4}$)

T_{d5} = design strength of bearing of the nut on the support (= $\phi_5 T_{n5}$)

The design strength of the tension rod is therefore given by:

$$T_d = \min [T_{d1}, T_{d2}, T_{d3}, T_{d4}, T_{d5}] \qquad (7.12.3)$$

For a rod of given material and diameter, the characteristics of the nut (material, length, cross-sectional area) can be chosen such that failure modes 3 and 4 do not control the design. Such nuts are said to be **matched nuts.** Also, the dimensions of the washer and/or bearing plate, should they be required by the specification, can be chosen such that the limit state of nut bearing on the support material does not control the design strength either. Thus, for a tension rod provided with a matching nut:

$$T_d = \min [T_{d1}, T_{d2}] \qquad (7.12.4)$$

Equations D1-1 and D1-2 of the LRFDS for the tension members, may be written with the help of Eq. 7.12.2 as:

$$T_{d1} = 0.9 F_y A_g = 0.9 F_y A_R \qquad (7.12.5)$$

$$T_{d2} = 0.75 F_u A_e = 0.75 F_u (0.75 A_R) = 0.75(0.75 F_u A_R) \qquad (7.12.6)$$

For uniform rods without upset ends, considered here:

$$A_R = \frac{\pi d_R^2}{4} \qquad (7.12.7)$$

where A_R is the cross-sectional area based on the nominal diameter of the rod (d_R).

For all steels presently used as tension rods $0.75(0.75 F_u) < 0.9 F_y$ (see Table 7.4.1). Thus, the design strength, T_{dR}, of a uniform tension rod reduces to the single relation:

$$T_{dR} = T_{d2} = 0.75(0.75 F_u) A_R = F_{dR} A_R \qquad (7.12.8)$$

where F_{dR} is the design stress for tension rods without upset ends. Values of F_{dR} are tabulated in Table 2 of the LRFDS for threaded rods of A36, A242,

A572, A588, A992, and A514 steels. The design tensile strength of uniform threaded rods is given by LRFDS Section J3.6 and Table J3.2.

7.12.4   Sag Rods

Sag rods are tension members used to provide additional intermediate support to purlins perpendicular to their weak axis (Fig. 3.4.2). They are often considered necessary for roofs with slopes steeper than 3 in./ft, especially when the purlins are torsionally weak, as is the case for steel channels. Sag rods placed at midpoints of purlins are usually satisfactory for light roof loads and moderate roof truss spacing. For larger spacings and or heavier loads, sag rods are placed at third points of each purlin. Each end of each sag rod is threaded and passed through holes punched in the webs of consecutive purlins. A nut is used at each end of the rod for anchorage. Adjacent sag rods are offset in plan view about 6 in. to facilitate installation. Note that if a metal deck is used and properly attached to the purlins, sag rods may not be necessary.

Sag rods are generally designed to support the simple beam reactions for the components of the factored gravity loads (snow, ice, roofing, and purlins) parallel to the roof surface. Wind forces are assumed to act normal to the roof surface and therefore do not influence the forces in sag rods. The maximum tensile force will be present in the sag rod nearest the ridge, because it must support the sum of forces in the lower sag rods. Since individual sag rods are placed between successive pairs of purlins, they may be designed individually, each to carry the tangential component from all the purlins below it. Thus, it is theoretically possible to use smaller rods for the lower sag rods, but the reduction in size is generally not worthwhile.

Anchorage must be provided to the sag rods at the ridge. One method is shown in Fig. 3.4.2b. In that connection, the tangential components from the two sides of the roof provide anchorage for each other via a horizontal tie rod between the two ridge purlins. The tensile force in this horizontal tie rod has as one of its components, the force in the upper sag rod. A free body diagram of one of the ridge purlins is shown in Fig. 3.4.2b.

Design of a Threaded Rod

EXAMPLE 7.12.1

A 30-ft-long walkway suspender in a hotel plaza is subjected to a dead load of 30 kips and a live load of 30 kips. Use A36 steel and the LRFD Specification to select a standard threaded rod. The elongation of the rod under service dead and live loads is to be limited to $\frac{1}{2}$ in.

Solution

For structures carrying live loads that induce impact, the nominal live load is to be increased as stipulated in ASCE Section 4.7.2. For hangers supporting floors and balconies, impact is 33 percent.

[continues on next page]

Example 7.12.1 continues ...

So, service live load, $L = 1.33(30) = 40$ kips
Service dead load, $D = 30$ kips
The required tensile strength of the member from the load combinations LC-1 to LC-7 is:

$$T_u = \max [1.4D, 1.2D + 1.6L] = \max [1.4(30), 1.2(30) + 1.6(40)]$$
$$= \max [42, 100] = 100 \text{ kips}$$

From LRFDM Table 2-3, it is observed that A36 steel is the preferred material specification for threaded rods. Also, $F_y = 36$ ksi, $F_u = 58$ ksi. The design strength of a uniform tension rod is given by Eq. 7.12.8. Thus:

$$T_{dR} = 0.75(0.75)(F_u A_R) \geq T_u$$

or

$$A_R \geq \frac{T_u}{0.5625F_u} = \frac{100}{0.5625(58)} = 3.07 \text{ in.}^2$$

So, the diameter of the standard threaded rod, $d_R \geq \sqrt{\frac{4(3.07)}{\pi}} = 1.98$ in.

Tentatively select a 2-in.-dia. standard threaded rod of A36 steel.
Check for elongation:
Cross-sectional area, $A_R = 3.14$ in.^{2}
Length of rod, $L_R = 30$ ft $= 360$ in.
Service load to be carried, $T = 1.0D + 1.0L = 30 + 40 = 70$ kips
From Eq. 7.3.1, elongation,

$$\Delta = \frac{TL_R}{A_R E} = \frac{70.0(360)}{3.14(29,000)} = 0.277 \text{ in.} < 0.5 \text{ in. O.K.}$$

So, provide a 2-in.-dia. standard threaded rod of A36 steel. Specify matching nuts and check bearing of the nuts on the supports.   (Ans.)

EXAMPLE 7.12.2

Design of a Sag Rod for a Roof

Design sag rods to support the purlins of an industrial building roof having a span of 96 ft and a rise of 24 ft. Roof trusses are spaced at 30 ft on center (see Fig. 3.4.2). Sag rods are to be placed at third points of each purlin. Assume roofing and purlin weigh 12 psf of roof surface. The snow load to be carried is 30 psf. Use standard threaded rods and A36 steel.

Solution

Span of roof truss, $L = 96$ ft
Rise, $h = 24$ ft

Inclination of the roof, $\theta = \tan^{-1}\left(\frac{24}{48}\right) = 26.6°$

Spacing of trusses, $S_T = 30$ ft

Spacing of sag rods, $S_S = \dfrac{30}{3} = 10$ ft

a. Loads
Tributary area of the roof surface per line of sag rods,

$$A = \frac{L/2}{\cos\theta} S_S = \left[\frac{48}{\cos 26.6°}\right](10) = 537 \text{ ft}^2$$

Weight of roofing and purlins = 12 psf

Dead load, $D = \dfrac{537(12)}{1000} = 6.44$ kips

Horizontal projection of the roof area, $A_H = 48(10) = 480$ ft^2
Snow load per unit area = 30 psf

Snow load, $S = \dfrac{480(30)}{1000} = 14.4$ kips

As there are only dead loads and snow loads, combination LC-3 controls. Factored vertical load on the roof, $V = 1.2D + 1.6S = 1.2(6.44) + 1.6(14.4) = 30.8$ kips.

b. Top sag rod
The top sag rod parallel to the roof carries only that component of the vertical load, V, parallel to the roof. Thus, tension in the top sag rod:

$$T_S = V \sin\theta = 30.8(\sin 26.6°) = 13.8 \text{ kips}$$

Required area of the threaded rod for the sag rod (Eq. 7.12.8):

$$A_R \geq \frac{T_S}{0.75(0.75F_u)} = \frac{13.8}{0.75(0.75)(58)} = 0.423 \text{ in.}^2$$

From Table 7-10 of the LRFDM, select a $\frac{3}{4}$-in.-diameter threaded rod having a cross-sectional area, 0.442 in.^{2} > 0.423 in.^{2} O.K.   (Ans.)

c. Horizontal tie rod
The horizontal tie rod between ridge purlins has to carry a tensile force, T_T:

$$T_T = \frac{T_S}{\cos\theta} = \frac{13.8}{\cos 26.6°} = 15.4 \text{ kips}$$

Required area of the horizontal tie rod:

$$A_R \geq \frac{T_T}{0.75(0.75F_u)} = \frac{15.4}{0.75 \times 0.75 \times 58} = 0.472 \text{ in.}^2$$

From Table 7-10 of the LRFD Manual, select a $\frac{7}{8}$-in.-diameter threaded rod of A36 steel having a cross-sectional area, 0.601 in.^{2} > 0.472 in.^{2} O.K.
(Ans.)

References

7.1 Birkemoe, P. C., and Gilmor, M. I. [1978]: "Behavior of Bearing-Critical Double-Angle Beam Connections," *Engineering Journal,* AISC, vol. 15, no. 4, 4th quarter, pp. 109–115.

7.2 Chesson, E., and Munse, W. H. [1957]: "Behavior of Riveted Connections in Truss Type Members," *Journal of the Structural Division,* ASCE, 83, ST1, January, pp. 1–61.

7.3 Cochrane, V. H. [1922]: "Rules for Rivet Hole Deductions in Tension Members," *Engineering News-Record,* vol. 89, November 16, pp. 847–848.

7.4 Easterling, W. S., and Gonzales, L. [1993]: "Shear Lag Effects in Steel Tension Members," *Engineering Journal,* AISC, vol. 30, no. 3, pp. 77–89.

7.5 Epstein, H. I. [1992]: "An Experimental Study of Block Shear Failure of Angles in Tension," *Engineering Journal,* AISC, vol. 29, no. 2, pp. 75–84.

7.6 Fisher, J. W., and Struik, J. H. A. [1974]: *Guide to Design Criteria for Bolted and Riveted Joints,* John Wiley and Sons, New York.

7.7 Hardash, S. G., and Bjorhovde, R. [1985]: "New Design Criteria for Gusset Plates in Tension," *Engineering Journal,* AISC, vol. 22, no. 2, 2nd quarter, pp. 77–94.

7.8 McGuire, W. [1968]: *Steel Structures,* Prentice Hall Inc., New Jersey.

7.9 Munse, W. H., and Chesson, J. H. [1963]: "Riveted and Bolted Joints: Net Section Design," *Journal of the Structural Division,* ASCE, 89, ST2, February, pp. 107–126.

7.10 Ricles, J. M. and Yura, J. A. [1983]: "Strength of Double-Row Bolted Web Connections," *Journal of the Structural Division,* ASCE, vol. 109, no. ST1, January, pp. 126–142.

P R O B L E M S

Note: 1. Assume STD punched holes, unless otherwise stated.
2. Assume A992 steel for W- and WT-shapes, unless otherwise stated.
3. Assume A36 steel for S-, C-, MC-, L-shapes and plates, unless otherwise stated.

Analysis

P7.1. Determine the net area for the following tension members:

(a) PL $\frac{1}{2} \times 8$ with holes for $\frac{3}{4}$-in.-dia. bolts, as shown in Fig. 7.6.1a.

(b) W12×120 with STD drilled holes for $\frac{7}{8}$-in.-dia. bolts, as shown in Fig. 7.6.1b.

(c) PL $\frac{1}{2} \times 8$ with holes for $\frac{3}{4}$-in.-dia. bolts with a gage, $g = 2.75$ in., pitch, $p = 3$ in., and a staggered pitch, $s = 1.5$ in., as shown in Fig. 7.6.2a.

(d) L4 × 3$\frac{1}{2}$ × $\frac{1}{2}$ with holes for $\frac{3}{4}$-in.-dia. bolts with pitch, $p = 2.5$ in., and a staggered pitch, $s = 1.25$ in., as shown in Fig. 7.6.3a.

(e) C8×18.75 with holes for $\frac{3}{4}$-in.-dia. bolts with pitch, $p = 2.5$ in, staggered pitch, $s = 1.25$ in., gage, g_a = 1.5 in., and gage, g_b = 2.5 in. as shown in Fig. 7.6.4a.

P7.2. Determine the reduction coefficient, U, using Eq. 7.7.2 for the following tension members:

(a) A L4×4×½ connected to a gusset by two bolts with a pitch, p = 3 in. (Fig. 7.7.4b).
(b) A W12×65 connected by two lines of bolts in each flange to gussets using a pitch, p = 3 in. (Fig. 7.7.4c). There are three bolts in each line.
(c) A W12×35 connected by two lines of bolts in each flange to gussets using a pitch, p = 3 in. (Fig. 7.7.4e). There are three bolts in each line.
(d) A WT7×45 connected by two lines of bolts in the flange to a gusset using a pitch, p = 3 in. (Fig. 7.7.4d). There are three bolts in each line.
(e) A PL½×6 connected to a gusset by two 9-in.-long longitudinal welds, as shown in Fig. 7.7.5a.
(f) A L6×4×½ with its long leg connected to a gusset by a 6-in.-long transverse weld only (Fig. 7.7.5b).
(g) A L6×4×½ with its long leg connected to a gusset by a transverse weld and two longitudinal welds having L_{w1} = 8 in. and L_{w2} = 4 in. (Fig. 7.7.5c).

P7.3. A 14 × ½ in. A242 steel plate is used as a tension member. The end connection to the gusset plate is by six ⅞-in.-dia. bolts, arranged with staggered pitches of 2 in. and 1½ in., and gage distances of 2½ in. as shown in Fig. P7.3. Determine the design tensile strength of the member. Neglect block shear failure.

P7.4. The web of a MC13×35 steel tension member is connected to a gusset plate by four rows of ¾-in.-dia. bolts using a gage of 3 in. and a pitch of 3 in. as shown in Fig. P7.4. Assume A588 steel and determine the design tensile strength of the member. Neglect block shear failure.

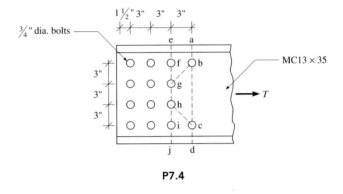

P7.4

P7.5. A L4×4×½ of A572 Grade 50 steel is used as a tension member. What is the design tensile strength of the member:

(a) If only a transverse weld is used to connect the member to the gusset, as shown in Fig. P7.5?

(b) If, in addition, two 6-in.-long longitudinal welds are provided?

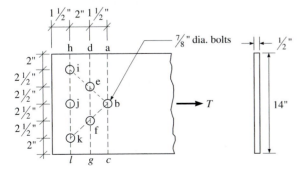

P7.3

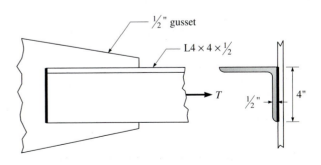

P7.5

P7.6. The web of a C12×25 steel tension member is connected to a gusset by three rows of $\frac{7}{8}$-in.-dia. A490-X bolts using a gage of 3 in. and a staggered pitch of $1\frac{1}{2}$ in. as shown in Fig. P7.6. Assume A588 steel and determine the design tensile strength of the member. Neglect block shear rupture.

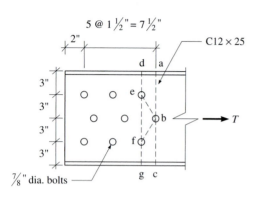

P7.6

P7.7. A C12325 channel shape is used as a tension member. At each end, the web of the channel is connected by six $\frac{3}{4}$-in.-dia. bolts (in two rows, using a pitch of 4 in. and a gage of 6 in.) to a thick gusset plate. In addition, there are two holes in each flange (located on the normal gage line) to connect bracing members in the perpendicular plane. The holes in the flanges are staggered 2 in. with respect to the holes in the web, as shown in Fig. P7.7. Determine the design strength of the member. What is the maximum recommended length of the member as per LRFDS?

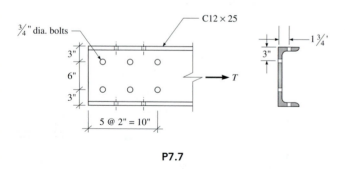

P7.7

P7.8. Determine the design tensile strength of a pair of $6 \times 4 \times \frac{1}{2}$ in. angles with their long legs connected to a $\frac{5}{8}$-in. gusset plate (Fig. P7.8). Use A36 steel and $\frac{7}{8}$-in.-dia.

bolts. The bolts are arranged on standard gages. The tensile force T is transmitted to the gusset plate by the six bolts on lines A and B. Assume open holes in the outstanding legs (for connection to braces in the plane perpendicular to the plane of the paper).

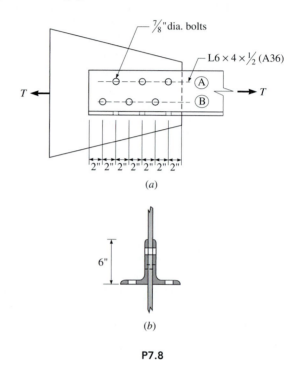

P7.8

P7.9. The end connection for a W14×82 tension member has two lines of $\frac{7}{8}$-in.-dia. bolts in each flange (Fig. P7.9). There are 3 bolts in each line arranged with a pitch, $p = 4$ in.; end distance, $L_e = 2$ in.; and side distance $L_s = 2$ in. Determine the block shear rupture strength.

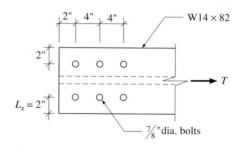

P7.9

P7.10. Determine the block shear rupture strength of the 2L4 ×3½ ×⅜ LLBB tension member shown in Fig. P7.10. The angles are connected to a gusset by three ⅞-in.-dia. bolts, using a pitch, $p = 3$ in.; end distance, $L_e = 1½$ in. Assume A 572 Grade 60 steel.

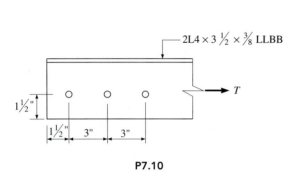

P7.10

P7.11. A ½ × 6 in. plate is welded to a ⅜-in.-thick gusset plate of A36 steel by ¼-in.-fillet welds using E70 electrodes. Two 6-in.-long longitudinal welds are provided in addition to a 6-in.-long transverse weld (Fig. P7.11). Determine the design tensile strength of the member. Include block shear strength of the gusset plate.

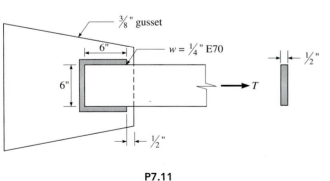

P7.11

P7.12. A MC6×18 channel tension member of A242 corrosion resistant steel is welded to a ⅜-in.-thick gusset plate of A36 steel, using a 6-in.-long transverse weld and two 4-in.-long longitudinal welds as shown in Fig. P7.12. Assume

E70 fillet welds of ¼-in leg size. Determine the design strength of the member and the gusset plate. Include block shear rupture strength.

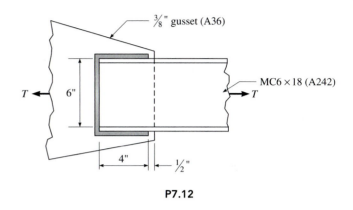

P7.12

P7.13. A ½ × 8 in. plate tension member is connected to a ⅜-in.-thick gusset plate by five A325-N bolts of ⅜-in.-diameter (Fig. P7.13). Assume A36 steel for the member and the gusset. Determine the tensile strength of the member. Include block shear rupture strength of the member and the gusset.

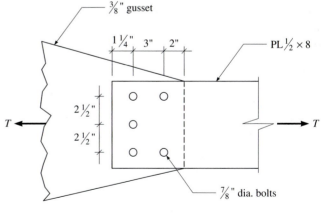

P7.13

P7.14. (a) Determine the design tensile strength of a C15×40 truss member of A572 Grade 42

steel shown in Fig. P7.14. The end connection to the gusset is by ten $\frac{7}{8}$-in.-dia. A490-X type bolts. They are arranged in four rows with a gage of 3 in., pitch of $2\frac{1}{2}$ in., and an end distance of $1\frac{1}{2}$ in. Include block shear rupture strength.

(b)   Redesign the end connection to increase the design strength of the member to the extent possible, if there are no architectural restrictions

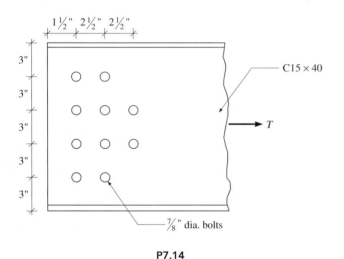

P7.14

Design

P7.15.   The axial tensile forces in the vertical member of a pedestrian truss bridge, from the code specified service loads, have been calculated as: 40 kips from dead load, 50 kips from live load, and 30 kips from wind. Select a 1-in.-thick plate of A36 steel of suitable width. Assume a single line of bolts of $\frac{3}{4}$-in.-diameter.

P7.16.   The web member of a roof truss is subjected to tensile forces of 40 kips and 60 kips under service dead load and live load, respectively. Select the lightest WT5, if

the end connection to gusset is by two lines of $\frac{3}{4}$-in.-dia. bolts in the flange. Assume that there will be more than three bolts in each line. Neglect block shear failure.

P7.17.   The factored tensile load on a truss bottom chord is 480 kips. Select the lightest W10 with its flanges connected to gusset plates at each end, with two lines of $\frac{3}{4}$-in.-dia. bolts in each flange. The bolts are A490-N type. Neglect block shear failure.

P7.18.   A W16 section is used as a tension member to carry a factored tensile load of 500 kips. The end connection is by $\frac{7}{8}$-in.-dia. A490-X bolts, in two rows, in each flange, using standard punched holes. The gusset plates are $\frac{5}{8}$-in.-thick. Find the lightest W16 section, and determine the number of bolts required. What is the maximum allowable length of such a member?

P7.19.   A 16-ft-long tension member is subjected to a factored axial load of 100 kips. Select a suitable, single, equal-leg angle of 5-in. leg size and an even number of $\frac{3}{4}$-in.-dia. bolts of A325-N type. The bolts are to be arranged in two lines, without stagger, with a pitch of 3 in. and an end distance of 2 in. Use standard gages.

P7.20.   Design the tension diagonal of an all welded Warren truss for a pedestrian bridge, in which the chords are made from WT7×19 sections. The factored load in the diagonal member is 125 kips, and its length is 15 ft. Use 2Ls of A36 steel LLBB, with their 4-in.-long legs welded to the stem of the WT. The weld consists of a transverse weld and two longitudinal welds.

P7.21.   Design a single-angle tension member 13 ft long and its connection at each end to a $\frac{3}{8}$-in. gusset plate. The tension is 24 kips dead load and 52 kips live load. The angle cannot extend more than 15 in. on the gusset. Use LRFD specification, A36 steel, $\frac{7}{8}$-in.-dia. bolts, slip critical connection. Include block shear rupture strength.

P7.22.   A tension member is composed of two C12×30 channels placed 12 in. back-to-back and tied together as shown in Fig. 7.2.2c (*ii*). The member is connected at ends to $\frac{1}{2}$-in.-thick gusset plates, by $\frac{7}{8}$-in-dia. A325-N bolts provided in three rows in the webs. If A572 Grade 42 steel is used, find the design strength as per LRFDS. Assume a gage of 3 in. a pitch of 3 in., and an end distance of 2 in. If the force in the member from dead load on the structure is 200 kips, what live load and wind load forces can be applied on the member as per ASCES? What is the maximum recommended length of such a member as per LRFDS?

P7.23. A tension member in a covered pedestrian bridge is subjected to a dead load of 50 kips, live load of 70 kips, and snow load of 38 kips. Select two angles with long legs back-to-back and separated by $\frac{3}{8}$ in. for end connections to gusset plates. Assume a single line of holes for $\frac{3}{4}$-in.-dia. A325-X bolts. Assume A242 Grade 50 steel. The member is 20 ft long. Determine the spacing of spacer plates, if necessary.

P7.24. Select a pair of American standard channels for a tension member subjected to a dead load of 68 kips and a live load of 140 kips. The channels are placed back-to-back and connected to a $\frac{1}{2}$-in. gusset plate by $\frac{7}{8}$-in.-dia. A490-N bolts [see Fig. 7.2.2c (i)]. Assume A588 Grade 50 steel for the member and the gusset. The member is 20 ft long. The bolts are arranged in two lines parallel to the length of the member.

P7.25. A steel truss is used to transfer loads from the upper floor columns of an office building across an arcade at street level. The lower chord of the truss is made up of two angles, with their vertical legs 8 in. back-to-back, connected to gussets as shown in Fig. 7.2.2a (ii). In addition, the two angles are connected by tie plates. The factored load to be carried by the member is 240 kips and the member length is 18 ft. Use $\frac{3}{4}$-in.-dia. A325-N bolts. Select suitable angles. Show the end connection details, including tie-plate. Consider block shear failure.

P7.26. A pinned member is to consist of four equal leg angles arranged as shown in Fig. 7.2.2b (i). The factored tensile load is 324 kips. Two $\frac{5}{8}$-in. bolts will be used in each angle. The member length is 40 ft. For architectural reasons the cross-sectional dimensions should not exceed 12 in. by 12 in. Include the design of tie plates.

P7.27. Two channels with their webs 6 in. back-to-back, as shown in Fig. 7.2.2c (iii), are used as a tension member to carry a factored load of 480 kips. The member is 40 ft long. At the ends, the flanges are connected by tie plates, while the webs are connected to gusset plates. Using A572 Grade 50 steel, LRFD Specification, and $\frac{3}{4}$-in.-dia. bolts, select the channels. Design the tie plates.

Rods

P7.28. A floor beam hanger is subjected to tensile forces of 30 kips dead load and 48 kips live load. Select a suitable round steel threaded rod. Assume A36 steel. (Note: Use impact factor from ASCES 7-98, Section 4.7.2.)

P7.29. Roof trusses of an industrial building, with a span of 80 ft and a rise of 20 ft, are spaced on 27-ft centers and have purlins spaced at 6-ft intervals (see Fig. 3.4.2). To support the purlins, sag rods are provided at $\frac{1}{3}$ points between the trusses. The estimated weight of the roof covering and of the purlins is 7 psf and 3 psf of the roof surface respectively. Use 35 psf of ground snow load. Design sag rods using A36 steel and LRFDS. (Assume: exposure factor, $C_e = 1.0$; thermal factor, $C_t = 1.0$; and importance factor $I = 1.0$, in ASCES 7-98, Eq. 7-1.)

P7.30. Steel trusses with a span of 100 ft and a pitch of 1:5 are spaced at 30 ft on centers to support the roof of an industrial building. The roof covering is lightweight cement boards on S-shape purlins, 6 ft on centers. The dead load of the purlins and the attachments may be assumed to be 3 psf. For the snow load recommended by ASCES 7-98 for Green Bay, Wisconsin, design sag rods at the $\frac{1}{3}$ points between trusses. Use LRFDS. (Assume: exposure factor, $C_e = 1.0$; thermal factor, $C_t = 1.0$; and importance factor $I = 1.0$, in ASCES 7-98, Eq. 7-1.)

P7.31. The height to eaves of an industrial building is 32 ft. The columns are spaced 24 ft on centers. The side wall is covered with lightweight cement boards weighing

5 psf. Aluminum and glass windows, weighting 8 psf, occupy 50 percent of the wall area. Girts are C5×9 spaced 5 ft on centers. Design sag rods located at the ¼ points between columns (see Fig. 3.4.3). Use A36 steel and LRFDS.

P7.32.   A three-hinged parabolic tied-arch has a span of 60 ft and a rise of 20 ft. It carries a uniform horizontally distributed vertical load of 3 klf. Determine the diameter of an A36 steel tie rod, and select suitable clevises and turn buckle. The load given is a factored load.

DowElanco Project, Harbor Beach, MI.

Lateral loads on this complex biochemical-process support-structure are transferred by using Chevron bracing. All steel was galvanized.

Photo courtesy Douglass Steel Fabricating Corp., Lansing MI, www.douglassteel.com.

CHAPTER 8

Axially Loaded Columns

8.1 Introduction

A *column* is a structural member used to transmit a compressive force along a straight path in the direction of the longitudinal axis of the member. Originally, only vertically-upright compression members were referred to as columns. Compression members are sometimes called *posts,* and the inclined member at the end of a through-bridge truss is usually called an *end post.* Some types of compression members in roof trusses are called *struts.* In particular, a member which connects adjacent trusses at the eave of an industrial building is called an *eave strut,* while the member that connects adjacent trusses at the ridge of an industrial building is called a *ridge strut* (Fig. 3.4.1). In this text, any compression member, whether horizontal, vertical, or inclined, is termed a *column* if the compressive force it transmits is the primary force determining its structural behavior. Members subjected to combined axial compression and flexure are studied in Chapter 11.

Steel columns can be broadly classified as either short columns, long columns, or intermediate columns, based on the manner in which they would fail when subjected to axial loads. A compression member may be considered a *short column,* if its length is of the same order of magnitude as its overall cross-sectional dimensions. Such a member will usually fail by the squashing of its material (Fig. 8.1.1a). If the material is ductile, a gradual bulging of the member will manifest, and failure occurs without bending. The internal stresses that develop are distributed uniformly over all cross sections throughout the length of the member, and throughout the loading process. The compressive stress in an axially loaded column is given by:

$$f = \frac{P}{A_g} \qquad (8.1.1)$$

where A_g = gross area of the column cross section
       P = axial load on the column

For short columns, a load producing yield stress in the material is usually considered as the limit load. We have:

$$P_y = A_g F_y \qquad (8.1.2)$$

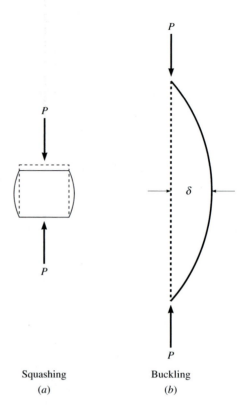

Squashing

(*a*)

Buckling

(*b*)

Figure 8.1.1: Short and long compression members.

where P_y = yield load of the column section
             F_y = yield stress of the material

Thus, the load carrying capacity of a short stocky column is independent of the length of the member.

   A ***long column*** is a compression member having great length relative to its least lateral dimension. Failure of a long column under axial load consists of a sudden lateral bending or buckling, as shown in Fig. 8.1.1*b*. For each column there is a limiting axial load ($P = P_{cr}$), known as the *buckling* or *critical* load. When this critical load, P_{cr}, has been reached, the column fails suddenly by buckling, even though the axial load does not produce an average unit stress as high as the yield stress of the material. Long columns will buckle elastically and the buckling stress remains below the proportional limit. ***Intermediate columns*** will also fail by buckling, but at buckling some of the fibers will reach the yield stress and some will not. Their behavior is said to be inelastic. Elastic buckling of long columns is studied in Sections 8.4 and 8.5, while inelastic buckling of intermediate columns is studied in Section 8.6.

8.2   Types of Sections for Columns

Several rolled shapes used as columns are shown in Fig. 8.2.1. The rolled W-section is the most widely used shape for columns in buildings, especially tall buildings. Connections of other members to W-shapes are easy to make, and fabrication costs are generally low. T-sections are suitable as compression chord members of welded roof trusses, as the web members may be welded directly to the stem of the tee. Single angles and channels are used as compression members in towers, and as bracing members in light trusses and frames. Square and rectangular HSS and steel pipe have more favorable distribution of material in their cross section compared to I-shapes, and often result in economical columns. Rods, bars, and plates have cross sections that result in members that are too slender to be used efficiently as compression members unless they are very short.

A column load may be so great that sufficient area cannot practically be provided by a single rolled shape. Also, the slenderness ratio of a column

Laced column built-up from two I-shapes.
Photo by S. Vinnakota.

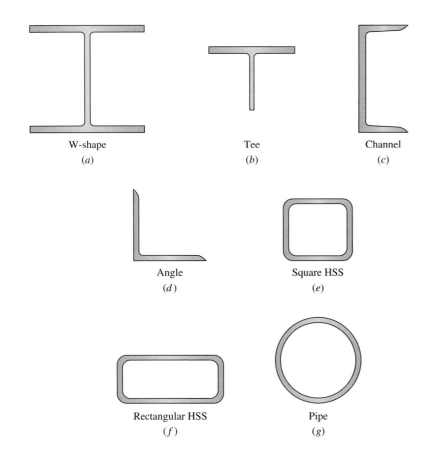

W-shape	Tee	Channel
(*a*)	(*b*)	(*c*)

Angle	Square HSS
(*d*)	(*e*)

Rectangular HSS	Pipe
(*f*)	(*g*)

Figure 8.2.1: Rolled shapes used as columns.

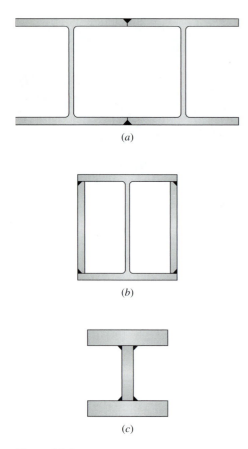

(a)

(b)

(c)

Figure 8.2.2: Built-up column sections made using continuous welds.

composed of a single shape may become so great that it reduces the design compressive stress to an undesirably low value. In both of these situations, the problem can be solved by using a built-up section consisting of several elements (rolled shapes and/or plates) bolted or welded together. The elements of a built-up section may be continuously welded (Fig. 8.2.2), closely spaced and interconnected by stitch plates in intervals (Figs. 8.2.3a and b), or widely spaced and interconnected by a continuous system of lacing (Figs. 8.2.3c, d, e). Stitch plates or lacing (shown dotted in Fig. 8.2.3), either bolted or welded, hold together the elements of the cross section and keep them in their proper positions so that they will act together as a single unit.

A built-up section consisting of two angles connected back to back is widely used as a chord or web member of bolted roof trusses and open web steel joists (Fig. 8.2.3a). End connections are made to gusset plates between the vertical legs. This section is also suitable as a web member of welded trusses where tee sections are used as chord members. A pair of channels is sometimes used as a web member in large roof and bridge trusses (Fig. 8.2.3b). The four-angle section shown in Fig. 8.2.3c, is often found in crane booms or on small masts and guyed towers. Generally, built-up sections that do not need lacing, such as those shown in Fig. 8.2.2, result in more economical sections than those shown in Fig. 8.2.3.

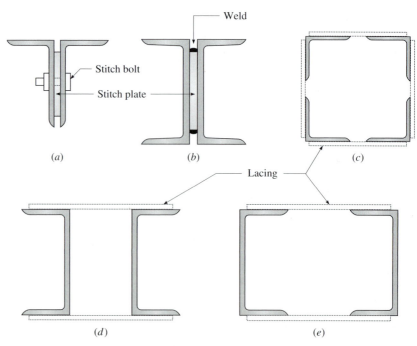

Figure 8.2.3: Built-up column sections made using stitch plates or lacing.

8.3   Stable, Neutral, and Unstable States of Equilibrium

Consider a structure in equilibrium. Upon this structure a small disturbing force is applied and then removed. Based upon the three possible responses of the structure, we can classify its equilibrium. We say that the equilibrium of the structure is **stable** if it tends to return to its original position. On the other hand, if the structure continues to move away from its original position, we say that its equilibrium is **unstable.** Finally, if the structure simply remains in its disturbed position, neither moving towards or away from its original position, we say that the equilibrium is **neutral.** In the first case, an addition of energy is required to produce the disturbance; in the second case, energy is released as the disturbance takes place; and in the third case, no change in the energy of the system occurs.

As a first example, consider the equilibrium of a spherical ball on three types of rigid surfaces shown in Figs. 8.3.1a, b, and c. It can be seen that the ball on the concave spherical surface (Fig. 8.3.1a) is in stable equilibrium, while the ball on the convex spherical surface (Fig. 8.3.1c) is in unstable equilibrium. The ball on the horizontal surface (Fig. 8.3.1b) is in neutral equilibrium. In the stable case, any displacement of the ball from its position of equilibrium will raise its center of gravity. A certain amount of work is needed to produce such a displacement. Thus the potential energy of the ball increases for any small displacement from the equilibrium position. In the unstable case, any displacement from the equilibrium position will lower the center of gravity of the ball and hence will decrease its potential energy. In the neutral case, there is no change in energy during a displacement. Note that in the case of stable equilibrium, the potential energy of the system prior

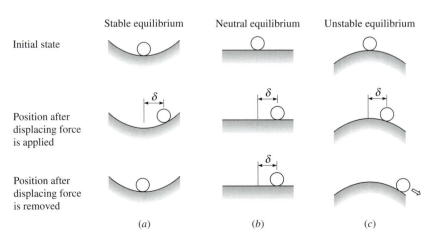

Figure 8.3.1: Representations of stable, neutral, and unstable equilibrium.

to a disturbance is a local minimum, while in the case of unstable equilibrium it is a local maximum.

As a second example, consider the long, pin-ended straight column under increasing axial load, P, shown in Figs. 8.3.2a, b, and c. When the applied load is small, the member maintains its linear shape and continues to remain in its straight equilibrium position. If at any time during this stage it is forced into a bent position (shown dotted in Fig. 8.3.2a), and the forcing agent is then removed, the column will return to the straight position. Thus the column is in a state of *stable equilibrium*. As the axial load is increased, a state will eventually be reached at which the member, if again forced into a slightly bent configuration, will retain its displaced shape upon removal of the forcing agent. At this load, the column is able to maintain equilibrium in both the straight and slightly bent configurations. Thus the column is in a state of *neutral equilibrium*. The phenomenon in which a perfectly straight member under compression may either assume a deflected position or may remain undeflected is known as **bifurcation**. The corresponding load P_{cr} is known as the **critical load, bifurcation load,**

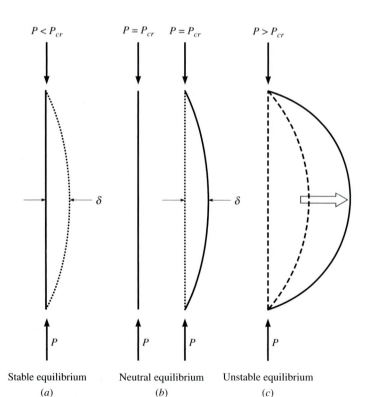

Figure 8.3.2: Equilibrium positions of a long column under axial load.

or ***buckling load*** of the column. If the axial load applied to a straight column is in excess of its buckling load, the member equilibrium is no longer stable. Thus, to determine the critical load of a column, one must find the load under which the member can be in equilibrium both in the straight and slightly bent configurations.

8.3.1   Limit States of Buckling

There are two general modes by which axially loaded steel columns can fail. These are ***member buckling*** and ***plate local buckling.*** Member buckling is characterized by no distortion of column section. On the other hand, local buckling is characterized by distortion of the cross section. Member buckling may take the form of ***flexural buckling, torsional buckling,*** or ***flexural-torsional buckling.***

In ***flexural buckling,*** the buckling deformations (deflections) all lie in one of the principal planes of the column cross section. No twisting of the cross section occurs for flexural buckling. In ***torsional buckling,*** the buckling deformations consist only of rotations of the cross sections about the longitudinal axis of the member. In ***flexural-torsional buckling,*** the buckling deformations consist of a combination of twisting and bending about the two flexural axes of the member. ***Plate buckling*** occurs when the compressive elements of the cross section of a member are so thin that they buckle locally before other modes of member buckling can occur.

The simplest type of buckling is flexural buckling. The limit state of flexural buckling is applicable for axially loaded columns with doubly symmetric sections such as bars, HSS, round HSS, and I-shapes, and singly symmetric sections, such as T- and C-shapes. The limit state of torsional buckling is applicable to axially loaded columns with doubly symmetric open sections with very slender cross-sectional elements. Examples of these sections include I-shapes, cruciform sections, and sections consisting of four angles placed back to back (Fig. X8.4.1*b*). The limit state of flexural-torsional buckling is applicable to columns with singly symmetric shapes, such as double-angle, T-, and C-shapes, and asymmetric cross sections. Flexural-torsional buckling analysis of such members will not be considered in this text.

The cross-sectional configurations and proportions of typical hot-rolled doubly symmetric column shapes such as wide flange sections, pipes, and square and rectangular HSS, are such that, under usual conditions of axial compression, failure by torsional buckling is nearly always preceded by failure by flexural buckling. The behavior of columns with thin sections that may fail by local plate buckling and singly symmetric or asymmetric hot-rolled shapes, such as channel, tee, and angle shapes that may fail by torsional or flexural-torsional buckling, are not treated in this chapter.

8.4   Elastic Flexural Buckling of a Pin-Ended Column

The following assumptions are made in deriving the basic differential equation for flexural buckling of a pin-ended column:

1. The column is prismatic and has a doubly symmetric cross section.
2. The column is perfectly straight.
3. The compressive load is applied along the centroidal axis of the column.
4. There are no transverse loads.
5. The ends of the member are ideally pinned. Thus, the lower end of the column is provided with an immovable hinge, while the upper end is supported so that it can rotate freely and move vertically but not horizontally.
6. The material is homogeneous and obeys Hooke's Law.
7. Plane sections before deformation remain plane after deformation.
8. The deformations of the member are small. Thus, the curvature can be approximated by the second derivative of the lateral displacement.
9. Influence of shear on deformations is neglected.
10. No twisting or distortion of the section occurs.

The initially straight pin-ended column ACB of length L, shown in Fig. 8.4.1a, is subjected to an axial thrust P applied through the centroid of the end section. The member is oriented with its longitudinal axis along the z axis of the coordinate system. In accordance with the criterion of neutral equilibrium discussed earlier, the critical load is that load for which the column is indifferent as to whether it is straight or deflected. That is, equilibrium in the slightly bent configuration, shown as AC′B in Fig. 8.4.1b is possible. We will assume that flexural buckling occurs about the member's y axis (i.e., buckling deformations lie in the xx plane).

The deflection at D, a distance z from A, is denoted by u relative to the straight line AB along the z axis. Figure 8.4.1c shows a free body diagram of the column segment AD′. Moment equilibrium about D′ of the free body in the buckled state gives:

$$M - Pu = 0 \rightarrow M = Pu \tag{8.4.1}$$

where M is the internal moment at section D. This relation states that for the column to retain its deflected shape, the internal moment must equal the external moment.

Assuming that the column has a uniform flexural rigidity, EI, about the y axis and that deflections are sufficiently small for $\left(\frac{du}{dz}\right)^2$ to be neglected, the resisting moment is given by:

$$M = EI\Phi = -EI\frac{d^2u}{dz^2} \tag{8.4.2}$$

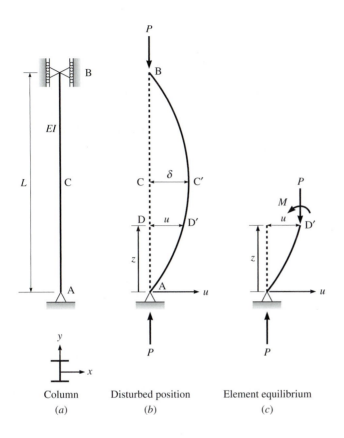

Column   Disturbed position   Element equilibrium
 (a)   (b)   (c)

Figure 8.4.1:   Pin-ended column under axial load.

where, E is the modulus of elasticity, I is the moment of inertia with respect to the axis of buckling, and Φ is the radius of curvature at section z. Substituting the results of Eq. 8.4.2 into the equilibrium expression of Eq. 8.4.1 and rearranging, we obtain the differential equation governing the deflected form of the column as follows:

$$EI\frac{d^2u}{dz^2} + Pu = 0 \rightarrow \frac{d^2u}{dz^2} + \frac{P}{EI}u = 0 \qquad (8.4.3)$$

or

$$\frac{d^2u}{dz^2} + \alpha^2 u = 0 \qquad (8.4.4)$$

where

$$\alpha^2 = \frac{P}{EI} \qquad (8.4.5)$$

Equation 8.4.4 is a second-order, linear, homogeneous differential equation with constant coefficients. The general solution of this equation is its homogeneous solution, namely,

$$u = A \sin \alpha z + B \cos \alpha z \qquad (8.4.6)$$

where A and B are constants of integration. This can easily be checked by computing $\frac{d^2u}{dz^2}$ from Eq. 8.4.6, and substituting for u and $\frac{d^2u}{dz^2}$ in Eq. 8.4.4. Equation 8.4.6 has three unknowns, A, B, and α (or P). To determine them, we employ the two boundary conditions (or end conditions). The end conditions of the column state that the deflections of the column are zero at both ends A and B.

$$u = 0 \quad \text{at} \quad z = 0; \quad u = 0 \quad \text{at} \quad z = L \qquad (8.4.7)$$

Substituting the end condition $u = 0$ at $z = 0$ into Eq. 8.4.6 gives $B = 0$, and the deflected curve becomes:

$$u = A \sin \alpha z \qquad (8.4.8)$$

Next, applying the end condition $u = 0$ at $z = L$, we obtain

$$A \sin \alpha L = 0 \qquad (8.4.9)$$

This relation can be satisfied in one of two ways, either:

$$A = 0 \qquad (8.4.10a)$$

or

$$\sin \alpha L = 0 \qquad (8.4.10b)$$

If $A = 0$, α (and consequently P) can have any value, and from Eq. 8.4.8, $u = 0$ for all z. This result is known as a *trivial solution,* as it simply states that a column is in equilibrium under any axial load P as long as the member remains perfectly straight. *Nontrivial solutions* describing the equilibrium position of the column in a slightly bent configuration are possible for particular or characteristic values of α, and hence P. They are obtained by solving Eq. 8.4.10b, resulting in:

$$\alpha L = n\pi \rightarrow \alpha = \frac{n\pi}{L} \qquad (8.4.11)$$

where $n = 0, 1, 2, 3, \ldots$ The value $n = 0$ is unimportant, because for that value the load P is also zero, meaning that the column is unloaded. For nonzero values of n, substitution of this expression into Eq. 8.4.5 gives the loads for which nontrivial solutions of Eq. 8.4.9 can be obtained:

$$P = P_{crn} = \frac{n^2 \pi^2 EI}{L^2} \qquad (8.4.12)$$

By substituting the value of α from Eq. 8.4.11 into Eq. 8.4.8, the corresponding displacements are given by:

$$u = u_n = A_n \sin \frac{n\pi z}{L} \qquad (8.4.13)$$

For the axial loads given by Eq. 8.4.12, called the **critical loads** or **buck-ling loads,** the column can be in equilibrium in a slightly bent shape as defined by Eq. 8.4.13.

Of the critical loads given by Eq. 8.4.12, the most important is the small-est one corresponding to $n = 1$. It is known as the **Euler load of a pin-ended column,** named after the Swiss mathematician Leonhard Euler (1707–1783). Thus, the Euler load of a pin-ended column is:

$$P_E = P_{cr1} = \frac{\pi^2 EI}{L^2} \qquad (8.4.14)$$

With the help of Eqs. 8.4.5 and 8.4.14, we observe that for $0 < P < P_E$ (or $0 < \alpha L < \pi$), the condition $\sin \alpha L = 0$ (Eq. 8.4.10b) cannot be satisfied, and the solution given by Eq. 8.4.13 does not exist. For this condition $A = 0$, and the only configuration possible for the column is a straight one (Fig. 8.4.2). Thus, for $P < P_E$ the straight configuration is stable. If $P = P_E$, the deformed shape of the buckled column given by Eq. 8.4.13 is valid, but its amplitude is inde-terminate. From Eq. 8.4.9, A can have any value when $\sin \alpha L = 0$, that is, at buckling load, P_E (Fig. 8.4.2).

Critical loads corresponding to $n > 1$ occur at larger loads and are pos-sible only if the column is braced at intermediate points between the ends of

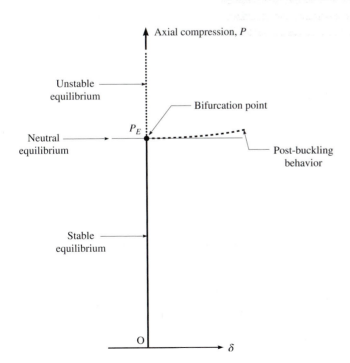

Figure 8.4.2: Stable, neutral, and unstable equi-librium of an ideal long column.

the column. The coefficient n indicates the number of half-sine waves that the deflected form of the column assumes. The configurations of the column for the first three critical loads ($n = 1, 2, 3$) are plotted in Fig. 8.4.3. Because small deflections are assumed, the curvature of the column can be represented by the second derivative of the displacement. Thus, by differentiating Eq. 8.4.13 twice, we obtain the following:

$$u'' = -\frac{n^2\pi^2}{L^2} A_n \sin \frac{n\pi z}{L}$$

From this equation and Eq. 8.4.13, we can see that the curvature is zero when z is equal to any multiple of L/n. Therefore, we may conclude that the points marked "X" in Figures 8.4.3a, b, and c are locations of zero curvature and are, therefore, inflection points of the buckled configuration. Recalling that curvature is also a function of internal bending moment, it can be concluded that the internal moment in the buckled column is zero at these points.

Values of n larger than 1 are not possible unless the pin-ended column is physically restrained from deflecting at the points where the reversal of

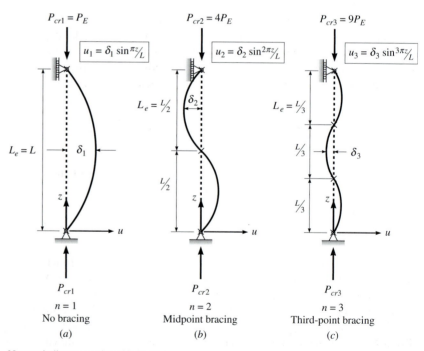

Figure 8.4.3: First three buckling modes of a pin-ended column.

Note: × indicates a point of inflection

curvature would occur. Thus, providing bracing at the mid-height (Fig. 8.4.3*b*) forces the buckled configuration into the second mode and results in a critical load multiplied by a factor of four. Providing bracing at the one-third points forces the buckled configuration into the third mode and results in a buckling load nine times that of the first mode (Fig. 8.4.3*c*). The distance between two successive inflection points of the buckled shape of a column is known as the **effective length of the column.** Indicating this length by L_e, we can rewrite Eqs. 8.4.12 and 13 as:

$$P_{crn} = \frac{\pi^2 EI}{(L/n)^2} = \frac{\pi^2 EI}{L_e^2} \tag{8.4.15}$$

$$u_n = A_n \sin\left(\frac{\pi z}{L/n}\right) = \delta_n \sin\left(\frac{\pi z}{L_e}\right) \tag{8.4.16}$$

where δ_n is the amplitude of the sine curve of the *n*th critical mode.

The stability of a pin-ended column could also be studied by considering the variation of the stiffness of that member as a function of the axial load, *P*. The critical load, P_{cr}, corresponds to the load at which the stiffness of the member reduces to zero. *For further information on stiffness of a pin-ended column as a function of the axial load, refer to Section W8.1 on the website* http://www.mhhe.com/Vinnakota.

As indicated above, for a pin-ended, axially loaded column, the critical load at which the column buckles is given by the Euler load given in Eq. 8.4.14. By dividing both sides of Eq. 8.4.14 by the cross-sectional area of the column, the value of the stress corresponding to the critical load can be determined. This stress is termed the **critical stress** and is denoted by f_{cr}.

$$f_{cr} = \frac{P_{cr}}{A} = \frac{P_E}{A} = \frac{\pi^2 EI}{AL^2} \tag{8.4.17}$$

Noting that $I = Ar^2$, where *r* is the radius of gyration of the column section with respect to the buckling axis, we obtain:

$$f_{cr} = F_E = \frac{\pi^2 E(Ar^2)}{AL^2} = \frac{\pi^2 E}{(L/r)^2} \tag{8.4.18}$$

where, F_E is the elastic buckling stress of a pin-ended column (Euler stress), and *L/r* is known as the *slenderness ratio* of the pin-ended column. A graphical interpretation of Eq. 8.4.18 is shown in Fig. 8.4.4, where the critical stress of the pin-ended column is plotted as a function of the slenderness ratio. The resulting curve is a hyperbola, known as the *Euler hyperbola*. However, since Eq. 8.4.18 is derived on the basis of elastic behavior, f_{cr} determined by this equation cannot exceed the yield stress of the material, F_y. So, the hyperbola

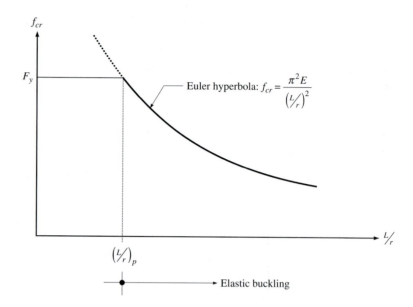

Figure 8.4.4: Variation of critical stress for an ideal pin-ended column.

shown in Fig. 8.4.4 is shown dashed beyond the material yield stress, and this portion of the curve cannot be used.

The nominal strength of a tension member (studied in Chapter 7) is given as $T = AF_y$, while the strength of a long, pin-ended, axially loaded column is shown to be given by $P_E = \pi^2 EI/L^2$. The load carrying capacity of a long column thus is not dependent on just the amount of material present in the cross section, as is the case with tension members, but also on its distribution, as evidenced by the presence of the moment of inertia term, I. A given column section has two principal moments of inertia, I_x and I_y. Associated with each is a load that will cause the member to buckle about the corresponding axis, indicated by P_{Ex} and P_{Ey}. The load that causes the pin-ended column to buckle is the smaller of the two values. Thus, for a pin-ended column, we can write:

$$P_{Ex} = \frac{\pi^2 EI_x}{L_x^2} ; \quad P_{Ey} = \frac{\pi^2 EI_y}{L_y^2} \tag{8.4.19}$$

$$P_E = \min [P_{Ex}, P_{Ey}] \tag{8.4.20}$$

Assuming that the member possesses no intermediate bracing about either axis, it will tend to buckle about the weaker axis—the one associated with the smaller moment of inertia. It follows that a round or square cross section would not have a preferential direction for buckling.

In the derivation of the critical load for the pin-ended column, we assumed that the column is perfectly straight. However, practical steel members almost always contain small, unavoidable, initial geometrical imperfections as a result of rolling, fabrication, and erection. *For information on the influence of such geometrical imperfections on the elastic behavior of a pin-ended column, refer to Section W8.2 on the website http://www.mhhe.com/ Vinnakota.*

Influence of Cross-Sectional Geometry

EXAMPLE 8.4.1

Determine the Euler stress and Euler load for the four pin-ended columns having a length L of 30 ft and a modulus of elasticity E of 29,000 ksi shown in Fig. X8.4.1. The cross sections are (a) a square bar of 4.36 in., (b) four $6 \times 4 \times \frac{1}{2}$ angles bolted back to back with $\frac{3}{4}$ in. spacing for stitch plates, (c) the same four angles welded together to form a box shape, and (d) the same four angles connected by lacing to form an open box of 12 in. by 16 in. The data for a single $L6 \times 4 \times \frac{1}{2}$ are shown in Fig. X8.4.1e.

Solution
All the sections considered are doubly symmetric. So the principal axes are the planes of symmetry, and the center of gravity, G, is located at the intersection of the two planes of symmetry as shown.

a. Square bar
Size 4.36 by 4.36 in.
$$A = bd = 4.36 \,(4.36) = 19.0 \text{ in.}^2$$

$$I = \frac{1}{12} bd^3 = \frac{1}{12}(4.36)(4.36)^3 = 30.1 \text{ in.}^4$$

$$r = \sqrt{\frac{I}{A}} = \sqrt{\frac{30.1}{19.0}} = 1.26 \text{ in.}$$

Slenderness ratio, $L/r = \dfrac{30(12)}{1.26} = 286$

Euler stress, $F_E = \dfrac{\pi^2 E}{(L/r)^2} = \dfrac{\pi^2(29,000)}{286^2} = 3.50 \text{ ksi}$ (Ans.)

Euler load, $P_E = F_E A = 3.50(19.0) = 66.5$ kips   (Ans.)

b. Cruciform section
For a single $6 \times 4 \times \frac{1}{2}$ angle, we have from LRFDM Table 1-7:
$$A_L = 4.72 \text{ in.}^2, \ I_{xL} = 17.3 \text{ in.}^4, \ \bar{y}_L = 1.99 \text{ in.}$$
$$I_{yL} = 6.21 \text{ in.}^4, \ \bar{x}_L = 0.986 \text{ in.}$$
For the built-up section shown in Fig. X8.4.1b:
$$A = 4(4.72) = 18.9 \text{ in.}^2$$
$$I_x = 4 \,[I_o + Ad^2]_x = 4 \,[17.3 + 4.72 \,(\tfrac{3}{8} + 1.99)^2] = 175 \text{ in.}^4$$

[continues on next page]

Example 8.4.1 continues ...

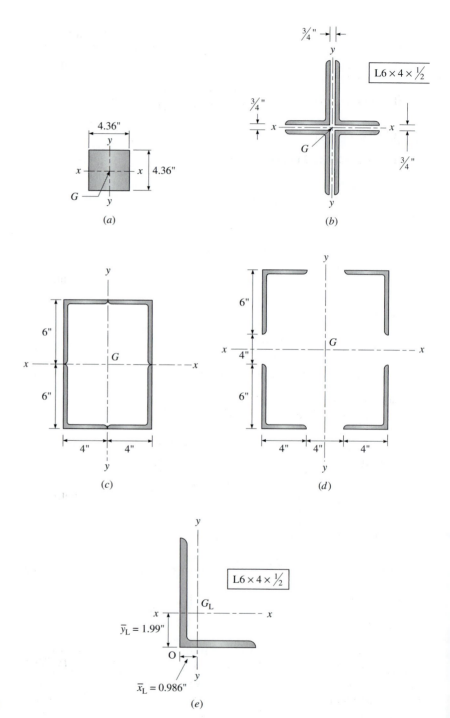

Figure X8.4.1

$$I_y = 4[I_o + Ad^2]_y = 4[6.21 + 4.72 \,(\% + 0.986)^2] = 59.8 \text{ in.}^4$$

$$r_x = \sqrt{\frac{175}{18.9}} = 3.04 \text{ in.}; \quad r_y = \sqrt{\frac{59.8}{18.9}} = 1.78 \text{ in.} \leftarrow \text{controls}$$

Slenderness ratio of column, $\dfrac{L}{r} = \dfrac{L}{r_y} = \dfrac{30(12)}{1.78} = 202$

Euler stress, $F_E = \dfrac{\pi^2(29,000)}{202^2} = 7.01 \text{ ksi}$ \hfill (Ans.)

Euler load, $P_E = 7.01(18.9) = 132 \text{ kips}$ \hfill (Ans.)

c. Box shape
For the compound section shown in Fig. X8.4.1c,

$A = 18.9 \text{ in.}^2$
$I_x = 4 [I_o + Ad^2]_x = 4 [17.3 + 4.72 \,(6.00 - 1.99)^2 \,] = 373 \text{ in.}^4$
$I_y = 4 [I_o + Ad^2]_y = 4 [6.21 + 4.72 \,(4.00 - 0.986)^2] = 196 \text{ in.}^4$

$$r_x = \sqrt{\frac{373}{18.9}} = 4.44 \text{ in.}; \quad r_y = \sqrt{\frac{196}{18.9}} = 3.22 \text{ in.} \leftarrow \text{controls}$$

Slenderness ratio of column, $\dfrac{L}{r} = \dfrac{L}{r_y} = \dfrac{30(12)}{3.22} = 112$

Euler stress, $F_E = \dfrac{\pi^2(29,000)}{112^2} = 22.8 \text{ ksi}$ \hfill (Ans.)

Euler load, $P_E = 22.8(18.9) = 431 \text{ kips}$ \hfill (Ans.)

d. Laced column
For the laced column shown in Fig. X8.4.1d,

$A = 18.9 \text{ in.}^2$
$I_x = 4[17.3 + 4.72(8.00 - 1.99)^2] = 751 \text{ in.}^4$
$I_y = 4[6.21 + 4.72(6.00 - 0.986)^2] = 500 \text{ in.}^4$

$$r_x = \sqrt{\frac{751}{18.9}} = 6.31 \text{ in.}; \quad r_y = \sqrt{\frac{500}{18.9}} = 5.14 \text{ in.} \leftarrow \text{controls}$$

Slenderness ratio, $\dfrac{L}{r} = \dfrac{L}{r_y} = \dfrac{30(12)}{5.14} = 70.0$

Euler stress, $F_E = \dfrac{\pi^2(29,000)}{70.0^2} = 58.4 \text{ ksi}$ \hfill (Ans.)

Euler load, $P_E = 58.4(18.9) = 1100 \text{ kips}$ \hfill (Ans.)

The results are summarized in the following table:

	Section	A (in.²)	L/r	F_E (ksi)	P_E (kips)
(a)	square bar	19.0	286	3.50	66.5
(b)	cruciform	18.9	202	7.01	133
(c)	box	18.9	112	22.8	431
(d)	laced section	18.9	70.0	58.4	1100

[*continues on next page*]

Example 8.4.1 continues ...

Remarks

1. All of the columns considered are of the same material, and have the same length, end conditions, and cross-sectional area. So, the increase in elastic critical strength from columns (*a*) to (*d*) is due to the improved arrangement of the material in the cross section. As the resistance to buckling of a column is mainly due to flexural resistance, long columns should be designed with as much column material as far from the buckling axis as is practical. Because of this, HSS provide greater buckling strength per unit volume of material, compared to I-shapes.

2. This simple example shows that the cross-shape, an open section built-up from four angles, is very inefficient to transmit compression loads (it can be shown that it is equally weak in torsion). Yet, this was the type of section used for top-chord (compression) members of the ill-fated 300 ft by 360 ft Hartford Civic Center Arena roof that collapsed in January 1978 [for details see Levy and Salvadori, 1992].

3. Attainment of the Euler load for the laced box assumes that all the fibers of the column continue to behave elastically under a uniform stress of 58.4 ksi. Influence of yielding on column strength is considered in Section 8.6.

8.5 Effective Length of Columns in Structures

The elastic buckling load of a column is influenced by the end restraint provided to the column. For example, if one end of the pin-ended column studied in Section 8.4 is fixed, the elastic buckling load increases two fold. *For details on the influence of end fixity on elastic buckling of such a column, refer to Section W8.3 on the website http://www.mhhe.com/Vinnakota.* More generally, the elastic buckling load of a column that is part of a structure can be written as:

$$P_e = \frac{\pi^2 EI}{(KL)^2} \tag{8.5.1}$$

Here,

$$
\begin{aligned}
L &= \text{length of the column} \\
KL &= \text{effective length of the column} \\
K &= \text{effective length factor} \\
P_e &= \text{elastic flexural buckling load of a column}
\end{aligned}
$$

Determination of the effective length factor, *K*, will be considered in detail in this section.

8.5.1 Introduction

In steel building frames, columns are generally fabricated in lengths of two stories or more. These lengths are spliced together rigidly. The bottom end of each column in the first tier is either bolted through connection angles or directly welded to the base plate on which the column rests. The unsupported length, L, of each column is taken as the distance between the centerline of the girders at successive floors. For the lowest story, it is the distance from the bottom end of that column to the centerline of the second floor girder.

Column bases can be designed to be either pinned or fixed connections (Fig. 8.5.1). Typically, pinned column base connections attach the base plate to the foundation via four anchor rods as shown in Fig. 8.5.1a. Fixed column base connections usually feature heavy base plates with four or more heavy anchor rods located outside the flange tips as far away from the buckling axes of the column as practical (Fig. 8.5.1b).

Simple, semi-rigid, and rigid girder-to-column connections (see Fig. 3.2.2) are all used in frames. Simple or shear type girder-to-column connections only attach the girder web to the column and are incapable of providing any significant rotational restraint (Figs. 8.5.1c and 3.2.2b). They are modeled as pinned connections. Rigid girder-to-column connections attach the girder web as well as the girder flanges rigidly to the column and are capable of providing rotational restraint to the column (Figs. 8.5.1d and 3.2.2a).

LRFDS Section C2 categorizes frames as braced and unbraced.

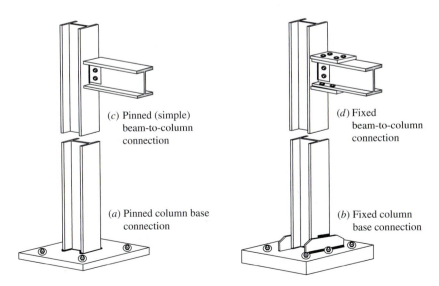

(c) Pinned (simple) beam-to-column connection

(d) Fixed beam-to-column connection

(a) Pinned column base connection

(b) Fixed column base connection

Figure 8.5.1: Column end connections.

- A **braced frame** is one in which the resistance to lateral loads or frame instability is provided by diagonal bracing, shear walls, or other equivalent means. If a column is part of an adequately braced frame, lateral translation of the top end of the column with respect to its bottom end does not occur when the frame buckles (Fig. 8.5.2a). That is, during the buckling process the joints of a braced frame remain in their original positions. The effective length of a column in a braced frame is at most equal to its actual length between supports (floors). Rotational restraint at the top and/or bottom end of a column due to girder and/or base plate connections reduces the effective length of the

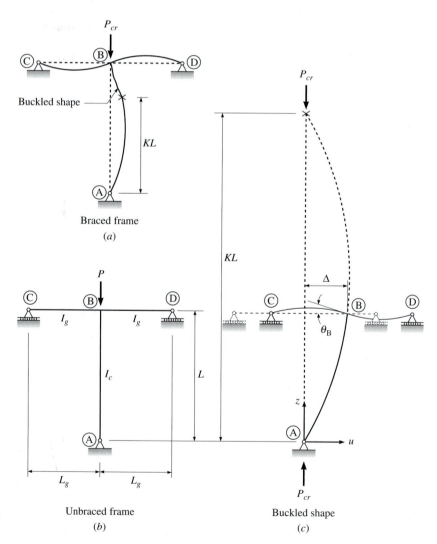

Figure 8.5.2: Buckling of braced and unbraced frames.

column to a value below its actual length. Thus, values of K for columns in braced frames are less than or equal to 1.

- On the other hand, an ***unbraced frame*** is one in which the resistance to lateral loads and frame buckling is provided entirely by the flexural resistance of its own members (girders and columns) and the rigidity of their connections. When buckling of such a frame occurs, the top end of a typical column translates laterally with respect to its bottom end (Figs. 8.5.2*b* and *c*). The effective length of a column in an unbraced frame always exceeds its actual length L between supports (floors). Thus, values of K for columns in unbraced frames are greater than 1.

The lateral deflection of the top end of a column with respect to its bottom end is known as ***sway.*** Thus, braced frames are also known as ***sidesway inhibited frames,*** while unbraced frames are known as ***sidesway uninhibited frames.***

The range of effective length factors for columns in portal frames is shown in Fig. 8.5.3.

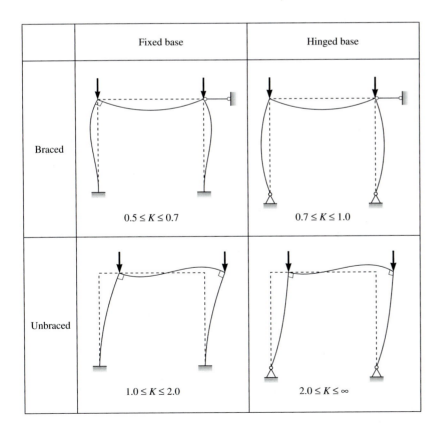

Figure 8.5.3: Range of effective length factors for columns in portal frames.

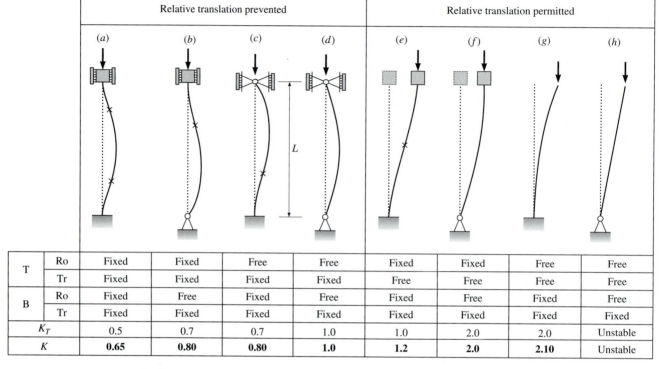

		Relative translation prevented				Relative translation permitted			
		(a)	*(b)*	*(c)*	*(d)*	*(e)*	*(f)*	*(g)*	*(h)*
T	Ro	Fixed	Fixed	Free	Free	Fixed	Fixed	Free	Free
	Tr	Fixed	Fixed	Fixed	Fixed	Free	Free	Free	Free
B	Ro	Fixed	Free	Fixed	Free	Fixed	Free	Fixed	Free
	Tr	Fixed	Fixed	Fixed	Fixed	Fixed	Fixed	Fixed	Fixed
K_T		0.5	0.7	0.7	1.0	1.0	2.0	2.0	Unstable
K		**0.65**	**0.80**	**0.80**	**1.0**	**1.2**	**2.0**	**2.10**	Unstable

T = Top   Ro = Rotation K_T = Theoretical K value
B = Bottom   Tr = Translation K = Recommended or design K value

Figure 8.5.4:   Effective length factors for isolated columns.

8.5.2   Influence of End Conditions on Isolated Columns

Fig. 8.5.4 gives effective length factors for eight isolated columns that are identical in all respects except for their end conditions. Idealized end conditions in which the rotational and translational end restraints are either fully realized or nonexistent are considered. For all columns, the base is fixed against translation. The dashed lines in Fig. 8.5.4 show the original position of the eight columns while the continuous lines represent the buckled shape when the axial load reaches the critical load for each column. Location of inflection points, or points of contraflexure where the curvature changes sign and the internal moment equals zero, are shown by an X on the buckled shape. Columns *a*, *b*, *c*, and *d* represent columns that are part of braced frames, as there is no translation of the top ends of these columns. Columns *e*, *f*, *g*, and *h* represent columns that are part of unbraced frames, as the top end of each of these columns is free to translate relative to the bottom end of the column.

Column d represents the standard pin-ended column. Column g has one end fixed and the other completely free for both rotation and translation. It is called a **cantilever column.** Flag poles and some bridge towers are structures which closely approximate this type of idealized column. The condition of full fixity at a column base can be approached only when a column is anchored securely by adequate embedment in a concrete footing or pile cap. It can also be approached when the column is attached by a moment-resisting bolted connection to a footing that is designed to resist overturning moment and for which the rotation is negligible. Rotational fixity of the top of a column is approached when the top of a column is rigidly connected to a heavy truss or a heavy girder having a rotational stiffness many times that of the column under consideration. Rotationally free ends are obtained by using a simple shear connection between the column and the girder web or by using a very flexible girder rigidly connected to the column.

In addition to the theoretical values for the effective length factors, Fig. 8.5.4 gives (shown in bold face) the **design values for K** suggested by the Structural Stability Research Council (SSRC). The values are also given in LRFDC Table C-C2.1. These values are modifications of the idealized values that take into account the fact that full fixity against rotation is, practically speaking, impossible to attain. Hinged column base connections designed only for vertical loads (Fig. 8.5.1a) will provide substantial restraint against rotation [Galambos, 1960]. Consequently, the effective length of column d of Fig. 8.5.4 may in reality be somewhat less than the actual length. In the interest of being conservative, however, LRFDC Table C-C2.1 recommends a value equal to 1.0. In a related situation, if the base of column f were truly pinned, K would actually exceed 2.0, because the flexibility of the adjoining girders would prevent realization of full fixity at the top of the column. However, the moment restraint of the base connection can be substantial. As a result, LRFDC recommends a design K value of 2.0 for column f.

The use of K values given in Fig. 8.5.4 (LRFDC Table C-C2.1) is often satisfactory for designing isolated columns and for making preliminary design of columns in frames. The end conditions shown in Fig. 8.5.4 represent theoretical extremes, while the end conditions in actual practice often fall somewhere in between. The effect of real end conditions that exist in practical structures is often taken into account by simply modifying the effective length factor K of the member or by using the alignment charts described in Section 8.5.4.

8.5.3   Influence of Intermediate Bracing

To reduce effective lengths of columns and thus increase their load carrying capacity, columns are frequently braced at one or more points along their length. The deflection of a buckled column at a brace point is zero. The bracing itself is generally part of the structural system for the rest of the building and often serves other functions. For example, in industrial buildings the

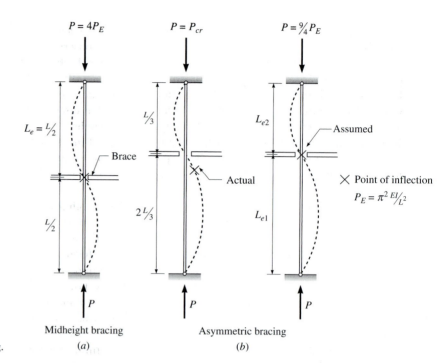

$P = 4P_E$

$P = P_{cr}$

$P = \frac{9}{4}P_E$

$L_e = \frac{L}{2}$

$\frac{L}{3}$

L_{e2}

Brace

Assumed

$\frac{L}{2}$

$2\frac{L}{3}$

L_{e1}

Actual

$\times$ Point of inflection

$P_E = \pi^2 \frac{EI}{L^2}$

P

P

P

Midheight bracing

Asymmetric bracing

Figure 8.5.5:   Influence of intermediate bracing.

(a)

(b)

vertical wall panels are attached to horizontal elements called *girts*, which in turn frame into the columns located along the wall (Fig. 3.4.3). In addition to supporting the wall panels, the girts serve to brace the column. The girts are usually fastened to the column web by simple shear type connections that are assumed to be incapable of providing any rotational restraint. The girts will therefore prevent the column from displacing laterally at brace points, but will not keep the column from rotating at those points.

Fig. 8.5.5*a* shows a pin-ended column of length, L, braced laterally at mid-height, resulting in two segments each having an unsupported length of $\frac{1}{2}L$. When the column buckles, it will deform into an S-shape as shown by the dotted line of Fig. 8.5.5*a*. Due to the symmetry of the column about the brace, the point of inflection coincides with the brace location. The shape of the column between the brace point and the column end is similar to that of a pin-ended column. The effective length of the column shown has thus been halved, thereby increasing its elastic critical load by a factor of four.

If, instead, the bracing were placed two-thirds of the way up from the bottom (Fig. 8.5.5*b*), two segments of unsupported lengths $\frac{1}{3}L$ and $\frac{2}{3}L$ would result. The longer (and hence more slender) segment of length $\frac{2}{3}L$ tends to buckle prior to the shorter segment. If the critical load of the column is reached, it would again buckle into a modified S-shape (Fig. 8.5.5*b*) with a point of inflection located in the longer segment near the brace point. Typically, however, the location of the inflection point is conservatively assumed to coincide with the brace point, as shown in the second schematic of Fig. 8.5.5b. Accordingly, the effective length of the column will be $\frac{2}{3}L$, and the critical load of the column will be $\frac{9}{4}$ times the critical load of the unbraced column. Thus, bracing the column at this point is not as effective at increasing the load carrying capacity as is bracing it at mid-height.

Finally, if braces are provided at each of the third points of the column, the points of inflection coincide with the brace points, making the effective length of the column as $\frac{1}{3}L$. The elastic critical load in this case is multiplied by a factor of nine (see Fig. 8.4.3c).

So, it can be concluded that the use of more girts reduces both the span of the wall panel and the effective length of columns in the wall. However, adding girts increases the fabrication and labor costs due to the increase in the number of connections and the number of pieces to be handled. The larger the loads on the column, and the longer it is, the more important it is to pay first attention to optimizing the column itself. Typically, however, the spacing of girts is based on the optimum span for the most economical wall paneling or siding available. This factor, in turn, establishes where the column is to be braced. Column properties are then selected on the basis of these brace points. The goal of the designer should always be to optimize the total cost of the column, bracing, and wall system together rather than the column alone or the wall system alone.

Columns are sometimes located in brick or masonry walls where the wall can serve as continuous lateral bracing in one plane ($KL = 0$) (see Example 8.5.1f). In such cases, I-shapes can be oriented so that the strong axis (x) lies in the plane of the wall. Thus $KL = 0$ for buckling about the weak axis (y). Wide flange sections having large r_x/r_y ratios are most economical for such columns.

8.5.4 Effective Length of Columns in Frames and Alignment Charts

The concept of the *effective length factor* permits the designer to find an equivalent pinned-end braced column of length KL that has the same buckling load as the actual column of length L that is part of a frame with given end and bracing conditions. The magnitude of K depends upon the rotational restraint supplied at the ends of the column and upon the resistance to lateral movement provided by the frame. The effective length of a framed column can be determined by carrying out a stability analysis of the entire frame [Bleich, 1952; Timoshenko and Gere, 1961; Galambos, 1968; Vinnakota and Badoux, 1970]. Such an analysis gives the magnitude of the external loads at the instant when the frame becomes unstable. From these results the buckling loads of individual compression members, and hence their effective lengths, can be evaluated. However, a stability analysis of an entire frame is often too

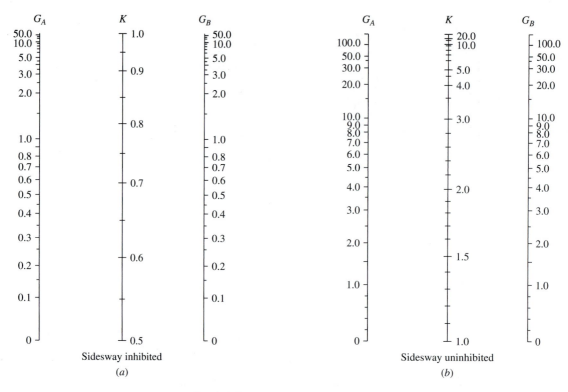

Figure 8.5.6: Alignment charts for effective length factors for columns in continuous frames. From Load and Resistance Factor Design Specification for Structural Steel Buildings, December 27, 1999. Copyright © American Institute of Steel Construction, Inc. Reprinted with permission. All rights reserved.

involved for routine design, so approximate methods using relatively simple procedures are needed.

Most engineers use the two alignment charts originally developed by Julian and Lawrence [1959] to determine the K factors for columns in rectangular frames (LRFDC Figs. C-C2a and C-C2b). Figure 8.5.6a shows the alignment chart for determining the effective length factor K for columns in multistory frames with sidesway inhibited (columns in braced frames). Figure 8.5.6b shows the alignment chart for columns in rectangular frames with sidesway uninhibited (columns in unbraced frames). A subassemblage consisting of the column in question and only the girders and columns that frame directly into that column is considered in the development of the alignment charts [Kavanagh, 1962; Galambos, 1968; Chen and Lui, 1987]. The model used for the determination of the effective length factor for column AB of the braced frame of Fig. 8.5.7a is shown in Fig. 8.5.7b. The model used for the determination of the effective length factor for the same column AB, but within the unbraced frame of Fig. 8.5.8a, is shown in Fig. 8.5.8b.

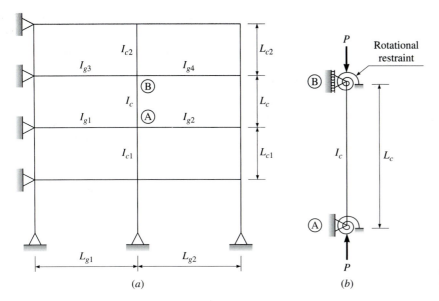

(a)

(b)

Figure 8.5.7: Buckling model for a column in a braced frame.

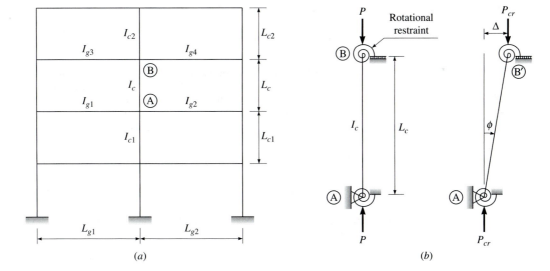

(a)

(b)

Figure 8.5.8: Buckling model for a column in an unbraced frame.

The alignment charts are based upon assumptions of idealized conditions that seldom are completely satisfied in real structures. These assumptions are as follows:

1. The structure consists of regular rectangular frames.
2. All members have constant cross section.

3. All girder-to-column connections are rigid connections.
4. The stiffness parameter $\alpha L(=\pi\sqrt{P/P_E})$ is the same for all columns.
5. All columns reach their buckling loads simultaneously.
6. At a joint, the restraining moment provided by the girders is distributed to the column above and below the joint considered, in proportion to the I/L ratios of the two columns.
7. No significant axial compression force exists in the girders.
8. Material behavior is linear elastic.
9. For braced frames, at the onset of buckling, rotations at opposite ends of girders are equal in magnitude and opposite in sense, producing symmetric single curvature bending. Note that the rotational stiffness of such a girder is $(2EI/L)_g$.

   For unbraced frames, at the onset of buckling, the rotations at the opposite ends of the girders are equal in magnitude and act in the same sense, thus producing reverse-curvature bending. Note that the rotational stiffness of such a girder is $(6EI/L)_g$.

   In order to use the alignment charts, it is necessary to know the preliminary sizes of the girders and columns framing into the column in question. Evaluation of the relative stiffness factors of the frame members connected to the ends of the column under consideration must then be made. The **relative stiffness factor,** also known as the G factor, is equal to the ratio of the summation of the rotational stiffnesses of the destabilizing members at a joint to the summation of the rotational stiffnesses of the stabilizing members at that joint. Generally speaking, columns are destabilizing members, while girders are stabilizing members. Thus:

$$G = \frac{\Sigma_c(2E_cI_c)/L_c}{\Sigma_g(2E_gI_g)/L_g} \quad \text{for a column in a braced frame} \quad (8.5.2a)$$

$$G = \frac{\Sigma_c(6E_cI_c)/L_c}{\Sigma_g(6E_gI_g)/L_g} \quad \text{for a column in an unbraced frame} \quad (8.5.2b)$$

or simply,

$$G = \frac{\Sigma_c(E_cI_c)/L_c}{\Sigma_g(E_gI_g)/L_g} \quad \text{for a column in a braced or unbraced frame}$$

$$(8.5.2c)$$

Here, E_c, I_c, L_c are the modulus of elasticity, the moment of inertia, and the length of a column, and E_g, I_g, L_g are the corresponding quantities for a girder. Σ_c indicates a summation for all columns, and Σ_g for all girders rigidly connected to the joint considered, and lying in the plane in which buckling of the column is being considered. The moments of inertia refer to the cross-sectional axes about which buckling takes place. Thus, if buckling occurs in the plane of the paper (i.e., if the deflections due to buckling are in the plane of the paper), the moments of inertia are about axes perpendicular to the plane of the paper. In a steel structure, the columns and girders generally have the same elastic modulus, so the terms E_c, E_g in Eqs. 8.5.2 usually cancel. For the specific case of column AB in Figs. 8.5.7 and 8.5.8, the expressions for G at ends A and B of the column can be expressed as follows:

$$G_A = \frac{\dfrac{I_c}{L_c} + \dfrac{I_{c1}}{L_{c1}}}{\dfrac{I_{g1}}{L_{g1}} + \dfrac{I_{g2}}{L_{g2}}}; \quad G_B = \frac{\dfrac{I_c}{L_c} + \dfrac{I_{c2}}{L_{c2}}}{\dfrac{I_{g3}}{L_{g3}} + \dfrac{I_{g4}}{L_{g4}}} \qquad (8.5.3)$$

In these expressions, the quantities I_c and L_c are the moment of inertia and the length of the column AB under investigation. The subscripts $c1$ and $c2$ refer to the adjacent columns, and the subscripts $g1$, $g2$, $g3$, and $g4$ refer to the adjacent girders. Also, for rectangular frames considered here, $L_{g3} = L_{g1}$ and $L_{g4} = L_{g2}$.

Using the slope-deflection method and the assumptions mentioned earlier (especially assumption 9 regarding girder deformations), the following buckling expressions can be obtained for column AB of the braced frame of Fig. 8.5.7 [Galambos, 1968]:

$$\frac{G_A G_B}{4}\left(\frac{\pi}{K}\right)^2 + \left(\frac{G_A + G_B}{2}\right)\left[1 - \frac{\left(\dfrac{\pi}{K}\right)}{\tan\left(\dfrac{\pi}{K}\right)}\right] + \frac{\tan\left(\dfrac{\pi}{2K}\right)}{\left(\dfrac{\pi}{2K}\right)} - 1 = 0$$

$$(8.5.4)$$

The nomograph of Fig. 8.5.6a is a graphical solution to this transcendental equation. Similarly, using the slope-deflection method and the assumptions listed earlier (especially assumption 9 regarding girder deformations), the following characteristic equation can be obtained for column AB of the unbraced frame of Fig. 8.5.8 [Galambos, 1968]:

$$\frac{G_A G_B\left(\dfrac{\pi}{K}\right)^2 - 36}{6\,(G_A + G_B)} - \frac{\left(\dfrac{\pi}{K}\right)}{\tan\left(\dfrac{\pi}{K}\right)} = 0 \qquad (8.5.5)$$

The nomograph given in Fig. 8.5.6b is a graphical solution to this transcendental equation.

When G_A and G_B have been calculated for a particular column, the appropriate alignment chart is entered with these two parameters, and a straight line is drawn between the two points representing G_A and G_B on the outer scales. The value of K then is read from the central scale at the point at which it crosses the line drawn. For example, for a column in an unbraced frame, with $G_A = 10$ and $G_B = 1.5$, $K = 2.0$ from the alignment chart for sidesway uninhibited case. Also, for a column in a braced frame, with $G_A = 1.0$ and $G_B = 0.35$, $K = 0.7$ from the alignment chart for sidesway inhibited case.

Note that when very stiff members such as deep girders or trusses are rigidly connected to flexible columns, G tends to zero, and the K factor tends to be small. On the other hand, if long, flexible girders are connected to short, heavy columns, G tends to infinity, and the K value tends to be large. Calculated K factors between 2.0 and 3.0 are quite common, and even larger values are occasionally obtained.

Equations 8.5.4 and 8.5.5, and hence the alignment charts in Fig. 8.5.6, were developed under the idealized conditions and assumptions listed earlier. When actual conditions differ from those assumed, care must be taken to avoid unrealistic designs. Methods are available to enable engineers to make simple modifications in the use of the alignment charts when some of the underlying assumptions are violated [Galambos, 1960; Le-Wu Lu, 1965; Galambos, 1988; Yura, 1971; Bjorhovde, 1984; ASCE, 1991]. A few of them are described below.

When the lower end of a column is attached to a footing, some judgement must be exercised in selecting the value for G. For a column pin-connected to a footing or foundation cap ($I_g \to 0$), G is theoretically infinity. However, unless a true friction-free pin is actually provided, the LRFDC recommends a G value of 10 for practical designs. If the column end is rigidly attached to a properly designed footing or foundation that provides full rotational restraint to the column base ($I_g \to \infty$), G is theoretically zero. However, the LRFDC recommends a G value of 1.0 to be used to account for the practical impossibility of achieving complete fixity (zero end slope). Smaller values may be used if justified by analysis. See the discussion accompanying the alignment charts in the LRFD Commentary (page 16.1–191 of the LRFDM). Thus, we use:

$$G = 10 \quad \text{for a hinged base} \tag{8.5.6a}$$

$$G = 1.0 \quad \text{for a fixed base} \tag{8.5.6b}$$

One of the assumptions made in the derivation of Eqs. 8.5.4 and 8.5.5 is that girder ends are rigidly connected to columns. However, if girder ends are hinged or clamped, the stiffness of the girder used in the determination of G factors must be modified. This is done by introducing a stiffness modifier α for the girders as follows:

$$G_A = \frac{\dfrac{I_c}{L_c} + \dfrac{I_{c1}}{L_{c1}}}{\alpha_{g1}\dfrac{I_{g1}}{L_{g1}} + \alpha_{g2}\dfrac{I_{g2}}{L_{g2}}} \; ; \quad G_B = \frac{\dfrac{I_c}{L_c} + \dfrac{I_{c2}}{L_{c2}}}{\alpha_{g3}\dfrac{I_{g3}}{L_{g3}} + \alpha_{g4}\dfrac{I_{g4}}{L_{g4}}} \tag{8.5.7}$$

For all frames:

1. If both ends of a girder are rigidly connected to columns, the modifier α_g for that girder equals 1. This is the standard case considered for all girders in Eq. 8.5.3.
2. If the near end of a girder is pinned, the girder cannot provide any restraint to the column end in question. Thus, the modifier α_g for that girder equals 0.

For braced frames:

1. If the end of a girder away from the joint being considered is hinged, such as girder BC of Fig. 8.5.2a, its rotational stiffness is $3EI_g/L_g$ instead of the value $2EI_g/L_g$ used in the derivation of Eq. 8.5.4 (also see Eq. 8.5.2a). Hence, to adjust for the far end of a girder being hinged, the corresponding I_g/L_g term in the G factor expression is multiplied by (3/2). Thus, the coefficient α_g for that girder is 1.5.
2. If the far end of a girder is fixed (i.e., clamped), its rotational stiffness is $4EI_g/L_g$ instead of the value $2EI_g/L_g$ used in the derivation of Eq. 8.5.4 (also see Eq. 8.5.2a). Hence, to adjust for a fixed far end of a girder, its I_g/L_g term is multiplied by (4/2). Thus, the coefficient α_g for that girder is 2.0.

For unbraced frames:

1. If the far end of a girder is hinged, such as girder BC of Fig. 8.5.2b, its rotational stiffness is $3EI_g/L_g$ instead of the value $6EI_g/L_g$ used in the derivation of Eq. 8.5.5 (also see Eq. 8.5.2b). Hence, to adjust for a hinged far end of a girder its I_g/L_g term is multiplied by (3/6). Thus, the coefficient α_g for that girder is 0.5.
2. If the far end of a girder is fixed, its rotational stiffness is $4EI_g/L_g$ instead of the value $6EI_g/L_g$ used in the derivation of Eq. 8.5.5 (also see Eq. 8.5.2b). Hence, to adjust for a fixed far end of girder, its I_g/L_g term in the G factor expression is multiplied by (4/6). Thus, the coefficient α_g for that girder is 0.67.

As an alternative to the use of the alignment chart of Fig. 8.5.6a, an approximate solution to the transcendental equation (Eq. 8.5.4) for columns in braced frames may be used [Dumonteil, 1992]:

$$K = \frac{3G_AG_B + 1.4(G_A + G_B) + 0.64}{3G_AG_B + 2.0(G_A + G_B) + 1.28} \qquad (8.5.8)$$

For the special case of a column hinged at A, that is, when G_A is infinitely large, the above approximation results in the following expression:

$$K = \frac{3G_B + 1.4}{3G_B + 2.0} \qquad (8.5.9)$$

or, by using the recommended value of 10 for G_A for practical hinged bases, we have:

$$K = \frac{31.4G_B + 14.6}{32.0G_B + 21.3} \qquad (8.5.10)$$

If the column is fixed at A, $G_A = 0$, theoretically, and Eq. 8.5.8 results in:

$$K = \frac{1.4G_B + 0.64}{2.0G_B + 1.28} \qquad (8.5.11)$$

or, by using the recommended value of 1.0 for G_A:

$$K = \frac{4.4G_B + 2.04}{5.0G_B + 3.28} \qquad (8.5.12)$$

In the particular case where $G_A = G_B = G$, we have:

$$K = \frac{G + 0.4}{G + 0.8} \qquad (8.5.13)$$

Again, as an alternative to the use of the alignment chart of Fig. 8.5.6b, the following approximate solution to the transcendental Eq. 8.5.5 may be used, to determine the effective length of columns in unbraced frames [Dumonteil, 1992]:

$$K = \sqrt{\frac{G_A(1.6G_B + 4.0) + (4G_B + 7.5)}{G_A + G_B + 7.5}} \qquad (8.5.14)$$

For a hinge at A, G_A is infinitely large, resulting in:

$$K = \sqrt{1.6G_B + 4.0} \qquad (8.5.15)$$

or, by using the recommended value of 10 for G_A for practical hinged base connections:

$$K = \sqrt{\frac{20.0G_B + 47.5}{G_B + 17.5}} \qquad (8.5.16)$$

If the column is fixed at A, $G_A = 0$, theoretically, resulting in:

$$K = \sqrt{\frac{4.0G_B + 7.5}{G_B + 7.5}} \qquad (8.5.17)$$

or, by using the recommended value of 1.0 for G_A:

$$K = \sqrt{\frac{5.6G_B + 11.5}{G_B + 8.5}} \qquad (8.5.18)$$

Finally, when $G_A = G_B = G$:

$$K = \sqrt{0.8G + 1.0} \qquad (8.5.19)$$

Equations 8.5.8 to 8.5.19 are well suited for use with hand held programmable calculators.

Compressive axial load in a girder reduces its rotational stiffness which, in turn, has an adverse effect on the effective length of the column (see assumption 7). To account for any compressive axial load, the girder stiffness parameter $(I/L)_g$ in Eq. 8.5.7 should be multiplied by the factor

$$1 - \frac{P_{ug}}{P_{crg}} \tag{8.5.20}$$

where P_{ug} is the axial compressive force in the girder under factored loads, and P_{crg} is the in-plane flexural buckling load of the girder based on a value of $K = 1.0$ [Galambos, 1998]. Tensile axial load in the girders can be ignored when determining G.

8.5.5 Compression Members in Trusses

In triangulated trusses, loads are generally applied only at the joints, and joints are assumed to be pinned. Thus, only axial forces exist in the members. Relative translation of the ends of a truss compression member in the direction normal to its longitudinal axis results from the aggregate axial deformations of all other members of the truss under load, and is relatively small. If the joints are rigidly connected, as by welding or by using heavily bolted gusset plates, some secondary bending develops. The effect of these secondary distortions on the buckling of truss frameworks is usually small and may be neglected in buckling analysis [Kavanagh, 1962; Galambos, 1998]. Hence, the study of a truss compression member reduces to that of a column partially restrained against rotation at each end and fully restrained against transverse movement at each end, both in the plane of the main truss and plane of the lateral truss.

In an ordinary truss, optimally designed for a load system fixed in position, buckling loads in compression members and yield loads in tension members will be reached at about the same level of applied live load. Hence, the SSRC Guide [Galambos, 1988] recommends that, unless an exact analysis is made, K is to be taken as unity. That is, for buckling in the plane of the truss, the effective length of a compression chord is to be taken as the full distance between panel points. However, in a roof truss of nearly constant depth, in which a single member of constant cross section is used throughout the full length of the truss, K may be taken as 0.9 and L as the distance between panel points. Web members in trusses that are designed for moving live load systems may be designed with $K = 0.85$. In such a design, the member closest to the buckling load will be restrained by the less heavily loaded adjacent members. Similar rules can be applied to evaluate the effective length of chords for buckling in the plane normal to the main truss, and for cases when the chord becomes a component of top plane lateral bracing truss system.

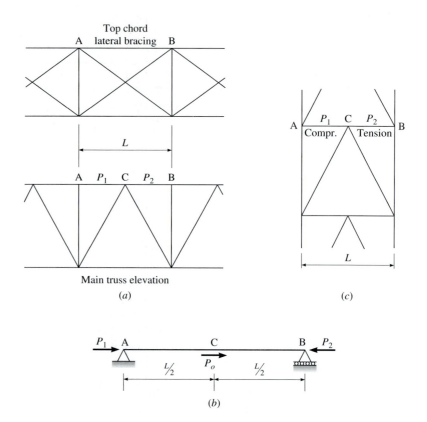

Figure 8.5.9: Compression members with axial load applied between supports.

If the magnitude of compressive force changes at a sub-panel point that is not braced normal to the plane of the main truss (Figs. 8.5.9a and b), the effective length factor for buckling normal to the plane of the main truss may be determined by the relation:

$$K = 0.75 + 0.25 \frac{P_1}{P_2} \qquad (8.5.21)$$

where P_2 = larger of the two compressive forces
$\quad\quad\quad$ P_1 = smaller of the two compressive forces
$\quad\quad\quad$ K = effective length factor for buckling normal to the plane of the main truss

In the case of horizontal web members of K-braced trusses (shown in Fig. 8.5.9c), where P_1 is compression (assigned a positive sign) and P_2 is tension (assigned a negative sign), the effective length factor K for buckling out of the plane of the K-truss is again given by Eq. 8.5.21. The effective length is KL where L is the full depth of the truss. When P_1 and P_2 are numerically equal but opposite in sign, the above relation yields a value of $K = 0.5$.

8.5.6 Buckling About x Axis and y Axis of a Column

Quite often a column has different support conditions with respect to its two principal axes. For example, it can be pin-ended with respect to one axis and fixed with respect to the other. Further, in many situations intermediate bracing is provided in one plane (usually the plane of a wall or a permanent partition), but no such bracing is provided in the perpendicular plane for functional or aesthetic reasons. In these situations, the column section ideally should be oriented such that the bracing lies in the plane perpendicular to the weak axis. In other cases, a column might be part of a braced frame in one direction, while being part of an unbraced frame in the other (perpendicular) direction. In situations of this type, wide flange sections and rectangular box sections can be used economically when oriented such that the strong axis of the column section (x axis) lies in the plane of the braced frame and the weak axis of the column section (y axis) lies in the plane of the unbraced frame. Note that, for flexural buckling about the x axis, the buckling deformations lie in the yy plane and vice-versa. Hence, we evaluate two effective lengths, $K_x L_x$ and $K_y L_y$, and arrive at two different elastic buckling loads: one for buckling about the x axis and one for buckling about the y axis. The controlling elastic buckling load for the entire column is the smaller of the two. The results can be summarized as follows:

$$P_{ex} = \frac{\pi^2 E I_x}{(K_x L_x)^2}; \quad P_{ey} = \frac{\pi^2 E I_y}{(K_y L_y)^2} \tag{8.5.22}$$

$$P_e = \min[P_{ex}, P_{ey}] \tag{8.5.23}$$

or in terms of stresses:

$$F_{ex} = \frac{\pi^2 E}{\left(\dfrac{K_x L_x}{r_x}\right)^2}; \quad F_{ey} = \frac{\pi^2 E}{\left(\dfrac{K_y L_y}{r_y}\right)^2} \tag{8.5.24}$$

$$F_e = \min[F_{ex}, F_{ey}] \tag{8.5.25}$$

where P_e = elastic flexural buckling load of the column
P_{ex} = elastic flexural buckling load about the column x axis
P_{ey} = elastic flexural buckling load about the column y axis
E = modulus of elasticity of column material
I_x = moment of inertia of the column cross section about its x axis
I_y = moment of inertia of the column cross section about its y axis
L_x = unsupported length of the column for buckling about its x axis
L_y = unsupported length of the column for buckling about its y axis
$K_x L_x$ = effective length of the column for buckling about its x axis
$K_y L_y$ = effective length of the column for buckling about its y axis

F_{ex} = elastic flexural buckling stress about its x axis
F_{ey} = elastic flexural buckling stress about its y axis
r_x = radius of gyration of the column cross section about its x axis
r_y = radius of gyration of the column cross section about its y axis
F_e = elastic flexural buckling stress of the column
A = cross-sectional area of the column

Care must be taken to associate the correct effective length with the appropriate radius of gyration in order to arrive at the correct effective slenderness ratios.

EXAMPLE 8.5.1

Influence of Support Conditions

A W8×31 column is 16 ft long and is braced at the top and bottom against sidesway in both the xx and yy planes of the column cross section. Assume that the material is elastic and has a modulus of elasticity, $E = 29{,}000$ ksi. Determine the elastic flexural buckling stress and elastic flexural buckling load, for the following support conditions:

a. Pinned at both ends about both axes.
b. Pinned about both axes at the top; pinned about the major axis and fixed about the minor axis at the base.
c. Pinned at both ends about both axes and laterally supported perpendicular to the weak axis at midheight.
d. Pinned at both ends about both axes and laterally supported perpendicular to the weak axis at a height 10 ft from the base.
e. Pinned about both axes at the top; pinned about the major axis and fixed about the minor axis at the base; and laterally supported perpendicular to the weak axis at a height 10 ft from the base.
f. Pinned about both axes at both ends and built into a wall such that it may be considered continuously supported for buckling about its weak axis.

Solution
Data:

Length of column, $L = 16$ ft
From Table 1-1 of the LRFDM, we obtain the following for a W8×31 section:

$A = 9.12$ in.^{2}
$I_x = 110$ in.^{4}; $r_x = 3.47$ in.
$I_y = 37.1$ in.^{4}; $r_y = 2.02$ in.

a. For the column shown in Fig. X8.5.1a, end conditions given in Fig. 8.5.4d apply for buckling about the x- and y-axes. Thus, $K_x = K_y = 1.0$.
$L_x = L = 16$ ft; $L_y = L = 16$ ft
$K_x L_x = 1.0(16) = 16.0$ ft; $K_y L_y = 1.0(16) = 16.0$ ft

$$\frac{K_xL_x}{r_x} = \frac{16.0(12)}{3.47} = 55.3; \quad \frac{K_yL_y}{r_y} = \frac{16.0(12)}{2.02} = 95.0 \leftarrow \text{controls}$$

Elastic flexural buckling stress, $F_e = \dfrac{\pi^2 E}{(KL/r)^2} = \dfrac{\pi^2(29,000)}{95.0^2}$

$$= 31.7 \text{ ksi} \qquad \text{(Ans.)}$$

Elastic flexural buckling load, $P_e = F_e A = 31.7\,(9.12)$

$$= 289 \text{ kips} \qquad \text{(Ans.)}$$

b. For the column shown in Fig. X8.5.1b, end constraints for buckling about its x axis correspond to Fig. 8.5.4d, while those for the y axis are represented by Fig. 8.5.4c. Thus, $K_x = 1.0$ and $K_y = 0.80$.
$L_x = L = 16$ ft; $L_y = L = 16$ ft
$K_xL_x = 1.0(16.0) = 16.0$ ft; $K_yL_y = 0.80(16.0) = 12.8$ ft

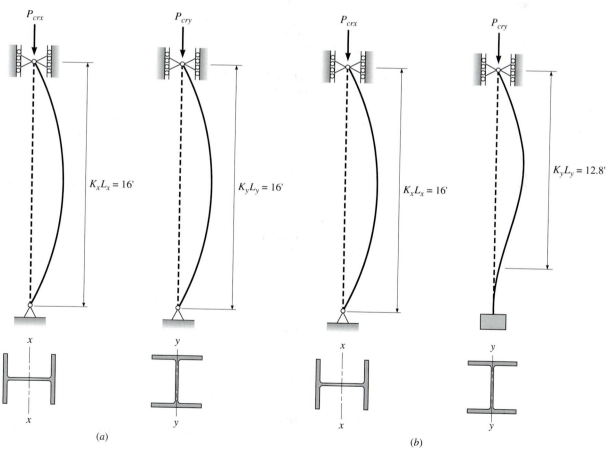

(a)   (b)

Figure X8.5.1

[*continues on next page*]

Example 8.5.1 continues ...

$$\frac{K_x L_x}{r_x} = \frac{16.0(12)}{3.47} = 55.3; \quad \frac{K_y L_y}{r_y} = \frac{12.8(12)}{2.02} = 76.0 \leftarrow \text{controls}$$

Elastic flexural buckling stress, $F_e = \dfrac{\pi^2(29,000)}{76.0^2} = 49.6$ ksi (Ans.)

Elastic buckling load, $P_e = 49.6(9.12) = 452$ kips   (Ans.)

c. For the column shown in Fig. X8.5.1c, $K_x = 1.0$, while Fig. 8.5.5a gives an effective length of $L/2$ for buckling about its y axis, or $K_y = 0.5$.

$L_x = L = 16$ ft; $L_{y1} = L_{y2} = 8$ ft

$K_x L_x = 1.0(16.0) = 16.0$ ft

$K_y L_y = (K_y L_y)_1 = (K_y L_y)_2 = 8.00$ ft

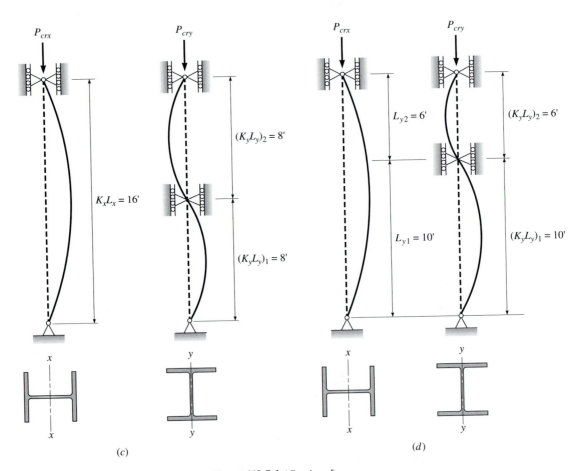

(c)

(d)

Figure X8.5.1 (*Continued*)

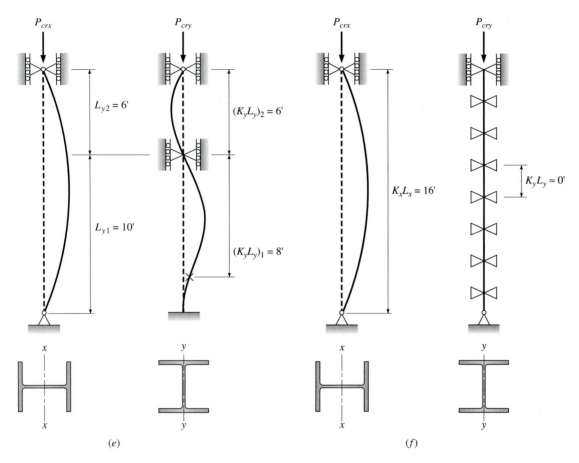

(e)   (f)

Figure X8.5.1 (*Continued*)

$$\frac{K_x L_x}{r_x} = \frac{16.0(12)}{3.47} = 55.3 \leftarrow \text{ controls;} \quad \frac{K_y L_y}{r_y} = \frac{8.0(12)}{2.02} = 47.5$$

Elastic flexural buckling stress, $F_e = \dfrac{\pi^2(29{,}000)}{55.3^2} = 93.6 \text{ ksi}$ (Ans.)

Elastic flexural buckling load, $P_e = 93.6(9.12) = 854$ kips   (Ans.)

d. For the column shown in Fig. X8.5.1d, we have:

$L_x = L = 16$ ft; $L_{y1} = 10$ ft; $L_{y2} = 6$ ft

$K_x L_x = 1.0(16.0) = 16.0$ ft

$(K_y L_y)_1 = 1.0(10.0) = 10.0$ ft; $(K_y L_y)_2 = 1.0(6.00) = 6.00$ ft

$K_y L_y = \max[10.0, 6.00] = 10.0$ ft

$$\frac{K_x L_x}{r_x} = \frac{16.0(12)}{3.47} = 55.3; \quad \frac{K_y L_y}{r_y} = \frac{10.0(12)}{2.02} = 59.4 \leftarrow \text{ controls}$$

[*continues on next page*]

Example 8.5.1 continues ...

Elastic flexural buckling stress, $F_e = \dfrac{\pi^2(29{,}000)}{59.4^2} = 81.1$ ksi (Ans.)

Elastic flexural buckling load, $P_e = 81.1(9.12) = 740$ kips (Ans.)

e. For the column shown in Fig. X8.5.1e, we have
$L_x = L = 16$ ft; $L_{y1} = 10$ ft; $L_{y2} = 6$ ft
$K_x L_x = 1.0(16.0) = 16.0$ ft
$(K_y L_y)_1 = 0.8(10.0) = 8.00$ ft; $(K_y L_y)_2 = 1.0(6.00) = 6.00$ ft

$\dfrac{K_x L_x}{r_x} = 55.3 \leftarrow$ controls; $\dfrac{K_y L_y}{r_y} = \dfrac{8.00(12)}{2.02} = 47.5$

Elastic flexural buckling stress, $F_e = 93.6$ ksi (Ans.)
Elastic flexural buckling load, $P_e = 854$ kips (Ans.)

f. For the column shown in Fig. X8.5.1f, we have:
$K_y L_y \approx 0$; $K_x L_x = 1.0(16.0) = 16.0 \leftarrow$ controls

$\dfrac{KL}{r} = \dfrac{K_x L_x}{r_x} = \dfrac{16.0(12)}{3.47} = 55.3$

Elastic flexural buckling stress, $F_e = 93.6$ ksi (Ans.)
Elastic flexural buckling load, $P_e = 854$ kips (Ans.)

| | \multicolumn{2}{c}{x} | \multicolumn{2}{c}{y} | | KL | Buckling | F_e | P_e |
	B	T	B	T	L_{y1}	r	Axis	(ksi)	(kips)
a	P	P	P	P	–	95.0	y axis	31.7	289
b	P	P	F	P	–	76.0	y axis	49.6	452
c	P	P	P	P	8′	55.3	x axis	93.6	854
d	P	P	P	P	10′	59.4	y axis	81.1	740
e	P	P	F	P	10′	55.3	x axis	93.6	854
f	P	P	P	P	C	55.3	x axis	93.6	854

x = buckling about x axis;   B = base;   T = top
y = buckling about y axis;   P = pinned;   F = fixed
L_{y1} = intermediate brace from base; C = continuous brace

Remarks

1. End rotational restraints raise the buckling load, as evidenced by the results of cases (*a*) and (*b*) shown in the table where the elastic critical load was increased 57 percent by fixing the base of the column.
2. It is also seen, by comparing the results of (*a*) and (*c*), that the critical load can be raised considerably by adding lateral bracing at mid-height. The

lateral force P_{br} required to gain this additional strength (force for which the brace is to be designed) is generally less than 2 percent of the axial load P in the column.

3. Increases in axial capacity depend not only on the presence but also on the location of the brace. This is seen by comparing the results of cases (*a*), (*c*) and (*d*). Ideally the braces should be located such that the axial capacity of the column in each of the segments is the same.

4. Attainment of the elastic buckling load, P_e tabulated requires that all the fibers of that column continue to behave elastically under a uniform stress of F_e. Influence of yielding on column strength is considered in Section 8.6.

Effective Length Factors, Unbraced Frame

EXAMPLE 8.5.2

The rectangular steel frame shown in Fig. X8.5.2 is used to support a footbridge in a chemical plant. The ends of the bridge are supported on rollers, so the frame is permitted to sway. All members are oriented with their webs in the plane of the frame. All columns are assumed to be adequately braced in the weak direction (i.e., no out-of-plane buckling occurs). All girder-to-column connections are rigid, unless otherwise shown. All girders are W16×50. The central column is a W12×58, while the other two columns are W8×48 shapes. Assume elastic behavior, and determine the effective length factors of the three columns, using:

1. The alignment chart.
2. The approximate equations suggested by Dumonteil [1992].

Solution

Data

W8×48: $I_x = 184$ in.^{4}; W12×58: $I_x = 475$ in.^{4}; W16×50: $I_x = 659$ in.^{4}
Note: All of the columns considered are members of an unbraced frame. So, the effective length factors for buckling in the plane of the frame will be greater than 1.0.

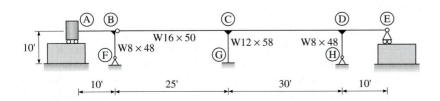

Figure X8.5.2

[*continues on next page*]

Example 8.5.2 continues ...

1. Use the alignment chart for unbraced (sidesway uninhibited) frames shown in Fig. C-C2.2*b* of the LRFDC and presented in Fig. 8.5.6*b*.

   a. Column BF

   The lower end F of the column BF is hinged, so we use the recommended value of 10.0 for G_F. At the top end of the column, two girders are connected, BA and BC. End B of girder BA is rigidly connected to the column, while the far end A is clamped, resulting in a value for $\alpha = (4/6)$. The near end B of girder BC is pin-connected to the column, resulting in a value of $\alpha = 0$.

   $$G_B = \frac{\dfrac{184}{10}}{\left(\dfrac{4}{6}\right)\dfrac{659}{10} + (0)\dfrac{659}{25}} = 0.419; \quad G_F = 10.0$$

   Connecting points $G_F = 10.0$ and $G_B = 0.419$ on the alignment chart for unbraced (sidesway uninhibited) frames, $K \approx 1.76$.
   So, for column BF: $K_x \approx 1.76$ (Ans.)

   b. Column CG

   The lower end G of column CG is fixed. So, we use the recommended value of 1.0 for G_G. At the top end of the column, two girders are connected, CB and CD. The near end C of girder CB is rigidly connected to the column, while the far end B is pinned, resulting in a value for $\alpha = (3/6)$. Both ends of girder CD are rigidly connected to columns, so $\alpha = 1.0$ for girder CD.

   $$G_C = \frac{\dfrac{475}{10}}{\dfrac{3}{6}\left(\dfrac{659}{25}\right) + 1.0\left(\dfrac{659}{30}\right)} = 1.35; \quad G_G = 1.0$$

   Connecting points $G_C = 1.35$ and $G_G = 1.0$ on the alignment chart for unbraced (sidesway uninhibited) frames, $K \approx 1.38$.
   So, for column CG: $K_x \approx 1.38$ (Ans.)

   c. Column DH

   The lower end H of the column DH is hinged. So, we use the recommended value of 10.0 for G_H. At the top end of the column, two girders DC and DE are rigidly connected to the column. The far end C of girder DC is rigidly connected ($\alpha = 1.0$), while the far end E of girder DE is hinged ($\alpha = 3/6$). Thus:

   $$G_D = \frac{\dfrac{184}{10}}{(1.0)\dfrac{659}{30} + \left(\dfrac{3}{6}\right)\dfrac{659}{10}} = 0.335; \quad G_H = 10.0$$

A value of $K \approx 1.75$ is read on the alignment chart for unbraced frames, corresponding to $G_D = 0.335$ and $G_H = 10.0$.
So, for column DH: $K_x \approx 1.75$ (Ans.)

2. Use the approximate equations suggested by Dumonteil [1992].
   Values of G factors obtained in part 1 are still valid.
   a. For column BF, $G_F = 10.0$ and $G_B = 0.419$. From Eq. 8.5.16:

$$K = \sqrt{\frac{20.0 G_B + 47.5}{G_B + 17.5}} = \sqrt{\frac{20.0(0.419) + 47.5}{0.419 + 17.5}} = 1.77$$

   So, for column BF: $K_x \approx 1.77$ (Ans.)
   b. For column CG, $G_C = 1.35$ and $G_G = 1.0$. So, from Eq. 8.5.18:

$$K = \sqrt{\frac{5.6 G_C + 11.5}{G_C + 8.5}} = \sqrt{\frac{5.6(1.35) + 11.5}{1.35 + 8.5}} = 1.39$$

   So, for column CG: $K_x \approx 1.39$ (Ans.)

   c. For column DH, $G_D = 0.335$ and $G_H = 10.0$. So from Eq. 8.5.16:

$$K = \sqrt{\frac{20.0 G_D + 47.5}{G_D + 17.5}} = \sqrt{\frac{20.0(0.335) + 47.5}{0.335 + 17.5}} = 1.74$$

   So, for column DH: $K_x \approx 1.74$ (Ans.)

Effective Length Factors, Braced Frame

EXAMPLE 8.5.3

Rework the previous example for the case in which the translation of the top ends of columns with respect to their bases is prevented by building the left end, A, of the girder AB into the abutment as shown in Fig. X8.5.3, using:

1. The alignment chart.
2. The approximate equations suggested by Dumonteil [1992].

Solution

Note: The columns are now part of a braced frame. So, the K-factor for buckling in the plane of the frame, for all the columns will be less than or equal to 1.

1. Use the alignment chart for braced (sidesway inhibited) frames shown in Fig. C-C2.2a of the LRFDC and presented in Fig. 8.5.6a.

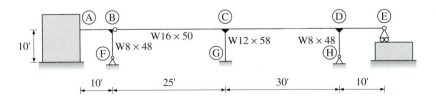

Figure X8.5.3

[continues on next page]

Example 8.5.3 continues ...

a. Column BF

Two girders BA and BC are connected at the top end of the column. End B of girder BA is rigidly connected to the column, while the far end A is clamped, resulting in a value of $\alpha = (4/2)$. The near end B of girder BC is pin-connected to the column, resulting in a value of $\alpha = 0$.

$$G_B = \frac{\dfrac{184}{10}}{\dfrac{4}{2}\left(\dfrac{659}{10}\right) + 0\left(\dfrac{659}{25}\right)} = 0.140; \quad G_F = 10.0$$

From the alignment chart for braced, that is, sidesway inhibited frames shown in Fig. C-C2.2a of the LRFDC, for $G_B = 0.140$ and $G_F = 10.0$, read $K \approx 0.73$.

So, for column BF: $K_x \approx 0.73$ (Ans.)

b. Column CG

At the top end of the column, two girders are connected, CB and CD. End C of girder CB is rigidly connected to the column, while the far end is pinned, resulting in a value for $\alpha = (3/2)$. Both ends of girder CD are rigidly connected to columns, so $\alpha = 1.0$ for girder CD. Thus:

$$G_C = \frac{\dfrac{475}{10}}{\dfrac{3}{2}\left(\dfrac{659}{25}\right) + 1.0\left(\dfrac{659}{30}\right)} = 0.772; \quad G_G = 1.0$$

From the alignment chart for braced, that is, sidesway inhibited frames, for $G_C = 0.772$ and $G_G = 1.0$, K ≈ 0.75 is read.

So, for column CG: $K_x \approx 0.75$ (Ans.)

c. Column DH

At the top end of this column, two girders are rigidly connected, DC and DE. The far end of girder DC is rigidly connected ($\alpha = 1.0$), while the far end of girder DE is hinged ($\alpha = 3/2$). Thus:

$$G_D = \frac{\dfrac{184}{10}}{1.0\left(\dfrac{659}{30}\right) + \dfrac{3}{2}\left(\dfrac{659}{10}\right)} = 0.152; \quad G_H = 10.0$$

Again, from the alignment chart for braced frames, with $G_D = 0.152$ and $G_H = 10.0$, K ≈ 0.73 is read.

So, for column DH: $K_x \approx 0.73$ (Ans.)

2. Use the approximate equations suggested by Dumonteil [1992]. Values of G factors obtained in part 1 are still valid.

a. For column BF, $G_B = 0.140$ and $G_F = 10.0$. So, from Eq. 8.5.10:

$$K = \frac{31.4G_B + 14.6}{32.0G_B + 21.3} = \frac{31.4(0.140) + 14.64}{32.0(0.140) + 21.28} = 0.74$$

So, for column BF: $K_x \approx 0.74$ (Ans.)

b. For column CG, $G_C = 0.772$ and $G_G = 1.0$. So, from Eq. 8.5.12:

$$K = \frac{4.4G_C + 2.04}{5.0G_C + 3.28} = \frac{4.4(0.772) + 2.04}{5.0(0.772) + 3.28} = 0.76$$

So, for column CG: $K_x \approx 0.76$ (Ans.)

c. For column DH, $G_D = 0.152$ and $G_H = 10.0$. So, from Eq. 8.5.10:

$$K = \frac{31.4G_D + 14.6}{32.0G_D + 21.3} = \frac{31.4(0.152) + 14.6}{32.0(0.152) + 21.3} = 0.74$$

So, for column DH: $K_x \approx 0.74$ (Ans.)

8.5.7 Leaning Columns

The framing plan for a low-rise steel building is shown in Fig. 8.5.10. In the east-west direction, lateral loads such as wind are resisted by the rigid-jointed frames along column lines A and D from line 2 to line 4. In the north-south direction, lateral loads are resisted by the braced frames along column lines 1 and 5 between lines B and C. If the lateral resistance of the remainder of the

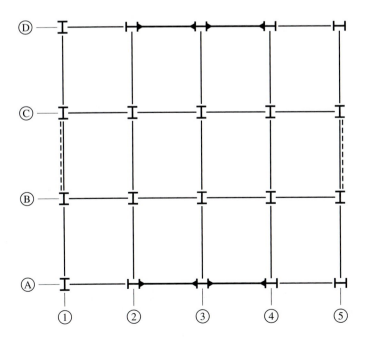

Figure 8.5.10: Framing plan for a low-rise building.

structure is considered negligible, then each rigid frame must be designed to resist the gravity load applied to its columns plus one-half of the total lateral load. Columns that are not part of the rigid frames are designed only to support their part of the gravity load, since they are not part of the lateral load resisting system. Direct application of the procedure developed in Section 8.5.4 indicates that for these gravity-only columns the coefficients α in Eq. 8.5.7 are zero, and thus the G factors are infinity. This results in an effective length factor of infinity from the alignment chart, and hence zero axial strength. Such pin-ended columns that are part of an unbraced rigid jointed frame, having zero stiffness against drift, are called **leaning columns.** [Geschwindner, 1995; LeMessurier, 1976, 1977; Vinnakota and Badoux, 1970; Yura, 1971].

Next consider an unbraced frame ABCD composed of a cantilever column AB, a leaning column CD, and a simple beam BD (Fig. 8.5.11a). The frame is subjected to gravity loads P and Q at joints B and D, respectively. Equilibrium of the structure in the undeflected configuration indicates that the columns AB and CD are subjected to axial forces P and Q, respectively, as shown in Fig. 8.5.11a. If the roof joints are permitted to translate an amount Δ, equilibrium in this displaced configuration will be as indicated in Fig. 8.5.11b. It is seen that a horizontal restraining force, $Q\Delta/L$, must be applied at D for the leaning column CD to be in moment equilibrium. This horizontal force is to be equilibrated by an equal and opposite force at joint B. Thus, when the frame buckles, column AB will be subjected to a destabilizing moment of $(P\Delta + Q\Delta)$ at its base. It is observed that this is the same second-order moment that would result if the individual column AB were to buckle under an axial load of $(P + Q)$. The assumption that the buckling load is $(P + Q)$ is slightly conservative for the individual column AB, since the deflected shape due to an axial load and that due to a lateral load differ somewhat. Nevertheless, in order to ensure sufficient lateral restraint to preclude

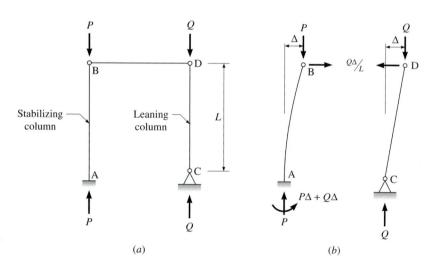

Figure 8.5.11: Behavior of an unbraced frame with a leaning column.

buckling of the leaning column CD, the restraining column AB must be designed to carry a fictitious axial load $(P + Q)$.

One can consider the restraining column in either one of two ways: design it for load $(P + Q)$ using the effective length factor as determined by the sidesway uninhibited nomograph of the LRFDC (Figure 8.5.6b) or design it for load P using a modified effective length factor that accounts for the destabilizing effect of the leaning column. Let K_o be the effective length factor of column AB that would be determined from the nomograph, which does not account for the leaning column ($K_o = 2$ in the example shown). Let K_m be the modified effective length factor that will account for the presence of the leaning column. We have:

$$P + Q = \frac{\pi^2 EI}{K_o^2 L^2} \tag{8.5.26}$$

and

$$P = \frac{\pi^2 EI}{K_m^2 L^2} \tag{8.5.27}$$

Dividing Eq. 8.5.26 by Eq. 8.5.27 and solving for K_m,

$$K_m = K_o \sqrt{\frac{P + Q}{P}} \tag{8.5.28}$$

Thus, if the restraining column AB from Fig. 8.5.11a were designed to carry the load P using the modified effective length factor K_m, it would provide sufficient lateral restraint to the leaning column CD to permit it to be designed to carry the load Q using $K = 1.0$. For frames with more than one restraining column and more than one leaning column, P and Q will be replaced by ΣP_i and ΣQ_j. Thus,

$$K_m = K_o \sqrt{\frac{\Sigma P_i + \Sigma Q_j}{\Sigma P_i}} \tag{8.5.29}$$

It should be noted that this approach maintains the assumption that all restraining columns in a story buckle in a sidesway mode simultaneously.

Sidesway buckling is a total story phenomenon. An individual column cannot fail by sidesway without all of the columns in the same story also buckling in a sway mode. However, each column's nonsway buckling load is reasonably independent of the buckling load of other columns. The alignment charts, both for sway inhibited and sway uninhibited, were developed based on the assumption that all of the individual columns in a story buckle simultaneously under their proportional share of the total gravity load (assumption 5 in Section 8.5.4). For the case of the sway uninhibited frame, when total story buckling commences, no individual column can offer any lateral restraint to another, as each column's total capacity has been spent simply in supporting its own individual gravity load. That is, there is no reserve strength available which could provide a bracing force (shear resistance) for other columns.

Consider the symmetric unbraced frame shown in Fig. 8.5.12a. The column bases are fixed, and the tops are hinged to the girders. The effective length of each column, therefore, equals 2.0. The outer columns A and C are proportioned for an ultimate axial load of 200 kips, and the interior column B is proportioned for an ultimate axial load of 400 kips. When sidesway occurs, $P\Delta$ moments of 200Δ, 400Δ, and 200Δ at bases A, B, and C, respectively, will be produced as shown in Fig. 8.5.12b.

Note that if the frame under discussion were a braced frame, the effective length factors for each of these columns would decrease to 0.7. Common practice, however, is to consider the columns to be pinned at both ends as seen in Fig. 8.5.12c, thus changing the effective length factor from 2.0 to 1.0. As a result, the strengths of the columns would then be four times as great as their unbraced counter parts, namely 800, 1600, and 800 kips for columns A, B, and C, respectively.

Refer now to the unbraced frame of Fig. 8.5.12d, which differs from that shown in Fig. 8.5.12a in that the loads applied to columns A and B are only 100 kips and 250 kips respectively, instead of their maximum strengths of 200 kips and 400 kips, respectively. Columns A and B will not buckle until the moments developed at their bases reach 200Δ and 400Δ, respectively. Thus, columns A and B can sustain horizontal destabilizing forces of $(200 - 100)(\Delta/L)$ and $(400 - 250)(\Delta/L)$, respectively, at their tops. These two columns together can, therefore, provide a total stabilizing force of $(100 + 150)(\Delta/L)$ at the top of column C. Thus, column C can carry its original load of 200 kips plus—as a result of the bracing available from columns A and B—an additional 250 kips, for a total of 450 kips. The maximum level to which horizontal bracing can increase the axial strength of column C is the capacity this member will have when its ends cannot translate with respect to one another; that is, 800 kips.

The results discussed above may be summarized as follows [Yura 1971]: *"In general, the total gravity load that produces sidesway can be distributed among the columns in a story in any manner. Sidesway will not occur until the total frame load on a story reaches the sum of the potential individual column loads for the unbraced frame. There is one limitation; the maximum load an individual column can carry is limited to the load permitted on that column for the braced case, K = 1.0."**

A more general approach to determining the effective length of columns in sway frames with and without leaning columns was presented by LeMessurier [1977]. In this method, the contribution of each column to the lateral resistance is accounted for individually. The effective length factor

* From "The Effective Length of Columns in Unbraced Frames," by J. A. Yura, *Engineering Journal*, AISC, Vol. 8, No. 2, pp. 27–42. Copyright © American Institute of Steel Construction, Inc. Reprinted with permission. All rights reserved.

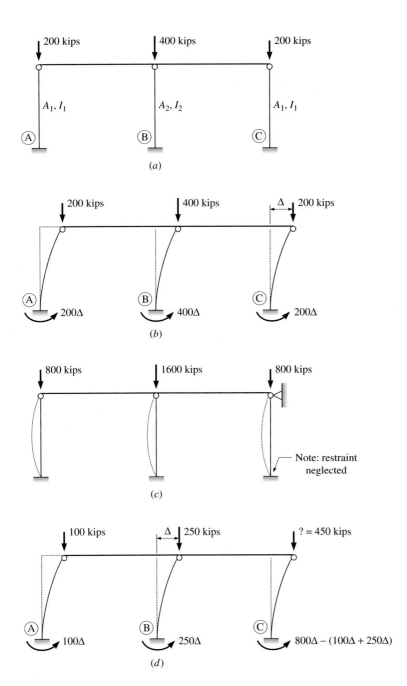

Figure 8.5.12: Bracing of heavily loaded columns by lightly loaded columns of an unbraced frame.

for each column that participates in resisting sidesway buckling is given by:

$$K_{mi} = \sqrt{\frac{P_{Ei}}{P_{ui}} \frac{(\Sigma_i P_{ui} + \Sigma_j Q_{uj})}{\Sigma_i P_{eoi}}} \qquad (8.5.30)$$

where

K_{mi} = modified effective length factor of a column providing sidesway resistance

P_{ui} = factored axial load on the restraining column i

$\Sigma_i P_{ui}$ = factored axial load on all the restraining columns of a story

$\Sigma_j Q_{uj}$ = factored axial load on all the leaning columns of a story

P_{Ei} = Euler load of the restraining column i

= $(\pi^2 EI_i)/L^2$

L = length of column (story height)

I_i = moment of inertia of the restraining column i

P_{eoi} = elastic flexural buckling load of the restraining column i based on the nomograph

= $(\pi^2 EI_i)/(K_{oi}L)^2$

K_{oi} = effective length factor of restraining column i obtained from the nomograph

Design of columns in a frame with leaning columns is considered in Example 8.10.5.

8.6   Inelastic Stability of Axially Loaded Columns

Euler developed his theory of elastic buckling in 1744 well before the use of steel columns in building construction [Euler, 1744]. As we remarked in Examples 8.4.1 and 8.5.1, for intermediate columns (members with small slenderness ratios), the buckling stress given by the Euler equation far exceeds the proportional limit of the material, and hence the Euler equation based on linear elastic behavior can no longer be used. In 1889 Engesser modified the Euler theory to make it applicable to inelastic buckling of straight columns, and in 1947 Shanely clarified certain doubts that had arisen regarding the correctness of Engesser's original work [Shanely, 1947]. These theories were developed for columns made from materials having a linear elastic stress-strain diagram in the elastic region followed by a continuously nonlinear curve in the inelastic region. The critical load obtained by the Engesser-Shanely model is known as the **tangent modulus load,** P_t, and is given by [Galambos, 1968]:

$$P_{cr} = P_t = \frac{\pi^2 E_t I}{L^2} = \frac{E_t}{E} P_E \qquad (8.6.1)$$

Here, E_t is the tangent modulus of the material defined as the slope of the tangent to the stress-strain curve at the point of the applied stress, for values of the stress between the proportional limit f_{pl} and the yield stress F_y.

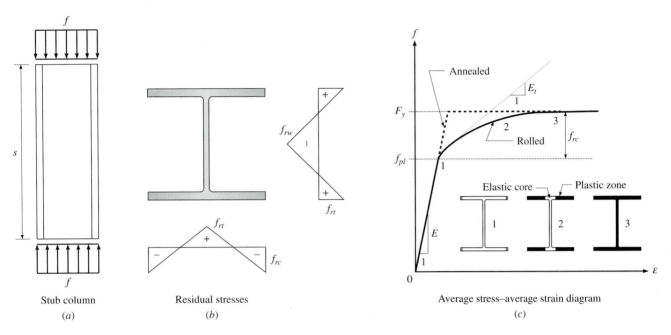

Figure 8.6.1:   Stub column test.

To determine the axial strength of columns of hot-rolled steel shapes, a test called a ***stub column test*** is often used (Fig. 8.6.1a). In this test, a short length of a hot-rolled W-shape is cut, its ends are carefully machined, and it is placed between the plates of a testing machine in which it is compressed concentrically and the load is increased gradually [Galambos, 1998]. The average stress-strain relationship for a stub column is influenced by residual stresses (see Section 2.6.2). It would be linear elastic in the elastic region followed by a continuously nonlinear curve in the inelastic region (Fig. 8.6.1c). With sufficient straining the average stress $(f = P/A)$ would eventually reach the yield stress of the material (F_y). If a compressive residual stress of magnitude f_{rc} were present at the flange tips (Fig. 8.6.1b), then yielding of the stub column would commence when the applied average stress f equals $(F_y - f_{rc}) = f_{pl}$. Thus, the influence of residual stresses in steel columns is to make the yielding over the cross section a gradual process and to make the stress-strain relationship nonlinear above the proportional limit, as shown in Fig. 8.6.1c. The slope of this stress-strain curve is the tangent modulus E_t of the member. Research at Lehigh University during the 1950s showed that the critical load for rolled steel columns can also be determined from Eq. 8.6.1 using the tangent modulus E_t of the average stress-strain diagram from a stub column test.

The maximum strength of an axially loaded, pin-ended steel column is a function of the length of the column, geometry of the cross section, yield stress of the material, residual stress distribution in the cross section, and

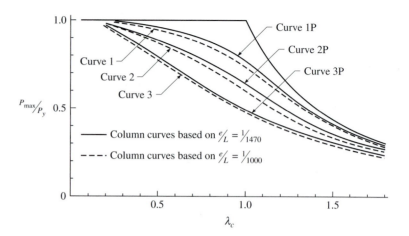

Figure 8.6.2: SSRC column strength curves. From *Guide to Stability Design Criteria for Metal Structures*, 5E, edited by Theodore V. Galambos. This material is used by permission of John Wiley & Sons, Inc.

initial geometry of the unloaded column. In a major study conducted at Lehigh University, Bjorhovde [1972] determined the maximum strength of steel columns using measured residual stress distributions, and assumed initial crookedness having a sinusoidal shape with a maximum amplitude of 1/1000 of the column length. The results were presented as three curves known as the ***SSRC column strength curves 1, 2, and 3*** (Fig. 8.6.2). Bjorhovde also developed multiple column curves where the initial out-of-straightness (*e*) was equal to its mean value of 1/1470 of the column length. Those curves are known as the ***SSRC column strength curves 1P, 2P, and 3P.*** The results are given in a nondimensional format using the ***column-slenderness parameter,*** $\lambda_c = (KL/r\pi)\sqrt{F_y/E}$. The value $\lambda_c = 1$ corresponds to the value of KL/r for which the elastic flexural buckling stress is equal to the yield stress of the material.

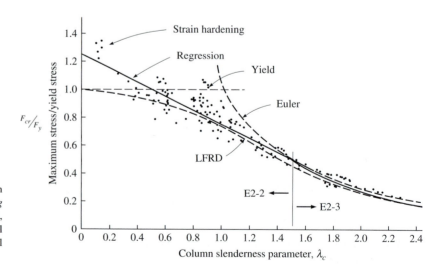

Figure 8.6.3: LRFDS column curve superposed on available test data. From *Elements for Teaching Load Resistance Factor Design* by Joseph A. Yura, April 1983. Copyright © American Institute of Steel Construction, Inc. Reprinted with permission. All rights reserved.

The single column curve that is adopted by the LRFD Specification for the design of steel columns (discussed in Section 8.7.1) is identical to the SSRC curve 2P [Galambos, 1998]. Figure 8.6.3 shows a plot of the LRFDS column curve superposed on the available test data [Yura, 1988; Hall, 1981]. *For further details on inelastic stability of axially loaded steel columns, refer to Section W8.4 on the website http://www.mhhe.com/Vinnakota.*

WWW

8.7   Design Strength of Axially Loaded Columns

8.7.1   Design Strength for Flexural Buckling of Axially Loaded Columns

The equations governing the design strength for flexural buckling of axially loaded steel columns are found in Chapter E of the LRFD Specification. The relation between design strength and factored loads now takes the form:

$$P_d \equiv \phi_c P_n \geq P_{\text{req}} = P_u \qquad (8.7.1)$$

where P_n = nominal axial compressive strength of the column, kips
ϕ_c = resistance factor for compression = 0.85
P_u = axial load in the column under factored loads, kips
P_d = design axial compressive strength of the column, kips
P_{req} = required axial compressive strength of the column, kips

The nominal strength P_n of rolled shape compression members is given by (LRFDS Eq. E 2-1):

$$P_n = F_{cr} A_g \qquad (8.7.2)$$

where A_g = cross-sectional area of the column, in.²
F_{cr} = critical stress of member, ksi

The LRFDS gives two formulae for F_{cr}, one for inelastic flexural buckling and the other for elastic flexural buckling, as functions of the column slenderness parameter λ_c (LRFDS Eq. E2-4):

$$\lambda_c = \frac{KL}{r\pi} \sqrt{\frac{F_y}{E}} \qquad (8.7.3)$$

where K = effective length factor
L = laterally unbraced length of member, in.
r = governing radius of gyration about the axis of buckling, in.
F_y = yield stress of the column material, ksi
E = modulus of elasticity of the material, ksi

The LRFDS considers that the transition from elastic to inelastic buckling of columns occurs at $\lambda_c = 1.5$. Thus:

- For $\lambda_c \leq 1.5$, buckling is inelastic, and the nominal critical compressive stress is given by LRFDS Eq. E2-2:

$$F_{cr} = \left(0.658^{\lambda_c^2}\right) F_y \qquad (8.7.4)$$

- For $\lambda_c > 1.5$, buckling is elastic, and the nominal critical compressive stress is given by LRFDS Eq. E2-3:

$$F_{cr} = 0.877 \left(\frac{1}{\lambda_c^2}\right) F_y \qquad (8.7.5)$$

The design compressive strength for axially loaded columns failing in flexural buckling is therefore given by:

$$P_d = \phi_c \left(0.658^{\lambda_c^2}\right) F_y A_g \quad \text{for} \quad \lambda_c \leq 1.5 \qquad (8.7.6)$$

$$P_d = \phi_c \frac{0.877}{\lambda_c^2} F_y A_g \qquad \text{for} \quad \lambda_c > 1.5 \qquad (8.7.7)$$

Columns with slenderness parameters less than or equal to 1.5 are known as *intermediate columns,* and their axial compressive strength is limited by inelastic buckling (Eq. 8.7.6). Columns with slenderness parameters greater than 1.5 are known as *long columns,* and their axial strength is limited by elastic buckling (Eq. 8.7.7). It is further seen that the *short column,* discussed in Section 8.1, does not actually exist as a separate category in the LRFDS strength equations. The ratio of the design strength to yield load, $P_y = F_y A_g$, as a function of the slenderness parameter, λ_c, is plotted in Fig. 8.7.1.

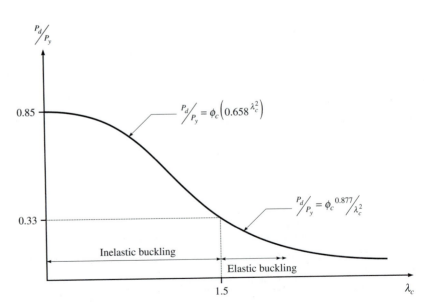

Figure 8.7.1: Design compressive strength of axially loaded columns as per LRFD Specification.

Note that a single set of equations is recommended for all steels, regardless of yield stress, and for all possible column cross-sectional shapes. Further note that the same set of equations is used for buckling about both the major and minor axes. These two relations, which closely fit SSRC Curve 2P, account for the presence of residual stresses and also include the influence of an initial crookedness, sinusoidal in shape, and having an amplitude of $L/1470$ at mid-height [Bjorhovde, 1972, Galambos, 1998].

Alternative expressions for the column slenderness parameter are given below:

$$\lambda_c = \frac{KL/r}{\pi\sqrt{E/F_y}} = \sqrt{\frac{F_y}{F_e}} = \sqrt{\frac{P_y}{P_e}} \qquad (8.7.8)$$

with

$$F_e = \frac{\pi^2 E}{(KL/r)^2} \qquad (8.7.9)$$

where F_e = elastic buckling stress, ksi
  P_y = yield (squash) load of the column section, kips
  P_e = elastic buckling load of the column, kips

Also, KL/r, represents the effective slenderness ratio of the column, while $\pi\sqrt{E/F_y}$, represents that particular value of the KL/r, for which the elastic flexural buckling stress equals the yield stress of the material.

An alternative format of Eqs. 8.7.4 and 8.7.5, expressed in terms of the effective slenderness ratio, KL/r, is:

$$F_{cr} = \left[0.658^{\frac{F_y}{F_e}}\right] F_y \quad \text{for} \quad \frac{KL}{r} \le 4.71\sqrt{\frac{E}{F_y}} \qquad (8.7.10)$$

$$F_{cr} = 0.877\, F_e \qquad \text{for} \quad \frac{KL}{r} > 4.71\sqrt{\frac{E}{F_y}} \qquad (8.7.11)$$

The factor 0.877 in Eqs. 8.7.5, 8.7.7, and 8.7.11 accounts for the effects of any unintended but potential initial crookedness on the strength of long columns. Eqs. 8.7.10 and 8.7.11 show clearly that column strength is a function of the yield stress of the material for intermediate length columns ($\lambda_c \le 1.5$) but is independent of the yield stress of the material for long columns ($\lambda_c > 1.5$).

8.7.2 Design Tables for Axially Loaded Columns

The LRFD Specification Numerical Values Table 3-36 and Table 3-50 give design compressive stresses, $\phi_c F_{cr}$, for flexural buckling of columns of various KL/r values for the two most commonly used grades of steel, namely those with $F_y = 36$ ksi and 50 ksi, respectively. They are applicable for all cross-sectional shapes. LRFD Specification Numerical Values Table 4 gives design compressive stress ratios, $\frac{\phi_c F_{cr}}{F_y}$, for flexural buckling for various values of the slenderness parameter, λ_c. Table 4 is applicable for all cross-sectional shapes and for all grades of steel. LRFDS Section B7 recommends a maximum value of 200 for the effective slenderness ratio, KL/r, for members in which the design is based on compression. Consequently, Tables 3-36 and 3-50 stop at the recommended upper limit of $KL/r = 200$. Members for which design is dictated by tension loading but must also account for compression under other load conditions are not to be subjected to the slenderness limit of 200.

Column Load Tables in Part 4 of the LRFD Manual give design axial compressive strengths, $\phi_c P_n$, for columns of various shapes. Tabular loads are computed in accordance with the LRFDS Section E2 for axially loaded members having unsupported lengths KL indicated at the left of each table, in ft. For example, LRFDM Table 4-2 provides design axial compressive strengths for each of the W-shapes normally used as columns. These include W14-, W12-, and W10-series members having a yield stress of 50 ksi. LRFDM Table 4-6 provides design axial compressive strengths for rectangular HSS shapes of 46 ksi yield stress steel. LRFDM Table 4-7 provides design axial compressive strengths for round HSS shapes, for 42 ksi yield stress steel. LRFDM Table 4-8 provides design axial compressive strengths for steel pipes, for 35 ksi yield stress steel. All design strengths are tabulated in kips. The heavy horizontal lines within the Column Load Tables indicate $KL/r = 200$, the maximum recommended effective slenderness ratio for columns. Design strengths are not listed for effective slenderness ratios beyond 200. Some useful properties of the shapes, used in the design of columns, are listed at the bottom of each Column Load Table.

The numerical values of the limiting effective slenderness ratio $4.71 \sqrt{E/F_y}$ for steels of different yield strengths are given in Table 8.7.1. Other parameters given in Table 8.7.1 will be discussed in Section 8.9.

8.7.3 Equivalent Effective Length $(K_x L_x)_y$

Design strengths for axial compression for I-shapes and rectangular HSS are given in the Column Load Tables for effective lengths with respect to the minor axis, $KL = K_y L_y$. That is, the tabulated strengths are P_{dy} values. When the end conditions of the column are different about the two axes, and/or when the minor axis is braced at closer intervals than the major axis, the strength of the column must be investigated with reference to both the major (x) and minor (y) axes. The ratio r_x/r_y furnished in the bottom part of the Column Load Tables facilitates the determination of the design strength of a column with respect to its major axis, P_{dx}. To this end, let us define the term *equivalent effective length, $(K_x L_x)_y$*, as the effective length with respect to

TABLE 8.7.1

Design Parameters for Compression Members And Plate Elements As a Function of F_y

	F_y (ksi)							
	36	42	46	**50**	60	65	90	100
$0.85F_y$	30.6	35.7	39.1	**42.5**	51.0	55.3	76.5	85.0
$\pi\sqrt{\dfrac{E}{F_y}}$	89.2	82.6	78.9	**75.7**	69.1	66.4	56.4	53.5
$4.71\sqrt{\dfrac{E}{F_y}}$	134	124	118	**114**	104	100	84.6	80.3
$0.45\sqrt{\dfrac{E}{F_y}}$	12.8	11.8	11.3	**10.8**	9.89	9.51	8.0	7.6
$0.56\sqrt{\dfrac{E}{F_y}}$	15.9	14.7	14.1	**13.5**	12.3	11.8	10.0	9.5
$0.75\sqrt{\dfrac{E}{F_y}}$	21.3	19.7	18.8	**18.1**	16.5	15.8	13.4	12.7
$1.49\sqrt{\dfrac{E}{F_y}}$	42.3	39.2	37.4	**35.9**	32.8	31.5	26.7	25.3

E = modulus of elasticity = 29,000 ksi
F_y = yield stress of material, ksi

the minor axis (y) equivalent in load carrying capacity to the actual effective length for buckling about the major axis (x). Thus:

$$\frac{(K_xL_x)_y}{r_y} = \frac{K_xL_x}{r_x} \qquad (8.7.12)$$

or

$$(K_xL_x)_y = \frac{K_xL_x}{(r_x/r_y)} \qquad (8.7.13)$$

where K_xL_x = major axis effective length (ft)
$(K_xL_x)_y$ = effective length with respect to the minor axis, equivalent in load carrying capacity to the actual effective length about the major axis (ft)

So, to obtain the equivalent effective length factor $(K_xL_x)_y$, divide the given major axis effective length, K_xL_x, by the (r_x/r_y) ratio of the section. The design strength P_{dx} can then be obtained by reentering the Column Load Table for that section with the effective length $KL = (K_xL_x)_y$. The smaller of the two strengths obtained, namely, P_{dy} or P_{dx}, will be the design strength, P_d, for the given column. For most of the W-shapes included in the Column Load Tables r_x/r_y values lie between 1.6 and 2.5.

EXAMPLE 8.7.1

Design Strength

Calculate the design strength of a W12×79 column with pinned ends and a length of 22 ft. Use (a) A992 steel, (b) A572 Grade 65 steel, and (c) A514 Grade 100 steel.

Solution

For a W12×79 section: $A = 23.2$ in.^{2}, $r_x = 5.34$ in., $r_y = 3.05$ in. For a pin-ended column of length, $L = 22$ ft,

$$\frac{K_x L_x}{r_x} = \frac{1.0(22.0)(12)}{5.34} = 49.4$$

$$\frac{K_y L_y}{r_y} = \frac{1.0(22.0)(12)}{3.05} = 86.6 \quad \leftarrow \text{ controls}$$

a. A992 steel: $F_y = 50$ ksi

   (i) For any hot-rolled steel column with a bisymmetrical cross section, the design axial compressive stress and the design axial compressive strength can be calculated by using Eqs. E2-1, 2, 3 and 4 of the LRFD Specification.
   The slenderness parameter of the A992 steel column is:

$$\lambda_c = (KL/r) \div \left[\pi\sqrt{\frac{E}{F_y}}\right] = (86.6) \div \left[\pi\sqrt{\frac{29,000}{50}}\right]$$

$$= 1.14 < 1.5$$

   So, buckling is inelastic and the buckling stress is given by Eq. 8.7.4 (Eq. E2-2 of the LRFDS):

$$F_{cr} = \left(0.658^{\lambda_c^2}\right) F_y = \left(0.658^{1.14^2}\right) 50 = 29.0 \text{ ksi}$$

   Design compressive stress, $\phi_c F_{cr} = 0.85(29.0) = 24.7$ ksi

   Design compressive strength, $P_d = \phi_c F_{cr} A_g = 24.7(23.2) = 573$ kips
   (Ans.)

   (ii) For any hot-rolled steel compression member, the design stress can be obtained from Table 4 of the LRFDS for a given value of the slenderness parameter, λ_c.

Thus, for $\lambda_c = 1.14$, we obtain: $\dfrac{\phi_c F_{cr}}{F_y} = 0.493$

Design compressive stress, $\phi_c F_{cr} = 0.493(50) = 24.7$ ksi
Design compressive strength, $P_d = \phi_c F_{cr} A_g = 24.7(23.2) = 573$ kips   (Ans.)

(iii) For a A992 steel compression member, the design compressive stress can also be read from Table 3-50 of the LRFDS for any given value of *KL/r*. Thus, for $(KL/r) = 86.6$, we obtain:
Design compressive stress, $\phi_c F_{cr} = 24.6$ ksi
Design compressive strength, $P_d = \phi_c F_{cr} A_g = 24.6(23.2) = 571$ kips   (Ans.)

(iv) The Column Load Tables for W-shapes (LRFDM Table 4-2) can be used to obtain the column design strength of rolled steel W-shapes commonly used as columns made of Grade 50 steels. As the end conditions are the same about both axes and there are no intermediate braces, we know that $K_y L_y$ controls the column strength and there is no need to calculate $(K_x L_x)_y$.
Entering the Column Load Table for W12-Shapes (LRFDM Table 4-2) with a W12×79 section and an effective length $KL = K_y L_y = 22$ ft, the design axial compressive strength of the column is given as 570 kips.   (Ans.)

b. A572 Grade 65 steel: $F_y = 65$ ksi

From part (*a*) we have: $KL/r = 86.6$
The slenderness parameter of the A572 Grade 65 steel column ($F_y = 65$ ksi) is:

$$\lambda_c = (KL/r) \div \left[\pi \sqrt{\frac{E}{F_y}} \right] = (86.6) \div \left[\pi \sqrt{\frac{29,000}{65}} \right]$$

$$= 1.30 < 1.5$$

The design compressive stress ratio, corresponding to a slenderness parameter λ_c of 1.30, is obtained from Table 4 of the LRFDS, as: $\phi F_{cr}/F_y = 0.419$.
So, design compressive stress, $\phi_c F_{cr} = 0.419(65) = 27.2$ ksi
Design strength of column, $P_d = \phi_c F_{cr} A_g = 27.2(23.2) = 631$ kips
                    (Ans.)

Thus, changing from A992 steel to A572 Grade 65 steel results in:

Increase in yield stress $= \dfrac{65 - 50}{50} \times 100 = 30.0\%$

Increase in column strength $= \dfrac{631 - 570}{570} \times 100 = 10.7\%$

[continues on next page]

Example 8.7.1 continues ...

c. A514 Grade 100 steel: $F_y = 100$ ksi

From LRFDM Table 2-1: Availability of Various Structural Shapes, we observe that rolled W-shapes are not available in this grade of steel. For the purpose of this problem we will assume that the member is obtained by welding together three plates having similar width to thickness ratios as the flange and web elements of a rolled W12×79 section.

The slenderness parameter of the A514 Grade 100 steel column ($F_y = 100$ ksi) is:

$$\lambda_c = (KL/r) \div \left[\pi \sqrt{\frac{E}{F_y}}\right] = (86.6) \div \left[\pi \sqrt{\frac{29,000}{100}}\right]$$

$$= 1.62 > 1.5$$

So, buckling is elastic and the critical stress is given by Eq. 8.7.5 (Eq. E2-3 of the LRFDS):

$$F_{cr} = \left[\frac{0.877}{\lambda_c^2}\right] F_y = \left(\frac{0.877}{1.62^2}\right) 100 = 33.4 \text{ ksi}$$

Design compressive stress, $\phi_c F_{cr} = 0.85(33.4) = 28.4$ ksi

Design compressive strength, $P_d = \phi_c F_{cr} A_g = 28.4(23.2) = 659$ kips

(Ans.)

Increase in yield stress $= \dfrac{100 - 50}{50} \times 100 = 100\%$

Increase in column strength $= \dfrac{659 - 570}{570} \times 100 = 15.6\%$

Remarks

From parts (b) and (c) of the above problem, we observe that an increase in yield stress does not result in a proportional increase in column strength.

EXAMPLE 8.7.2

Use of Column Load Tables

A 28-ft axially loaded W10×77 is laterally supported perpendicular to its y axis at midheight. Determine the design axial compressive strength of this column, if A992 steel is used and if the column is pinned about both axes at both ends. Use the Column Load Tables in Part 4 of the LRFDM.

Solution

Section: W10×77; $F_y = 50$ ksi

Length of column, $L = 28$ ft

For buckling about x axis: $L_x = 28$ ft; $K_x = 1.0$; $K_x L_x = 28.0$ ft
For buckling about y axis: $L_y = 14$ ft; $K_y = 1.0$; $K_y L_y = 14.0$ ft
Entering the Column Load Table for W10-shapes (LRFDM Table 4-2)
with $KL = K_y L_y = 14.0$ ft, it is observed that for a W10×77 the design
strength, $\phi_c F_{cr} = 708$ kips $= P_{dy}$. From the bottom part of the table we
note that, for a W10×77 shape, $r_x/r_y = 1.73$. So, the equivalent effective
length of the given column for buckling about the x axis is:

$$(K_x L_x)_y = \frac{K_x L_x}{r_x/r_y} = \frac{28}{1.73} = 16.2 \text{ ft}$$

Since $(K_x L_x)_y = 16.2$ ft $> (K_y L_y) = 14$ ft, buckling about the x axis con-
trols. Reentering the Column Load Table for W10-shapes with $KL = (K_x L_x)_y = 16.2$ ft, it is observed that the W10×77 has a design strength
$\phi_c F_{cr} = 638$ kips $= P_{dx}$.
Design compressive strength of the column is thus:
$P_d = \min [P_{dx}, P_{dy}] = \min [638, 708] = 638$ kips   (Ans.)

Built-Up Column

EXAMPLE 8.7.3

A built-up column in the ground floor of a 60-story building is composed of
five plates welded together as shown in Fig. X8.7.3. Assuming that A572
Grade 42 steel is used, determine the axial load capacity if $K_x L_x = 36$ ft and
$K_y L_y = 30$ ft.

Solution

From Table 2-2 of the LRFDM, we observe that plates up to 6-in. thickness are available in A572 Grade 42 steel.

Figure X8.7.3

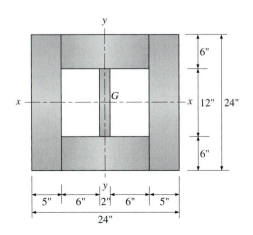

[*continues on next page*]

Example 8.7.3 continues ...

For the built-up section, we have with the help of Fig. X8.7.3:

$$A = 24(24) - 2(6)(12) = 432 \text{ in.}^2$$

$$I_x = \frac{1}{12}(24)(24^3) - 2\left(\frac{1}{12}\right)(6)(12^3) = 25{,}900 \text{ in.}^4$$

$$I_y = \frac{1}{12}(24)(24^3) - \frac{1}{12}(12)(14^3) + \frac{1}{12}(12)(2^3) = 24{,}900 \text{ in.}^4$$

$$r_x = \sqrt{\frac{25{,}900}{432}} = 7.74 \text{ in.}; \quad r_y = \sqrt{\frac{24{,}900}{432}} = 7.59 \text{ in.}$$

For the column:

$$\frac{K_x L_x}{r_x} = \frac{36.0(12)}{7.74} = 55.8 \leftarrow \text{controls}; \frac{K_y L_y}{r_y} = \frac{30.0(12)}{7.59} = 47.4$$

Slenderness parameter,

$$\lambda_c = \frac{K_x L_x}{r_x} \div \left[\pi \sqrt{\frac{E}{F_y}}\right] = 55.8 \div \left[\pi \sqrt{\frac{29{,}000}{42}}\right] = 0.676$$

From Table 4 of the LRFDS, for $\lambda_c = 0.676$, the design compressive stress ratio is $\phi_c F_{cr}/F_y = 0.702$.

Design compressive stress, $\phi_c F_{cr} = 0.702(42) = 29.5$ ksi
Design compressive strength of the column,

$$P_d = \phi_c F_{cr} A_g = 29.5(432) = 12{,}700 \text{ kips} \qquad \text{(Ans.)}$$

8.8   Inelastic Effective Length Factors

The discussion in Section 8.5.4 concerning the evaluation of effective length factors of columns in rectangular frames was restricted to the buckling of perfectly elastic frames. However, in reality, instability of steel frames is more likely to take place after the stresses at some parts of the frame have reached the yield stress.

The rotational stiffness of a column axially loaded in the elastic range is proportional to EI. However, its stiffness in the inelastic range can be more accurately taken as $E_t I$, where E_t is the tangent modulus. The parameter G used in the development of the alignment chart expression involves the rotational stiffnesses of the adjacent columns and girders and is therefore a function of the modulus of elasticity of the columns and girders (Eqs. 8.5.2a and b). For elastic behavior, E cancels out of the equation, resulting in the familiar expressions for the coefficients G_A and G_B given in Eqs. 8.5.3. When the KL/r ratio of a column is less than the limiting value $4.71 \sqrt{E/F_y}$ axial column behavior is inelastic. If the elastic E still applies for the girder members, but the inelastic E_t applies for the columns, this can be accounted for by adjusting the G values as follows:

$$G_i = \frac{\Sigma(E_t I/L)_c}{\Sigma(EI/L)_g} = \frac{E_t}{E} G_e = \tau G_e \tag{8.8.1}$$

where G_e = elastic G factor assuming that both the columns and the girders behave elastically (Eq. 8.5.3)
G_i = inelastic G factor assuming that the girders behave elastically
while the columns behave inelastically
τ = *stiffness reduction factor*

Hence, as the plastification of the column section increases under increased
load, the elastic girder offers more restraint to the column relative to the elastic case [Yura, 1971]. Thus, the value of G normally used in the alignment
chart is reduced by the stiffness reduction factor, τ. A reduced value for G
results in a smaller value for the effective length of the column and, in turn,
a higher value for the design axial compressive strength.

Since the critical buckling stress is directly proportional to the ratio of the
moduli (Eq. 8.6.1), the stiffness reduction factor may also be expressed as:

$$\tau = \frac{E_t}{E} = \frac{F_{cr,\,inelastic}}{F_{cr,\,elastic}} \tag{8.8.2}$$

In estimating the stiffness reduction factor in a design problem, Yura [1971]
assumed that a column is designed very close to its capacity. In the context
of the LRFD, this is equivalent to the assumption that:

$$P_u \approx \phi_c P_n = \phi_c F_{cr,\,inelastic} A_g \tag{8.8.3}$$

in which $\phi_c = 0.85$, P_u is the factored axial load in the column, and A_g is the
cross-sectional area of the column. Rearranging Eq. 8.8.3 we obtain:

$$F_{cr,\,inelastic} = \frac{P_u}{\phi_c A_g} \tag{8.8.4}$$

The LRFD expression for the inelastic critical stress for a column of slenderness λ_c is:

$$F_{cr,\,inelastic} = \left[0.658^{\lambda_c^2}\right] F_y \tag{8.8.5}$$

The LRFD expression for the elastic critical stress is:

$$F_{cr,\,elastic} = \frac{0.877}{\lambda_c^2} F_y \tag{8.8.6}$$

By solving for λ_c in Eqs. 8.8.4 and 8.8.5 and substituting this value into
Eq. 8.8.6 (overlooking for the moment that λ_c might be less than 1.5), we
obtain a value for the *elastic* critical stress, that is, the value from the Euler
hyperbola corresponding to the value of λ_c. Then, by using the expression for

stiffness reduction factor from Eq. 8.8.2, we obtain (LRFDC Eq. C-C2-3):

$$\tau = 1.0 \qquad\qquad \text{for} \quad \frac{P_u}{P_y} \le \frac{1}{3} \qquad (8.8.7a)$$

$$= -7.38\left(\frac{P_u}{P_y}\right) \log\left(\frac{P_u}{\phi_c P_y}\right) \quad \text{for} \quad \frac{P_u}{P_y} > \frac{1}{3} \qquad (8.8.7b)$$

where P_y is the squash load of the column cross section ($F_y A_g$).

Values of the stiffness reduction factor, τ, for different values of P_u/A_g are presented in Table 4-1 of the LRFDM for steels with $F_y = 35, 36, 42, 46,$ and 50 ksi. For values of P_u/A_g smaller than those with entries in this table, the column behaves elastically, and the reduction factor is 1.0. Note that, if the base of a column is fixed ($G = 1.0$) or pinned ($G = 10$), the value of G at that end should not be multiplied by the stiffness reduction factor τ.

EXAMPLE 8.8.1

Inelastic Effective Length Factor

An interior column of a building is subjected to an axial compressive load of 1660 kips under factored loads. The column is part of a braced frame for buckling about its minor axis. For buckling about the major axis, the column is part of an unbraced rigid jointed frame as shown in Fig. X8.8.1. All girders are W18×40 sections. Assume that the column cross section remains constant above and below the level considered. Determine if a W14×159 A992 steel column is adequate. $L_g = 32$ ft and $L_c = 12$ ft.

Solution

a. Data
    From Table 1-1 of the LRFDM, we obtain for:

$$\text{W18×40:} \quad I_x = 612 \text{ in.}^4$$

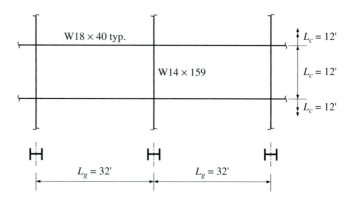

Figure X8.8.1

$$W14 \times 159: \quad A = 46.7 \text{ in.}^2; \quad r_y = 4.00 \text{ in.}$$

$$I_x = 1900 \text{ in.}^4; \quad r_x = 6.38 \text{ in.}$$

Factored axial load on the column, $P_u = 1660$ kips

b. Strength based on elastic G-factors

$$G_A = \frac{\left(\dfrac{1900}{12}\right)2}{1.0\left(\dfrac{612}{32}\right)(2)} = 8.28 = G_B = G$$

Using the alignment chart for the sway uninhibited case (Fig. 8.5.6b):

$$K_x \approx 2.76$$

Thus,

$$\frac{K_y L_y}{r_y} = \frac{1.0(12)(12)}{4.00} = 36.0$$

$$\frac{K_x L_x}{r_x} = \frac{2.76(12)(12)}{6.38} = 62.3 \leftarrow \text{controls}$$

From Table 3-50 of the LRFDS, we have for:

$$\frac{KL}{r} = 62.3 \rightarrow \phi_c F_{cr} = 32.0 \text{ ksi}$$

Design strength of the column, $P_{dc} = 32.0(46.7) = 1490$ kips which is less than the required strength of 1660 kips.

c. Strength based on the inelastic G factors

$$\frac{P_u}{P_y} = \frac{1660}{46.7(50)} = 0.711$$

Since $P_u/P_y = 0.711 > \frac{1}{3}$, we observe that the column partially plastifies under the given load. Using Eq. 8.8.7b, the stiffness reduction factor,

$$\tau = -7.38(0.711) \log\left(\frac{0.711}{0.85}\right) = 0.407$$

Alternatively:

Axial stress, $f_a = \dfrac{P_u}{A_g} = \dfrac{1660}{46.7} = 35.6$ ksi

From Table 4-1 of the LRFDM, for:

$$f_a = 35.6 \text{ ksi, and } F_y = 50 \text{ ksi} \rightarrow \tau = 0.408$$

[*continues on next page*]

Example 8.8.1 continues ...

$G_i = \tau G_e = 0.408(8.28) = 3.38$, resulting in a modified effective length factor, $K_x \approx 1.91$. So,

$$\frac{K_x L_x}{r_x} = \frac{1.91(12)(12)}{6.38} = 43.1$$

From Table 3-50 of the LRFDS, for $KL/r = 43.1$, $\phi_c F_{cr} = 37.1$ ksi
Design axial compressive strength of the column:

$$P_d = 37.1(46.7) = 1730 \text{ kips} > P_{req} = P_u = 1660 \text{ kips}$$

The W14×159 of A992 steel is therefore adequate. (Ans.)

EXAMPLE 8.8.2

Inelastic, Unbraced About Both Axes

A W14×233 edge column is a member of a rigid frame unbraced about both axes. W21×83 girders, 25 ft long, are connected to both the flanges of the column, and 22 ft long W18×55 beams frame into the web from one side, at the top and bottom of the column. Story height is 12 ft. Use A992 steel and assume that the column cross section remains the same in the floors above and below the column under consideration. Verify the safety of the column for a factored axial load of 2220 kips.

Solution

a. Data

W14×233: $A = 68.5$ in.^{2}
$I_x = 3010$ in.^{4}; $r_x = 6.63$ in.
$I_y = 1150$ in.^{4}; $r_y = 4.10$ in.
W21×83: $I_x = 1830$ in.^{4}; W18×55: $I_x = 890$ in.^{4}

Factored load, $P_u = 2220$ kips

b. Strength based on elastic G factors

For buckling about the major axis, $G_A = \dfrac{2\left(\dfrac{3010}{12}\right)}{2\left(\dfrac{1830}{25}\right)} = 3.43 = G_B = G$

From the alignment chart for unbraced frames, $K_x \approx 1.93$.

For buckling about the minor axis: $G_A = \dfrac{2\left(\dfrac{1150}{12}\right)}{1\left(\dfrac{890}{22}\right)} = 4.74 = G_B = G$

From the alignment chart for unbraced frames, $K_y \approx 2.19$.
Effective slenderness ratios of the column:

$$\frac{K_x L_x}{r_x} = \frac{1.93(12.0)(12)}{6.63} = 41.9$$

$$\frac{K_y L_y}{r_y} = \frac{2.19(12.0)(12)}{4.10} = 76.9 \leftarrow \text{controls}$$

From Table 3-50 of the LRFDS, we obtain, $\phi_c F_{cr} = 27.6$ ksi.
Substituting,

$$P_d = 27.6(68.5) = 1890 \text{ kips} < 2220 \text{ kips}$$

The section is therefore unsatisfactory when elastic G factors are used
to determine the effective length factor for the column.   (Ans.)

c. Strength based on inelastic G-factors

Axial compressive stress, $f_a = \dfrac{P_u}{A_g} = \dfrac{2220}{68.5} = 32.4$ ksi

From Table 4-1 of the LRFDM for Grade 50 steel and $f_a = 32.4$ ksi, the
stiffness reduction factor is $\tau = 0.564$.

Alternatively calculate, $\dfrac{P_u}{P_y} = \dfrac{2220}{68.5(50)} = 0.648$

Using Eq. 8.8.7b, $\tau = -7.38(0.648) \log\left(\dfrac{0.648}{0.85}\right) = 0.564$

So, for buckling about the major axis

$$G_i = \tau \times G_e = 0.564(3.43) = 1.93$$

From the alignment chart for unbraced frames, $K_x \approx 1.59$.

For buckling about the minor axis:

$$G_i = \tau \times G_e = 0.564(4.74) = 2.67$$

From the alignment chart for unbraced frames, $K_y \approx 1.77$.

So,

$$\frac{K_x L_x}{r_x} = \frac{1.59(144)}{6.63} = 34.5$$

$$\frac{K_y L_y}{r_y} = \frac{1.77(144)}{4.10} = 62.2 \leftarrow \text{controls}$$

[*continues on next page*]

Example 8.8.2 continues ...

From Table 3-50 of the LRFDS, $\phi_c F_{cr} = 32.1$ ksi, resulting in:
$P_d = 32.1(68.5) = 2200$ kips $< P_u = 2220$ kips N.G.
So, the W14×233 of A992 is not satisfactory (although it is within
1.0 percent of the required strength).   (Ans.)

8.9   Local Buckling of Rectangular Plate Elements

A rectangular plate is a flat structural element characterized by having its
length and width dimensions much greater than its thickness. It is supported
on all four (and sometimes three) edges and may also be flexurally (rotationally) restrained along some or all of those edges. The geometry and edge
loading used for flat plates is shown in Fig. 8.9.1. The length of this plate,
measured in the direction of main longitudinal stress in the plate, is a. The
width of the plate is b, and the thickness is t.

8.9.1   Behavior of a Compressed Plate

Let us consider a perfectly flat rectangular plate, simply supported on all four
edges (Fig. 8.9.2) subjected to edge compression in one direction only (i.e.,
$f_y = f_b = f_{xy} = 0$). The load is applied through rigid end blocks, thus forcing

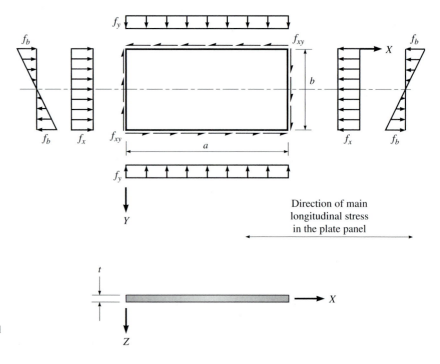

Figure 8.9.1: Rectangular plate under general
loading.

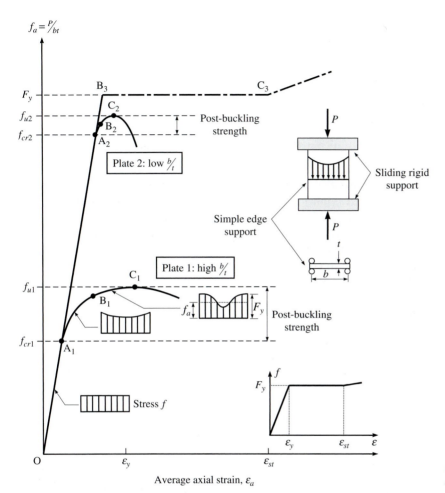

Figure 8.9.2: Behavior of rectangular plates under edge compression.

the edges to remain straight in the process of loading. The plate is made of linearly elastic, plastic, strain-hardening material containing no residual stresses. A diagram of plate behavior is obtained by plotting the average compressive stress f_a versus the average strain ε_a, where

$$f_a = \frac{P}{bt} \tag{8.9.1}$$

The heavy solid line $OA_1B_1C_1$ is typical for a plate with a high width-thickness ratio b/t. Several stages can be observed. Initially, when loaded by in-plane compressive forces, the perfectly flat plate develops a uniform stress field. Thus, if f_x is the stress intensity at a particular point, $f_x = f_a$ across the width, and no lateral deflection of the plate occurs. As the magnitude of the loading is increased, the intensity of the stress field increases proportionally until $f_a = f_{cr1}$. At that point, the flat plate is no longer stable, and out-of-plane deflection occurs. The magnitude of this loading is called the ***critical load*** or ***buckling load of the plate,*** and its value depends both on the plate geometry and the boundary conditions present [Timoshenko and Gere, 1961].

For columns, buckling is often synonymous with failure. This is not always the case for plates. Provided that the material is elastic, the buckling is restrained by the plate material spanning in the transverse direction. The plate can then continue to carry increased loads in its deflected form, and the plate is said to have stable post-buckled or ***post-critical behavior.*** After the point of buckling, the stress is no longer uniformly distributed along the width of the plate (i.e., $f_x \neq f_a$). See the detail for portion A_1B_1 of the curve in Fig. 8.9.2 for clarification. Due to the out-of-plane deformation, a redistribution of stress takes place with the maximum value located at the edges. Note that the stress distribution becomes nonuniform even though the load is applied through ends that are rigid and perfectly straight. Further increases in the load lead to greater stress redistribution. The stress along the edges gradually increases until the edges start yielding at point B_1 on the curve. The yielded region then spreads quickly inward until the ultimate load is reached at point C_1 on the curve. Generally, a localized plastic mechanism is formed at this point. The average stress at this load $(\frac{P_u}{bt})$ is defined as the ultimate stress, f_{u1}. The increase in the average stress beyond buckling $(f_{u1} - f_{cr1})$ can be quite substantial for high b/t ratios.

For plates with lower b/t ratios, the critical stress is closer to F_y, and yielding starts almost immediately after buckling. In this case, the ultimate stress f_{u2} is then only slightly higher than the critical stress f_{cr2}, as indicated by the curve $OA_2B_2C_2$ in Fig. 8.9.2. In plates having a width-thickness ratio below some specific value, the average stress f_a reaches the yield point F_y without buckling. The plate may then undergo further straining at this stress level as indicated by the line segment OB_3C_3 in Fig. 8.9.2. Eventually, the plate will buckle and fail at a strain that usually occurs somewhere between B_3 and C_3. However, for plates having very low b/t ratios, failure can occur in the strain-hardening region that lies beyond C_3. Note that relatively thick plates that buckle at a high stress and begin to yield soon thereafter do not exhibit a significant amount of post-buckling strength, whereas relatively thin plates that buckle at a low stress but do not yield until much later can be expected to have considerable post-buckling strength.

Two aspects of plate manufacture and fabrication influence the strength of a plate. The first is associated with residual stresses that exist in the plate prior to loading that result from rolling and welding processes (see Fig. 2.6.4). Residual stresses reduce the maximum allowable load, especially if the stress pattern is coherent, as would be the case if it is induced by welding along the edges of the plate. The other aspect is associated with the great difficulty of making a perfectly flat plate. Thus, a plate will inevitably have some initial deflection or imperfection, even at zero load. This, too, has a tendency to reduce the maximum load capacity. The behavior of a

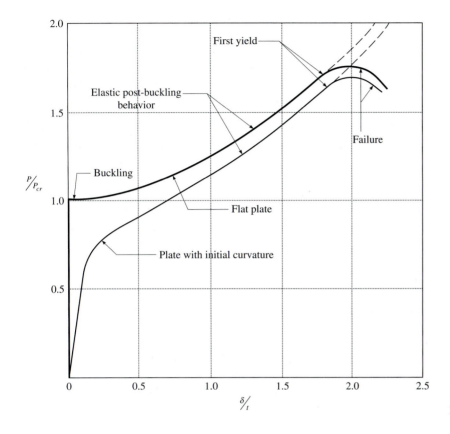

Figure 8.9.3: Post-buckling behavior of thin plates.

plate with initial curvature as compared to that of a flat plate is schematically shown in Fig. 8.9.3. Note that a plate with initial curvature has a continuously rising load-deformation curve until the ultimate strength is reached. That is, plates possessing inherent initial deflection do not exhibit buckling behavior.

Structural shapes such as I-shapes, channels, tees, and angles are composed of plate elements. In general, when one element buckles, only that element and not necessarily the entire member becomes deformed. Buckling of a plate element of a steel member is generally referred to as *local buckling*. When local buckling occurs, the buckled element of the section no longer supports its proportional share of any additional load that the member is subjected to. In other words, the efficiency of the cross section is reduced.

8.9.2   Limiting Width-to-Thickness Ratios, λ_r, Plate Elements in Compression

The *critical stress for a rectangular plate* of width b and length a ($a > b$), simply supported on all four edges and uniformly compressed in the longitudinal direction is given by [Bleich, 1952; Timoshenko and Gere, 1961]:

$$f_{cr} = \frac{\pi^2 E k_c}{12(1 - \mu^2)(b/t)^2} \tag{8.9.2}$$

where E = modulus of elasticity ($= 29{,}000$ ksi, for steel)
   μ = Poisson's ratio ($= 0.3$ for steel)
   t = thickness of the plate
   b = width of the loaded edge of the plate
   f_{cr} = critical stress

The ratio a/b is called the ***aspect ratio*** of the plate. The term k_c is generally referred to as the ***plate buckling coefficient.*** It depends upon the plate aspect ratio. For a long plate simply supported on all four edges,

$$k_c = 4 \quad \text{for} \quad a/b \geq 4 \tag{8.9.3}$$

For information on elastic buckling of a uniformly compressed rectangular plate with simply supported edges, refer to Section W8.5 on the website http://www.mhhe.com/Vinnakota.

For plates having other combinations of boundary conditions (e.g., simple, fixed, or free edges), other expressions for k_c for use in Eq. 8.9.2 have been developed, all of which depend upon the aspect ratio of the plate. The effects of some combinations of simply supported, clamped (fixed), and free edge conditions upon the buckling of plates with large aspect ratios are shown in Fig. 8.9.4 [Galambos, 1998]. Except for Case 5 in Fig. 8.9.4, the plate buckles in one or more half-sine waves in the direction of compressive stress. The plate in Case 5 buckles in only one half-sine wave in one direction, and one quarter-sine wave in the other direction.

Generally speaking, all of the rolled shapes shown in Fig. 8.2.1, with the exception of the round HSS, are composed of connected elements that, for the purposes of analysis and design, can be treated as long, flat, rectangular plates. When such a flat plate element is subjected to direct compression along the short edges, the plate can buckle locally before the member as a whole becomes unstable, or before the yield stress of the material is reached. Such behavior, known as ***local buckling*** of plate elements, is characterized by out-of-plane deflection of the plate element and consequent distortion of the cross section of the compression member.

Now, the plate buckling coefficient, k_c, of a flat rectangular plate is a function of the boundary conditions along the four edges of the plate and of the plate aspect ratio, a/b. For column plate elements, the dimension, a, represents the length of the column, indicating that the aspect ratio for these plates is quite large. For such plates with large aspect ratios, studies have shown that the most significant parameter that influences the value of k_c is the support conditions of the unloaded edges. That is, for these plates, the value of k_c is, practically speaking, independent of the support conditions of

Axial Compression	Conditions at Nonloaded Edges	k_c
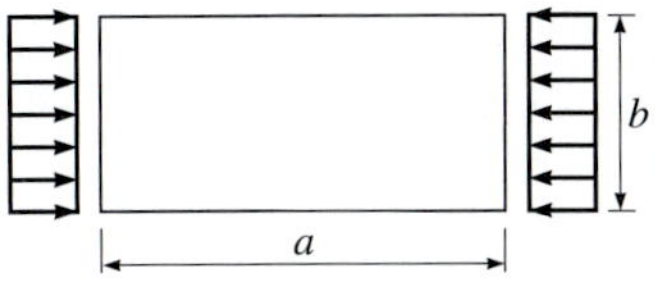	**Case 1:** Both fixed	6.97
All loaded edges simply supported	**Case 2:** One fixed, one simply supported	5.42
$\frac{a}{b} > 4.0$	**Case 3:** Both simply supported	4.00
	Case 4: One fixed, one free	1.277
	Case 5: One simply supported, one free	0.425

Figure 8.9.4: Approximate buckling coefficients for long rectangular plates in axial compression [Adapted from Galambos, 1998].

the loaded edges and can be considered to be strictly a function of the support conditions of the unloaded edges. Column plate elements can, therefore, be classified into two broad categories: stiffened elements and unstiffened elements.

- A *stiffened element* is a plate element supported along both edges parallel to loading. For example, the web of a W-shape column is supported, by the flanges, along its two longitudinal edges parallel to the loading. Thus, the web of an I-shape is an example of a stiffened plate.
- An *unstiffened element* is a plate element with one free edge parallel to the direction of loading. For example, each half-flange of an I-shape has one free edge and another edge supported by the web. So, each half-flange of an I-shape is an example of an unstiffened element.

From Eq. 8.9.2, we observe that the compressive stress at which any rectangular plate element of a member cross section buckles locally is inversely proportional to $(b/t)^2$, much the same as the buckling stress of a column is inversely proportional to $(L/r)^2$. The **width-to-thickness ratio,** (b/t), of a plate element is a measure of the plate slenderness and is indicated by λ. The value of $\lambda = \lambda_o$, for which the theoretical plate buckling stress equals the yield stress of the material, may be obtained from Eq. 8.9.2, by substituting $f_{cr} = F_y$ and rearranging, as:

$$\lambda_o = \sqrt{\frac{\pi^2 E k_c}{12(1 - \mu^2) F_y}} = 0.951 \sqrt{\frac{E k_c}{F_y}} \qquad (8.9.4)$$

To prevent premature failure of compression members by local buckling, a cross section should be selected such that the resistance to local buckling offered by the individual plate elements should be the same (or greater) than the resistance offered by the whole member to primary buckling. To this end, design limits in the United States are generally simplified to assure that compression elements of a column can reach the yield stress of the material before local buckling occurs, even though the slenderness ratio of the column as a whole would probably prevent any element from reaching the yield stress. However, the provisions prescribed are not necessarily conservative for design, since residual stresses and initial imperfections will have their greatest strength reducing influence precisely at b/t ratios equal to λ_o. Hence, to prevent premature failure of compression members by local buckling, the LRFD Specification limits the b/t values of individual plate elements to 0.7 of the theoretical value λ_o. That is:

$$\left(\frac{b}{t} = \lambda\right) \le \left(\lambda_r = 0.7\lambda_o = 0.666 \sqrt{\frac{E k_c}{F_y}}\right) \qquad (8.9.5)$$

So, if members are to function at stresses that are commensurate with the strengths of the steels of which they are made, the plate elements of high-strength steel members will have to be more stocky than those of low-strength steels.

Figure 8.9.5 shows several examples of unstiffened and stiffened elements in various cross-sectional shapes. In each case the width, b, and the thickness, t, of the elements considered are shown. Note that, for determining the width-to-thickness ratio of a plate element in a section, the width b of each element is explicitly defined in LRFDS Section B5, as follows:

Unstiffened Elements

1. For flanges of I-shaped members and tees, the width b is half of the full flange width, b_f.
2. For flanges of channels and zees, the width b is the full flange width, b_f.
3. For angles, the width b is the full length of the longer leg, l_1.
4. For stems of tees, the width b is the full nominal depth, d.
5. For plates, the width b is the distance from the free edge to the first row of bolts or lines of welds.

Stiffened Elements

6. For webs of rolled I-, C- and Z-shapes, the width b equals h, the clear distance between flanges less the fillet radius at each flange.

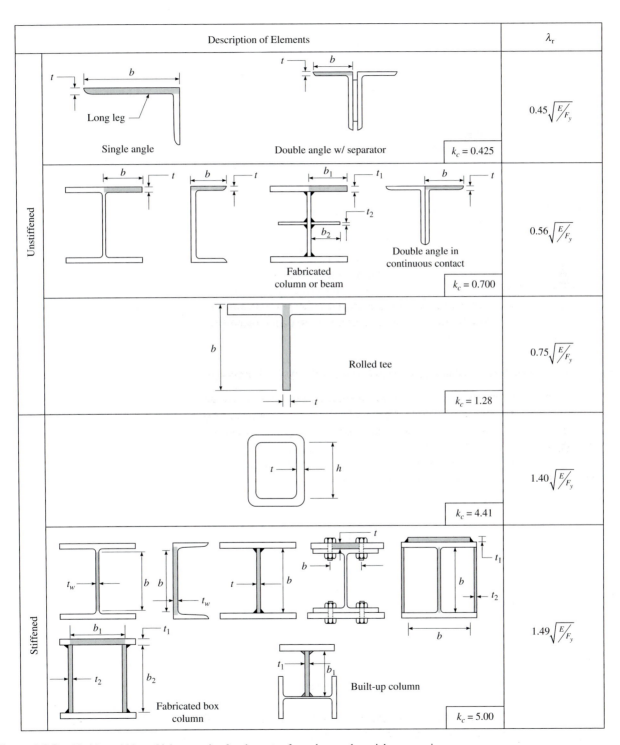

Figure 8.9.5: Limiting width-to-thickness ratios for elements of members under axial compression.

7. For webs of built-up welded sections, the width b equals h, the clear distance between flanges.
8. For cover plates in built-up sections, the width b is the distance between adjacent lines of bolts or welds.
9. For flanges of rectangular structural tubes, the width b is the clear distance between webs less the inside corner radius on each side. If the corner radius is not known, the width may be taken as the total section width minus three times the thickness.

A lower bound for the critical stress (or, equivalently for k_c) of column plate elements can be determined by hypothetically assuming that any element edge that is attached to another element is simply supported, or is free if not so attached. The theoretical value of critical stress so calculated for an element can in actuality be exceeded because elements are actually rotationally restrained at their connections. For example, at the junction of a web and the flange of an I-shape, the web edge is somewhere between the hinged and fully fixed conditions ($k_c = 4.00$ and $k_c = 6.97$, from Fig. 8.9.4). An approximate value of $k_c = 5.00$, a value about $\frac{1}{3}$ of the way between hinged and fixed conditions, is used in the LRFD Specification (see Fig. 8.9.5). Similarly, the k_c value for a flange of an I-shape (an unstiffened element) is taken to be 0.700, again about $\frac{1}{3}$ of the way between the hinged and fixed conditions ($k_c = 0.425$ and $k_c = 1.28$, from Fig. 8.9.4). Now, one leg of an angle derives little edge support (restraint) from the other leg when local buckling begins. If one leg starts to buckle, it will carry the other leg with it. On the other hand, in the case of a tee, the tendency of the stem to buckle locally is restrained by the much thicker flange plate which is in no danger of buckling at that same stress. These factors are taken care of by using k_c values of 0.425, 0.700 and 1.28 for the leg of an angle, the half-flange of an I and the stem of a tee, respectively, in Eq. 8.9.5 (see Fig. 8.9.5). Thus:

For unstiffened elements:

$$\text{Single angle:} \quad k_c = 0.425 \longrightarrow \lambda_{ra} = 0.45\sqrt{E/F_y} \quad (8.9.6a)$$

$$\text{Flange half:} \quad k_c = 0.700 \longrightarrow \lambda_{rf} = 0.56\sqrt{E/F_y} \quad (8.9.6b)$$

$$\text{Stem of tee:} \quad k_c = 1.28 \longrightarrow \lambda_{rs} = 0.75\sqrt{E/F_y} \quad (8.9.6c)$$

For stiffened elements:

$$\text{Web of an I- or C-shape:} \quad k_c = 5.00 \longrightarrow \lambda_{rw} = 1.49\sqrt{E/F_y} \quad (8.9.7a)$$

$$\text{Side of a tube:} \quad k_c = 4.41 \longrightarrow \lambda_{rt} = 1.40\sqrt{E/F_y} \quad (8.9.7b)$$

When the width-to-thickness ratio of all elements in an axially compressed column are less than their corresponding values of λ_r, local buckling does not occur before overall buckling of the member. In such cases, the nominal critical compressive stress of the column is that given by LRFDS Section E2 (i.e., Eqs. 8.7.4 and 8.7.5).

In this text, the following nomenclature will be used for plate elements of certain shapes often used as columns:

Flange of an I- or T-shape: $\lambda_f = \dfrac{b_f}{2t_f}$

Web of an I- or C-shape: $\lambda_w = \dfrac{h}{t_w}$

Values of the ratios $b_f/2t_f$ and h/t_w for I-shapes, and h/t ratios for HSS are tabulated in the Dimensions and Properties Tables in Part 1 of the LRFD Manual. Values of λ_r for different steels and plate elements are given in Table 8.7.1.

When the b/t ratio of one or more elements in an axially compressed column exceeds the corresponding λ_r value, local buckling precedes overall member buckling. As previously mentioned, the buckled stiffened elements will develop post-buckling strength, and the column will continue to resist additional compressive load. However, once local buckling occurs, stress is no longer distributed uniformly over the cross section. As a result, the column cannot remain straight, and will necessarily begin to bend. Eventually, the column assumes a buckled shape of considerable magnitude—which for all intents and purposes constitutes failure—at a load that is less than the buckling strength of the overall member. The reduced axial design compressive strength of the column—as influenced by plate local buckling—may be determined from Appendix B5.3 of the LRFDS. Since these cases are not encountered frequently in practice, they are not covered in this text.

Local Buckling Checks

EXAMPLE 8.9.1

A 22-ft long column in an existing building is part of a braced frame in both *xx* and *yy* planes of the cross section. It is fixed at the base and pinned at the top about both axes. In addition, the column is laterally supported perpendicular to its weak axis at a height 12 ft from the base. The column is a W8×35 section of A572 Grade 65 steel. Determine the permissible factored load on the column. Make all appropriate *b/t* checks.

Solution

a. Design strength
For a W8×35:

$A = 10.3$ in.^{2}, $r_x = 3.51$ in., $r_y = 2.03$ in.

$L = 22$ ft; $L_x = L = 22$ ft; $L_{y1} = 12$ ft; $L_{y2} = 10$ ft

$K_x L_x = 0.80(22) = 17.6$ ft

[*continues on next page*]

Example 8.9.1 continues ...

$(K_yL_y)_1 = 0.8(12) = 9.60$ ft; $(K_yL_y)_2 = 1.0(10) = 10.0$ ft
$K_yL_y = \max\,[(K_yL_y)_1,\ (K_yL_y)_2] = 10.0$ ft

The slenderness ratios are:

$$\frac{K_xL_x}{r_x} = \frac{17.6(12)}{3.51} = 60.2 \longleftarrow \text{controls}; \qquad \frac{K_yL_y}{r_y} = \frac{10.0(12)}{2.03} = 59.1$$

Slenderness parameter of the A572 Grade 65 steel column is:

$$\lambda_c = \left(\frac{KL}{r}\right) \div \left[\pi\sqrt{\frac{E}{F_y}}\right] = 60.2 \div \left[\pi\sqrt{\frac{29{,}000}{65}}\right] = 0.907$$

From Table 4 of the LRFDS, the design compressive stress ratio corresponding to this slenderness parameter is, $\phi_cF_{cr}/F_y = 0.602$
Design strength of the column, $P_d = \phi_cF_{cr}A_g = 0.602(65)(10.3) = 403$ kips

b. Width-to-thickness ratio checks of plate elements
We observe that the flange is an unstiffened element and the web is a stiffened element. Also, from LRFDM Table 1-1, we have for a W8×35:

$$\lambda_f = \frac{b_f}{2t_f} = 8.10; \qquad \lambda_w = \frac{h}{t_w} = 20.5$$

Figure X8.9.1

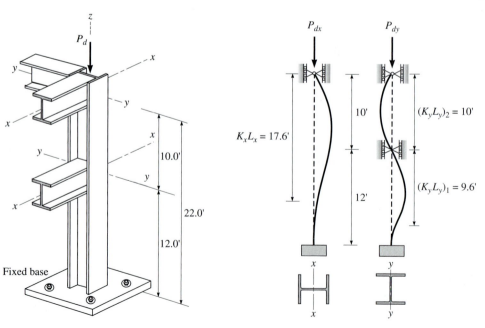

From the relations given in Eqs. 8.9.6a and 8.9.7a, we can write:

$$\text{flange} \longrightarrow \lambda_{rf} = 0.56\sqrt{\frac{E}{F_y}} = 0.56\sqrt{\frac{29{,}000}{65}} = 11.8$$

$$\text{web} \longrightarrow \lambda_{rw} = 1.49\sqrt{\frac{E}{F_y}} = 1.49\sqrt{\frac{29{,}000}{65}} = 31.5$$

So, $\lambda_f < \lambda_{rf}$ and $\lambda_w < \lambda_{rw}$, for the given section. As the width to thickness ratio checks are satisfied, neither flange local buckling nor web local buckling will precede member buckling.

So, the design axial compressive strength of the column is 403 kips.

(Ans.)

Local Buckling Checks

EXAMPLE 8.9.2

The cross section of a column in an existing multistory building is composed of a W16×100 to the web of which a PL½×20 and a PL1×16 are welded, forming a built-up section symmetric about its *x-x* axis, as shown in Fig. X8.9.2. Evaluate the design axial strength of the column if $K_x L_x$ = 24 ft and $K_y L_y$ = 30 ft. All material is of A36 steel. Use LRFDS. **Note:** The Inland Steel Building in Chicago has columns composed of sections similar in shape to the built-up section considered in this example.

Solution

For the W16×100 section

$$A = 29.7 \text{ in.}^2; \quad t_w = 0.585 \text{ in.}$$
$$I_x = 1{,}500 \text{ in.}^4; \quad I_y = 186 \text{ in.}^4$$

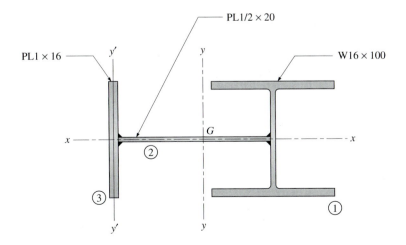

Figure X8.9.2

[continues on next page]

Example 8.9.2 continues ...

a. Section properties

The monosymmetric built-up section considered is composed of three elements: namely, the W-section and the two plates. We have:

Sectional area, $A = 29.7 + 20(\frac{1}{2}) + 1(16) = 55.7$ in.^{2}

The horizontal (*x-x*) axis of symmetry is a principal axis of the compound section. Let the center of gravity of the built-up section, *G*, located on the *x* axis, be at a distance $\bar{x}$ from the center of gravity of the vertical plate. We have:

$$\bar{x} = \frac{29.7(0.500 + 20.0 + 0.293) + 10.0(0.500 + 10.0)}{55.7} = 13.0 \text{ in.}$$

$$I_x = 1500 + \left[\frac{1}{12}(20.0)\left(\frac{1}{2}\right)^3\right] + \left[\frac{1}{12}(1)16.0^3\right] = 1840 \text{ in.}^4$$

$$I_y = [186 + 29.7(20.8 - 13.0)^2]$$
$$+ \left[\frac{1}{12}\left(\frac{1}{2}\right)(20.0^3) + 10.0(13.0 - 10.5)^2\right]$$
$$+ \left[\frac{1}{12}(16.0)(1^3) + 16.0(13.0^2)\right] = 5090 \text{ in.}^4$$

$$r_x = \sqrt{\frac{1840}{55.7}} = 5.75 \text{ in}; \quad r_y = \sqrt{\frac{5090}{55.7}} = 9.56 \text{ in.}$$

b. Design axial strength

$$K_x L_x = 24 \text{ ft}, \quad K_y L_y = 30 \text{ ft}$$

$$\frac{K_x L_x}{r_x} = \frac{24.0(12)}{5.75} = 50.1 \leftarrow \text{controls}; \quad \frac{K_y L_y}{r_y} = \frac{30.0(12)}{9.56} = 37.7$$

The design compressive stress corresponding to a *KL/r* = 50.1 and A36 steel can be obtained from Table 3-36 of the LRFDS: $\phi_c F_{cr} = 26.8$ ksi. Design axial compressive strength of the column,

$$P_d = 26.8(55.7) = 1490 \text{ kips} \qquad \text{(Ans.)}$$

c. Width-to-thickness checks

For W16×100 shape: From Table 1-1 of the LRFDM and Eq. 8.9.6*b*, we obtain the following:

$$\lambda_f = \frac{b_f}{2t_f} = 5.29 < \lambda_{rf} = 0.56\sqrt{\frac{E}{F_y}} = 0.56\sqrt{\frac{29,000}{36}} = 15.9 \text{ O.K.}$$

As the horizontal plate is welded to the midpoint of the web of the W-shape, the web acts as two stiffened elements:

$$\lambda = \frac{b}{t} = \frac{h/t_w}{2} = \frac{23.2}{2} = 11.6$$

$$\lambda_r = 1.49\sqrt{\frac{E}{F_y}} = 1.49\sqrt{\frac{29{,}000}{36}} = 42.3$$

As $\lambda < \lambda_r$ the web is not a slender element.
For the PL ½×20, acting as a stiffened element:

$$\lambda = \frac{b}{t} = \frac{20.0}{0.5} = 40.0 < \lambda_r = 1.49\sqrt{\frac{E}{F_y}} = 42.3 \text{ O.K.}$$

For the PL1×16, acting as two unstiffened elements:

$$\lambda = \frac{b}{t} = \frac{8}{1} = 8.00 < \lambda_r = 0.56\sqrt{\frac{E}{F_y}} = 15.9 \text{ O.K.}$$

As the width-to-thickness ratio checks for all plate elements are satisfied, local buckling of plate elements will not occur before the flexural buckling of the member.
Therefore, the design axial compressive strength of the member, $P_d = 1490$ kips   (Ans.)

8.10   Design of Axially Loaded Columns

From the geometry of a given structural arrangement, the length L and unsupported lengths L_x and L_y of a column to be designed are generally known. The required axial compressive strength of that column is obtained from the analysis of the structure under factored loads using the load combinations stipulated in the ASCES Section 2.3.2 (load combinations LC-1 to LC-6 of Section 4.10 in this text). Using this information, the designer must select a rolled shape or built-up section capable of carrying the required axial load.

Several criteria that influence which structural shape is chosen (HSS, W-, WT-, 2L-, etc.) are listed below:

- Most columns are used in frames in which buckling of the member in two perpendicular planes must be considered. For a column that has identical support conditions about both axes at both ends, and that has no intermediate bracing in either plane, the most efficient cross-sectional shape is a hollow circular pipe. Next in efficiency is the square HSS. Although both of these sections will have equal radii of gyration about both of the two axes about which buckling is considered, the circular pipe will require less material than the square HSS to support the same load. Next in efficiency, after the circular pipe and square HSS, is the rectangular HSS. Note that the torsional rigidities of all HSS are much greater than those of open sections having similar area and width-to-thickness ratios. Therefore, torsional buckling is not a problem for tubular shapes. A practical objection to the use of pipe and HSS is the difficulty of connecting beams to such sections. Also, the ends of pipe and tube columns should be sealed against moisture and air penetration to prevent corrosion from inside.

- For columns having identical end restraint conditions in both principal planes, I-shapes make for uneconomical columns because for such members the radius of gyration about the weak axis is significantly less than the radius of gyration about the strong axis. The added stiffness associated with the strong axis is not utilized, since buckling is controlled by the slenderness associated with the weak axis.

- Given the same cross-sectional area and end restraints, an I-shape having less web depth but greater flange width (a wide flange shape compared to an American standard beam) makes for a more efficient column.

- Often times, columns are laterally supported at one or more intermediate points in the plane in which weak axis buckling deformations would occur. This plane may be the plane of a braced frame, an exterior wall, or a *permanent* interior partition. In such cases, an I- or W-shape oriented such that its weak axis is perpendicular to the plane of bracing or wall becomes very economical.

- W-shapes of a given nominal size are divided into groups that are rolled with the same set of rollers. Because of the fixed dimensions of each set of rollers, the clear distance between the inner faces of the flanges is constant for each shape in that group, while the overall depths may vary considerably. For example, the inside dimension for each of the shapes from a W12×65 to a W12×336 is approximately 10 ⅞ in, but the overall depths vary from 12.1 to 16.8 in. By selecting shapes from one series for as many stories of a building as possible, the splice details can be simplified, resulting in overall economy.

The slenderness parameter of the column that determines the design compressive stress cannot be computed until r is known, and r is not known until the cross section has been selected. Consequently, column design is generally an iterative procedure. A recommended procedure is as follows:

1. Assume a reasonable value for the design compressive stress $(\phi_c F_{cr})$, and compute the approximate required cross-sectional area. If from the given length and load to be carried, it appears that the column will be relatively stocky, a stress near the maximum ($\approx 0.85F_y$) should be assumed. If it appears that the column will be slender, a stress close to $0.55F_y$ should be assumed.

2. Select a trial section having the area determined in Step 1. The section selected should be consistent with clearance requirements, connection feasibility (especially for splicing with columns below and above), and other practical considerations. Determine the K factors from the appropriate alignment chart, if necessary, and compute the maximum effective slenderness ratio for the selected section.

3. Determine the design axial compressive strength of the column using Table 3-36, 3-50, or Table 4 of the LRFDS.

4. If the design compressive strength of the selected section is less than the required compressive strength, select another trial section, using the design compressive stress for the first section as a guide in selecting the second.

   If the design compressive strength is within ± 1 to $\pm 2\%$ of the required compressive strength, the design would generally be considered acceptable. If P_d exceeds P_u by more than 5 percent, it might be possible to select a more economical section.

5. Check local stability (width-to-thickness ratios) of plate elements. Revise selections if necessary.

6. Repeat the above steps as many times as necessary to select a column section having a slenderness ratio that satisfies the stability requirements of the overall member and individual plate elements that satisfy the width-to-thickness ratios required to prevent local buckling.

Member selection can be expedited considerably by using the Column Selection Tables given in Part 4 of the LRFD Manual. For W-shapes and rectangular HSS, the procedure is as follows:

1. Enter the appropriate Column Load Table for W-shapes with $KL = K_y L_y$, and move horizontally from right to left until a section with $\phi_c P_n$ greater than or equal to the required strength P_u is obtained. The section tentatively selected is adequate for buckling about the minor axis.

2. Read the r_x/r_y value given in the bottom part of the table for the selected shape.

3. Calculate the equivalent effective length factor $(K_x L_x)_y$ using Eq. 8.7.13.

4. If $(K_x L_x)_y$ is greater than $K_y L_y$, the major axis buckling controls the design strength. For such cases, reenter the Column Load Table for the section tentatively selected with $KL = (K_x L_x)_y$, and move horizontally to the left if necessary, until a section with $\phi_c P_n$ greater than or equal to the required strength P_u is obtained.

5. Make all appropriate checks for the section selected.

Numerical values for design compressive stresses given in LRFDS Tables 3-36, 3-50, and 4 are based on flexural buckling and LRFDS Eqs. 2-2 and 2-3. It is assumed that plate local buckling does not precede member buckling. All W-shapes in the Column Load Tables satisfy both the flange and web width-to-thickness ratio limits. However, for some other shapes (such as certain WT-, and 2L-shapes), the limits are exceeded. In those cases, the tabulated values have been calculated according to the requirements of LRFDS Appendix B, and no further reduction is needed. For these reasons, if a compression member to be analyzed can be found in the Column Load Tables, then tabular strength values should be used. The LRFD Manual also provides Column Load Tables for a variety of double-angle configurations.

EXAMPLE 8.10.1

Column Design Using LRFDS Tables

Select the lightest W12 column of A572 Grade 60 steel to support a dead load of 100 kips and a live load of 300 kips. The column is 14 ft long and $K_x = K_y = 1$.

Solution

Dead load, D = 100 kips
Live load, L = 300 kips
Required axial capacity, $P_u = 1.2(100) + 1.6(300) = 600$ kips
As $K_x = 1.0$, $K_y = 1.0$, and $L = 14$ ft, we have $K_x L_x = 14.0$ ft and $K_y L_y = 14.0$ ft, indicating that minor axis buckling controls the design for all selections.

The LRFDS requires that:

$$P_d \geq P_u \quad \rightarrow \quad \phi_c F_{cr} A_g \geq 600 \text{ kips}$$

Trial 1

A trial section could be selected by assuming a suitable value for design compressive stress, $\phi_c F_{cr}$. Usually a value between $0.55F_y$ and $0.85F_y$ may be assumed. As the column ends are pinned at both ends, about both axes, and there is no intermediate lateral bracing, the column is likely to be somewhat slender. So, we will assume a design stress of $0.6F_y$. We have:

$$\phi_c F_{cr} = 0.6F_y = 0.6(60) = 36.0 \text{ ksi}$$

$$\text{Required area, } A_g \geq \frac{600}{36.0} = 16.7 \text{ in.}^2$$

So, from Table 1-1 in Part 1 of the LRFDM, select a W12×58 with $A = 17.0$ in.^{2} > 16.7 in.^{2} and $r_y = 2.51$ in., resulting in:

$$\frac{KL}{r} = \frac{K_y L_y}{r_y} = \frac{14.0(12)}{2.51} = 66.9$$

The slenderness parameter of the A572 Grade 60 steel column ($F_y = 60$ ksi) is:

$$\lambda_c = (KL/r) \div \left[\pi \sqrt{\frac{E}{F_y}} \right]$$

$$= 66.9 \div \left[\pi \sqrt{\frac{29,000}{60}} \right] = 0.969 < 1.5$$

The design compressive stress ratio, corresponding to a slenderness parameter λ_c of 0.969, could be obtained from Table 4 of the LRFDS, as: $\frac{\phi F_{cr}}{F_y} = 0.573$. Thus,

Design compressive stress, $\phi_c F_{cr} = 0.573(60) = 34.4$ ksi

Design strength of column, $P_d = \phi_c F_{cr} A_g = 34.4(17.0) = 585$ kips < 600 kips

As the design strength provided by the W12×58 is less than the required strength, this section is inadequate.

Revised value for required gross area, $A_g \geq \frac{600}{34.4} = 17.4$ in.^{2}

Trial 2

Select the next heavier W12-shape, namely, W12×65 with $A = 19.1$ in.^{2} and $r_y = 3.02$ in. Proceeding as before, we obtain:

$$\frac{KL}{r} = 55.6; \quad \lambda_c = 0.805; \quad \frac{\phi_c F_{cr}}{F_y} = 0.646; \quad \phi_c F_{cr} = 38.8 \text{ ksi}$$

Design axial strength of the column,

$$P_d = 38.8(19.1) = 741 \text{ kips} > 600 \text{ kips} \quad \text{O.K.}$$

Check for Local Buckling

$$\lambda_f = \frac{b_f}{2t_f} = 9.92 < \lambda_{rf} = 0.56 \sqrt{\frac{E}{F_y}} = 0.56 \sqrt{\frac{29,000}{60}} = 12.3 \quad \text{O.K.}$$

$$\lambda_w = \frac{h}{t_w} = 24.9 < \lambda_{rw} = 1.49 \sqrt{\frac{E}{F_y}} = 1.49 \sqrt{\frac{29,000}{60}} = 32.8 \quad \text{O.K.}$$

So, select a W12×65 of A572 Grade 60 steel.   (Ans.)

Lightest W10

Select a suitable W10-shape for the interior column located at the ground level of the building structure shown in Fig. 5.1.1 for the factored axial load determined in Example 5.2.4. Use A992 steel. The building is braced in both directions.

EXAMPLE 8.10.2

[*continues on next page*]

Example 8.10.2 continues ...

Solution

From Example 5.2.4, the factored load on the column, P_u = 335 kips.
Length of column, L = 12 ft
As the structure is adequately braced in both perpendicular directions using braced bents in the end walls, and as there are no intermediate braces provided, $L_x = L_y$ = 12.0 ft.
Assume that the column base is hinged about both axes.
Since the beam-to-column connections are simple and the continuity of the columns at floor levels (restraint provided by adjacent lightly loaded columns) is neglected, $K_x = K_y$ = 1.0.
So, $K_x L_x = K_y L_y$ = 12.0 ft.
As the effective lengths are the same about both axes, it follows that minor axis buckling will determine the design strength of the column.
Entering the Column Load Table for W10-shapes (LRFDM Table 4-2) with a value for $KL = K_y L_y$ = 12 ft, we observe that a W10×45 provides a design strength P_d of 388 kips > P_{req} = 335 kips.
So, select a W10×45 shape of A992 steel for the interior column. (Ans.)

EXAMPLE 8.10.3

Column Design Using Column Load Tables

Select a suitable W14 column 24 ft in length to support a factored axial load of 850 kips in the interior of an industrial building. The column base is rigidly fixed to a rigid pile foundation, and the top of the column is rigidly framed to very stiff trusses. Sidesway in the *yy* plane of the column cross section is permitted. However, bracing is provided to prevent sidesway in the *xx* plane as shown in (Fig. X8.10.3). Assume A588 Grade 50 steel.

Solution

Factored axial load, P_u = 850 kips
Length, L = 24.0 ft = $L_x = L_y$
For buckling about the major axis: bottom fixed, top rotationally fixed, sidesway permitted.
For buckling about the minor axis: bottom fixed, top rotationally fixed, sidesway prevented.
From Table C-C2.1 of the LRFDC (or from Fig. 8.5.4), the recommended values for the effective length factors are: K_x = 1.20 and K_y = 0.65, resulting in:

$$K_x L_x = 1.2(24.0) = 28.8 \text{ ft}; \quad K_y L_y = 0.65(24.0) = 15.6 \text{ ft}$$

Entering the Column Load Table (LRFDM Table 4-2) for W14 series with a value for $KL = K_y L_y$ = 15.6 ft and F_y = 50 ksi, we observe that

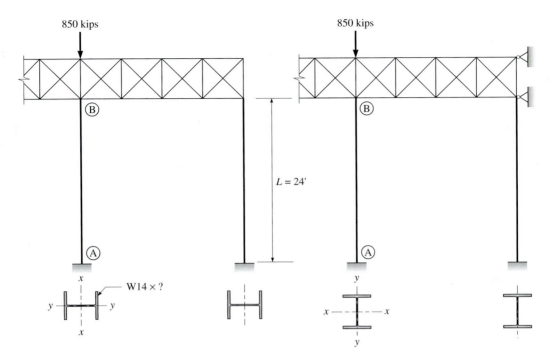

Figure X8.10.3

a W14×90 provides a minor axis design strength P_{dy} of about 934 kips and has a r_x/r_y value of 1.66.
Equivalent effective length for major axis buckling is:

$$(K_x L_x)_y = 28.8/1.66 = 17.3 \text{ ft}$$

Since $(K_x L_x)_y = 17.3 \text{ ft} > K_y L_y = 15.6 \text{ ft}$, major axis buckling controls. Reentering the Column Load Table with an effective length $KL = (K_x L_x)_y = 17.3$ ft, we observe that the W14×90 column has a major axis buckling strength, P_{dx}, of 892 kips. Hence:

$$P_d = \min [P_{dx}, P_{dy}] = \min [892, 934]$$

$$= 892 \text{ kips} > P_{req} = 850 \text{ kips} \text{ O.K.}$$

So, use a W14×90 column of A588 Grade 50 steel.   (Ans.)

Lightest Column Shape

EXAMPLE 8.10.4

Select the lightest wide flange column section of A992 steel for a factored axial load of 690 kips. The 14 ft long column is part of a frame unbraced about both of its principal axes, such that $K_x L_x = 36.0$ ft and $K_y L_y = 15.0$ ft.

[continues on next page]

Example 8.10.4 continues ...

Solution

Required strength, P_{req} = 690 kips

$K_x L_x$ = 36.0 ft; $K_y L_y$ = 15.0 ft; F_y = 50 ksi

a. Lightest W10

Entering the Column Load Table for W10-Shapes (LRFDM Table 4-2) with $KL = K_y L_y$ = 15.0 ft and P_{req} = 690 kips, we observe that a W10×88 has a minor axis strength, P_{dy}, of 782 kips. As r_x/r_y = 1.73 for this shape, we obtain:

$$(K_x L_x)_y = \frac{36.0}{1.73} = 20.8 \text{ ft} > K_y L_y = 15.0 \text{ ft}$$

Reentering the Column Load Table for W10-Shapes (LRFDM Table 4-2) with $KL = (K_x L_x)_y$ = 20.8 ft, we observe that a W10×112 with $P_d = P_{dx}$ = 742 kips is the lightest W10-shape that provides the required strength of 690 kips for buckling about both the x and y axes.

b. Lightest W12

From the Column Load Table for W12-Shapes, we observe that W12-shapes of weight comparable to, or slightly below, the W10×112 section just selected have r_x/r_y ratios of about 1.75. Consequently, $(K_x L_x)_y = \frac{36.0}{1.75}$ = 20.6 ft.

Entering the Column Load Table for W12 series (LRFDM Table 4-2), we observe that for $KL = (K_x L_x)_y$ = 20.6 ft, a W12×96 provides a design strength of 751 kips.

c. Lightest W14

Entering the Column Load Table for W14-Shapes with $KL = K_y L_y$ = 15.0 ft, we observe that the design compressive strength, P_{dy} = 694 kips for a W14×82. As r_x/r_y = 2.44, $(K_x L_x)_y$ = 36/2.44 = 14.8 ft $< K_y L_y$ = 15.0 ft, indicating that the major axis buckling does not control the column strength. So, $P_d = P_{dy}$ = 697 kips > 690 kips.

d. Lightest W

To summarize, we have:

W10×112:	P_d = 742 kips
W12×96:	P_d = 752 kips
W14×82:	P_d = 694 kips

So, try the lightest section, W14×82, for which:

$$A = 24.0 \text{ in.}^2; \quad r_x = 6.05 \text{ in.}; \quad r_y = 2.48 \text{ in.}$$

$$\frac{K_x L_x}{r_x} = \frac{36.0(12)}{6.05} = 71.4; \quad \frac{K_y L_y}{r_y} = \frac{15.0(12)}{2.48} = 72.6 \leftarrow \text{controls}$$

From Table 3-50 of the LRFDS corresponding to a $KL/r = 72.6$ design compressive stress, $\phi_c F_{cr} = 28.9$ ksi. Design strength, $P_d = 28.9(24.0) = 694$ kips $> P_{req} = 690$ kips, which confirms the results provided by using the Column Load Table.

e. Width-to-thickness ratio checks of plate elements

$$\lambda_f = \frac{b_f}{2t_f} = 5.92 < \lambda_{rf} = 0.56 \sqrt{\frac{29,000}{50}} = 13.5 \quad \text{O.K.}$$

$$\lambda_w = \frac{h}{t_w} = 22.4 < \lambda_{rw} = 1.49 \sqrt{\frac{29,000}{50}} = 35.9 \quad \text{O.K.}$$

We see that W14×82 of A992 steel is the lightest satisfactory section.

(Ans.)

Frame with Leaning Columns

EXAMPLE 8.10.5

ABCD is a rigid jointed, hinged base steel frame with adjacent leaning columns EF and GH connected to it as shown in Fig. X8.10.5. Also shown are the factored gravity loads acting on the frame. Sidesway is possible in the plane of the frame. However, the columns are braced top and bottom against sidesway out of the plane of the frame. In addition, the exterior columns are provided with bracing at mid-height. Select suitable W14-shapes of A992 steel for the columns.

Solution

a. Design of exterior column EF
   Factored load, $P_u = 500$ kips
   As the column is braced against sidesway out of the plane of the frame and is provided with bracing at mid-height, $K_y L_y = 1.0(8.0) = 8.00$ ft.

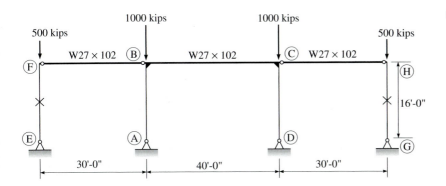

Figure X8.10.5

[continues on next page]

Example 8.10.5 continues ...

In the plane of the frame, the leaning column is designed as a pin-ended column. Using an effective length factor $K_x = 1.0$, we have:

$$K_x L_x = 1.0(16.0) = 16.0 \text{ ft}$$

Entering Column Load Tables for W-Shapes (LRFDM Table 4-2) with $KL = K_y L_y = 8.00$ ft, we observe that a W14×53 has a $P_{dy} = 552$ kips, which is greater than the required strength of 500 kips. Table 4-2 also gives $r_x/r_y = 3.07$.

$$(K_x L_x)_y = \frac{16.0}{3.07} = 5.20 \text{ ft} < K_y L_y = 8.00 \text{ ft}$$

indicating that major axis buckling will not control. Hence, $P_d = P_{dy} = 552$ kips.

Select a W14×53 of A992 steel for the two exterior columns.

b. Design of interior column AB

The two interior columns must brace the entire frame.

(i) Yura's method:

For out-of-plane buckling the column has to be designed for an axial load of 1000 kips. As the column is braced top and bottom against sidesway out of the plane of the frame, we have:

$$K_y = 1.0; \quad K_y L_y = 16.0; \quad P_{uy} = 1000 \text{ kips}$$

For in-plane buckling the column has to carry the load applied directly to it and its share of the load applied to the leaning columns. We obtain:

$$P_{ux} = 1000 + 500 = 1500 \text{ kips}$$

The K_x factor to use for the design of columns AB and CD will depend on the rigidity of these columns. Try a W14×132 ($A = 38.8$ in.²; $I_x = 1530$ in.⁴; $r_x = 6.28$ in.).

$$G_B = \frac{1530/16}{3620/40} = 1.06; \quad G_A = 10 \quad \text{(hinged base)}$$

From the alignment chart for unbraced frames (Fig. 8.5.6b), or from Eq. 8.5.16: $K_x \approx 1.92$.

The effective slenderness ratio of the column for in-plane buckling is:

$$\frac{K_x L_x}{r_x} = \frac{1.92(16.0)(12)}{6.28} = 58.7$$

From Table 3-50 of the LRFDS, we obtain, $\phi_c F_{cr} = 33.1$ ksi. So,

$$P_{dx} = 33.1(38.8) = 1280 \text{ kips} < P_{ux} = 1500 \text{ kips}$$

The section is therefore unsatisfactory when elastic G factors are used to determine the effective length factor for the column.

Required area: $A_{\text{req}} \geq \dfrac{1500}{33.1} = 45.3$ in.^{2}

Select a W14×159 ($A = 46.7$ in.^{2}; $I_x = 1900$ in.^{4}; $r_x = 6.38$ in.). By repeating the calculations we observe that this section provides a design strength of 1520 kips. O.K.   (Ans.)

(ii) LeMessurier's method

In this approach, the actual load $P_u = 1000$ kips is used for the design of the stabilizing column AB, but the effective length factor K_o must be modified to a value K_m given by Eq. 8.5.30 to take into account the presence of leaning columns in the frame.

Assume W14×132.

The effective length factor K_x (now called K_o) is calculated in part i above to be 1.92.

Note that there are two stabilizing columns and two leaning columns and that the loads and the structure are symmetric. Equation 8.5.30 with the help of Eqs. 8.5.1 and 8.4.19 gives:

$$K_{m,AB} = \sqrt{\frac{P_{E,AB}}{P_{u,AB}} \frac{(P_{u,AB} + Q_{u,EF})}{P_{eo,AB}}} = K_{o,AB} \sqrt{\frac{(1000 + 500)}{1000}}$$

$$= 1.92 \sqrt{1.5} = 2.35$$

$$\frac{K_{mx} L_x}{r_x} = \frac{2.35(16.0)(12)}{6.28} = 71.8$$

From Table 3-50 of the LRFDS, $\phi_c F_{cr} = 29.2$ ksi, resulting in:

$$P_{dx} = 29.2(38.8) = 1130 \text{ kips} > P_{\text{req}} = 1000 \text{ kips} \quad \text{O.K.}$$

$$\frac{K_y L_y}{r_y} = \frac{1.0(16.0)(12)}{3.76} = 51.1$$

From Table 3-50 of the LRFDS, $F_{dc} = 35.1$ ksi, resulting in:
$P_d = 35.1(38.8) = 1360$ kips $> P_{\text{req}} = 1000$ kips   O.K.
So, select a W14×132 of A992 steel for the interior columns AB and CD.   (Ans.)

References

8.1   ASCE [1991]: *Effective Length and Notional Load Approaches for Assessing Frame Stability: Implications for American Steel Design*, Task Committee on Effective Length, American Society of Civil Engineers, Reston, VA.

8.2 Bjorhovde, R. [1972]: Deterministic and Probabilistic Approaches to the Strength of Steel Columns, Ph.D. dissertation, Lehigh University, Bethlehem, PA, May.

8.3 Bjorhovde, R. [1984]: "Effect of End Restraint on Column Strength—Practical Applications," *Engineering Journal,* AISC, vol. 22, no. 1, pp. 1–13.

8.4 Bleich, F. [1952]: *Buckling Strength of Metal Structures,* Engineering Societies Monographs, McGraw-Hill, New York, NY.

8.5 Chen, W. F., and Lui, E. M. [1987]: *Structural Stability Theory and Implementation,* Elsevier.

8.6 Dumonteil, P. [1992]: "Simple Equations for Effective Length Factors," *Engineering Journal,* AISC, vol. 29, no. 3, pp. 111–115.

8.7 Euler, L. [1744]: *De Curvis Elasticis,* M. M. Bousquet, Lausanne and Geneva, pp. 267–268.

8.8 Galambos, T. V. [1960]: "Influence of Partial Base Fixity on Frame Stability," *Journal of the Structural Division,* ASCE, vol. 86, no. 5, pp. 85–108, May.

8.9 Galambos, T. V. [1968]: *Structural Members and Frames,* Prentice Hall, Englewood Cliffs, NJ.

8.10 Galambos, T. V., Ed. [1998]: *Guide to Stability Design Criteria for Metal Structures,* 4th ed. Structural Stability Research Council, John Wiley & Sons, New York, NY.

8.11 Geschwindner, L. F. [1995]: "A Practical Approach to the Leaning Column," *Engineering Journal,* AISC, vol. 32, no. 2, pp. 63–72.

8.12 Hall, D. H. [1981]: "Proposed Steel Column Design Criteria," *Journal of the Structural Division,* ASCE, vol. 107, no. ST4, pp. 649–670.

8.13 Julian, O. G., and Lawrence, L. S. [1959]: Notes on J and L Nomograms for Determination of Effective Lengths, Unpublished.

8.14 Kavanagh, T. C. [1962]: "Effective Length of Framed Columns," *Transactions of the ASCE,* vol. 127, pp. 81–101.

8.15 LeMessurier, W. J. [1976]: "A Practical Method of Second Order Analysis, Part 1–Pin Jointed Systems," *Engineering Journal,* AISC, vol. 13, no. 4, pp. 89–96.

8.16 LeMessurier, W. J. [1977]: "A Practical Method of Second-Order Analysis, Part 2—Rigid Frames," *Engineering Journal,* AISC, vol. 14, no. 2, April, pp. 49–67.

8.17 Levy, M., and Salvadori, M. [1992]: *Why Buildings Fall Down,* W. W. Norton & Company, New York.

8.18 Le-Wu Lu [1965]: "Effective Length of Columns in Gable Frames," *Engineering Journal,* AISC, vol. 2, no. 1, pp. 6–7, January.

8.19 Shanley, F. R. [1947]: "Inelastic Column Theory," *Journal of Aeronautical Sciences,* vol. 14, no. 5, pp. 261–267, May.

8.20 Timoshenko, S. P., and Gere, J. M. [1961]: *Theory of Elastic Stability,* 2nd ed., Engineering Societies Monographs, McGraw-Hill, New York, NY.

8.21 Vinnakota, S., and Badoux, J. C. [1970]: "Elastic Buckling of Rectangular Frames" (in French), *Bulletin Technique de la Suisse Romande,* Lausanne, no. 23, pp. 335–348, November.

8.22 Yura, J. A. [1971]: "The Effective Length of Columns in Unbraced Frames," *Engineering Journal,* AISC, vol. 8, no. 2, pp. 37–42.

8.23 Yura, J. A. [1988]: *Elements for Teaching Load & Resistance Factor Design,* AISC, Chicago, IL.

▌P R O B L E M S

Analysis

P8.1. Plot the LRFDS column curve $\phi F_{cr}/F_y$ vs. λ_c. Show all the salient points.

P8.2. Use the LRFD Specification and plot the design axial compressive stress ϕF_{cr} vs. effective slenderness ratio KL/r for steel columns. Include the following steels.

 (a) A36   (b) A992
 (c) A572 Grade 65   (d) A514 Grade 90

P8.3 and P8.4. Determine the elastic flexural buckling stress and elastic flexural buckling load for the pin-ended columns using the Euler equation. Assume $E = 29,000$ ksi.

P8.3. A solid square bar 2.0 in. by 2.0 in.

 (a) $L = 6.0$ ft   (b) $L = 12$ ft

P8.4. A W12×96

 (a) $L = 50$ ft   (b) $L = 25$ ft

P8.5. Determine the effective lengths of each of the columns of the frame shown in Fig. P8.5. The members are oriented so that the webs are in the plane of the frame. The structure is unbraced in the plane of the frame. All connections are rigid, unless indicated otherwise. In the direction perpendicular to the frame, the frame is braced at the joints. The connections at these points of bracing are simple connections (no rotational restraint).

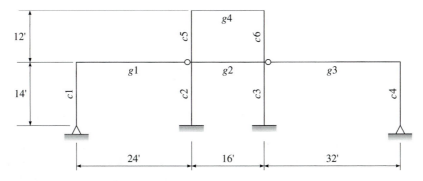

P8.5

(a) Use the alignment charts.
(b) Use the equations given in Section 8.5.

The sections used for different members are listed below:

c.1: W10×39; c2, c3, c5, c6: W10×49;
    c4: W10×45
g1: W16×40; g2: W14×34; g3: W16×50;
    g4: W14×30

P8.6. Repeat Problem P8.5, assuming the structure is braced in the plane of the frame also, by diagonal members in the middle bay.

P8.7. Determine the factored axial compressive strength of an 18-ft-long W8×48 column fixed about both axes at its ends. Assume A242 Grade 50 steel. Solve using specification equations only. Check calculations by using various tables in the LRFD Manual.

P8.8. A W12×65 column 30 ft long is fixed at both ends about both axes. In addition, it is braced in the weak direction at the one-third points. A242 Grade 42 steel is used. Determine the design axial compressive strength of the column.

P8.9. A S10×35 standard I-shape is used as a column in an industrial building. The column considered is 21 ft long. It can be assumed to be hinged about both axes at the top and fixed about both axes at the base. In addition, transverse braces (girts) are connected to the web of the column. What factored axial load is permitted by the LRFDS, if the bracing is positioned at:

(a) points 7 ft apart   (b) points 5 ft 3 in apart?

P8.10. A W12×72 of A588 Grade 50 steel is used as a column 16 ft long. The column ends are pinned, and the weak axis is braced 10 ft from the lower end. The structure is braced in both the xx and yy planes of the column. Determine the design axial compressive strength of the column.

P8.11. A W12×58 steel shape, strengthened by welding two $\frac{3}{8}$ × 8 in. plates to the outside face of the flanges, is used as a column. The member has a length of 15 ft and is assumed to be pin-connected at both ends. It is braced at mid-length to prevent movement in the x-direction only. Determine the factored axial compressive strength of the column if A572 Grade 60 steel is used.

P8.12. A 20-ft-long, A572 Grade 42 steel column is obtained by welding a $\frac{1}{2}$ × 12 in. plate to two MC13×50 channel shapes, to form a doubly symmetric section, as shown in Fig. P8.12. Determine the factored axial compressive strength of the column as per the LRFDS. Assume $K_x = 1.0$ and $K_y = 1.4$.

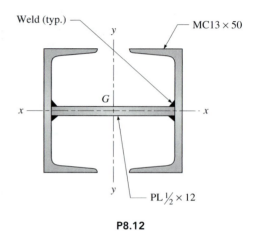

P8.12

P8.13. The cross-section of a structural steel column is a builtup section obtained by welding the web of a W12×26 shape to the flange of a W8×48 to form a monosymmetric section shown in Fig. P8.13. The 20-ft-long column is fixed

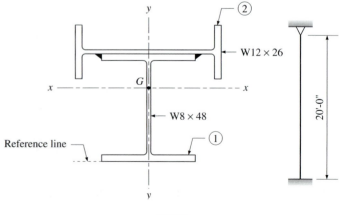

P8.13

at the base and pinned at the top about both axes. The column is part of a braced frame in both the *xx* and *yy* planes. Assume A572 Grade 65 steel. Determine the design axial compressive strength of the column.

P8.14. A cantilever column (fixed at the base and free at the top, about both axes) is obtained by welding two WT9 ×53's to the web of a W24×76 section to form a doubly-symmetric compound section shown in Fig. P8.14. The column is 20 ft long. Assume A572 Grade 60 steel and determine the design axial compressive strength of the column, as per the LRFD Specification.

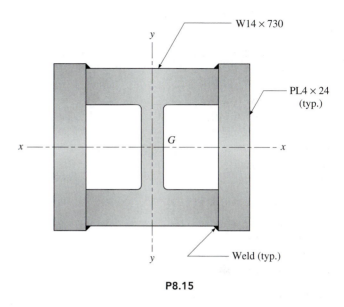

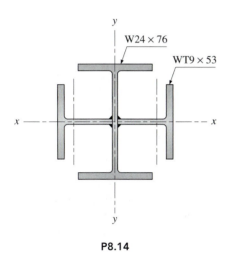

P8.14

P8.15. A heavy column used in the ground floor of a highrise building is formed by welding two 4×24 in. plates to the flanges of a W14×730 section as shown in Fig. P8.15. Assume A572 Grade 42 steel and determine the design axial compressive strength of the column, if $K_x L_x =$ 32 ft and $K_y L_y = 28$ ft. Use the LRFD Specification.

P8.16. Determine the design axial compressive strength of the columns considered in the Problem P8.5, assuming A992 steel is used.

P8.17. Determine the design axial compressive strength of the columns considered in the Problem P8.6, assuming A992 steel is used.

Design

P8.18. An eight-story office building has a bay size of 40 ft by 36 ft. The structure is designed for a roof dead load of 80 psf, a roof live load of 20 psf, a floor dead load of 80 psf, and a floor live load of 50 psf. Make a preliminary design, and select a suitable W14-shape of A992 steel for an interior column at ground level. Assume that $K_x = K_y = 1.0$ and $L_x = L_y = 14$ ft.

P8.19. Select the lightest W-section for a column supporting a factored axial load of 620 kips. Support conditions, applicable for both principal axes, are full fixity at the bottom and a pinned connection at the top. $L_x = L_y =$ 17 ft 6 in. Assume A992 steel is used.

P8.20. The interior column of an industrial building supports a roof dead load of 55 psf and a snow load of 35 psf. The column is 25 feet long. The bottom end is fixed, and the top end pinned. In addition, its weak axis is braced by struts at a point 6 feet down from the column top as shown in Figure P8.20. The column has a tributary area of 1800 ft². The structure is braced in both the *xx* and *yy* planes of the column. Select a suitable W-shape of A588 Grade 50 steel.

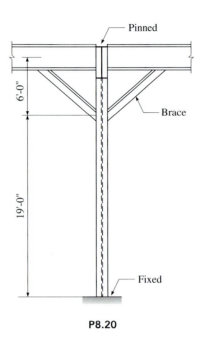

P8.20

P8.21. Select the lightest W12 of A588 Grade 50 steel to carry an axial compressive load of 500 kips dead load and 800 kips live load. The 30-ft-long trussmember is assumed to be pinned at both ends about both axes. Make all checks. Repeat the problem if, in addition, the member is provided with weak direction support at mid-length.

P8.22. Select the lightest W14 of A572 Grade 50 steel to carry an axial compression load of 100 kips dead load and 360 kips live load. The 28-ft-long column is pinned at both ends about both axes and in addition has weak direction support at mid-height. Make all checks.

P8.23. Design the lightest W-shape of A992 steel to support a factored axial compression load of 960 kips. The effective length with respect to its minor axis is 14 ft, and the effective length with respect to its major axis is 30 ft.

P8.24. A column is built into a wall so that it may be considered as continuously braced in the weak direction. The column is 18 ft long and may be considered hinged at both ends with respect to strong axis buckling. The factored axial load to be carried is 700 kips. Use A992 steel. Select the lightest W-shape used as columns.

P8.25. Repeat Problem P8.24, if the selection is extended to W-shapes up to 18-in. nominal depth.

P8.26. A steel column is built into a wall so that it may be considered as continuously braced for buckling about its y-axis. If the factored load on the column is 1000 kips and $K_x L_x = 22$ ft, select the lightest W-section for the column. The depth of the section provided is to be limited to 18 in. (nominal).

P8.27. Select the lightest W12-shape of A588 Grade 50 steel to carry an axial compressive force of 600 kips under factored loads. Assume the member to be part of a braced frame in both planes. The idealized support conditions are that the member is hinged in both principal directions at the top of a 30 ft height, supported in the weak direction at 12 ft and 22 ft from the bottom, and fixed in both directions at the base.

P8.28. The roof structure of an industrial building is supported by A992 steel columns, located between the roof beam and floor girder as shown in Fig. P8.28. The columns are connected in such a manner that the ends may be assumed pinned against rotation about the strong (x) axis, and fixed with respect to the weak (y) axis. If the length of the column is 24 ft and the factored axial load is 242 kips, find the lightest W10-shape.

P8.29.   (*a*)   Find the most economical W12-section to carry an 800 kip factored axial compressive load. The 22-ft-long column is pinned at both ends about both axes. In addition, it is braced at mid-height. Use A992 steel and the LRFD Specifications.

      (*b*)   After the column sections were ordered, the architect of the project decided not to provide the lateral bracing at mid-height of the column as assumed in (*a*). The fabricator strengthened the shapes received, by welding two $\frac{3}{8} \times 14$ in. plates of A572 Gr 50 steel in stock. Which of the two arrangements shown in Fig. P8.29 did he use? Substantiate your results.

P8.30. An interior column of a building is to support a factored axial load of 640 kips. It is 22 ft long and is to be of

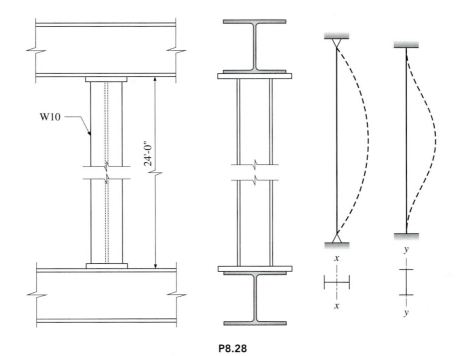

P8.28

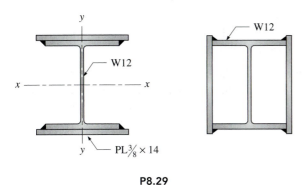

P8.29

A992 steel. The column base is rigidly fixed to the footing, and the top of the column is rigidly connected to stiff trusses. Assume that bracing is provided to the structure to prevent sidesway in the *xx* (strong) plane of the column, but the sidesway in the *yy* (weak) plane of the column is not prevented. Select the lightest column shape.

P8.31. Select the lightest W-shape of A588 Grade 50 steel for a 18-ft-long column to support an axial dead load of 290 kips and a live load of 280 kips. The column is assumed to be hinged at the base, while the top of the column is rigidly framed to very stiff girders. Assume that the structure is braced to prevent sidesway in the *xx* plane of the column section, while sidesway in the *yy* plane is uninhibited.

P8.32. Design a 12-ft-long W12 interior column of a building structure shown in Fig. P8.32, to support a factored axial load of 820 kips. The column is part of an unbraced, rigid jointed frame in the plane of its web. W27×94 girders, 28 ft long, are rigidly connected to each flange, at the top and bottom of the column under consideration. Assume the same sections are used for columns just above and below. The column considered is braced normal to its web at the top and bottom so that sidesway is prevented in that plane. Assume A992 steel.

P8.33. Select a suitable W12-shape for the columns of a fixed-base, rigid-jointed, unbraced, portal frame shown in Fig. P8.33, using A992 steel and the LRFD Specification. The horizontal girder is a W18×76. The webs of

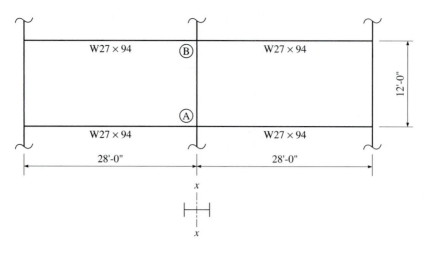

P8.32

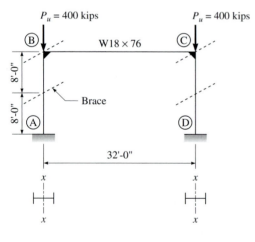

P8.33

these rolled sections lie in the plane of the frame. In the plane perpendicular to the frame, bracing (girts) are provided at top and mid-height of columns using simple flexible beam-to-column connections. The factored axial load on the columns is 400 kips. (Hint: As a starting point assume that I_x of the column section lies between 300 and 500 in.⁴).

P8.34. The column AB of a two-bay, single-story, rigid-jointed steel frame is subjected to a factored axial compressive load of 1930 kips (Fig. P8.34). The W30×90

girders are 30 ft long, and the columns are 16 ft long. The bases are hinged. Sidesway is possible in the plane of the frame. In the perpendicular direction $K_y = 0.8$. Assume A242 Grade 46 steel and check the adequacy of the W14×193. Include inelastic behavior of the column.

P8.35. Check the adequacy of a W12×152 for column AB of the rigid-jointed steel frame shown in Fig. P8.35. The columns are 12 ft long, and the columns above and below AB may be assumed to be approximately of the same size as AB. The adjacent girders are all of W18×50 and 32 ft long. The webs of all columns and girders shown lie in the plane of the paper. For buckling in the perpendicular plane, take $K = 0.9$. All members are of A992 steel. Take inelastic behavior of columns into account. The factored axial load in the member is 1540 kips.

P8.36. A W14×193 column is part of a rigid-jointed, unbraced frame in both xx and yy planes. Two W24×84 girders, 36 ft long, are rigidly connected to the flanges, while two W18×60 beams, 18 ft long, are rigidly connected to the web (Fig. P8.36). A992 steel is used for all members. The column in the tier above is also a W14×193, while the column in the tier below is a W14×211. The story height is 12 ft 6 in. A particular load combination results in a factored axial compressive load of 2000 kips in the column under consideration. Is the section safe according to the LRFDS?

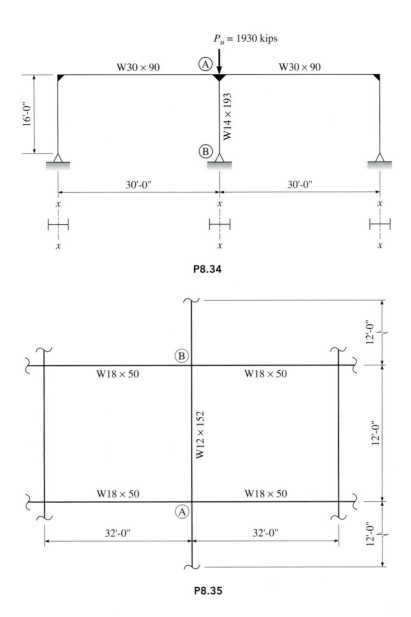

$P_u = 1930$ kips

W30 × 90   Ⓐ   W30 × 90

W14 × 193

16'-0"

Ⓑ

30'-0"   30'-0"

x   x   x
x   x   x

P8.34

12'-0"

Ⓑ
W18 × 50   W18 × 50

W12 × 152

12'-0"

W18 × 50   W18 × 50
Ⓐ

32'-0"   32'-0"

12'-0"

P8.35

P8.37. Select suitable W12-shapes for columns AB and CD of the axially loaded, unbraced frame with leaning column shown in Fig. P8.37. The girder BC is rigidly connected to the left column but has only a simple connection to the right column. The columns are braced top and bottom against sidesway out of the plane of the frame. In addition, the columns are braced at midheight by girts. Assume A992 steel. The loads given are factored loads.

P8.38. Design the columns AB and CD of the axially loaded, unbraced frame with leaning columns shown in Fig. P8.38. The girders are rigidly connected to the exterior column, while the other connections are simple. The columns are braced top and bottom against sidesway out of the plane of the frame. In addition, the exterior columns are braced at midheight by girts. Assume A992 steel. Use W12-shapes. The loads given are factored loads.

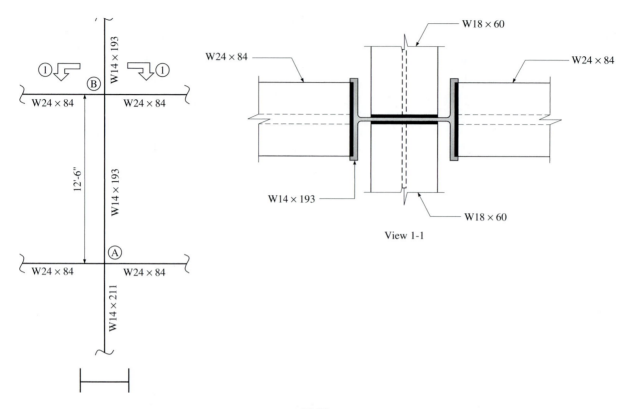

P8.36

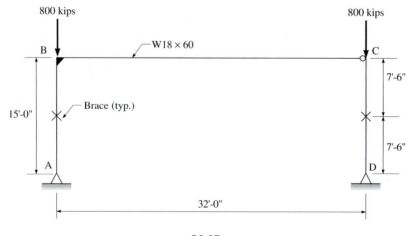

P8.37

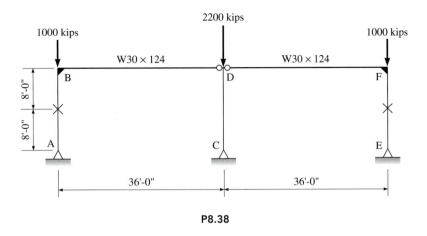

P8.38

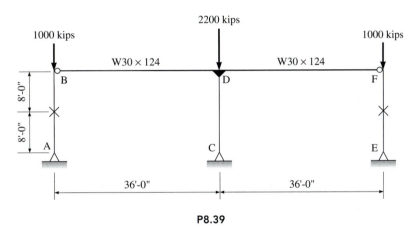

P8.39

P8.39. Design the columns AB and CD of the axially loaded, unbraced frame with leaning columns shown in Fig. P8.39. The girders are rigidly connected to the interior columns, while the connections to the exterior columns are simple. The columns are braced top and bottom against side-sway out of the plane of the frame. In addition, the exterior columns are braced at midheight by girts. Assume A992 steel. Use W12-shapes. The loads given are factored loads.

Super Foods Warehouse, Flint, Michigan.

910 tons of structural steel and steel joists were used for the framing of this 250,000 sq ft warehouse.
Photo courtesy Douglass Steel Fabricating Corp., Lansing MI, www.douglassteel.com

C H A P T E R 9

Adequately Braced Compact Beams

9.1 Introduction

A **beam** is a structural member designed to carry loads applied transversely to its longitudinal axis, and to transfer these loads to designated points on the beam called **supports**. The supports for a beam may consist of bearing walls, columns, or other beams (sometimes referred to as **girders**) which the beam frames. Beams considered here are long, straight members having a constant cross-sectional area. They are primarily subjected to bending, which is usually accompanied by shear. Less frequently, they are also subjected to torsion.

9.1.1 Classification of Beams

Beams may be classified or grouped in various ways based on how they are supported, their location, the function they serve in a building, and their physical properties such as web slenderness, and so on, as further described below.

Based on the position, kind, or number of supports in the plane of bending, a beam may be classified as one of several types:

- A **simply supported beam** (or **simple beam**) is hinged at one end and roller-supported at the other end (Fig. 9.1.1a).
- A **cantilever beam** is fixed (or built-in or clamped) at one end and free at the other end (Fig. 9.1.1b).
- An **overhanging beam** rests on two supports in such a way that it extends freely beyond the support at one or both ends (Fig. 9.1.1c).
- A **propped cantilever** is fixed (or built-in) at one end and roller supported at the other end (Fig. 9.1.1d).
- A **fixed-ended beam** has both ends fixed against rotation (Fig. 9.1.1e).
- A **continuous beam** is supported on three or more supports (Fig. 9.1.1f).
- A **restrained beam** is partially fixed at one or both ends.

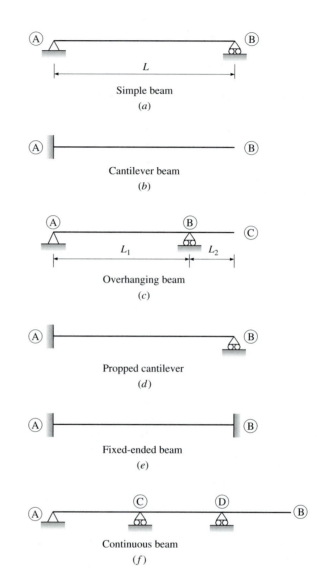

Figure 9.1.1: Classification of beams.

From a structural analysis standpoint, simple, cantilever, and overhanging beams are statically determinate, while propped cantilevers, fixed-ended, and continuous beams are examples of indeterminate beams.

Based on the function served and/or location in a building, beams may also be known by other names.
Thus:

- A *floor beam* directly supports a floor deck or a floor slab (Fig. 3.3.1).
- A *roof beam* directly supports a roof deck or a roof slab.

- A *joist* is a closely spaced beam that directly supports the roofing or flooring (metal deck or concrete slab) of a building.
- A *girder* is generally a large-sized beam in a structure that supports smaller beams. Girders usually have relatively wide spacing (Fig. 3.3.2).
- A *stringer* is an inclined beam that supports stair steps.
- A *header* is a beam framed at right angles to two other (usually parallel) beams that supports joists or floor beams on one side only. Headers are used to frame openings in the floor, such as those required for stairwells or elevator shafts. Note that torsion may be a significant consideration when designing headers.
- A *trimmer* is one of the beams that supports a header.
- A *spandrel beam,* in the exterior wall of a building, supports the weight of the wall from floor to floor, in addition to its share of the loads from the adjacent floor (Fig. 3.3.1).
- A *purlin* is a roof beam that spans adjacent trusses or frames of an industrial building (Fig. 3.4.1).
- A *girt* is a horizontal wall beam that spans the wall columns of an industrial building. Girts transfer the wind load on the wall system to the columns and, sometimes, also the weight of the wall system (Fig. 3.4.1).
- A *rafter* is a roof beam that supports purlins or joists (Fig. 3.2.1*d*), and which is usually sloping.
- A *lintel* spans above a window, door, or other wall opening and supports the wall immediately above it.
- A *transfer beam* is a horizontal framing member commonly used in tall buildings. In such structures, it is often desirable (perhaps for architectural or commercial reasons) to alter the plan arrangement of some of the lower story columns to obtain more spacious, column-free floor areas. To achieve this, heavy beams (i.e., transfer beams) are used to support the typical floor columns and transfer their loads to the larger but less numerous columns below. The transfer beam is usually very large, sometimes a full story in depth.

Traditionally, rolled steel I-shapes used as flexural members are called *beams,* while I-shapes that are built-up from plate elements are called *plate girders.* However, the LRFD Specification classifies I-shaped flexural members as *beams* or *plate girders* based on the value of the web slenderness ratio h/t_w, where the dimension h is the depth of the web between the toes of the flange fillets and t_w, the web thickness [Yura et al., 1978]. Thus, regardless of whether it is a rolled I-shape or built-up I-shape, the flexural member is treated as a *beam,* if:

$$\lambda_w \leq \lambda_{rw} \tag{9.1.1}$$

where

$$\lambda_{rw} = 5.70\sqrt{\frac{E}{F_y}} \tag{9.1.2}$$

with

λ_w = slenderness of the web plate (h/t_w)
λ_{rw} = limiting web slenderness separating beams and plate girders
h = height of the web
   = distance between the toes of the flange fillets for rolled I-shapes (see Fig. 9.1.2a)
   = clear distance between flanges less the fillet at each flange for a built-up I-shape (see Fig. 9.1.2f)
t_w = thickness of the web
F_y = yield stress of the web material

Also, regardless of whether it is a rolled I-shape or built-up I-shape, the flexural member is treated as a **_plate girder_**, if:

$$\lambda_w > \lambda_{rw} \tag{9.1.3}$$

For steels with $F_y = 50$ ksi, λ_{rw} equals 137. From values of h/t_w for rolled I-shapes given in LRFDM Tables 1-1, 1-2 and 1-3, observe that all of the standard hot-rolled I-shapes tabulated in Part 1 of the LRFDM satisfy Eq. 9.1.1; that is, when used as flexural members they are considered beams (and not as plate girders).

9.1.2 Beam Cross Sections

The most commonly used rolled beams are I-shapes, which are doubly symmetric sections. These shapes are highly efficient when loaded in the plane of the web and supported laterally. The advantage of these shapes is that a large percentage of the metal is placed in the flanges where it is highly stressed and has a large effective lever arm. Three types of I-beams are currently rolled: wide flange (W-) shapes (Fig. 9.1.2a), American standard (S-) shapes (Fig. 9.1.2b), and miscellaneous (M-) shapes. W-shapes range in depth from 4 to 44 in., while S-shapes range in depth from 3 to 24 in., and M-shapes range in depth from 4 to 12 in. Unlike S-shapes that have flanges of varying thickness, W-shapes have flanges of uniform thickness. Also, as their name implies, W-shapes have wider flanges than S-shapes. The wider flange width of W-shapes, as compared to S-shapes, results in beams with greater lateral stability. Miscellaneous sections are of the same shape as W-sections but are lighter in weight for the same depth. Rectangular HSS are doubly symmetric beam sections that result in torsionally stiff, laterally stable beams (Fig. 9.1.2c). They are recommended for long spans and design situations involving incomplete lateral support. Tees (Fig. 9.1.2d) are rarely used as beams, since they are relatively inefficient in bending. Channels (Fig. 9.1.2e) are used to carry light loads, and are often used as purlins, girts, lintels, trimmers, and headers.

When the loads to be carried exceed the capacity of available rolled beams, or when more efficient sections to carry lighter loads over longer

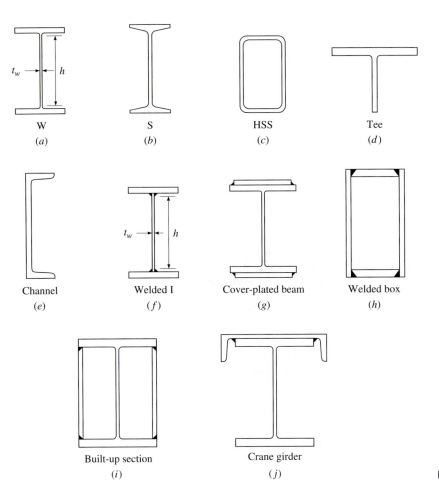

Figure 9.1.2: Beam cross sections.

spans are desired, built-up members are often used instead of rolled shapes (Figs. 9.1.2*f, g, h, i,* and *j*). In the prefabricated metal building industry, it is common practice to weld three plates together to form efficient, I-shaped beam sections (Fig. 9.1.2*f*). The strength and stiffness of an available rolled S- or W-shape may be augmented by welding plates, known as ***cover plates,*** to the flanges (Fig. 9.1.2*g*). Four plates welded together to form a torsionally stiff box shape are used for heavily loaded, laterally unsupported beams such as crane girders (Fig. 9.1.2*h*). As mentioned earlier, a plate girder is an I-section with a slender web and is often built up by welding three plates together.

Cross-sectional shapes having at least one axis of symmetry are often chosen for steel beams. Furthermore, the beam is usually positioned so that all loads and reactions lie in a plane of symmetry. This results in bending with shear, but precludes torsion from occurring. The member is then said to be in ***simple bending,*** and all deformations occur in that plane of symmetry. ***Pure bending*** is obtained when the beam is loaded in a plane of symmetry by equal end couples. Even though pure bending is not a common loading case for real structures, it is often used as a reference loading case for discussions and examples, as it often produces the most severe loading on a beam. Figure 9.1.3 shows a simply supported beam of a W-shape under a set of concentrated loads. Also shown are the shear force diagram, bending moment diagram, and the deflected shape of the beam. It is assumed that the reader is familiar with the construction of such shear force and bending moment diagrams for beams shown in Fig. 9.1.1 for different loads. The problem of beam design consists mainly in providing adequate bending and

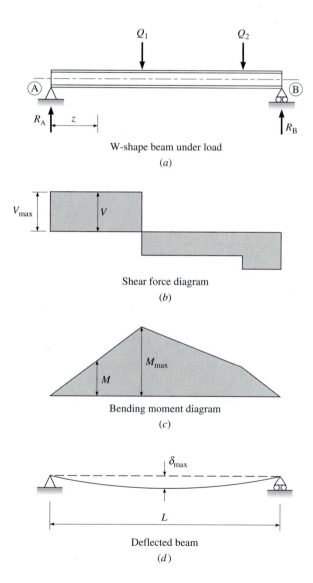

W-shape beam under load

(a)

Shear force diagram

(b)

Bending moment diagram

(c)

Deflected beam

(d)

Figure 9.1.3: Simply supported beam.

shear strengths, and in assuring that the deflections are less than certain allowable values. Behavior and design of beams subject to uniaxial flexure, without torsion and axial forces, is treated in Chapter F of the LRFD Specification. Behavior and design of such beams is covered in Chapters 9 and 10 of this text, while biaxially bent beams are considered in Section 11.16. Distribution of flexural stresses in elements subjected to pure bending will be considered in Section 9.2, while distribution of shear stresses in beam elements will be considered in Section 9.3. Plastic analysis of beams will be introduced in Section 9.4. Section 9.5 gives a brief introduction to various buckling phenomenon in beams, indicating in particular various means to prevent such buckling from reducing the bending and shear strength of the members. Serviceability limit states for beams will be considered in Section 9.6. Design of adequately braced compact section beams will be presented in Section 9.7. Open web steel joists and joist girders are considered in Section 9.8. Beam bearing plates are considered in Section 9.9, while column base plates are studied in Section 9.10.

9.2   Yield Moment, M_y, and Plastic Moment, M_p

In this section, the relations between bending moments and bending deformations are obtained for a steel beam element of unit length subjected to bending moments M, as shown in Fig. 9.2.1a. In deriving these relations, the following assumptions will be made.

1. The moments are applied in a plane of symmetry of the cross section.
2. There are no axial loads acting on the element.
3. Transverse sections of the element, originally plane and perpendicular to the longitudinal axis of the element before loading, remain plane and perpendicular after bending has commenced.
4. The stress-strain diagram for steel is idealized by two straight lines as shown in Fig. 9.2.1b. Up to the yield stress, F_y, the material is linearly

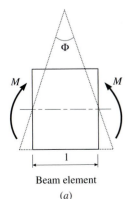

Beam element

(a)

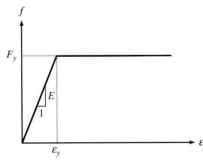

Idealized stress-strain diagram for steel

(b)

Figure 9.2.1: Unit-length beam element under uniform moment.

elastic. After the yield stress has been reached, the strain increases greatly without any further increase of stress.

5. The material in the beam has the same stress-strain characteristics in compression as in tension.

The beam elements considered in this section are a rectangular section, an I-section, and a monosymmetric section.

Rectangular Section

A rectangular steel beam element, of unit length and subjected to bending moments of magnitude M, is considered first (Fig. 9.2.2). Figures 9.2.2a, b, c, and d show the strain and stress distributions at four different stages, as the bending moment M is increased progressively from zero. It is assumed that each fiber behaves as shown in Fig. 9.2.1b and that the Bernoulli hypothesis, namely, plane sections before bending remain plane after bending, is valid for all values of M. For the applied bending moments shown, the top fibers will be in compression and the bottom fibers in tension. The *curvature, Φ,* is the angle change between the original and deformed positions of the end sections of the unit-length element. As long as the member behaves elastically, from mechanics of materials [Timoshenko and Young, 1962]:

$$\Phi = \frac{M}{EI} = \frac{\varepsilon}{y} \tag{9.2.1}$$

From the stress-strain diagram of Fig. 9.2.1b, if the maximum strain in the cross section is less than the yield strain, ε_y, the stress diagram is also linear as shown in Fig. 9.2.2a. From mechanics:

$$M = S f_{max} \tag{9.2.2}$$

$$\Phi = \frac{f_{max}}{Ec} \tag{9.2.3}$$

where S, denotes the elastic section modulus; c, the extreme fiber distance (d/2); and f_{max}, the extreme fiber stress. As the moment is increased, the strains and stresses likewise increase, and their distribution over the depth, at least initially, remains linear. When the maximum strain just equals ε_y, the extreme fiber stress f_{max} just equals F_y, and the stress diagram will continue to have a linear elastic distribution, as shown in Fig. 9.2.2b. The corresponding moment, M_y, for which the extreme fibers of the section just plastify, is known as the **yield moment** of the cross section. Using the flexure formula (Eq. 9.2.2), the yield moment may be written as:

$$M_y = SF_y \tag{9.2.4}$$

The corresponding curvature is given by:

$$\Phi_y = \frac{F_y}{Ec} \tag{9.2.5}$$

For the rectangular section considered here, M_y can be determined with the help of Fig. 9.2.2b. As there are no axial loads acting on the section, the

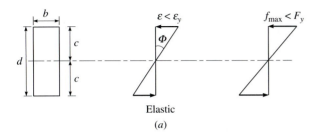

Elastic

(a)

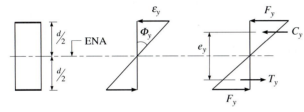

Yield moment, M_y

(b)

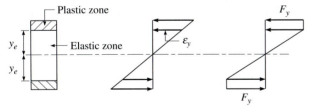

Partially plastic

(c)

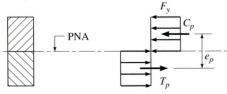

Plastic moment, M_p

(d)

Figure 9.2.2: Elastic and inelastic behavior of a rectangular section beam element.

horizontal summation of forces over the section must be zero. Hence, the resultant C_y of compressive stresses in the upper half of the cross section is equal to the resultant T_y of tensile stresses in the lower half. The value of each of these forces is equal to the product of the average stress multiplied by the area upon which the stress acts. Since the average stress in a linear stress distribution is one-half the maximum stress, we obtain:

$$|T_y| = |C_y| = \left(\frac{bd}{2}\right)\left(\frac{F_y}{2}\right) \tag{9.2.6}$$

The resisting moment of the section consists of the couple composed of the equal, oppositely directed forces C_y and T_y. The forces C_y and T_y act through the centroids of the triangular stress distributions at a distance $\bar{y}$ from the N.A. Since $\bar{y} = \frac{2}{3}c = \frac{2}{3}\left(\frac{d}{2}\right)$, the moment arm of the resisting couple is $e_y = 2\bar{y} = \frac{2}{3}d$. Equating the applied moment to the resisting moment of the section, we have:

$$M_y = |C_y|\, e_y = |T_y|\, e_y = \left(\frac{1}{6}bd^2\right)F_y = SF_y \tag{9.2.7}$$

where

$$S = bd^2/6 \tag{9.2.8}$$

is the elastic section modulus of the rectangular section.

For further increases of the moment above M_y, the extreme fiber strain exceeds ϵ_y; however, the stress cannot increase above F_y (Fig. 9.2.1b). So, the additional moment causes the outer fibers to yield or undergo additional strains, while the stresses remain at the limiting level of F_y as shown in Fig. 9.2.2c. At this stage the section is elastic over a depth $2y_e$, but plastic outside this depth. The stress is constant at F_y over a portion of the cross section called the **plastic zone** and varies linearly over the remaining portion called the **elastic zone** or **elastic core.** The process of successive yielding of the fibers as the bending moment is increased is called **plastification of the cross section.** With additional moment, this process will continue until the stress diagram approaches a nearly bi-rectangular distribution. As a convenient approximation, it is generally assumed that the bi-rectangular distribution shown in Fig. 9.2.2d represents the maximum bending strength of the cross section. Theoretically, such a stress distribution necessarily requires an infinite strain. The bending moment associated with this bi-rectangular stress distribution, wherein the entire cross section has plastified, is referred to as the **fully plastic moment,** or simply **plastic moment,** M_p, of the cross section. The moment cannot be increased any further as the axial stiffness of the plastified fibers and consequently the bending stiffness of the plastified section are zero, and the cross section will undergo large plastic rotation at the constant value of the moment, M_p. For the rectangular section considered here, M_p can be determined directly, with the help of Fig. 9.2.2d. Here, the tensile and compressive stress resultants T_p and C_p are both equal to $\left(\frac{bd}{2}F_y\right)$. As the stress distribution is uniform, the stress resultants act at the centroids of the tensile and compressive zones, respectively. So, the lever arm $e_p = \frac{d}{4} + \frac{d}{4} = \frac{d}{2}$. The resisting moment of these forces is:

$$M_p = |C_p|e_p = |T_p|e_p = \left(\frac{bd}{2}F_y\right)\frac{d}{2} = \frac{bd^2}{4}F_y = ZF_y \quad (9.2.9)$$

Here, Z is defined as the **plastic section modulus.** It is a geometrical characteristic of the cross section analogous to the elastic section modulus and represents the resistance to bending of a completely yielded cross section. Thus, for a rectangular section the plastic section modulus is given by:

$$Z = bd^2/4 \quad (9.2.10)$$

In Fig. 9.2.2, ENA represents the elastic neutral axis and PNA, the plastic neutral axis.

For the partially plastic section in Fig. 9.2.2c, the stress is constant at F_y over the plastic zone and varies linearly over the elastic core of height $2y_e$. The resisting moment of the elastic zone, as determined by the flexure formula (Eq. 9.2.2), is:

$$M_{ez} = \frac{1}{6}b(2y_e)^2 F_y = \frac{2}{3}by_e^2 F_y \quad (9.2.11)$$

For the plastic region, which is symmetrical about the neutral axis, the resisting moment is:

$$M_{pz} = 2\int_{y_e}^{c} y\,(b\,dy)F_y = 2bF_y\left[\frac{y^2}{2}\right]_{y_e}^{c} = b\left[c^2 - y_e^2\right]F_y \quad (9.2.12)$$

The total resisting moment of a partially plastified rectangular section is therefore given by:

$$M_{ep} = M_{ez} + M_{pz} = \left[\frac{2}{3}y_e^2 + c^2 - y_e^2\right]bF_y$$

$$= \frac{bd^2}{4}F_y\left[1 - \frac{1}{3}\left(\frac{y_e}{c}\right)^2\right] = M_p\left[1 - \frac{1}{3}\left(\frac{y_e}{c}\right)^2\right] \quad (9.2.13)$$

The usual elastic formula for curvature (Eq. 9.2.3) must still hold for the central elastic core of depth $2y_e$ so that the curvature for the elastic-plastic section can be written as:

$$\Phi = \frac{F_y}{Ey_e} \quad (9.2.14)$$

From Eqs. 9.2.5, 9.2.13, and 9.2.14, we obtain:

$$\frac{\Phi}{\Phi_y} = \frac{c}{y_e}; \quad \frac{M_{ep}}{M_p} = \left[1 - \frac{1}{3}\left(\frac{\Phi_y}{\Phi}\right)^2\right] \text{ for } \frac{\Phi}{\Phi_y} \geq 1.0 \quad (9.2.15)$$

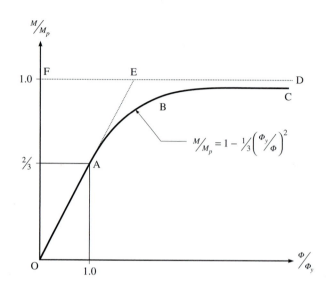

Figure 9.2.3: Moment-rotation response of a rectangular section beam element.

The curve OABC in Fig. 9.2.3 shows the moment-curvature relation for a rectangular shape of mild steel. It is seen from Eq. 9.2.15 that the bending moment will asymptotically approach the value M_p as Φ tends to infinity but will not attain it for any finite curvature. For example, when the extreme fiber strain is $2\varepsilon_y$, $\Phi/\Phi_y = 2$ and $M_{ep} = 0.917M_p$; and when the extreme fiber strain is $6\varepsilon_y$, $\Phi/\Phi_y = 6$ and $M_{ep} = 0.991M_p$. Thus, although the assumed bi-rectangular stress distribution corresponding to M_p cannot physically exist (as it requires infinite curvature), it represents a practical limiting value with negligible error. As structural steels exhibit a certain amount of strain harden-ing, the plastic moment can be attained in tests and even exceeded at a finite curvature. Thus, for structural steel shapes, it is certainly reasonable to assume that the plastic moment can actually be attained. As a matter of standard con-vention, the **plastic moment** is taken to be the maximum attainable moment possible; that is, the benefits of strain hardening are neglected.

I-Section
The behavior of an I-section beam element bent about its major axis is shown by the bending moment-curvature (M vs. Φ) diagrams shown in Fig. 9.2.4. The solid line curve pertains to a wide flange beam that is free of residual stresses. As the moment is gradually increased from zero, the beam passes through four stages of behavior. Stage I represents linear elastic behavior; Stage II represents the domain in which yielding begins at the outer fibers of the section and progresses through the flanges; Stage III represents the domain in which yielding starts at the outer fibers of the web and progresses through the web; and Stage IV represents the state in which the entire sec-tion has yielded. Zones of yielding and stress distributions over the beam

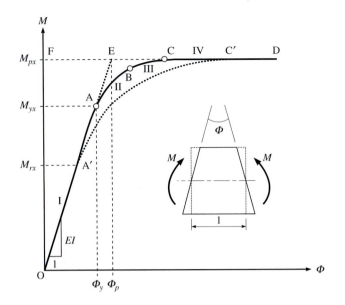

Stage	I	II	III	IV
Plastic zones				
Stress distribution				

Figure 9.2.4: Moment-rotation response of an I-section beam element.

cross section during each of the stages are also shown in this figure. Note that the neutral axis, which is the transverse axis on which bending stresses are zero, remains at the midheight of the section in all stages of loading due to the symmetry of the section. The bending moment that causes first yielding, M_{yx}, is the moment at the end of Stage I and for which Φ_y is the corresponding curvature. The constant moment during Stage IV is the plastic moment, M_{px}. In actual beams, the presence of residual stresses will contribute to yielding, and as indicated by the dashed line in the figure, inelastic behavior will occur at a **reduced yield moment** M_{rx} before the yield moment M_{yx} is reached. However, the residual stresses will not reduce the plastic bending moment of the section; the presence of residual stresses does not reduce the ultimate capacity of the section, which remains at M_{px}. Thus, we have for an I-shape bent about its major axis:

$$M_{rx} = S_x (F_y - f_{rc}) \tag{9.2.16}$$

$$M_{yx} = S_x F_y \tag{9.2.17}$$

$$M_{px} = Z_x F_y \tag{9.2.18}$$

where f_{rc} is the residual compressive stress in the flange.

In **plastic analysis** of beams and frames (discussed later in Section 9.4), the actual nonlinear moment-curvature diagram of a beam element (such as the curve OA'C'D of the I-shape shown in Fig. 9.2.4) is often replaced by the bilinear relationship OAED wherein the beam is assumed to behave linearly-elastically with a constant flexural rigidity EI until the fully plastic moment of magnitude M_{px} is reached; thereafter, the beam rotates plastically under a constant moment. Such a relationship between moment and curvature is said to be **elastic-perfectly plastic.** This corresponds to the behavior of an ideal beam section for which (1) the web is sufficiently strong to carry all necessary shear and maintain a fixed distance between the flanges, but has no bending strength and (2) the flanges are sufficiently thin that the strain, and hence the stress, within the flange may be considered a constant. For such an idealized section, yield moment coincides with the plastic moment, and the shape factor (to be defined later in this section) equals one. An alternative idealization of the moment-curvature relationship of a cross section is given by OFD (Fig. 9.2.4). In this case, no change of curvature occurs for moments less than M_{px}, whereas indefinitely large curvatures are possible once M_{px} is reached. This latter type of moment-curvature relationship is said to be **rigid-perfectly plastic.**

Monosymmetric Sections

Up to this point, only doubly symmetric shapes, that is, sections that possess two axes of symmetry, were considered. When such sections are bent in the plane of the web, the neutral axis passes through the centroid for all states of stress (elastic, elastic-plastic, and fully plastic).

For beams that are monosymmetrical in section, such as the T-beam bent within its web-plane shown in Fig. 9.2.5a, the location of the neutral axis depends upon the state of stress in the member. For these sections, the neutral

Figure 9.2.5: Shift of neutral axis with plastification of a monosymmetric section.

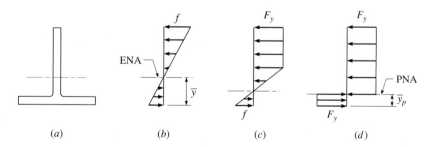

(a) (b) (c) (d)

axis begins to shift with the moment once the extreme outer fibers become plastified (see Figs. 9.2.5b, c, and d). In Fig. 9.2.5, ENA represents the elastic neutral axis and PNA, the plastic neutral axis.

The fully plastic moment of the monosymmetric section shown in Fig. 9.2.6 may be determined as follows. In the fully plastic stage the stress is constant at F_y over the tensile and compressive plastic zones. The condition that the resultant axial force on the section be zero, which must be satisfied to establish equilibrium, may be written with the help of Fig. 9.2.6 as:

$$\Sigma(F)_z = 0 \;\rightarrow\; T_p - C_p = 0 \;\rightarrow\; A_t F_y - A_c F_y = 0$$
$$\rightarrow A_t = A_c \qquad (9.2.19)$$

where
T_p = resultant of tensile stresses
C_p = resultant of compressive stresses
A_t = area of the tensile plastic zone
A_c = area of the compressive plastic zone

This relation shows that the **plastic neutral axis** (PNA) divides the cross section into two equal areas. Thus, for monosymmetric cross sections under plastic moment, the neutral axis does not pass through the centroid of the section, but divides the cross section into two equal areas. The plastic moment M_p equals the sum of the moments of the stresses in the section. Thus:

$$M_p = T_p \bar{y}_t + C_p \bar{y}_c = T_p e_p = (A_t \bar{y}_t + A_c|\bar{y}_c|)F_y = ZF_y$$
$$(9.2.20)$$

where
$\bar{y}_t$ = distance of the centroid of A_t from the PNA
$\bar{y}_c$ = distance of the centroid of A_c from the PNA

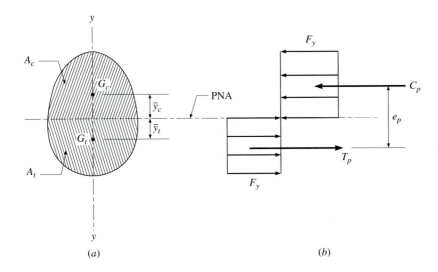

(a)   (b)

Figure 9.2.6: Plastic moment of a monosymmetric section.

e_p = lever arm between the forces C_p and $T_p = \bar{y}_t + \bar{y}_c$

M_p = plastic moment of the section

Z = plastic section modulus of the section

Shape Factor

The ratio of the plastic moment, M_p, to the yield moment, M_y, is a geometric characteristic of the cross section and is referred to as the **shape factor, α,** and its value is always greater than one. That is:

$$\alpha = \frac{M_p}{M_y} = \frac{ZF_y}{SF_y} = \frac{Z}{S} \tag{9.2.21}$$

The value of the **shape factor** is an indication of the extra moment capacity, beyond first yielding, that a cross section of a ductile material can develop due to plastic redistribution of stresses in a cross section. Shape factors of most rolled I-shapes bent about their strong axis vary between 1.09 and 1.20. The most frequent value for all wide flange shapes is 1.12. Thus, the plastic moment of wide flange sections is approximately 10 percent greater than the moment at first yield. Other shape factor values are 1.5 for a rectangular section; 1.70 for a solid circular section; 2.00 for a diamond-shaped section; 1.27 for a thin walled circular tube (thickness, $t <<$ diameter, d); and approximately 1.20 for a thin-walled rectangular tube ($b = d/2$) bent about its major axis. For an I-shape bent about its minor axis, the value is slightly greater than 1.5 since in this orientation, the section essentially consists of two rectangles separated by a distance.

EXAMPLE 9.2.1

Normal Stress Distribution

Compare the bending stress distribution and bending force distribution in a W24×104 with that in a $1\frac{1}{4}''$ by 24″ rectangular section subjected to a bending moment of 3100 in.-kips. Assume A572 Grade 50 steel.

Solution

a. PL1¼×24 rectangular section (Fig. X9.2.1a)

$$d = 24.0 \text{ in.}; \qquad t = 1.25 \text{ in.}; \qquad c = 12.0 \text{ in.}$$

$$A = 30.0 \text{ in.}^2; \qquad I = \frac{1}{12}(1.25)(24.0^3) = 1440 \text{ in.}^4$$

In the case of the rectangular section, the extreme fiber stress will be:

$$f_{\max} = \frac{Mc}{I} = \frac{3100(12.0)}{1440} = 25.8 \text{ ksi} \qquad \text{(Ans.)}$$

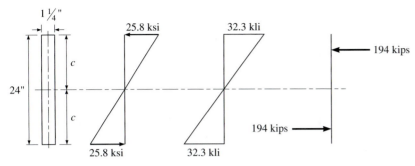

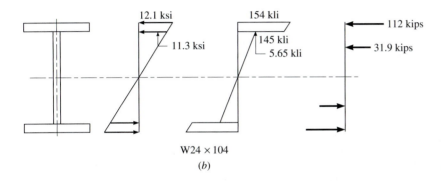

Figure X9.2.1

The force carried by the fibers per in. of section height, at the extreme fiber is then:

$$U = f_{max} \, b = 25.8(1.25) = 32.3 \text{ kli}$$

The compressive stress resultant is:

$$C = \frac{1}{2} U \frac{d}{2} = \frac{1}{2} (32.3)(12.0) = 194 \text{ kips}$$

Lever arm $= \frac{2}{3} (24.0) = 16.0$ in.

Resisting moment, $M = 194(16.0) = 3100$ in.-kips

b. W24×104 bent about major axis

$$\rightarrow \quad d = 24.1 \text{ in.}; \quad t_w = 0.500 \text{ in.};$$
$$b_f = 12.8 \text{ in.}; \quad t_f = 0.750 \text{ in.}$$
$$A = 30.6 \text{ in.}^2; \quad I_x = 3100 \text{ in.}^4$$

Note that both the rectangular section and the W-shape selected have essentially the same cross-sectional area.

[continues on next page]

Example 9.2.1 continues ...

In the wide flange section of Fig. X9.2.1b, the extreme fiber stress will be:

$$f_{max} = \frac{Mc}{I} = \frac{3100(12.1)}{3100} = 12.1 \text{ ksi} \qquad \text{(Ans.)}$$

The force carried by the fibers per in. of section height, at the extreme fiber, is then:

$$U_1 = f_{max} b_f = 12.1(12.8) = 155 \text{ kli}$$

At the bottom edge of the top flange, the fiber stress will be:

$$f = \frac{My}{I} = \frac{3100(11.3)}{3100} = 11.3 \text{ ksi} \qquad \text{(Ans.)}$$

The force carried by the fibers per in. of section height, at the bottom edge of the top flange:

$$U_2 = f b_f = 11.3(12.8) = 145 \text{ kli}$$

The force carried by the fibers per in. of section height at the top of the web is:

$$U_3 = f t_w = 11.3(0.5) = 5.65 \text{ kli}$$

Stress resultant in the compression flange:

$$C_f = \frac{1}{2}(U_1 + U_2) t_f = \left(\frac{155 + 145}{2}\right)(0.75) = 112 \text{ kips}$$

Stress resultant in the top half of the web:

$$C_w = \frac{1}{2} U_3 \left(\frac{1}{2} d_w\right) = \frac{5.65}{2}(11.3) = 31.9 \text{ kips}$$

Resisting moment contributed by flanges:

$$= C_f e_f = 112(23.4) = 2620 \text{ in.-kips}$$

Resisting moment contributed by web:

$$= C_w e_w = (31.9)(2)\left(\frac{2}{3}\right)(11.3) = 481 \text{ in.-kips}$$

Total resisting moment, $M = 2620 + 481 = 3100$ in.-kips

Remarks

This example shows that the flanges of an I-shape resist the major portion of any applied bending moment. Thus, in a flanged beam with a very thin web and thick flanges, the stresses in the web make a relatively small contribution to the bending resistance of the beam and thus can be ignored in some of the discussions concerning the behavior of wide flange beams. Also, the stress in the flange at any particular section can be taken as uniform over its thickness. It is, in fact, very nearly so if the flange is wide and not too thick, and the section not too shallow.

Plastic Moments of Sections   EXAMPLE 9.2.2

Determine the yield moment M_y, plastic moment M_p, plastic section modulus Z_x, and shape factor α, for (a) $1\frac{1}{4}$ by 24 in. rectangular section and (b) W24×104 section. Assume A572 Grade 50 steel in both cases.

Solution

a. Rectangular section

Elastic section modulus, $\quad S_x = \dfrac{I_x}{c} = \dfrac{1440}{12} = 120$ in.^{3}

Yield moment, $\quad M_y = S_x F_y = 120(50) = 6000$ in.-kips $\qquad$ (Ans.)
The distribution of stresses at plastic moment are as shown in Fig. X9.2.2a. We have:

$$T_p = \tfrac{1}{2}\,dt F_y = \tfrac{1}{2}(24.0)(1.25)(50) = 750 \text{ kips}$$

$$e_p = \tfrac{1}{2}d = 12.0 \text{ in.}$$

Plastic moment,

$$M_p = T_p e_p = 750(12) = 9000 \text{ in.-kips} \qquad \text{(Ans.)}$$

Plastic section modulus,

$$Z_x = \frac{M_p}{F_y} = \frac{9000}{50} = 180 \text{ in.}^3 \qquad \text{(Ans.)}$$

Shape factor,

$$\alpha = \frac{M_p}{M_y} = \frac{9000}{6000} = 1.5 \qquad \text{(Ans.)}$$

b. W24×104
   From LRFDM Table 1-1:

$\quad d = 24.1$ in.; $\qquad t_w = 0.500$ in.; $\quad A = 30.6$ in.^{2}
$\quad b_f = 12.8$ in.; $\qquad t_f = 0.750$ in.
$\quad I_x = 3100$ in.^{4}; $\qquad S_x = 258$ in.^{3}

Yield moment, $M_y = S_x F_y = 258(50) = 12,900$ in.-kips $\qquad$ (Ans.)

[continues on next page]

Example 9.2.2 continues ...

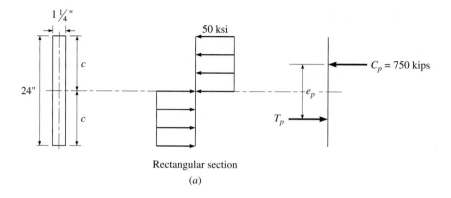

Rectangular section

(a)

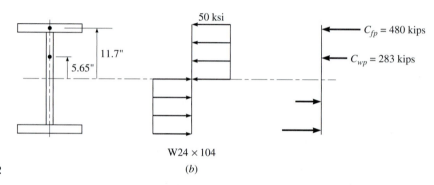

W24 × 104

(b)

Figure X9.2.2

Noting that from Example 9.2.1, the compressive stress resultant in the top flange is 112 kips, corresponding to an applied moment of 3100 in.-kips, its value at the yield moment may be evaluated as:

$$\frac{C_{fy}}{C_f} = \frac{M_y}{M} \rightarrow C_{fy} = \frac{12{,}900}{3100}(112) = 466 \text{ kips}$$

Similarly, from Example 9.2.1, the compressive stress resultant in the top half of the web is:

$$C_{wy} = \frac{12{,}900}{3100}(31.9) = 133 \text{ kips}$$

The distribution of stresses at plastic moment is as shown in Fig. X9.2.2*b*. Since this section is doubly symmetric, the PNA will remain at midheight. By letting C_{fp} be the stress resultant in the flange and C_{wp} the stress resultant in the top-half of the web, we observe:

$$C_{fp} = b_f t_f F_y = 12.8(0.750)(50) = 480 \text{ kips}$$

$$C_{wp} = \frac{1}{2} d_w t_w F_y = \frac{1}{2}[24.1 - 2(0.750)](0.500)(50) = 283 \text{ kips}$$

$$\bar{y}_{fp} = \frac{1}{2}d - \frac{1}{2}t_f = 11.7 \text{ in.}; \quad \bar{y}_{wp} = \frac{1}{4}[24.1 - 2(0.75)] = 5.65 \text{ in.}$$

Plastic moment,

$$M_p = 2(C_{fp}\,\bar{y}_{fp} + C_{wp}\,\bar{y}_{wp}) = 2(480)(11.7) + 2(283)(5.65)$$

$$= 14,400 \text{ in.-kips} \qquad \text{(Ans.)}$$

Plastic section modulus,

$$Z_x = \frac{M_p}{F_y} = \frac{14,400}{50} = 288 \text{ in.}^3 \qquad \text{(Ans.)}$$

This corresponds closely with the value of $Z_x = 289$ in.³ given in Table 1-1 of the LRFDM for a W24×104. However, the tabulated values include the influence of the fillets at the junction of the flange and web and hence are more accurate.

Shape factor of the W-shape for major axis bending is:

$$\alpha = \frac{M_p}{M_y} = \frac{14,400}{12,900} = 1.12 \qquad \text{(Ans.)}$$

Plastic Moment of a Monosymmetric Section

EXAMPLE 9.2.3

Calculate the values of S, Z, and α about the x axis for the built-up, monosymmetric I-shape shown in Fig. X9.2.3. The shape is obtained by welding flange plates (PL2×10 and PL2×6) to a web plate (PL1×18).

Solution

Total area, $A = 10.0(2.0) + 1.0(18.0) + 6.0(2.0) = 50.0$ in.²

a. Elastic properties
   To satisfy the condition $\int_A f\,dA = 0$, the elastic neutral axis must pass through the center of gravity of the cross section.

$$\bar{y} = \frac{\Sigma A_i \bar{y}_i}{\Sigma A_i} = \frac{[10.0(2.0)(1.0)] + [1.0(18.0)(11.0)] + [6.0(2.0)(21.0)]}{50.0}$$

$$= 9.40 \text{ in. below the top of the upper flange}$$

$$I_x = \Sigma \left[\frac{bh^3}{12} + Ad^2 \right]$$

$$= \left[\frac{10(2^3)}{12} + 20.0(9.40 - 1.00)^2 \right] + \left[\frac{1(18^3)}{12} + 18.0(11.0 - 9.40)^2 \right]$$

$$+ \left[\frac{6(2^3)}{12} + 12.0(21.0 - 9.40)^2 \right] = 3570 \text{ in.}^4$$

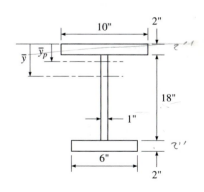

Figure X9.2.3

[continues on next page]

Example 9.2.3 continues ...

$$S_{xb} = \frac{I_x}{c_b} = \frac{3570}{12.6} = 283 \text{ in.}^3 \qquad \text{(Ans.)}$$

$$S_{xt} = \frac{I_x}{c_t} = \frac{3570}{9.40} = 380 \text{ in.}^3 \qquad \text{(Ans.)}$$

b. Plastic properties

To satisfy the condition $\int_A F_y \, dA = 0$, for the unequal flanged I cross section, the PNA divides the section into two equal areas. As $A/2 = 50/2 = 25.0 \text{ in.}^2$ is greater than the area of the top flange, it follows that the PNA will be in the web. To locate the PNA, find the distance y_p, as measured from the top of the top flange, such that the area above the PNA is equal to the area below.

$$10(2) + (1)(\bar{y}_p - 2) = (1)(20 - \bar{y}_p) + 6(2) \rightarrow \bar{y}_p = 7.00$$

Plastic section modulus,

$$Z_x = \sum A_i |\bar{y}_{pi}| = 2(10)(6.0) + 1(5.0)(2.5)$$
$$+ 1(13.0)(6.5) + 2(6)(14.0)$$
$$= 385 \text{ in.}^3 \qquad \text{(Ans.)}$$

$$\text{Shape factor, } \alpha = \frac{M_p}{M_y} = \frac{Z_x}{S_{xb}} = \frac{385}{283} = 1.36 \qquad \text{(Ans.)}$$

Note the shift of the neutral axis from the ENA to PNA, a distance of 2.4 in., due to the asymmetry of the section about the x axis.

9.3   Shear Stresses in Beam Elements

Assuming that the shear stress is uniformly distributed across the cut of width b, the horizontal shear stress, τ, in a longitudinal plane located at a distance y from the neutral axis of the section (Fig. 9.3.1) may be obtained from [Timoshenko and Young, 1962]:

$$\tau = \frac{VQ'}{Ib} = \frac{VA'\bar{y}'}{Ib} = \frac{q_{sv}}{b} \qquad (9.3.1)$$

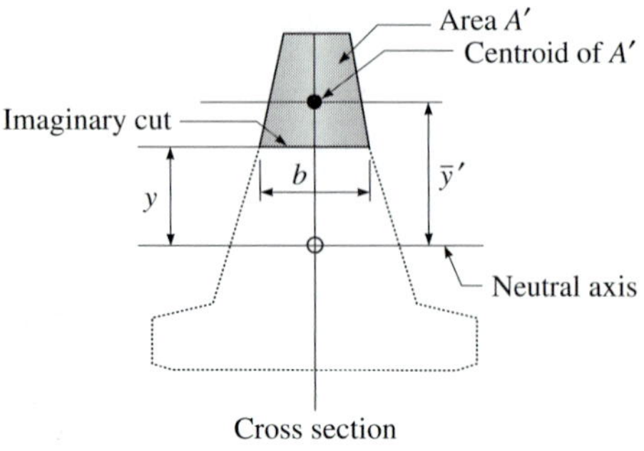

Here,

τ = shear stress at a point on a given cross section of the beam, located at a distance y from the ENA

V = shear force at the section considered

I = moment of inertia of the entire cross section with respect to neutral axis of the cross section

b = width of the beam at the point on the cross section where τ is to be calculated

A' = area of that portion of the section lying outside and beyond the line, defined by distance y, along which τ is desired

Figure 9.3.1: Parameters A' and b for calculating shear stresses in a beam.

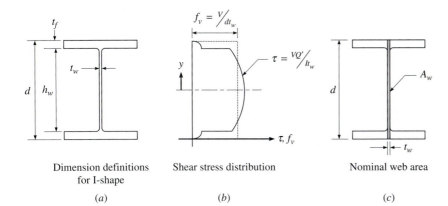

Dimension definitions
for I-shape

(*a*)

Shear stress distribution

(*b*)

Nominal web area

(*c*)

Figure 9.3.2: Shear stress distribution in an I-shape.

$\bar{y}'$ = distance from the neutral axis of the beam section to the centroid of A'

Q' = static moment, about the neutral axis of the section, of those areas of the section farther removed from the neutral axis than the point in question ($= A' \bar{y}'$)

q_{sv} = shear flow

A shear stress intensity diagram for an I-shape is shown in Fig. 9.3.2*b*. In the flange area, the value of *b* is the width of the flange, b_f, whereas for the web, the value of *b* is the thickness of the web, t_w. The maximum shear stress occurs at the neutral axis (midheight) of the beam section. For rolled steel shapes, it is easier to compute the average shear stress on the basis of a nominal web area based on the overall beam depth ($A_w = dt_w$). This is also the value used in general practice for purposes of design (Fig. 9.3.2*c*). Thus, the nominal shear stress in an I-shape for design purposes is given by:

$$f_v = \frac{V}{A_w} \tag{9.3.2}$$

Yielding of the web represents one of the shear limit states. Taking the shear yield stress as 60 percent of the tensile yield stress of the material, the nominal shear strength of a rolled steel I-shape, corresponding to the limit state of yield, is:

$$V_n = F_{yv} A_w = 0.6 F_y A_w = 0.6 F_y dt_w \tag{9.3.3}$$

This relation is valid as long as the web is stable (see Section 9.5.4).

Shear Stress Distribution

EXAMPLE 9.3.1

Compare the shear stress distribution in a W24×104 with that in a PL $1\frac{1}{4}$×24, each subjected to a shear force of 310 kips.

[continues on next page]

Example 9.3.1 continues ...

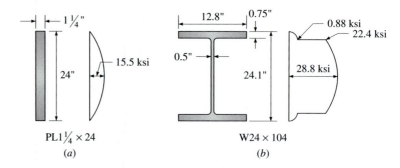

Figure X9.3.1

(a)   (b)

Solution

a. PL1¼ × 24

$$d = 24.0 \text{ in.}; \qquad t = 1.25 \text{ in.}; \qquad c = d/2 = 12.0 \text{ in.}$$
$$A = 30.0 \text{ in.}^2; \qquad I = 1440 \text{ in.}^4$$

For the extreme fibers the value of $Q' = 0$, hence $\tau = \tau_1 = 0$.
The value of Q' for a plane at a distance y from the neutral axis is:

$$Q' = (c - y)t\,\frac{(c + y)}{2} = \tfrac{1}{2}t\,(c^2 - y^2) = 90.0 - 0.625y^2$$

The shear stress at a distance y from the neutral axis is:

$$\tau = \frac{VQ'}{It} = \frac{310(90.0 - 0.625y^2)}{1440(1.25)}$$

Value of Q' is a maximum at the neutral axis and equals 90.0 in.³. Shear stress at the neutral axis is:

$$\tau_{\max} = \frac{VQ'}{It} = \frac{310(90.0)}{1440(1.25)} = 15.5 \text{ ksi} \qquad\qquad \text{(Ans.)}$$

The average intensity of shear is:

$$\tau_{\text{avg}} = \frac{V}{dt} = \frac{310}{30.0} = 10.3 \text{ ksi}$$

We have:

$$\frac{\tau_{\max}}{\tau_{\text{avg}}} = \frac{15.5}{10.3} = 1.50$$

The distribution of shear stress on a rectangular section is therefore parabolic in shape, having a maximum intensity of 3/2 of the average shear intensity.

b. W24×104

$$d = 24.1 \text{ in.}; \qquad t_w = 0.500 \text{ in.};$$
$$b_f = 12.8 \text{ in.}; \qquad t_f = 0.750 \text{ in.}$$
$$I_x = 3100 \text{ in.}^4; \qquad h_w = d - 2t_f = 22.6 \text{ in.}$$

For the extreme fibers, the value of $Q' = 0$, hence $\tau = \tau_1 = 0$.
For a horizontal plane in the flange located at a distance y from the neutral axis,

$$Q' = \tfrac{1}{2} b_f (c - y)(c + y) = \tfrac{1}{2} b_f(c^2 - y^2)$$

And the shear stress is:

$$\tau = \frac{VQ'}{Ib} = \frac{V}{Ib_f}\left[\tfrac{1}{2} b_f (c^2 - y^2)\right]$$

For a horizontal plane in the flange, located at the junction of the flange and web, we obtain:

$$Q' = \tfrac{1}{8} b_f(d^2 - h_w^2) = \tfrac{1}{8} \times 12.8 \times (24.1^2 - 22.6^2) = 112 \text{ in.}^3$$

$$\tau = \tau_2 = \frac{310(112)}{3100(12.8)} = 0.875 \text{ ksi} \qquad \text{(Ans.)}$$

For a horizontal plane in the web, located at a distance y from the neutral axis,

$$Q' = \tfrac{1}{8} b_f(d^2 - h_w^2) + \tfrac{1}{2} t_w \left(\tfrac{1}{2} h_w - y\right)\left(\tfrac{1}{2} h_w + y\right)$$
$$= \tfrac{1}{8} b_f (d^2 - h_w^2) + \tfrac{1}{2} t_w \left(\tfrac{1}{4} h_w^2 - y^2\right)$$

For a horizontal plane in the web, located at the flange-to-web junction (i.e., for $y = h_w/2$), the second term is zero and we obtain:

$$Q' = \tfrac{1}{8} b_f (d^2 - h_w^2) = 112 \text{ in.}^3$$

and the shear stress is:

$$\tau = \tau_3 = \frac{310(112)}{3100(0.5)} = 22.4 \text{ ksi} \qquad \text{(Ans.)}$$

For a horizontal plane at the neutral axis ($y = 0$), we have:

$$Q' = 112 + \tfrac{1}{2} (0.5)(11.3^2) = 144 \text{ in.}^3$$

and the shear stress is:

$$\tau = \tau_4 = \tau_{max} = \frac{310(144)}{3100(0.5)} = 28.8 \text{ ksi} \qquad \text{(Ans.)}$$

The average shear stress in the web is:

$$= \tfrac{1}{2} (\tau_3 + \tau_4) = \tfrac{1}{2} (22.4 + 28.8) = 25.6 \ \text{ksi}$$

The part of the shear load resisted by the two flanges is approximately:

$$V_f = 2(12.8)(0.750)\left(\tfrac{1}{2}\right)(0.875) = 8.40 \text{ kips}$$

while the shear resisted by the web is:

$$V_w = 25.6(0.500)(22.6) = 289 \text{ kips}$$

[continues on next page]

Example 9.3.1 continues ...

Note

1. From these results and those in Example 9.2.1, it is seen that the web of an I-shape carries the major part of the shear force at a section, while the flanges carry the major part of the bending moment at that section.
2. For design for shear, the web area A_w is defined as the overall depth d times the web thickness t_w (LRFDS Section F2-1). Thus, the nominal shear stress in the web is:

$$f_v = \frac{V}{A_w} = \frac{V}{dt_w} = \frac{310}{24.1(0.5)} = 25.7 \text{ ksi}$$

9.4   Plastic Analysis of Beams

Consider a statically indeterminate rigid-jointed, planar, steel structure of indeterminacy, I, subjected to a set of proportional loads. A set of **proportional loads** is a set in which all loads are kept in constant proportion to one another. Quite simply, proportional loading occurs when all loads are multiplied by the same (load) factor. As the loads increase, plastic hinges appear in succession at sections where the absolute value of the bending moment has a local maximum, equal to the plastic moment. If the structure does not carry any distributed loads, the only possible locations of plastic hinges are at the end sections of the members and at sections where concentrated loads are applied. Once a plastic hinge forms at a section, the magnitude of the bending moment at this section remains constant at the known value $M_p = ZF_y$, and the degree of redundancy of the structure is reduced by one. The structure therefore becomes statically determinate when the Ith plastic hinge forms. The next plastic hinge transforms this statically determinate system into a mechanism with one degree of freedom which can deform under virtually constant load. Thus, the formation of the $(I + 1)$th hinge represents the collapse of the structure. The load at which the $(I + 1)$th hinge appears is known as the **plastic limit load** or **collapse load.**

The plastic limit load is greater than the elastic limit load because complete plastification of a cross section requires more load than what is merely needed to initiate yielding at the extreme fibers. Moreover (assuming that a mechanism has not yet developed), redistribution of the bending moment within the structure occurs as each plastic hinge develops and transforms the structure into one which possesses one less degree of indeterminancy. This redistribution process is dependent upon the overall geometry and support conditions of the structure, the geometry and material properties of member cross sections, and the particular arrangement of the applied loads. Therefore, the ratio of collapse load to yield load varies from structure to structure, and for a given structure from load to load, and must be determined by performing a structural analysis, known as *plastic analysis.*

For an introduction to plastic analysis of beams, refer to Section W9.1 on the website http://www.mhhe.com/Vinnakota.

....

9.5   Introduction to Buckling Phenomenon in Beams

9.5.1   Schematics of Lateral Buckling of Beams

Consider a simply supported beam of span, L (Fig. 9.5.1a). The beam is subjected to a pair of equal and opposite end-moments M^o applied in the plane of the web (Figs. 9.5.1a and b). The beam is therefore in a state of pure bending about its major axis. For low values of M^o the beam deflects in the plane of the web as shown in Fig. 9.5.1c. As the moment is constant over length L of the beam, the resultant force of the compressive stresses in the top flange ($= C_f$) is also constant over the length L of the compression flange (Fig. 9.5.1d). As the moments M^o on the beam are increased, the flange compressive force C_f also increases. Assume that in addition to the vertical supports to the beam, the compression flange is also provided with lateral supports at end sections A and B as shown in Figs. 9.5.1a and e.

As a first approximation, the beam compression flange could be visualized as an isolated column pinned about both axes at both ends (Fig. 9.5.1f), and having a length L; rectangular cross section b_f by t_f; and yield stress F_y and subjected to an axial compressive force C_f. The horizontal and vertical axes of the rectangular section are denoted as 1-1 and 2-2, respectively (Fig. 9.5.1g). Let C_{fcr1} and C_{fcr2} be the critical axial compressive strengths of the flange column for buckling about 1-1 and 2-2 axes. The tendency of the compression flange to buckle about axis 1-1 (i.e., buckling in the plane of the beam web) at C_{fcr1} is restricted, as the compression flange is not an isolated free member as assumed in the above discussion, but is continuously supported vertically by the web of the beam, tying it to the tension flange which has no tendency to buckle. When the applied moments M° on the beam are increased to a level to produce a compressive force C_{fcr2} in the top flange, it buckles about axis 2-2, in the plane 1-1. Thus, if a long I-shaped beam with its compression flange supported at ends only is gradually loaded until it fails, it is probable that the failure will take the form of a sidewise or lateral buckling of the compression flange (Fig. 9.5.1h). This tendency of an I-beam bent about the major axis to buckle about its minor axis is known *as lateral-torsional buckling* or *lateral buckling of the beam.*

Lateral buckling of a long unsupported beam will occur before the moment in the critical section reaches the yield moment, M_y. That is, lateral buckling of the beam occurs in the elastic domain. As in the case of any other column, the buckling strength of the compression flange-column of the W-shape could be increased by providing bracing in plane 1-1. Thus, if the beam is provided with an additional lateral brace at mid-length of the beam, the compression flange-column buckles in the second mode.

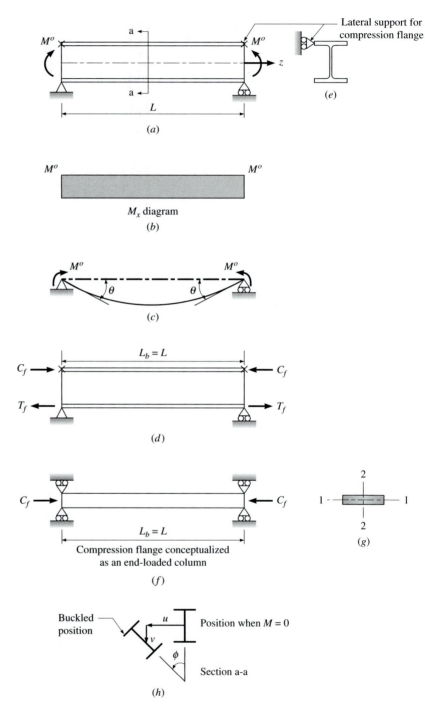

Figure 9.5.1: Lateral buckling of an I-shaped beam under uniform moment.

The compression flange of a beam acting as a column differs from the columns studied in Chapter 8 in two important respects:

* First, the compression flange is not an isolated member, but is significantly restrained by the web throughout its length.
* Second, the compression force in the flange is not, generally speaking, constant along its length; rather it varies with the bending moment. Both of these conditions are favorable, and so the stability limit load of the compression flange of a beam is considerably greater than the stability limit load of an isolated column having the same cross-sectional area.

As noted above, when an I-shaped beam is bent about its strong axis, it may buckle out of the plane of bending. The buckling deformation basically consists of a lateral movement of the compression flange. The tension flange, however, has no such tendency to move laterally, except due to its connection to the compression flange via the web. Since the tendencies of the two flanges to move laterally are not identical, twisting of the section inevitably occurs. That is, the reluctance of the tension flange to move laterally is transmitted to the compression flange via the web and results in a twisting of the cross section and the torsional resistance of the section. The torsional resistance of an I-shape is made up of two parts: (1) the torsional resistance that would be obtained under uniform torsion, also known as **pure torsion** or **St. Venant's torsion** and (2) torsional resistance due to coupled differential bending of the flanges, which induces shear forces in the flanges and thus creates a torsion couple. This is known as **warping torsional resistance.**

The compression flange-column of the beam under uniform moment discussed above, and the columns discussed in Chapter 8, have their axial loads applied entirely at the column ends, and thus experience a constant axial force along their entire length. In contrast, the compressive force in the flange of a transversely loaded beam is nonuniform along its length (Fig. 9.5.2). The value of the axial force is zero at points of zero bending moment and increases gradually as the point of maximum moment in the beam is approached. The type of transverse loading on the beam affects the type of axial load to which the flange is subjected. For example, for a simply supported beam with a single concentrated load at the center of the span, the shear is constant from one end to the center of the beam (Fig. 9.5.2b). Thus, for this condition of loading, a beam with a constant cross section will have uniform increments of axial load transferred from the web to the flange for each unit length of the beam. The amount of this increment will be the shear flow, $q_{sv} = VQ'/I$ in which V is the shear force in the beam at a section, Q' is the static moment of the compression flange about the neutral axis of the beam, and I is the moment of inertia of the beam about its neutral axis. This flow is depicted in Fig. 9.5.2c. The compression flange consequently acts as a column with compressive force varying from zero at the ends to a maximum at the middle. Now the critical axial load of a column loaded by distributed axial forces in the manner of a beam compression flange is always greater than that of a column of equal

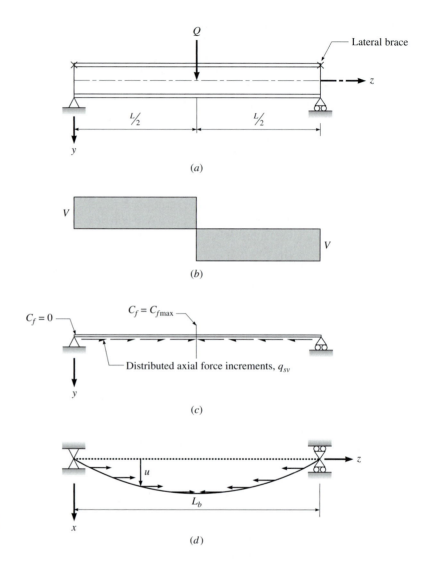

Figure 9.5.2: Compression flange-column of a simply supported beam under a central concentrated load.

dimensions and support conditions loaded entirely at its ends. The plan view of the centerline of the compression flange-column, before and after buckling, are shown in Fig. 9.5.2d.

The simple models considered here show that a beam compression flange tends to act as a column and buckle if it is not continuously supported laterally or if the distance between the adjacent lateral supports is several times the width of the compression flange. This column-like behavior of the compression flange is influenced by the end restraint conditions of the compression flange, the position and type of intermediate lateral supports to the compression flange, the torsional restraint which the section offers (through the geometric properties), the geometric and material properties of the flange itself, and the type of loading on the beam (which defines the distribution of axial load on the compression flange). When the distance between the lateral supports exceeds about nine times the flange width, for a $F_y = 50$ ksi steel W-shape bent about its major axis, its flexural strength is reduced below the plastic moment of the section due to lateral buckling of the compression flange. Determination of this reduction in strength and the phenomenon of lateral buckling of beams will be considered in detail in Chapter 10.

9.5.2   Lateral Bracing of Beams

The single most important parameter in preventing the lateral buckling of a beam is the spacing, L_b, of the lateral bracing. For the bracing to be fully effective, both the stiffness and the strength of the bracing must be adequate to prevent both twisting and lateral deflection of the beam cross section at the brace points.

Bracing for individual beams may be divided into two categories (Fig. 9.5.3):

1. *Point* or *discrete bracing* where lateral supports are provided at intervals by members such as purlins, joists, beams, or other framing element, transverse to the member being braced.
2. *Continuous bracing* where the lateral support is provided in a continuous manner by an element such as a concrete slab or a steel deck.

The beam of Fig. 9.5.3*a* is laterally supported by purlins having connections of appreciable depth and having sufficient stiffness to prevent twisting of the beam. For such a detail, lateral support can be assumed when the connection is made near the compression flange. If, instead, the connection is close to the tension flange, it is doubtful whether adequate lateral support can be achieved. Lateral bracing to the compression flange may also be provided by purlins resting on the top flange of the beam and connected to it by bolts (Fig. 9.5.3*b*), or by open web steel joists welded or bolted to the top flange (Fig. 9.5.3*c*). Positive lateral bracing is also obtained with cast-in-place concrete floor slabs, if the top flange of the beam is recessed into the bottom of the floor slab (Fig. 9.5.3*d*), or connected to it by an appropriate number of studs welded to the top flange of the beam as shown in Fig. 9.5.3*e*. Another method relies upon using an appropriate number of puddle welds and studs to connect the top flange of the beam to a steel deck and place concrete in the metal deck as shown in Fig. 9.5.3*f*.

For the purlins to be effective as lateral supports they must act to induce a point of inflection in the beam at the point of connection. Thus the designer should make certain that the purlins themselves are prevented from moving in their axial direction. For example, in Fig. 9.5.4*a*, simultaneous lateral buckling of the entire system of beams may occur, as shown by the dotted lines, as the purlins permit simultaneous movement of the connection points without offering any lateral resistance. The unbraced length L_b of the beams

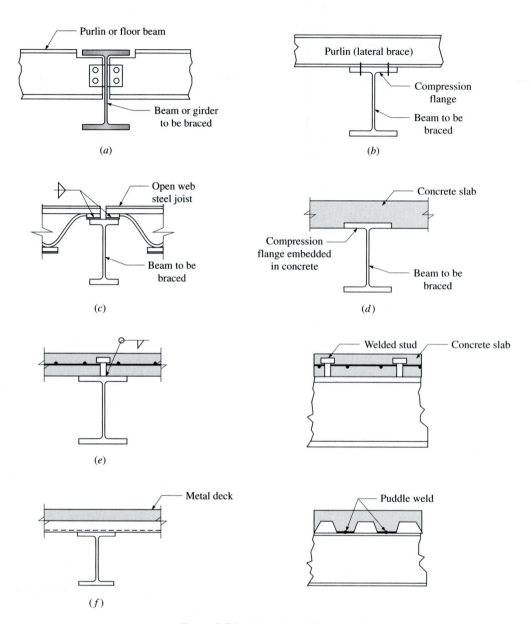

Figure 9.5.3: Examples of discrete and continuous lateral bracing of beams.

therefore equals the span length L. To prevent this type of system buckling, it is necessary to anchor one end of the purlins to a wall (Fig. 9.5.4b), or to provide diagonal bracing in one or more of the bays (Fig. 9.5.4c). The resulting horizontal truss, with its high stiffness, prevents longitudinal movement of the purlins. The diagonal bracing need not be placed in all bays, since

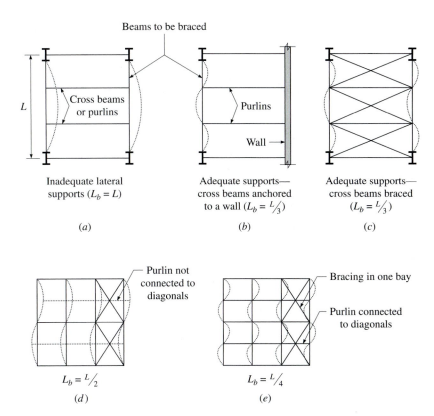

Figure 9.5.4: Lateral bracing systems for roof and floor beams.

movement in one bay requires movement in the other bays as well. Figures 9.5.4*d* and 9.5.4*e* illustrate that bracing which spans two bays can be as effective as bracing each bay individually, provided that a positive connection can be established between the intermediate purlin and bracing at their common point of intersection. The concrete slab or metal deck connected to the beams of Figs. 9.5.3*d*, *e*, and *f* could be considered as doing the job of diagonal bracing if that material were rigid enough.

Theoretically, the lateral force required to prevent buckling of a perfectly straight beam is zero. Even for beams having initial crookedness equal to the full mill tolerance, the required lateral force is still very small [Yura and Helwig, 1998]. LRFD Specification Section C3.4 gives rigorous rules to determine the strength and stiffness of lateral bracing for beams. Empirical rules have been used to determine the strength and stiffness requirements for braces for preliminary design. A typical conservative rule of thumb is to use a brace having an axial compressive strength equal to 2 percent of the factored compressive force in the compression flange being braced. When a beam supports slab construction, a typical rule of thumb is to assume the beam to be effectively restrained laterally if the connection of the slab to the beam is capable of resisting a lateral force of

2.5 percent of the maximum force in the compression flange of the beam, considered as distributed uniformly along the beam. For further details refer to Part 2 of the LRFDM.

It is sometimes difficult to judge whether the connecting members and other parts of the structure actually do provide lateral support to the compression flange of a beam. For example, metal deck roofs that are connected to the purlins with metal straps or by using puddle welds at intervals, probably furnish partial lateral support only. Such cases of partial support are usually transformed to an equivalent case of full support by using for L_b a multiple of the actual spacing of the connectors.

Lateral bracing available to the beams in the finished structure is often more than adequate to prevent lateral and torsional displacement at the points of support. However, during the intermediate stages of construction, the amount of bracing provided may be quite low, and therefore it is essential to check the adequacy of the beams under construction loads with the bracing actually available during those stages.

9.5.3 Plate Local Buckling

Classification of Beam Shapes Based on Plate Local Buckling

Since beam shapes, such as the I-shape considered in earlier sections, are composed of plate elements and some of these elements are in compression when the shape is subjected to bending, the flexural strength of the section based on its overall behavior shown in Fig. 9.2.4 can only be achieved if the plate elements do not buckle locally. For example, one half of the compression flange of a beam subjected to uniform moment acts as an unstiffened plate element under edge compression, as shown in Figs. 9.5.5a, b, and c, and will buckle as shown in Figs. 9.5.5d and e. The web plate acts as a stiffened plate element under flexural compression, and may buckle locally. Such local buckling of plate elements can cause premature failure of the entire section, or at the very least cause stresses to become nonuniform and reduce the overall strength.

The LRFD Specification classifies cross-sectional shapes as compact sections, noncompact sections, or slender-element sections (LRFDS Section B5.1):

- *Compact sections* can develop a fully plastic stress distribution (plastic moment) and possess a rotational capacity of approximately three times the yield rotation capacity associated with the onset of local buckling.
- *Noncompact sections* can develop the yield stress in compression elements before local buckling occurs, but will not resist inelastic local buckling at the strain levels required to develop a fully plastic stress distribution.
- *Slender-element sections* develop local buckling elastically before the yield stress is achieved.

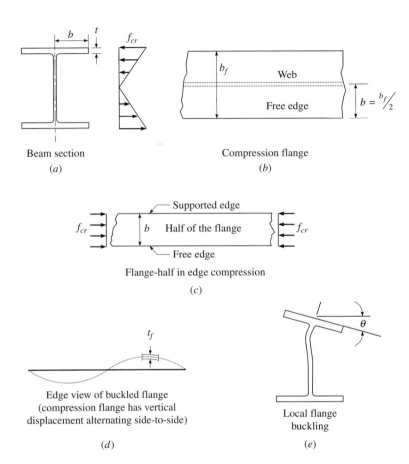

Beam section
(a)

Compression flange
(b)

Flange-half in edge compression
(c)

Edge view of buckled flange
(compression flange has vertical
displacement alternating side-to-side)

(d)

Local flange
buckling

(e)

Figure 9.5.5: Local buckling of compression flange of a rolled steel beam.

The dividing line between compact and noncompact compression elements is the limiting width-to-thickness ratio λ_p, while the dividing line between noncompact and slender compression elements is the limiting width-to-thickness ratio λ_r. For a section to qualify as a **compact section,** all of its compression elements must have width-to-thickness ratios smaller than the limiting ratios, λ_p. On the other hand, if the width-to-thickness ratio of any one compression element of a section exceeds λ_r, the section is referred to as a **slender-element section.** If the width-to-thickness ratio of one or more compression elements exceeds λ_p, but none exceeds λ_r, the section is a **noncompact section.**

Note that one-half of the compression flange of an I-shape acts as an unstiffened element under uniform edge compression, whereas the web acts as a stiffened element under flexural compression. To ensure the required hinge rotations of three or more, the width, b_f, and thickness, t_f, of the flange of a compact section must be such that the flange can compress plastically to strain hardening, ε_{st}, without buckling. An I-shape is compact, if:

$$\lambda_f \leq \lambda_{pf} \quad \text{and} \quad \lambda_w \leq \lambda_{pw} \qquad (9.5.1)$$

It is a slender-element section if:

$$\lambda_f > \lambda_{rf} \quad \text{or} \quad \lambda_w > \lambda_{rw} \qquad (9.5.2)$$

and it is a noncompact section in all other cases. From LRFDS Table B5.1, for rolled steel I-shapes:

$$\lambda_{pf} = 0.38\sqrt{\frac{E}{F_y}} ; \quad \lambda_{pw} = 3.76\sqrt{\frac{E}{F_y}} \qquad (9.5.3)$$

$$\lambda_{rf} = 0.83\sqrt{\frac{E}{F_L}} ; \quad \lambda_{rw} = 5.70\sqrt{\frac{E}{F_y}} \qquad (9.5.4)$$

where $F_L = F_y - F_r$ and F_r is the compressive residual stress in the flange, taken as 10 ksi for rolled shapes. The limits λ_{pf}, λ_{pw}, λ_{rf}, and λ_{rw} are given in Table 9.5.1 for rolled I-shapes used as beams, for different steels.

Remarks

- For all rolled I-shapes, the webs are compact for all structural steels.
- None of the rolled steel I-shapes in the LRFD Manual are slender-element shapes for any available structural steels.

TABLE 9.5.1

Limiting Width-to-Thickness Ratios for Elements of Rolled I-Shapes Used As Beams

Element	λ	F_y (ksi)	36	42	46	50	60	65
Compression flange	λ_{pf}	$0.38\sqrt{E/F_y}$	10.8	9.99	9.54	**9.15**	8.35	8.03
	λ_{rf}	$0.83\sqrt{E/F_L}$	27.7	24.9	23.6	**22.3**	20.0	19.1
Web in flexure	λ_{pw}	$3.76\sqrt{E/F_y}$	107	98.8	94.4	**90.6**	82.7	79.4
	λ_{rw}	$5.70\sqrt{E/F_y}$	162	150	143	**137**	125	120
Web in shear	λ_{pv}	$2.45\sqrt{E/F_y}$	69.5	64.4	61.5	**59.0**	53.9	51.7

λ_{pf} = Limiting b/t ratio for the flange of compact hot-rolled I-shape
λ_{pw} = Limiting b/t ratio for the web of compact hot-rolled I-shape
λ_{rf} = Limiting b/t ratio for the flange of noncompact hot-rolled I-shape
λ_{rw} = Limiting b/t ratio for the web of noncompact hot-rolled I-shape
λ_{pv} = Limiting b/t ratio for inelastic shear buckling of the web of a hot-rolled I-shape
$F_L = F_y - 10$ ksi

$$\lambda_f = \frac{b_f}{2t_f}; \quad \lambda_w = \frac{h}{t_w}$$

Note: Values for $F_y = 50$ ksi steel are shown in boldface.

- There are only five noncompact sections among $F_y = 50$ ksi steel W-shapes used as beams. They are the W21×48, W14×99, W14×90, W12×65, and W10×12. This is indicated in the Beam Tables 5-2, 5-3, 5-4, 5-6 and 5-7 of the LRFDM, with a [††].

9.5.4   Shear Buckling of Web

A simply supported rolled steel I-beam with a set of concentrated loads is shown in Fig. 9.5.6a. The I-beam is oriented so that bending is about the major principal axis. At a distance z from the left support, the section is subjected to a shear V (Figs. 9.5.6b and c). The state of stress on a small element at the neutral axis of the cross section, in a region where the bending stresses

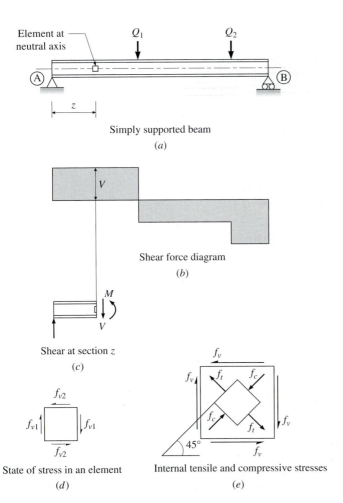

Element at neutral axis

Simply supported beam
(a)

Shear force diagram
(b)

Shear at section z
(c)

State of stress in an element
(d)

Internal tensile and compressive stresses
(e)

Figure 9.5.6: Shear buckling of the web of a rolled steel I-beam.

are zero, is shown in Fig. 9.5.6d. On the right and left sides equal unit shearing stresses f_{v1} occur, which are produced by the external shear V in the beam at the section considered. To maintain rotational equilibrium of the element, equal shearing stresses f_{v2} occur at right angles to f_{v1} and in such direction that the moment couple formed by stresses f_{v2} is equal and opposite to that by f_{v1}. Thus, when a unit shearing stress occurs in one plane, an equal unit shearing stress exists in a plane perpendicular to the first for any element of material from a web.

As illustrated by the sketch in Fig. 9.5.6e, an element subjected to uniform shear stress along all four edges develops internal tensile and compressive stresses (f_t, f_c), that are a maximum on planes at 45° to the edges and equal, in magnitude, to the edge shear stress, f_v. Thus, when the shear stresses on the element are increased by increasing the loads on the beam, the internal compressive stresses increase too. If the compressive stresses along the diagonal are large enough, the beam web will buckle forming a series of wrinkles or waves perpendicular to the direction of the compressive stress. Whether shear buckling of the web precedes the shear yielding will depend on the width-to-thickness ration (h/t_w) of the web. If this ratio is large—that is, if the web is too slender—it can buckle in shear, either inelastically or elastically. Shear buckling can be avoided by selecting a section with a relatively large web area and low slenderness. Shear buckling of the web plate does not occur before shear yielding of the web provided the web slenderness satisfies the relation (LRFDS Section F2.2):

$$\lambda_w \le \lambda_{pv} \tag{9.5.5}$$

with

$$\lambda_{pv} = 2.45 \sqrt{\frac{E}{F_y}} \tag{9.5.6}$$

where λ_w = slenderness of the web $(= h/t_w)$
       λ_{pv} = limiting slenderness of the web below which shear yielding rather than shear buckling controls the design

9.6   Serviceability Limit States

In addition to strength limit states, such as bending strength and shear strength, beams should satisfy certain serviceability limit states. The two primary serviceability considerations for beams are deflections and vibrations.

9.6.1   Deflections

With the development of steels of higher strength, and the increase in the demand for large column free floor areas requiring beams of long spans, control of deflections has become more important. Service load deflections of beams are limited for various functional and/or structural reasons, some of which are listed below:

- Large deflections in beams do not generate confidence in the persons using a structure, although it may be completely safe from a strength viewpoint.
- Excessive deformation in one component of a structure may trigger failure in another component. Thus floor beams that deflect too much could produce cracks in plaster ceilings attached to them. If a spandrel beam over a window sags too much it could break the brittle glass. Similarly, excessive deflections of floor beams may produce cracks in partition walls below them or cause doors to jam.
- If the beam or joist is part of a flat roof, excessive deflections may result in poor drainage, thus increasing the probability of ponding failure.
- Excessive deflection of floor beams supporting sensitive laboratory equipment or machinery may result in misalignments as well as dangerous vibrations.
- Excessive member deflection might produce distortion in connections.
- Excessive deflections are indications of a lack of stiffness of the element and/or the structure, which might lead to premature instability failure.

If any one of these situations occurs, even if the beam does not actually collapse, the structure is deemed to have failed to serve its function properly.

In general, beam deflection is a function of the span length, end restraints, modulus of elasticity of the material, moment of inertia of the cross section, and loading. For a simple beam with constant cross section, the maximum central deflection depends upon the loading.

$$\text{For a uniformly distributed load: } \delta = \frac{5}{384} \frac{qL^4}{EI} \qquad (9.6.1)$$

$$\text{For a single concentrated load at the center: } \delta = \frac{1}{48} \frac{QL^3}{EI} \quad (9.6.2)$$

$$\text{For two equal concentrated loads at the third points: } \delta = \frac{23}{648} \frac{QL^3}{EI} \quad (9.6.3)$$

where δ = central deflection, in.
  L = span, in.
  q = uniformly distributed load, kips/in.
  Q = concentrated load, kips
  I = moment of inertia, in.^{4}

The deflection of a simply supported beam under uniformly distributed load is given by Eq. 9.6.1; for the same member with both ends fixed, the central deflection is reduced by 80 percent. Thus the deflection of a partially restrained beam will be between these two limits.

The allowable deflection, δ_{all}, of a beam depends on the purpose for which the beam is designed. It is usually specified relative to the beam span, L. Thus, for example, the service live load deflection of beams supporting plastered ceilings is generally limited to $L/360$. This is the limit beyond which cracks are likely to open up in ceilings with brittle finishes such as plaster. Though no structural failure occurs, it is found that it takes only a small crack in a plastered ceiling to spoil the appearance and to alarm the occupants. Generally, only the live load deflection need be limited to the pre-scribed value, since deflection due to dead load takes place before plastering. Where the beam supports some types of precision machinery, such as satel-lite tracking equipment, the deflection may be limited to as little as $L/2000$. Sometimes, the limiting deflection is given as an absolute value, independ-ent of the span length. For example, there is not much yield to a glass block wall or plate glass window. Hence, if a lintel is used to clear an opening for a glass window, the fully loaded lintel-beam should be designed to provide a prescribed clearance, say $\frac{1}{2}$ in., regardless of its span. For additional infor-mation on serviceability design considerations refer to Galambos and Ellingwood [1986] and Fisher and West [1990].

Cambering of Beams

Cambering is the process of predeforming a steel beam so that its deflection under load is less noticeable to the eye. Cambering is accomplished by intro-ducing a slight convexity into a member and is usually expressed in terms of the maximum ordinate at midspan. Beams may be cambered for an amount equal to the dead load deflection (usual practice), or the dead load deflection plus part of the live load deflection, at the discretion of the engineer. He or she may be influenced by the relative percentages of dead load and live load, the probable frequency and intensity of live load, aesthetics, and the per-formance history of similar members.

Cambering can be performed at either the rolling mill or the fabricating shop. Beams can be cambered by the application of brute force using presses and/or offset rollers *(cold cambering)* or by the application of heat at various points along their length *(heat cambering)*. Heat cambering should be limited to low carbon steels. Usually the cambering, if performed by the fabricator, is done after the member has been cut to length and punched or drilled. Beams that require square and parallel ends (such as shear end-plate connections, end-plate moment connections, or welded moment connections studied in Chapter 13) must be cut after cambering.

Members of uniform cross section with simple end connections, such as beams, girders, and composite floor beams, lend themselves to cambering. Spandrel beams (especially those supporting fascia materials), short beams (less than 25 ft in length), shallow beams (wide flange shapes less than 14 in. nominal depth and standard beams less than 12 in. nominal depth), and beams with fully restrained moment connections or significant semi-rigid moment connections do not lend themselves to cambering. The minimum camber that will remain permanent in a 25-ft beam is about $\frac{1}{2}$ in. A $\frac{1}{4}$-in. camber would be elastic and would disappear as soon as the bending force is removed.

Maximum camber is limited in order to avoid over stressing the steel during the cambering operation. If the expense of cambering can be eliminated by selecting a stiffer (but usually heavier and thus more costly) section, so as to limit the service dead and live load deflection to permissible values, an overall economical savings may result.

9.6.2   Ponding

Structural failures, including sudden and total collapse, have occurred in some poorly designed flat roofs during heavy downpours. In effect, the deflection curve produced by the increasing weight of accumulated rain water may form a shallow basin in a roof. If the roof drains do not empty the basin of water faster than rain water flows in, the roof becomes unstable, and in the worse case, an intense, continuous rainfall may cause failure of the flat roof. This failure is termed **ponding** failure.

As a rough safeguard against ponding, for nominally flat roofs designed with less than the minimum recommended slope of $\frac{1}{4}$ in. per ft, the sum of the deflections of the supporting deck, purlins, girders, or trusses under a 1-in. depth of water (5 psf load) should not exceed $\frac{1}{2}$ in. Otherwise, the designer should present computations substantiating the safety of the roof slope used.

Marino [1966] has given an extensive treatment of ponding that forms the basis for the LRFDS provisions in Section K2 and the charts provided in Appendix K2. Ruddy [1986] has applied the procedure for ponding to study the failure of metal decking supported by steel beams and girders under a load of wet concrete.

....

9.7   Design of Adequately Braced Compact Section Beams

9.7.1   Design Bending Strength

The bending strength of a laterally braced compact section beam is the plastic moment M_p. If the shape has a large shape factor, α, significant inelastic deformation may occur at service load if the bending moment is allowed to reach M_p at factored loads. So, LRFDS Section F1-1 imposes a limit of $1.5M_y$ at factored load to control the amount of plastic deformation for sections with shape factors greater than 1.5. Thus the bending strength of adequately braced beams of compact shapes, according to LRFDS Section F1-1, may be written as:

$$M_d = \phi_b M_n = \min\left[\phi_b M_p, \quad \phi_b(1.5)M_y\right]$$

$$= \min\left[\phi_b Z F_y, \quad \phi_b(1.5S)F_y\right] \qquad (9.7.1)$$

where M_d = design bending strength
      M_n = nominal bending strength
      ϕ_b = resistance factor (= 0.90)
      M_p = plastic moment
      M_y = moment corresponding to the onset of yielding at the extreme
          fiber for an elastic stress distribution
      Z = plastic section modulus with respect to the bending axis
      S = elastic section modulus with respect to the bending axis

The first term within the brackets, in Eq. 9.7.1, represents the limit state of strength corresponding to the formation of a plastic hinge, while the second term is introduced to prevent excessive working load deformation for sections with high shape factors. Elastic and plastic section moduli for various rolled steel structural shapes are tabulated in Part 1 of the LRFDM. The Manual also has formulas for the plastic section moduli of various geometric shapes (Table for Properties of Geometric Sections, in Part 17 of the LRFDM).

I-Shape Bent about the Major Axis

For an I-shape bent about the strong axis, Z/S will always be less than 1.5. So, for adequately braced, compact, I-shaped beams bent about the major axis, the design flexural strength is given by the simplified relation:

$$M_{dx} = \phi_b M_{px} = \phi_b Z_x F_y \tag{9.7.2}$$

For a beam to be considered adequately braced, its compression flange should be either continuously braced, or the distance L_b between adjacent lateral braces should satisfy the relation (LRFDS Eq. F1-4):

$$L_b \leq L_p \tag{9.7.3}$$

where

$$L_p = 1.76 r_y \sqrt{\frac{E}{F_y}} \tag{9.7.4}$$

with

      L_p = limiting unbraced compression flange length for full plastic
          moment capacity, for the particular case where the moment is
          uniform over L_b, in.
      r_y = radius of gyration of the section about the minor axis, in.

For I-shapes the value of r_y varies from 0.22 to 0.25b_f. Hence, for I-shapes with F_y = 50 ksi, an approximate value for L_p is obtained as [1.76 (0.22b_f) $\sqrt{29,000}$]/ $\sqrt{50}$ = 9.3b_f. This relation could be used as a preliminary guide in the selection of suitable sections to arrive at adequately braced beams under uniform moment.

For the section of a beam to be compact, LRFDS Section B5.1 requires that the flange plate (an unstiffened element) and the web plate (a stiffened element) satisfy the following conditions:

$$\lambda_f \leq \lambda_{pf}; \quad \lambda_w \leq \lambda_{pw} \qquad (9.7.5a \text{ and } b)$$

with

$$\lambda_f = \frac{b_f}{2t_f}; \quad \lambda_w = \frac{h}{t_w} \qquad (9.7.6)$$

$$\lambda_{pf} = 0.38 \sqrt{\frac{E}{F_y}}; \quad \lambda_{pw} = 3.76 \sqrt{\frac{E}{F_y}} \qquad (9.7.7)$$

where b_f = width of the flange, in.
 t_f = thickness of the flange, in.
 h = clear distance between flanges less the fillet at each flange, in.
 t_w = thickness of the web, in.
 λ_f = b/t ratio of the flange plate
 λ_w = b/t ratio of the web plate
 λ_{pf} = limiting b/t ratio for the flange of a compact hot-rolled I-shape
 λ_{pw} = limiting b/t ratio for the web of a compact hot-rolled I-shape

For all standard hot-rolled I-shapes listed in the LRFD manual, the webs are compact for all structural steels. So, only the flange ratio needs to be checked. Most of these shapes will also satisfy the flange requirement and will therefore be classified as compact. As mentioned previously, for $F_y = 50$ ksi steel, there are only five **noncompact W-shapes** used as beams, namely, W21×48, W14×99, W14×90, W12×65, and W10×12.

I-Shape Bent about the Minor Axis

For an I-shape bent about the minor axis, the shape factor Z/S will always be greater than 1.5. So, for a compact, I-shaped beam bent about its minor axis, the design flexural strength, given by Eq. 9.7.1 simplifies to:

$$M_{dy} = \phi_b M_n = \phi_b (1.5S_y)F_y \qquad (9.7.8)$$

As will be seen in Chapter 10, lateral buckling of a beam will not occur if the moment of inertia of the section about the bending axis is equal to or less than the moment of inertia out of plane. Thus, the limit state of lateral-torsional buckling need not be checked for I-shapes bent about their minor axis. Further, as the web stresses are zero for an I-shape bent about its minor axis, only the flange b/t ratio needs to be checked for compactness, using Eq. 9.7.5a.

9.7.2   Design Shear Strength

For I-shapes bent about their major axis, it is assumed that only the web resists the shear and that the intensity of shear stress is uniform throughout the depth. Shear strength of an I-shaped beam is a function of the shear yield stress of the web material, the web area, and the web slenderness. The shear yield stress is generally taken as $0.6F_y$. According to LRFDS Section F2.1, the web area is to be taken as the overall depth times the web thickness for rolled shapes. The design shear strength for rolled I-shapes for the limit state of shear yielding of the web is given by (LRFDS Eq. F2-1):

$$V_d = \phi_v V_n = \phi_v (0.6F_y) A_w \quad \text{for } \lambda_w \leq \lambda_{pv} \tag{9.7.9}$$

where

$$\lambda_{pv} = 2.45 \sqrt{\frac{E}{F_y}} \tag{9.7.10}$$

with

V_d = design shear strength, kips
V_n = nominal shear strength, kips
ϕ_v = resistance factor (= 0.90)
F_y = yield stress of the web material, ksi
F_{yv} = shear yield stress of the web material (= $0.6F_y$), ksi
A_w = area of the web (= dt_w) in.^{2}
λ_w = slenderness of the web (= h/t_w)
λ_{pv} = limiting slenderness of the web below which shear yielding, rather than shear buckling, controls the design
h = clear distance between flanges less the fillet at each flange
t_w = web thickness

Values of λ_w are given in LRFDM Tables 1-1, 1-2 and 1-3 for W-, M-, and S-shapes, respectively.

The rolled W-shape with the most slender web, as given in the LRFDM Table 1-1, is a W30×90 with λ_w = 57.5. The limits λ_{pv} for webs of rolled I-shapes used as beams are given in Table 9.5.1 for different steels. When $\lambda_w > \lambda_{pv}$, the web shear strength is based on the shear buckling strength of the web. For materials with F_y = 50 ksi, λ_{pv} equals 59.0 from Table 9.5.1, indicating that the shear buckling will not be a design criteria for any of the rolled W-shapes of A992 steel. Values of V_d (= $\phi_v V_n$) are given in several beam selection tables such as LRFDM Tables 5-2, 5-3, and 5-4.

9.7.3 Deflections

LRFDS Section L3.1 stipulates that deformations in structural members and systems due to service loads shall not impair the serviceability of the structure. However, no specific limits are furnished. Deflections are likely to control the size of long span shallow beams under heavy loads, and of beams with severe deflection limitations. In many practical situations, if the span-to-depth exceeds 20, deflection is likely to be critical. Formulas for maximum deflections for a variety of beams and loading conditions are

TABLE 9.7.1

Suggested Deflection Limits

Beams carrying plaster or other brittle finish	*L*/360
Roof members supporting nonplaster ceiling	*L*/240
Roof members not supporting ceiling	*L*/180
Cantilevers	*L*/180

given in Table 5-17: Shears, Moments and Deflections, in Part 5 of the LRFD Manual.

Serviceability limit states are checked under unfactored loads. Thus, for roof and floor beams the maximum deflection is checked for the load combination: $1.0D + 1.0(L, L_r, S, \text{ or } R)$. Usually, the beam is provided with a camber by the fabricator, corresponding to the dead load D; the deflection is then controlled for service live load only. Some suggested values for deflection limits are given in Table 9.7.1.

9.7.4   Required Strengths of Indeterminate Beams Using Elastic Analysis

LRFDS Section A5.1 permits either elastic analysis or plastic analysis in the design of adequately braced, indeterminate beams with compact sections. Further, when elastic analysis is used, recognition is given to the reserve strength of the beam beyond first yield due to plastic redistributon of moments mentioned in Section 9.4. Thus, LRFDS Section A5.1 permits a 10 percent reduction in maximum negative moments produced by factored gravity loading at support points of fixed or continuous beams, providing that the maximum positive moment is increased by one-tenth of the average negative moments. Thus, if M_1 and M_2 are the negative moments at the ends of a beam such that $|M_2| \geq |M_1|$ and M_3 is the maximum positive moment of the beam obtained from an elastic analysis, the required bending strengths are:

$$M_{\text{req}}^{-} = \frac{9}{10} M_2 \tag{9.7.11}$$

$$M_{\text{req}}^{+} = M_3 + \frac{1}{10} \left| \frac{M_1 + M_2}{2} \right| \tag{9.7.12}$$

This reduction is not permitted for moments produced by loading on cantilevers, for members made of A514 steel, or for hybrid beams.

9.7.5   Required Strengths of Cantilevers and Beams with Overhangs

Most cantilevers are formed by using moment connections to column flanges or by overhanging beams over supports. The first situation occurs, for example, when balconies are fixed at an intermediate point between column ends or when equipment is held on brackets in industrial buildings.

Loading Cases

As mentioned in Chapter 4, live loading is a random type of load and, hence, should be imposed on the structure such that maximum or critical moments, shears, and forces are induced on each individual element. This will be illustrated with the help of a beam ABC having a main span AB (length, L_1; rigidity, EI_1) and overhang BC (length, L_2; rigidity, EI_2), shown in Fig. 9.1.1c. For this structure, the following load cases are suggested.

1. Maximum dead load and live load ($1.2D + 1.6L$) on the main span with dead load only ($1.2D$) on the cantilever. This case will give the maximum span moment and deflection, and the maximum downward reaction and shear at support A.
2. Maximum dead load and live load ($1.2D + 1.6L$) on the cantilever with dead load only ($1.2D$) on the main span. This case will give the maximum cantilever moment and the maximum deflection at the cantilever tip.
3. Both span and cantilever fully loaded with dead load and live load ($1.2D + 1.6L$). This case will give the maximum reaction at B together with the maximum shear in the beam on the span side of B.
4. Maximum dead load and live load ($1.2D + 1.6L$) on the cantilever with dead load only ($0.9D$) on the main span. This case will give the maximum uplift force at support A for which anchor rods are to be designed if necessary.

Deflection

Cantilevers are often subjected to large deflections. For a cantilever beam with a constant cross section, the maximum deflection at the free end is:

For a uniformly distributed load: $\quad \delta = \dfrac{1}{8} \dfrac{qL^4}{EI}$ (9.7.13)

For a concentrated load at the tip: $\quad \delta = \dfrac{1}{3} \dfrac{QL^3}{EI}$ (9.7.14)

These values for cantilever deflection are based on the assumption that the support of the cantilever is absolutely rigid—a situation not normally encountered in practice. The additional movement of the tip due to the support rotation must therefore be added to the values given above to assess the total tip deflection. The service live load deflection at the cantilever tip is usually limited to $L/180$. The deflection at the free end of an overhanging beam (point C of the beam ABC, shown in Fig. 9.1.1c), may be calculated from the relations:

For a concentrated load Q at the free end C:

$$\delta_C = \frac{QL_2^3}{3EI_2} + \left[\frac{QL_2L_1}{3EI_1} \right] L_2 \qquad (9.7.15)$$

For a uniformly distributed load q over BC:

$$\delta_C = \frac{qL_2^4}{8EI_2} + \left[\frac{qL_2^2L_1}{6EI_1}\right]L_2 \qquad (9.7.16)$$

For a uniformly distributed load q over span AB:

$$\delta_C = -\left[\frac{qL_1^3}{24EI_1}\right]L_2 \qquad (9.7.17)$$

9.7.6   Design of Adequately Braced Rolled Steel I-Beams

The successful design of steel beams involves satisfying many different limit states. Some of these are flexural strength, lateral buckling strength, shear strength, flange local buckling (FLB) strength, web local buckling (WLB) strength, deflection, web local yielding, web crippling, vertical buckling, and ponding. The design procedure generally consists of proportioning the beam to satisfy the bending requirements first, and then if it appears necessary, checking the chosen section for some or all of the above mentioned limit states. The experienced designer can tell by inspection when limit states other than bending need be considered.

Let M_u be the required moment capacity of the beam under factored loads. Since most rolled shapes are compact shapes, initially assume this condition to be true and later perform checks for FLB and WLB. Also, assume that the beam is adequately braced laterally. Using the specified steel yield stress F_y, the required plastic section modulus $Z_{x\,req}$ is determined from Eq. 9.7.2:

$$Z_{x\,req} \geq \frac{12M_u}{\phi_b F_y} \qquad (9.7.18)$$

where M_u = required bending strength under factored loads, ft-kips
         ϕ_b = resistance factor (= 0.90)
         F_y = yield stress, ksi
         $Z_{x\,req}$ = required plastic section modulus, in.^{3}

The LRFD Manual has several tables and charts that greatly facilitate the design of I-shapes used as beams. Thus:

- LRFDM Table 1-1: W-Shapes—Dimensions and Properties provides values of strong and weak axis moments of inertia, I_x, I_y; elastic section moduli, S_x, S_y; and plastic section moduli, Z_x, Z_y for W-shapes.
- LRFDM Table 5-17: Shears, Moments and Deflections provides beam diagrams and formulas for cases frequently encountered in design. Here, shear force and bending moment diagrams for various loading conditions on various beams with particular end support conditions are given. Formulas for reactions, shears, moments, and deflections are

also provided for these cases. The diagrams and formulas are based on elastic analysis.

- LRFDM Table 5-3: W-Shapes—Selection by Z_x lists W-shapes normally used as beams within groups, in descending order of strong axis plastic section modulus, Z_x. The top member in each group, printed in boldface, has the least weight, the largest Z_x, the largest depth, and the largest $\phi_b M_{px}$ in that group for a given steel. The table also includes, for steels with $F_y = 50$ ksi, several design strengths ($\phi_b M_{px}$, $\phi_b M_{rx}$, $\phi_v V_n$, and $\phi_b M_{py}$), and several valuable design parameters (L_p, L_r, BF, I_x, Z_y, and I_y). Here, $\phi_b M_{px}$ represents the maximum major axis flexural design strength and L_p is the limiting laterally unbraced compression flange length for full plastic moment capacity and uniform moment. The other terms will be discussed in Chapter 10.

- LRFDM Table 5-2: W-Shapes—Selection by I_x gives values of I_x for all W-shapes normally used a beams, listed within groups in descending order of magnitude. Here again, boldface type identifies the shapes that are the lightest and the deepest in their respective groups. This table facilitates the selection of beams having adequate stiffness for controlling deflections.

- LRFDM Table 5-4: Maximum Total Factored Uniform Load Tables is applicable to W-shapes having adequate lateral support . These tables provide, for steels having $F_y = 50$ ksi, the maximum factored uniform loads (in kips) for various practical span lengths. Note that the total load (in kips) is given—*not* the maximum distributed load per foot. The latter may be obtained by dividing the tabulated value by the span length of the beam. Note also that the values in these tables include the (factored) self-weight of the beam; to obtain the net load, the self-weight must be subtracted. The design shear strength, plastic section modulus, and other useful design properties are also provided for each shape. LRFDM Part 5 also includes similar tables for A36 steel S-shapes (Table 5-8), C-shapes (Table 5-9), and MC-shapes (Table 5-10) used as simple beams subjected to strong axis bending.

LRFDM Table 5-5. W-Shapes—Plots of $\phi_b M_{nx}$ vs. L_b can be used for the selection of suitable W-shapes with $F_y = 50$ ksi for continuously braced ($L_b = 0$) or adequately braced ($L_b \leq L_p$) beams. The use of this table will be explained in Section 10.4.4.

Because steel sections cost so many cents per pound, it is often desirable to select the lightest possible shape that satisfies all the limit states. However, the lightest sections in the beam selection table are also the deepest in their respective groups. Now, 6 in. of head room saved in each story of a tier building, achieved by selecting a more shallow but heavier section than the lightest section tabulated, will reduce the required height of a 20-story building by an entire story. This saving often represents an important economy. In such situations, where the beam size is limited to a certain (nominal) depth, for architectural of clearance reasons or to match the depth of adjacent beams, proceed up the column headed "shape" in the Beam Selection Table (LRFDM Table 5-3), until a beam within the required depth is reached.

There are two common approaches to allowing for self-weight of a beam in its design, as described below:

1. The weight of the beam is neglected in the original calculation of the maximum factored bending moment, and a section is tentatively selected from the LRFDM Beam Selection Table. Then the weight of that tentative section (or a little bit more) can be used as the estimated beam weight. The additional required bending strength for the factored weight of the beam is then found, and the sum of the two computations is checked against the design bending strength furnished for the choosen section. The section tentatively selected is then revised if necessary as demonstrated in Example 9.7.1.
2. The weight of the beam is guessed in advance and is included in the original computation for the required bending strength and then revised if necessary after selecting a section from the Beam Selection Table (see Example 9.7.3).

The first method is probably the easier for beginners.

Web shear should be considered in the design of every beam. For typical span lengths, the limit state of flexure will control the shape selection, and the limit state of shear is merely checked for satisfaction. For the design of short and heavily loaded beams, the reverse of the above scenario will probably prevail. That is, for short and heavily loaded beams, the limit state of shear will control the section selection, and the limit state of flexure is merely checked. Thus, when the length of a beam is less than six times its depth for a W-shape, or nine times its depth in the case of an S-shape, shear is likely to control the design. However, beams with large concentrated loads near either or both supports must always be investigated for shear. Shear strength should also be investigated for all beams where the web area has been reduced by a hole cut for a service pipe or a ventilation duct, or if the end of the beam has been coped to provide end connections. Spandrel beams might also be governed by shear since the vertical shear stresses in those beams are sometimes augmented by torsional shear stresses induced by the connecting floors.

The design procedure used for the design of a rolled steel I-beam with continuous lateral bracing can be summarized as follows:

1. Draw a sketch of the beam with its factored loads. Determine the maximum shear force V_{max} and the maximum bending moment M_{max} either by drawing shear force and bending moment diagrams under appropriate factored load combinations or by using appropriate beam formulas and diagrams such as those given in LRFDM Table 5-17. We have:

$$\text{Required bending strength, } M_u = M_{max}$$

$$\text{Required shear strength, } V_u = V_{max}$$

If F_y is other than 50 ksi, determine $Z_{x\text{req}}$ using Eq. 9.7.18.

2. The Beam Selection Table (LRFDM Table 5-3) is directly applicable to adequately braced beams. It can be used either by entering the column headed Z_x with $Z_{x\,req}$, or (when $F_y = 50$ ksi) by entering the corresponding $\phi_b M_{px}$ column with M_u. Select a section in the shape column having values equal to or greater than $Z_{x\,req}$ or M_u. That shape and all shapes above it are strong enough for the loading considered. The first section appearing in boldface type opposite to or above the $Z_{x\,req}$ or (M_u when $F_y = 50$ ksi) is the lightest section that will satisfy the flexural limit state. Select a suitable section that satisfies, in addition to the strength criteria, any architectural constraints on depth, nominal size, availability, and so on. Keep in mind that the design strength, $\phi_b M_{px}$, of the chosen beam must be sufficient to include the effect of the self-weight of the beam. Get the values of L_p, $\phi_v V_n$, I_x for the shape selected.

3. Check now the starting assumption that the selected shape will be compact, by comparing the width-to-thickness ratios for the flange and the web of the section selected.

$$\lambda_f \leq \lambda_{pf}; \quad \lambda_w \leq \lambda_{pw}$$

4. If the beam is not continuously laterally braced, the lateral bracing interval, L_b, should be checked against the L_p value for the section selected.

$$L_b \leq L_p$$

If this condition is not satisfied, and if the moment is not uniform over the segment L_b, refer to Chapter 10. Otherwise, reduce the spacing of lateral bracing or select a heavier section.

5. Check the shear strength of the section selected at the point of maximum shear.

$$V_d = \phi_v V_n \geq V_u$$

If the design shear strength V_d is less than V_u, another beam with greater web area may be chosen, or the web may be reinforced with doubler plates or stiffeners.

6. Deflection must be checked against allowable deflection.

$$\delta \leq \delta_{all}$$

7. Web local yielding and web crippling must be checked at points of support (Sections 9.9.1 and 9.9.2) and concentrated loading.

EXAMPLE 9.7.1

Design for Flexure

Select the lightest W-section of A242 Grade 42 corrosion resistant steel for a simple beam to carry a uniformly distributed dead load of 0.2 klf, roof live load of 0.8 klf, and a snow load of 1.2 klf. The simply supported span is 32 ft. The compression flange is fully supported laterally by the roof deck. Make b/t checks. Check for shear.

Solution

a. Required strength
Span, $L = 32$ ft
At this stage, neglect self-weight of the beam.

$$D = 0.2 \text{ klf}; \quad L_r = 0.8 \text{ klf}; \quad S = 1.2 \text{ klf}$$

Load combinations:

$$(\text{LC-1}) \rightarrow 1.4D \rightarrow 1.4(0.2) = 0.28 \text{ klf}$$

$$(\text{LC-3}) \rightarrow 1.2D + 1.6(L_r \text{ or } S) = 1.20(0.2) + 1.6(1.2)$$

$$= 2.16 \text{ klf}$$

So, the factored load $q_u = 2.16$ klf. For the simply supported beam:

$$M_u = M_{\text{max}} = \frac{q_u L^2}{8} = \frac{2.16(32^2)}{8} = 277 \text{ ft-kips}$$

$$V_u = V_{\text{max}} = \frac{q_u L}{2} = \frac{2.16(32)}{2} = 34.6 \text{ kips}$$

b. Beam selection
The beam is continuously laterally supported ($L_b = 0$). So, assuming a compact section,

$$Z_{x\text{req}} \geq \frac{M_u}{\phi_b F_y} = \frac{277(12)}{0.9(42)} = 87.9 \text{ in.}^3$$

Enter LRFDM Table 5-3: W-Shapes Selection by Z_x, with $Z_{x\text{req}}$ of 87.9 in.^{3} and observe that a W21×44 in boldface with Z_x value of 95.8 in.^{3} is the lightest suitable section. Using a value of 50 plf for self-weight, the additional moment and shear due to factored self-weight (dead load) are:

$$\Delta M_{\text{max}} = \frac{1.2(50)}{1000}\left(\frac{32^2}{8}\right) = 7.68 \text{ ft-kips}$$

$$\Delta V_{\text{max}} = \frac{1.2(50)}{1000}\left(\frac{32}{2}\right) = 0.960 \text{ kips}$$

resulting in the revised values for M_{max}, V_{max}, and Z_x of:

$$M_{\text{max}} = 277 + 7.68 = 285 \text{ ft-kips}$$

$$V_{\text{max}} = 34.6 + 0.960 = 35.6 \text{ kips}$$

$$Z_{x\text{req}} = \frac{285}{277}(87.9) = 90.4 \text{ in.}^3 < 95.8 \text{ in.}^3 \quad \text{O.K.}$$

[continues on next page]

Example 9.7.1 continues ...

Hence, the W21×44 of A242 Grade 42 steel selected earlier is adequate for flexure.

c. Checks

For A242 Grade 42 steel, with $F_y = 42$ ksi, the limiting slenderness ratios for plate buckling in flexure and shear are given in Table 9.5.1 as:

$$\lambda_{pf} = 0.38 \sqrt{\frac{E}{F_y}} = 0.38 \sqrt{\frac{29000}{42}} = 9.99$$

$$\lambda_{pw} = 3.76 \sqrt{\frac{E}{F_y}} = 3.76 \sqrt{\frac{29000}{42}} = 98.8$$

$$\lambda_{pv} = 2.45 \sqrt{\frac{E}{F_y}} = 2.45 \sqrt{\frac{29000}{42}} = 64.4$$

From Table 1-1 of the LRFDM for a W21×44,

$$d = 20.7 \text{ in.}; \qquad t_w = 0.350 \text{ in.}$$

$$\lambda_f = \frac{b_f}{2t_f} = 7.22; \quad \lambda_w = \frac{h}{t_w} = 53.6$$

As $\lambda_f < \lambda_{pf}$ and $\lambda_w < \lambda_{pw}$, the section is compact as assumed, indicating that FLB and WLB will not reduce the flexural strength calculated above. Also, as $\lambda_w < \lambda_{pv}$, shear buckling of the web will not occur. Shear strength of the web is therefore given by Eq. 9.7.9 (LRFDS Eq. F2.1), namely:

$$V_d = \phi_v V_n = 0.9(0.6F_y) \, dt_w = 0.9(0.6)(42)(20.7)(0.350)$$

$$= 164 \text{ kips} > V_{max} = 35.6 \text{ kips} \quad \text{O.K.}$$

Select a W21×44 of A242 Grade 42 steel.   (Ans.)

EXAMPLE 9.7.2

Design for Shear

A 6-ft simply supported beam is to carry a factored uniformly distributed load of 30 klf. Assume adequate continuous lateral bracing and neglect self-weight of beam. Select:

a. The lightest W-shape of A992 steel.
b. The lightest W12-shape of A992 steel.

Solution

$$L = 6 \text{ ft}; \quad q_u = 30 \text{ klf}; \quad F_y = 50 \text{ ksi}$$

As the loading is rather heavy and the span is short, shear might control the member selection.

$$M_u = M_{max} = \frac{q_u L^2}{8} = \frac{1}{8}(30.0)(6.0^2) = 135 \text{ ft-kips}$$

$$V_u = V_{max} = \frac{q_u L}{2} = \frac{1}{2}(30.0)(6.0) = 90.0 \text{ kips}$$

For A992 steel, with $F_y = 50$ ksi, the limiting web slenderness ratio for plate yielding in shear is given in Table 9.5.1 as:

$$\lambda_{pv} = 2.45\sqrt{\frac{E}{F_y}} = 2.45\sqrt{\frac{29000}{50}} = 59.0$$

The limiting slenderness ratios for plate buckling in flexure are given in Table 9.5.1 as:

$$\lambda_{pf} = 0.38\sqrt{\frac{E}{F_y}} = 0.38\sqrt{\frac{29000}{50}} = 9.15$$

$$\lambda_{pw} = 3.76\sqrt{\frac{E}{F_y}} = 3.76\sqrt{\frac{29000}{50}} = 90.6$$

a. Lightest W-shape
    Enter LRFDM Table 5-3: W-Shapes Selection by Z_x with $M_{req} = 135$ ft-kips, and select a W12×26 section in bold face, for which:

$$\phi_b M_{px} = 140 \text{ ft-kips} > M_u = 135 \text{ ft-kips} \qquad \text{O.K.}$$

From Table 1-1 of the LRFDM for W12×26:

$$d = 12.2 \text{ in.,} \quad t_w = 0.230 \text{ in.,} \quad \lambda_w = \frac{h}{t_w} = 47.2$$

As $\lambda_w < \lambda_{pv}$ for the web, the design shear strength of the web is given by Eq. 9.7.9 (LRFDS Eq. F2-1), namely:

$$V_d = \phi_v V_n = 0.9(0.6 F_y d t_w) = 0.9(0.6)(50)(12.2)(0.230)$$

$$= 75.9 \text{ kips} < V_u = 90 \text{ kips} \qquad \text{N.G.}$$

indicating that the W12×26 section is inadequate for the limit state of shear strength. (Instead of calculating, the design shear strength $\phi V_n = 75.9$ kips could have been read from LRFDM Table 5-3: W-Shapes Selection by Z_x, corresponding to a W12×26 shape).
Select a heavier and/or deeper section, from one of the sections located above the W12×26 shape, in that table. For example, select a W14×26 shape that is also in boldface, for which:

$$\phi_v V_n = 95.7 \text{ kips} > V_u = 90 \text{ kips} \quad \text{O.K.}$$

$$\phi_b M_{px} = 150 \text{ kips} > M_u = 135 \text{ ft-kips} \quad \text{O.K.}$$

[continues on next page]

Example 9.7.2 continues ...

For a W14×26 section: $\lambda_f = \dfrac{b_f}{2t_f} = 5.98;$ $\lambda_w = \dfrac{h}{t_w} = 48.1$

As $\lambda_f < \lambda_{pf}$ and $\lambda_w < \lambda_{pw}$, the section is compact as assumed, indicating that FLB and WLB will not reduce the flexural strength calculated above. So, provide a W14×26 section of A992 steel.   (Ans.)

b. Lightest W12-shape

Enter the bottom part of LRFDM Table 5-4: W-Shapes—Maximum Total Factored Uniform Load Tables corresponding to a W12×26 that has adequate flexural strength as seen in Part (*a*). Move horizontally to the left until the design shear strength $\phi_v V_n$ tabulated is greater than the required shear strength V_u of 90.0 kips. Observe that a W12×35 with $\phi_v V_n = 101$ kips is the lightest W12-shape that is adequate in shear. From LRFDM Table 5-3: W-Shapes Selection by Z_x, the W12×35 has a bending strength of 192 ft-kips > 135 ft-kips required. So, select a W12×35 of A992 steel.   (Ans.)

E X A M P L E   9 . 7 . 3   Design for Deflection

Select the lightest W-section to carry a uniform dead load of 0.6 klf and a uniform live load of 1.1 klf on a simply supported span of 40 ft. Adequate lateral bracing is provided to the compression flange. The live load deflection is to be limited to $L/360$. Use A992 steel.

Solution

Span, $L = 40$ ft; $F_y = 50$ ksi
Estimate the beam weight at 70 plf.
$q_u = 1.20q_D + 1.60q_L = 1.2(0.60 + 0.070) + 1.6(1.1) = 2.56$ klf
Required bending strength,

$$M_u = M_{max} = \frac{q_u L^2}{8} = \frac{2.56(40^2)}{8} = 512 \text{ ft-kips}$$

Required shear strength,

$$V_u = V_{max} = \frac{q_u L}{2} = \frac{2.56(40)}{2} = 51.2 \text{ kips}$$

Enter the LRFDM Table 5-3: W-Shapes Selection by Z_x, with $M_u = 512$ ft-kips and select a W21×62 in bold face,

$$\phi_b M_{px} = 540 \text{ ft-kips} > M_u = 512 \text{ ft-kips}   \text{O.K.}$$

The section is compact for A992 steel.

Allowable deflection, $\delta_{all} = \dfrac{L}{360} = \dfrac{40(12)}{360} = 1.33$ in.

Maximum service live load deflection at the center (Eq. 9.6.1),

$$\delta_L = \frac{5q_L L^4}{384EI} = \frac{5(1.1)(40^4)(12^3)}{384(29,000)I} = \frac{2185}{I} \text{ in.}$$

where I is the moment of inertia of the beam selected. For the W21×62 beam selected from strength considerations, $I = 1330$ in.^{4}, resulting in:

$$\delta_L = \frac{2185}{1330} = 1.64 \text{ in.} > \delta_{all} = 1.33 \text{ in.} \qquad \text{N.G.}$$

indicating that the deflection controls the design. The required moment of inertia to satisfy the deflection limit may be calculated from the relation:

$$\frac{2185}{I} \leq \delta_{all} = 1.33 \rightarrow I \geq \frac{2185}{1.33} = 1640 \text{ in.}^4$$

Enter the LRFDM Table 5-2: W-Shapes Selection by I_x, with $I_{req} = 1640$ in.^{4} and select the first section in boldface, namely, W24×68 for which:

$$I = 1830 \text{ in.}^4 > I_{req} = 1640 \text{ in.}^4 \qquad \text{O.K.}$$

Dead load deflection may be calculated by proportion of the loads as:

$$\delta_D = \frac{2185}{1830}\left(\frac{0.67}{1.1}\right) = 0.727 \text{ in.}$$

Select a W24×68 of A992 steel and provide a camber of ¾ in. As the weight of the beam, 68 plf, is less than the assumed value of 70 plf, the selection is O.K. **(Ans.)**

Design for Deflection

EXAMPLE 9.7.4

Select the lightest W-section of A588 Grade 50 corrosion resistant steel for the 20-ft-long cantilever beam shown in Fig. X9.7.4. The unfactored live loads are $Q_1 = 10$ kips and $Q_2 = 20$ kips. Assume continuous lateral bracing and neglect self-weight of the beam. The maximum service load deflection is to be limited to $L/800$.

Solution
Span, $L = 20$ ft

a. Serviceability limit state
   As the loads on the beam are relatively heavy and the deflection criteria rather stringent, design the beam to satisfy the limit state of deflection first, and then check the section obtained for flexure and shear.

[continues on next page]

Example 9.7.4 continues ...

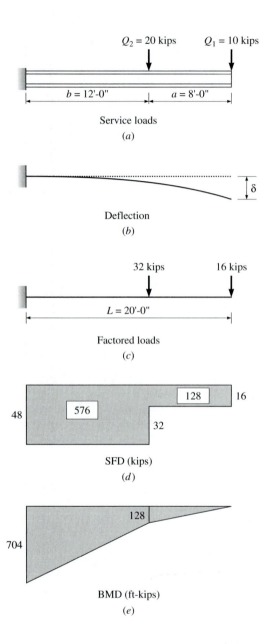

Figure X9.7.4

Maximum permissible deflection, $\delta_{all} = \dfrac{L}{800} = \dfrac{20(12)}{800} = 0.300$ in.

The maximum deflection of the cantilever occurs at the free end. Let I be the moment of inertia of the beam. Using expressions for deflection in Case 22 and 21 of the LRFDM Table 5-17: Shears, Moments and

Deflections, the deflection of the free end of the cantilever beam is given by:

$$\delta = \frac{Q_1 L^3}{3EI} + \frac{Q_2 b^2}{6EI}(3L - b)$$

$$= \frac{1}{I}\left\{\frac{10(20^3)}{3} + \frac{20(12^2)}{6}[3(20) - 12]\right\}\frac{12^3}{29{,}000}$$

$$= \frac{2962}{I}$$

For serviceability limit state of deflection, we require:

$$\delta_L \leq \delta_{all} \longrightarrow \frac{2962}{I} \leq 0.300 \longrightarrow I \geq 9870 \text{ in.}^4$$

From the LRFDM Table 5-2: W-Shapes Selection by I_x, find that a W40×167 with an I of 11,600 in.^{4} is the lightest section that satisfies this requirement.

b. Strength limit states
    From the bending moment diagram, the required bending strength

$$M_u = M_{max} = 704 \text{ ft-kips}$$

From LRFDM Table 5-3: W-Shapes Selection by Z_x, for a continuously braced W40×167 beam of Grade 50 steel:

$$M_d = \phi_b M_{px} = 2600 \text{ ft-kips} > M_u = 704 \text{ ft-kips} \quad \text{O.K.}$$

From the shear force diagram, the required shear strength is:

$$V_u = V_{max} = 48 \text{ kips}$$

From LRFDM Table 5-2, the design shear strength of a W40×167 of Grade 50 steel,

$$V_d = \phi_v V_n = 677 \text{ kips} > 48 \text{ kips} \quad \text{O.K.}$$

So, select a W40×167 of A588 Grade 50 steel.   (Ans.)

Design

EXAMPLE 9.7.5

Design an interior floor beam and an interior floor girder for the braced building shown in Fig. 5.1.1 for the required strengths determined in Examples 5.2.2 and 5.2.3, respectively. Use W-shapes and A992 steel. Assume continuous lateral support, both for the beams and the girders from the steel decking.

[*continues on next page*]

Example 9.7.5 continues ...

Solution

a. Floor beam

The following data is obtained from Example 5.2.2:

$$L = 25.0 \text{ ft}$$

Required bending strength, $M_u = 122$ ft-kips
Required shear strength, $V_u = 20$ kips
Service live load, $q_L = 0.40$ klf;   service dead load, $q_D = 0.76$ klf

From LRFDM Table 5-3: W-Shapes Selection by Z_x, observe that a W14×22 (in bold face) has adequate flexural strength. Select the W14×22. Design bending strength,

$$M_d = \phi_b M_{px} = 123 \text{ ft-kips} > 122 \text{ ft-kips} \qquad \text{O.K.}$$

From LRFDM Table 5-3 the design shear strength for a W14×22:

$$V_d = \phi_v V_n = 85.1 \text{ kips} > 20 \text{ kips} \qquad \text{O.K.}$$

As the ceiling is suspended from the floor beams, the maximum allowable deflection,

$$\delta_{\text{all}} = \frac{L}{360} = \frac{25(12)}{360} = 0.833 \text{ in.}$$

Moment of inertia, $I_x = 199$ in.^{4}
Maximum service live load deflection,

$$\delta_L = \frac{5}{384} \frac{q_L L^4}{EI} = \frac{5(0.4)(25^4)(12^3)}{384(29,000)(199)}$$

$$= 0.609 \text{ in.} < 0.833 \text{ in.} \qquad \text{O.K.}$$

Dead load deflection,

$$\delta_D = \delta_L \frac{q_D}{q_L} = 0.609\left(\frac{0.76}{0.40}\right) = 1.16 \text{ in.}$$

So, select a W14×22 of A992 steel and provide a camber of $1\frac{1}{4}$ in. As the weight of the selected beam, 22 plf, is less than the assumed value of 40 plf (see Example 5.2.2), the selection is O.K.   (Ans.)

b. Floor girder

The following data is obtained from Example 5.2.3:
Span of the girder, $L = 24$ ft
Required bending strength, $M_u = 290$ ft-kips
Required shear strength, $V_u = 37.0$ kips
The girder is essentially subjected to concentrated loads transferred by the floor beams at the third points.
Service live load, $Q_L = 8$ kips;   service dead load, $Q_D = 19$ kips

From LRFDM Table 5-3: W-Shapes Selection by Z_x, observe that a W18×40 (in bold face) has a design bending strength,

$$M_d = \phi_b M_{px} = 294 \text{ ft-kips} > 290 \text{ ft kips} \quad \text{O.K.}$$

From LRFDM Table 5-3 the design shear strength for a W18×40:

$$V_d = \phi_v V_n = 152 \text{ kips} > 37 \text{ kips} \quad \text{O.K.}$$

As the ceiling is suspended from the floor girders, the maximum allowable deflection is,

$$\delta_{\text{all}} = \frac{L}{360} = \frac{24(12)}{360} = 0.800 \text{ in.}$$

Moment of inertia, $I_x = 612 \text{ in.}^4$

The girder is subjected to two equal concentrated loads at the third points, by the floor beams framing into it. The maximum service live load deflection which occurs at midspan (from Eq. 9.6.3) is,

$$\delta_L = \frac{23}{648} \frac{Q_L L^3}{EI} = \frac{23(8)(24^3)(12^3)}{648(29,000)(612)}$$

$$= 0.382 \text{ in.} < 0.800 \text{ in.} \quad \text{O.K.}$$

Service dead load deflection at the center,

$$\delta_D = \delta_L \frac{Q_D}{Q_L} = 0.382 \left(\frac{19}{8} \right) = 0.907 \text{ in.}$$

So, select a W18×40 of A992 steel and provide a camber of 1 in. As the weight of the selected girder, 40 plf, is less than the assumed value of 60 plf (see Example 5.2.3), the selection is O.K.   (Ans.)

Design Using Reduced Moments   EXAMPLE 9.7.6

A 24-ft-long beam AB is fixed at end A and roller supported at end B with a 60 kip factored concentrated load at the center C. Assume that the compression flange can be laterally braced such that $L_b < L_p$. Use elastic analysis and select the lightest W-shape of A992 steel. Shear and deflection need not be checked. Neglect self-weight of the beam.

Solution

Span, $L = 24$ ft
Factored central concentrated load, $Q_u = 60$ kips

[*continues on next page*]

Example 9.7.6 continues ...

From Case 13 in LRFDM Table 5-17: Shears, Moments and Deflections:

$$M_A = \frac{3Q_uL}{16} = \frac{3(60)(24)}{16} = 270 \text{ ft-kips} = M^-_{max}$$

$$M_C = \frac{5Q_uL}{32} = \frac{5(60)(24)}{32} = 225 \text{ ft-kips} = M^+_{max}$$

These results are from an elastic analysis of the indeterminate beam. LRFDS Section A5.1 permits the beam to be designed for a reduced moment.

Reduction in the moment at support A = 10% M_A = 27.0 ft-kips
The moment at support B cannot be changed (hinge).
Increase in moment at C = ½ (27.0 + 0.0) = 13.5 ft-kips
The adjusted moment diagram consists of:

$$M_A = 270 - 27.0 = 243 \text{ ft-kips}; \quad M_B = 0.0$$

$$M_C = 225 + 13.5 = 239 \text{ ft-kips}$$

So, the required flexural strength of the beam is M_u = 243 ft-kips.
Enter LRFDM Table 5-3 with a required bending strength of 243 ft-kips and select a W18×35 with:

$$M_d = \phi_b M_{px} = 249 \text{ ft-kips} > 243 \text{ ft-kips}$$

The section is compact as required by the LRFDS Section A5.1. Also, for this section L_p = 4.31 ft.
So, select a W18×35 of A992 steel and provide bracing at 4 ft intervals.
(Ans.)

9.8   Open Web Steel Joists and Joist Girders

Open web steel joists are standardized, prefabricated, welded steel trusses which are used as simply supported beams. Briefly introduced in Section 3.5.2 (Fig. 3.5.2), these trusses are particularly well suited for single-story structures with high ceilings such as gymnasiums, factories, and shopping centers, where fire proofing and acoustic needs are minimal.

WWW   *For information on open web steel joists and joist girders, refer to Section W9.2 on the website http://www.mhhe.com/Vinnakota.*

9.9   Beam Bearing Plates

Beams supported by masonry or concrete walls generally rest on steel bearing plates (Fig. 9.9.1). The primary purpose of a **bearing plate** is to distribute the beam reaction over an area large enough to keep the average pressure on the support material within the limits of its design stress. Another purpose of the bearing plate is to help seat the beam at its required elevation. Steel

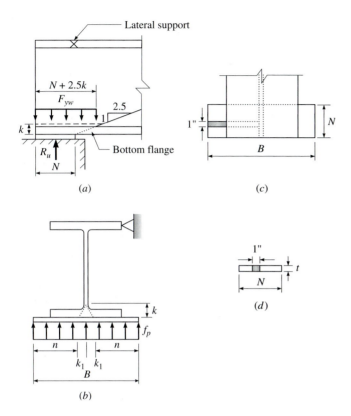

Figure 9.9.1:   Beam bearing plate.

bearing plates are usually shipped separately to the construction site and grouted in place before the beam is set. The bearing plate need not be attached to the beam. However, the bearing plate could also be attached (welded) to the beam during fabrication. In this latter case, a setting plate will be embedded in mortar in the wall.

Let R_u be the concentrated reaction, under factored loads, transmitted from the beam to the support. Also, let:

N = length of the bearing plate, parallel to the length of the beam

B = width of the bearing plate, perpendicular to the length of the beam

t = thickness of the bearing plate

F_{ypl} = yield stress of the bearing plate material

Limit states of strength that influence bearing plate design are:

- Limit state of beam web local yielding, R_{dy}.
- Limit state of beam web crippling, R_{dc}.
- Limit state of bearing of plate on concrete or masonry, R_{dp}.
- Limit state of flexural strength of bearing plate, R_{df}.

The dimension N is determined so that web local yielding and web crippling are prevented. The width B is calculated next so that the bearing area is sufficient to prevent the supporting material from being crushed in bearing. Finally, the thickness t is evaluated so that the bearing plate has sufficient flexural strength. Note that the compression flange of the beam must be properly braced laterally at the support, as indicated in Figs. 9.9.1a and b.

9.9.1   Beam Web Local Yielding

Beam web local yielding at a support is the result of high stress concentration at the junction of the flange and web, where the transfer of pressure from the relatively wide flange to a narrow web takes place. The section at the web toe of the fillet is the most critical location for failure. The pressure is assumed to spread out from the edge of the bearing plate at a slope of 2.5 horizontal to 1 vertical. At the limit state, the stress at the critical section is assumed uniform and equal to the yield stress of the web material. The design strength corresponding to localized yielding of the beam web due to the reaction at the member end is given by (LRFDS Eq. K1-3):

$$R_{dy} = \phi R_n = \phi(2.5k + N)t_w F_{yw} \tag{9.9.1}$$

where ϕ = resistance factor ($= 1.0$)
$\quad\quad\ t_w$ = thickness of the beam web, in.
$\quad\quad\ k$ = distance from the bottom of the beam flange to web toe of fillet, in.
$\quad\quad F_{yw}$ = yield stress of the beam web, ksi

Let $N = N_1$ be the length of bearing required for a beam end reaction R_u corresponding to the limit state of web local yielding. By equating R_{dy} from Eq. 9.9.1 to R_u, and solving for N we obtain:

$$N_1 = \frac{R_u - \phi R_1}{\phi R_2} \tag{9.9.2}$$

where

$$\phi R_1 = \phi(2.5kt_w F_{yw}); \quad \phi R_2 = \phi t_w F_{yw} \tag{9.9.3}$$

Values of k are tabulated in Part 1 of the LRFDM for hot-rolled, steel I-shapes. Also, the beam constants ϕR_1 and ϕR_2 are given in the LRFDM Table 9-5: Beam End Bearing Constants for steels having $F_y = 50$ ksi and typical beam shapes.

9.9.2   Beam Web Crippling

From LRFD Specification Section K1-4, the crippling design strength of the beam web at the member end, when $N/d \le 0.2$, is given by:

$$R_{dc} = \phi_r\left(0.40t_w^2\right)\left[1 + 3\left(\frac{N}{d}\right)\left(\frac{t_w}{t_f}\right)^{1.5}\right]\sqrt{\frac{t_f}{t_w}EF_{yw}} \tag{9.9.4}$$

or, when $N/d > 0.2$, by:

$$R_{dc} = \phi_r(0.40t_w^2)\left[1 + \left(\frac{4N}{d} - 0.2\right)\left(\frac{t_w}{t_f}\right)^{1.5}\right]\sqrt{\frac{t_f}{t_w}EF_{yw}} \quad (9.9.5)$$

Here,

d = overall depth of the beam, in.

t_f = thickness of the beam flange, in.

ϕ_r = resistance factor (= 0.75)

Let $N = N_2$ be the length of bearing required for a beam end reaction R_u corresponding to the limit state of web crippling. By equating R_{dc} from Eq. 9.9.4 or 9.9.5 to R_u, solving for N, and rearranging, we obtain:

$$N_2 = \frac{R_u - \phi_r R_3}{\phi_r R_4} \quad \text{when } \frac{N}{d} \le 0.2 \quad (9.9.6)$$

$$N_2 = \frac{R_u - \phi_r R_5}{\phi_r R_6} \quad \text{when } \frac{N}{d} > 0.2 \quad (9.9.7)$$

Here,

$$\phi_r R_3 = 0.75\left[0.4t_w^2\sqrt{\frac{t_f}{t_w}EF_{yw}}\right] \quad (9.9.8a)$$

$$\phi_r R_4 = 0.75\left[0.4t_w^2\left(\frac{3}{d}\right)\left(\frac{t_w}{t_f}\right)^{1.5}\sqrt{\frac{t_f}{t_w}EF_{yw}}\right] \quad (9.9.8b)$$

$$\phi_r R_5 = 0.75\left\{0.4t_w^2\left[1 - 0.2\left(\frac{t_w}{t_f}\right)^{1.5}\right]\sqrt{\frac{t_f}{t_w}EF_{yw}}\right\} \quad (9.9.8c)$$

$$\phi_r R_6 = 0.75\left[0.4t_w^2\left(\frac{4}{d}\right)\left(\frac{t_w}{t_f}\right)^{1.5}\sqrt{\frac{t_f}{t_w}EF_{yw}}\right] \quad (9.9.8d)$$

Values of beam constants $\phi_r R_3$, $\phi_r R_4$, $\phi_r R_5$, and $\phi_r R_6$ are also given in the LRFDM Table 9-5: Beam End Bearing Constants for steels having $F_y = 50$ ksi and typical beam shapes.

Often bearing length as computed above is very small or even negative. In such cases, the bearing length used is determined by practical considerations, such as the length required for safety during erection of the beam, the need for holes in the bearing plate, and so on. Wall thickness, clearance requirements, or other constraints may also place limits on N. However, at beam end reactions the minimum value of N should not be less than the k value of the beam. As a rule, a length of 3 in. is considered as the practical minimum. The bearing length N required for a beam end reaction is then:

$$N \ge \max[N_1, N_2, k, 3 \text{ in.}] \quad (9.9.9)$$

If the value of N provided is less than N_1 or N_2, bearing stiffeners should be provided.

9.9.3 Bearing Strength of Support Material

The bearing plate is assumed to distribute the beam end reaction uniformly to the area of the support material under the bearing plate. The design bearing pressure is dependent on the type of support material (for example, concrete, masonry, etc.). When the support material is concrete, the design bearing pressure depends on the compressive strength of unconfined concrete, f'_c and on the degree of confinement of the concrete below the bearing plate. The nominal bearing pressure of concrete is assumed to vary linearly with its compressive strength f'_c. This is a reasonable assumption for compressive strengths up to 6 ksi. The degree of confinement affects the triaxial state of stress in the concrete below the bearing plate. Thus, when the concrete support is wider on all sides than the loaded area $A_1 (= B \times N)$, the bearing strength also increases in proportion to the square root of the ratio A_2/A_1. Here, A_2 is the area of the concrete support. If A_2 is not concentric with A_1, then A_2 should be taken as the largest concentric area that is geometrically similar to A_1 (for details, see Fig. 9.10.2). The limiting ratio of A_2/A_1 above which no increase in bearing strength occurs is assumed to be 4.0. The LRFDS relation for bearing strength of concrete, given in Section J9, is essentially based on the ACI 318 Code [1999] requirements. The design bearing load on concrete, for a beam bearing plate, can be written with the help of LRFDS Eq. J9-2 as:

$$R_{dp} = \phi_c R_{np} = \phi_c F_p A_1 = \phi_c (0.85 \, f'_c \, \beta) BN \qquad (9.9.10)$$

with

$$\beta = \min \, [\sqrt{\rho}, 2]; \quad \rho = \frac{A_2}{A_1} \qquad (9.9.11a \text{ and } b)$$

where
R_{dp} = design bearing load on concrete, kips
R_{np} = nominal bearing load on concrete, kips
ϕ_c = resistance factor for bearing on concrete (= 0.60)
f'_c = 28-day compressive strength of concrete, ksi
F_p = nominal bearing pressure, ksi
A_1 = area of steel bearing plate, in.^{2} = BN
A_2 = effective support area, below the bearing plate and within the body of the support, that is geometrically similar to and concentric with the area BN, in.^{2}
β = confinement coefficient

By equating R_{dp} from Eq. 9.9.10 to R_u, and solving for B:

$$B = \frac{R_u}{\phi_c (0.85 f'_c \, \beta) N} \qquad (9.9.12)$$

The value of B is rounded to the next higher inch.

9.9.4   Bending Strength of Bearing Plate

The beam end reaction R_u is assumed to be transferred from the beam web to the bearing plate over an area equal to $2k_1N$. The bearing plate, in turn is assumed to distribute this load uniformly onto the concrete surface over an area BN as shown in Fig. 9.9.1b. The required bearing plate thickness is determined by considering a unit strip (a 1-in.-wide strip) of the bearing plate and assuming it to act as a cantilever beam of length n where $n = \frac{1}{2} B - k_1$, and which is subjected to a uniformly distributed upward pressure f_p. In reality, the end slope of the loaded beam increases the stress at the inner face of the wall (Fig. 9.9.1a). The bending of the bearing plate and the bottom flange of the beam, in the transverse direction, will reduce the bearing pressure at the edges of these members. Thus, the moment computed from the assumed (uniform) pressure curve is appreciably greater than that computed from the probable pressure curve. The contribution of the beam flange overlapping the bearing plate in resisting bending is ignored in the design, another conservative assumption. The critical section for bending is taken at the flange toe of the fillet, a distance k_1 from the center of the web. The following relations can be written with the help of Figs. 9.9.1b and d:

Required moment capacity, $\quad M_u = f_p \, (1 \times n) \dfrac{n}{2} = \dfrac{R_u}{BN} \dfrac{n^2}{2}$

Design bending strength, $\quad M_d = \phi_b M_p = \phi_b Z F_{ypl} = 0.90 \left[\dfrac{1}{4} (1)(t^2) \right] F_{ypl}$

The limit state of bending for the bearing plate, $M_d \geq M_u$, therefore results in:

$$t \geq \sqrt{\dfrac{2n^2 R_u}{0.90 BN F_{ypl}}} \tag{9.9.13}$$

where t = thickness of bearing plate, in.
$\quad\quad\quad N$ = length of bearing plate, in.
$\quad\quad\quad B$ = width of bearing plate, in.
$\quad\quad\quad F_{ypl}$ = yield stress of bearing plate material, ksi
$\quad\quad\quad R_u$ = beam reaction under factored loads, kips
$\quad\quad\quad n$ = $(B/2) - k_1$
$\quad\quad\quad k_1$ = distance from the mid-thickness of the web to the flange toe of the flange-to-web fillet, in.

High allowable bearing pressures and/or relatively light loads result in small required bearing areas, and/or bearing plate thickness. For such cases, the beam may rest (without bearing plates) directly on the support. However, the beam's bottom flange should be checked for transverse bending such that:

$$t_f \geq \sqrt{\frac{2n_f^2 R_u}{0.90 b_f N F_{yf}}} \qquad (9.9.14)$$

where $n_f = \frac{1}{2}b_f - k_1$ and F_{yf} is the yield stress of the flange material, which in this case acts like a bearing plate.

EXAMPLE 9.9.1

Design; Beam Bearing Plate

Determine the size of bearing plate for W24×62 A992 steel beam for an end reaction of 120 kips, under factored loads. The beam rests on a concrete wall with a 28-day compressive strength of 4.0 ksi.

Solution

a. Data

From Table 1-1 of the LRFDM for a W24×62:

$$d = 23.7 \text{ in.}; \qquad b_f = 7.04 \text{ in.}$$

$$k = 1.19 \text{ in.}; \qquad k_1 = \frac{17}{16} = 1.06 \text{ in.}$$

Beam reaction under factored loads, $R_u = 120$ kips

b. Length of bearing plate, N

Assume $\dfrac{N}{d} > 0.2$, which requires $N > 0.2(23.7) = 4.74$ in.

From LRFDM Table 9-5: Beam End Bearing Constants, for a W24×62 of A992 steel:

$$\phi_v V_n = 275 \text{ kips} > R_u = 120 \text{ kips} \qquad \text{O.K.}$$

$$\phi R_1 = 63.9 \text{ kips}; \qquad \phi R_2 = 21.5 \text{ kli}$$

$$\phi_r R_5 = 68.5 \text{ kips}; \qquad \phi_r R_6 = 8.22 \text{ kli}$$

Minimum length of bearing plate to prevent web local yielding,

$$N_1 = \frac{R_u - \phi R_1}{\phi R_2} = \frac{120 - 63.9}{21.5} = 2.61 \text{ in.}$$

Minimum length of bearing plate to prevent web crippling,

$$N_2 = \frac{R_u - \phi_r R_5}{\phi_r R_6} = \frac{120 - 68.5}{8.22} = 6.27 \text{ in.}$$

Try $N = 8$ in.

Check: $N/d = 8.00/23.7 = 0.338 > 0.2$, as assumed.   O.K.

c. Width of bearing plate, B

Assume conservatively that the bearing plate covers the full area of the support. Hence, the confinement factor $\beta = 1$ and the required plate width can be found from:

$$\phi_c(0.85f_c'\beta)\,BN = 0.6(0.85)(4.0)(1.0)[B(8.00)] \geq 120$$

$$B \geq 7.35 \text{ say 8 in., rounding to the nearest inch.}$$

The flange width of a W24×62 is 7.04 in., making the bearing plate slightly wider than the beam flange.

d. Thickness of the bearing plate, t

The span of the cantilever strip is:

$$n = \frac{B - 2k_1}{2} = \frac{8.00 - 2(1.06)}{2} = 2.94 \text{ in.}$$

The required thickness is:

$$t \geq \sqrt{\frac{2n^2R_u}{\phi_b BN F_{ypl}}} = \sqrt{\frac{2(2.94^2)(120)}{0.9(8)(8)(36)}} = 1.00 \text{ in.}$$

Use a PL1×8 ×0'-8 of A36 steel.   (Ans.)

9.10   Column Base Plates

Steel columns usually rest on concrete footings or pedestals in order to transmit the load to the foundation. Compressive stresses in steel columns are considerably higher than the design compressive stress of concrete. Hence, a rolled steel plate, called a *base plate,* is generally inserted between the column and the concrete footing to distribute the column load over a sufficient area of the footing, thereby preventing crushing of the concrete (Figure 9.10.1). Since the base plate projects outside of the column dimensions, the bearing pressure from the foundation below produces flexure in the base plate. However, in contrast to beam bearing plates in which bending is in one direction, column base plates are subjected to two-way bending, and each direction must be considered separately. The column must be anchored to the base plate and the plate in turn anchored to the concrete foundation.

For small columns, base plates can be shop welded directly to the column. To prepare for erection of such columns, leveling plates—¼-in.-thick steel plates sheared to the same size as the base plates, sent to the field in advance of the main column, and grouted in place at the correct elevation—are generally used (Fig 9.10.1a). Anchor rods, to secure the columns in position, are embedded in the concrete footing and project above the base plate to engage washers and nuts. For large columns, the base plates are usually shipped loose, set to proper level, and grouted in place before the erection of columns.

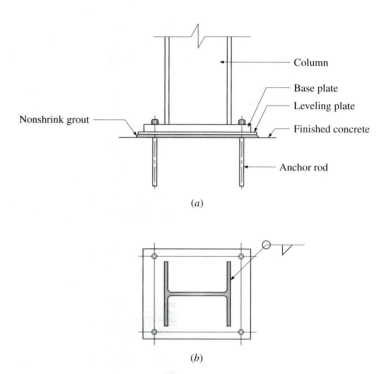

Figure 9.10.1:   Column base plate details.

In general, base plates of columns in braced frames with simple beam-to-column connections (Fig. 3.3.5) are only subjected to axial compressive forces under gravity loads (load combinations LC-1, LC-2, and LC-3). Base plates of columns within braced bays of such structures are also generally subjected to shear forces in addition to axial forces due to lateral loads (load combinations LC-3 and LC-4) transmitted by the diagonal member connected to the column base. Base plates of columns in unbraced frames are subjected to bending moments in addition to axial compressive forces and shear forces, if the bases are fixed. Under certain conditions, base plates may also be subject to axial tension, or uplift forces. The maximum uplift force, if any, on the base is generally given by the load combination LC-6 or LC-7. Simple column base details, appropriate for columns assumed to be pinned at the base, are illustrated in Figs. 9.10.1a and b. Actually, though, these base plates do carry some moment and are useful in resisting frame instability, as discussed in Chapter 8.

9.10.1   Axially Loaded Base Plates

Let B be the width of the base plate (dimension parallel to column flange), N the length (dimension parallel to column web), and t the thickness (Figs. 9.10.2a and b). Design of axially loaded column base plates is based on the assumption that the factored axial load, P_u, from the column is distributed uniformly over an equivalent rectangular area on the top surface of the base

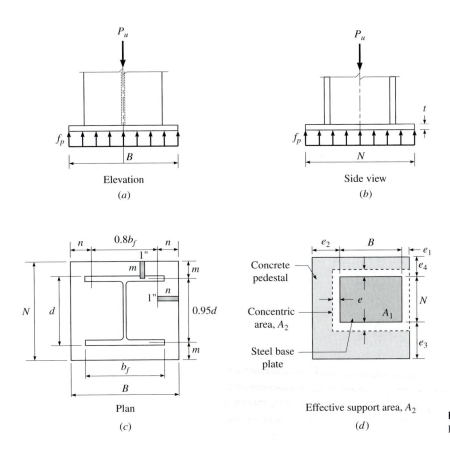

Figure 9.10.2: Base plate design for an axially loaded column

plate. The recommended analysis of the base plates for I-shaped columns, given in Part 14 of the LRFDM, assumes this equivalent rectangular area to be $0.80b_f$ by $0.95d$ where b_f and d are the flange width and depth of the column section, respectively (Fig. 9.10.2c). Within this area the plate is assumed fixed (rigid), and outside this area it is assumed to bend into a concave saucer-shaped surface. The distribution of bearing pressure on the bottom of the base plate from the foundation depends on the relative stiffness of the plate and the foundation. It is nonuniform and somewhat bell-shaped with greater pressure within the rectangular (stiffer) portion defined above and less pressure in the cantilevered (flexible) portions. Even if this distribution were known, the resulting stresses in the plate could not be determined easily, since bending occurs in two directions. So, the analysis of steel base plates is generally based on the following additional assumptions:

- The bearing pressure is uniformly distributed over the plate area BN.
- Those portions of the plate which project from the equivalent rectangular area act as inverted cantilever beams. That is, bending is considered to be in one direction only.

Note that both these assumptions are generally conservative. The LRFD Specification does not give a particular method for designing column base plates. The method presented below follows the one given in Part 14 of the LRFDM.

The design axial load on a column base plate, corresponding to the limit state of bearing on concrete, may be written (with the help of LRFDS Eq. J9-2) as:

$$P_{dp} = \phi_c F_{np} A_1 = \phi_c [0.85 f_c' \beta] BN \qquad (9.10.1)$$

with

$$\beta = \min[\sqrt{\rho}, \ 2] \qquad (9.10.2a)$$

$$\rho = \frac{A_2}{A_1} \qquad (9.10.2b)$$

$$A_2 = (B + 2e)(N + 2e) \qquad (9.10.2c)$$

$$e = \min(e_1, e_2, e_3, e_4) \qquad (9.10.2d)$$

where P_{dp} = design bearing load on concrete, kips
ϕ_c = resistance factor for bearing of concrete (= 0.60)
F_{np} = nominal bearing pressure, ksi
A_1 = area of steel base plate, in.^{2} = BN
f_c' = specified 28-day compressive strength of concrete, ksi
e_1, e_2, e_3, e_4 = projections, if any, of the pedestal beyond the base plate, in plan view, as shown in Fig. 9.10.2d, in.
A_2 = effective support area, below the base plate and within the body of the footing, that is geometrically similar to and concentric with the area A_1, in.^{2}
β = confinement coefficient

The general relation for LRFD of a base plate, corresponding to the limit state of bearing on concrete, is:

$$P_{dp} \geq P_u \qquad (9.10.3)$$

where P_{dp} = design bearing load on concrete, kips
P_u = factored axial load on the column, kips

When the plate dimensions B and N are significantly larger than the dimensions b_f and d of the steel shape, the conventional method described below is used.

The overhangs m and n of the base plate parallel and perpendicular to the web, respectively, are:

$$m = \frac{(N - 0.95d)}{2}; \quad n = \frac{(B - 0.80 b_f)}{2} \qquad (9.10.4)$$

By considering the overhangs of unit width and of length m and n to bend as inverted cantilevers, two limit states of bending can be written for the two critical sections. Thus, for the cantilever strip parallel to the column web, the required bending strength is:

$$M_u = f_p (1) m \frac{m}{2} = f_p \frac{m^2}{2} \tag{9.10.5}$$

where f_p is the actual bearing pressure on the contact surface ($= P_u / BN$). The design bending strength of this cantilever (a rectangular plate of unit width, thickness t and yield stress F_{ypl}) is

$$\phi_b M_p = \phi_b Z F_{ypl} = \phi_b \frac{1}{4} (1) t^2 F_{ypl} \tag{9.10.6}$$

with the resistance factor $\phi_b = 0.90$. The plate thickness is then obtained from the relation:

$$\phi_b M_p \geq M_u \rightarrow t_1 \geq m \sqrt{\frac{2P_u}{\phi_b F_{ypl} BN}} \tag{9.10.7}$$

Similarly, considering a cantilever of unit width perpendicular to the column web:

$$t_2 \geq n \sqrt{\frac{2P_u}{\phi_b F_{ypl} BN}} \tag{9.10.8}$$

Columns with factored axial loads less than their full axial design strengths, and/or foundations with high values of F_{np}, generally require small base plates. For example, columns of low-rise conventional construction and columns in preengineered metal buildings often result in lightly loaded base plates. A **lightly loaded base plate** is one where the required base plate area A_1 is approximately equal to or even less than the column flange width times the column depth. It is seen from Eqs. 9.10.7 and 9.10.8 that, if the values of m and n for a particular base plate are close to zero, the plate thickness computed from the above expressions will also be close to zero. Such a value for thickness is not realistic. Lightly loaded base plates were studied by several researchers [Murray, 1983; Thornton, 1990a, b]. DeWolf and Ricker [1990] presented a compilation of existing information on the design of base plates. For lightly loaded base plates, the critical moment in the base plate occurs at the face of the column web halfway between the sides of the flanges. Also, thin base plates lift off the foundation during loading. Thus, the assumption of uniform stress distribution at the interface, made in the design of thick plates, is no longer valid. The currently accepted procedure given in Part 14 of the LRFDM considers a portion of the lightly loaded base plate contained within the column depth and width as two rectangular plates of dimensions $\frac{b_f}{2}$ by d. Each plate is assumed to be completely fixed to the column web along the side of length d, simply supported along the sides of length $b_f/2$, and free along the fourth edge. The required plate thickness for a lightly loaded base plate is given by [Thornton, 1990b] and Part 14 of the LRFDM:

$$t_3 \geq n^* \sqrt{\frac{2P_u}{\phi_b F_{ypl} BN}} \tag{9.10.9}$$

where

$$n^* = \lambda n'; \quad n' = \frac{1}{4}\sqrt{b_f d}$$

$$\lambda = \min\left[\frac{2\sqrt{x}}{1 + \sqrt{1 - x}}, 1\right] \tag{9.10.10}$$

$$x = \left[\frac{4b_f d}{(b_f + d)^2}\right]\frac{P_u}{P_{dp}}$$

Since both the terms in brackets and the ratio of P_u to P_{dp} are always less than or equal to one, the value of x will always be less than or equal to one.

Finally, the required thickness of the base plate is determined as the maximum thickness obtained through the preceding analyses.

$$t \geq \max[t_1, t_2, t_3] \tag{9.10.11}$$

Design Procedure

The steps in the design of base plates for axially loaded steel I-shaped columns are as follows:

1. Determine the factored axial compressive load P_u on the base plate using appropriate load combinations.
2. If the area of the pedestal A_2 is given, determine the required plate area A_{1req} from:

$$A_{1req} = \max\left\{\frac{1}{A_2}\left[\frac{P_u}{\phi_c(0.85f'_c)}\right]^2, \frac{P_u}{2\phi_c(0.85f'_c)}, b_f d\right\} \tag{9.10.12}$$

If the ratio $A_2/A_1 = \rho$ is given, determine the required plate area A_1 from:

$$A_{1req} = \max\left[\frac{P_u}{\phi_c(0.85 f'_c \beta)}, \frac{P_u}{\phi_c(0.85 f'_c)}, b_f d\right] \tag{9.10.13}$$

Here b_f and d are the dimensions of the column section, f'_c is the cylinder strength of concrete and $\phi_c = 0.6$.

3. Establish B and N (width and length of base plate) such that the projections m and n are approximately equal and $BN > A_{1req}$. this condition is

achieved by letting:

$$B \approx \sqrt{A_{1req}} - 0.5(0.95d - 0.80b_f) > b_f \qquad (9.10.14)$$

Round the value of B to the nearest inch. Next calculate:

$$N = A_{1req}/B > d \qquad (9.10.15)$$

rounding the value to the next higher integer number. Often, at least one of the dimensions, B or N, is made an even number. When either value (B or N) is limited by other considerations, such as interference with a gusset plate (for connecting to a diagonal brace), length of a shear lug, and availability of plates of a certain width, that dimension is established first and the other is obtained from:

$$BN \geq A_{1req} \qquad (9.10.16)$$

Calculate the actual value of area A_1 provided ($= BN$).
4. Next determine the projections m and n.
5. Evaluate the parameters β, P_{dp}, x, λ, n', and the projection n^*.
6. Determine $l = \max (m, n, n^*)$.
7. Calculate:

$$t_{req} = l\sqrt{\frac{2P_u}{0.9 F_{ypl} BN}} \qquad (9.10.17)$$

This is the design thickness of the base plate and is rounded off to the nearest higher $\frac{1}{8}$ in. for thicknesses up to $1\frac{1}{4}$ in. and to the nearest higher $\frac{1}{4}$ in. thereafter.
8. Base plates not greater than 2 in. thick need not be milled if satisfactory contact in the bearing is present. Base plates greater than 2 in. thick, but not greater than 4 in. thick, must be straightened by either pressing or milling to obtain satisfactory contact in the bearing, at the option of the fabricator. Base plates greater than 4 in. thick must be finished if the bearing area does not meet flatness tolerance. Also, finishing of base plates is not required in the following cases: (1) bottom surfaces of base plates when grout is used to ensure full contact on foundations, and (2) top surfaces of base plates when complete-joint-penetration groove welds are provided between the column and the base plate. The thickness of the base plate to be ordered is obtained by adding a finishing allowance (t_{fin}) to the design thickness t. That is:

$$t_o = t + t_{fin} \qquad (9.10.18)$$

LRFDM Table 14-1 provides finish allowances for carbon steel base plates based on the width, thickness, and whether one or two sides are to be finished.

EXAMPLE 9.10.1   **Design; Lightly Loaded Column Base Plate**

A W10×45 column of A992 steel supports a factored axial load of 340 kips. Design a base plate for the column if the supporting concrete has a cylinder strength $f'_c = 3$ ksi and the base plate is of A36 steel. Assume that the area of the concrete support is large compared to that of the base plate.

(Note: the column considered in this example is the interior column C of the building shown in Fig. 5.1.1, for which the required axial strength is calculated in Example 5.2.4 and the member selected in Example 8.10.2).

Solution

1. Data:

   Factored axial load, $P_u = 340$ kips

   W10×45 column: $b_f = 8.02$ in.; $d = 10.1$ in.

   Concrete: $f'_c = 3.0$ ksi

   Base plate: A36 steel $F_y = 36.0$ ksi

   $A_2 >> A_1$ hence $\beta = 2.0$

   $F_{dp} = \phi_c(0.85 f'_c \beta) = 0.6(0.85)(3.0)(2.0) = 3.06$ ksi

   $$A_{1req} = \frac{P_u}{F_{dp}} = \frac{340}{3.06} = 111 \text{ in.}^2 > b_f d = 81.0 \text{ in.}^2 \qquad \text{O.K.}$$

2. Plan dimensions of base plate

   Optimize plan dimensions.

   $\Delta = 0.5(0.95d - 0.80 b_f) = 0.5[(0.95)(10.10) - 0.80(8.02)] = 1.59$ in.

   $$B \approx \sqrt{A_{1req}} - \Delta = \sqrt{111} - 1.59 = 8.95$$

   say 10.0 in. $> b_f = 8.02$ in. \qquad O.K.

   $$N \geq A_{1req}/B = \frac{111}{10.0} = 11.1 \text{ say } 12 \text{ in.} > d = 10.10 \text{ in.} \qquad \text{O.K.}$$

   So, the plate dimensions are 10 by 12 in.

   $$A_1 = 10.0(12.0) = 120 > 111 \text{ in.}^2 \qquad \text{O.K.}$$

3. Plate thickness

   $$m = \frac{1}{2}(N - 0.95d) = \frac{1}{2}[12.0 - 0.95(10.1)] = 1.20 \text{ in.}$$

   $$n = \frac{1}{2}(B - 0.80 b_f) = \frac{1}{2}[10.0 - 0.80(8.02)] = 1.79 \text{ in.}$$

$$P_{dp} = F_{dp} A_1 = 3.06(120) = 367 > 340 \text{ kips} \qquad \text{O.K}$$

$$x = \frac{4b_f d}{(b_f + d)^2} \frac{P_u}{P_{dp}} = \frac{4(81.0)}{(18.12)^2} \left(\frac{340}{367.2}\right) = 0.916$$

$$\lambda = \frac{2\sqrt{x}}{1 + \sqrt{1 - x}} = \frac{2\sqrt{0.916}}{1 + \sqrt{1 - 0.916}} = 1.48 > 1.0 \text{ set } \lambda = 1.0$$

$$n' = \frac{1}{4} \sqrt{b_f d} = \frac{1}{4} \sqrt{81.0} = 2.25 \text{ in.}$$

$$n^* = \lambda n' = 1.0(2.25) = 2.25 \text{ in.}$$

$$l = \max [m, n, n^*] = \max [1.20, 1.79, 2.25] = 2.25 \text{ in.}$$

$$t_{\text{req}} = l \sqrt{\frac{2P_u}{0.9F_y A_1}} = 2.25 \sqrt{\frac{2(340)}{0.9(36)(120)}} = 0.941 \text{ in.}$$

Provide $t = 1$ in. > 0.941 in.
Use a PL1×10×1′-0 of A36 steel as base plate.   (Ans.)

References

9.1   ACI [1999]: *Building Code Requirements for Structural Concrete,* American Concrete Institute, ACI 318–99. Detroit, MI.

9.2   DeWolf, J. T. and Ricker D. T. [1990]: *Column Base Plates, Steel Design Guide Series 1,* Publ. No. D801, AISC, Chicago, IL.

9.3   Fisher, J. M. and West, M. A. [1990]: *Serviceability Design Considerations for Low-Rise Buildings, Steel Design Guide Series 3,* AISC, Chicago, IL.

9.4   Galambos, T. V. and Ellingwood, B. [1986]: "Serviceability Limit States: Deflection," *Journal of Structural Engineering,* ASCE, vol. 112, no. 1, January, pp. 67–84.

9.5   Marino, F. J. [1966]: "Ponding of Two-Way Roof Systems," *Engineering Journal,* AISC, vol. 3, no. 3, July, pp. 93–100.

9.6   Murray, T. M. [1983]: "Design of Lightly Loaded Steel Column Base Plates," *Engineering Journal,* AISC, vol. 20, no. 4, pp. 143–152.

9.7   Ruddy, J. L. [1986]: "Ponding of Concrete Deck Floors," *Engineering Journal,* AISC, vol. 23, no. 3, pp. 107–115.

9.8   Thornton, W. A. [1990a]: "Design of Small Base Plates for Wide Flange Columns," *Engineering Journal,* AISC, vol. 27, no. 3, pp. 108–110.

9.9   Thornton, W. A. [1990b]: "Design of Base Plates for Wide Flange Columns—A Concatenation of Methods," *Engineering Journal,* AISC, vol. 27, no. 4, pp. 173–174.

9.10 Timoshenko, S. and Young, D. H. [1962]: *Elements of Strength of Materials,* 4th Ed., Van Nostrand Co., Princeton.

9.11   Yura, J. A. and Helwig, T. A. [1998]: *Bracing for Stability,* Structural Stability Research Council, USA.

9.12   Yura, J. A., Galambos, T. V., and Ravindra, M. K. [1978]: "The Bending Resistance of Steel Beams," *Journal of the Structural Division,* ASCE, vol. 104, ST9, pp. 1355–1370.

PROBLEMS

P9.1.   Determine the value of S_x, Z_x, α_x of the following shapes, using the dimensions given in the LRFDM.

  (*a*) W35×230

  (*b*) W16×36

  (*c*) WT18×115

  (*d*) WT8×18

P9.2.   Determine the value of S_x, Z_x, α_x of the following built-up sections:

  (*a*) A W16×36 with one ½ by 12 in. plate welded to each flange.

  (*b*) A W16×36 with one ½ by 12 in. plate welded to the top of the flange

  (*c*) A W16×36 with a C12 by 20.7 with its web welded to the top flange.

P9.3.   Determine the values of M_{yx} and M_{px} for the shapes given in Problem P9.1. Assume A992 steel.

P9.4.   Determine the values of M_{yx} and M_{px} for the built-up shapes given in Problem P9.2. Assume A36 steel for all elements.

P9.5.   (*a*)   A W18×35 beam of A992 steel spans 30 ft and is connected to columns at either end by means of simple shear connections. Compute the uniformly distributed factored load that the member can resist. Assume continuous lateral bracing for the compression flange.

  (*b*)   Find the maximum spacing of lateral supports for the design to still hold good.

P9.6.   Determine the uniformly distributed factored load that can be carried by a W21×62 beam of A572 Grade 42 steel over a simply supported span of 24 ft with lateral supports at 6 ft.

P9.7.   A W24×176 of A572 Grade 60 steel is used for a simple span of 36 ft. If the only dead load present is the weight of the beam, what is the largest service concentrated live load that can be placed at midspan? Assume continuous lateral bracing. Deflection need not be checked.

P9.8.   A triangular opening in the floor of an industrial building results in the factored loads on a W18×35 simple beam as shown in Fig. P9.8. The beam weight is not included. Use A588 Grade 50 steel. Assume full lateral support for the compression flange. Check the adequacy of the beam.

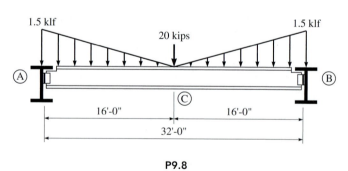

P9.8

P9.9.   A 34-ft-long W27×94 shape of A992 steel is used as a simply supported beam. It is subjected to a factored concentrated load of 80 kips at 12 ft from each support. In addition, it is subjected to a factored moment of 340 ft-kips, one at its left-end (anticlockwise) and one at its right-end (clockwise). Neglect self-weight of the beam in the calculation, and check if the beam is safe. Assume full lateral support for the compression flange.

P9.10. Repeat Problem 9.9 with the end moment acting at the left end only.

P9.11. A wide flange beam AB frames across an open well in a building and is subjected to the factored loads shown in Figure P9.11. Determine if a W16×50 of $F_y = 50$ ksi steel is sufficient under these conditions. Assume continuous lateral support to the compression flange.

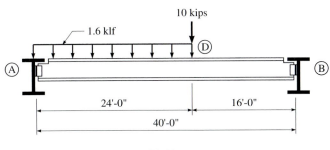

P9.11

P9.12. A simple beam consists of a W18×55 with a PL1×12 cover plate welded to each flange. Determine the factored uniform load the beam can support in addition to its own weight for a 28-ft simple span. Assume A572 Grade 50 steel and full lateral support.

P9.13. A 36-ft-long simple beam consists of a built-up section obtained by welding three A36 steel plates as shown in Fig. P2.6. Determine the factored uniformly distributed load the beam can support, in addition to its self-weight. The beam is continuously laterally supported.

P9.14. A W16×57 of A992 steel is used for the overhanging beam ABCD (Fig. P9.14) under the service loads (50 percent dead and 50 percent live) shown. Assume continuous lateral support, and check the adequacy of the beam for bending and shear.

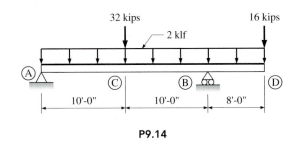

P9.14

Deflections

P9.15. A W14×30 of A992 steel is used as a simply supported, uniformly loaded beam with a span length of 28 ft. The service dead load (not including the self-weight) is 0.4 klf, and the service live load is 0.78 klf. Assume that the beam is continuously laterally supported by the floor deck. Check the adequacy of the beam for bending and shear. Calculate the maximum dead load and live load deflections of the beam.

P9.16. A W30×148 of $F_y = 50$ ksi steel has been selected for the simply supported beam shown in Figure P9.16. Loads Q consist of 50 kips dead load and 80 kips live load. Check the adequacy of the beam for bending and shear. Calculate the maximum dead load and live load deflections. Assume continuous lateral support for the compression flange.

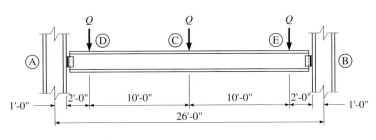

P9.16

Shear

P9.17. (a) A W12×30 of A992 steel is used as a simply supported beam with a span length of 10 ft. It is subjected to two concentrated loads of 80 kips each, located 1 ft from each support. The loads given are service live loads. Neglect self-weight of the beam in the calculations. Check the adequacy of the beam for bending and shear, assuming that the beam is continuously laterally supported.

(b) Redesign, if necessary (limit the selection to W12's only).

Design

P9.18. A large steel beam simply spans 60 ft and carries an applied load of 1.5 klf ($D = 0.5$ klf and $L = 1.0$ klf). Select the lightest adequate W-shape of A992 steel. Include the effect of member self-weight. Assume that the beam is continuously laterally supported.

P9.19. Select the lightest W-section to carry a uniform dead load of 0.3 klf and a live load of 0.6 klf on a simply supported span of 34 ft. The beam is continuously laterally supported. Assume no deflection limitations. Use A572 Grade 42 steel.

P9.20. A simply supported transfer beam AB with a span of 48 ft must carry three column loads of 100 kips each (Fig. P9.20). The loads are factored loads. Assume full lateral support for the compression flange. Select the most economical W-section assuming A992 steel, if the nominal depth is limited to 30 in. Include the effect of member self-weight.

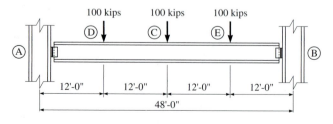

P9.20

P9.21. The 32-ft-long steel beams AB of Figure P9.21 are spaced 8 ft on center. They are simply supported at points D and E, and must carry a service floor load of 100 psf (50 percent dead load and 50 percent live load) and a wall load of 0.6 kips per linear foot of wall (100 percent dead load). Assume adequate lateral support, A992 steel, and select the lightest W-shape for flexure and shear. The maximum service load deflection of the beam should be within ±½ in.

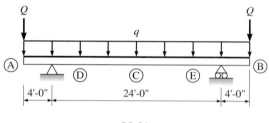

P9.21

P9.22. A simply supported beam with a span of 24 ft has a uniform load of 0.6 klf over the entire span. In addition, it has a 40 kip load 8 ft from the left end and a 30 kip load 8 ft from the other end of the beam. Select a W-shape of A572 Grade 60 steel that can safely support the given factored loads. Assume full lateral support.

P9.23. A simply supported beam AB, with a span of 28 ft, has a 40 kip load at a point C, 16 ft from end A. In addition, the beam is subjected to a uniformly distributed load of 1.6 klf over the portion AC. Select a W-shape of A572 Grade 50 steel that can safely support the given factored loads. Assume full lateral support to the compression flange.

P9.24. A 20-ft-long cantilever beam AB is fixed at end A, and free at end B. It is subjected to a concentrated factored load of 24 kips at point C, where AC = 12 ft. In addition, the beam is subjected to a uniformly distributed factored load of 2.5 klf over its entire length. Select a W-shape of A572 Grade 60 steel that can support the loads. Assume adequate lateral support to the beam flanges.

P9.25. A simply supported beam ABCD with an overhanging end is supported at points B and D such that

AB = 8 ft and BC = CD = 12 ft. The beam is subjected to a uniformly distributed, factored load of 3 klf over the entire length. In addition, it is subjected to concentrated, factored loads of 10 kips and 40 kips, at points A and C, respectively. Select a W-shape of A572 Grade 50 steel that can support the factored loads. Assume adequate lateral support for the compression flange.

P9.26. Select the most economical wide flange section for the 30-ft-long overhang beam, with supports at the left end and 10 ft from the right end. It is subjected to the following service loads: 1.5 klf dead load and 2 klf live load uniformly distributed over the entire length of the beam. In addition, the beam is subjected to concentrated live loads of 30 kips and 10 kips at 10 ft and 30 ft respectively from the left end. The beam is continuously laterally braced. Assume A572 Grade 42 steel.

P9.27. Select the lightest W-shape for a 18-ft-long cantilever beam of A572 Grade 42 steel. It is subjected to concentrated, service live loads of 16 kips and 10 kips acting at 12 ft and 18 ft, respectively, from the fixed end. The beam is continuously laterally braced. The maximum service load deflection is limited to 1/600 of its span length. Neglect beam weight in all the calculations.

P9.28. In a building, roof beams are spaced at 10-ft intervals and are connected to girder webs by simple connections. The span of the beams is 24 ft. The beam is designed for the following service loads: dead load (not including the self-weight), 65 psf; roof live load, 20 psf; snow load, 40 psf; and rain load, 25 psf. Deflection of the beam is limited to $L/240$. Select a suitable W-shape of A242 Grade 42 steel. Assume the beam is continuously laterally supported by the roof decking. Specify any camber provided.

P9.29. A simple beam in a research lab is to support a dead load of 1 klf and a live load of 1.5 klf. In addition, it is to carry a concentrated live load (movable) of 6 kip. The load consists of precision machinery which requires that the deflection be limited to a maximum of $L/800$. The beam has an effective span of 40 ft and is continuously laterally supported. Use A992 steel and select a suitable section.

P9.30. A W-shape is used to support a factored load of 12 klf on a 20-ft simple span. The architect specifies that the beam be no more than 18 in. in depth. Assume full

lateral support to the compression flange, and select a suitable A992 steel beam. Is the deflection acceptable if the beam carries a plastered ceiling? Live load is two-thirds of the total load on the beam.

P9.31. A simply supported beam with continuous lateral support for the entire span of 30 ft is subjected to a factored, uniformly distributed load of 2.4 klf. The beam depth is limited to not more than 14 in. (nominal), and the deflection is limited to $L/360$. Select a suitable A992 shape.

P9.32. To save construction depth with a precast plank system, a structural tee is used for the overhang beam ABC with supports at A and B, as shown in Fig. P9.32. The beam is subjected to a dead load of 0.6 klf and a live load of 0.3 klf. The dead load includes provision for self-weight of the beam. Assume continuous lateral support, and select a suitable WT6 of A992 steel.

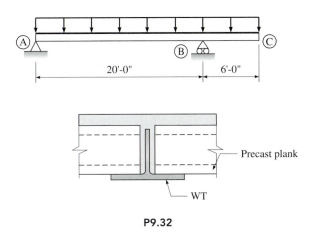

P9.32

P9.33. The bay size for a shopping center is 32′ by 30′. The beams are spaced at 8 ft centers and have a span of 30 ft. All the beams and girders are simply supported. The floor deck consists of 5-in-thick slab of lightweight concrete (100 pcf) directly supported by the beams. Dead load from flooring and ceiling is 10 psf. The floor live load, including provision for partitions, is 120 psf. Design an interior beam without counting on composite action between the slab and the steel beam, but assuming that lateral buckling of the beam is prevented. Limit the live load deflection to $L/360$. Use A992 steel.

P9.34.   Design a spandrel beam of the shopping center given in Problem P9.33. The spandrel beam supports in addition to floor loads, a fascia panel weighing 0.6 klf. There are no clearance limitations, but the total service load deflection is to be limited to $\frac{1}{2}$ in. Assume continuous lateral support.

P9.35.   Design an interior girder of the shopping center described in Problem 9.33. Due to clearance limitations, the nominal depth of the girder is limited to 24 in. Limit the live load deflection to $L/360$. Determine the camber to be specified. Assume continuous lateral support.

P9.36.   The floor framing system of an office building has bay sizes of 27 by 34 ft. The floor beams are at 9 ft centers and have a span of 34 ft. Assume that the beams and girders are simply supported. The floor deck consists of a 5-in.-thick reinforced concrete "one-way" slab of ordinary concrete, directly supported by the beams, and carries a live load of 100 psf. Design the beam without counting on composite action between the slab and the steel beam, but assuming that lateral buckling of the beams is prevented. Check deflection, and specify camber. Use A992 steel.

P9.37.   Design an interior girder of the floor framing system of Problem P9.36. Due to clearance limitations, the depth of girder cannot exceed 27 in. (nominal). Assume continuous lateral support.

P9.38.   Design a spandrel girder of the floor framing system of Problem P9.36. The spandrel girder supports, in addition to beam reactions, a fascia panel weighing 0.8 klf. There are no clearance limitations. Assume continuous lateral support.

P9.39.   Select the most economical W12-section that can carry a concentrated dead load of 40 kips and a live load of 30 kips at the third point of a 6-ft simple span. Check for bending, shear and local buckling. Neglect self-weight of beam. Assume adequate lateral bracing to the compression flange.

P9.40.   Select the lightest beam section to carry a uniformly distributed, factored load of 20 klf on a simple span of 8 ft. Compression flange is adequately supported. Use A992 steel.

P9.41.   Select the lightest W-section to support a factored uniformly distributed load of 2.6 klf over a simple span of 25 ft. Assume A572 Grade 60 steel and continuous lateral support. Make all checks. The given load does not include self-weight of the beam.

P9.42.   Determine the size of the bearing plate required for an end reaction, under a factored load of 90 kips, for a W16×45 A992 steel beam. The beam rests on a concrete wall with a 28-day compressive strength of 3.0 ksi.

P9.43.   A W14×426 column of A992 steel supports a factored axial load of 5310 kips. Design a base plate for the column if the supporting concrete has a cylinder strength $f'_c = 3$ ksi. Assume that (*a*) The full area of the concrete support is covered by the base plate, (*b*) The support will be a 4 ft by 4 ft concrete pier.

P9.44.   A W10×45 column of A992 steel supports a factored axial load of 565 kips. Design a base plate for the column if the supporting concrete has a cylinder strength $f'_c = 3.0$ ksi. Assume that (*a*) The full area of the concrete support is covered by the base plate, (*b*) The footing size is 20 in. by 20 in.

Troy Community Center, Troy, Michigan.

The 625 ton steel skeleton incorporated many curved, rolled steel beams and sweeping members to accentuate the variety of functions the building will serve.

Photo courtesy Douglass Steel Fabricating Corp., Lansing MI, www.douglassteel.com

C H A P T E R

Unbraced Beams

10

10.1   Introduction

Beams are often provided with lateral bracing at intervals, in the form of purlins, girts, cross beams, cross frames, ties, or struts, framing in laterally, where the lateral system is itself adequately stiff and braced. Chapter 9 considered the behavior and design of adequately braced compact I-beams. Such a member, when loaded in the plane of its web, will deflect only in that plane until the maximum bending strength is attained. However, if a beam does not have sufficient lateral stiffness and/or lateral support to ensure such in-plane behavior, the beam will buckle out of the plane of loading and twist. The load at which this lateral-torsional buckling occurs may be substantially lower than the beam's in-plane load carrying capacity considered in Chapter 9. For idealized, perfectly straight, compact I-beams having long unbraced lengths, lateral-torsional buckling occurs in the elastic domain. A perfectly straight beam of intermediate unbraced length may yield before the elastic critical load is reached (due to the combined effects of in-plane bending stresses and residual stresses), and may subsequently buckle inelastically.

Elastic lateral buckling of beams with various support conditions and loading types will be studied in Section 10.2. Inelastic lateral-torsional buckling of beams of I-shaped sections is considered in Section 10.3, while load and resistance factor design of compact I-shaped beams is considered in Section 10.4. Beams of noncompact sections are considered in Section 10.5. Finally, the flexural strength of compact beams having box, channel, and tee cross-sectional shapes is considered in Section 10.6.

10.2   Elastic Lateral-Torsional Buckling of Beams of I-Shaped Sections

Torsionally Simple I-Beam under Uniform Moment

Consider a simply supported, prismatic, doubly symmetric I-beam AB of length L. Let us say that this beam is subjected to a pair of equal and opposite end moments $M°$ applied in the plane of the web (Figs. 10.2.1a and b). The beam is therefore in a state of pure bending about the Z axis. The beam

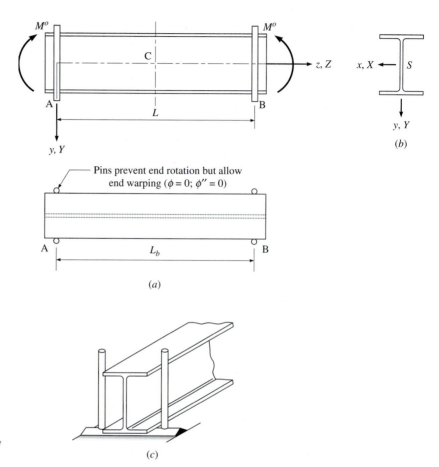

Figure 10.2.1: Laterally and torsionally simply supported I-beam under uniform moment.

is supported laterally at both ends in such a way that lateral deflection and twist of the end sections about the longitudinal axis of the beam are prevented. However, the ends of the beam are free to rotate about the two principal axes of the cross section, the X and Y axes, and the beam ends are free to warp. The support conditions assumed here are shown schematically in Figs. 10.2.1a and 10.2.1c. Here, the distance between the lateral supports is indicated as L_b. In practice, any beam end-connection in which restraint is provided only to the web of the beam approximates the support conditions assumed here. An example is a beam connected to a column by a pair of flexible web (cleat) angles (see Fig. 3.2.2b).

Under the action of the applied moment, the beam will initially bend in the plane of the web. As the moment M^o is increased gradually, a limit state will be reached at which the in-plane bending of the beam becomes unstable and a slightly deflected and twisted form of the beam becomes possible. Such neutral equilibrium behavior is called *lateral-torsional buckling (LTB)* or simply **lateral buckling** of the beam. The lowest moment M_{cr}^o at which this condition occurs defines the **critical moment** or **lateral buckling load of the beam.**

We define two sets of coordinate systems. The X-Y-Z axes are fixed with respect to the original, undeformed position of the member; the x-y-z axes are local coordinate axes that are fixed with respect to each cross section but move with the deformed position of the member (Fig. 10.2.2c). The x and y axes coincide with the principal axes of the cross section. The z axis is always tangent to the longitudinal z axis of the deflected member.

The buckled position of the beam at any section z is defined by the in-plane displacement, v, and the out-of-plane displacement, u, of the shear center, S, and the angle of twist of the cross section, ϕ. We will now develop three equilibrium equations for the beam in its deformed position, in terms of the three displacement variables u, v, and ϕ.

From Fig. 10.2.2, the components of the external moments acting on a cross section at a distance Z from the origin with respect to X-Y-Z coordinates are:

$$(M_X)_{ext} = M^o; \qquad (M_Y)_{ext} = 0; \qquad (M_Z)_{ext} = 0 \qquad (10.2.1)$$

The external bending and twisting moments about the x, y and z axes are next found by taking the components of M^o about each axis, as shown in Figs. 10.2.2b and c. We have, for small θ and ϕ:

$$(M_x)_{ext} = (M_X)_{ext} \cos\phi \cos\theta \approx (M_X)_{ext} \approx M^o$$

$$(M_y)_{ext} = (M_X)_{ext} \sin\phi \cos\theta \approx \phi(M_X)_{ext} \approx \phi M^o \qquad (10.2.2)$$

$$(M_z)_{ext} = (M_X)_{ext} \sin\theta \approx (M_X)_{ext} \tan\theta \approx \frac{du}{dz}(M_X)_{ext} \approx u'M^o$$

Here ϕM^o is the lateral bending moment induced by the twisting ϕ, and $u'M^o$ is the torque induced by the lateral deflection, u, of the beam. This tendency of the beam to twist is resisted by a combination of St. Venant torsion and warping torsion.

Since the displacements u and v, and their respective slopes are assumed to be small, curvatures in the principal planes of the deflected cross section may be taken as u'' and v''. The internal bending and torsional resistances are therefore given by:

$$(M_x)_{int} = -EI_x v'' \qquad (10.2.3a)$$

$$(M_y)_{int} = -EI_y u'' \qquad (10.2.3b)$$

$$(M_z)_{int} = GJ\phi' - EC_w \phi''' \qquad (10.2.3c)$$

Here, each prime (') indicates differentiation with respect to z once. Also, E is the modulus of elasticity, G is the shear modulus, I_x is the moment of inertia about the x-axis, I_y is the moment of inertia about the y-axis, J denotes the

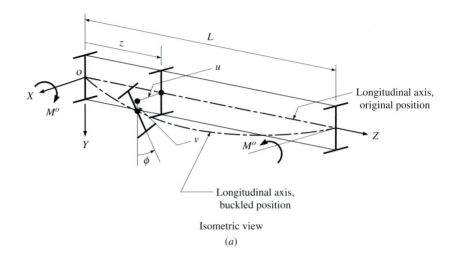

Isometric view

(a)

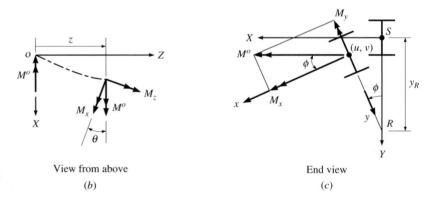

View from above

(b)

End view

(c)

Figure 10.2.2: Lateral-torsional buckling of a simply supported beam.

St. Venant torsional constant, and C_w represents the warping torsional constant of the cross section. Thus EI_x is the major axis flexural rigidity, EI_y is the minor axis flexural rigidity, GJ is the torsional rigidity, and EC_w is the warping rigidity of the beam cross section. The minus sign in Eq. 10.2.3a indicates that a positive deflection in the y-z plane will result in negative curvature.

By equating the internal moments to the external moments, we obtain the three equilibrium equations:

$$EI_x v'' + M^o = 0 \tag{10.2.4a}$$

$$EI_y u'' + M^o \phi = 0 \tag{10.2.4b}$$

$$GJ\phi' - EC_w \phi''' - M^o u' = 0 \tag{10.2.4c}$$

These are the three governing differential equations for the lateral-torsional buckling of a symmetric I-section beam under uniform moment.

The first equation contains only the variable v and thus is uncoupled from the other two equilibrium equations. It describes the in-plane behavior of the beam before lateral buckling occurs. The vertical deflection in the plane of the web, v, can be obtained by integrating Eq. 10.2.4a twice, and using the boundary conditions $v = 0$ at both $z = 0$ and $z = L$, we have:

$$v = \frac{M^o L^2}{2EI_x}\left[\left(\frac{z}{L}\right) - \left(\frac{z}{L}\right)^2\right] \tag{10.2.5}$$

The maximum deflection v_C, which occurs at the center of the beam, is obtained by setting $z = L/2$ in this expression, resulting in:

$$v_C = \frac{M^o L^2}{8EI_x} \tag{10.2.6}$$

Thus, for a given beam of length L and elastic rigidity EI_x there is initially a linear relationship between M^o and v_C (Fig. 10.2.3a).

Equations 10.2.4b and 10.2.4c are coupled, since both contain the deformation variables u and ϕ. Because the buckling deformations are coupled, the analysis of the lateral buckling problem is inherently more complex than that presented in Chapter 8 for flexural (planar) buckling of columns. By differentiating Eq. 10.2.4c once with respect to z, and substituting the value for u'' from Eq. 10.2.4b into it, we obtain a single relation:

$$EC_w \phi'''' - GJ\phi'' - \frac{(M^o)^2}{EI_y}\phi = 0 \tag{10.2.7}$$

or, dividing both sides by EC_w:

$$\phi'''' - \frac{GJ}{EC_w}\phi'' - \frac{(M^o)^2}{EI_y EC_w}\phi = 0 \tag{10.2.8}$$

This is a fourth-order differential equation with constant coefficients that describes the lateral-torsional buckling behavior of the beam under consideration. Introducing the notations α and β such that:

$$2\alpha = \frac{GJ}{EC_w}; \quad \beta = \frac{(M^o)^2}{EI_y EC_w} \tag{10.2.9a and b}$$

Eq. 10.2.8 may be rewritten as:

$$\phi'''' - 2\alpha\phi'' - \beta\phi = 0 \tag{10.2.10}$$

Assume a solution of the form:

$$\phi = Ce^{mz} \tag{10.2.11}$$

resulting in:

$$\phi'' = Cm^2 e^{mz}; \quad \phi'''' = Cm^4 e^{mz} \tag{10.2.12}$$

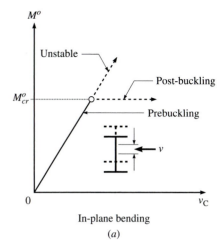

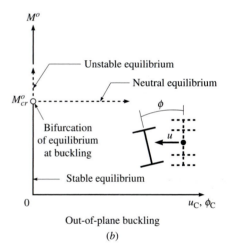

Figure 10.2.3: Moment vs. deformation curves for a simply supported I-beam.

Substitution of Eqs. 10.2.11 and 10.2.12 into Eq. 10.2.10 results in:

$$Ce^{mz}(m^4 - 2\alpha m^2 - \beta) = 0 \qquad (10.2.13)$$

As e^{mz} cannot be zero: either $C = 0$, or the term within the brackets equals zero. If $C = 0$, $\phi = 0$ in view of Eq. 10.2.11, indicating that the beam remains straight (no buckling), a trivial solution. So, at buckling,

$$m^4 - 2\alpha m^2 - \beta = 0 \qquad (10.2.14)$$

which has the solution:

$$m = \pm \sqrt{\alpha \pm \sqrt{\alpha^2 + \beta}} \qquad (10.2.15)$$

Since $\beta > 0$ from Eq. 10.2.9b, $\sqrt{\alpha^2 + \beta} > \alpha$, and we observe that m has two real roots, ω and $-\omega$, and two complex roots, $i\Omega$ and $-i\Omega$, where:

$$\omega = +\sqrt{\alpha + \sqrt{\alpha^2 + \beta}} \qquad \text{(real root)} \qquad (10.2.16a)$$

$$\Omega = +\sqrt{-\alpha + \sqrt{\alpha^2 + \beta}} \quad \text{(real part of complex root)} \qquad (10.2.16b)$$

and i is the imaginary number. Using these four values for m, the expression for ϕ from Eq. 10.2.11 may be written as:

$$\phi = C_1 e^{\omega z} + C_2 e^{-\omega z} + C_3 e^{i\Omega z} + C_4 e^{-i\Omega z} \qquad (10.2.17)$$

The exponential functions may be expressed in terms of circular and hyperbolic functions using the Euler formulas:

$$e^{\omega z} = \cosh \omega z + \sinh \omega z; \quad e^{-\omega z} = \cosh \omega z - \sinh \omega z$$
$$e^{i\Omega z} = \cos \Omega z + i \sin \Omega z; \quad e^{-i\Omega z} = \cos \Omega z - i \sin \Omega z \qquad (10.2.18)$$

By substituting these values for the exponential functions and defining new constants A_1, A_2, A_3 and A_4 such that:

$$A_1 = (C_1 - C_2); \quad A_2 = (C_1 + C_2) \qquad (10.2.19)$$
$$A_3 = (C_3 - C_4)i; \quad A_4 = (C_3 + C_4)i$$

we obtain:

$$\phi = A_1 \sinh \omega z + A_2 \cosh \omega z + A_3 \sin \Omega z + A_4 \cos \Omega z \qquad (10.2.20)$$

where A_1, A_2, A_3 and A_4 are known as constants of integration and ω and Ω are positive real quantities defined by Eq. 10.2.16.

Differentiating Eq. 10.2.20 twice, we obtain:

$$\phi'' = A_1 \omega^2 \sinh \omega z + A_2 \omega^2 \cosh \omega z - A_3 \Omega^2 \sin \Omega z$$
$$- A_4 \Omega^2 \cos \Omega z \qquad (10.2.21)$$

The four constants of integration are determined from the boundary conditions at the ends of the beam. For the torsionally simple end conditions assumed (Fig. 10.2.1), the ends of the beam cannot rotate about the z axis and the ends are free to warp. That is:

$$\phi = 0; \quad \phi'' = 0 \quad \text{at } z = 0 \text{ and } z = L = L_b \qquad (10.2.22)$$

From the two conditions at $z = 0$, we obtain:

$$A_2 + A_4 = 0; \quad \omega^2 A_2 - \Omega^2 A_4 = 0 \qquad (10.2.23)$$

resulting in:

$$A_2 = 0; \quad A_4 = 0 \qquad (10.2.24)$$

Consequently, Eqs. 10.2.20 and 10.2.21 reduce respectively to:

$$\phi = A_1 \sinh \omega z + A_3 \sin \Omega z \qquad (10.2.25)$$

$$\phi'' = A_1 \omega^2 \sinh \omega z - A_3 \Omega^2 \sin \Omega z \qquad (10.2.26)$$

From the two end conditions at $z = L$, we therefore obtain:

$$A_1 \sinh \omega L + A_3 \sin \Omega L = 0 \qquad (10.2.27)$$

$$A_1 \omega^2 \sinh \omega L - A_3 \Omega^2 \sin \Omega L = 0$$

which can be rewritten as:

$$\begin{bmatrix} \sinh \omega L & \sin \Omega L \\ \omega^2 \sinh \omega L & -\Omega^2 \sin \Omega L \end{bmatrix} \begin{bmatrix} A_1 \\ A_3 \end{bmatrix} = \begin{bmatrix} 0 \\ 0 \end{bmatrix} \qquad (10.2.28)$$

This system is always satisfied for $A_1 = A_3 = 0$ in which case from Eq. 10.2.25, $\phi = 0$ everywhere, indicating that the beam is in the stable pre-buckling position. The system has nontrivial solutions when its determinant equals zero. By expanding the determinant and rearranging, we obtain the following characteristic equation or buckling condition:

$$(\omega^2 + \Omega^2) \sinh \omega L \sin \Omega L = 0 \qquad (10.2.29)$$

Since ω and Ω are positive nonzero quantities and as $\sinh \omega L$ is zero only at $\omega L = 0$, that is, at $M^o = 0$, it follows that for nontrivial solutions we must have:

$$\sin \Omega L = 0 \qquad (10.2.30)$$

This relation will be satisfied for certain specific values of Ω, known as the *characteristic* or *eigenvalues* of the system of equations. We have:

$$\Omega L = n\pi \qquad (10.2.31)$$

in which n is an integer (1, 2, 3, . . .). Substituting the value of Ω from Eq. 10.2.31, and using the definitions of α and β from Eq. 10.2.9, in

Eq. 10.2.16b results in:

$$-\frac{GJ}{2EC_w} + \sqrt{\left(\frac{GJ}{2EC_w}\right)^2 + \frac{(M^o)^2}{EI_y\,EC_w}} = \frac{n^2\pi^2}{L^2} \tag{10.2.32}$$

Solving this relation for M^o gives the nth mode of the critical moment as:

$$M^o_{crn} = \frac{\pi}{(L/n)}\sqrt{EI_y\,GJ}\sqrt{1 + \left(\frac{\pi}{L/n}\right)^2\frac{EC_w}{GJ}} \tag{10.2.33}$$

The lowest critical moment corresponds to the lowest value of the integer n ($n = 1$), and is given by:

$$M^o_{cr} = \frac{\pi}{L}\sqrt{EI_y\,GJ}\sqrt{1 + \frac{\pi^2}{L^2}\frac{EC_w}{GJ}} \tag{10.2.34}$$

This expression gives the elastic **lateral-torsional buckling moment** or **critical moment** of a doubly symmetric I-shaped beam with torsionally simple end conditions, under the action of constant moment in the plane of the web.

If we substitute Eq. 10.2.30 into Eq. 10.2.27, we find that $A_1 \sinh \omega L = 0$. As $\sinh \omega L$ is zero only at $\omega L = 0$, it follows that $A_1 = 0$. Thus, from Eqs. 10.2.25 and 10.2.31 we observe that the twist of the buckled beam at the lowest critical moment is given by:

$$\phi = A_3 \sin \Omega z = A_3 \sin \frac{\pi z}{L} \tag{10.2.35a}$$

or

$$\phi = \phi_C \sin \frac{\pi z}{L} \tag{10.2.35b}$$

in which $A_3 = \phi_C$ represents the twist of the beam at midspan C ($z = L/2$) and is of indeterminate magnitude. The lateral deflection at buckling can be obtained by substituting Eq. 10.2.35a into Eq. 10.2.4b, integrating twice, and substituting the boundary conditions $u = 0$ at $z = 0$ and $z = L$. We obtain:

$$u = \frac{M^o_{cr}\,L^2}{(\pi^2 EI_y)}A_3 \sin \frac{\pi z}{L} = u_C \sin \frac{\pi z}{L} \tag{10.2.36}$$

in which u_C represents the indeterminate, central lateral deflection of the beam at buckling. For an idealized perfectly straight elastic beam considered here, there are no out-of-plane deformations until the applied moment reaches the critical value M^o_{cr}, at which point the beam buckles by deflecting laterally a distance, u_C, and twisting through an angle ϕ_C at the center, as shown in Fig. 10.2.3b.

During buckling, a section rotates about a point R in the plane of the web, located a distance y_R below the shear center, S (Fig. 10.2.2c). For small values of ϕ, we have with the help of Eqs. 10.2.35a and 10.2.36:

$$y_R = \frac{u}{\phi} = \frac{M_{cr}^o L^2}{\pi^2 E I_y} \qquad (10.2.37)$$

Substitution of the value of M_{cr}^o from Eq. 10.2.34 in this relation results in:

$$y_R = \left(\frac{L}{\pi}\right) \sqrt{\frac{GJ}{EI_y}} \sqrt{\left(1 + \frac{\pi^2}{L^2}\frac{EC_w}{GJ}\right)} \qquad (10.2.38a)$$

or, rewriting:

$$y_R = \sqrt{\frac{C_w}{I_y}} \sqrt{\left(1 + \frac{L^2}{\pi^2}\frac{GJ}{EC_w}\right)} \qquad (10.2.38b)$$

The second radical in Eq. 10.2.38a is nearly equal to unity for very long members, and the second radical in Eq. 10.2.38b is nearly equal to unity for very short members. Also, as $C_w = I_y(d - t_f)^2/4$ for I-shapes, we can write:

$$y_R \approx \frac{L}{\pi}\sqrt{\frac{GJ}{EI_y}} \quad \text{for long beams} \qquad (10.2.39a)$$

$$\approx \frac{(d - t_f)}{2} \quad \text{for short beams} \qquad (10.2.39b)$$

So, for an I-shape beam under uniform moment at the instant of lateral buckling, the **center of rotation,** R, is at the center of the tension flange for short beams and moves downwards from this point as the member becomes longer. Also, y_R varies directly with L for long beams.

Influence of Various Parameters on Elastic Lateral-Torsional Buckling of Beams

The elastic lateral-torsional buckling moment of a doubly symmetric I-shaped beam is influenced by several parameters, such as the I_y/I_x ratio of the beam section, the end support conditions, the presence of lateral supports to the compression flange, transverse loads and their level of application in the web plane, and continuity in the major- and minor-axis planes of the beam. For example, it can be shown that lateral-torsional buckling cannot occur if the moment of inertia about the bending axis is equal to or less than the moment of inertia out of plane. Thus, for shapes bent about the minor axis, and shapes for which $I_x = I_y$ (such as square or circular shapes), the limit state of lateral-torsional buckling is not applicable and yielding controls if the section is compact.

The critical maximum moment in a segment of unbraced length L_b, of a beam of doubly symmetric I-section, when such a member is loaded by end couples in the plane of the web and/or by transverse loads applied at the shear center axis in the plane of the web, may be closely approximated by:

$$M_{cr} = C_b\left(\frac{\pi}{K_b L_b}\right)\sqrt{EI_y\,GJ}\sqrt{1 + \frac{\pi^2}{(K_b L_b)^2}\frac{EC_w}{GJ}} \qquad (10.2.40)$$

Here C_b is a coefficient that depends on the load distribution and end conditions, L_b is the distance between adjacent lateral supports, and K_b is the notional effective length factor of the beam. An alternate form of this relation is given by:

$$M_{cr} = C_b\left(\frac{\pi}{K_b L_b}\right)^2\sqrt{EI_y\,EC_w}\sqrt{1 + \frac{(K_b L_b)^2}{\pi^2}\frac{GJ}{EC_w}} \qquad (10.2.41)$$

For hot-rolled sections and plate girders of normal proportions, the two terms under the second square root of Eq. 10.2.40 are of comparable magnitude. The value of the second radical in Eq. 10.2.40 approaches unity in the case of short beams and girders of very shallow or thick-walled sections, indicating that the St. Venant torsional resistance is dominant. On the other hand, for long beams and girders of deep or thin-walled sections, the second radical in Eq. 10.2.41 approaches unity; thus the resistance of the compression flange to buckling (warping torsion) governs.

For information on the influence of the various parameters relating to the elastic lateral-torsional buckling of beams of I-section, refer to Section W10.1 on the website http://www.mhhe.com/Vinnakota.

The support conditions of a cantilever beam differ from those of simply supported beams in that a cantilever is usually completely fixed at one end and completely free at the other. Consequently, when a cantilever is subjected to concentrated or distributed gravity loads, the tension flange moves farther during buckling, as compared to the compression flange in the case of simple beams.

For information on the elastic lateral-torsional buckling of cantilever beams, refer to Section W10.2 on the website http://www.mhhe.com/ Vinnakota.

10.3 Inelastic Stability of Beams of I-Shaped Sections

10.3.1 Introduction

The results obtained in Section 10.2 are only valid while the beam remains elastic, which is only true for a slender long-span beam. For a short-span steel beam, the yielding of fibers may start before the elastic buckling load can be attained. Such yielding reduces the stiffness of the beam and hence reduces the resistance of the beam to lateral buckling. As a consequence, the inelastic buckling load may be substantially lower than the elastic buckling load. Yielding of a straight, I-shaped steel beam is affected by its in-plane loading and support conditions, and by the residual stresses present in such sections.

Figure 10.3.1 shows the behavior of an I-section beam subjected to end moments M^o that cause symmetric single curvature bending to occur about the x axis of the cross section. An initially straight short-span steel beam initially behaves elastically as the applied moments are gradually increased. For an annealed beam (that is, one which contains no residual stresses), the extreme fibers of both flanges will simultaneously plastify at some level of applied loading. Yielding spreads through more of the cross section as the applied moments are further increased. The inelastic lateral-torsional buckling moment, M^o_{crI}, of such a partially plastified, initially straight beam can be evaluated by using the elastic core (that is, the effective cross section) as is done for columns. The elastic core, as it pertains to flexure, is described by the effective rigidities $(EI_y)_e$ and $(EC_w)_e$. The presence of residual stresses in rolled steel sections causes yielding to begin prior to the point at which the yielding would initiate in an annealed beam. Unlike columns and annealed beams, however, the elastic core is not doubly symmetrical for rolled steel beams. In effect, for beams, yielding of the compression flange starts at the tips where the residual stresses are compressive, while the tension flange begins yielding at its junction with the web where the residual stress is tensile (Fig. 10.3.2). Yielding need not necessarily begin simultaneously in each flange. The presence of residual stresses in conjunction with the action of bending reduces the double symmetry of the unloaded section to a mono-symmetry which exists about the web only. The center of gravity, G_e, and the shear center, S_e, of the elastic core (a section which now possesses monosymmetry only) no longer coincide. Thus the inelastic lateral-torsional buckling moment, M^o_{crI}, is highly dependent upon the magnitude of the residual stresses, especially the compressive residual stresses which occur at the flange tip. Also, yielding of the flange tip causes considerable reduction in the effective rigidities $(EI_y)_e$ and $(EC_w)_e$ of the elastic core. The inelastic lateral buckling

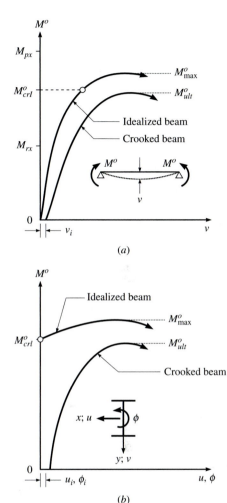

(a)

(b)

Figure 10.3.1: Inelastic stability of I-shaped beams.

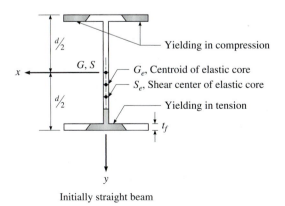

Initially straight beam

Figure 10.3.2: Yielding in the cross section of an I-shaped beam.

strength can be evaluated by using the properties of the elastic core [Galambos, 8.9].

When the major axis bending moment varies along the length of an inelastic beam, the distribution of yield zones, and hence the effective cross section and its properties, also vary along the length. The beam, in effect, acts as a nonprismatic member. The distribution of yielding along a beam due to in-plane bending and residual stresses can be determined by an in-plane bending analysis. From this, the variation along the beam of the out-of-plane section properties, such as $(EI_y)_e$, $(GJ)_e$, and $(EC_w)_e$, can be determined.

Real beams are not perfectly straight, but have small initial curvatures and twists. Unavoidable load eccentricities influence the inelastic behavior of the beams as well. Since lateral buckling involves both lateral displacements and twist, both initial sweep, u_i, and the initial twist, ϕ_i, influence this behavior. For geometrically imperfect beams lateral deflection and twist increase continuously from the start of loading, and tend to become very large as the applied moment approaches the critical moment (Fig. 10.3.1b). These additional deformations produce additional stresses (lateral bending stresses and warping stresses). The maximum stress in the beam now occurs at the tip of one flange only. If the combined effects of plasticity, residual stresses, and geometrical imperfections are considered, the elastic core of a partially plastified section will be nonsymmetric. The center of gravity and shear center of the elastic core shift both horizontally and vertically with respect to the centroidal axes of the original section.

Vinnakota [1977] used the finite difference method to study such inelastic flexural behavior of beams with residual stresses and geometrical imperfections. The loads were increased steadily until the solutions for deformations diverged, indicating that the maximum load capacity M_{ult}^o had been reached.

10.3.2   Lateral Buckling Strength vs. Unbraced Length

Figure 10.3.3 shows the relationship between the applied major axis moment, M^o, and the resulting end rotation, θ, for four beams. The beams are all of the same compact I-shape, flexurally and torsionally simply supported at the ends. Lateral bracing is provided at the ends of the member only, so that the unbraced length, L_b, equals the span length, L. Depending on the unbraced length, L_b, lateral-torsional buckling may occur at any level of loading: after the member has reached the plastic moment, M_{px}, as shown by curves 1' and 1 in Fig. 10.3.3; between the reduced yield moment, M_{rx}, and the plastic moment, M_{px}, as shown by curve 2; or even at moments below M_{rx} as shown by curve 3. Thus, very slender beams fail more or less elastically by excessive lateral deformation at loads that are close to M_{crE}^o, as discussed in Section 10.2. Stocky beams will attain M_{px} with negligible lateral deformation. Finally, beams of intermediate slenderness fail inelastically by excessive lateral deformation at loads that are close to M_{crI}^o, as discussed in Section 10.3.1.

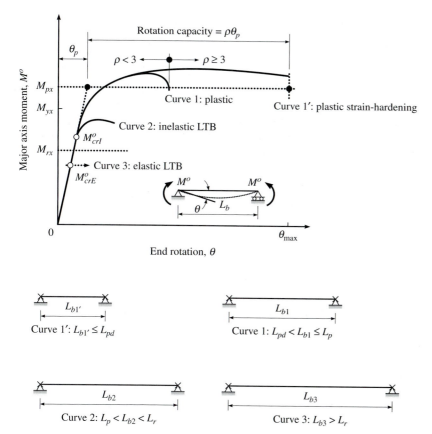

Figure 10.3.3: Behavior of compact I-shaped beams of different unbraced lengths.

Curve 1′ represents the ideal response where the moment continues to rise above the plastic moment M_{px} due to strain-hardening, and the beam after undergoing a significant amount of inelastic deformation becomes unstable. Such instability leads to a gradual decrease in moment capacity as the maximum deformation, θ_{max}, is approached. Beams with unbraced lengths shorter than a limiting value L_{pd} show such behavior. The increase in bending strength beyond M_{px} is not taken into account in design calculations, however. Curve 1 represents inelastic behavior where the plastic moment M_{px} is attained but little rotation capacity is exhibited because of inadequate lateral support to resist lateral-torsional buckling, or inadequate flange and/or web plate stiffness to resist plate local buckling. Beams with unbraced lengths greater than L_{pd} but shorter than another limiting unbraced length L_p show this behavior. Curve 2 represents inelastic behavior where the reduced yield moment M_{rx} is reached or exceeded. However, inelastic lateral-torsional buckling prevents the member from attaining the plastic moment strength M_{px}. Beams with unbraced length greater than L_p but less than another limiting

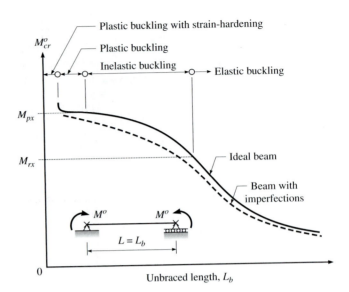

Figure 10.3.4: Lateral buckling strength vs. unbraced length.

unbraced length L_r manifest this behavior. Curve 3 represents elastic behavior where the moment capacity M^o_{cr} is controlled by elastic lateral-torsional buckling. Beams with unbraced lengths greater than the limiting unbraced length L_r show such behavior.

The variation of lateral buckling strength, M^o_{cr}, as a function of unbraced length, L_b, is shown in Figure 10.3.4 for compact I-shaped steel beams. The solid curve in this figure represents schematically the variation of the critical load when bifurcation-type instability occurs for a perfectly straight beam. The dashed curve represents the case when initial deformations are included. There are essentially three ranges of behavior: (1) plastic behavior, where the unbraced length is short enough that buckling occurs after the plastic moment is reached; (2) inelastic lateral buckling, when instability occurs after some portions of the beam have yielded; and (3) elastic lateral buckling, which governs for long beams. The first two ranges are of importance for the beams within a completed structure, while the third range is generally of concern during construction, as there is often a brief but potentially dangerous period between the placing of the beam and the installation of bracing members, decking, and/or slab.

10.3.3   Limiting Unbraced Lengths L_r and L_p

The *limiting unbraced length, L_r,* is the unbraced length of a beam segment under uniform moment at which the elastic lateral-torsional buckling regime begins, while the *limiting unbraced length, L_p,* is the largest unbraced length of a beam segment subjected to a uniform moment for which the plastic moment M_{px} develops.

Unbraced Length, L_r

Unbraced length L_r is the value of L_b obtained from Eq. 10.2.40 by setting M_{cr} equal to $M_{rx} = S_x(F_y - F_r)$, beam effective length factor $K_b = 1$, and moment modification factor $C_b = 1$. After substituting $I_y = Ar_y^2$ and rearranging, the following expression is obtained:

$$L_r = \frac{r_y \, \pi}{(F_y - F_r) \, S_x} \sqrt{\frac{EAGJ}{2}} \sqrt{1 + \sqrt{1 + \frac{4C_w}{I_y}\left(\frac{S_x}{GJ}\right)^2 (F_y - F_r)^2}}$$

(10.3.1)

where L_r = limiting laterally unbraced length for elastic lateral buckling, for a beam segment under uniform moment, in.
       r_y = radius of gyration about minor axis, in.
       F_y = yield stress of material, ksi
       F_r = compressive residual stress in flange
            = 10 ksi, for rolled shapes
            = 16.5 ksi, for welded built-up shapes
       A = cross-sectional area, in.^{2}
       C_w = warping constant, in.^{6}
       J = St. Venant torsional constant, in.^{4}
       I_y = moment of inertia about the y axis, (Ar_y^2) in.^{4}
       S_x = elastic section modulus about major axis, in.^{3}
       E = modulus of elasticity of steel (29,000 ksi)
       G = shear modulus of elasticity of steel (11,200 ksi)

In the LRFD Specification, Eq. 10.3.1 is given in the form (LRFDS Eq. F1-6):

$$L_r = \frac{r_y X_1}{(F_y - F_r)} \sqrt{1 + \sqrt{1 + X_2 (F_y - F_r)^2}}$$

(10.3.2)

where X_1 and X_2 are known as **_beam buckling factors_**. From LRFDS Eqs. F1-8 and F1-9:

$$X_1 = \frac{\pi}{S_x} \sqrt{\frac{EAGJ}{2}}; \quad X_2 = \frac{4C_w}{I_y}\left(\frac{S_x}{GJ}\right)^2$$

(10.3.3)

Values of X_1 and X_2 are tabulated in Part 1 of the LRFD Manual for rolled steel I-shapes.

Unbraced Length, L_p

As the distance between the lateral supports will be relatively short to achieve the plastic moment capacity (and the rotation capacity), the value of the second radical in Eq. 10.2.41 for critical moment approaches unity. For a uniform moment over the unbraced length, C_b equals one, and neglecting any restraint offered by any adjacent segments ($K_b = 1$). Equation 10.2.41 then reduces to [Lay and Galambos, 1965, 1967]:

$$M_{cr}^o = \frac{\pi^2 E}{L_b^2} \sqrt{I_y C_w} \qquad (10.3.4)$$

Since the critical moment must reach the plastic moment, we substitute $M_{cr}^o = M_{px} = Z_x F_y$. Also, for a thin walled I-shape, we have $C_w = I_y h^2/4$, and $I_y = A r_y^2$. The maximum unbraced length to achieve M_p is then obtained from Eq. 10.3.4 as:

$$L_b \leq r_y \sqrt{\frac{E}{F_y}} \left[\sqrt{\frac{\pi^2}{2}\left(\frac{hA}{Z_x}\right)} \right] \qquad (10.3.5)$$

Using a conservative (low) value of 1.5 for hA/Z_x results in:

$$L_b \leq 2.72 r_y \sqrt{\frac{E}{F_y}} \qquad (10.3.6)$$

Test results presented by Bansal [1971] indicate that a lower limit than Eq. 10.3.6 is necessary to achieve adequate rotational capacity. Thus, the LRFDS has set the limiting value of an unbraced length as:

$$L_p = 1.76 r_y \sqrt{\frac{E}{F_y}} \qquad (10.3.7)$$

10.4 LRFD of Beams of Compact I-Shaped Sections

10.4.1 General

Criteria for the LRFD design of beams is based on the work of Galambos and Ravindra [1976] and Yura, Galambos, and Ravindra [9.12]. The strength requirement according to the LRFD may be written as:

$$M_d \equiv \phi_b M_n \geq M_u \qquad (10.4.1)$$

where M_u = required bending strength under factored loads
M_n = nominal bending strength
ϕ_b = resistance factor for bending $(= 0.90)$
M_d = design bending strength

The criteria discussed in this section are valid for beams when the moments are obtained from an elastic structural analysis using factored loads. The value of M_u to be used is the maximum factored moment in the unbraced segment considered. The nominal flexural strength of a beam is the lowest value obtained according to the limit states of:

- Yielding of the section.
- Lateral-torsional buckling of the member.
- Local buckling of the flange plate in axial compression.
- Local buckling of the web plate in flexural compression.

Only beams with prismatic, homogeneous, compact I-shaped sections, loaded in the plane of the web, are considered in this section. That is, sections considered here satisfy the following relations:

$$\lambda_f \le \lambda_{pf}; \quad \text{and} \quad \lambda_w \le \lambda_{pw} \tag{10.4.2}$$

where:

$$\lambda_{pf} = 0.38 \sqrt{E/F_y}; \quad \lambda_{pw} = 3.76 \sqrt{E/F_y} \tag{10.4.3a}$$

$$\lambda_f = \frac{b_f}{2t_f}; \quad \lambda_w = \frac{h}{t_w} \tag{10.4.3b}$$

Hence, flange local buckling and web local buckling will not control the design flexural strength of the members considered in this section (for non-compact sections, see Section 10.5.2).

The case of uniform moment along a laterally unbraced length of beam is used by the LRFDS as the basic loading case for lateral-torsional buckling. The uniform moment causes constant compression in one flange over the entire unbraced length. The basic relationship between the design flexural strength, M_d^o, and unbraced length, L_b, is shown in Fig. 10.4.1 for a compact I-section beam. There are three principal zones on the basic curve delimited by L_p and L_r. LRFDS Eq. F1-4 (or Equation 10.3.7) defines the maximum

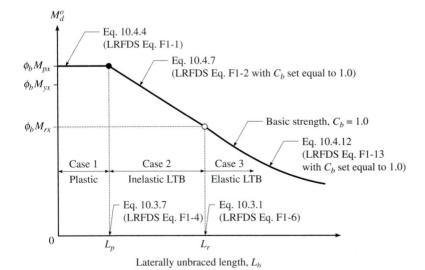

Figure 10.4.1: Basic design bending strength, M_d^o, of a compact I-beam under uniform moment.

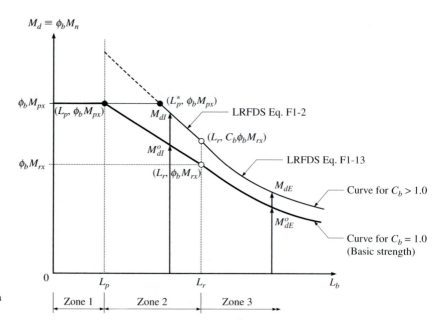

Figure 10.4.2: Design bending strength, M_d, as a function of unbraced length, L_b, and C_b factor.

unbraced length L_p capable of reaching M_{px} under a uniform moment. Elastic lateral-torsional buckling will occur when the unbraced length is greater than L_r as given by LRFDS Eq. F1-6 (or Eq. 10.3.1). LRFDS Eq. F1-2 with C_b set equal to 1.0 (or Equation 10.4.7) defines the inelastic lateral-torsional buckling strength M_{dI}^o, which has a straight line variation between point $(L_p, \phi_b M_{px})$ indicated by the solid circle ($\bullet$) in Fig. 10.4.1, and point $(L_r, \phi_b M_{rx})$, indicated by the open circle ($\bigcirc$). Lateral-torsional buckling strength in the elastic region M_{dE}^o, (when $L_b > L_r$) is given by LRFDS Eq. F1-13 with C_b set equal to 1.0 (Eq. 10.4.12).

When the bending moment varies over the unbraced length the axial force in the compression flange likewise varies along the segment. This results in a lower average compressive force in the flange, and thus less likelihood of lateral-torsional buckling. To account for this, the LRFDS allows the basic strength, M_d^o, to be multiplied by a moment modification factor C_b, wherever there is a nonuniform distribution of moments over the unbraced length. However, as shown in Fig. 10.4.2, the value of $C_b M_d^o$ is limited to a maximum value of $\phi_b M_{px}$.

10.4.2   Limit State of Yielding of the Section

For a compact I-shaped beam bent about its major axis, the design strength corresponding to the limit state of yielding of the section may be written as:

$$M_d = \phi_b M_{px} = \phi_b Z_x F_y \qquad (10.4.4)$$

where M_{px} = plastic moment, in.-kips
$\quad\quad\ Z_x$ = plastic section modulus, in.^{3}

10.4.3   Limit State of Lateral-Torsional Buckling

As discussed previously in Section 10.3, the behavior of a given I-shaped steel beam subjected to a uniform moment is defined by the two limiting unbraced lengths L_p and L_r. Here, L_p is the limiting laterally unbraced length for full plastic bending capacity for a beam segment under uniform moment (Eq. 10.3.7 or LRFDS Eqs. F1-4), and L_r is the limiting laterally unbraced length for elastic lateral buckling for a beam segment under uniform moment (Eq. 10.3.2 or LRFDS Eq. F1-6).

Let L_b be the distance between points braced against lateral displacement of the compression flange or the distance between points braced to prevent twist of the cross section. Depending on the value of L_b relative to L_p and L_r, the limit state of lateral buckling falls into three different zones.

Zone 1 ($L_b \leq L_p$)
For laterally braced compact I-shaped beam segments with $L_b \leq L_p$, the limit state of lateral-torsional buckling does not apply. The nominal major axis flexural strength of the segment can reach M_{px}. The design bending strength of the segment is therefore given by:

$$M_d = \phi_b M_n = \phi_b M_{px} \tag{10.4.5}$$

Zone 2 ($L_p < L_b \leq L_r$)
For laterally unbraced compact I-shaped beam segments with $L_p < L_b \leq L_r$, the governing limit state is the inelastic lateral-torsional buckling of the unbraced segment. The design bending strength is:

$$M_d = \phi_b M_n = M_{dI} = \min [C_b M_{dI}^o, \quad \phi_b M_{px}] \tag{10.4.6}$$

with

$$M_{dI}^o = \phi_b M_{px} - (\phi_b M_{px} - \phi_b M_{rx}) \frac{(L_b - L_p)}{(L_r - L_p)} \tag{10.4.7}$$

$$= \phi_b M_{px} - BF (L_b - L_p) \tag{10.4.8}$$

$$BF = (\phi_b M_{px} - \phi_b M_{rx})/(L_r - L_p) \tag{10.4.9}$$

$$M_{rx} = S_x F_L = S_x (F_y - F_r) \tag{10.4.10}$$

where M_{dI} = design bending strength of a given beam segment and moment
$\quad\quad\quad\quad$ diagram, for $L_p < L_b \leq L_r$
$\quad\ M_{dI}^o$ = design bending strength of the same segment under uniform
$\quad\quad\quad\quad$ moment (given by Eq. 10.4.7 or 10.4.8)

C_b = modification factor for nonuniform moment diagram over length L_b

= 1.0 for uniform moment over L_b

M_{rx} = value of M where yielding just commences due to applied compressive flexural stresses and compressive residual stresses in the compression flange, in.-kips

L_b = laterally unbraced length, in.

L_p = limiting laterally unbraced length for full plastic bending capacity, when the segment is under uniform moment (Eq. 10.3.7), in.

L_r = limiting laterally unbraced length for elastic lateral buckling (Eq. 10.3.1), when the segment is under uniform moment, in.

BF = beam factor (defined by Eq. 10.4.9)

S_x = elastic section modulus about major axis, in.^{3}

F_y = yield stress, ksi

F_r = compressive residual stress in flange

= 10 ksi, for rolled shapes

= 16.5 ksi, for welded built-up shapes

F_L = $F_y - F_r$

Zone 3 ($L_b > L_r$)

For laterally unbraced compact I-shaped beam segments with $L_b > L_r$, the governing limit state is the elastic lateral-torsional buckling of the unbraced segment. The design bending strength is now given by (LRFDS Eq. F1-13):

$$M_d = \phi_b M_n = M_{dE} = \min[C_b M_{dE}^o, \quad \phi_b M_{px}] \qquad (10.4.11)$$

with

$$M_{dE}^o = \phi_b M_{cr}^o \qquad (10.4.12)$$

$$M_{cr}^o = \frac{\pi}{L_b} \sqrt{EI_y\,GJ + \frac{\pi^2}{L_b^2} EC_w EI_y} \qquad (10.4.13a)$$

$$= \frac{S_x X_1 \sqrt{2}}{L_b/r_y} \sqrt{1 + \frac{X_1^2 X_2}{2(L_b/r_y)^2}} \qquad (10.4.13b)$$

where M_{dE} = design bending strength of a given beam segment and moment diagram for $L_b > L_r$

M_{dE}^o = design bending strength of the same segment under uniform moment

M_{cr}^o = critical buckling moment of the segment under uniform moment

The other variables are defined in Section 10.3.3.

Moment Modification Factor C_b

In a beam bent into single curvature, the entire unbraced length of one flange is in compression. However, if the same beam is bent in double curvature, only a portion of each flange length is in compression while the remainder of the length is in tension. Hence, for a given beam and set of end moments, the lateral-torsional buckling strength is higher when the beam segment is bent in double curvature than when the beam is bent in single curvature. As discussed in Section 9.5.1, the most severe loading case for lateral-torsional buckling of a simply supported beam braced only at its ends is that of equal and opposite end moments producing symmetric single curvature bending.

Variation of the major axis moment distribution over the unbraced length also has an influence on its resistance to inelastic lateral-torsional buckling. In effect, yielding takes place only in the high moment regions. That is, the reductions in the section rigidities due to plastification occur in the high moment regions. The resistance to inelastic lateral-torsional buckling depends heavily on the location and extent of these regions of reduced rigidity. For example, for a simple beam under reverse curvature moments, the high moment gradient results in yielding that is confined to short regions at the ends of the beam. In this case the central region, which is mostly responsible for the buckling resistance remains elastic and there is only a small reduction in buckling strength. Thus, when the applied loading produces a nonuniform distribution of moment over the unbraced length, the likelihood of lateral-torsional buckling is lower, as compared to the basic loading case of a uniform moment over L_b. The moment modification factor, C_b, is then applied to account for this reduced likelihood (see Web Section 10.1). Kirby and Nethercot [1979] proposed an expression for C_b which can be used for moment diagrams of any shape, and which results in good estimates of the critical moment M_{cr}. Their equation, which has been adopted by the LRFDS as Eq. F1-3, is

$$C_b = \frac{12.5M_{max}}{2.5M_{max} + 3M_A + 4M_B + 3M_C} \qquad (10.4.14)$$

where M_{max} = absolute value of the maximum moment within the unbraced segment (including the end points)

M_A = absolute value of the moment at the quarter-point of the unbraced segment

M_B = absolute value of the moment at the midpoint of the unbraced segment

M_C = absolute value of the moment at the three-quarter point of the unbraced segment

Observe that when the bending moment is uniform, the above relation results in a value of one for C_b. Also note that for all other distributions, Eq. 10.4.14 results in a value greater than one for C_b. Figure 10.4.3 gives the

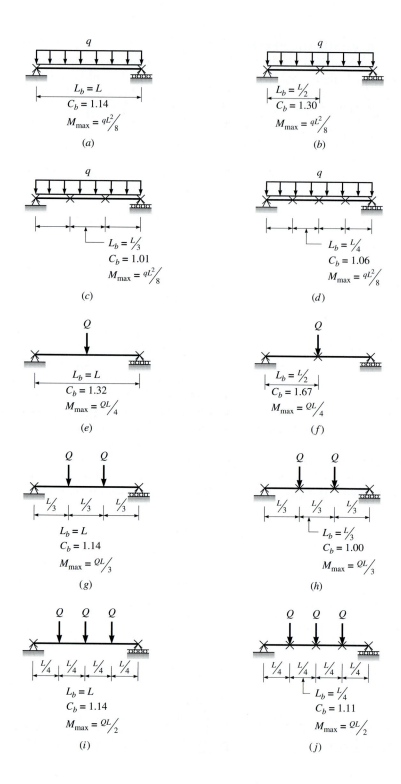

Figure 10.4.3: Values of C_b for simply supported beams.

values of L_b, C_b, and M_{max} to be used for the critical segment of simple beams for several common cases of loading and arrangement of lateral bracing, using this relation.

A loading case that frequently occurs in practice is the application of concentrated loads at the brace points of a beam. The bending moment within any such braced segment will therefore vary linearly from M_1 at one end to M_2 at the other end. Let:

$$r_M = \pm \frac{|M_1|}{|M_2|} \qquad (10.4.15)$$

Here, M_2 always designates the moment having the larger absolute value and M_1 the moment having the smaller absolute value. The ratio is assigned a positive sign when the moments tend to produce reverse curvature bending over the unbraced length L_b and a negative sign when they tend to cause single curvature bending. Since M_2 is always the larger moment, the parameter r_M always lies between -1 and $+1$. Equation 10.4.14 for C_b is valid for such straight line moment diagrams as well. Table 10.4.1 lists the values of C_b for several values of the end moment ratio, r_M between -1 and $+1$, using Eq. 10.4.14.

To sum up, the bending modifying factor C_b is a coefficient which is used to take advantage of the beneficial effect of any moment gradient which may occur between points of lateral support. According to LRFDS Section F1.2a, C_b may be taken as 1.0 for all cases; this is a conservative approach as it is equivalent to assuming the most severe loading of constant bending moment. Equation 10.4.14 is not applicable to cantilevers or overhangs, where the free end is unbraced. For these cases, C_b is prescribed to be taken as 1.0.

Parameters C_b^*, L_p^*

For values of $C_b > 1$, the maximum unbraced length, L_p^*, which enables the nominal moment M_n to reach M_{px}, that is, for M_d to equal $\phi_b M_{px}$, can be determined from Eq. 10.4.6 with the help of Eq. 10.4.8, as follows:

$$\phi_b M_{px} = C_b [\phi_b M_{px} - BF (L_p^* - L_p)]$$

$$L_p^* = L_p + (C_b - 1) \frac{\phi_b M_{px}}{BF} \le L_r \qquad (10.4.16)$$

For values of C_b equal to or greater than a limiting value C_b^*, the inelastic buckling region described by Eq. 10.4.6 disappears, that is for $C_b = C_b^*$, $M_d = \phi_b M_{px}$ for all $L_b \le L_r$. This value, C_b^*, may be obtained from Fig. 10.4.4 by observing that for $L_b = L_r$ and $C_b = C_b^*$, we have $M_d = \phi_b M_{px}$ and also $M_d = C_b M_{dl}^o = C_b^* \phi_b M_{rx}$, which results in:

$$C_b^* = \frac{\phi_b M_{px}}{\phi_b M_{rx}} \qquad (10.4.17)$$

From LRFDM Table 5-3, we have for a W21×44 section of steel with $F_y = 50$ ksi, $\phi_b M_{px} = 359$ ft-kips, $\phi_b M_{rx} = 246$ ft-kips. We then obtain

TABLE 10.4.1

Values of C_b and C_m Factors for Linear Variation of Bending Moment over the Unbraced Length, L_b

M_1/M_2	C_b	C_m	M_1/M_2	C_b	C_m
−1.00	1.00	1.00	0.0	1.67	0.60
−0.95	1.02	0.98	0.05	1.72	0.58
−0.90	1.04	0.96	0.10	1.79	0.56
−0.85	1.06	0.94	0.15	1.85	0.54
−0.80	1.09	0.92	0.20	1.92	0.52
−0.75	1.11	0.90	0.25	2.00	0.50
−0.70	1.14	0.88	0.30	2.08	0.48
−0.65	1.16	0.86	0.35	2.15	0.46
−0.60	1.19	0.84	0.40	2.16	0.44
−0.55	1.22	0.82	0.45	2.16	0.42
−0.50	1.25	0.80	0.50	2.17	0.40
−0.45	1.28	0.78	0.55	2.18	0.38
−0.40	1.32	0.76	0.60	2.19	0.36
−0.35	1.35	0.74	0.65	2.20	0.34
−0.30	1.39	0.72	0.70	2.21	0.32
−0.25	1.43	0.70	0.75	2.22	0.30
−0.20	1.47	0.68	0.80	2.23	0.28
−0.15	1.52	0.66	0.85	2.24	0.26
−0.10	1.56	0.64	0.90	2.25	0.24
−0.05	1.61	0.62	0.95	2.26	0.22
			1.00	2.27	0.20

Notes: 1.
$$C_b = \frac{12.5}{2.5 + 3\left|\dfrac{M_A}{M_2}\right| + 4\left|\dfrac{M_B}{M_2}\right| + 3\left|\dfrac{M_C}{M_2}\right|}$$

2. $C_m = 0.6 - 0.4\,(M_1/M_2)$
3. M_1 and M_2 are the end moments. M_1/M_2 is positive for reverse curvature bending. M_A, M_B, M_C are the moments at quarter points of the segment.
4. For definition and use of C_m factor see Section 11.9.1.

from Eq. 10.4.17, $C_b^* = 359/246 = 1.46$. The effect of C_b on the design flexural strength is illustrated in Fig. 10.4.4, for this section, for values of $C_b = 1.0$, 1.30, 1.46, and 2.27.

10.4.4 Design Aids

Part 1 of the LRFDM gives values of r_y, Z_x, X_1 and X_2 for all rolled steel W-, M-, and S-shapes (Tables 1-1, 1-2, and 1-3). In addition, values of torsion

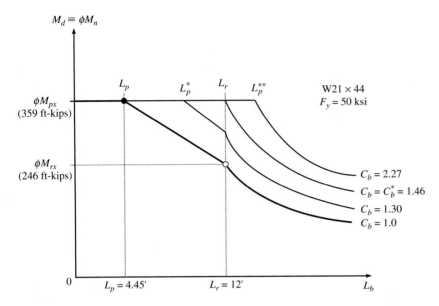

Figure 10.4.4: Influence of C_b on the design flexural strength of W21×44 beams.

constants J and C_w are given in Part 1 of the LRFDM for all hot-rolled W-, M-, and S-shapes (Tables 1-25, 1-26, 1-27). Part 5 of the LRFD Manual contains several beam selection tables for W-shapes used as beams: Table 5-2: W-Shape Selection by I_x; Table 5-3: W-Shape Selection by Z_x; Table 5-6: W-Shape Selection by I_y; and Table 5-7: W-Shape Selection by Z_y. These four tables list many beam design parameters, such as Z_x, I_x, L_p, L_r, $\phi_b M_{px}$, $\phi_b M_{rx}$, BF, $\phi_v V_n$, and Z_y, I_y, $\phi_b M_{py}$, for 50 ksi steels. For example, Table 5-3: W-Shape Selection by Z_x, lists the W-shapes used as beams, in descending order of strong-axis plastic section modulus, Z_x.

Beams that are not fully laterally supported cannot be designed conveniently using the beam selection tables, unless the interval of lateral support is less than the length L_p. The design moment plots given in LRFDM Table 5-5: W-Shapes—Plots of $\phi_b M_{nx}$ vs. L_b provide a simple design solution for beams that are not fully laterally supported, for $F_y = 50$ ksi steel. In these plots, the design moments for W-shapes, $\phi_b M_{nx}$ (in ft-kips) are plotted as functions of the unbraced length, L_b (in feet), for $C_b = 1.0$ and $F_y = 50$ ksi. As already mentioned, values corresponding to L_p are indicated in the charts by solid circles (●), and values corresponding to L_r are indicated by open circles (○). The curves do not extend beyond an arbitrarily set span/depth limit of 30. The curves are constructed without regard to deflection; therefore deflection checks, if necessary, must be made separately.

The design moment plots are constructed to assist in selecting the lightest available beam for a given combination of unbraced length and design moment. For any particular shape, the solid portion of the curve indicates that it is the most economical section by weight in that range. The dashed portion of the curve indicates ranges in which a lighter weight beam will satisfy the loading conditions. For beams of equal weight and sufficient loading capacity the deeper beam, having a lower design moment capacity than the more shallow beam, is indicated as a dashed curve to assist in making a selection for a limited depth condition. For example, observe that in LRFDM page 5-90, for $M_{req} = 500$ ft-kips and $L_b = 8$ ft, the W24×62 is shown dotted while the W21×62 is indicted as a continuous line.

To determine the design bending strength of a given I-shaped beam of unbraced length L_b and moment modification factor C_b, locate the design bending strength curve corresponding to that shape in the beam selection plots of LRFDM Table 5-5. Proceed vertically at the given unbraced length L_b (abscissa) to intersect the curve at point P. If point P is to the left of the solid circle on the curve, then $L_b < L_p$ and the ordinate of P gives the design bending strength of the beam ($\phi_b M_{px}$). If the point P is to the right of the solid circle, then $L_b > L_p$, and the ordinate of P represents the design strength of that beam under uniform moment over $L_b (=M_d^o)$. To obtain the beam design strength, M_d, for values of C_b other than 1.0, multiply the ordinate of P by C_b, keeping in mind though that M_d may never exceed $\phi_b M_{px}$.

EXAMPLE 10.4.1

Elastic Lateral-Buckling Loads

A W24×104 of A992 steel is to be used as a simply supported beam on a span of 36 ft with lateral supports at the end only. Compute the elastic critical moment and maximum bending stress for each of the loading conditions given below (Fig. X10.4.1):

a. Pure bending (uniform moment) about the major axis.
b. Constant moment gradient ($M_1 = +0.2M_2$).
c. Uniformly distributed load q, applied at the shear center axis.
d. Central concentrated load Q applied at the shear center.

Solution

Yield stress, $F_y = 50$ ksi
For a rolled steel section residual compressive stress, $F_r = 10$ ksi.

Therefore,

$$F_{yr} = F_y - F_r = 40 \text{ ksi}$$

a. Pure bending about major axis
From LRFDM Table 1-1: W-Shapes, for a W24×104:

$$S_x = 258 \text{ in.}^3; \quad I_y = 259 \text{ in.}^4$$

From LRFDM Table 1-25: W-Shapes Torsional Properties, for a W24×104:

$$J = 4.72 \text{ in.}^4; \quad C_w = 35{,}200 \text{ in.}^6; \quad \sqrt{\frac{EC_w}{GJ}} = 139$$

For the simply supported beam, of length $L = 36$ ft, we obtain:

$$\frac{\pi}{L}\sqrt{EI_y\, GJ} = \frac{\pi}{36(12)}\sqrt{29{,}000(259)(11{,}200)(4.72)}$$

$$= 4582.4 \text{ in.-kips} = 382 \text{ ft-kips}$$

$$\frac{\pi}{L}\sqrt{\frac{EC_w}{GJ}} = \frac{\pi}{36(12)}(139) = 1.01$$

For a simply supported beam braced at the supports and subjected to uniform major axis moment, the critical lateral buckling moment,

$$M_{cr}^o = \frac{\pi}{L}\sqrt{EI_y\, GJ}\sqrt{1 + \frac{\pi^2}{L^2}\frac{EC_w}{GJ}} = 382\sqrt{1 + (1.01)^2}$$

$$= 382(1.42) = 542 \text{ ft-kips} \qquad \text{(Ans.)}$$

Maximum stress in the critical section corresponding to this moment is:

$$f_{\max} = \frac{M_{cr}^o}{S_x} = \frac{542(12)}{258} = 25.2 \text{ ksi} < F_{yr} = 40 \text{ ksi} \qquad \text{(Ans.)}$$

So, buckling is elastic as assumed.

b. End moments $(0.2M_2, M_2)$ producing single curvature bending
Let A, B, C be the quarter points of the beam. We have from Fig. X10.4.1b.
$M_A = 0.4M_2$, $M_B = 0.6M_2$, $M_C = 0.8M_2$ and $M_{\max} = M_2$. The C_b factor may be obtained from Eq. 10.4.14:

$$C_b = \frac{12.5M_{\max}}{2.5M_{\max} + 3M_A + 4M_B + 3M_C}$$

$$= \frac{12.5M_2}{2.5M_2 + 3(0.4M_2) + 4(0.6M_2) + 3(0.8M_2)} = 1.47$$

From Eq. 10.4.15, end moment ratio,

$$r_M = -\frac{|0.2M_2|}{|1.0M_2|} = -0.20$$

Note that r_M is negative, as the moments produce single curvature bending. As a check, from Table 10.4.1, read $C_b = 1.47$ for this value of r_M. The critical value of the end moment M_2 is now given by:

$$M_{2cr} = C_b M_{cr}^o = 1.47(542) = 797 \text{ ft-kips} = M_{\max} \qquad \text{(Ans.)}$$

$$f_{\max} = \frac{M}{S_x} = \frac{797(12)}{258} = 37.1 \text{ ksi} < F_{yr} = 40 \text{ ksi} \qquad \text{(Ans.)}$$

So, buckling is elastic as assumed.   [*continues on next page*]

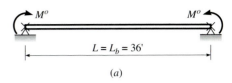

$L = L_b = 36'$

(a)

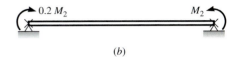

(b)

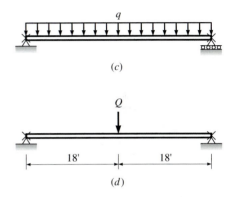

(c)

18'   18'

(d)

Figure X10.4.1

Example 10.4.1 continues ...

c. Uniformly distributed load

From the definition of C_b, we have: $M_{qcr} = C_b M^o_{cr}$
For a uniformly distributed load acting along the shear center axis on a simply supported beam, $C_b = 1.14$ (Fig. 10.4.3a), resulting in:

$$M_{qcr} = 1.14(542) = 618 \text{ ft-kips} = M_{max} = \frac{q_{cr} L^2}{8}$$

$$q_{cr} = \frac{8}{L^2} M_{qcr} = \frac{8}{36^2}(618) = 3.82 \text{ klf} \qquad \text{(Ans.)}$$

$$f_{max} = \frac{618}{258}(12) = 28.7 \text{ ksi} < F_{yr} = 40 \text{ ksi} \qquad \text{(Ans.)}$$

So, buckling is elastic as assumed.

d. Central concentrated load

From the definition of C_b, we have: $M_{Qcr} = C_b M^o_{cr}$
For a central concentrated load acting at the shear center, on a torsionally simply supported beam, we have (from Fig. 10.4.3e) $C_b = 1.32$ resulting in:

$$M_{Qcr} = 1.32(542) = 715 \text{ ft-kips} = M_{max} = \frac{Q_{cr} L}{4}$$

$$Q_{cr} = \frac{4 M_{Qcr}}{L} = \frac{4(715)}{36} = 79.6 \text{ kips} \qquad \text{(Ans.)}$$

$$f_{max} = \frac{715}{258}(12) = 33.3 \text{ ksi} < F_{yr} = 40 \text{ ksi} \qquad \text{(Ans.)}$$

The beam therefore undergoes lateral buckling in the elastic domain.

EXAMPLE 10.4.2

Use of LRFDM Tables 5-3 and 5-5

Determine the design bending strength of a W24×104 beam segment of A992 steel, if:

a. $L_b = 10$ ft
b. $L_b = 20$ ft and $C_b = 1.0$
c. $L_b = 20$ ft and $C_b = 1.10$
d. $L_b = 20$ ft and $C_b = 2.00$
e. $L_b = 30$ ft and $C_b = 1.0$
f. $L_b = 30$ ft and $C_b = 1.50$
g. $L_b = 30$ ft and $C_b = 2.20$

Solution
From LRFDM Table 5-3, observe that for a W24×104 of $F_y = 50$ ksi:

$$\phi_b M_{px} = 1080 \text{ ft-kips}; \quad L_p = 10.3 \text{ ft}; \quad BF = 18.6 \text{ kips}$$

$$\phi_b M_{rx} = 774 \text{ ft-kips}; \quad L_r = 26.9 \text{ ft}$$

a. $L_b = 10$ ft

As $L_b = 10 \text{ ft} < L_p = 10.3$ ft

$$M_d = \phi_b M_n = \phi_b M_{px} = 1080 \text{ ft-kips} \tag{Ans.}$$

Alternatively, from the beam selection plots (LRFDM Table 5-5), for a W24×104 of $F_y = 50$ ksi steel, and $L_b = 10$ ft, read $M_d = M_d^o = 1080$ ft-kips
(Ans.)

b. $L_b = 20$ ft and $C_b = 1.0$

As $L_p = 10.3 \text{ ft} < L_b = 20 \text{ ft} < L_r = 26.9 \text{ ft}$ and $C_b = 1.0$:

$$M_d = M_d^o = \phi_b M_{px} - BF(L_b - L_p)$$
$$= 1080 - 18.6 (20.0 - 10.3) = 900 \text{ ft-kips} \tag{Ans.}$$

Alternatively, from the beam selection plots (LRFDM Table 5-5), for a W24×104 of $F_y = 50$ ksi steel, $C_b = 1.0$ and $L_b = 20$ ft, read:

$$M_d = M_d^o = 900 \text{ ft-kips} \tag{Ans.}$$

c. $L_b = 20$ ft and $C_b = 1.10$

As $L_p = 10.3 < L_b = 20 \text{ ft} < L_r = 26.9 \text{ ft}$ and $C_b > 1.0$:

$$M_d = \phi_b M_n = \min [C_b M_d^o, \phi_b M_{px}]$$

As $M_d^o = 900$ ft-kips as calculated in part (b):

$$M_d = \min [1.10(900), 1080] = 990 \text{ ft-kips} \tag{Ans.}$$

d. $L_b = 20$ ft and $C_b = 2.00$

As $L_p = 10.3 \text{ ft} < L_b = 20 \text{ ft} < L_r = 26.9 \text{ ft}$ and $C_b > 1.0$:

$$M_d = \min [C_b M_d^o, \phi_b M_{px}]$$
$$= \min [2.00(900), 1080] = 1080 \text{ ft-kips} \tag{Ans.}$$

e. $L_b = 30$ ft and $C_b = 1.0$

As $L_b = 30 \text{ ft} > L_r = 26.9 \text{ ft}$ and $C_b = 1.0$:

$$M_d = \phi_b M_n = M_d^o = \phi_b M_{cr}$$

The value of M_d^o can conveniently be read from the beam selection plots (LRFDM Table 5-5), corresponding to a W24×104 shape of $F_y = 50$ ksi steel, $C_b = 1.0$ and $L_b = 30$ ft, as 650 ft-kips. So,

$$M_d = 650 \text{ ft-kips} \tag{Ans.}$$

[continues on next page]

Example 10.4.2 continues ...

f. $L_b = 30$ ft and $C_b = 1.50$
As $L_b = 30$ ft $> L_r = 26.9$ ft and $C_b > 1.0$:

$$M_d = \phi_b M_n = \min[C_b M_d^o, \phi_b M_{px}]$$

As $M_d^o = 650$ ft-kips from part (*e*) above:

$$M_d = \min[1.50(650), 1080] = \min[975, 1080] = 975 \text{ ft-kips} \quad \text{(Ans.)}$$

g. $L_b = 30$ ft and $C_b = 2.20$
As $L_b = 30$ ft $> L_r = 26.9$ ft and $C_b > 1.0$:

$$M_d = \phi_b M_n = \min[C_b M_d^o, \phi_b M_{px}]$$

$$= \min[2.20(650), 1080] = \min[1430, 1080] = 1080 \text{ ft-kips}$$

(Ans.)

EXAMPLE 10.4.3

Analysis of Beam under Distributed Loads

A W16×45 of A992 steel is used as a simple beam of 33-ft span, as shown in Fig. X10.4.3. Determine the factored, uniformly distributed load q_u that the beam can support if lateral supports are provided: (*a*) at 5 ft 6 in. intervals; (*b*) at the ends and at the third points of the span; and (*c*) at the ends and at the center only. Shear and deflection need not be checked.

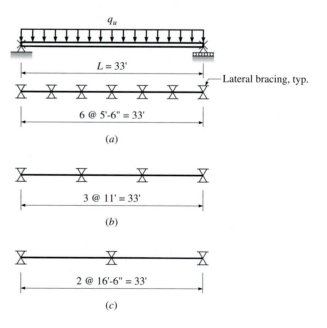

Figure X10.4.3

Solution

Span, $L = 33$ ft

From LRFDM Table 5-3, for a W16×45 of $F_y = 50$ ksi steel:

$$\phi_b M_{px} = 309 \text{ ft-kips}; \quad \phi_b M_{rx} = 218 \text{ ft-kips}; \quad BF = 9.45 \text{ kips}$$

$$L_p = 5.55 \text{ ft}; \quad\quad L_r = 15.1 \text{ ft}; \quad\quad I_y = 32.8 \text{ in.}^3$$

a. When lateral supports are provided at 5 ft 6 in. intervals
   As $L_b = 5.5$ ft $< L_p = 5.55$ ft,

$$M_d = \phi_b M_{px} = 309 \text{ ft-kips} = M_{max} = \frac{q_{u1} L^2}{8}$$

$$q_{u1} = \frac{309(8)}{(33)^2} = 2.27 \text{ klf}$$

b. When lateral supports are provided at the ends and at the third-points of
   the span
   As $L_p = 5.55$ ft $< L_b = 11$ ft $< L_r = 15.1$ ft,

$$M_{dI}^o = \phi_b M_{px} - BF(L_b - L_p)$$

$$= 309 - 9.45(11.0 - 5.55) = 257 \text{ ft-kips}$$

(Alternatively, the value of M_{dI}^o can be read from LRFDM Table 5-5, for
a W16×45 with $L_b = 11$ ft, as 257 ft-kips.)
From Fig. 10.4.3, for a simple beam under uniformly distributed load
with lateral bracing at the supports and at third-points only, $C_b = 1.01$.
Design bending strength is therefore given by Eq. 10.4.6:

$$M_d = \min[C_b M_{dI}^o, \phi_b M_{px}] = \min[1.01(257), 309]$$

$$= 260 \text{ ft-kips} = \frac{q_{u2} L^2}{8}$$

$$q_{u2} = \frac{260(8)}{(33)^2} = 1.91 \text{ klf}$$

c. When lateral supports are provided at the ends and at the center only
   As $L_b = 16.5$ ft $> L_r = 15.1$ ft, the limit state of elastic lateral buckling
   controls. From the Torsional Properties Table for W-Shapes (LRFDM
   Table 1-25), for a W16×45:

$$\text{Torsion constant, } J = 1.11 \text{ in.}^4; \quad \sqrt{\frac{EC_w}{GJ}} = 68.1 \text{ in.}$$

From Eq. 10.4.13a,

$$M_{cr}^o = \frac{\pi}{L_b} \sqrt{EI_y GJ} \sqrt{1 + \left(\frac{\pi}{L_b}\right)^2 \frac{EC_w}{GJ}}$$

[continues on next page]

Example 10.4.3 continues ...

$$= \frac{1}{12}\left[\frac{\pi}{16.5(12)}\right]\sqrt{29{,}000(32.8)(11{,}200)(1.11)}\sqrt{1 + \left[\frac{\pi(68.1)}{16.5(12)}\right]^2}$$

$$= 212 \text{ ft-kips}$$

$$M_{dE}^o = \phi_b\, M_{cr}^o = 0.9(212) = 191 \text{ ft-kips}$$

From Fig. 10.4.1, or from Table 5.1 of the LRFDM, for a simple beam under uniformly distributed load with lateral bracing at the supports and midspan only, $C_b = 1.30$. So:

$$M_d = \min\ [C_b\, M_{dE}^o,\ \phi_b\, M_{px}] = \min\ [1.30(191),\ 309]$$

$$= \min\ [248,\ 309] = 248 \text{ ft-kips}$$

Thus,

$$248 \text{ ft-kips} = \frac{q_{u3}\, L^2}{8}$$

$$q_{u3} = \frac{248(8)}{33^2} = 1.82 \text{ klf} \qquad \text{(Ans.)}$$

EXAMPLE 10.4.4

Analysis of Simple Beam under Central Concentrated Load

A W24×76 A588 Grade 50 steel beam is used for a simple span of 30 ft (Fig. X10.4.4). Determine the largest single concentrated service live load, applied at the shear center of the cross section, if lateral supports are provided: (*a*) at the ends and at the centerline, as for the finished structure; (*b*) at the ends only, as during construction. Neglect beam weight. Deflection and shear need not be checked.

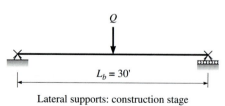

Lateral supports: finished structure
(*a*)

Lateral supports: construction stage
(*b*)

Figure X10.4.4

Solution

Span, $L = 30$ ft

From LRFDM Table 5-3 for a W24×76 of A588 Grade 50 steel:

$$\phi_b\, M_{px} = 750 \text{ ft-kip}; \quad BF = 19.8 \text{ kips}$$

$$L_p = 6.78 \text{ ft}; \qquad L_r = 18.0 \text{ ft}$$

From LRFDM Table 1-1 for a W24×76:

$$X_1 = 1760; \quad X_2 = 18{,}600 \times 10^{-6}; \quad S_x = 176 \text{ in.}^3; \quad r_y = 1.92 \text{ in.}$$

a. Design load on finished structure

In the finished structure, lateral braces are provided at the ends and at the center. So, $L_b = 15$ ft. As $L_p = 6.78 < L_b = 15 < L_r = 18.0$ ft,

$$M_d^o = \phi_b M_{px} - BF(L_b - L_p)$$

$$= 750 - 19.8(15.0 - 6.78) = 587 \text{ ft-kips}$$

From Fig. 10.4.3, or from LRFDM Table 5-1, for a simple beam under central concentrated load and lateral supports at the ends and the center, $C_b = 1.67$.

As $L_p = 6.78 < L_b = 15 < L_r = 18.0$ ft, the design strength is given by:

$$M_d = \min[C_b M_d^o, \phi_b M_{px}] = \min[1.67(587), 750]$$

$$= \min[980, 750] = 750 \text{ ft-kips}$$

Let Q_f represent the service live load at the center in the finished structure, then:

$$M_u = \frac{Q_u L}{4} = \frac{(1.6 Q_f)(30)}{4} \le 750$$

$$Q_{f\,max} = 62.5 \text{ kips} \qquad \text{(Ans.)}$$

b. Design load, construction stage

During construction, lateral braces are provided at the ends only. So, $L_b = 30.0$ ft.

From Fig. 10.4.3, or from LRFDM Table 5-1, $C_b = 1.32$ for a simple beam under central concentrated load with lateral supports at the ends only.

Slenderness ratio $L_b/r_y = 30(12)/1.92 = 188$

As $L_b = 30.0 > L_r = 18.0$ ft, the limit state of elastic lateral buckling applies.

$$M_{cr}^o = \frac{S_x X_1 \sqrt{2}}{L_b/r_y} \sqrt{1 + \frac{X_1^2 X_2}{2(L_b/r_y)^2}}$$

$$= \frac{176(1760) \sqrt{2}}{188(12)} \sqrt{1 + \frac{1760^2 (18,600 \times 10^{-6})}{2(188)^2}} = 262 \text{ ft-kips}$$

$$M_d^o = 0.9(262) = 236 \text{ ft-kips}$$

The design strength of the beam is given by:

$$M_d = \min[C_b M_d^o, \phi_b M_{px}] = \min[1.32(236), 750]$$

$$= \min[312, 750] = 312 \text{ ft-kips}$$

Let Q_c represent the concentrated construction load at the center, then:

$$M_u = \frac{Q_u L}{4} = \frac{(1.6 Q_c)(30)}{4} \le 312$$

$$Q_{c,max} = 26.0 \text{ kips} \qquad \text{(Ans.)} \qquad \qquad \textit{[continues on next page]}$$

Example 10.4.4 continues ...

Remark

This example clearly shows the importance of checking the safety of structures during construction stages, under factored construction loads, when all the bracing may not yet be in place.

EXAMPLE 10.4.5

Analysis of Simple Beam under Restraining Moment at One End

A W18×76 beam AB is subjected to three factored concentrated loads at points D, C, E, and to a factored external moment at the support A, as shown in Fig. X10.4.5. Lateral supports are provided at the ends and at the load points D and E, 24 ft from each end. Verify if the section is adequate for flexure. Shear and deflection need not be checked. Neglect bending moment due to self-weight. Assume A242 Grade 50 steel.

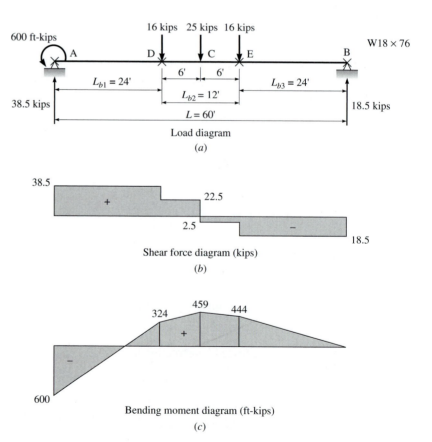

Load diagram
(a)

Shear force diagram (kips)
(b)

Bending moment diagram (ft-kips)
(c)

Figure X10.4.5

Solution

a. Data

From LRFDM Table 1-1 for a W18×76 shape, $\lambda_f = 8.11$; $\lambda_w = 37.8$.
From Table 9.5.1, for a rolled W-shape of A242 Grade 50 steel: $\lambda_{pf} = 9.15$, $\lambda_{pw} = 90.6$.
As $\lambda_f < \lambda_{pf}$ and $\lambda_w < \lambda_{pw}$, the W18×76 is a compact section, indicating that FLB and WLB will not precede the attainment of member flexural strength.
From LRFDM Table 5-3, for a W18×76 of Grade 50 steel,

$$\phi_b M_{px} = 611 \text{ ft-kips}; \quad L_p = 9.22 \text{ ft}; \quad L_r = 24.8 \text{ ft}; \quad BF = 11.1 \text{ kips}$$

For the simple beam under the factor loads and end moment given, support reactions are calculated and the shear force and bending moment diagrams are drawn, as shown in Fig. X10.4.5.

b. Check Segment 1, AD

$L_{b1} = 24.0$ ft

The maximum moment in Segment 1 occurs at the left end of the segment, and

$$M_{max} = 600 \text{ ft-kips}$$

Absolute values of the bending moments at quarter-points of the segment are 369, 138, and 93.1 ft-kips respectively. The moment modification factor, using Eq. 10.4.14, is:

$$C_b = \frac{12.5(600)}{2.5(600) + 3(369) + 4(138) + 3(93.1)} = 2.18$$

As $L_p = 9.22 < L_b = 24.0 < L_r = 24.8$ ft:

$$M_d^o = \phi_b M_{px} - BF (L_b - L_p) = 611 - 11.1(24.0 - 9.22)$$

$$= 447 \text{ ft-kips}$$

$$M_d = \min[C_b M_d^o, \phi_b M_{px}] = \min[2.18 (447), 611]$$

$$= \min[974, 611]$$

$$= 611 \text{ ft-kips}$$

As $M_{max} = 600$ ft-kips $< M_d = 611$ ft-kips, Segment 1 is O.K.

c. Check Segment 2, DE

$L_{b2} = 12.0$ ft; $M_{max} = 459$ ft-kips
Absolute values of the bending moments at the quarter-points of the segment are 392, 459, and 452 ft-kips, respectively.

$$C_b = \frac{12.5(459)}{2.5(459) + 3(392) + 4(459) + 3(452)} = 1.04$$

[continues on next page]

Example 10.4.5 continues ...

As $L_p = 9.22 < L_b = 12 < L_r = 24.8$ ft:

$$M_d^o = \phi_b M_{px} - BF(L_b - L_p) = 611 - 11.1(12.0 - 9.22)$$

$$= 580 \text{ ft-kips}$$

$$M_d = \min[C_b M_d^o, \phi_b M_{px}] = \min[1.04(580), 611]$$

$$= \min[603, 611] = 603 \text{ ft-kips}$$

As $M_{max} = 459$ ft-kips $< M_d = 603$ ft-kips, Segment 2 is O.K.

d. Check Segment 3, EB

$L_b = 24.0$ ft; $M_{max} = 444$ ft-kips

As the bending moment varies linearly from zero at one end to a maximum at the other end ($r_M = 0.0$), $C_b = 1.67$ from Table 10.4.1. From the calculations for Segment 1, $M_d^o = 447$ ft-kips.

$$M_d = \min[C_b M_d^o, \phi_b M_{px}] = \min[1.67(447), 611]$$

$$= \min[746, 611]$$

$$= 611 \text{ ft-kips}$$

As $M_{max} = 444$ ft-kips $< M_d = 611$ ft-kips, lateral-buckling strength of Segment 3 is adequate too.

Thus, the W18×76 of A242 Grade 50 is safe under the given factored loads.   (Ans.)

10.4.5   Design Procedure

To choose an appropriate W-shape of $F_y = 50$ ksi steel for a given beam segment of unbraced length L_b, and a factored major-axis moment M_u^o uniform over the unbraced length, enter an appropriate beam selection plot in LRFDM Table 5-5 at the required resisting moment (ordinate) and proceed to the right to meet the vertical line corresponding to the unbraced length (abscissa) at point P. Any beam that passes through or is located above and to the right of the intersection point P is a satisfactory solution. Of these, beams indicated by broken lines satisfy the requirements of unbraced length and uniform moment but are not the lightest sections available for the specified problem. The first beam shown in solid line immediately to the right and above the intersection point P is the lightest satisfactory section.

For unbraced lengths L_b greater than L_p and C_b values greater than one, the design requirement (Eq. 10.4.1) for a required bending strength M_u can be rewritten with the help of Eqs. 10.4.6 and 10.4.11 as:

$$\min[C_b M_d^o; \phi_b M_{px}] \geq M_u \tag{10.4.18}$$

or, equivalently, by:

$$M_d^o \geq \frac{M_u}{C_b} \equiv M_{ueq}^o \qquad (10.4.19a)$$

and

$$\phi_b M_{px} \geq M_u \qquad (10.4.19b)$$

where M_{ueq}^o = equivalent required factored uniform moment for the segment considered.

So, to choose an appropriate beam section for a given unbraced length L_b, required bending strength M_u, and a moment modification factor C_b, the following procedure may be used:

1. Calculate the equivalent required factored uniform moment M_{ueq}^o from Eq. 10.4.19a.
2. Locate the point P corresponding to L_b (abscissa) and M_{ueq}^o (ordinate) on the appropriate page of the beam selection plots (LRFDM Table 5-5).
3. Select a few sections that are above and to the right of point P. All these shapes therefore satisfy Eq. 10.4.19a.
4. From this set, select the subset of sections that also satisfy Eq. 10.4.19b. The values of $\phi_b M_{px}$ for the selected sections can be read from the beam selection plots as the ordinate corresponding to $L_b = 0$ for the chosen section, or can be found in the W-Shape Selection Tables (LRFDM Tables 5-2, 5-3, 5-4).
5. Select the lightest of this subset of sections as solution.
6. Make checks for the limit states of shear, deflection, and so on. if warranted.

Design of Simple Beam under Concentrated Loads

EXAMPLE 10.4.6

Select a A588 Grade 50 wide flange shape for a simply supported beam having a span length of 28 ft to support two concentrated live loads of 40 kips, each load placed 8 ft from a support (Fig. X10.4.6a). Lateral supports are provided at the ends of the beam and the load points. The maximum live load deflection must not exceed $L/360$.

Solution

a. Preliminary selection
   Neglect contribution of self-weight (dead load) in the preliminary calculations.
   Live load, $Q_L = 40.0$ kips
   Factored load, $Q_u = 1.6Q_L = 64.0$ kips
   Support reaction, $V_A = 64.0$ kips

[continues on next page]

Example 10.4.6 continues ...

$M_B = M_C = M_{max} = 64.0(8.0) = 512$ ft-kips

For segment AB: $L_{b1} = 8$ ft; $M_{max} = 512$ ft-kips

As the bending moment varies linearly, $C_b > 1.0$

For segment BC: $L_{b2} = 12$ ft; $M_{max} = 512$ ft-kips

As the moment is constant over BC, $C_b = 1.0$

It is evident that segment BC is more critical. Enter the beam selection plots for W-Shapes (LRFDM Table 5-5) with $M_{req} = 512$ ft-kips and $L_b = 12.0$ ft, and select a W24×68 corresponding to the first continuous line to the right and above. It provides a design strength $\phi_b M_{n(C_b=1)} = 563$ ft-kips.

b. Check strength limit states

From the LRFDM Table 5-3 for a W24×68 of Grade 50 steel:

$$\phi_b M_{px} = 664 \text{ ft-kips}; \quad BF = 18.6 \text{ kips}; \quad \phi_v V_n = 266 \text{ kips}$$

$$L_p = 6.61 \text{ ft}; \qquad L_r = 17.4 \text{ ft}; \qquad I_x = 1830 \text{ in.}^4$$

As $L_p = 6.61 < L_{b2} = 12 < L_r = 17.4$ ft, we have:

$$M_d^o = \phi_b M_{px} - BF(L_b - L_p) = 664 - 18.6(12.0 - 6.61)$$

$$= 564 \text{ ft-kips}$$

$$M_d = \min[C_b M_d^o; \phi_b M_{px}] = \min[1.0(564), 664]$$

$$= 564 \text{ ft-kips}$$

Contribution of self-weight to bending moment: $\dfrac{[1.2(0.07)](28^2)}{8} = 8.23$ ft-kips

As $M_d = 564 > M_{req} = (512 + 8.23) = 520$, flexural strength is O.K. Also, $V_d = \phi_v V_n = 266$ kips $> V_{req} = 64.0 + 1.20$ kips. So, shear is O.K.

c. Check serviceability limit state

From Case 9 in the LRFDM Table 5-17: Shears, Moments and Deflection, the maximum deflection at the center of the simply supported beam, under unfactored service live loads is:

$$\delta = \frac{Qa}{24EI}(3L^2 - 4a^2) = \frac{40.0(8.0)[3(28^2) - 4(8^2)]}{24(29,000)(1830)}(12^3) = 0.913 \text{ in.}$$

Maximum permissible deflection, $\delta_{all} = \dfrac{L}{360} = \dfrac{28(12)}{360} = 0.933$ in.

As $\delta < \delta_{all}$ the W24×68 of A588 Grade 50 steel satisfies all design requirements. (Ans.)

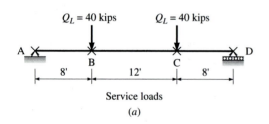

$Q_L = 40$ kips $\qquad Q_L = 40$ kips

A — B — C — D

8' | 12' | 8'

Service loads

(a)

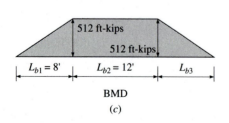

$Q_u = 64$ kips $\qquad Q_u = 64$ kips

8' | 12' | 8'

Factored loads

(b)

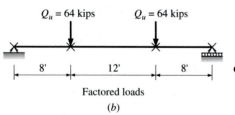

512 ft-kips

512 ft-kips

$L_{b1} = 8'$ | $L_{b2} = 12'$ | L_{b3}

BMD

(c)

Figure X10.4.6

Design of Simple Beam under Distributed Loads   E X A M P L E   1 0 . 4 . 7

Select the lightest W-section for a simply supported beam of 30-ft span.
Lateral supports are provided at the beam ends and midspan only. The dead
load is 0.50 klf (includes self-weight), and live load is 2.50 klf. Use A992
steel. Deflection of the beam under service live load is to be limited to $\frac{5}{4}$ in.

Solution

Required strength
$L = 30.0$ ft; $L_b = 15.0$ ft
$D = 0.50$ klf; $L = 2.50$ klf
$q_u = 1.2\,(0.50) + 1.6(2.50) = 4.60$ klf

$$M_{max} = \frac{q_u L^2}{8} = \frac{4.60(30^2)}{8} = 518 \text{ ft-kips}$$

From Fig. 10.4.3, or from LRFDM Table 5-1, for a simple beam with lateral supports at the ends and midspan only, and subjected to uniformly
distributed load, $C_b = 1.30$.
As the unbraced length L_b is large, assume that $L_b > L_p$ for the section to
be selected.
$$M_d = \min \left[C_b M_d^o, \phi_b M_{px} \right] \geq M_u = 518$$

So, for safe design we should have, $\phi_b M_{px} \geq M_u = 518$

and, simultaneously, $M_d^o \geq \dfrac{M_u}{C_b} = \dfrac{518}{1.30} = 398 \text{ ft-kips} = M_{ueq}^o$

Enter the beam selection plots for W-Shapes (LRFDM Table 5-5), with
$M_{ueq}^o = 398$ and $L_b = 15.0$ ft, and observe that a W21×62 (with
$M_d^o = 406$ ft-kips) and a W18×71 (with $M_d^o = 420$ ft-kips) are likely to
be satisfactory. From the beam selection plots also observe that
$\phi_b M_{px} = 540$ ft-kips for a W21×62 and 548 ft-kips for a W18×71. So,
both of the sections tentatively selected satisfy the strength requirements.
Serviceability limit state for live load deflection requires:

$$\delta = \frac{5}{384} \frac{2.5(30^4)(12^3)}{29000I} \leq \frac{5}{4} \text{ in.} \rightarrow I \geq 1260 \text{ in.}^4$$

From LRFDM Table 5-2, observe that a W21×62 has an $I_x = 1330$ in.^{4},
while the value is 1170 in.^{4} for the W18×71. So, select the W21×62 section: the only section from the set that satisfies both the strength and
serviceability limit states. The design bending strength is:

$$M_d = \min \left[1.30(406), 540 \right] = 528 \text{ ft-kips} > 518 \text{ ft-kips}   \text{O.K.}$$

Select a W21×62 of A992 steel.   (Ans.)

EXAMPLE 10.4.8

Design of Beam under Distributed and Concentrated Loads

Select the lightest W-section for the simply supported beam AB shown in Fig. X10.4.8a. The beam is subjected to a central concentrated load of 40 kips and a uniformly distributed load of 2 klf. The loads are factored loads. Lateral supports are provided at the ends and at the midspan only. Use A992 steel. Deflection need not be checked.

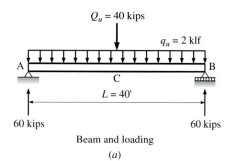

Beam and loading
(a)

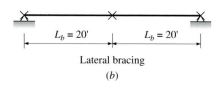

Lateral bracing
(b)

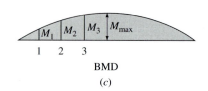

BMD
(c)

Figure X10.4.8

Solution

Span, $L = 40$ ft

Reaction, $R_A = \dfrac{40}{2} + \dfrac{2(40)}{2} = 60$ kips

Required shear strength, $V_u = V_{max} = 60$ kips
Bending moment at a distance z from end A,

$$M_z = 60z - 2\frac{z^2}{2} = 60z - z^2 \quad \text{for } z \leq 20 \text{ ft}$$

Unbraced length, $L_b = 20$ ft
Let 1, 2, 3 be the quarter-points of the unbraced segment AC.

$$M_1 = M_{(z=5')} = 60(5) - 5^2 = 275 \text{ ft-kips}$$

$$M_2 = M_{(z=10')} = 60(10) - 10^2 = 500 \text{ ft-kips}$$

$$M_3 = M_{(z=15')} = 60(15) - 15^2 = 675 \text{ ft-kips}$$

$$M_{max} = M_{(z=20')} = 60(20) - 20^2 = 800 \text{ ft-kips}$$

$$C_b = \frac{12.5(800)}{2.5(800) + 3(275) + 4(500) + 3(675)} = 1.46$$

Required bending strength, $M_u = M_{max} = 800$ ft-kips
As the unbraced length, 20 ft, is quite large, and likely to be greater than L_p of the section to selected, design flexural strength may be written as:

$$M_d = \min [C_b M_d^o, \phi_b M_{px}]$$

So, for safe design we should have, $\phi_b M_{px} \geq M_u = 800$ ft-kips

and, simultaneously, $M_d^o \geq \dfrac{M_u}{C_b} = \dfrac{800}{1.46} = 548$ ft-kips $= M_{ueq}^o$

Observe from the beam selection plots for W-Shapes (LRFDM Table 5-5) that, for a beam segment with $L_b = 20$ ft, $F_y = 50$ ksi and $M_{ueq}^o = 548$ ft-kips, a W27×84 with a M_d^o value of 600 ft-kips is a possible choice. Also for this section, from LRFDM Table 5-3:

$$\phi_b M_{px} = 915 \text{ ft-kips}; \phi_b M_{rx} = 639 \text{ ft-kips}; L_r = 19.3 \text{ ft}$$

$$\phi_v V_n = 332 \text{ kips}$$

As $L_b = 20$ ft $> L_r = 19.3$ ft, the flexural strength is determined by the limit state of elastic lateral-torsional buckling.

$$M_d = \min[C_b M_d^o, \phi_b M_{px}] = \min[1.46(600), 915]$$

$$= \min[876, 915] = 876 \text{ ft-kips}$$

As $M_d = 876$ ft-kips $> M_u = 800$ ft-kips, the section is O.K. Also, the section selected is compact for Grade 50 steel. Further, the design shear strength,

$$V_d = \phi_v V_n = 332 \text{ kips} > V_u = 60 \text{ kips} \qquad \text{O.K.}$$

Use a W27×84 of A992 steel.   (Ans.)

Design of Beam with Overhangs

EXAMPLE 10.4.9

Select the lightest W-section for the overhanging beam AB shown in Fig. X10.4.9a. It has a 32-ft main span and 12-ft overhangs on each side. It is subjected to concentrated loads Q at the free ends and a load $2Q$ at the center C. Each concentrated load Q represents 10 kips dead load and 20 kips live load. Lateral support is provided at the vertical supports and at the load points. Assume A572 Grade 50 steel. Neglect the self-weight of the beam in the calculations. Shear and deflection need not be checked.

Solution

a. Loading

Dead load = 10 kips;   Factored dead load = $1.2D$ = 12 kips
Live load = 20 kips;   Factored live load = $1.6L$ = 32 kips

Two loading cases, and their corresponding moment diagrams are shown in Fig. X10.4.9. Segment 1 of Loading Case I (Fig. X10.4.9b) has the highest absolute value of moment ($|+560| = 560$), but the value of C_b for this segment will be relatively high due to the occurrence of reverse curvature bending between the ends of the segment. On the other hand, Segment 1 of Loading Case II (Fig X10.4.9c) has a slightly lower absolute value for maximum moment ($|-528| = 528$), but will most

[continues on next page]

Example 10.4.9 continues ...

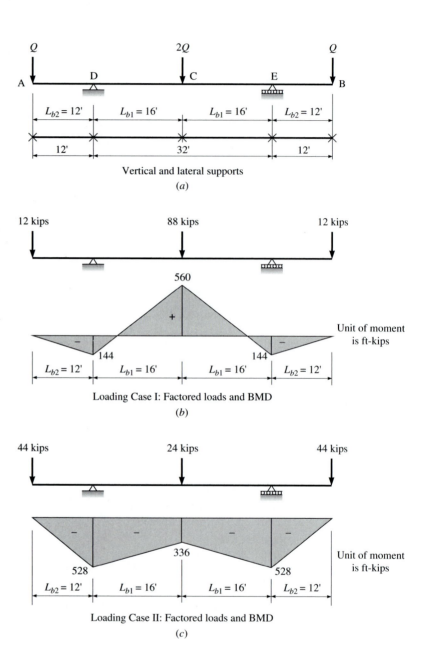

Figure X10.4.9

likely have a relatively low value of C_b since single curvature bending is maintained throughout the segment. Based upon this reasoning, consider Segment 1 of Loading Case II first, and select a section that satisfies the requirements for this segment. Then check to verify that all segments of both loading cases are satisfied.

b. Check Segment 1, Loading Case II (Fig X10.4.9c)

$$M_{max} = 528 \text{ ft-kips}; L_b = 16 \text{ ft}; r_M = -\frac{|M_1|}{|M_2|} = -\frac{336}{528} = -0.64$$

From Table 10.4.1, $C_b = 1.17$ for $r_M = -0.64$.

$$M_{ueq}^o = \frac{M_{max}}{C_b} = \frac{528}{1.17} = 451 \text{ ft-kips}$$

Enter beam selection plots for W-Shapes, $F_y = 50$ ksi, and $C_b = 1.0$ (LRFDM Table 5-5) with $L_b = 16$ ft, $M_{ueq}^o = 451$ ft-kips, and observe that a W24×68 with $M_d^o = 488$ ft-kips is a possible choice. Also, from LRFDM Table 5-3 for a W24×68 of Grade 50 steel:

$$\phi_b M_{px} = 664 \text{ ft-kips}; \quad L_p = 6.61 \text{ ft}$$

$$L_r = 17.4 \text{ ft}; \qquad BF = 18.6 \text{ kips}$$

For the segment considered, with $L_b = 16$ ft:

$$M_d^o = \phi_b M_{px} - BF(L_b - L_p) = 664 - 18.6(16.0 - 6.61)$$

$$= 489 \text{ ft-kips}$$

and for $C_b = 1.17$,

$$M_d = \min[C_b M_d^o; \phi_b M_{px}] = \min[1.17(489), 664]$$

$$= \min[572, 664]$$

$$= 572 \text{ ft-kips}$$

As $M_d = 572$ ft-kips $> M_{max} = 528$ ft-kips, the W24×68 shape is O.K. for Segment 1 under Loading Case II.

c. Check Segment 1, Loading Case I (Fig. X10.4.9b)

We have: $M_{max} = 560$ ft-kips; $\quad L_b = 16$ ft; $r_M = \dfrac{|M_1|}{|M_2|} = \dfrac{144}{560} = +0.26$

From Table 10.4.1, $C_b = 2.02$ for $r_M = +0.26$.
The design bending strength for this segment is:

$$M_d = \min[C_b M_d^o, \phi_b M_{px}] = \min[2.02(489), 664]$$

$$= \min[988, 664]$$

$$= 664 \text{ ft-kips}$$

As $M_d = 664$ ft-kips $> M_{max} = 560$ ft-kips, the W24×68 shape selected is also O.K. for Segment 1, Loading Case I.   [*continues on next page*]

Example 10.4.9 continues ...

d. Check Segment 2

Upon inspection, it is seen that only Load Case II needs to be checked, as it clearly controls. Thus, for Segment 2, Loading Case II,

$$M_{max} = 528; \quad L_b = L_{b2} = 12 \text{ ft}; \quad r_M = \frac{M_1}{M_2} = \frac{0}{528} = 0$$

From Table 10.4.1, $C_b = 1.67$ for $r_M = 0.0$.

$$M_d^o = \phi_b M_{px} - BF(L_b - L_p) = 664 - 18.6(12.0 - 6.61)$$

$$= 564 \text{ ft-kips}$$

$$M_d = \min[C_b M_d^o, \phi_b M_{px}] = \min[1.67(564), 664]$$

$$M_d = 664 \text{ ft-kips} > M_{max} = 528 \text{ ft-kips} \qquad \text{O.K.}$$

Thus, a W24×68 section of A572 Grade 50 steel satisfies the flexural strength requirements of all segments of both loading cases.   (Ans.)

EXAMPLE 10.4.10

Design of Simple Beam under Moving Loads

A 18-ft-long, crane runway beam carries the two end wheels of a crane. The wheels are spaced 12 ft on centers, and each transfers a maximum vertical live load of 24 kips to the beam. Neglect the lateral force and longitudinal force on the member. The crane beam is without lateral support except at the columns. The specified impact is 25 percent of the live load. Design the beam, using A242 Grade 50 steel. The maximum service live load deflection is to be limited to $L/500$.

Solution

Span, $L = 18$ ft;   Unbraced length, $L_b = 18$ ft

a. Required strengths

Wheel load, $Q = 24$ kips;   Wheel spacing, $a = 12$ ft
Impact factor, $I = 25\%$
Equivalent statically applied concentrated wheel load, $Q_L = 1.25(24.0) = 30$ kips.
The maximum live load moment will occur at midspan, with one of the crane wheels at the center of the span (Fig. X10.4.10b).
Assume a 90 lb rail, weighing 90 lb/yard or 30 lb/ft. Also, assume the weight of the crane beam to be 90 lb/ft.

Bending moment due to dead load, $M_D = \dfrac{q_D L^2}{8} = \dfrac{(30 + 90)}{1000} \dfrac{18^2}{8}$

$$= 4.86 \text{ ft-kips}$$

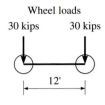

Wheel loads

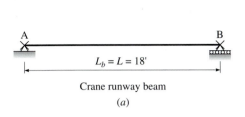

Crane runway beam

(a)

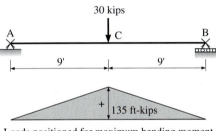

Loads positioned for maximum bending moment

(b)

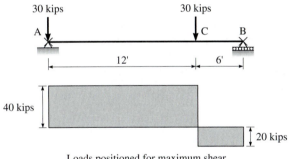

Loads positioned for maximum shear

(c)

Figure X10.4.10

Bending moment due to live load, $M_L = \dfrac{Q_L L}{4} = \dfrac{30(18)}{4}$

$$= 135 \text{ ft-kips}$$

Required bending strength, $M_u = 1.2M_D + 1.6M_L = 1.2(4.86) + 1.6(135) = 222$ ft-kips

The maximum live load shear will occur at a support, when one of the wheels is at that support (Fig. X10.4.10c).

Required shear strength,

$$V_u = 1.2(0.05)(9) + 1.6\left[30 + 30\left(\frac{6}{18}\right)\right] = 64.5 \text{ kips}$$

[continues on next page]

Example 10.4.10 continues ...

b. Serviceability limit state
Maximum allowable deflection,

$$\delta_{all} = \frac{L}{500} = \frac{18(12)}{500} = 0.432 \text{ in.}$$

Maximum deflection occurs at the center, with the service live load (i.e., the wheel load of 31.25 kips) located at the center. If I is the moment of inertia of the crane beam, the maximum deflection is given by:

$$\delta = \frac{QL^3}{48EI} = \frac{30(18^3)(12^3)}{48(29,000)(I)} = \frac{217}{I}$$

The serviceability limit state for deflection may be expressed as:

$$\delta \leq \delta_{all} \rightarrow \frac{217}{I} \leq 0.432 \rightarrow I \geq 502 \text{ in.}^4$$

From LRFDM Table 5-2: W-Shape Selection by I_x, observe that W18×40, W16×45, and W14×53 shapes with $I_x > 502$ are the lightest of the W18, W16, and W14 series that satisfy the deflection criteria.

c. Strength limit states
As the bending moment is essentially due to the concentrated load at midspan, use $C_b = 1.32$ (value from Fig. 10.4.3 or from the LRFDM Table 5-1, given for a simple beam with lateral bracing at the supports only). So,

$$M^o_{ueq} = \frac{M_u}{C_b} = \frac{222}{1.32} = 168 \text{ ft-kips}$$

Entering the beam selection plots for W-Shapes, $C_b = 1.0$, $F_y = 50$ ksi (LRFDM Table 5-5) with $\phi_b M_n = M^o_{ueq} = 168$ ft-kips and $L_b = 18$ ft, observe that W18×50, W16×50 and W14×43 are possible sections. From the beam selection plots, also observe that for these three sections $\phi_b M_{px}$ is greater than the required strength of 222 ft-kips.
Thus, W18×50 and W16×50 satisfy both the strength and serviceability requirements. So, select the W18×50 section. For this section, $M^o_d = 212$ ft-kips and $\phi_b M_{px} = 379$ ft-kips, resulting in:

$$M_d = \min [C_b M^o_d, \phi_b M_{px}] = \min [1.32(212), 379] = 280 \text{ ft-kips}$$

As $M_d = 280 > M_u = 222$ ft-kips, the section is O.K.
From LRFDM Table 5-3, the design shear strength of a W18×50 of Grade 50 steel is:

$$V_d = \phi_v V_n = 173 \text{ kips} > 64.5 \text{ kips}   \text{O.K.}$$

As the weight of the beam selected (50 lb) is less than the assumed value of 90 lb used in the dead load calculations and as the contribution of self-weight to bending moment is small, no revision is necessary. The section satisfies the requirements for compactness as there is no footnote in the beam selection tables stating otherwise.

Therefore use a W18×50 of A242 Grade 50 steel.   (Ans.)

10.5   Additional Topics

10.5.1   Holes in Beam Flanges

Ideally, holes in beam flanges should be placed at points of low bending moments. This, however, may not always be possible. For example, when flange-plated bolted semi-rigid or rigid beam-to-column end connections are used, bolt holes are required in the flanges of the beam.

To ascertain the influence of bolt holes on the moment capacity of a section, the flange is considered as an isolated tension member. The design tensile strength corresponding to yield on gross area of the flange is $T_{dfg} (= 0.9 F_y A_{fg})$, and the design strength corresponding to fracture on the net area of the flange is $T_{dfn} (= 0.75 F_u A_{fn})$. When the limit state of yield controls, that is, when $T_{dfg} \leq T_{dfn}$, no deduction for bolt holes need be made in either flange. On the other hand, if the limit state of fracture controls, when $T_{dfn} < T_{dfg}$, the member flexural properties tabulated in Part 1 of the LRFDM, which are based on the gross area of the flange, must be reevaluated based on an effective tension flange area,

$$A_{fe} = \frac{0.75}{0.90} \frac{F_u}{F_y} A_{fn} = \frac{5}{6} \frac{F_u}{F_y} A_{fn} \qquad (10.5.1)$$

This is LRFDS Eq. B10.3. In this relation:

A_{fg} = gross area of tension flange $(= b_f t_f)$
A_{fn} = net area of tension flange
F_u = specified minimum tensile strength
F_y = specified minimum yield stress

LRFDS Section B10 requires further that the maximum flexural strength be based on the elastic section modulus. That is,

$$M_d = \phi_b S_{xe} F_y \qquad (10.5.2)$$

where

$$S_{xe} = \frac{I_{xe}}{c} = \frac{I_{xe}}{d_b/2} \qquad (10.5.3)$$

and I_{xe} is the effective moment of inertia, which is obtained by subtracting from the gross moment of inertia $2(A_{fg} - A_{fe}) \bar{y}_h^2$. Here $\bar{y}_h$ is the distance from the neutral axis to the center of the bolt hole and d_b is the depth of the beam.

EXAMPLE 10.5.1

Holes in Beam Tension Flange

Determine the design flexural strength of an adequately braced W24×94 section of A992 steel, having two lines of holes in each flange for 1-in.-dia. bolts.

Solution

A992 steel: $F_y = 50$ ksi; $F_u = 65$ ksi

From LRFDM Table 1-1 for a W24×94:

$$b_f = 9.07 \text{ in.}; \quad t_f = 0.875 \text{ in.}; \quad I_x = 2700 \text{ in.}^4$$

$$d_b = 24.3 \text{ in.}; \quad c = d_b/2 = 12.2 \text{ in.}$$

$$\bar{y}_h = (d - t_f)/2 = (24.3 - 0.875)/2 = 11.7 \text{ in.}$$

From LRFDM Table 5-3 for a W24×94: $\phi_b M_{px} = 953$ ft-kips

Bolts: $d = 1.00$ in.

Assume standard punched holes; $d_e = d + \frac{1}{8} = 1.13$ in.

$A_{fg} = b_f t_f = 9.07(0.875) = 7.94$ in.^{2}

$A_{fn} = A_{fg} - nd_e t_f = 7.94 - 2(1.13)(0.875) = 5.96$ in.^{2}

Design yield strength of the tension flange,

$$T_{dfg} = 0.9A_{fg}F_y = 0.9(7.94)(50) = 357 \text{ kips}$$

Design fracture strength of the tension flange,

$$T_{dfn} = 0.75A_{fn}F_u = 0.75(5.96)(65) = 291 \text{ kips}$$

As the fracture strength is less than the yield strength, the flexural properties must be based on effective tension flange area given by Eq. 10.5.1,

$$A_{fe} = \frac{5}{6}\frac{F_u}{F_y}A_{fn} = \frac{5}{6}\left(\frac{65}{50}\right)(5.96) = 6.46 \text{ in.}^2$$

Effective moment of inertia, $I_{xe} = I_x - 2(A_{fg} - A_{fe})\bar{y}_h^2$

$$= 2700 - 2(7.94 - 6.46)(11.7^2) = 2300 \text{ in.}^4$$

Effective section modulus, $S_{xe} = \dfrac{I_{xe}}{c} = \dfrac{2300}{12.2} = 189 \text{ in.}^3$

Design bending strength,

$$M_d = \phi_b F_y S_{xe} = \frac{0.9(50)(189)}{12} = 709 \text{ ft-kips} \qquad \text{(Ans.)}$$

Reduction in strength due to the presence of holes in the tension flange

$$= \frac{(\phi_b M_{px} - M_d)}{\phi_b M_{px}} = \frac{(953 - 709)(100)}{953} = 25.6\%$$

10.5.2   Noncompact Sections

As mentioned in Section 9.5.3, if the width-to-thickness ratio of one or more compression elements exceeds λ_p, but none exceeds λ_r, the section is non-compact. A noncompact section is one for which the reduced yield moment, M_{rx}, can be reached but is not capable of reaching a fully plastic moment, M_{px}. Beams with noncompact elements must be checked against the reduced bending strength as influenced by local buckling of the plate elements. The local buckling design strength of a noncompact section is given by the LRFDS Eq. A-F1-3, as:

$$M_d = \phi_b M_n = \phi_b M_{px} - (\phi_b M_{px} - \phi_b M_{rx}) \frac{(\lambda - \lambda_p)}{(\lambda_r - \lambda_p)}$$

$$\text{for } \lambda_p < \lambda \le \lambda_r \qquad\qquad (10.5.4)$$

where the slenderness parameters λ, λ_p, and λ_r are defined for I-shaped members loaded in the plane of the web, as follows:

- For the limit state of flange local buckling (FLB):

$$\lambda = \lambda_f = \frac{b_f}{2t_f}; \quad \lambda_{pf} = 0.38 \sqrt{\frac{E}{F_y}}; \quad \lambda_{rf} = 0.83 \sqrt{\frac{E}{F_L}} \qquad (10.5.5)$$

where $F_L = F_y - F_r$
$\quad\quad F_r$ = residual compressive stress in flange (10 ksi, for rolled shapes)

- For the limit state of web local buckling (WLB):

$$\lambda = \lambda_w = \frac{h}{t_w}; \quad \lambda_{pw} = 3.76 \sqrt{\frac{E}{F_y}}; \quad \lambda_{rw} = 5.70 \sqrt{\frac{E}{F_y}} \qquad (10.5.6)$$

The slenderness limits λ_{pf}, λ_{rf}, λ_{pw} and λ_{rw} are given for various yield stresses in Table 9.5.1.

The variation of design bending strength, M_d, with the generalized slenderness ratio, λ, is shown in Fig. 10.5.1 for the limit states of LTB (for $C_b = 1.0$), FLB, and WLB. The smallest of M_d as determined by lateral-torsional buckling ($M_{d,\text{LTB}}$ with appropriate C_b), flange local buckling ($M_{d,\text{FLB}}$), or web local buckling ($M_{d,\text{WLB}}$) determines the design strength of the beam. That is:

$$M_d = \min [M_{d,\text{LTB}}, M_{d,\text{FLB}}, M_{d,\text{WLB}}] \qquad (10.5.7)$$

λ	Limit State
L_b/r_y	LTB, $C_b = 1.0$
$b_f/2t_f$	FLB
h/t_w	WLB

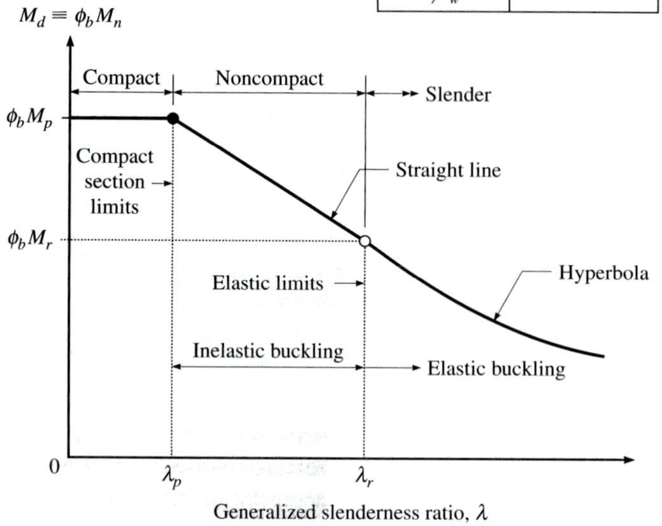

Figure 10.5.1: Design bending strength vs. generalized slenderness ratio, λ, for limit states of LTB, FLB, and WLB.

The rolled I-shaped section with the most slender web, as tabulated in Part 1 of the LRFDM, is a M12×10 with $\lambda_w = h/t_w = 74.7$. For $F_y = 70$ ksi (highest available grade of structural steel for rolled I-shapes, from LRFDM Table 2-1), λ_{pw} from Eq. 10.5.6 equals $3.76(29,000/70)^{0.5} = 76.5$, indicating that web local buckling will not be a design criteria for any of the rolled steel I-sections. Hence, for rolled steel I-sections used as beams, Eq. 10.5.7 reduces to:

$$M_d = \min [M_{d,\,\text{LTB}}, M_{d,\,\text{FLB}}] \tag{10.5.8}$$

There are five sections listed within the steel beam selection tables (LRFDM Tables 5-2, 5-3, 5-4, 5-6, 5-7, and 5-8) that are noncompact when F_y is 50 ksi. These sections, which are identified in those tables with a superscript ††, are the W21×48, W14×99, W14×90, W12×65, and W10×12. (The following I-shapes are noncompact: W8×10 and W6×15; however, they are not used as beams and as such are not listed in the beam selection tables). For the five noncompact shapes included in the LRFDM beam selection tables, the tabulated values of $\phi_b M_{px}$ are actually values of the design strength based on flange local buckling $M_{d,\,\text{FLB}}$ (Fig. 10.5.2). Also, the tabulated values of L_p are the values of unbraced length L'_p at which the nominal bending strength based on inelastic lateral-torsional buckling (with the unbraced segment under uniform moment) equals the nominal strength based

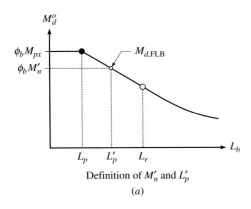

Definition of M'_n and L'_p

(a)

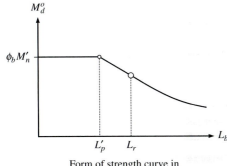

Form of strength curve in
the beam selection plots

(b)

Figure 10.5.2: M^o_d vs. L_b curve for an I-beam with noncompact flange.

on flange local buckling. Thus, from Eq. 10.4.8:

$$\phi_b M'_n = \phi_b M_{px} - BF(L'_p - L_p) = M_{d,\,\text{FLB}}$$

$$L'_p = L_p + \frac{(\phi_b M_{px} - M_{d,\,\text{FLB}})}{BF} \qquad (10.5.9)$$

For noncompact shapes, values tabulated as L_p in the beam selection tables for I-shapes (LRFDM Tables 5-2, 5-3, 5-4, 5-6, 5-7, and 5-8) and beam selection plots (LRFDM Table 5-5) are, in reality, values of L'_p. Thus the designer need have no additional concerns with noncompact shapes when F_y equals 50 ksi. One may use the tables or plots, because even though the section may be noncompact, the necessary changes in the flexural strength ($\phi_b M_{px}$ to $\phi_b M'_{px}$) and the limiting unbraced length L_p to L'_p have already been made to the tabulated values. The designer has to use the formulas presented in this section, however, for shapes with F_y other than 50 ksi.

EXAMPLE 10.5.2

Noncompact Beam under Major Axis Moment

A W12×65 beam of A992 steel has an unbraced segment of length 11 ft. Calculate the values of L_p, L'_p, $\phi_b M_{px}$, and $\phi_b M'_{px}$. Also, determine the maximum uniform factored major-axis bending moment that may be applied to this beam segment as per LRFD.

Solution

From LRFDM Table 1-1 for a W12×65 shape:

$$Z_x = 96.8 \text{ in.}^3; \quad r_y = 3.02 \text{ in.}$$

$$\lambda_f = \frac{b_f}{2t_f} = 9.92; \quad \lambda_w = \frac{h}{t_w} = 24.9$$

The limiting slenderness parameters for the flange and web of a rolled steel I-section are as follows:

$$\lambda_{pf} = 0.38\sqrt{\frac{29,000}{50}} = 9.15; \quad \lambda_{rf} = 0.83\sqrt{\frac{29,000}{(50-10)}} = 22.3$$

$$\lambda_{pw} = 3.76\sqrt{\frac{29,000}{50}} = 90.6; \quad \lambda_{rw} = 5.70\sqrt{\frac{29,000}{50}} = 137$$

As $\lambda_{pf} < \lambda_f < \lambda_{rf}$, the flange is noncompact. And as $\lambda_w < \lambda_{pw}$, the web is compact.

The limiting unbraced length for full plastic bending capacity, for $C_b = 1.0$, is given by:

$$\lambda_p = 1.76 r_y \sqrt{\frac{E}{F_y}} = \frac{1.76(3.02)}{12}\sqrt{\frac{29,000}{50}} = 10.7 \text{ ft}$$

Plastic moment, $M_{px} = Z_x F_y = 96.8(50)/12 = 403$ ft-kips.

$$\phi_b M_{px} = 0.9(403) = 363 \text{ ft-kips}$$

From LRFDM Table 5-3, for a W12×65 of $F_y = 50$ ksi steel:

$$\phi_b M_{rx} = 264 \text{ ft-kips}; \quad L_r = 31.7 \text{ ft}; \quad BF = 5.01$$

The flange local buckling strength of the noncompact section is obtained with the help of Eq. 10.5.4, as:

$$M_{d,\text{FLB}} = \phi_b M_{px} - (\phi_b M_{px} - \phi_b M_{rx})\frac{(\lambda_f - \lambda_{pf})}{(\lambda_{rf} - \lambda_{pf})}$$

$$= 363 - (363 - 264)\frac{(9.92 - 9.15)}{(22.3 - 9.15)} = 357 \text{ ft-kips} = \phi_b M'_{px}$$

(Ans.)

Also, from Eq. 10.5.9:

$$L_p' = L_p + \frac{(\phi_b M_{px} - M_{d,\text{FLB}})}{BF} = 10.7 + \frac{(363 - 357)}{5.01} = 11.9 \text{ ft}$$

(Ans.)

Note that the values of $\phi_b M_{px}$ and L_p tabulated in LRFDM beam selection tables, for the noncompact shape considered, are actually $\phi_b M'_{px}$ and L_p' calculated.
Unbraced length for the beam segment, $L_b = 11.0$ ft
As $L_b < L_p' = 11.9$ ft, the design bending strength of the beam segment,

$$M_d = \phi_b M'_{px} = 357 \text{ ft-kips}$$

(Ans.)

Noncompact Beam under Minor Axis Moment

EXAMPLE 10.5.3

A simply supported W12×65 beam of A992 steel spans 10 ft. Determine the maximum uniform factored minor-axis bending moment that may be applied to this beam as per LRFD.

Solution

From LRFDM Table 1-1 for a W12×65 shape:

$$Z_y = 44.1 \text{ in.}^3; \quad S_y = 29.1 \text{ in.}^3$$

The nominal plastic moment about the minor axis is:

$$M_{py} = \min[Z_y F_y, 1.5 S_y F_y] = \min[44.1(50), 1.5(29.1)(50)]$$

$$= \min[2210, 2180] = 2180 \text{ in.-kips}$$

$$\phi_b M_{py} = 0.9(2180)/12 = 164 \text{ ft-kips}$$

From Example 10.5.2, the flange is noncompact.
The nominal flexural strength for a slenderness parameter of λ_r is:

$$M_{ry} = S_y(F_y - F_r) = 29.1(50 - 10)/12 = 97.0 \text{ ft-kips}$$

$$\phi_b M_{ry} = 0.9(97.0) = 87.3 \text{ ft-kips}$$

From Example 10.5.4, $\lambda_{pf} = 9.15$; $\lambda_f = 9.92$; and $\lambda_{rf} = 22.3$.
The flange local buckling strength of the noncompact section is again obtained with the help of Eq. 10.5.1, as:

$$M_{d,\text{FLB}} = \phi_b M_{py} - (\phi_b M_{py} - \phi_b M_{ry}) \frac{(\lambda_f - \lambda_{pf})}{(\lambda_{rf} - \lambda_{pf})}$$

$$= 164 - (164 - 87.3) \frac{(9.92 - 9.15)}{(22.3 - 9.15)} = 160 \text{ ft-kips} = \phi_b M'_{py}$$

[continues on next page]

Example 10.5.3 continues ...

For an I-shape bent about its minor axis, the limit state of lateral buckling is not applicable. So, the maximum uniform factored minor-axis bending moment that can be applied to the W12×65 beam equals 160 ft-kips.

(Ans.)

10.6   Flexural Strength of Compact Beams Other Than I-Shapes

Doubly symmetric, I-sections are the most commonly used shapes for beams as they are easier to produce, and more importantly, are much easier to connect to other members. However, in situations where the beam is to be used in a laterally unsupported state over long spans such as crane booms, the use of box sections having high torsional stiffness (and hence, high flexural-torsional buckling strength) should be seriously considered. Channels are often used as purlins, girts, eave struts, lintels, trimmers and headers for stairwells, lift shafts, and other openings. For rolled channels used as beams, the LRFD Specification permits the design strength to be calculated by equations for doubly symmetric beams, as given in Section 10.4. The effect of eccentricity of load should be considered in the strength evaluation if the loads are not applied through the shear center of the channel.

For information on the flexural strength of compact beams other than I-shapes such as bars, rectangular HSS, C-, monosymmetric I-, and T-shapes, refer to Section W10.3 on the website http://www.mhhe.com/Vinnakota.

WWW

References

10.1 Bansal, J. [1971]: "The Lateral Instability of Continuous Beams," AISI Report No. 3, American Iron and Steel Institute, New York, August.

10.2 Galambos, T. V., and Ravindra, M. K. [1976]: "Load and Resistance Factor Design Criteria for Steel Beams," *Research Report No. 27*, Civil Engineering Department, Washington University, St. Louis, Mo., February.

10.3 Kirby, P. A., and Nethercot, D. A. [1979]: *Design for Structural Stability,* New York, Wiley.

10.4 Lay, M. G., and Galambos, T. V. [1965]: "Inelastic Steel Beams under Uniform Moment," *Journal of the Structural Division,* ASCE, vol. 91, ST6, December, pp. 67–93.

10.5 Lay, M. G., and Galambos, T. V. [1967]: "Inelastic Steel Beams Under Moment Gradient," *Journal of the Structural Division,* ASCE, vol. 93, ST1, February, pp. 381–399.

10.6 Vinnakota, S. [1977]: "Finite Difference Method for Plastic Beam-Columns," Chapter 10 of *Theory of Beam-Columns,* vol. 2, eds. W. F. Chen and T. Atsuta, McGraw-Hill, New York, pp. 451–503.

PROBLEMS

Analysis

P10.1. Plot to scale the design bending strength M_d versus the laterally unsupported length, L_b, using C_b of 1.0, 1.2, 1.75, and 2.3 for a W14×82 beam. Assume: (*a*) A992 steel, (*b*) A572 Grade 42 steel.

P10.2. Plot to scale the design bending strength M_d versus the laterally unsupported length, L_b, using C_b of 1.0, 1.2, 1.75, and 2.3 for a W24×76 beam. Assume: (*a*) A992 steel, (*b*) A572 Grade 65 steel.

P10.3. Determine the design bending strength of a W18×35 beam segment of A992 steel, if:

(*a*) $L_b = 4$ ft

(*b*) $L_b = 8$ ft and $C_b = 1.0$

(*c*) $L_b = 8$ ft and $C_b = 1.1$

(*d*) $L_b = 8$ ft and $C_b = 1.5$

(*e*) $L_b = 12$ ft and $C_b = 1.0$

(*f*) $L_b = 12$ ft and $C_b = 1.3$

(*g*) $L_b = 12$ ft and $C_b = 2.3$

P10.4. Determine the design bending strength of a W14×68 beam segment of A992 steel, if:

(*a*) $L_b = 8$ ft

(*b*) $L_b = 16$ ft and $C_b = 1.0$

(*c*) $L_b = 16$ ft and $C_b = 1.2$

(*d*) $L_b = 16$ ft and $C_b = 1.67$

(*e*) $L_b = 30$ ft and $C_b = 1.0$

(*f*) $L_b = 30$ ft and $C_b = 1.3$

(*g*) $L_b = 12$ ft and $C_b = 2.27$

P10.5. A W24×176 of A992 steel is used for a simple span of 36 ft. If the only dead load present is the weight of the beam, what is the largest service concentrated live load that can be placed at midspan? Assume: (*a*) Lateral bracing is provided at the beam ends and at midspan only; (*b*) Lateral bracing is provided at the beam ends only; (*c*) Recalculate, if, in addition, the service live load deflection of the beam is to be limited to $L/480$.

P10.6. Determine the concentrated live load that a W18×50 of A992 steel can carry if the span is 32 ft with the load located at the midspan. Lateral bracing is provided: (*a*) at the beam ends and at quarter points only, (*b*) at the beam ends and at the midspan only. Shear and deflection need not be checked.

P10.7. A W30×108 of A992 steel is used for a simple span of 36 ft. If the only dead load present is the weight of the beam, what is the largest service uniformly distributed live load that can be placed on the beam? Lateral bracing is provided at the beam ends and quarter points only.

P10.8. Verify the adequacy of the W16×50 beam in Problem P9.11 if continuous lateral bracing is provided over the length AD only.

P10.9. Check the adequacy of the beam in Problem P9.16, if lateral bracing is provided at the supports and at the concentrated load points only.

P10.10. A 34-ft-long W27×94 shape of A572 Grade 50 steel is used as a simply supported beam. It is subjected to a factored concentrated load of 90 kips at 12 ft from each support. In addition, the beam is subjected to a factored moment of 340 ft-kips at its left end (anticlockwise). Neglect self-weight of the beam in the calculations, and check if the beam is safe as per the LRFDS. Lateral bracing is provided at the end points and load points only.

P10.11. A W18×86 is used for a partially restrained beam, supporting the factored loads shown in Fig. P10.11. Lateral supports are provided at the ends and at load points only. Using A992 steel and neglecting the dead weight of the beam, find if the beam is overloaded. Shear and deflection need not be checked.

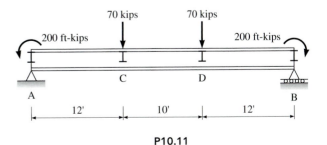

P10.11

P10.12. Determine the concentrated live load that a W18×50 of A572 Grade 60 steel can carry if the span is 40 ft with the load located at the midspan. Lateral bracing is provided: (*a*) at the beam ends and at quarter points only, (*b*) at the beam ends and at the midspan only, and (*c*) at the beam ends only.

P10.13. Repeat Problem P10.5 if the shape is of A572 Grade 60 steel.

....

Design

P10.14. Select the lightest available W-section of A992 steel for a factored moment of 530 ft-kips, if (*a*) L_b = 6 ft; (*b*) L_b = 10 ft; and (*c*) L_b = 19 ft. Assume C_b = 1.0.

P10.15. (*a*) A 30-ft-long simply supported beam is loaded at the third-points of the span with concentrated loads of 10 kips (5 kips dead load and 5 kips live). Lateral bracing is provided at the supports and load points. The self-weight of the beam may be neglected. Select the lightest W-shape of A572 Grade 50 steel.

(*b*) Redesign if the deflections are to be limited to *L*/360, under combined service dead and live loads.

P10.16. Select the lightest W-section to carry a uniform dead load of 0.3 klf and a live load of 0.6 klf on a simply supported span of 34 ft. Assume no deflection limitations. Use A572 Grade 50 steel. The beam is laterally supported at: (*a*) the beam ends and midspan only, (*b*) the beam ends only.

P10.17. A 24-ft-long simply supported beam is loaded with a uniform dead load of 1.6 klf, including the beam weight and a uniform of snow load of 2.5 klf. Lateral bracing is provided at beam ends and at 8 ft intervals. Select the lightest W-shape of A992 steel.

P10.18. Repeat Problem P9.23 if lateral bracing is provided at beam ends and point C only.

P10.19. A simple beam supporting elevator machinery carries a concentrated load of 6 kips at midspan. It is

laterally unsupported and has an effective span of 14 ft. The beam deflection is to be limited to *L*/1000. Allowance is to be made for a 10 percent impact, which is to be included in computing the deflection. Select a suitable A992 steel beam.

P10.20. Redesign the beam in Problem 9.18, if lateral bracing is provided at: (*a*) intervals of 12 ft; (*b*) the ends and midspan only.

P10.21. Redesign the beam in Problem 9.20, if lateral bracing is provided at the support points A and B and at the load point C only.

P10.22. A simple beam carries a uniformly distributed dead load of 2.1 klf plus its self-weight, and three 12 kip concentrated live loads at the quarter points of 40-ft span. Determine the lightest W-section to carry the loads, if lateral bracing is provided at the supports and the load points only. Use A992 steel.

P10.23. Select the lightest W-shape of A572 Grade 50 steel for a simple beam of span 36 ft. The beam is subjected to a uniformly distributed dead load of 1 klf acting over the entire length. In addition, the beam receives concentrated loads (6 kips dead and 14 kips live, respectively) from purlins located at quarter points of the beam. Lateral bracing is provided at beam ends and quarter points only. The loads given are service loads.

P10.24. A W27×146 of A992 steel has four bolt holes in each flange for $^7/_8$-in.-dia. bolts.

(*a*) Determine the reduced elastic section modulus S_x.

(*b*) Determine the reduced plastic section modulus Z_x. What is the percentage reduction in Z_x?

P10.25. A W24×104 of A572 Grade 60 steel has two bolt holes in each flange for $1^1/_8$-in.-dia. bolts.

(*a*) Determine the reduced elastic section modulus S_x.

(*b*) Determine the reduced plastic section modulus Z_x. What is the percentage reduction in Z_x?

Plate Local Buckling and Shear Buckling

P10.26. Select the lightest W-section to support a factored uniformly distributed load of 2.6 klf over a simple span of 25 ft. Assume A572 Grade 60 steel and continuous lateral support. Make all checks. The given load does not include self-weight of the beam.

P10.27. A W8×10 beam with a yield stress of 50 ksi has an unbraced segment length of 6 ft and is subjected to a uniform bending moment. Calculate the values of L_p, M_p, L'_p, and M'_p. Also determine the maximum uniform factored bending moment that may be applied to the beam segment.

P10.28. A M10×8 beam is simply supported over a span of 12 ft. Adequate lateral support is provided to the beam, which supports a uniformly distributed factored load, including its own weight, of 0.5 klf and a factored live load consisting of two concentrated loads of 5 kips each acting at the third-points of the span. Assume A572 Grade 50 steel, and check the adequacy of the beam for bending and shear.

P10.29. A propped cantilever, with a span $L = 30$ ft, is subjected to a single concentrated load, Q, at midspan. The load consists of 50 kips dead load and 40 kips snow load. The self-weight of the beam may be neglected. Assume full lateral support, and select the lightest W24 shape. Use elastic analysis with redistribution of moments taken into account.

P10.30. Determine if the section selected in Problem P10.29 is satisfactory, if the beam is laterally supported at the supports and at the point of application of the concentrated load.

P10.31. A fixed ended beam, with a span $L = 30$ ft, is subjected to a single concentrated load, Q, at a distance $a = 12$ ft from the left support. The load Q consists of 40 kip dead load and 80 kip live load. The self-weight of the section may be neglected. Assume full lateral support and select the lightest W24 shape. Use elastic analysis with redistribution of moments taken into account.

P10.32. Determine if the section selected in Problem P10.31 is satisfactory, if the beam is laterally supported at the supports and at the point of application of the concentrated load.

P10.33. Select the lightest W-section required for a two span continuous beam ABC with $L_1 = L_2 = L = 32$ ft. The beam is hinged at A and roller-supported at B and C. It supports a factored distributed load, including its self-weight, of 12 klf. Assume full lateral support to the beam. Use elastic analysis with redistribution of moments taken into account. Check for shear.

P10.34. Select the lightest W-section required for a three span continuous beam ABCD with $L_1 = L_2 = L_3 = L = 24$ ft. The beam is hinged at A and roller supported at B, C, and D. It supports a factored distributed load, including its self-weight of 10 klf. Assume full lateral support to the beam. Use elastic analysis with redistribution of moments taken into account. Check for shear.

Steel Skeleton, St. Joseph's Community Hospital, West Bend, WI.
Photo courtesy Michael Henke, Construction Supply & Erection, Inc., Germantown, WI, www.cseconstruction.com

11

Members under Combined Forces

11.1 Introduction

Members in steel structures are often subjected to axial force—either tension or compression—and flexure about one or both axes of symmetry. The member subjected to axial compression and bending is known as a **beam-column,** and is the major element treated in this chapter and in Chapter H of the LRFD Specification. The bending moments in beam-columns may result from:

- Transverse loads acting between the ends of a compression member, as shown in Fig. 11.1.1a.
- Eccentricity of the longitudinal force at one or both ends, as shown in Fig. 11.1.1b.
- Flexure of the connecting members, as shown in Figs. 11.1.1c and d.

Beam-columns in framed structures are generally subjected to end forces only. The axial force in a beam-column typically results from the axial load transferred by the column above the particular member, and from the end-shears on the adjacent beams and girders framing into its ends. The bending moments at the ends of the beam-column represent resistance to the bending moments imparted at the ends by the girders and beams which frame it. Beam-columns in steel structures are often subjected to bending moments acting in two principal planes. These biaxial bending moments result from the space action of the framing system. The column shape is usually oriented so as to produce significant bending about the major axis of the member, but the minor-axis bending moments may become relatively significant as well, as the minor-axis bending resistance of an I-shaped section is relatively small compared to the major-axis bending resistance. Figure 11.1.2 shows columns that are part of a braced frame in the xx plane, and part of an unbraced frame in the yy plane of its cross section. A column of an industrial building supporting wind loads acting on a wall, transferred by girts as shown in Fig. 3.4.1, is an example of a beam-column with transverse loads. The top chord of a bridge truss subjected to wind loads is another example.

In a beam-column subjected to end-moments and/or in-span transverse loads, in addition to axial loads, the bending moments that develop in the member with these axial loads removed, are called **primary moments** or

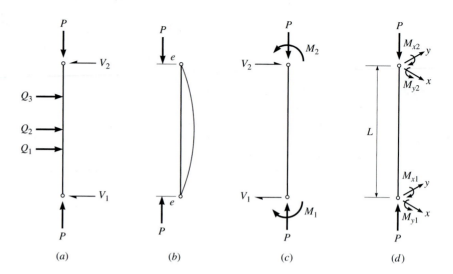

Figure 11.1.1: Typical beam-columns.

(a)   (b)   (c)   (d)

first-order moments. The additional moments induced by the interaction of axial force with the deflections are known as *secondary moments.*

As the bending moment on a beam-column approaches zero, the member tends to become a centrally loaded column, a problem that has been treated in Chapter 8 on columns. As the axial force on a beam-column approaches zero, the problem becomes that of a beam, which has been treated in Chapter 9 for the adequately laterally supported case and in Chapter 10 for the laterally unsupported case. All of the parameters that affect the behavior of a beam or a column—such as length of the member, geometry and material

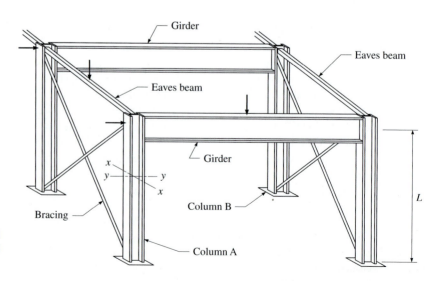

Figure 11.1.2: Beam-columns as part of a braced frame in the *xx* plane and as part of an unbraced frame in the *yy* plane.

properties of the cross section, magnitude and distribution of transverse loads and moments, presence or absence of lateral bracing, and whether the member is part of an unbraced or braced frame—will also influence the behavior, strength, and design of beam-columns. The design of beam-columns is treated in Chapters H, C and B of the LRFD Specification. Strength of steel sections under axial compression and uniaxial, or biaxial bending is considered in Section 11.2, wherein the reduced plastic moment, M_{pc}, is introduced. Second-order moments in elastic beam-columns are determined in Section 11.3, wherein moment amplification factor, B_1, and equivalent moment factor, C_m, are introduced, Lateral torsional buckling of beam-columns is described in Section 11.4. Planar strength of steel beam-columns is reviewed in Section 11.5, while inelastic lateral-torsional buckling of steel beam-columns is discussed in Section 11.6. Strength of biaxially bent steel beam-columns is considered in Section 11.7. The moment amplification factor, B_2, is introduced in Section 11.8. Use of LRFD interaction formulas for beam-column design is explained in detail in Section 11.9. Limiting slenderness ratios for local buckling of web and flange plates are given in Section 11.10. Some building structures consisting of members subjected to axial compression and transverse loads are described in Section 11.11, while members subjected to axial tension and bending are considered in Section 11.12. Use of LRFD interaction formulas for beam-columns is illustrated in Section 11.13 with the help of several analysis examples. Two alternate methods for member selection for steel beam-columns are explained in Section 11.14, while their use is illustrated in Section 11.15 with the help of a large number of beam-column design examples. Finally, biaxial bending and asymmetrical bending of beams are treated in Section 11.16, as a special case of the biaxially bent beam-column problem.

11.2   Strength of Sections under Combined Loading

11.2.1 Strength of Sections under Axial Compression and Uniaxial Bending

If a short, theoretically, zero length beam-column segment is subjected to the combined action of bending moment and axial compressive force, the available flexural strength of the section is reduced due to the presence of axial load from the full value of M_p to a lesser value that will be denoted as M_{pc}. In what follows, we will assume that the bending moment acts in a plane of symmetry and that the material behavior is elastic perfectly plastic. That is, the fibers have a stiffness E in the elastic domain and zero stiffness in the plastic domain when stressed to the yield stress F_y.

Let us consider a rectangular section of width b and depth d subjected to an axial compressive force P and a bending moment M in its vertical plane

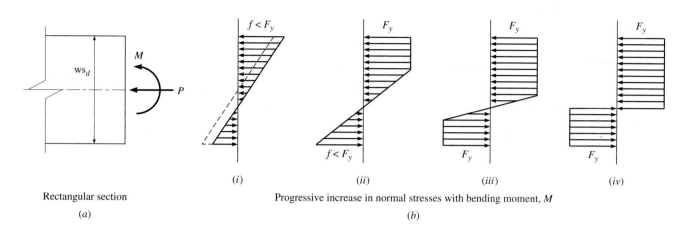

Rectangular section

(a)

Progressive increase in normal stresses with bending moment, M

(b)

Figure 11.2.1: Stress distributions in a rectangular section under axial compression P and moment M.

of symmetry (Fig. 11.2.1). The axial load P is maintained constant while the moment M is progressively increased. Figure 11.2.1b shows the stress distribution at various stages (*i, ii, iii, iv*) of loading. Due to the presence of axial compressive force, yielding starts first in the outer fiber on the compression side of the section (distribution *ii*). Only after a portion of this side has yielded will the tension side fibers begin to yield (distribution *iii*). The bending moment tends to an ultimate value M_{pc}, which corresponds to the distribution *iv* in Fig. 11.2.1b. When the cross section is entirely plastic, the stress distribution has the bi-rectangular form shown here. However, the stress block no longer contains equal compression and tension areas (compare Fig. 11.2.1b(*iv*) to Fig. 9.2.2d). That is, the plastic neutral axis (PNA) no longer coincides with the centroidal axis. The rectangular section under the action of the axial load P and the moment M_{pc} Fig. 11.2.1b(*iv*) is reconsidered in Fig. 11.2.2.

Let y_o be the distance of the PNA from the axis of symmetry. The bi-rectangular stress distribution of Fig. 11.2.2b(*i*) may be decomposed into the symmetric and antisymmetric distributions shown in Figs. 11.2.2b(*ii*) and (*iii*). The stresses in Fig. 11.2.2b(*ii*) have as their resultant the axial force P acting over the area of height $2y_o$, whereas the stresses in Fig. 11.2.2b(*iii*) have as their resultant the reduced plastic moment M_{pc}. We have:

$$P = 2y_o b F_y \qquad (11.2.1)$$

Noting that the compressive and tensile stress resultants of Fig. 11.2.2b(*iii*) act at the centroids of the areas above and below y_o, respectively

$$M_{pc} = 2\left[b\left(\frac{d}{2} - y_o\right)F_y\right]\left[y_o + \frac{1}{2}\left(\frac{d}{2} - y_o\right)\right] \qquad (11.2.2)$$

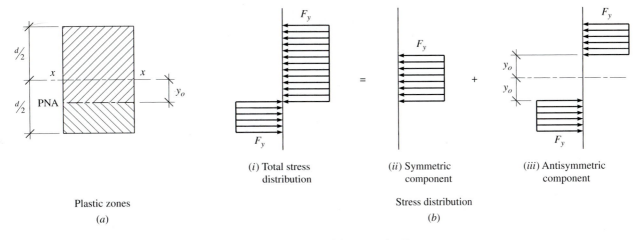

(*i*) Total stress
distribution

(*ii*) Symmetric
component

(*iii*) Antisymmetric
component

Plastic zones

Stress distribution

(*a*)

(*b*)

Figure 11.2.2: Reduced plastic moment M_{pc} of a rectangular section under axial compression P.

Substituting the value of y_o from Eq. 11.2.1 into Eq. 11.2.2 and noting that

$$bdF_y = P_y \qquad (11.2.3a)$$

and

$$\frac{bd^2}{4} F_y = ZF_y = M_p \qquad (11.2.3b)$$

where P_y is the yield load and M_p the fully plastic moment of the rectangular cross section, we obtain:

$$M_{pc} = M_p\left[1 - \left(\frac{P}{P_y}\right)^2\right] \qquad (11.2.4)$$

The reduced plastic moment M_{pc} of a rectangular section may therefore be expressed in the nondimensional form:

$$\frac{M_{pc}}{M_p} = \left[1 - \left(\frac{P}{P_y}\right)^2\right] \qquad (11.2.5)$$

Using a similar procedure, the reduced plastic moment M_{pcx} of an I-shape bent about its major axis, in the presence of an axial compressive load P, may be shown to satisfy the approximate relations [Driscoll et al., 1965]:

$$\frac{M_{pcx}}{M_{px}} = 1 \qquad \text{for } 0 < \frac{P}{P_y} \le 0.15 \quad (11.2.6a)$$

$$\frac{M_{pcx}}{M_{px}} = 1.18\left(1 - \frac{P}{P_y}\right) \quad \text{for } 0.15 < \frac{P}{P_y} \le 1.0 \quad (11.2.6b)$$

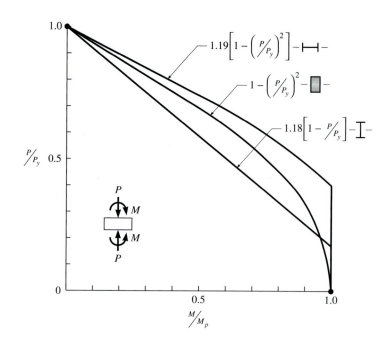

Figure 11.2.3: Interaction curves for rectangular and I-sections under axial compression and uniaxial bending.

Similarly, the reduced plastic moment M_{pcy} of an I-shape bent about its minor axis, in the presence of an axial load P, may be expressed by the simplified relation:

$$\frac{M_{pcy}}{M_{py}} = 1.19\left[1 - \left(\frac{P}{P_y}\right)^2\right] \leq 1.0 \qquad (11.2.7)$$

For further details on the strength of an I-section under axial compression and *uniaxial bending, refer to Section W11.1 on the website http://www.mhhe. com/Vinnakota.*

The interaction curves that result from Eqs. 11.2.5, 11.2.6, and 11.2.7 for rectangular sections and I-sections under axial compression and uniaxial bending are plotted in Fig. 11.2.3. Observe that when the axial force is zero, $M = M_p$, and when the axial force reaches the squash load P_y, the moment capacity is zero.

11.2.2   Strength of Sections under Axial Compression and Biaxial Bending

For information on the strength of an I-section under axial compression and *biaxial bending, refer to Section W11.2 on the website http://www.mhhe. com/ Vinnakota.*

Second-Order Moments
in Beam-Columns

11.3.1 Beam-Column with End Moments M_A, M_B

Consider a pin-ended member AB of length L and flexural rigidity EI. It is subjected to end moments M_A and M_B, with $M_B > M_A$, in addition to an axial compressive force P, as shown in Fig. 11.3.1a. The member is of uniform section, and the applied forces bend the member in a plane of symmetry. The maximum second-order moment is given by [Chen and Atsuta, 1977]:

$$M^*_{max} = |M_B| \left[\sqrt{\frac{1 + 2r_M \cos \phi + r_M^2}{\sin^2 \phi}} \right] \qquad (11.3.1)$$

where

$$r_M = \pm \frac{|M_A|}{|M_B|} \qquad (11.3.2)$$

$$\phi^2 = \frac{PL^2}{EI} = \frac{1}{\pi^2} \left(\frac{P}{P_E} \right) \qquad (11.3.3)$$

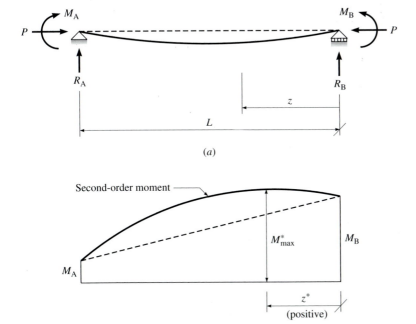

(a)

(b)

Figure 11.3.1: Beam-column with end moments M_A, M_B.

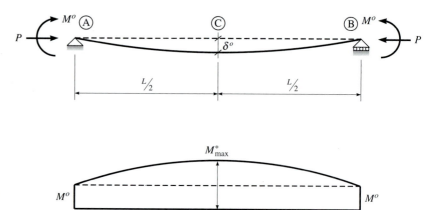

Figure 11.3.2: Beam-column under uniform moment.

Here, the moment ratio r_M is assigned a positive sign if the beam-column is bent in double-curvature and a negative sign if it is bent in single-curvature. Also, P_E is the Euler load of the column. The absolute value for the larger of the two end moments, namely of M_B, is used in Eq. 11.3.1 because, in design, we are interested only in the magnitude, not the direction of M^*_{max}.

For the particular case where the beam-column is subjected to a uniform moment that produces symmetric single curvature bending, that is, with $M_A = M_B = M^o$ as shown in Fig. 11.3.2, the maximum second-order moment occurs at midspan. Its magnitude is obtained by setting $r_M = -1$ in Eq. 11.3.1 as:

$$M^*_{max} = M^o \sqrt{\frac{2(1 - \cos\phi)}{\sin^2\phi}} = M^o \sec\frac{\phi}{2} \qquad (11.3.4)$$

For further details on the elastic behavior of a beam-column with axial compression and end moments only, refer to Section W11.3 on the website http://www.mhhe.com/Vinnakota.

11.3.2 Beam-Columns with Lateral Loads

For details on the elastic behavior of a beam-column with axial compression and lateral loads, refer to Section W11.4 on the website http://www.mhhe.com/Vinnakota.

11.3.3 Moment Amplification Factor, B_1, and Equivalent Moment Factor, C_m

Member sizes of beam-columns are generally based on the magnitude of the maximum moment, while the location of this maximum moment does not enter in the design process. Hence, the concept of equivalent moment, schematically shown in Fig. 11.3.3 for a beam-column under unequal end

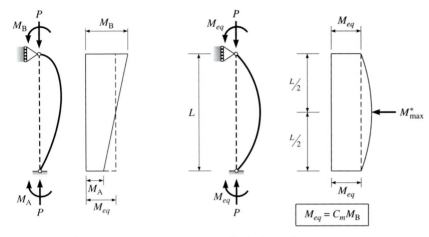

Unequal end moments
(a)

Equivalent end moment
(b)

$$M_{eq} = C_m M_B$$

Figure 11.3.3: Equivalent uniform moment for a beam-column under an axial load P and end moments M_A, M_B.

moments, is generally followed in design specifications. In simple terms, the maximum second-order moment of a beam-column under given axial load P and bending moment diagram (given end moments M_A and M_B with $|M_B| > |M_A|$ in Fig. 11.3.3a) is numerically equal to the maximum second-order moment of the same member and axial load P, under a pair of equal and opposite end moments M_{eq} known as ***equivalent end moments*** (Fig. 11.3.3b). Note that the axial load P and the second-order moment M^*_{max} are the same for both members. So, to determine the magnitude of the equivalent end moment M_{eq}, we let Eq. 11.3.4, with M^o replaced by M_{eq}, equal Eq. 11.3.1:

$$M_{eq}\left[\frac{\sqrt{2(1 - \cos\phi)}}{\sin\phi}\right] = |M_B|\left[\frac{\sqrt{1 + 2r_M\cos\phi + r_M^2}}{\sin\phi}\right] \quad (11.3.5)$$

Solving for M_{eq} results in:

$$M_{eq} = \left[\sqrt{\frac{1 + 2r_M\cos\phi + r_M^2}{2(1 - \cos\phi)}}\right]|M_B| = C_m|M_B| \quad (11.3.6)$$

This relation applies when the magnified moment within the span exceeds the end moment M_B. Here, C_m is known as the ***equivalent moment factor*** or ***moment reduction factor***. From Eq. 11.3.6 it is seen that C_m is a function of the moment ratio r_M and also the axial load ratio $\frac{P}{P_E}$ (see Eq. 11.3.3). Simplified expressions for C_m, as a function of moment ratio only, have been proposed by Campus and Massonnet [1956], Austin [1961], and others to approximate the exact value of C_m given by Eq. 11.3.6. For the design of beam-columns, the LRFD Specification (LRFDS Eq. C1-3) adopted the simplified linear expression suggested by Austin, namely:

$$C_m = 0.6 - 0.4 r_M \tag{11.3.7}$$

Values of C_m for different values of r_M are given in Table 10.4.1.

Thus, to evaluate the maximum second-order moment for a nonsway beam-column subjected to end moments only, we need to multiply the larger of the two end moments by C_m. Once the equivalent moment M_{eq} is obtained, the maximum second-order moment M^*_{max} in the beam-column can be determined using Eq. 11.3.4 with M^o replaced by M_{eq}. Thus,

$$M^*_{max} = M_{eq} \sec \frac{\phi}{2} = \left(C_m \sec \frac{\phi}{2} \right) |M_B| \tag{11.3.8}$$

This relation can be rewritten in the general format:

$$M^*_{max} = B_1 |M_B| = B_1 M_{max} \tag{11.3.9}$$

where M_{max} is the maximum first-order moment acting on the beam-column, and B_1 is a **moment amplification factor.** It is a measure of the second-order moment that develops due to the interaction of the axial force in a column with the maximum chord deflection ($P\delta$ effect). For a pin-ended member subjected to end moments only, we have:

$$B_1 = C_m \sec \frac{\phi}{2} \approx \frac{C_m}{\left(1 - \dfrac{P}{P_e} \right)} \tag{11.3.10}$$

For the result to have physical meaning, the moment magnification factor must be greater than or equal to unity, otherwise the end moment M_B itself will be taken as M^*_{max}.

For design purposes, the exact relations for maximum second-order moment(s) for transversely loaded pinned-pinned, fixed-pinned, or fixed-fixed beam-columns, obtained by the differential equation approach described in Section W11.4 on the website, could also be approximated by the single formula:

$$M^*_{max} = B_1 M_{max} = \frac{C_m}{\left(1 - \dfrac{P}{P_e} \right)} M_{max} \tag{11.3.11}$$

where M^*_{max} = maximum second-order moment in the member
M_{max} = maximum first-order moment in the member
B_1 = moment magnification factor
P = axial compression in the member
P_e = elastic critical load of the beam-column about the bending axis, consistent with the end support conditions = $\dfrac{\pi^2 EI}{(KL)^2}$
C_m = equivalent moment factor

For example, for a simply supported beam-column under a central concentrated load:

$$C_m = \left[1 - 0.2\frac{P}{P_e}\right]; \quad P_e = P_E = \frac{\pi^2 EI}{L^2} \qquad (11.3.12)$$

This is the value for C_m given in LRFDC Table C-C1.1.

For further details on the equivalent moment factors, C_m, for beam-columns with lateral loads, refer to Section W11.5 on the website http://www. mhhe.com/Vinnakota.

11.4   Elastic Lateral-Torsional Buckling of Beam-Columns

Let us consider a perfectly straight beam-column of a uniform, doubly symmetric I-shape and of length L. The ends of the beam-column are assumed to be flexurally and torsionally simply supported and free to warp. The member is bent about its major axis by equal and opposite end moments M^o and loaded by an axial force P, with the applied moments acting in the plane of symmetry (yz plane) as shown in Fig. 11.4.1a. The beam-column will buckle at an elastic critical moment $M^o = M^o_{ccrE}$ when a deflected and twisted equilibrium position, such as that shown in Fig. 11.4.1, is possible. Deformation of the member at any section can be broken down into three distinct motions: a vertical displacement v in the y direction, a lateral displacement u in the x direction, and a rotation ϕ about the z axis.

The elastic critical moment for such a flexurally and torsionally simply supported beam-column of doubly symmetric I-shape subjected simultaneously to an axial load P is obtained as [Galambos, 1968]:

$$M^o_{ccrE} = \sqrt{\frac{(I_x + I_y)}{A}(P_{Ey} - P)(P_{Ez} - P)} \qquad (11.4.1)$$

wherein

$$P_{Ey} = \frac{\pi^2 EI_y}{L^2}; \quad P_{Ez} = \frac{1}{r_o^2}\left[\frac{\pi^2 EC_w}{L^2} + GJ\right] \qquad (11.4.2)$$

Here,

M^o_{ccrE} = elastic critical moment of a beam-column of a doubly symmetric I-shape loaded by an axial compressive force P and uniform major axis moment, M^o

P = axial compression in the member

P_{Ey} = flexural buckling load of the member about the weak axis

P_{Ez} = torsional buckling load of the member

L = length of the member

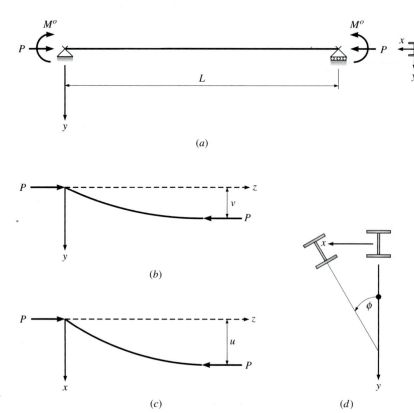

Figure 11.4.1: Lateral-torsional buckling of a beam-column of doubly symmetric I-shape.

$$
\begin{aligned}
EI_y &= \text{bending stiffness about the } y \text{ axis} \\
GJ &= \text{St. Venant torsional stiffness} \\
EC_w &= \text{warping stiffness} \\
I_x &= \text{moment of inertia of the section about the } x \text{ axis} \\
A &= \text{area of cross section} \\
\bar{r}_o^2 &= \frac{(I_x + I_y)}{A}
\end{aligned}
$$

The elastic critical moment for a flexurally and torsionally simply supported beam of a doubly symmetric I-shape under uniform moment M^o only, given by Eq. 10.2.34, may now be rewritten with the help of Eq. 11.4.2 as:

$$
M^o_{crE} = \sqrt{\frac{\pi^2 EI_y}{L^2}\left(\frac{\pi^2}{L^2}EC_w + GJ\right)} = \sqrt{\left(\frac{I_x + I_y}{A}\right)P_{Ey}P_{Ez}} \quad (11.4.3)
$$

The critical moment for lateral-torsional buckling of a flexurally and torsionally simply supported beam-column under uniform major axis moment, for the doubly symmetric I-shaped beam-column, given by Eq. 11.4.1, may now be rewritten in the nondimensional form:

$$\frac{M^o_{ccrE}}{M^o_{crE}} = \sqrt{\left(1 - \frac{P}{P_{Ey}}\right)\left(1 - \frac{P}{P_{Ez}}\right)} \qquad (11.4.4)$$

For further information on the elastic lateral-torsional buckling of beam-columns of singly or bisymmetric I-sections, refer to Section W11.6 on the website http://www.mhhe.com/Vinnakota.

11.5   Planar Strength of Beam-Columns

Wide-flange steel beam-columns bent, in a symmetric single curvature, about their major axis by end moments M^o_x and an axial force P are considered in this section (Fig. 11.5.1a). We will assume that the material is elastic perfectly plastic. The applied axial load, P, is smaller than the maximum force P_{cr} which the member can support as a column, and thus there exists some reserve of capacity to support bending moments. We assume that P is applied first and then held constant. The moments M^o_x are subsequently applied and increased monotonically until failure occurs. The variation of the central deflection δ with M^o_x is shown as Curve A in Fig. 11.5.2. The applied moment at a section z from one of the supports consists of the primary moment M^o_x and the secondary moment Pv, where v is the deflection of the member at that section (Fig. 11.5.1a). At low levels of loading, when all the sections of the member are elastic, the bending resistance of the section is given by $EI_x v''$ where EI_x is the elastic rigidity (stiffness) of the section. As the load level increases, the summation of compressive stresses and bending

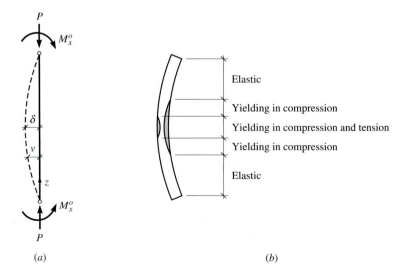

(a)   (b)

Figure 11.5.1: I-section beam-column under an axial load P and major axis moments M^o_x.

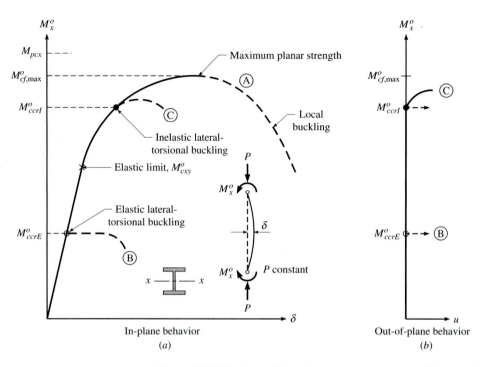

Figure 11.5.2: Behavior of I-section beam-columns under axial compression P and major axis moments M_x^o.

stresses at the outer fiber on the inside face of the critical section, located at the midheight of the member, reaches the compressive yield stress of the material and yielding occurs. We designate this moment at first yield as M_{cxy}^o. As the load increases above M_{cxy}^o, yielding progresses through the midsection to the adjacent sections, as shown in Fig. 11.5.1*b*. Since these yielded fibers have zero stiffness, the stiffness of the member as a whole is reduced. Subsequently each successive transverse deflection increment results in additional yielding and further reduction in stiffness. The moment increment required to produce each successive deflection increment thus becomes progressively smaller until a value of zero is attained at the ***maximum flexural strength of the beam-column,*** designated here as $M_{cf,\text{max}}^o$. Beyond this point, any further increase in deflection must be accompanied by a decrease in applied moment M_x^o as the rate of increase in the $P\delta$ moment becomes much faster than the rate of increase of the internal moment of the cross section. Failure of the member, by the formation of a plastic hinge at the midsection, occurs when the second-order moment at that point equals the moment carrying capacity, M_{pcx}, of the section. The beam-column is then considered to have failed through instability caused by excessive bending in the plane of applied moments. The maximum moment $M_{cf,\text{max}}^o$ of the beam-column is always less than or at most equal to the reduced plastic moment M_{pcx} of the cross section, as defined in Eq. 11.2.6.

Residual stresses, which cause premature yielding and consequently a premature reduction in stiffness, may reduce both the initial yield and the maximum strength of beam-columns. Similarly, an initial crookedness, which increases the secondary bending moment caused by the axial load, reduces both the initial yield load and the maximum strength.

Failure as delineated by Curve A is assumed to take place in the plane of bending. This will truly be the case if the member is bent about its weak axis. If, however, the member is bent about its strong axis, as in the foregoing discussion, failure in the plane of bending is possible only if the member is adequately braced against lateral-torsional buckling. Lateral-torsional buckling (LTB) of the beam-column may occur, however, in the elastic range corresponding to an elastic critical moment M_{ccrE}^{o} (Curve B), defined in Eq. 11.4.1. For shorter unbraced lengths, LTB of the beam-column may occur in the inelastic range corresponding to an inelastic bifurcation moment M_{ccrI}^{o} (Curve C). Another form of failure which may occur in the member is flange- or web-local buckling (FLB or WLB), which is the buckling of the plate elements of the cross section. Like LTB, plate local buckling may occur in the elastic or inelastic range. Also, like LTB, plate local buckling may reduce the load carrying capacity of the member.

For further details on the planar strength of steel beam-columns, refer to Section W11.7 on the website http://www.mhhe.com/Vinnakota.

11.6   Inelastic Lateral-Torsional Buckling of Steel Beam-Columns

If a beam-column has significantly different bending stiffnesses in the principal directions, is laterally unsupported, and is subjected to external moments about the strong axis, it may fail due to lateral-torsional buckling before it would fail due to excessive bending in the plane of the moment. Traditionally, the strength of such members is considered from the standpoint of bifurcation (Fig. 11.6.1), assuming an initially straight member. Thus, the out-of-plane deformations (u, θ) at any point are assumed to remain zero until the inelastic critical moment M_{ccrI}^{o} is reached. The in-plane behavior of the beam-column up to the critical load can therefore be analyzed independently of the out-of-plane buckling behavior. The solution from this in-plane behavior can then be substituted in the flexural-torsional equations which govern the out-of-plane buckling of the beam-columns. [Galambos, 1968; Massonnet, 1976]. The occurrence of lateral-torsional buckling reduces the maximum load carrying capacity of the member below the in-plane flexural strength $M_{cf,max}^{o}$ to a value designated as M_{clmax}^{o}. As mentioned earlier, real beam-columns will have initial imperfections (displacements u_i, v_i, and initial twist, θ_i). When such a beam-column is loaded by the axial force P and

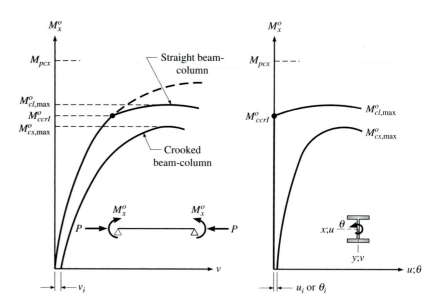

Figure 11.6.1: Inelastic lateral-torsional buckling of steel beam-columns.

major axis moments M_x^o the member exhibits the nonbifurcation type of instability, in which the deformations increase (from u_i, v_i, and θ_i), until a maximum moment is reached, beyond which static equilibrium can only be sustained by decreasing the moment. The maximum strength based on the spatial behavior of such initially crooked beam-columns $M_{cs,max}^o$ could be lower than the lateral buckling load M_{ccrl}^o of the corresponding initially straight beam-column, as shown schematically in Figs. 11.6.1a and b [Vinnakota, 1977].

For information on the inelastic lateral-torsional buckling of steel beam-columns, refer to Section W11.8 on the website http://www.mhhe.com/Vinnakota.

11.7 Strength of Biaxially Bent Steel Beam-Columns

Columns in structures are frequently subjected to bending moments in two perpendicular directions in addition to an axial load. These biaxial moments may result from an axial load eccentrically applied with respect to the principal axes of the column cross section (isolated column), or they may be the result of the space action of the entire structure (restrained column). This latter case is the more realistic and more frequently encountered in practice, but it is also more difficult to resolve. For a given plane, the end restraints for a column considered will be rotational only if the structure is braced, or rotational and directional if the structure is unbraced. The corner column of a tier building is an example of a biaxially bent beam-column.

For information on the behavior and strength of biaxially bent steel beam-columns, refer to Section W11.9 on the website http://www.mhhe.com/Vinnakota.

11.8   Moment Amplification Factor, B_2

Let us consider a cantilever column AB of length L that is simultaneously acted on by a transverse load H and axial force P, as shown in Fig. 11.8.1a. We will assume that the loads are applied in a plane of symmetry and that the member is laterally braced, if necessary, so that it can only bend in the plane of the applied loads. Let Δ represent the deflection of the free end, that is, the relative translation of the top of the column with respect to its base (Fig. 11.8.1b). We will assume that the member is prismatic with a flexural rigidity, EI, and that the behavior is elastic.

If the coordinate axes are taken as shown in Fig. 11.8.1c, the external moment at a distance z from the origin, taken as the free end, is:

$$M^* = Hz + Pv \tag{11.8.1}$$

Equating this expression to the internal resisting moment $-EIv''$ gives:

$$EI\frac{d^2v}{dz^2} + Pv = -Hz \tag{11.8.2}$$

The general solution of Eq. 11.8.2 is:

$$v = C_1 \sin \alpha z + C_2 \cos \alpha z - \frac{Hz}{P} \tag{11.8.3}$$

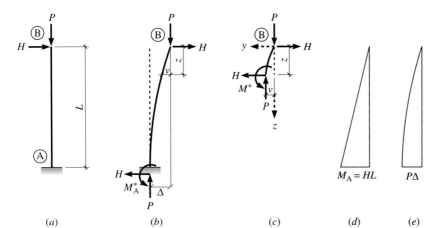

(a) (b) (c) (d) (e)

Figure 11.8.1: An example of a beam-column in a sway frame.

where C_1 and C_2 are the integration constants and the parameter α^2 equals P/EI. The constants C_1 and C_2 can be determined by using the boundary conditions. From the boundary condition, $v_{(z=0)} = 0$, we obtain:

$$C_2 = 0 \qquad (11.8.4)$$

and from the boundary condition, $v'_{(z=L)} = 0$, we have:

$$C_1 = \frac{H}{P\alpha} \frac{1}{\cos \alpha L} \qquad (11.8.5)$$

Substitution of these results in Eq. 11.8.3 gives:

$$v = \frac{H}{P\alpha} \left[\frac{\sin \alpha z}{\cos \alpha L} - \alpha z \right] \qquad (11.8.6)$$

Letting $z = L$ in this relation, we obtain the sway or drift of the cantilever column as:

$$\Delta = \frac{H}{P\alpha} [\tan \alpha L - \alpha L] \qquad (11.8.7)$$

Multiplying and dividing the above expression by $L^3/(3EI)$ gives:

$$\Delta = \frac{HL^3}{3EI} \frac{3(\tan \phi - \phi)}{\phi^3} = \Delta_o \frac{3(\tan \phi - \phi)}{\phi^3} \qquad (11.8.8)$$

where

$$\Delta_o = \frac{HL^3}{3EI} \qquad (11.8.9)$$

$$\phi = \alpha L = L \sqrt{\frac{P}{EI}} = \frac{\pi}{2} \sqrt{\frac{P}{P_e}} \qquad (11.8.10)$$

with

$$P_e = \frac{\pi^2 EI}{(2L)^2} = \frac{\pi^2 EI}{(KL)^2} \qquad (11.8.11)$$

Here Δ_o is the deflection that would exist if the transverse load H were acting by itself (i.e., first-order deflection), and P_e is the elastic buckling load of the cantilever column.

Differentiating the expression 11.8.6 twice and multiplying by $-EI$, we obtain the second-order moment at any point z as:

$$M^* = H\left(\frac{EI}{P}\right) \alpha \frac{\sin \alpha z}{\cos \alpha L} = HL\left[\frac{\sin \alpha z}{\alpha L \cos \alpha L} \right] \qquad (11.8.12)$$

The maximum bending moment occurs at the base of the column, point A. Its value is obtained by letting $z = L$ in Eq. 11.8.12, as:

$$M_A^* = M_{max} = M_A \frac{\tan \phi}{\phi} \tag{11.8.13}$$

Here, $M_A (= HL)$ is the first-order moment at the base of the cantilever beam under transverse load H (Fig. 11.8.1d). The expressions for the maximum moment can be simplified by using the power series expansion for $\tan \phi$, namely:

$$\tan\phi = \phi + \frac{1}{3}\phi^3 + \frac{2}{15}\phi^5 + \frac{17}{315}\phi^7 + \cdots \cdots \tag{11.8.14}$$

Upon substituting Eq. 11.8.14 into Eqs. 11.8.13 and simplifying, we obtain:

$$M_A^* \approx \left[\frac{1 - 0.18(P/P_e)}{1 - (P/P_e)} \right] M_A \tag{11.8.15}$$

Equations 11.8.13, 11.8.15, and Eq. A5.1.17 of Appendix A5.1 are different expressions for the second-order moment at the base of a cantilever beam-column in terms of the corresponding first-order moment. Considering that beam-columns will generally be part of a total structure, these equations have to be modified to include the total load on the columns within a story and the total lateral load within a story. We can generalize Equations 11.8.13, 11.8.15, and A5.1.17 in the form:

$$M_{II} = B_2 M_I \tag{11.8.16}$$

where B_2 is a moment amplification factor for columns in unbraced multistory frames. It is a measure of the second-order moment that develops in a column due to the interaction of the axial force in a column and the relative translation, Δ, of the ends of that column ($P\Delta$ effect).

For a derivation of the amplification factor B_2 for columns in unbraced multistory frames, based on the story stiffness concept, refer to Section W11.10 on the website http://www.mhhe.com/Vinnakota.

<hr/>

11.9   LRFD Interaction Formulas for Beam-Column Design

In present practice, the approach used to design steel frames is to modify an individual member design in a manner that approximately accounts for continuity or frame action. This is accomplished as follows:

1. First, we isolate a compression member plus its adjacent members at both ends (see Figs. 8.5.7 and 8.5.8) and determine the bifurcation load of the subassemblage which is expressed in terms of the effective length factor, K. This is conveniently done by calculating the G-factors at each end of

the member and making use of the alignment charts given in the LRFD Manual, or using equations given in Section 8.5.

2. Next, the distribution of bending moments and internal forces throughout the structure, under factored loads, is determined for design purposes. The LRFD interaction equations for beam-column design, given in Chapter H, are based on the premise that the moments used are the maximum second-order elastic moments over the length of the member. Two approaches are possible:

- The LRFD Specification encourages the designer to determine the internal forces and bending moments from a direct second-order elastic analysis of the structure under factored loads. In a second-order analysis, the equilibrium equations are formulated on the deformed structure. Such calculations account for the second-order moments produced by the gravity loads acting on the displaced structure (i.e., the ***P*Δ *effect,*** also known as ***chord rotation effect***). The second-order analysis must also account for the reduction in the stiffness of the individual columns by their axial loads (i.e., the ***P*δ *effect,*** also known as the ***member curvature effect***).

- However, member forces and moments are often obtained from a first-order elastic analysis of the frame under factored loads. This type of analysis is based on the initial geometry of the structure and disregards the influence of axial force on the member stiffness. Next, the second-order moments are calculated approximately by making use of the ***B*₁ *factor*** (*P*δ effect) and the ***B*₂ *factor*** (*P*Δ effect), if applicable.

3. The member is then designed as a beam-column by a simplified interaction equation which accounts in an approximate manner for continuity, inelasticity, and second-order effects. In this manner, member design is substituted for frame design.

11.9.1   Linear Interaction Formulas

The LRFD Specification gives a set of interaction formulae (Eqs. H1-1*a* and H1-1*b*) to design prismatic members subject to axial compression and flexure about one or both axes of symmetry. The interaction equations have originally been derived for the no-joint translation case where the effective lengths are taken as the actual lengths. However, it has been found that these same interaction equations can be used for columns in unbraced frames. They are (from LRFDS Section H1):

$$\frac{P_u}{\phi_c P_n} + \frac{8}{9}\left[\frac{M_{ux}}{\phi_b M_{nx}} + \frac{M_{uy}}{\phi_b M_{ny}}\right] \leq 1 \quad \text{for } \frac{P_u}{\phi_c P_n} \geq 0.2 \quad (11.9.1a)$$

$$\frac{1}{2}\frac{P_u}{\phi_c P_n} + \frac{M_{ux}}{\phi_b M_{nx}} + \frac{M_{uy}}{\phi_b M_{ny}} \leq 1 \quad \text{for } \frac{P_u}{\phi_c P_n} < 0.2 \quad (11.9.1b)$$

where P_u = required compressive strength
$\quad\quad M_u$ = required flexural strength including second-order effects
$\quad\quad P_n$ = nominal axial compressive strength
$\quad\quad M_n$ = nominal flexural strength
$\quad\quad \phi_c$ = resistance factor for compression (= 0.85)
$\quad\quad \phi_b$ = resistance factor for flexure (= 0.90)
$\quad\quad x$ = subscript relating symbol to strong axis bending
$\quad\quad y$ = subscript relating symbol to weak axis bending

Note, particularly, that M_u is the maximum second-order moment in the member under consideration. In the calculation of this moment, inclusion of the detrimental second-order effects of axial compression, the $P\delta$ and $P\Delta$ moments, is required. The LRFD Specification permits three alternative methods to calculate this required flexural strength:

1. For members in structures designed on the basis of plastic analysis, the required flexural strength is determined from a second-order plastic analysis of the structure (LRFDS Section C1.1).
2. For members in structures designed on the basis of elastic analysis, the required bending strength for members and connections is determined from a second-order elastic analysis of the structure under factored loads. Thus if M_u^* is the maximum bending moment in a beam-column from such an analysis,

$$M_u = M_u^* \tag{11.9.2}$$

3. For members in structures designed on the basis of elastic analysis and where bending moments are obtained from first-order elastic analyses of the structure, the required second-order flexural strength of a beam-column is obtained from the following approximate relation (LRFDS Eq. C1-1):

$$M_u \equiv M_u^* = B_1 M_{nt} + B_2 M_{lt} \tag{11.9.3}$$

where B_1, B_2 = moment amplification factors to be described later
$\quad\quad M_{nt}$ = **no-translation moment** or maximum first-order moment in the beam-column, assuming there is no lateral translation of the frame, whether the frame is actually braced or not (the subscript "*nt*" stands for no lateral translation).
$\quad\quad M_{lt}$ = **lateral translation moment** or maximum first-order moment in the beam-column, as a result of lateral translation of the frame only (the subscript "*lt*" stands for lateral translation).

In practice, for each member the no lateral translation moments M_{nt} corresponding to a given factored load combination are determined from a first-order analysis of the given frame provided with fictitious supports against lateral translation at each floor level and at the roof level (Fig. 11.9.1*a*). Also determined from this analysis are the fictitious horizontal reactions R_i at each floor level and roof level to prevent their lateral translation. The lateral translation moments M_{lt} in each member are next determined from a first-order sway analysis of the frame acted on only by the reverse of the reactions R_i at the fictitious lateral supports (Fig. 11.9.1*a*).

There is no joint translation when symmetric rectangular frames are subjected to symmetric gravity loads. Determination of M_{nt} and M_{lt} moments for such structures could be simplified by the following procedure. Make a

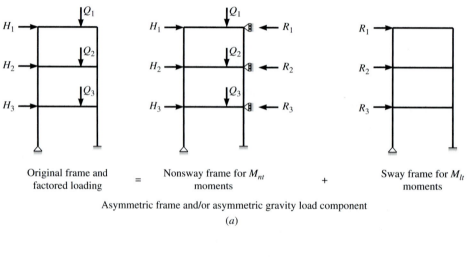

Asymmetric frame and/or asymmetric gravity load component

(*a*)

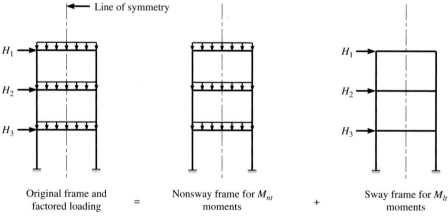

Symmetric frame with symmetric gravity load component

(*b*)

Figure 11.9.1: M_{nt} and M_{lt} moments for rectangular frames.

first-order analysis of the given frame under only the gravity components (D, L, L_r, S, R) of the factored load combination (Fig. 11.9.1b). This gives the M_{nt} moments. Next, make a first-order analysis of the given frame under only the lateral components (W, E) of that factored load combination (Fig. 11.9.1b). This results in M_{lt} moments. It follows that for symmetric rectangular frames subjected to a load combination consisting of symmetric gravity loads only, M_{lt} moments in all members will be zero. Similarly, if the frame is braced, M_{lt} moments will be zero for all load combinations.

The amplification factor B_1 and B_2 are especially important when the columns are slender.

Moment Magnification Factor, B_1

The **moment magnifier B_1** is a $P\delta$ moment amplification factor. That is, it accounts for the amplification of the first-order nt moments associated with the member curvature effects. So, it applies to columns in braced and unbraced frames as well as trusses. The magnifier B_1 is given by LRFD Specification Eq. C1-2, namely:

$$B_1 = \max\left[\frac{C_m}{1 - \dfrac{P_u}{P_{e1}}}, 1.0\right] \qquad (11.9.4)$$

Here:

B_1 = magnifier of M_{nt} moment
C_m = equivalent moment factor
P_u = factored axial compression load on the beam-column
P_{e1} = elastic buckling load, using the moment of inertia I in the plane of bending and the effective length factor K_{nt} (≤ 1.0) in the plane of bending based on the assumption that sidesway is prevented (that is, treat as a braced frame, even when the column under consideration is part of an unbraced frame)

Thus:

$$P_{e1} = \frac{\pi^2 EI}{(KL)_{nt}^2} = \frac{\pi^2 EA}{(KL/r)_{nt}^2} = F_e A \qquad (11.9.5)$$

Values of the elastic buckling stress F_e for different values of the effective slenderness ratio, $(KL/r)_{nt}$ may be obtained from Table 11.9.1.

For beam-columns loaded with end moments only, the actual distribution of the second-order moment along the length of the member may be either Type 1 (Fig. 11.9.2a), where the maximum second-order moment occurs between the ends of the member, or Type 2, where the maximum second-order moment equals the larger end moment (Fig. 11.9.2b). The C_m factor adjusts the moment amplification factor to account for bending moment distributions other than uniform bending. Since a uniform moment

TABLE 11.9.1

Elastic Buckling Stress, F_e, for Steel of Any Yield Stress

$\dfrac{KL}{r}$	F_e (ksi)	$\dfrac{KL}{r}$	F_e (ksi)	$\dfrac{KL}{r}$	F_e (ksi)	$\dfrac{KL}{r}$	F_e (ksi)
21	649	41	170	61	76.9	81	43.6
22	591	42	162	62	74.5	82	42.6
23	541	43	155	63	72.1	83	41.6
24	497	44	148	64	69.9	84	40.6
25	458	45	141	65	67.7	85	39.6
26	423	46	135	66	65.7	86	38.7
27	393	47	130	67	63.8	87	37.8
28	365	48	124	68	61.9	88	37.0
29	340	49	119	69	60.1	89	36.1
30	318	50	114	70	58.4	90	35.3
31	298	51	110	71	56.8	91	34.6
32	280	52	106	72	55.2	92	33.8
33	263	53	102	73	53.7	93	33.1
34	248	54	98.2	74	52.6	94	32.4
35	234	55	94.6	75	51.0	95	31.7
36	221	56	91.3	76	49.6	96	31.1
37	209	57	88.1	77	48.3	97	30.4
38	198	58	85.1	78	47.0	98	29.8
49	188	59	82.2	79	45.9	99	29.2
40	179	60	79.5	80	44.7	100	28.6

$$F_e = P_e/A = \frac{\pi^2 E}{(KL/r)^2}$$

Note: For P_{e1} use $(KL/r)_{nt}$ and for P_{e2} use $(KL/r)_{lt}$.

is the worst case loading of a beam-column, C_m is always less than or equal to 1. C_m is to be taken as follows:

1. For compression members not subject to transverse loading between their supports in the plane of bending, C_m converts the linearly varying primary bending moment into an equivalent uniform moment. The equivalent moment factor is given by Eq. C1-3 of the LRFD Specification, namely:

$$C_m = 0.6 - 0.4 r_M \qquad (11.9.6)$$

where

$$r_M = \pm \frac{|M_1|}{|M_2|} \qquad (11.9.7a)$$

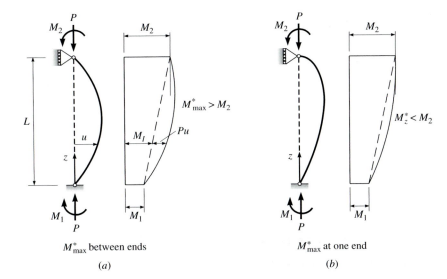

M_{max}^* between ends

(a)

M_{max}^* at one end

(b)

Figure 11.9.2: Maximum moment in a beam-column under axial load P and end moments M_1, M_2.

Here, M_1 and M_2 are the moments at the ends of that portion of the member unbraced in the plane of bending under consideration. The end moment with the larger absolute value of the two end moments is always designated as moment M_2, and the moment at the other end as M_1. The factor r_M is assigned a positive sign when the segment is bent in reverse curvature, and a negative sign when the segment is bent in a single curvature. Thus,

$$-1.0 \leq r_M \leq +1.0 \qquad (11.9.7b)$$

2. For compression members subjected to transverse loading between their supports, the value of C_m can be determined by a rational analysis (see Section 11.3.3). Values of C_m for various cases of end conditions and transverse loads are given in LRFDM Table C-C1.1. For the restrained end cases, the values of B_1 will be most accurate if values of $K_{nt} < 1.0$ corresponding to the end boundary conditions are used in calculating P_{e1}. In lieu of using the referenced equations and the table, $C_m = 1.0$ can be used conservatively for transversely loaded beam-columns with unrestrained ends and $C_m = 0.85$ for members with restrained ends.

In summary, the maximum first-order moment $M_{1,max}$ of a beam-column when multiplied by the appropriate C_m factor transforms the various loading conditions to the case of the same beam-column under an equivalent uniform moment $M_{eq} = C_m M_{1,max}$.

Moment Magnification Factor, B_2

The **moment magnifier B_2** is a $P\Delta$ moment amplification factor; it amplifies the member end moments associated with lateral translation of the story.

This factor is therefore applied to columns in unbraced frames only and is defined by the LRFDS formula C1-4, namely:

$$B_2 = \cfrac{1}{1 - \cfrac{\Delta_{oh}}{L} \cfrac{\displaystyle\sum_{i=1}^{n} P_{ui}}{\displaystyle\sum_{j=1}^{m} H_j}} \tag{11.9.8}$$

or alternatively by the LRFDS formula C1-5, namely:

$$B_2 = \cfrac{1}{1 - \cfrac{\displaystyle\sum_{i=1}^{n} P_{ui}}{\displaystyle\sum_{i=1}^{n} P_{e2}}} \tag{11.9.9}$$

where $\sum P_{ui}$ = required axial strength of all columns in a story, kips
Δ_{oh} = lateral interstory deflection, in.
$\sum H_j$ = sum of all story horizontal forces producing Δ_{oh}, kips
L = story height, in.
P_{e2} = elastic buckling load of the member i, using the effective length factor K_{lt} (≥ 1.0) in the plane of bending for the unbraced frame and the moment of inertia I in the plane of bending

Thus:

$$P_{e2} = \frac{\pi^2\, EI}{(KL)^2_{lt}} = \frac{\pi^2\, EA}{(KL/r)^2_{lt}} = F_e A \tag{11.9.10}$$

In a framed building, sidesway instability cannot occur in an individual column; rather, the effect must develop, at the very least, within an entire bay because of the diaphragm action of the floors. Consequently, all columns in given story of a frame have the same B_2 factor.

The parameter Δ_{oh}/L is known as the drift index. In tier buildings the drift index is usually limited at service loads to a value between 0.0015 and 0.0030, and at factored loads to about 0.004.

Summary
In this text, the LRFD interaction equations (Eqs. 11.9.1a and b) for beam-columns are rewritten as follows:

$$\frac{P_u}{P_d} + \frac{8}{9}\left[\frac{M^*_{ux}}{M_{dx}} + \frac{M^*_{uy}}{M_{dy}}\right] \leq 1.0 \quad \text{for} \quad \frac{P_u}{P_d} \geq 0.2 \tag{11.9.11a}$$

$$\frac{1}{2}\frac{P_u}{P_d} + \frac{M_{ux}^*}{M_{dx}} + \frac{M_{uy}^*}{M_{dy}} \le 1.0 \quad \text{for} \quad \frac{P_u}{P_d} < 0.2 \qquad (11.9.11b)$$

where P_u = required compressive strength
M_u^* = required flexural strength including second-order effects
P_d = design compressive strength = $\phi_c P_n$
M_d = design flexural strength = $\phi_b M_n$
x = subscript relating symbol to strong axis bending
y = subscript relating symbol to weak axis bending

The applicable interaction equation is based on the first term that represents the axial load ratio. The design compressive strength P_d must be based on the larger of the two effective slenderness ratios, namely, $\frac{K_x L_x}{r_x}$, $\frac{K_y L_y}{r_y}$.

The beam-column interaction equations have a linear form in which a direct relationship exists between the axial load and amplified bending moments. In particular, for bending about one axis only, the interaction equations have the bilinear form shown in Fig. 11.9.3. Note that in the interaction equations, the K factor occurs six times. The member amplification factor B_1 is especially important when the column is slender and becomes more so at higher axial load ratios for a given member. The story amplification factor B_2 is especially important when the columns in a story are slender as a group and becomes more so at higher levels of gravity loads on the floor.

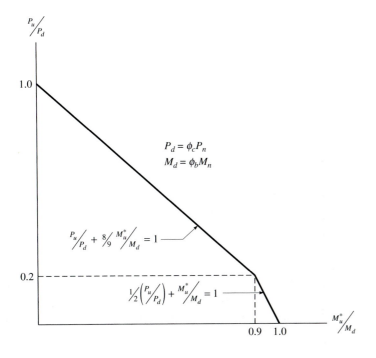

Figure 11.9.3: Graphical representation of LRFDS interaction Eqs. H1-1a and H1-1b.

Equation 11.9.3 often results in a conservative estimate of the maximum second-order moment in the beam-column. This is because the amplified moment resulting from the $P\delta$ effect (the term $B_1 M_{nt}$) and the amplified moment resulting from the $P\Delta$ effect (the term $B_2 M_{lt}$) do not generally occur at the same location. The $P\Delta$ effect always magnifies the end moments. The maximum second-order elastic no-translation moment, $B_1 M_{nt}$, might occur within the length of the member, as evidenced by the parameter B_1 being greater than 1. In effect, as the axial force becomes large or the member becomes more slender, it is more likely that the moment, $B_1 M_{nt}$, does not occur at the end of the member.

Design Aids

Table 4-2 of the LRFDM gives the design axial compressive strength for W-shapes used as columns having effective lengths KL, with respect to the minor axis, given in the left most column for material with $F_y = 50$ ksi. The effective lengths are in ft, and all the tabulated loads are in kips. To obtain an effective length with respect to the minor axis equivalent in load carrying capacity to the actual effective length about the major axis, $K_x L_x$ should be divided by r_x/r_y. The larger of the two effective lengths (with respect to the minor axis) is used to enter the column load tables to obtain the design axial compressive strength. The value of r_x/r_y is listed at the bottom portion of the column load tables together with other section and member properties (A, I_x, I_y, r_y, L_p, L_r, etc.). Values of the parameters $[P_{ex}(KL)^2/10^4]$ and $[P_{ey}(KL)^2/10^4]$ are also listed at the bottom of the column load tables. Here K is the effective length factor, and L is the actual unbraced length in the plane of bending.

Additional design tables for beam-columns are given in Part 6 of the LRFD Manual and will be discussed in Section 11.14.2.

11.10 Plate Local Buckling in Beam-Columns

The application of Eq. 11.9.11 for the design of beam-columns implies that the component plates of the cross section are sufficiently compact; the section can thus resist a moment equal to the reduced plastic moment capacity M_{pc} of the section. The web plate of an I-shaped beam-column subjected to a major-axis moment and axial load is strained nonuniformly across its height (Fig. 11.10.1). Now, the critical b/t ratio of a plate depends to a large extent on the magnitude of the maximum strain. A maximum strain ε_{max}, equal to four times the yield strain ε_y, is assumed to be sufficient for developing adequate rotation capacity of a plastic hinge. The dashed curve shown in Fig. 11.10.1 indicates the theoretical relationship between the maximum d/t_w ratio and the magnitude of the axial compressive force P/P_y, presented

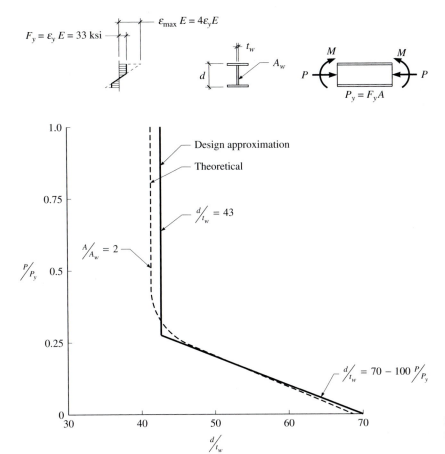

Figure 11.10.1: Width-to-thickness requirements for webs of beam-columns (Courtesy of ASCE, 1971).

by Haaijer and Thurlimann [1958]. The results are for a steel with $F_y = 33$ ksi, and total section area A equal to twice the web area A_w. A design approximation suggested by Haaijer and Thurlimann [1958] is also shown by the solid line in Fig. 11.10.1. To be applicable for steels with other yield stresses, this may be written as:

$$\frac{d}{t_w} \le 2.42 \sqrt{\frac{E}{F_y}}\left[1 \, - \, 1.4\frac{P}{P_y}\right] \quad \text{for } \frac{P}{P_y} \le 0.27 \quad (11.10.1a)$$

$$\frac{d}{t_w} \le 1.45 \sqrt{\frac{E}{F_y}} \quad \text{for } \frac{P}{P_y} \ge 0.27 \quad (11.10.1b)$$

For rolled shapes, a small upward variation of the b/t ratio is permissible to account for the restraint provided by the flanges.

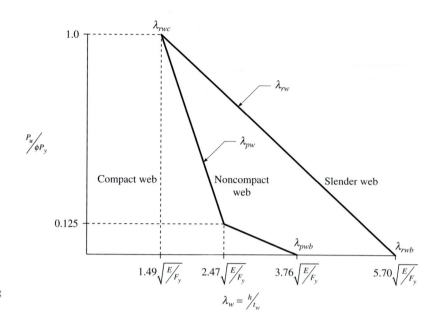

Figure 11.10.2: Influence of axial load on limiting web slenderness ratios for beam-column.

Let

$$\lambda_f = \frac{b_f}{2t_f}; \quad \lambda_w = \frac{h}{t_w} \tag{11.10.2}$$

where λ_f and λ_w are the b/t ratios of the flange and web of the I-shape used as a beam-column. The LRFDS design rules for web local buckling of W-shape beam-columns may be summarized as follows (Fig. 11.10.2):

$$\lambda_w \leq \lambda_{pw} \qquad \text{the shape is compact}$$

$$\lambda_{pw} < \lambda_w < \lambda_{rw} \quad \text{the shape is noncompact} \tag{11.10.3}$$

$$\lambda_w > \lambda_{rw} \qquad \text{the shape is slender}$$

The LRFD b/t limitations for rolled steel beam-columns are based on the following observations:

1. The requirements, loading, and support conditions for the compression flange of a beam-column are essentially the same as those for the compression flange of a beam. So, the same limits on the flange width-to-thickness ratios should be used.

2. If the axial force in the beam-column is relatively small, the PNA will be close to the mid-depth of the cross section. So, when the axial force is relatively small, the limitation on the web depth-to-thickness ratio for beam-columns should be the same as that for beams.

3. If the axial force in the beam-column is large, the PNA will be in the tension flange, and the entire web plate will be subjected to compression.

Hence, for beam-columns subjected to relatively large axial forces, the limiting web depth-to-thickness ratio should be the same as that for an axially loaded column.

4. For intermediate values of the axial force, the PNA will lie between the centroid and the tension flange. As P_u/P_y increases, the portion of the web depth that is subjected to compressive stress increases. So, the limit on the web slenderness ratio should become more restrictive.

LRFDS Table B5-1 prescribes the following limits for the web of a beam-column:

$$\lambda_{pw} = \lambda_{pwb}\left[1 - 2.75\frac{P_u}{\phi_b P_y}\right] \quad \text{for } \frac{P_u}{\phi_b P_y} \le 0.125 \quad (11.10.4)$$

$$= \max\left[1.12\sqrt{\frac{E}{F_y}}\left(2.33 - \frac{P_u}{\phi_b P_y}\right); \lambda_{rwc}\right] \quad \text{for } \frac{P_u}{\phi_b P_y} > 0.125 \quad (11.10.5)$$

$$\lambda_{rw} = \lambda_{rwb}\left[1 - 0.74\frac{P_u}{\phi_b P_y}\right] \quad (11.10.6)$$

with

$$\lambda_{pwb} = 3.76\sqrt{\frac{E}{F_y}}; \ \lambda_{rwb} = 5.70\sqrt{\frac{E}{F_y}} \quad (11.10.7a)$$

$$\lambda_{rwc} = 1.49\sqrt{\frac{E}{F_y}} \quad (11.10.7b)$$

where P_y is the yield load of the section (= AF_y); λ_{pwb} is the limiting slenderness ratio of compact webs in pure bending; λ_{rwb} is the limiting slenderness ratio of noncompact webs in pure bending; and λ_{rwc} is the limiting slenderness ratio of webs in pure compression.

The limiting values of b/t for the limit state of local buckling of flange of rolled I-shapes are:

$$\lambda_{pf} = 0.38\sqrt{\frac{E}{F_y}}; \ \lambda_{rf} = 0.83\sqrt{\frac{E}{F_L}} \quad (11.10.8)$$

where $F_L = F_y - F_r = (F_y - 10)$ ksi.

A majority of the rolled shapes satisfy the worst case limit of $1.49\sqrt{E/F_y}$, which means that these shapes have compact webs no matter what the axial load. Shapes listed in the column load tables in the LRFDM Part 4 that do not satisfy this criterion are marked by ††, and only these need be checked for compactness of the web. Shapes whose flanges are not compact are also marked ††, so if there is no indication to the contrary, shapes found in the column load tables are compact for $F_y = 50$ ksi, for use as beam-columns. Thus, we observe that the flanges of W14×99, W14×90 and W12×65 are noncompact for $F_y = 50$ ksi, while the web of a W14×43 may be noncompact for combined axial and flexural loading for $F_y = 50$ ksi. For noncompact and slender sections, the design bending strength is defined in LRFDS Appendix F.

11.11   Structures with Transversely Loaded Members

11.11.1   Trusses with Transversely Loaded Members

In calculating the forces in truss members, the loads are assumed to be applied only at the joints. In most trusses this assumption is true because the joists, purlins, or beams—which carry the roof or floor—are attached to the trusses at the joints. (Recall that the spacing of joists equals the panel length of a truss.) A few exceptions exist. A situation may arise where, owing to the required joist spacing, it is not possible to locate the joists over the panel points. Also, the bottom chords of roof trusses, which are in tension due to truss action, may be subjected to bending by the hanging of hardware such as light fixtures and duct work between the truss joints. Finally, the transverse load on a truss member may simply be the self-weight of a heavy, horizontal or inclined chord member of a truss.

If transverse loads are applied directly to some of the truss members between the panel points, these members act as beams to transmit the transverse loads to the truss joints. If the joints are frictionless pin joints, these flexural effects are confined to the members that are subjected to the loads between joints. However, the ends of bolted or welded truss members are in reality much closer to being rigidly jointed than to being hinged. Also, for chords of a truss that are continuous through several panel points, the flexural effects in the transversely loaded members are transmitted through the joints to other members so that significant bending moments can be produced in all members of the truss structure. These members should therefore be designed as beam-columns to provide flexure for the beam action under the transverse load, as well as the axial force from the truss action under joint loads.

11.11.2   Columns Supporting Crane Girders

In an industrial building, machinery and products are quite often moved about by rectilinear traveling cranes. The most common of all rectilinear cranes is the ***overhead traveling crane*** in which there is, in addition to the lifting motion, provision for two horizontal movements at right angles to one another; the load can thus be deposited at any point within the rectangle covered by the movement of the crane. Traveling cranes consist of a ***bridge,*** generally spanning the width of a shop or foundry, which moves longitudinally on overhead rails provided at the end of the bridge, giving the straight line motion in one direction. On this bridge is mounted a ***trolley*** or cab which

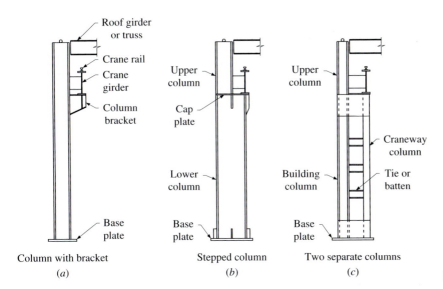

Figure 11.11.1: Crane girder columns.

moves transversely along the bridge, thus giving the straight line motion in the other direction. The cranes travel on rails fastened to *runway girders,* also called *crane girders,* that are supported by the columns. Crane girders are simple beams resting on brackets welded or bolted to the columns. They are laterally unsupported except at the columns.

Three common types of crane girder columns in industrial buildings are shown in Fig. 11.11.1.

1. Figure 11.11.1a shows a column of uniform section for the entire length with the crane runway girders supported by column brackets. This arrangement is suitable only for light loads, as the bracket will induce severe bending in the column for larger crane loads.

2. Figure 11.11.1b shows a stepped column, with the lower shaft a single heavy wide flange section. The upper shaft, supporting the roof structure, is a lighter wide flange shape. The crane girder is seated on the step so formed: centered over the inner flange of the enlarged section. A small bracket, usually stiffened, is used to support the overhanging flange of the crane girder. A stepped column may be used for intermediate loads (20 to 400 kips). Bending below the step is resisted by the deep section of the lower shaft. The upper shaft usually supports the gravity and wind loads from the roof. To transmit the bending moments from the upper shaft, the junction of the two shafts must be strongly spliced and stiffened.

3. For very heavy loads, separate columns under the crane girders and connected to the main building columns by means of diaphragms are used (Fig. 11.11.1c). The two sections are designed independently of each other. The building column is proportioned to take the gravity and wind loads from the building and the lateral thrust from the crane. The crane

column is designed to support the vertical loads from the crane girder and a horizontal longitudinal force at the top in the direction of the crane girder. The two sections are separated far enough to provide the necessary crane clearance. Use of separate columns has several advantages for heavy crane loads: the crane loads can be applied concentrically on the crane column; the designer can select any type and size of column for this member. Regardless of the orientation of the building column, the crane column can be placed with the web parallel to the plane of the girder to obtain maximum resistance to longitudinal forces. End stiffeners of the crane girder can be placed over the flanges of the crane column. The building column can be independently oriented so that its strong axis can resist wind loads on the building and lateral forces from the crane.

11.12 Members Subjected to Tension and Bending

Bending moments in tension members may be the result of connection eccentricity, due to loads such as the self-weight of a nonvertical member or the transverse loads such as wind acting along the length of the member. As discussed in Appendix A5.1, the effect of tension load is always to decrease the primary bending moment. Hence, secondary bending effects can conservatively be ignored in the design of a member subjected to an axial tension force and bending.

The strength design requirements for members with doubly or singly symmetric shapes subjected to axial tension force and bending are given in LRFDS Section H1.1 as Eqs. H1-1a and b. They may be rewritten as:

$$\frac{T_u}{T_d} + \frac{8}{9}\left[\frac{M_{ux}}{M_{dx}} + \frac{M_{uy}}{M_{dy}}\right] \leq 1 \quad \text{for } \frac{T_u}{T_d} \geq 0.2 \qquad (11.12.1a)$$

$$\frac{1}{2}\frac{T_u}{T_d} + \left[\frac{M_{ux}}{M_{dx}} + \frac{M_{uy}}{M_{dy}}\right] \leq 1 \quad \text{for } \frac{T_u}{T_d} < 0.2 \qquad (11.12.1b)$$

where T_u = required tensile strength under factored loads, kips
M_u = required flexural strength under factored loads, ft-kips
T_d = design tensile strength of the member, kips
M_d = design flexural strength of the member, ft-kips

The required flexural strength M_u is the maximum moment in the member. As discussed in Appendix A5, for a member subjected to combined tension and flexure the secondary moments reduce the primary moments. So, in the calculation of the required strengths M_{ux} and M_{uy}, inclusion of beneficial second-order effects of tension is optional. That is M_{ux}, M_{uy}, and T_u can be obtained directly from an elastic first-order analysis and B_1 and B_2 equal 1.0. T_d is based on the fracture of the effective net area, or yielding of the gross area, whichever is less.

11.13   Analysis Examples

Column Analysis (Type: *ntx*, *nty*; Load: *P*, M_x)   E X A M P L E   1 1 . 1 3 . 1

A W14×109 of A992 is used as a beam-column. The member is part of a
braced frame in both the *xx* and *yy* planes. The 12-ft-long member is subjected to an axial load of 610 kips. It is bent in single curvature with equal
and opposite end moments of 360 ft-kips about its major axis and is not subjected to any intermediate transverse loads. These forces are obtained from a
first-order analysis of the structure under factored loads. Assume $K_x = K_y =$
1.0, and check the adequacy of the member.

Solution

Section: W14×109; $F_y = 50$ ksi
Factored axial load, $P_u = 610$ kips
Factored maximum first-order moment, $M_{ux} = 360$ ft-kips
Member length, $L = 12$ ft
$K_x = 1.0; K_x L_x = 1.0(12) = 12.0$ ft
$K_y = 1.0; K_y L_y = 1.0(12) = 12.0$ ft ← Controls
From LRFDM Table 4-2, for a W14×109 $F_y = 50$ ksi steel column with
$KL = K_y L_y = 12$ ft, the design axial compressive strength is $P_d = 1220$
kips.

Axial load ratio, $\dfrac{P_u}{P_d} = \dfrac{610}{1220} = 0.500 > 0.2$

So, LRFDS Eq. H1-1*a* governs. As bending is about the major axis only
($M_{uy}^* = 0$), this equation becomes:

$$\frac{P_u}{P_d} + \frac{8}{9}\frac{M_{ux}^*}{M_{dx}} \le 1.0$$

From LRFDM Table 5-3, for a W14×109, $L_p = 13.2$ ft, and $\phi_b M_{px} =$
720 ft-kips.
As $L_b < L_p$, the design bending strength, $M_d = \phi_b M_{px} = 720$ ft-kips.
As the column is part of a braced frame, there are no M_{lt} moments. So,

$$M_{ux}^* = B_{1x} M_{ux}$$

From LRFDM Table 4-2, corresponding to a W14×109, we obtain P_{ex}
$(KL)^2 = 35,500 \times 10^4$. So,

$$P_{e1x} = \frac{35,500 \times 10^4}{[12.0(12)]^2} = 17,100 \text{ kips}$$

[*continues on next page*]

Example 11.13.1 continues ...

As the member is subjected to uniform moment, $C_{mx} = 1.0$ and

$$B_{1x} = \frac{C_{mx}}{1 - \dfrac{P_u}{P_{e1x}}} = \frac{1}{1 - \dfrac{610}{17,100}} = 1.04$$

$$M_{ux}^* = 1.04(360) = 374 \text{ ft-kips}$$

Substitution of these values in the interaction equation results in:

$$\text{LHS} = 0.500 + \frac{8}{9}\left(\frac{374}{720}\right) = 0.962 < 1.0 \quad \text{O.K.}$$

So, the W14×109 section is adequate. (Ans.)

EXAMPLE 11.13.2

Column Analysis (Type: *ntx*, *nty*; Load: *P*, *M$_x$*, *M$_y$*)

A W12×79 shape of A992 steel is used as a column in a braced frame. Under factored loads, the 21-ft-long pin-ended column is subjected to an axial load of 419 kips. In addition, moments of 200 ft-kips about the strong axis and 20 ft-kips about the weak axis are applied at the top end only. The member is braced at the ends about both axes. Additionally, lateral support occurs in the

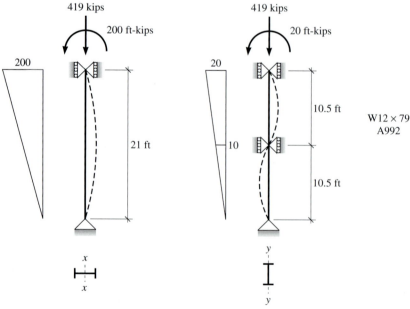

Bending about the *x* axis   Bending about the *y* axis

Figure X11.13.2   (a)   (b)

weak direction at midheight. Check the beam-column for compliance with the LRFD specification.

Solution

a. W12×79 section

$$I_x = 662 \text{ in.}^4; \qquad I_y = 216 \text{ in.}^4; \qquad A = 23.2 \text{ in.}^2$$

$$r_x = 5.34 \text{ in.}; \qquad r_y = 3.05 \text{ in.}$$

A 992 steel: $F_y = 50$ ksi

b. Design axial strength

From Fig. X11.13.2, $K_x L_x = 21.0$ ft; $K_y L_y = 10.5$ ft

$$\frac{K_x L_x}{r_x} = \frac{21.0(12)}{5.34} = 47.2; \qquad \frac{K_y L_y}{r_y} = \frac{10.5(12)}{3.05} = 41.3$$

Major-axis buckling controls the axial strength of the column. From LRFDS Table 3-50, design axial compressive stress, $F_{dc} = \phi_c F_{cr} = 36.1$ ksi.

Axial strength of the column, $P_d = F_{dc}A = 36.1(23.2) = 838$ kips

Axial load ratio, $\dfrac{P_u}{P_d} = \dfrac{419}{838} = 0.500 > 0.2$

So, we have to use Eq. 11.9.11a (or LRFDS Eq. H1-1a) for verification of member strength.

c. Design flexural strength

For major-axis bending, the laterally unbraced length L_b is 10.5 ft.
From LRFDM Table 5-3, for a W12×79 section of $F_y = 50$ ksi steel, read:

$$\phi_b M_{px} = 446 \text{ ft-kips}; \quad L_p = 10.8 \text{ ft}; \quad \phi_b M_{py} = 201 \text{ ft-kips}$$

Since $L_b = 10.5 < L_p = 10.8$ ft,

Major-axis bending strength, $M_{dx} = \phi_b M_{px} = 446$ ft-kips

Minor-axis bending strength, $M_{dy} = \phi_b M_{py} = 201$ ft-kips

d. Second-order moments

As the column is part of a braced frame about both axes, the M_{lt} moments are zero.

$$M_{ux}^* = B_{1x} M_{ntx}; \quad M_{uy}^* = B_{1y} M_{nty}$$

For bending about the major axis, the maximum first-order moment,

$$M_{ntx} = M_{2x} = 200 \text{ ft-kips}$$

[*continues on next page*]

Example 11.13.2 continues ...

The linear variation of major axis moments over the height of the column results in:

$$r_M = \pm \frac{|M_1|}{|M_2|} = \frac{|0|}{|200|} = 0.0 \rightarrow C_m = 0.6 - 0.4r_M = 0.6$$

$$(KL)_{ntx} = K_x L_x = 21.0 \text{ ft}$$

$$P_{e1x} = \frac{\pi^2 E I_x}{(KL)_{ntx}^2} = \frac{\pi^2 (29,000)(662)}{[21.0(12)]^2} = 2980 \text{ kips}$$

$$B_{1x} = \max \left[\frac{C_{mx}}{1 - \dfrac{P_u}{P_{e1x}}}, 1.0 \right] = \max \left[\frac{0.6}{1 - \dfrac{419}{2980}}, 1.0 \right] = 1.0$$

So, $M_{ux}^* = B_{1x} M_{2x} = 1.0(200) = 200$ ft-kips

For minor-axis bending, the maximum first-order moment,

$$M_{nty} = M_{2y} = 20 \text{ ft-kips}$$

The linear variation of minor-axis bending moments over the upper half of the column results in:

$$r_M = -\frac{|M_1|}{|M_2|} = -\frac{|10|}{|20|} = -0.5 \rightarrow C_m = 0.6 - 0.4r_M = 0.8$$

Corresponding to $(KL)_{nty} = K_y L_y = 10.5$ ft,

$$P_{e1y} = \frac{\pi^2 E I_y}{(KL)_{nty}^2} = \frac{\pi^2 (29,000)(216)}{(10.5 \times 12)^2} = 3890 \text{ kips}$$

$$B_{1y} = \frac{C_{my}}{1 - \dfrac{P_u}{P_{e1y}}} = \frac{0.8}{1 - \dfrac{419}{3890}} = 0.897 < 1.0 \text{ use } 1.0$$

So:

$$M_{uy}^* = B_{1y} M_{nty} = 1.0(20.0) = 20.0 \text{ ft-kips}$$

e. Check interaction formula H1-1*a*

$$\frac{P_u}{P_d} + \frac{8}{9} \frac{M_{ux}^*}{M_{dx}} + \frac{8}{9} \frac{M_{uy}^*}{M_{dy}} \leq 1.0$$

$$\text{LHS} = 0.500 + \frac{8}{9} \left(\frac{200}{446} \right) + \frac{8}{9} \left(\frac{20.0}{201} \right)$$

$$= 0.500 + 0.399 + 0.089 = 0.988 < 1.0 \quad \text{O.K.}$$

f. Check local buckling of plates

$$\lambda_f = \frac{b_f}{2t_f} = 8.22; \quad \lambda_w = \frac{h}{t_w} = 20.7$$

Nominal yield load of the section, $P_y = AF_y = 23.2(50) = 1160$ kips
From Table B5.1 of the LRFDS, for $F_y = 50$ ksi steel, we have:

For flange, $\lambda_{pf} = 0.38\sqrt{\dfrac{E}{F_y}} = 0.38\sqrt{\dfrac{29{,}000}{50}} = 9.15$

Axial load ratio, $\dfrac{P_u}{\phi P_y} = \dfrac{419}{0.9(1160)} = 0.401 > 0.125$

So:

$$\lambda_{pw} = \max\left[1.12\sqrt{\frac{E}{F_y}}\left(2.33 - \frac{P_u}{\phi_b P_y}\right), 1.49\sqrt{\frac{E}{F_y}} \right]$$

$$= \max\left[1.12\sqrt{\frac{29{,}000}{50}}(2.33 - 0.401), 1.49\sqrt{\frac{29{,}000}{50}} \right]$$

$$= \max[52.0, 35.9] = 52.0$$

For the W12×79 section under consideration:

$$\lambda_f = 8.22 < \lambda_{pf} = 9.15 \qquad \text{O.K.}$$

$$\lambda_w = 20.7 < \lambda_{pw} = 52.0 \qquad \text{O.K.}$$

So, the section is compact.
Thus, the W12×79 of A992 steel is acceptable as a beam-column to carry
the given factored loads, according to the LRFDS.   (Ans.)

Column Analysis (Type: *ntx, nty;* Load: *P, q_y*)

EXAMPLE 11.13.3

A W18×76 beam-column of A992 steel is subject to an axial compressive
factored load of 100 kips (Fig. X11.13.3). The 24-ft-long member is pinned
at both ends about both axes. Determine the intensity of uniformly distributed factored load that could be applied in the plane of the web:

a. If the member is laterally braced at the ends only.
b. If the member is provided with an additional lateral brace at the midspan.

Make *b/t* checks.

Solution

Data:
   Factored axial load, $P_u = 100$ kips
   W18×76 section of A992 steel:

[*continues on next page*]

Example 11.13.3 continues ...

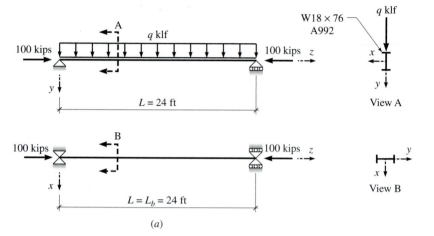

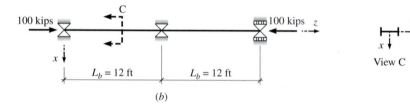

Figure X11.13.3

(b)

$$A = 22.3 \text{ in.}^2; \qquad r_x = 7.73 \text{ in.}$$
$$I_x = 1330 \text{ in.}^4; \qquad r_y = 2.61 \text{ in.}$$

b/t checks

$$\lambda_f = \frac{b_f}{2t_f} = 8.11; \qquad \lambda_w = \frac{h}{t_w} = 37.8$$

Limiting width-to-thickness ratio for a compact flange (from LRFDS Table B5.1):

$$\lambda_{pf} = 0.38\sqrt{\frac{E}{F_y}} = 0.38\sqrt{\frac{29,000}{50}} = 9.15$$

Squash load, $P_y = AF_y = 22.3(50) = 1120$ kips
We have:

$$\frac{P_u}{\phi P_y} = \frac{100}{0.9(1120)} = 0.099 < 0.125$$

Limiting width-to-thickness ratio for a compact web in combined flexure and axial compression (from LRFDS Table B5.1):

$$\lambda_{pw} = 3.76\sqrt{\frac{E}{F_y}}\left(1 - \frac{2.75P_u}{\phi P_y}\right) = 65.7$$

As $\lambda_f < \lambda_{pf}$ and $\lambda_w < \lambda_{pw}$ the section is compact and flange local buckling and web local buckling will not occur before the plastification of the section.

a. With lateral bracing at the member ends only (Fig. X11.3.3a)

Design axial strength

$$K_x L_x = 1.0(24.0) = 24.0 \text{ ft}; \qquad K_y L_y = 1.0(24.0) = 24.0 \text{ ft}$$

$$\frac{K_x L_x}{r_x} = \frac{24.0(12)}{7.73} = 37.3; \frac{K_y L_y}{r_y} = \frac{24.0(12)}{2.61} = 110 \leftarrow \text{controls}$$

From LRFDS Table 3-50, $F_{dc} = \phi_c F_{cr} = 17.6$ ksi.

Hence, $P_d = F_{dc} A = 17.6(22.3) = 392$ kips.

Axial load ratio, $\dfrac{P_u}{P_d} = \dfrac{100}{392} = 0.255 > 0.2.$

Therefore, LRFDS Eq. H1-1a governs.

Design bending strength

Unbraced length, $L_b = 24$ ft
From LRFDM Table 5-5, for a W18×76 and $L_b = 24$ ft, we have $M_{dx}^o = 447$ ft-kips. Also, from LRFDM Table 5-3, for a W18×76, $\phi_b M_{px} = 611$ ft-kips and $L_p = 9.22$ ft.
From LRFDM Table 5-1, for a simply supported beam under a uniformly distributed transverse load and with lateral supports at the ends only, we have $C_b = 1.14$. As $L_p = 9.22$ ft $< L_b = 24$ ft, we have:

$$M_{dx} = \min [C_b M_{dx}^o, \phi_b M_{px}] = \min [1.14(447), 611] = 510 \text{ ft-kips}$$

As there are no M_{lt} moments, we have: $M_{ux}^* = B_{1x} M_{ntx}$

From Table 11.9.1, corresponding to $KL/r = K_x L_x/r_x = 37.3$, we obtain:

$$F_e = 206 \text{ ksi} \rightarrow P_{elx} = F_e A = 206(22.3) = 4590 \text{ kips}$$

From LRFDC Table C-C1-1, for a simply supported beam-column under uniformly distributed transverse load, $C_m = 1.0$. So:

$$B_{1x} = \max \left[\frac{C_{mx}}{1 - \dfrac{P_u}{P_{elx}}}, 1.0 \right] = \max \left[\frac{1.0}{1 - \dfrac{100}{4590}}, 1.0 \right] = 1.02$$

$$M_{ux}^* = B_{1x} M_{ntx} = 1.02 M_{ntx}$$

Load calculation

As there are no minor axis moments, Eq. 11.9.11a (LRFDS Eq. H1-1a)

[continues on next page]

Example 11.13.3 continues ...

reduces to:

$$\frac{P_u}{P_d} + \frac{8}{9}\frac{M^*_{ux}}{M_{dx}} \le 1.0 \rightarrow 0.255 + \frac{8}{9}\left(\frac{1.02M_{ntx}}{510}\right) \le 1.0$$

$$M_{ntx} = \frac{q_{uy}L^2}{8} \le 419 \rightarrow q_{uy} \le \frac{8(419)}{24^2} = 5.82 \text{ klf}$$

The factored uniformly distributed load that could be applied on the beam, in addition to the self-weight, is therefore given by:

$$q_{uy1} = 5.82 - 1.2(0.076) = 5.73 \text{ klf} \qquad \text{(Ans.)}$$

b. With an additional brace at midspan (Fig. X11.3.3b)

Design axial strength

$$\frac{K_xL_x}{r_x} = 37.3; \quad K_yL_y = 12.0 \text{ ft}; \quad \frac{K_yL_y}{r_y} = 55.2 \leftarrow \text{still controls}$$

From Table 3-50 of the LRFDS, $F_{dc} = 34.0$ ksi and $P_d = 34.0(22.3) = 758$ kips

$$\frac{P_u}{P_d} = \frac{100}{758} = 0.132 < 0.2, \text{ therefore use Eq. 11.9.11}b \text{ (LRFDS Eq. H1-1}b\text{)}.$$

Design bending strength

From LRFDM Table 5-1, for a simply supported beam under a uniformly distributed transverse load and laterally braced at the ends and at midspan, $C_b = 1.30$

From LRFDM Table 5-5, for a W18×76 and $L_b = 12$ ft, we have $M^o_{dx} = 580$ ft-kips.

As, $L_p = 9.22$ ft $< L_b = 12.0$ ft

$$M_{dx} = \min [1.30(580), 611] = \min [754, 611] = 611 \text{ ft-kips}$$

Load calculation

From part (*a*), the maximum second-order moment,

$$M^*_{ux} = 1.02 \, M_{ntx}$$

$$\frac{1}{2}\frac{P_u}{P_d} + \frac{M^*_{ux}}{M_{dx}} \le 1.0 \rightarrow \frac{1}{2}(0.132) + \frac{1.02M_{ntx}}{611} \le 1.0$$

$$M_{ntx} = \frac{q_{uy}L^2}{8} \le 560 \rightarrow q_{uy} \le \frac{560(8)}{24^2} = 7.78 \text{ klf}$$

So, the factored uniformly distributed transverse load that could be applied on the beam-column, in addition to its self-weight, is:

$$q_{uy2} = 7.78 - 1.2(0.076) = 7.69 \text{ klf} \qquad \text{(Ans.)}$$

EXAMPLE 11.13.4

Column Analysis (Type: *ntx, nty*; Load: *T, M_{ux}, M_{uy}*)

A W14×22 tension member of A992 steel is subjected to a factored tensile load $T_u = 100$ kips and factored bending moments M_{ux} of 28 ft-kips and M_{uy} of 6 ft-kips. Is the member satisfactory if $L_b = 8$ ft and $C_b = 1.67$?

Solution

From LRFDM Table 1-1, for a W14×22, we obtain, $A_g = 6.49$ in.[2]
From beam selection tables for W-shapes (LRFDM Table 5-3, for example), for a W14×120 shape of $F_y = 50$ ksi, we have:

$$L_p = 3.67 \text{ ft}; \qquad \phi_b M_{px} = 123 \text{ ft-kips}; \qquad BF = 6.22 \text{ kips}$$

$$L_r = 9.65 \text{ ft}; \qquad \phi_b M_{py} = 15.8 \text{ ft-kips}$$

As $(L_p = 3.67 \text{ ft}) < (L_b = 8 \text{ ft}) < (L_r = 9.65 \text{ ft})$,

$$M_{dx} = \min [C_b\{\phi_b M_{px} - BF(L_b - L_p)\}, \ \phi_b M_{px}]$$
$$= \min\{1.67 [123 - 6.22(8.0 - 3.67)], 123\}$$
$$= \min [1.67(96.1), 123] = 123 \text{ ft-kips}$$

Design tensile strength of the member, corresponding to the limit state of yield on gross area:

$$T_d = T_{d1} = \phi_t A_g F_y = 0.9(6.49)(50) = 292 \text{ kips}$$

Axial load ratio, $\dfrac{T}{T_d} = \dfrac{100}{292} = 0.342 > 0.20$

So, use Eq. 11.9.11a (LRFDS Eq. H1-1a):

$$\frac{T_u}{T_d} + \frac{8}{9}\left[\frac{M_{ux}}{M_{dx}} + \frac{M_{uy}}{M_{dy}} \right] = 0.342 + \frac{8}{9}\left[\frac{28}{123} + \frac{6.0}{15.8} \right]$$

$$= 0.342 + 0.202 + 0.338 = 0.882 < 1.0 \text{ safe}$$

So, the W14×22 of A992 is adequate.   (Ans.)
Note that, for convenience, we have assumed that tensile yielding (T_{d1}) controls over tensile rupture (T_{d2}). If this is not the case, the design tensile strength should be based upon the tensile rupture design strength.

11.14   Design of Beam-Columns

The design of steel beam-columns is an iterative procedure. Thus to design a beam-column for a given set of factored loads, the designer typically chooses a trial section and then checks it for compliance with LRFDS Eq. H1-1a or b, as the case may be, to verify that the capacity of the member is just adequate for the axial load and moment it must support. If these calculations indicate that the capacity of the member is fully or almost fully utilized, the design is complete. On the other hand, if the analysis indicates that either the capacity of the member is exceeded (LHS $\gg$ 1.0) or that the section is substantially under stressed (LHS $\ll$ 1.0), and therefore uneconomical because excess material is provided, a new section is selected and the calculations repeated. Typically, several trial sections must be investigated before the most economical section is established.

11.14.1 Preliminary Member Selection: Method 1 [Uang et al., 1990]

A fast method for selecting a trial W-shape for beam-columns using an ***equivalent axial load*** was reported by Uang, Wattar, and Leet [1990]. The procedure uses column selection tables (such as Table 4-2 of the LRFDM) developed for concentrically loaded columns to design beam-columns subjected to axial load and bending moments. Their method is presented below.

Multiplying both sides of the beam column interaction equation 11.9.11*a* for $\dfrac{P_u}{P_d} \geq 0.2$ by P_d, we obtain:

$$P_u + \left[\frac{8}{9}\frac{P_d}{M_{dx}}\right]M_{ux}^* + \left[\frac{8}{9}\frac{P_d}{M_{dx}}\right]\left[\frac{M_{dx}}{M_{dy}}\right]M_{uy}^* \leq P_d \quad (11.14.1)$$

This relation can be rewritten as:

$$P_u + mM_{ux}^* + muM_{uy}^* \leq P_d \quad (11.14.2)$$

or

$$P_{ueq} \leq P_d \quad (11.14.3)$$

Here,

$$m = \frac{8}{9}\frac{P_d}{M_{dx}} = \frac{8}{9}\frac{\phi_c P_n}{\phi_b M_{nx}}; \quad u = \frac{M_{dx}}{M_{dy}} = \frac{M_{nx}}{M_{ny}} \quad (11.14.4)$$

and

$P_{ueq} =$ equivalent axial load on the member, equivalent to the given axial load and bending moments acting on the beam-column, used to enter the concentrically loaded column tables for preliminary member selection

For a first approximation, valid for all W-shapes of Grade 50 steel, values of m for all different values of KL are given in Table 11.14.1. More accurate values of m, for use in subsequent cycles, are also given in that table as a function of the nominal depth of the column section. The value of u, for a first approximation, can be taken as 2.0. The factor u is given in Table 11.14.2 for several W-shapes used as columns and $F_y = 50$ ksi [Uang et al., 1990; AISC, 1994].

TABLE 11.14.1

Factors, *m*, for Preliminary Beam-Column Design

Values of *m*

$F_y = 50$ ksi						
KL (ft)	10	12	14	16	18	20	22 and over

1st Approximation

All Shapes	1.9	1.8	1.7	1.6	1.4	1.3	1.2

Subsequent Approximation

	10	12	14	16	18	20	22 and over
W10	2.0	1.9	1.8	1.7	1.5	1.4	1.3
W12	1.7	1.6	1.5	1.5	1.4	1.3	1.2
W14	1.5	1.4	1.4	1.3	1.3	1.2	1.2

Adapted from AISC, LRFD Manual [AISC, 1994]. Copyright © American Institute of Steel Construction, Inc. Reprinted with permission. All rights reserved.

A suggested procedure for beam-column design using Method 1 is as follows.

1. For a given value of the effective length *KL*, select a first approximate value of *m* from the second row of the Table 11.14.1. If a moment about the weak axis is also present, assume $u = 2$. Assume $B_1 = 1.0$ and $B_2 = 1.0$, or use estimated values for these factors.

2. Evaluate:

$$P_{ueq} = P_u + mM_{ux}^* + muM_{uy}^* \qquad (11.14.5a)$$

where P_u = actual factored axial load, kips

M_{ux}^* = factored, second-order bending moment about the strong axis, ft-kips

M_{uy}^* = factored, second-order bending moment about the weak axis, ft-kips

m = factor taken from Table 11.14.1, for the controlling *KL*

u = factor taken from Table 11.14.2

3. Calculate $\dfrac{P_u}{P_{ueq}}$. If $\dfrac{P_u}{P_{ueq}} < 0.2$ modify the equivalent axial load as:

$$P_{ueq} = \frac{1}{2}P_u + \frac{9}{8}mM_{ux}^* + \frac{9}{8}muM_{uy}^* \qquad (11.14.5b)$$

▌ **TABLE 11.14.2**

▌ **Factors, _u_, for Preliminary Beam-Column Design**

W14×808	**2.03**	W14×90	1.94	W12×87	2.02
W14×730	2.03	W14× 82	2.68	W12×79	2.08
W14×665	2.02	W14×74	2.62	W12×72	1.98
W14×605	2.01	W14×68	2.56	W12×65	1.95
W14×550	2.01	W14×61	2.44	W12×58	2.22
W14×500	2.00	W14×53	2.7	W12×53	2.16
W14×455	1.99	W14×48	2.56	W12×50	2.51
W14×426	1.99	W14×43	2.37	W12×45	2.37
W14×398	1.98			W12×40	2.22
W14×370	1.97	**W12×336**	**2.17**		
W14×342	1.97	W12×305	2.16	**W10×112**	**2.02**
W14×311	1.96	W12×279	2.15	W10×100	2.01
W14×283	1.95	W12×252	2.14	W10×88	1.99
W14×257	1.94	W12×230	2.13	W10×77	1.96
W14×233	1.93	W12×210	2.13	W10×68	1.93
W14×211	1.93	W12×190	2.11	W10×60	1.90
W14×176	1.92	W12×170	2.11	W10×54	1.87
W14×159	1.92	W12×152	2.11	W10×49	1.83
W14×145	1.90	W12×136	2.09	W10×45	2.17
W14×132	1.99	W12×120	2.07	W10×39	2.04
W14×120	1.99	W12×106	2.06	W10×33	1.87
W14×109	1.97	W12×96	2.04		
W14×99	1.95				

Adapted from AISC, LRFD Manual [AISC, 1994]. Copyright © American Institute of Steel Construction, Inc. Reprinted with permission. All rights reserved.

4. From the appropriate column load table select a tentative section to support P_{ueq}. That is, select a section such that for the controlling KL:

$$P_d = \phi_c P_n \geq P_{req} = P_{ueq} \qquad (11.14.6)$$

5. Based on the nominal depth of the section selected in Step 4, select a subsequent approximate value of m from Table 11.14.1 and u value from Table 11.14.2.

6. With the new values selected, recalculate P_{ueq}.

7. Repeat Steps 2 to 6 until the value of P_{ueq} stabilizes.

8. Check the section obtained in Step 7 with the appropriate interaction formula Eq. 11.9.11a or b (LRFDS Eq. H1-1a or H1-1b).

9. Make $\frac{b}{t}$ checks for all plate elements.

In using the above procedure for design of beam-columns, it should be recalled that in Part 4 of the LRFDM, the axial strength values tabulated for W-shapes are for an effective length KL based on the least radius of gyration, r_y.

Therefore, if the column end-support conditions are different about the two axes, and/or if the column is braced between the end points, the table should be reentered with a modified effective length based on r_x, namely, $(K_x L_x)_y$ obtained by dividing $K_x L_x$ by the ratio r_x/r_y. Values of r_x/r_y are given in the column load table. They are about 2.1 for "square" column shapes ($b_f \approx d$) and about 1.7 for column shapes with $b_f \approx 0.8d$.

The tables for concentrically loaded columns in Part 4 of the LRFDM are limited to the W10, W12, and W14 column shapes only. When the moment is large in proportion to the axial load (that is, when the member more closely approximates a beam than a column), there will often be a much deeper and appreciably lighter beam shape that will satisfy the appropriate interaction equation. Yura [1988] suggests converting the combined loading to an equivalent bending moment about the x axis, using the approximate relation:

$$M_{ueq} = M_{ux}^* + P_u \frac{d}{2} \qquad (11.14.7)$$

Initially, a span-to-depth (L/d) value between 18 and 24 may be assumed. A trial shape can then be selected from the beam selection tables or beam selection plots in Part 5 of the LRFDM, for which:

$$M_{dx} \geq M_{ueq} \qquad (11.14.8)$$

11.14.2   Preliminary Member Selection: Method 2 [Aminmansour, 2000]

A new method for design of steel beam-columns was recently presented by Aminmansour [2000]. Aids for the design of beam-columns using this approach are included in Part 6 of the LRFDM, in the form of Tables 6-1 and 6-2. A discussion of his method is given below.

The beam-column interaction equation for $\dfrac{P_u}{P_d} \geq 0.2$, namely, Eq. 11.9.11$a$, can be rewritten as:

$$\left[\frac{1}{P_d}\right] P_u + \left[\frac{8}{9}\frac{1}{M_{dx}}\right] M_{ux}^* + \left[\frac{8}{9}\frac{1}{M_{dy}}\right] M_{uy}^*$$

$$\leq 1.0 \quad \text{for } \left[\frac{1}{P_d}\right] P_u \geq 0.2 \qquad (11.14.9)$$

Or equivalently as:

$$b P_u + m M_{ux}^* + n M_{uy}^* \leq 1.0 \quad \text{for } b P_u \geq 0.2 \qquad (11.14.10a)$$

where:

$$b = \frac{1}{P_d} = \frac{1}{\phi_c P_n} \qquad (11.14.11)$$

$$m = \frac{8}{9} \frac{1}{M_{dx}} = \frac{8}{9} \frac{1}{\phi_b M_{nx}} \qquad (11.14.12)$$

$$n = \frac{8}{9} \frac{1}{M_{dy}} = \frac{8}{9} \frac{1}{\phi_b M_{ny}} \qquad (11.14.13)$$

Similarly, the beam-column interaction equation for $\dfrac{P_u}{P_d} < 0.2$, namely Eq. 11.9.11b becomes:

$$\frac{1}{2} b P_u + \frac{9}{8} m M_{ux}^* + \frac{9}{8} n M_{uy}^* \leq 1.0 \text{ for } b P_u < 0.2 \quad (11.14.10b)$$

Values of b, m, and n for a large number of W-shapes for a range of effective lengths (KL) and unbraced lengths (L_b) for steels with $F_y = 50$ ksi are given in LRFDM Table 6-2 (W-Shapes: Values of b, m, and n for Beam-Columns). Also, to assist in making a trial selection of a W-shape to resist combined axial compression and flexure, median values of b, m, and n are given in LRFDM Table 6-1 (Median Values of b, m and n for Beam-Columns).

Remarks

1. The coefficient b is a function of the member design axial compressive strength. The limit states of elastic or inelastic flexural buckling and local buckling of the flange and web elements in axial compression are considered. Values of b listed in LRFDM Tables 6-1 and 6-2 are based on $K_y L_y$, the effective length with respect to the least radius of gyration. In cases where major axis buckling controls, $K_x L_x$ is critical and the designer must use its equivalent, $(K_x L_x)_y$, to look up b in these tables.
2. The coefficient m is a function of the member design flexural strength about the x axis. The limit states of yielding, lateral-torsional buckling, flange local buckling, and web local buckling in flexural compression were considered. Values of m are listed for $C_b = 1$ (uniform moment over the unbraced length L_b). Values of m are tabulated for all compact and noncompact sections with $L_b \leq L_r$. Cases for which $L_b > L_r$ are not listed due to the low structural effectiveness of such members.
3. The coefficient n is a function of the member design flexural strength about its weak axis. The limit states of yielding, and flange local buckling were included.
4. Values of b, m, and n listed in LRFDM Tables 6-1 and 6-2 are 1000 times their actual values, to avoid large numbers of decimal places. Therefore, all values of b, m, and n listed in these tables must be multiplied by 10^{-3} before use.

A suggested procedure for selecting a preliminary section for beam-columns with $b P_u \geq 0.2$ is as follows:

1. When axial effects appear to be more dominant, select an initial value for m from LRFDM Table 6-1 for the desired nominal depth and given unbraced length L_b. When major-axis flexural effects appear to dominate, it is recommended that the initial value for b be taken from LRFDM Table 6-1 for the desired nominal depth and effective length KL, instead. If minor-axis flexural effects are present in either case, also select a value of n from LRFDM Table 6-1.
2. Use Eq. 11.14.10a and m or b (and n, if applicable) found in Step 1 to solve for b or m.
3. From LRFDM Table 6-2, select a section with the approximate values of b, m, and n obtained in Step 2. Note that the KL value used to select b, and the L_b value used to select m need not be equal.
4. Use b, m, and n for the section selected in Step 3 to check Eq. 11.14.10a.
5. Repeat Steps 3 and 4 for different sections until the values of b, m, and n stabilize and the LHS of the interaction equation is close to or only slightly less than unity.
6. LRFDM Table 6-2 allows designers to select sections that are more efficient for a given set of conditions. Generally speaking, when selecting sections it is desirable to choose those with smaller values of b, m, and n. If a beam-column is subjected to a relatively large axial load, the designer may select a section with a smaller b value, though m and n may be slightly larger than those of another section. The dominating role of the first term in Eq. 11.14.10a, in such a case, may still lead to a smaller total for this combination.
7. For beam-columns with $bP_u < 0.2$, replace Eq. 11.14.10a with Eq. 11.14.10b in the above procedure.

11.15 Design Examples

In Part 5 of the LRFDM, any effects of local buckling on the beam design strength were already included in the preparation of the design tables and design plots for flexural strength. Therefore, we can use these tables to determine the parameter M_d without having to check if local buckling governs. Similarly, in Part 4 of the LRFDM, any effects of flange local buckling on the design axial strength of columns was accounted for in the preparation of design tables. Therefore, we can also use these tables to determine the parameter P_d directly; an analysis of local buckling is unnecessary. In Part 4 of the LRFDM, a foot note symbol on a section indicates that the section web must be verified to determine if web local buckling, under combined flexure and compression, governs the design bending strength. Any section that is used as a beam-column but not listed in Part 4 of the LRFDM, must also be checked for the governing of bending strength by local buckling.

EXAMPLE 11.15.1   **Column Design (Type: *ntx, nty*; Load: *P, M_x*)**

Select a W14-shape for a column to support, under factored loads, an axial load of 1600 kips and a major-axis moment at the top and bottom of 210 ft-kips, as determined by a first-order analysis of the structure. The moments are in the opposite direction resulting in single curvature. The interior column considered is part of a braced frame in both directions. Use A992 steel.

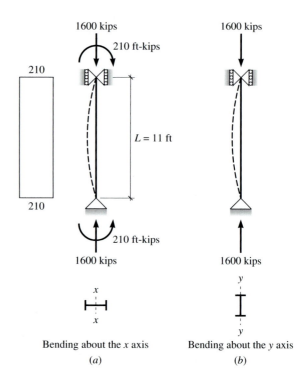

Figure X11.15.1

Solution

a. Data

Factored axial load, $P_u = 1600$ kips

For a column in a braced frame, $K = 1.0$ for design (LFRDS Section C2.1).

$$K_x L_x = K_y L_y = 1.0(11 \text{ ft}) = 11.0 \text{ ft}$$

Since there are no minor axis moments, M_{uy}^* does not need to be assessed. Because the column is part of a braced frame, $M_{ltx} = 0$, and Eq. 11.9.3 (LRFDS Eq. C1-1) reduces to $M_{ux}^* = B_{1x} M_{ntx}$

From Fig. X11.15.1, maximum first-order major axis moment, $M_{ntx} = 210$ ft-kips

b. Preliminary selection

Assume $B_{1x} = 1.1$

$$M_{ux}^* = B_{1x}M_{ntx} = 1.1(210) = 231 \text{ ft-kips}$$

Method 1

As bending is about the x axis only, we select a trial shape using:

$$P_{ueq} = P_u + mM_{ux}^*$$

For a W14 with $KL = 11$ ft, $m = 1.45$ from Table 11.14.1. Substitution into the above relation results in:

$$P_{ueq} = 1600 + 1.45(231) = 1940 \text{ kips}$$

Entering the column load table for W12-shapes (LRFDM Table 4-2) with $KL = 11$ ft and $P_{req} = 1940$ kips, we observe that a W14×176 has a design strength $P_d = 2030$ kips.
Check the W14×176.

Method 2

As bending is about the x axis only, $M_{uy}^* = 0$ and Eq. 11.14.10a reduces to

$$bP_u + mM_{ux}^* \leq 1.0$$

Since the axial load is relatively large, we get the median value of m from LRFDM Table 6-1. Corresponding to W14-shapes with $L_b = 11$ ft, we have $m = 0.826 \times 10^{-3}$ (ft-kips)$^{-1}$. So,

$$b(1600) + (0.826 \times 10^{-3})(231) \leq 1.0 \rightarrow b \leq 0.506 \times 10^{-3}$$

From LRFDM Table 6-2, try W14×176, which has:

$$b = 0.491 \times 10^{-3} < 0.506 \times 10^{-3} \quad \text{O.K.}$$

and

$$m = 0.741 \times 10^{-3} \text{ (ft-kips)}^{-1}$$

From Eq. 11.14.10a we obtain:

$$(0.491 \times 10^{-3})(1600) + (0.741 \times 10^{-3})(231) = 0.786 + 0.171$$

$$= 0.957 < 1.0 \quad \text{O.K.}$$

Also,

$$bP_u = 0.786 > 0.2 \quad \text{O.K.}$$

Check the W14×176. [*continues on next page*]

Example 11.15.1 continues ...

c. **Column effect**

For the W14×176 section selected, from Table 1-1 of the LRFDM:

$$A = 51.8 \text{ in.}^2; \qquad r_x = 6.43 \text{ in.}; \qquad I_x = 2140 \text{ in.}^4$$

$P_d = 2030$ kips

Axial load ratio, $\dfrac{P_u}{P_d} = \dfrac{1600}{2030} = 0.788 > 0.2$

So, use the interaction equation 11.9.11a (LRFDS Eq. H1-1a), which for $M_{uy}^* = 0$ reduces to

$$\frac{P_u}{P_d} + \frac{8}{9}\frac{M_{ux}^*}{M_{dx}} \le 1.0$$

We now need to examine the beam effect to determine the final values of M_{ux}^* and M_{dx}.

d. **Beam effect**

The laterally unbraced length, $L_b = L_y = 11$ ft

From LRFDM Table 5-3, $L_p = 14.2$ ft and $\phi_b M_{px} = 1200$ ft-kips, for a W14×176.

As $L_b < L_p$, we have, $M_{dx} = \phi_b M_{px} = 1200$ ft-kips

The moment magnification factor is calculated next using Eq. 11.9.4. For a pin-ended member under a uniform moment, $C_{mx} = C_m = 1.0$.

$$P_{e1x} = \frac{\pi^2 EI_x}{(KL)_{ntx}^2} = \frac{\pi^2 (29{,}000)(2140)}{[11.0(12)]^2} = 35{,}200 \text{ kips}$$

$$B_{1x} = \max\left[\frac{C_{mx}}{1 - \dfrac{P_u}{P_{e1x}}}, 1.0\right] = \max\left[\frac{1.0}{1 - \dfrac{1600}{35{,}200}}, 1.00\right] = 1.05$$

$$M_{ux}^* = 1.05(210) = 221 \text{ ft-kips}$$

e. **Check the interaction formula**

$$\frac{P_u}{P_d} + \frac{8}{9}\frac{M_{ux}^*}{M_{dx}} = 0.79 + \frac{8}{9}\left(\frac{221}{1200}\right) = 0.952 < 1.0 \qquad \text{O.K.}$$

So, the W14×176 of A992 steel selected is O.K.   (Ans.)

EXAMPLE 11.15.2

Column Design (Type: *ntx, nty*; Load: *P, M_x, M_y*)

Select the lightest W12 section of A992 steel for an interior column in a framed structure braced in both the *xx* and *yy* directions. The column is 14-ft long. Under factored gravity loads, the column is subjected to an axial load

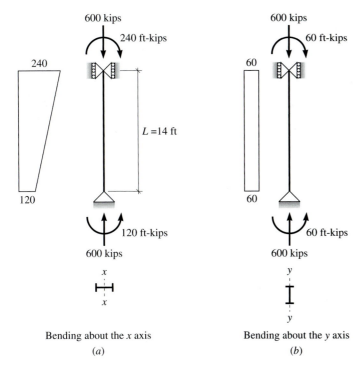

Bending about the *x* axis

(*a*)

Bending about the *y* axis

(*b*)

Figure X11.15.2

of 600 kips and the major and minor axis moments shown in Fig. X11.15.2.
Assume $K_x = K_y = 1.0$ and no lateral bracing between the floor levels.

Solution

a. Data

Column length, $L = 14$ ft; $K_x L_x = 14$ ft; $K_y L_y = 14$ ft
There are no M_{lt} moments for a braced frame, $M_{lt} = 0$

$$M_{x2} = 240 \text{ ft-kips}; \quad M_{x1} = 120 \text{ ft-kips}; \quad r_{Mx} = -\frac{120}{240} = -0.5$$

$$C_{mx} = 0.6 - 0.4(-0.5) = 0.8$$

$$M_{y2} = 60 \text{ ft-kips}; \quad M_{y1} = 60 \text{ ft-kips}; \quad r_{My} = -1.0$$

$$C_{my} = 0.6 - 0.4(-1) = 1.0$$

From Table 10.4.1, for $r_M = -0.5$, select $C_b = 1.25$.

b. Selection of trial section

The amplification factor B_{1x} is assumed to be 1.0 since a favorable
moment gradient exists over the column height. B_{1y} is assumed to be 1.1,
for the purpose of arriving at a trial section. For each of the two axes:

$$M_{ux}^* = B_{1x} M_{ntx} \approx 1.0(240) = 240 \text{ ft-kips}$$
$$M_{uy}^* = B_{1y} M_{nty} \approx 1.1(60) = 66 \text{ ft-kips}$$

[*continues on next page*]

Example 11.15.2 continues ...

For a W12 column of F_y = 50 ksi steel and KL = 14 ft, m = 1.5 from Table 11.14.1. Assume a u value of 2.0 for the first trial. So,

$$P_{ueq} = P_u + mM_{ux}^* + muM_{uy}^* = 600 + 1.5(240) + 1.5(2.0)(66)$$
$$= 600 + 360 + 198 = 1160 \text{ kips}$$

From column selection tables (LRFDM Table 4-2), for $KL = K_y L_y$ = 14 ft and F_y = 50 ksi steel, select:

$$\text{W12}\times120 \text{ with } P_d = 1220 \text{ kips} > P_{req} = 1160 \text{ kips}$$

So, the W12×120 is a potential trial shape. From Table 11.14.2, this shape has a u value of 2.07 resulting in the revised value:

$$P_{ueq} = 600 + 360 + 1.5(2.07)(66) = 1170 \text{ kips}$$

Reentering the column selection tables, it is seen that the W12×120 with an axial load capacity of 1220 kips appears to be an acceptable shape. So, let us try a W12×120 shape.

c. Design strengths
From Table 1-1 of the LRFDM, for a W12×120 shape:

$$A = 35.3 \text{ in.}^2; \quad I_x = 1070 \text{ in.}^4; \quad I_y = 345 \text{ in.}^4$$

Also, from column load tables, P_d = 1220 kips.

Axial load ratio, $\dfrac{P_u}{P_d} = \dfrac{600}{1220} = 0.492$

As the axial load ratio is greater than 0.2, the interaction equation 11.9.11a (LRFDS Equation H1-1a) applies. Unbraced length, $L_b = L = 14$ ft. Thus, from beam selection plots for W-shapes (LRFDM Table 5-5), for a W12×120 shape having F_y = 50 ksi steel, L_b = 14 ft and C_b = 1.0, we find that M_{dx}^o = 682 ft-kips. Finally, as $L_p < L_b < L_r$ and C_b = 1.25, we have:

$$M_{dx} = \min [C_b M_{dx}^o, \phi_b M_{px}] = \min [1.25(682), 698] = 698 \text{ ft-kips}$$
$$M_{dy} = \phi_b M_{py} = 315 \text{ ft-kips}$$

d. Second-order moments M_{ux}^*, M_{uy}^*
The magnification factors B_{1x} and B_{1y} are calculated from Eq. 11.9.4.

$$P_{e1x} = \frac{\pi^2 EI_x}{(KL)_{ntx}^2} = \frac{\pi^2(29,000)(1070)}{[14.0(12)]^2} = 10,900 \text{ kips}$$

$$P_{e1y} = \frac{\pi^2 EI_y}{(KL)_{nty}^2} = \frac{\pi^2(29,000)(345)}{[14.0(12)]^2} = 3500 \text{ kips}$$

$$B_{1x} = \max \left[\frac{0.8}{1 - \frac{600}{10,900}}, 1.0 \right] = \max [0.847, 1.0] = 1.0$$

$$B_{1y} = \max \left[\frac{1.0}{1 - \frac{600}{3500}}, 1.0 \right] = \max [1.21, 1.0] = 1.21$$

The second-order moments are therefore:

$$M_{ux}^* = B_{1x}M_{ntx} = 1.0(240.0) = 240 \text{ ft-kips}$$

$$M_{uy}^* = B_{1y}M_{nty} = 1.21(60.0) = 72.6 \text{ ft-kips}$$

The quantities necessary to perform a check of the interaction equation have now been determined. The LHS of the interaction equation 11.9.11a (LRFDS Eq. H1-1a) is:

$$\frac{P_u}{P_d} + \frac{8}{9}\frac{M_{ux}^*}{M_{dx}} + \frac{8}{9}\frac{M_{uy}^*}{M_{dy}} = 0.492 + \frac{8}{9}\left(\frac{240}{698}\right) + \frac{8}{9}\left(\frac{72.6}{315}\right)$$

$$= 0.492 + 0.306 + 0.205 = 1.00 \qquad \text{Accept}$$

So, select a W12×120 of A588 Grade 50 steel.   (Ans.)

Column Design (Type: *ntx, nty*; Load: *P, M$_x$, M$_y$*)

EXAMPLE 11.15.3

Select a suitable W-shape of A992 steel for a 12-ft-long, corner column at the ground level of a building structure. The column is hinged at the base, and the structure is braced in both perpendicular directions. The column is subjected to a factored axial load, P_u, of 147 kips, first-order major-axis moment, M_{ux}, of 5.2 ft-kips, and a first-order minor-axis moment, M_{uy}, of 2.2 ft-kips.

Solution
a. Data

Length of column, $L = 12$ ft
As the column is part of a braced frame, and the base is hinged, we have:

$$K_x L_x = L = 12.0 \text{ ft}; \quad K_y L_y = L = 12.0 \text{ ft}; \quad L_b = L = 12.0 \text{ ft}$$

Factored axial load, $P_u = 147$ kips
M_{lt} moments are zero, as the column is part of a braced frame in both directions. So

$$M_{ux}^* = B_{1x}M_{ux}; \quad M_{uy}^* = B_{1y}M_{uy}$$

[*continues on next page*]

Example 11.15.3 continues ...

b. Preliminary selection

Assume $B_{1x} = 1.0$ and $B_{1y} = 1.0$. From LRFDM Table 6-1, for W10-shapes with $KL = L_b = 12$ ft, we obtain: $m = 5.15 \times 10^{-3}$; $n = 13 \times 10^{-3}$

Substituting into Eq. 11.14.10a we obtain:

$$b\,(147) + (5.15 \times 10^{-3})(5.2) + (13 \times 10^{-3})(2.2)$$

$$= 1.0 \rightarrow b = 6.44 \times 10^{-3}$$

From LRFDM Table 6-2, we observe that for a W10×30: $b = 5.97 \times 10^{-3} < 6.44 \times 10^{-3}$ and

$$m = 8.24 \times 10^{-3}; \qquad n = 27.5 \times 10^{-3}$$

Substituting these values in Eq. 11.14.10a, we obtain:

$$\text{LHS} = (5.97 \times 10^{-3})(147) + (8.24 \times 10^{-3})(5.2) + (27.5 \times 10^{-3})(2.2)$$

$$= 0.875 + 0.043 + 0.061 = 0.979 < 1.0 \quad \text{O.K.}$$

Select a W10×30 and check.

c. Check W10×30

For a W10×30: $A = 8.84$ in.^{2}; $r_x = 4.38$ in.; $r_y = 1.37$ in.

$I_x = 170$ in.^{4}; $I_y = 16.7$ in.^{4}

$$\frac{K_x L_x}{r_x} = \frac{12.0(12)}{4.38} = 32.9; \quad \frac{K_y L_y}{r_y} = \frac{12.0(12)}{1.37} = 105$$

From LRFDS Table 3-50, for $KL/r = 105$, $F_{dc} = 19.0$ ksi, resulting in:

$$P_d = 19.0(8.84) = 168 \text{ kips}$$

$$\frac{P_u}{P_d} = \frac{147}{168} = 0.875 > 0.2$$

So, Eq. 11.9.11a governs. From beam design plots for W-shapes (LRFDM Table 5-5), for a W10×30 with $L_b = 12$ ft, we read $M_{dx}^o = 108$ ft-kips. Also, from these plots, $\phi_b M_{px} = 137$ ft-kips. For a member subjected to a bending moment that varies linearly from zero at one end to a maximum at the other end, $r_M = 0$ and from Table 10.4.1, $C_b = 1.67$. Hence,

$$M_{dx} = \min[1.67(108), 137] = 137 \text{ ft-kips}$$

Also, from LRFDM Table 5-3, for a W10×30, $\phi_b M_{py} = 32.3$. So, $M_{dy} = \phi_b M_{py} = 32.3$ ft-kips.

d. Second-order moments M_{ux}^*, M_{uy}^*
The magnification factors B_{1x} and B_{1y} are calculated from Eq. 11.9.4.

$$C_{mx} = 0.6 - 0.4r_{Mx} = 0.6$$
$$C_{my} = 0.6 - 0.4r_{My} = 0.6$$

$$P_{elx} = \frac{\pi^2 EI_x}{(KL)_{ntx}^2} = \frac{\pi^2 (29,000)(170)}{(12.0 \times 12)^2} = 2350 \text{ kips}$$

$$P_{ely} = \frac{\pi^2 EI_y}{(KL)_{nty}^2} = \frac{\pi^2 (29,000)(16.7)}{(12.0 \times 12)^2} = 231 \text{ kips}$$

$$B_{1x} = \max \left[\frac{0.6}{1 - \dfrac{147}{2350}}; 1.0 \right] = \max [0.640; 1.0] = 1.00$$

$$B_{1y} = \max \left[\frac{0.6}{1 - \dfrac{147}{231}}; 1.0 \right] = \max [1.65; 1.0] = 1.65$$

The second-order moments are therefore:

$$M_{ux}^* = B_{1x} M_{ntx} = 1.00 \times 5.2 = 5.20 \text{ ft-kips}$$

$$M_{uy}^* = B_{1y} M_{nty} = 1.65 \times 2.2 = 3.63 \text{ ft-kips}$$

e. Check for limit state of strength
Substituting all known values into Eq. 11.9.11a, we obtain:

$$\text{LHS} = 0.875 + \frac{8}{9} \left(\frac{5.20}{137} + \frac{3.63}{32.3} \right) = 0.875 + 0.034 + 0.100$$

$$= 1.01 \quad \text{O.K. Accept}$$

So, select a W10×30 of A992 steel.   (Ans.)

Column Design (Type: *ntx, nty*; Load: *P, q_x, q_y*)

EXAMPLE 11.15.4

A 30-ft-long chord member of a truss is subjected, under factored loads, to an axial compressive force of 690 kips and a uniformly distributed factored load of 0.3 klf causing bending about its weak axis. Use A992 steel and select the appropriate W14 section. Include the influence of the self-weight of the member which causes bending about the major axis.

[continues on next page]

Example 11.15.4 continues ...

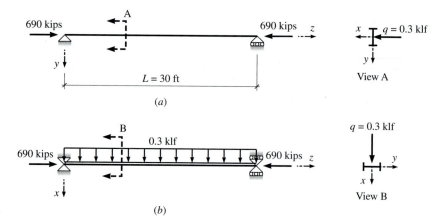

Figure X11.15.4

(b)

Solution

a. Data

$P_u = 690$ kips

$K_x L_x = K_y L_y = 1.0(30) = 30.0$ ft

From the column load tables for W14-shapes (LRFDM Table 4-2) for $F_y = 50$ ksi and $KL = K_y L_y = 30$ ft, $P_d = 762$ kips ($> P_u = 690$ kips) for a W14×120. So, we need a section heavier than a W14×120 to resist the combined effects of axial load and bending moments. As the distributed loads are light we choose the next heavier section, W14×132, and calculate the major-axis bending moment due to self-weight.

The factored distributed load due to self-weight, $q_{uy} = \dfrac{1.2(132)}{1000} = 0.16$ klf.

The maximum major-axis bending moment at the center of the beam,

$$M_{ntx} = \frac{q_{uy} L_x^2}{8} = \frac{0.16(30^2)}{8} = 18.0 \text{ ft-kips}$$

Factored uniformly distributed transverse load, $q_{ux} = 0.30$ klf

The maximum, minor-axis bending moment at the center of the beam is:

$$M_{nty} = \frac{q_{ux} L_y^2}{8} = \frac{0.3(30)^2}{8} = 33.8 \text{ ft-kips}$$

b. Preliminary selection

As the member is quite flexible, we will assume $B_{1x} = 1.2$ and $B_{1y} = 1.5$, resulting in:

$$M_{ux}^* = B_{1x} M_{ntx} = 1.20(18.0) = 21.6 \text{ ft-kips}$$

$$M_{uy}^* = B_{1y} M_{nty} = 1.50(33.8) = 50.7 \text{ ft-kips}$$

For the biaxially bent beam-column:

$$P_{ueq} = P_u + mM_{ux}^* + muM_{uy}^*$$

For a W14 with $KL = 30$ ft and $F_y = 50$ ksi, $m = 1.2$ from Table 11.14.1. Also, for W14 sections heavier than a W14×132, $u \approx 1.99$ from Table 11.14.2. Hence, we have:

$$P_{ueq} = 690 + 1.2(21.6) + 1.2(1.99)(50.7) = 837 \text{ kips}$$

In the column load tables for W-shapes (LRFDM Table 4-2) for $F_y = 50$ ksi and $KL = 30$ ft, $P_d = 844$ kips ($> P_{ueq} = 837$ kips) for the W14×132 selected.

So, try a W14×132.

c. Check the section
Axial strength of the column, $P_d = 844$ kips

The axial load ratio, $\dfrac{P_u}{P_d} = \dfrac{690}{844} = 0.818 > 0.2$

Therefore, Eq. 11.9.11a (LRFDS Eq. H1-1a) applies. Furthermore, the section is compact according to LRFDS B5, as there is no indication to the contrary in Table 4-2 of the LRFDM.

Major-axis bending
From beam selection tables for W-shapes (LRFDM Table 5-3, for example), for a W14×132 and $F_y = 50$ ksi:

$$\phi_b M_{px} = 878 \text{ ft-kips}; \qquad I_x = 1530 \text{ in.}^4; \qquad I_y = 548 \text{ in.}^4$$

$$L_p = 13.3 \text{ ft}; \qquad\qquad \phi_b M_{py} = 419 \text{ ft-kips}$$

From beam selection plots for W-shapes (LRFDM Table 5-5), for a W14×132 of $F_y = 50$ ksi steel, $C_b = 1.0$ and $L_b = 30$ ft, $M_{dx}^o = 763$ ft-kips. A simply supported beam under uniformly distributed load and laterally braced at the ends only has a C_b value of 1.14 from Table 5-1 of the LRFDM. As the unbraced length, L_b, is between L_p and L_r, the design major axis bending strength of the member is:

$$M_{dx} = \min[C_b M_{dx}^o; \phi_b M_{px}] = \min[1.14(763), 878]$$

$$= \min[870, 878] = 870 \text{ ft-kips}$$

We have, $M_{ux}^* = B_{1x} M_{ntx}$
For a beam-column simply supported at the ends and subjected to uniformly distributed lateral load, $C_m = 1.0$. Also,

$$P_{e1x} = \frac{\pi^2 EI_x}{(KL)_{ntx}^2} = \frac{\pi^2 (29{,}000)(1530)}{[30(12)]^2} = 3380 \text{ kips}$$

$$B_{1x} = \max\left[\frac{1.0}{\left(1 - \dfrac{690}{3380}\right)}, 1.00\right] = 1.26$$

[*continues on next page*]

Example 11.15.4 continues ...

$$M_{ux}^* = B_{1x} M_{ntx} = 1.26(18.0) = 22.7 \text{ ft-kips}$$

Minor-axis bending
$$M_{dy} = \phi_b M_{py} = 419 \text{ ft-kips}$$
For the member under consideration, with no lateral translation of ends possible, we have: $M_{uy}^* = B_{1y} M_{nty}$
For a beam-column with transverse loading, C_m may be obtained from LRFDC Table C-C1.1. For uniformly distributed load, $C_m = 1.0$. We have:

$$P_{e1y} = \frac{\pi^2 E I_y}{(KL)_{nty}^2} = \frac{\pi^2 (29,000)(548)}{[30(12)]^2} = 1210 \text{ kips}$$

$$B_{1y} = \frac{1.0}{1 - \frac{690}{1210}} = 2.33$$

$$M_{uy}^* = 2.33(33.8) = 78.8 \text{ ft-kips}$$

d. Check for limit state of strength

$$\frac{P_u}{P_d} + \frac{8}{9} \frac{M_{ux}^*}{M_{dx}} + \frac{8}{9} \frac{M_{uy}^*}{M_{dy}} = 0.818 + \frac{8}{9}\left(\frac{22.7}{870}\right) + \frac{8}{9}\left(\frac{78.8}{419}\right)$$

$$= 0.818 + 0.023 + 0.167 = 1.01$$

The W14×132 of A992 steel is therefore acceptable.   (Ans.)

EXAMPLE 11.15.5

Column Design (Type: *ntx, nty;* Load *P, Q_y*)

A W10-shape is used as a continuous top chord member of the 128-ft-long roof truss of an industrial building as shown in Fig. X11.15.5a. The eight-panel Pratt truss is 12-ft deep at the center and 8-ft deep at the supporting columns. The top chord supports purlins at the panel points and midway between the panel points. Under factored gravity loads, each purlin transmits a load of 35 kips (end purlin 17.5 kips). These loads include provision for the dead weight of the truss too. Use A992 steel and select the lightest section in the center panel.

Solution

a. General.
Due to the presence of purlins between truss panel points, the upper chord will be subjected to combined bending and direct compression. Since the upper chord is continuous over the joints, it should be analyzed

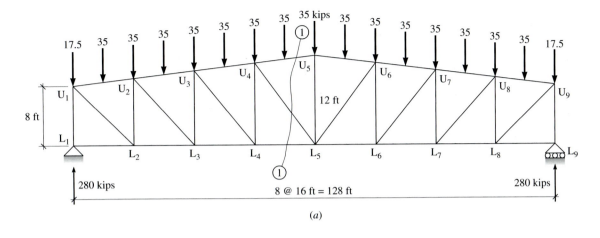

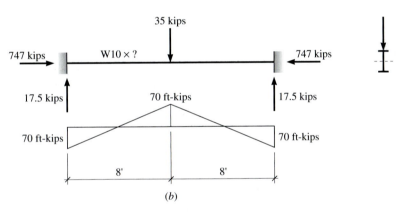

Figure X11.15.5

as a continuous beam for moments. The symmetrical arrangement of the loading and spans is such, however, that the upper chord may be considered a series of fixed-ended beams for moment considerations. From Case 16 in beam diagrams, given in Table 5-17 of the LRFDM, the fixed-end moment and the central moment, which are equal in magnitude, are:

$$M_{\max} = \frac{Q_u L}{8} = \frac{35(16)}{8} = 70.0 \text{ ft-kips} = M_u$$

These end moments and the corresponding reactions are shown in Fig. X11.15.5*b*. When the beam reactions are added to the loads that are directly applied to the panel points, the truss loading condition consists of 35 kips at U_1 and U_9, and 70 kips at the interior panel points U_2 to U_8. The maximum compressive force will occur in member $U_4 U_5$ (and in

[*continues on next page*]

Example 11.15.5 continues ...

U_5U_6) and can be obtained by considering the equilibrium of a free body of the portion of the truss to the left of section 1-1 about the panel point L_5 (neglecting the slight influence of the inclination of the top chord):

$$\sum (M)_{L5} = [280(4) - 35(4) - 70(3) - 70(2) - 70(1)]16 - 12P$$

$$= 0 \rightarrow P = 747 \text{ kips}$$

Hence the beam-column is to be designed for:

$$P_u = 747 \text{ kips}; \quad M_{ntx} = 70.0 \text{ ft-kips}; \quad M_{ltx} = M_{lty} = 0; \quad M_{nty} = 0$$

For a column in a structure braced against sidesway, such as the truss member under consideration, $K = 1.0$. We assume that lateral support is provided for this member at its ends and at the center by the purlins. Thus, $K_x L_x = 16.0$ ft; $K_y L_y = 8.0$ ft.

b. Trial section

We will assume a KL value of 10 ft, as major axis buckling may control. For W10-shape of A992 steel, $KL = 10$ ft, we read a m value of 2.0 from Table 11.14.1. So, assuming $B_{1x} = 1.1$ to start with:

$$P_{ueq} = P_u + mM^*_{ux} = 747 + 2.0(1.1)(70.0) = 901 \text{ kips}$$

Entering column selection tables for W10 series (LRFDM Table 4-2) with $KL = K_y L_y = 8$ ft,

W10×88 has $P_{dy} = 999$ kips > 901 kips required; $r_x/r_y = 1.73$

$$(K_x L_x)_y = \frac{16.0}{1.73} = 9.25 \text{ ft} \rightarrow P_{dx} = 965 \text{ kips} > 901 \text{ kips}$$

As the design strength about both axes is greater than the required value of 901 kips, we will consider the W10×88 shape.

c. Axial and bending strengths

W10×88: $A = 25.9$ in.²; $r_x = 4.54$ in.; $r_y = 2.63$ in.

$$\frac{K_x L_x}{r_x} = \frac{16(12)}{4.54} = 42.3; \quad \frac{K_y L_y}{r_y} = \frac{8(12)}{2.63} = 36.5$$

From LRFDS Table 3-50, for:

$$\frac{KL}{r} = 42.3 \rightarrow F_{dc} = \phi_c F_{cr} = 37.3 \text{ ksi}$$

$$\rightarrow P_d = 37.3(25.9) = 966 \text{ kips}$$

$$\frac{P_u}{P_d} = \frac{747}{966} = 0.773 > 0.2$$

So, use Eq. 11.9.11*a* (LRFDS Eq. H1-1*a*).

From beam selection tables for W-shapes (LRFDM Table 5-3, for example), for a

$\qquad$ W10×88, L_p = 9.29 ft; $\phi_b M_{px}$ = 424 ft-kips

$\qquad$ As L_b = 8 ft < L_p = 9.29 ft, M_{dx} = $\phi_b M_{px}$ = 424 ft-kips

d. Second-order moment, M_{ux}^*

For bending about the x axis, we have F_e = 160 ksi, from Table 11.9.1,

for $\left(\dfrac{KL}{r}\right)_{ntx}$ = $\dfrac{K_x L_x}{r_x}$ = 42.3.

P_{elx} = $F_e A$ = 160.0(25.9) = 4150 kips
From Table C-C1.1 of the LRFDC, we have for a fixed ended beam with a central transverse load:

$$C_{mx} = 1 - 0.2 \frac{P_u}{P_{elx}} = 1 - 0.2\left(\frac{747}{4150}\right) = 0.96$$

$$B_{1x} = \frac{C_{mx}}{1 - \dfrac{P_u}{P_{elx}}} = \frac{0.96}{1 - \dfrac{747}{4150}} = 1.17$$

$$M_{ux}^* = B_{1x} M_{ntx} = 1.17(70) = 81.9 \text{ ft-kips}$$

e. Check for limit state of strength
For the beam-column under major axis bending, we have:

$$\frac{P_u}{P_d} + \frac{8}{9}\frac{M_{ux}^*}{M_{dx}} = 0.773 + \frac{8}{9}\left(\frac{81.9}{424}\right) = 0.773 + 0.172 = 0.945$$

$$< 1.00 \quad \text{O.K.}$$

Select a W10×88 of A992 steel. $\qquad\qquad$ (Ans.)

Column Design (Type: *ltx, nty*; Load: *P, M$_x$*)

EXAMPLE 11.15.6

A W14 column, 14-ft-long, is part of an unbraced frame in the plane of the web and part of a braced frame in the perpendicular direction such that K_x = 1.5 and K_y = 1.0. First-order factored load analysis gives an axial load of 650 kips under (1.2D + 0.5L + 1.6W), the nonsway gravity moments M_{nt} under (1.2D + 0.5L), and the sway moments M_{lt} under 1.6W, as shown in Fig. X11.15.6. For the story under consideration, ΣP_u = 20,000 kips. The allowable story drift index is 0.002 due to total horizontal (unfactored) wind forces ΣH = 500 kips. Select the lightest section using A992 steel.

[continues on next page]

Example 11.15.6 continues ...

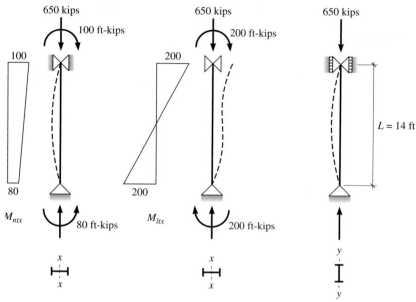

Figure X11.15.6

Bending about the x axis

(a)

Bending about the y axis

(b)

Solution

a. Data

Length of the column, $L = 14$ ft

$$K_x L_x = 1.5(14) = 21.0 \text{ ft}; \quad K_y L_y = 1.0(14.0) = 14.0 \text{ ft}$$

$P_u = 650$ kips; $M_{uy}^* = 0$

Maximum first-order no-translation moment, $M_{ntx} = 100$ ft-kips

Maximum first-order translation moment, $M_{ltx} = 200$ ft-kips

Story drift index, $\Delta_o/h = 0.002$

b. Preliminary selection

Assume $B_{1x} = B_{2x} = 1.0$ giving:

$$M_{ux}^* = B_{1x}M_{ntx} + B_{2x}M_{ltx} = 1.0(100) + 1.0(200) = 300 \text{ ft-kips}$$

To start with, we will assume that the axial capacity is determined by buckling about the minor axis. For a W14 section of A992 steel and $KL = K_y L_y = 14$ ft, value of m is 1.4 from Table 11.14.1. Hence:

$$P_{ueq} = P_u + mM_{ux}^* = 650 + 1.4(300) = 1070 \text{ kips}$$

Enter the column selection table for W14-shapes (LRFDM Table 4-2) with P_{req} = 1070 kips; KL = 14 ft; F_y = 50 ksi, to find that for a W14×109,

$$P_{dy} = 1170 \text{ kips} > P_{req} = 1070 \text{ kips and } \frac{r_x}{r_y} = 1.67$$

$$(K_x L_x)_y = \frac{21.0}{1.67} = 12.6 \text{ ft} < K_y L_y = 14 \text{ ft}$$

indicating that major-axis buckling will not control and $P_d = P_{dy} = 1170$ kips

c. Check the section
For a W14×109 section, A = 32.0 in.^{2}; I_x = 1240 in.^{4}

$$\frac{P_u}{P_d} = \frac{650}{1170} = 0.555 > 0.2$$

Hence use interaction Eq. 11.9.11a (LRFDS Eq. H1-1a).
The laterally unbraced length L_b is 14 ft.
From beam selection plots for W-shapes (LRFDM Table 5-5), we have for a W14×109 shape of F_y = 50 ksi steel, $\phi_b M_{px}$ = 720 ft-kips, and for L_b = 14 ft and C_b = 1.0, M^o_{dx} = 715 ft-kips.
The ratio of (total) end moments is:

$$r_M = \frac{|M_1|}{|M_2|} = + \frac{|-200 + 80.0|}{|200 + 100|} = 0.4$$

So, from Table 10.4.1, C_b = 2.16.
As $L_p < L_b$, we have:

$$M_{dx} = \min [C_b M^o_{dx}, \phi_b M_{px}] = \min [2.16(715), 720] = 720 \text{ ft-kips}$$

Next B_{1x} is calculated from Eq. 11.9.4. We have:

$$C_{mx} = 0.6 - 0.4\left(\frac{M_{nt1}}{M_{nt2}}\right) = 0.6 - 0.4(-0.8) = 0.92$$

$$P_{e1x} = \frac{\pi^2 EI_x}{(KL)^2_{ntx}} = \frac{\pi^2(29,000)(1240)}{[14.0(12)]^2} = 12,600 \text{ kips}$$

$$B_{1x} = \max \left[\frac{0.92}{1 - \dfrac{650}{12,600}}, 1.00 \right] = \max [0.970, 1.00] = 1.00$$

$$B_{2x} = \frac{1}{1 - \dfrac{\Delta_{oh}}{L} \dfrac{\Sigma P_u}{\Sigma H}} = \frac{1}{1 - 0.002\left(\dfrac{20,000}{500}\right)} = 1.09$$

[continues on next page]

Example 11.15.6 continues ...

Factored, second-order moment is:

$$M^*_{ux} = B_{1x}M_{ntx} + B_{2x}M_{ltx} = 1.0(100) + 1.09(200) = 318 \text{ ft-kips}$$

Substituting in the interaction Eq. 11.9.11a:

$$\text{LHS} = 0.555 + \frac{8}{9}\left(\frac{318}{720}\right) = 0.948 < 1.0 \quad \text{O.K.}$$

Hence adopt a W14×109 of A572 Grade 50 steel.   (Ans.)

EXAMPLE 11.15.7

Column Design (Type: *ltx, nty;* Load: *P, M_x, M_y*)

Select a W14 section of A992 steel for a 14-ft-long beam-column, part of an unbraced frame in the *y* plane and part of a braced frame in the *x* plane. $K_x = 1.4$ and $K_y = 1.0$. Factored axial load on column P_u is 840 kips. First-order, single curvature moments under gravity loads are $M_{ntx} = 280$ ft-kips and $M_{nty} = 40$ ft-kips. The first-order reverse-curvature moments under factored wind loads are $M_{ltx1} = 200$ ft-kips and $M_{ltx2} = 50$ ft-kips. The drift index is 1/400 under $\Sigma H = 100$ kips. Total factored gravity load above this story is 6800 kips.

Solution

a. Data
   W14-shape of A992 steel

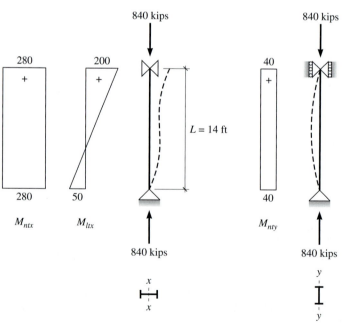

840 kips

280
+
280
M_{ntx}

200
+
50
M_{ltx}

$L = 14$ ft

840 kips

Bending about the *x* axis
(a)

840 kips

40
+
40
M_{nty}

840 kips

Bending about the *y* axis
(b)

Figure X11.15.7

$L = 14.0$ ft; $K_x = 1.4$; $K_y = 1.0$
$K_x L_x = 1.4(14) = 19.6$ ft; $K_y L_y = 1.0(14.0) = 14.0$ ft $= L_b$
$P_u = 840$ kips; $M_{ntx} = 280$ ft-kips; $M_{nty} = 40.0$ ft-kips
$M_{ltx1} = 200$ ft-kips; $M_{ltx2} = 50.0$ ft-kips; $M_{ltx} = 200$ ft-kips
$M_{lty} = 0$ (column part of braced frame in xx plane)

$\Delta_o/h = 1/400$; $\Sigma H = 100$ kips; $\Sigma P_u = 6800$ kips

b. Preliminary selection

Assume $B_{1x} = 1.05$; $B_{1y} = B_{2x} = 1.1$

$M^*_{ux} = B_{1x} M_{ntx} + B_{2x} M_{ltx} = 1.05(280) + 1.1(200) = 514$ ft-kips

$M^*_{uy} = B_{1y} M_{nty} = 1.1(40.0) = 44.0$ ft-kips

From Table 11.14.1, corresponding to $F_y = 50$ ksi, for a W14-shape with $KL = 14$ ft, select $m = 1.4$. Assume $u = 2.0$.

$$P_{ueq} = P_u + mM^*_{ux} + muM^*_{uy}$$

$$= 840 + 1.4(514) + 1.4(2.0)(44.0) = 1680 \text{ kips}$$

From column load tables for W-shapes (LRFDM Table 4-2), for a W14× 159 and $KL = K_y L_y = 14$ ft, $P_{dy} = 1740$ kips > 1680 kips. Also $r_x/r_y = 1.60$ for this section, resulting in:

$$(K_x L_x)_y = 19.6/1.6 = 12.3 < K_y L_y = 14 \text{ ft}$$

So, $K_y L_y$ controls the design. For the W14×159 section, $u = 1.92$ from Table 11.14.2, resulting in a revised value,

$$P_{ueq} = 840 + 1.4(514) + 1.4(1.92)(44.0) = 1680 \text{ kips}$$

The W14×159 is still valid.

c. Check the selected section
For the W14×159 of A992 steel column with $(K_x L_x)_y = 12.3$ ft and $K_y L_y = 14$ ft, $P_d = 1740$ kips.

$$\frac{P_u}{P_d} = \frac{840}{1740} = 0.483 > 0.2, \text{ so use Eq. 11.9.11}a.$$

For a member with no translation moments producing symmetric, single curvature bending, $C_{mx} = C_{my} = 1.0$.
P_{elx} and P_{ely} are the Euler buckling loads of the column considered, as part of braced frames. From column load table for W-shapes (LRFDM Table 4-2), for a W14×159:

$$P_{ex}(KL)^2 = 54,400 \times 10^4; \quad P_{ey}(KL)^2 = 21,400 \times 10^4$$

[*continues on next page*]

Example 11.15.7 continues ...

Conservatively take $(KL)_{ntx} = (KL)_{nty} = 1.0(14) = 14.0$ ft. Then,

$$P_{elx} = \frac{54{,}400 \times 10^4}{[14.0(12)]^2} = 19{,}300 \text{ kips}; \ P_{ely} = \frac{21{,}400 \times 10^4}{[14.0(12)]^2}$$

$$= 7580 \text{ kips}$$

$$B_{1x} = \max \left[\frac{1.0}{1 - \dfrac{840}{19{,}300}}, 1.0 \right] = \max[1.05, 1.0] = 1.05$$

$$B_{1y} = \max \left[\frac{1.0}{1 - \dfrac{840}{7580}}, 1.0 \right] = \max[1.12, 1.0] = 1.12$$

Also,

$$B_{2x} = \frac{1}{1 - \dfrac{\Sigma P_u}{\Sigma H}\left(\dfrac{\Delta_{oh}}{L}\right)} = \frac{1}{1 - \dfrac{6800}{100}\left(\dfrac{1}{400}\right)} = 1.20$$

We obtain

$$M_{ux}^* = B_{1x} M_{ntx} + B_{2x} M_{ltx} = 1.05(280) + 1.20(200) = 534 \text{ ft-kips}$$

$$M_{uy}^* = B_{1y} M_{nty} = 1.12(40.0) = 44.8 \text{ ft-kips}$$

From beam selection tables for W-shapes (LRFDM Table 5-3, for example), for a W14×159 and $F_y = 50$ ksi:

$$\phi_b M_{px} = 1080 \text{ ft-kips}; \quad L_p = 14.1 \text{ ft}; \quad \phi_b M_{py} = 541 \text{ ft-kips}$$

As $L_b = 14.0$ ft $< L_p = 14.1$ ft, $M_{dx} = \phi_b M_{px} = 1080$ ft-kips

$M_{dy} = \phi_b M_{py} = 541$ ft-kips

We therefore have:

$$\frac{P_u}{P_d} + \frac{8}{9}\left[\frac{M_{ux}^*}{M_{dx}} + \frac{M_{uy}^*}{M_{dy}}\right] = 0.483 + \frac{8}{9}\left[\frac{534}{1080} + \frac{44.8}{541}\right]$$

$$= 0.996 < 1.0 \qquad \text{O.K.}$$

So, the W14×159 of A992 steel selected is O.K.   (Ans.)

EXAMPLE 11.15.8   **Column Design (Type: *ltx, lty*; Load: *P, M_x, M_y*)**

Select a W12 section of A992 steel for a 16-ft-long beam-column. The member is part of a symmetric frame unbraced about either axes. Under factored loads,

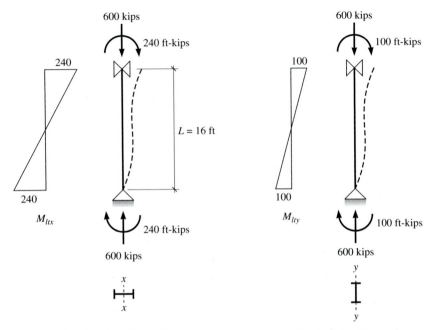

Bending about the *x* axis
(*a*)

Bending about the *y* axis
(*b*)

Figure X11.15.8

P_u = 600 kips, M_{ntx} = M_{nty} = 0, M_{ltx} = 240 ft-kips and M_{lty} = 100 ft-kips. Also K_x = 1.5 and K_y = 1.3. For all columns in the story under consideration, ΣP_{ui} = 10,800 kips, ΣP_{e2x} = 120,000 kips and ΣP_{e2y} = 82,000 kips.

Solution

a. Data

L = 16.0 ft; L_b = 16.0 ft

K_x = 1.5; $K_x L_x$ = 1.5(16.0) = 24.0 ft

K_y = 1.3; $K_y L_y$ = 1.3(16.0) = 20.8 ft

P_u = 600 kips; M_{ntx} = M_{nty} = 0

M_{ltx} = 240 ft-kips; M_{lty} = 100 ft-kips

ΣP_{ui} = 10,800 kips; ΣP_{e2x} = 120,000 kips; ΣP_{e2y} = 82,000 kips

From Eq. 11.9.9 (LRFDS Eq. C1-5):

$$B_{2x} = \cfrac{1}{1 - \cfrac{\Sigma P_{ui}}{\Sigma P_{e2x}}} = \cfrac{1}{1 - \cfrac{10,800}{120,000}} = 1.10$$

$$B_{2y} = \cfrac{1}{1 - \cfrac{\Sigma P_u}{\Sigma P_{e2y}}} = \cfrac{1}{1 - \cfrac{10,800}{82,000}} = 1.15$$

[*continues on next page*]

Example 11.15.8 continues ...

$$M^*_{ux} = B_{1x}M_{ntx} + B_{2x}M_{ltx} = 0 + 1.10(240) = 264 \text{ ft-kips}$$

$$M^*_{uy} = B_{1y}M_{nty} + B_{2y}M_{lty} = 0 + 1.15(100) = 115 \text{ ft-kips}$$

b. Preliminary selection

Entering Table 11.14.1 with $KL = K_yL_y = 20.8$ ft, $F_y = 50$ ksi and W12-shapes we obtain $m = 1.3$. Assume $u = 2.0$. We obtain:

$$P_{ueq} = P_u + mM^*_{ux} + mu\,M^*_{uy}$$

$$= 600 + 1.3(264) + 1.3(2.0)(115) = 1240 \text{ kips}$$

From LRFDM column load table for W-shapes, for $KL = K_yL_y = 20.8$ ft and $F_y = 50$ ksi, a W12×170 has $P_{dy} = 1370$ kips > 1240 kips. Also, $u = 2.11$ from Table 11.14.2, resulting in:

$$P_{ueq} = 600 + 1.3(264) + 1.3(2.11)(115) = 1260 \text{ kips}$$

So, a W12×170 is still O.K.

c. Check the selected section

W12×170 of A992 steel. From LRFDM Table 5-3:

$$L_p = 11.4 \text{ ft}; \qquad L_r = 68.9 \text{ ft}; \qquad A = 50.0 \text{ in.}^2$$

$$\phi_b M_{px} = 1030 \text{ ft-kips}; \quad BF = 5.67 \text{ kips}; \quad \phi_b M_{py} = 463 \text{ ft-kips}$$

With equal, reverse-curvature moments at the ends $C_b = 2.27$.
As $L_p = 11.4$ ft $< L_b = 16.0$ ft $< L_r = 68.9$ ft

$$M^o_{dx} = \phi_b M_{px} - BF\,(L_b - L_p)$$

$$= 1030 - 5.67(16.0 - 11.4) = 1000 \text{ ft-kips}$$

$$M_{dx} = \min\,[C_b M^o_{dx}, \phi_b M_{px}] = \min\,[2.27(1000), 1030] = 1030 \text{ ft kips}$$

The axial load ratio, $\dfrac{P_u}{P_d} = \dfrac{600}{1372} = 0.437 > 0.2$, so use Eq. 11.9.11a (LRFDS Eq. H1-1a).

$$\frac{P_u}{P_d} + \frac{8}{9}\left[\frac{M^*_{ux}}{M_{dx}} + \frac{M^*_{uy}}{M_{dy}}\right] \le 1.0$$

$$\text{LHS} = 0.437 + \frac{8}{9}\left(\frac{264}{1030} + \frac{115}{463}\right)$$

$$= 0.886 < 1.0 \qquad \text{O.K.}$$

So, the W12×170 of A992 steel selected is adequate.   (Ans.)

Column Design (Type: *ltx, lty*; Load: *P, M$_x$, M$_y$*)

Select a W14 section of A992 steel for a beam-column in an unbraced frame with factored loads: P_u = 400 kips; symmetric single curvature moments, M_{ntx} = 150 ft-kips, M_{nty} = 50 ft-kips; and reverse curvature moments M_{ltx} = 116 ft-kips and M_{lty} = 72 ft-kips. There are no transverse loads along the span. Story height is 15 ft. K_x = 1.4 and K_y = 1.2. The allowable story drift index is 1/500, due to unfactored horizontal wind forces in the *yy* plane of 120 kips (causing bending about *xx* axis) and 82 kips in *xx* direction (causing bending about *yy* axis). Total factored gravity load above this level is ΣP_u = 7200 kips.

Solution

a.   Data

$$L \;=\; 15 \text{ ft}; \qquad L_b \;=\; L = 15.0 \text{ ft}$$
$$K_x \;=\; 1.4; \qquad K_x L_x = 1.4(15) = 21.0 \text{ ft}$$
$$K_y \;=\; 1.2; \qquad K_y L_y = 1.2(15) = 18.0 \text{ ft}$$
$$P_u \;=\; 400 \text{ kips}$$
$$M_{ntx} \;=\; 150 \text{ ft-kips}; \quad M_{nty} \;=\; 50 \text{ ft-kips; symmetric single curvature}$$

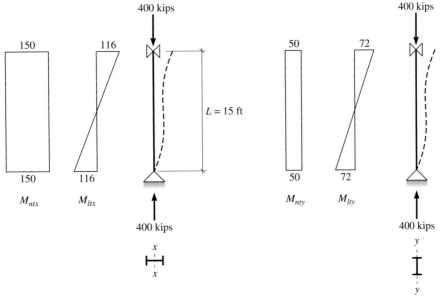

Bending about the *x* axis
(*a*)

Bending about the *y* axis
(*b*)

Figure X11.15.9

[*continues on next page*]

Example 11.15.9 continues ...

$M_{ltx} = 116$ ft-kips; $M_{lty} = 72$ ft-kips; reverse curvature
$\sum P_u = 7200$ kips

Drift index, $\dfrac{\Delta_{oh}}{L} = \dfrac{1}{500}$, corresponding to:

- a horizontal force $H = 120$ kips for bending about xx axis of the column, and
- a horizontal force $H = 82$ kips for bending about yy axis.

$$B_{2x} = \cfrac{1}{1 - \dfrac{\sum P_u}{\sum H}\left(\dfrac{\Delta_{oh}}{L}\right)} = \cfrac{1}{1 - \dfrac{7200}{120}\left(\dfrac{1}{500}\right)} = 1.14$$

$$B_{2y} = \cfrac{1}{1 - \dfrac{\sum P_u}{\sum H}\left(\dfrac{\Delta_{oh}}{L}\right)} = \cfrac{1}{1 - \dfrac{7200}{82.0}\left(\dfrac{1}{500}\right)} = 1.21$$

b. Preliminary selection
Assume $B_{1x} = B_{1y} = 1.0$. From LRFDS Eq. C1-1:

$$M^*_{ux} = B_{1x}M_{ntx} + B_{2x}M_{ltx} = 1.0(150) + 1.14(116) = 282 \text{ ft-kips}$$

$$M^*_{uy} = B_{1y}M_{nty} + B_{2y}M_{lty} = 1.0(50.0) + 1.21(72.0) = 137 \text{ ft-kips}$$

From Table 11.14.1, for a W14 column of $F_y = 50$ ksi steel and $KL = K_yL_y = 18$ ft, the coefficient $m = 1.3$. Assume $u = 2.0$.

$$P_{ueq} = 400 + 1.3(282) + 1.3(2.0)(137) = 1120 \text{ kips}$$

By interpolation of the tabulated values in the column load table for W-shapes (LRFDM Table 4-2), corresponding to $KL = K_yL_y = 18$ ft, we observe that a W14×120 has $P_d = 1180$ kips ($> P_{req} = 1120$ kips). This selection has $r_x/r_y = 1.67$, resulting in:

$$(K_xL_x)_y = \frac{K_xL_x}{r_x/r_y} = \frac{21.0}{1.67} = 12.6 \text{ ft} < K_yL_y = 18 \text{ ft}$$

indicating that x-axis buckling will not control the design. Also, $u = 1.99$ for a W14×120, from Table 11.14.2. The revised value of:

$$P_{ueq} = 400 + 1.3(282) + 1.3(1.99)(137) = 1120 \text{ kips}$$

Still select a W14×120.

c. Check selected section:
From LRFDM Table 4-2, for a W14×120 column with $F_y = 50$ ksi, $KL = K_yL_y = 18$ ft,

$$P_d = 1180 \text{ kips;} \qquad\qquad A = 35.3 \text{ in.}^2$$

$$P_{ex} (KL)^2/10^4 = 39,500; \qquad P_{ey} (KL)^2/10^4 = 14,200$$

The magnification factors B_{1x} and B_{1y} are determined, using conservatively, $(KL)_{ntx} = L = 15$ ft and $(KL)_{nty} = L = 15$ ft. Thus,

$$P_{e1x} = \frac{39{,}500 \times 10^4}{(180)^2} = 12{,}200 \text{ kips}; \quad P_{e1y} = \frac{14{,}200 \times 10^4}{(180)^2}$$
$$= 4380 \text{ kips}$$

For the symmetric, single-curvature, first-order, nt moments given, $C_{mx} = 1.0$ and $C_{my} = 1.0$. Thus,

$$B_{1x} = \max\left[\frac{C_{mx}}{1 - \dfrac{P_u}{P_{e1x}}}, 1.0\right] = \max\left[\frac{1.0}{1 - \dfrac{400}{12{,}200}}, 1.0\right] = 1.03$$

$$B_{1y} = \max\left[\frac{C_{my}}{1 - \dfrac{P_u}{P_{e1y}}}, 1.0\right] = \max\left[\frac{1.0}{1 - \dfrac{400}{4380}}, 1.0\right] = 1.10$$

The second-order factored moments are:

$$M_{ux}^* = 1.03(150) + 1.14(116) = 287 \text{ kips}$$
$$M_{uy}^* = 1.10(50.0) + 1.21(72.0) = 142 \text{ kips}$$

From LRFDM Table 5-3, for a W14×120 of A992 steel, $\phi_b M_{px} = 795$ ft-kips; $L_p = 13.2$ ft, $\phi_b M_{py} = 380$ ft-kips.
We will conservatively assume that $C_b = 1.0$.
From beam selection plots (LRFDM Table 5-5), for a W14×120 beam segment with $L_b = 15$ ft and $C_b = 1.0$, $M_{dx} = 783$ ft-kips.

$$M_{dy} = \phi_b M_{py} = 380 \text{ ft-kips}$$

As the axial load ratio, $\dfrac{P_u}{P_d} = \dfrac{400}{1180} = 0.339 > 0.2$, we use Eq. 11.9.11a

(LRFDS Eq. H1-1a). Substituting the various terms in the interaction formula , we obtain:

$$\text{LHS} = 0.339 + \frac{8}{9}\left(\frac{287}{783} + \frac{142}{380}\right) = 0.997 \qquad \text{O.K.}$$

Select a W14×120 of A992 steel. (Ans.)

Design of Crane Column (Type: *ntx, nty*; Load: *P, M_x*)

EXAMPLE 11.15.10

Design an interior crane column for a mill building. The column of A992 steel, 28-ft high, carries two crane girders in addition to the roof load. The crane runway on each side of the column is to be supported by means of a

[*continues on next page*]

Example 11.15.10 continues ...

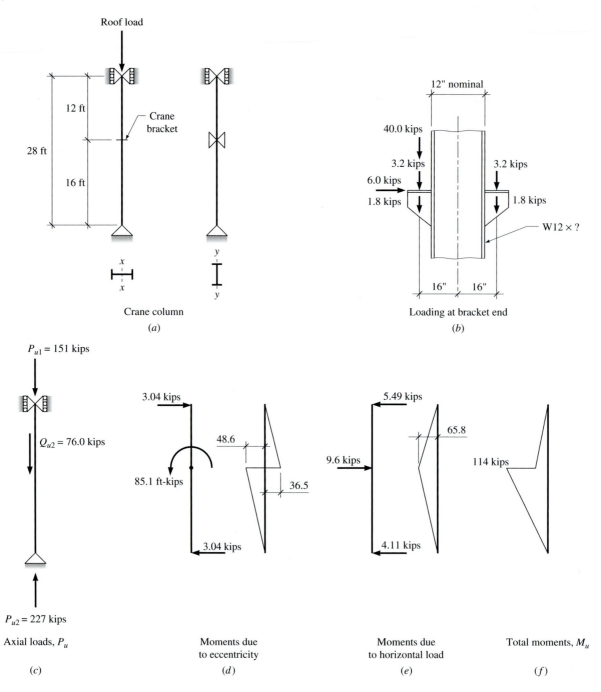

Roof load

12 ft

Crane
bracket

28 ft

16 ft

x

x

y

y

Crane column

(a)

12" nominal

40.0 kips

3.2 kips   3.2 kips

6.0 kips

1.8 kips   1.8 kips

W12 × ?

16"   16"

Loading at bracket end

(b)

P_{u1} = 151 kips

Q_{u2} = 76.0 kips

P_{u2} = 227 kips

Axial loads, P_u

(c)

3.04 kips

48.6

85.1 ft-kips

36.5

3.04 kips

Moments due
to eccentricity

(d)

5.49 kips

65.8

9.6 kips

114 kips

4.11 kips

Moments due
to horizontal load

(e)

Total moments, M_u

(f)

Figure X11.15.10

bracket welded to the flange of the column at a distance 12-ft below the top. The roof covering including decking, insulation, joists, truss, and piping is 30 psf of the roof surface, and the snow load is assumed to be 30 psf. The contributory area for the column under consideration is 1800 sq. ft. Maximum vertical load from wheels, including the impact of wheels, is given as 40 kips. The horizontal load from the crane wheels is 6 kips. The weight of the crane girder and rails is 3.2 kips, and the weight of each bracket is estimated as 1.8 kips. The eccentricity of the crane loads is 16 in. Select a suitable W12-shape. Assume the column is pinned at both ends about both axes and laterally supported at bracket level by the crane girders.

Solution

a. Factored loads
Roof load $= 1.2(30) + 1.6(30) = 84$ psf

Column axial load at roof level, $P_{u1} = \dfrac{84(1800)}{1000} = 151$ kips

Vertical load on left bracket $= 1.2(3.2 + 1.8) + 1.6(40) = 70.0$ kips
Vertical load on right bracket $= 1.2(3.2 + 1.8) = 6.00$ kips
By transferring these loads to the column center, we obtain:

Additional axial load at bracket level, $Q_{u2} = 70.0 + 6.00 = 76.0$ kips
Moment from eccentricity of crane loads $= 1.33(70.0 - 6.00) = 85.1$ ft-kips which produces a horizontal reaction at the column base of (Fig. X11.15.10):

$$A_1 = \frac{85.1}{28.0} = 3.04 \text{ kips}$$

and a bending moment just below the bracket of:
$M_{u1} = 3.04 \times 16 = 48.6$ ft-kips
The factored horizontal load from the crane girder, $H_u = 1.6(6.00) = 9.60$ kips

This horizontal force produces a reaction at the base, $A_2 = \dfrac{9.60(12)}{28} = 4.11$ kips

and a bending moment at bracket level, $M_{u2} = 4.11(16) = 65.8$ ft-kips
The net (maximum) factored bending moment,

$$M_u = 48.6 + 65.8 = 114 \text{ ft-kips}$$

The factored axial load in the bottom part of the column is:

$$P_u = 151 + 76.0 = 227 \text{ kips}$$

The column will be designed conservatively, for an axial load P_u of 228 kips applied at the roof level and a bending moment, $M_{ltx} = 114$ ft-kips. We thus have:

$K_x L_x = 28$ ft; $(K_y L_y)_1 = 16$ ft; $(K_y L_y)_2 = 12$ ft; $K_y L_y = 16$ ft

[*continues on next page*]

Example 11.15.10 continues ...

b. Preliminary member selection
The equivalent axial load on the beam-column is:

$$P_{ueq} = P_u + mM_{ux}^*$$

Assume minor-axis buckling controls the design axial strength. For a W12 column with $KL = K_y L_y = 16$ ft, $m = 1.5$ from Table 11.14.1. Assume $B_{1x} = 1.1$. So,

$$P_{ueq} = 228 + 1.5(1.1)(114) = 416 \text{ kips}$$

Using the column load table for W-shapes, select a W12×53, which for $KL = 16$ ft has $P_d = 428$ kips ($> P_{ueq} = 416$ kips) and $r_x/r_y = 2.11$. We have:

$$(K_x L_x)_y = \frac{K_x L_x}{(r_x/r_y)} = \frac{28.0}{2.11} = 13.3 \text{ ft} < K_y L_y = 16 \text{ ft}$$

as assumed. So, try W12×53:

$$A = 15.6 \text{ in.}^2; \qquad r_x = 5.23 \text{ in.}; \qquad r_y = 2.48 \text{ in.}$$

c. Column action

$$\frac{K_x L_x}{r_x} = \frac{28.0(12)}{5.23} = 64.2; \qquad \frac{K_y L_y}{r_y} = \frac{16.0(12)}{2.48} = 77.4 \leftarrow \text{controls}$$

From LRFDS Table 3-50, for $\dfrac{KL}{r} = 77.4$, $F_{dc} = 27.4$ ksi giving $P_d = 27.4(15.6) = 428$ kips and an axial load ratio, $\dfrac{P_u}{P_d} = \dfrac{228}{428} = 0.533 > 0.20$

So, the interaction formula to be checked is Eq. 11.9.11a (LRFDS Eq. H1.1a).

$$\frac{P_u}{P_d} + \frac{8}{9} \frac{M_{ux}^*}{M_{dx}} \leq 1.0$$

d. Beam action
From LRFDM Table 5-3, for a W12×53 beam of A992 steel:

$$L_p = 8.76 \text{ ft}; \qquad \phi_b M_{px} = 292 \text{ ft-kips}; \qquad BF = 4.78 \text{ kips}$$
$$L_r = 25.6 \text{ ft}; \qquad \phi_b M_{py} = 108 \text{ ft-kips}$$

The lower segment, which is longer and subjected to heavier axial load and higher maximum bending moment, is more critical. As $L_p < L_b = 16$ ft $< L_r$,

$$M_d = \min\{C_b[\phi_b M_{px} - BF(L_b - L_p)], \phi_b M_{px}\}$$

The variation of bending moment over the segment is linear, and as

$$\frac{M_1}{M_2} = \frac{0}{114} = 0, \; C_b = 1.67 \text{ from Table 10.4.1, and}$$

$$M_d = \min \{1.67[292 - 4.78(16 - 8.76)], 292\} = 292 \text{ ft-kips}$$

$$M^*_{ux} = B_{1x} M_{ntx}$$
with

$$B_{1x} = \frac{C_{mx}}{1 - \dfrac{P_u}{P_{elx}}} = \frac{1 - 0.2 \dfrac{P_u}{P_{elx}}}{1 - \dfrac{P_u}{P_{elx}}}$$

The moment reduction factor C_{mx} is conservatively obtained from LRFDC Table C-C1.1 corresponding to a pin-ended column with a central concentrated load. As the bending is about the major axis,

$$P_{elx} = \frac{\pi^2 EA}{(KL/r)^2_{ntx}} = \frac{\pi^2(29,000)(15.6)}{(64.2)^2} = 1080 \text{ kips}$$

$$B_{1x} = \frac{1 - 0.2\left(\dfrac{228}{1080}\right)}{1 - \dfrac{228}{1080}} = 1.21$$

$$M^*_{ux} = 1.21(114) = 138 \text{ ft-kips}$$

e. **Limit state of strength**

$$\frac{P_u}{P_d} + \frac{8}{9}\frac{M^*_{ux}}{M_{dx}} = 0.533 + \frac{8}{9}\left(\frac{138}{292}\right) = 0.533 + 0.420$$

$$= 0.953 < 1.0 \quad \text{O.K.}$$

So, select a W12×53 of A992 steel. (Ans.)

Design of a Portal Frame

EXAMPLE 11.15.11

Design the members of the hinged-base, rigid jointed portal frame ABCD shown in Fig. 5.1.2, studied in Example 5.3.1. The girder BC is subjected to a uniformly distributed service load of 2.5 klf (1 klf D + 1.5 klf S). In addition, the frame is subjected to a concentrated dead load of 30 kips at the column tops B and C. The wind load on the frame consists of a 15-kip horizontal force acting at the joint B. The dead loads given include provision for self-weight of members. Use A992 W-shapes with their webs in the plane of the frame.

[continues on next page]

Example 11.15.11 continues ...

The columns are braced at the top and bottom against y-axis displacement and at midheight against y-axis buckling. Lateral bracing for the girder is provided at the ends B and C, and at the quarter points F, E, and G. Limit the drift to $h/250$ and the live load deflection to $L/360$.

Solution

a. Data

Span, $L = 60$ ft

height, $h = 20$ ft

Assume the moment of inertia of girder BC to be twice that of the columns.

$$I_g = 2\, I_c$$

The internal forces in the members under the factored loads of the three load combinations (namely, LC-1, LC-2, and LC-3) were determined in Example 5.3.1. The results for column CD are summarized in Table X5.3.1.*b*. Observe that the load combination LC-3 controls the design of the columns.

b. Design of column CD (Fig. 5.1.2)

Length, $L = 20$ ft; $L_x = 20$ ft; $L_y = 10$ ft

For x-axis buckling:

$$G_C = \frac{I_c/h}{I_g/L} = 1.5; \quad G_D = 10.0 \text{ (hinged base)}$$

For these G-factors, the monograph for sidesway permitted case (Fig. 8.5.6*b*) yields $K_x = 2.0$. For y-axis buckling, the column is part of a braced structure, pinned at both ends and provided with a brace at mid-height. So,

$$K_y L_y = 1.0(10) = 10.0 \text{ ft}; \qquad K_x L_x = 2.0(20) = 40.0 \text{ ft}$$

Factored axial load on the column, $P_u = 148$ kips

There are no bending moments about the minor axis, $M_{uy}^* = 0$

The vertical component of the load combination LC-3, namely, symmetric gravity load acting on a symmetric structure, produces M_{nt} moments in the columns (see Figs. 5.1.2 and 11.9.1).

Further, the horizontal component of the load combination (wind load) produces M_{lt} moments. Thus, for column CD, we obtain from Table X5.3.1*b*:

$$M_{ntx} = 748 \text{ ft-kips}, \qquad M_{ltx} = 120 \text{ ft-kips}$$

Factored second-order major-axis moment, $M_{ux}^* = B_{1x} M_{ntx} + B_{2x} M_{ltx}$

Assume $B_{1x} = 1.05$, $B_{2x} = 1.20$. This results in:

$$M_{ux}^* = 1.05(748) + 1.20(120) = 929 \text{ ft-kips}$$

As the axial load is relatively light, we select a W14-shape for bending and then check it for interaction. The total first-order moment at C is

868 ft-kips. From LRFDM Table 5-3, the lightest W14-shape capable of resisting this moment is a W14×132 with $\phi_b M_{px} = 878$ ft-kips. In view of the simultaneously acting axial load on the column, let us tentatively select a W14×145. From LRFDM Table 5-3, for a W14×145:

$$\phi_b M_{px} = 975 \text{ ft-kips}; \quad L_p = 14.1 \text{ ft}; \quad I_x = 1710 \text{ in.}^4$$

As the unbraced length $L_b = 10$ ft $< L_p$, the design major-axis bending strength,

$$M_{dx} = \phi_b M_{px} = 975 \text{ ft-kips}$$

The upper segment of the column with larger end moments is more critical. From LRFDM Table 4-2, for a W14×145:

$$\frac{r_x}{r_y} = 1.59 \rightarrow (K_x L_x)_y = \frac{40}{1.59} = 25.2 \text{ ft}$$

and, from the same table, for $KL = 25.2$ ft, $P_d = 1190$ kips. Also,

$$P_{ex} (KL)^2 = 48900 \times 10^4$$

With $G_C = 1.5$ and $G_D = 10$, the alignment chart for sidesway prevented columns (Fig. 8.5.6a) gives $K_{ntx} = 0.88$. So,

$$P_{elx} = \frac{48{,}900 \times 10^4}{[0.88(20)(12)]^2} = 11{,}000 \text{ kips}$$

$$r_M = 0; \quad C_{mx} = 0.6 - 0.4 r_M = 0.6$$

$$B_{1x} = \max \left[\frac{C_{mx}}{1 - \dfrac{P_u}{P_{elx}}}, 1.0 \right] = \max \left[\frac{0.6}{1 - \dfrac{148}{11{,}000}}, 1.0 \right] = 1.0$$

$$P_{e2x} = \frac{48{,}900 \times 10^4}{[2.0(20)(12)]^2} = 2120 \text{ kips}$$

$$B_{2x} = \frac{1}{1 - \dfrac{\Sigma P_u}{\Sigma P_{e2x}}} = \frac{1}{1 - \dfrac{2(144)}{2(2120)}} = 1.07$$

The factored second-order major-axis moment,

$$M_{ux}^* = 1.0(748) + 1.07(120) = 876 \text{ ft-kips}$$

$$\frac{P_u}{P_d} = \frac{148}{1190} = 0.124 < 0.20 \quad \text{Use LRFDS Eq. H1-1}b:$$

$$\frac{1}{2} \frac{P_u}{P_d} + \frac{M_{ux}^*}{M_{dx}} = \frac{0.124}{2} + \frac{876}{975} = 0.96 < 1.0 \quad \text{O.K.}$$

[*continues on next page*]

Example 11.15.11 continues ...

The sidesway of the frame under a unit horizontal load at joint B (as calculated in Example 5.3.1): $\Delta_{(1)} = \dfrac{106}{I_c}$ in.

Drift under service wind load of 15 kips,

$$\Delta = \Delta_{(1)} Q_W = \frac{105.9}{I_c} Q_W = \frac{106}{1710}(15) = 0.930 \text{ in.}$$

$$\Delta_{all} = \frac{h}{250} = \frac{20(12)}{250} = 0.960 > 0.930 \text{ in.} \quad \text{O.K.}$$

So, the W14×145 shape selected is adequate for the column.

c. Design of girder BC (Fig. 5.1.2)

Moment at B and C under a factored vertical load of 3.6 klf from Table X5.3.1*b* is −748 ft-kips.

Moment at E under a factored vertical load of 3.6 klf is +242.3(3.6) = +872 ft-kips

Moment at F under a factored vertical load of 3.6 klf, noting that the shear at E is zero for the symmetrical load considered, is:

$$M_F = 872 - 3.6(15)\left(\frac{15}{2}\right) = +467 \text{ ft-kips}$$

Moment at F due to the factored horizontal load $= +\dfrac{120}{2} = +60$ ft-kips

Total moment at F due to vertical and horizontal loads = +527 ft-kips
Total moment at E due to vertical and horizontal loads = 872 + 0 = +872 ft-kips
Unbraced length of the segment EF, $L_b = 15$ ft
Variation of the total moment is assumed to be linear over the unbraced length FE.

Moment ratio, $r_M = -\dfrac{527}{872} = -0.60$

From Table 10.4.1, $C_b = 1.19$
Maximum moment over the segment, $M_u = 872$ ft-kips. Hence,

$$M_{u(C_b = 1)} = \frac{M_u}{C_b} = \frac{872}{1.19} = 733 \text{ ft-kips}$$

Entering the beam design plots (LRFDM Table 5-4) with $L_b = 15$ ft and $M_{req} = 733$ ft-kips, we observe that a W27×84 is the lightest section that satisfies this requirement. This section, which has a $\phi_b M_{px}$ value of 918 ft-kips > $M_u = 872$ ft-kips, is adequate for flexural strength.

The maximum vertical deflection under the uniformly distributed gravity load occurs at the midspan, E, of the girder. For a unit distributed load of 1 klf this value is calculated in Example 5.3.1, as $\delta_{E(1)} = \dfrac{4486}{I_g}$ in.

Maximum vertical deflection under the nominal snow load of 1.5 klf, is given by:

$$\delta_E = \delta_{(1)} q_S = \frac{4490}{I_g}(1.5) = \frac{6730}{I_g} \text{ in.}$$

Allowable deflection, $\delta_{all} = \dfrac{L}{360} = \dfrac{60(12)}{360} = 2.00 \text{ in.}$

The serviceability limit state for deflection becomes:

$$\frac{6730}{I_g} \le 2.00 \rightarrow I_g \ge 3370 \text{ in.}^4$$

So, the W27×84 with an I_x of 2850 in.[4] $<$ 3370 in.[4] is not adequate for serviceability. From LRFDM Table 5-2, it is seen that a W30×90 with $I_x = 3610$ in.[4] is the lightest section that satisfies the serviceability requirement. So, select a W30×90 for the beam BC. With a W14×145 selected for the columns and a W30×90 for the beam, we have:

$$\frac{I_g}{I_c} = \frac{3610}{1710} = 2.11$$

As this value is quite close to the value of 2 assumed to determine the internal forces, no redesign is necessary. Deflection of the beam under service dead load of 1 klf is:

$$\delta_D = \frac{6730}{3620}(1.0) = 1.86 \text{ in.}$$

So, specify a camber of 2 in. for the beam.

To summarize, select a W14×145 for the columns, a W30×90 for the beam, and provide a camber of 2 in. for the beam. As the roof is a flat roof, the member sizes selected have to be checked for possible ponding failure. The design of the corner connections is considered in Example X13.10.1.

11.16   Biaxial Bending and Asymmetrical Bending of Beams

11.16.1   Biaxial Bending

Simple bending, studied in Chapters 9 and 10, occurs for prismatic beams of any cross-sectional shape, as long as the plane of loading is parallel to a principal plane and produces no torsion. When the loading plane of a beam does not coincide with either of the principal planes (as in Fig. 11.16.1), and the loading causes no torsion, bending occurs along both principal axes and the beam is said to be in *biaxial bending.*

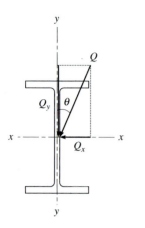

Figure 11.16.1: Beam subjected to biaxial bending.

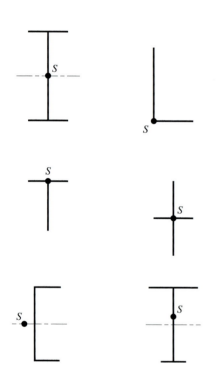

Figure 11.16.2: Shear centers of some cross sections.

The condition that no twisting takes place for simple and biaxial bending of beams requires that the line of action of an applied load must pass through the shear center of the cross section. The location of the shear center for several common cross sections is shown in Fig. 11.16.2, where the shear center is indicated by an S. For doubly symmetric sections, the shear center coincides with the centroid. For cross sections composed of rectangular plates meeting at a common point (e.g., L- and T-shapes), the shear center is at this point of intersection. For singly symmetric sections, such as C-shapes, the shear center lies on the axis of symmetry. Formulas for the shear-center location of other shapes and procedures are given in references.

Figure 11.16.1 shows the section of a rolled steel beam which is subjected to an inclined load Q. The load Q is located in a plane perpendicular to the longitudinal axis of the member and passes through the shear center of the cross section. It makes an angle θ with the vertical plane of symmetry. The load Q may be resolved into two components parallel to the principal axes, yy and xx, so that:

$$Q_y = Q \cos \theta; \quad Q_x = Q \sin \theta \qquad (11.16.1)$$

The component Q_y produces bending about the x axis, while the component Q_x produces bending about the y axis. Thus the beam is subjected to biaxial bending. Let M_{ux} and M_{uy} be the maximum bending moments about the x axis and y axis, respectively, under factored loads.

When a member is subjected to biaxial bending, the interaction formula given in LRFDS Eq. H1-1b, with the axial load term equated to zero, must be satisfied. The interaction formula then reduces to:

$$\frac{M_{ux}}{\phi_b M_{nx}} + \frac{M_{uy}}{\phi_b M_{ny}} = 0 \rightarrow \frac{M_{ux}}{M_{dx}} + \frac{M_{uy}}{M_{dy}} = 0 \qquad (11.16.2)$$

where
M_{ux} = factored bending moment about the x axis
M_{uy} = factored bending moment about the y axis
M_{nx} = nominal bending strength about the x axis
M_{ny} = nominal bending strength about the y axis
M_{dx} = design bending strength about the x axis
M_{dy} = design bending strength about the y axis
ϕ_b = resistance factor (=0.9)

For preliminary member selection, the procedure developed by Aminmansour [2000] discussed in Section 11.14.2 could be used. Noting that for biaxially bent beams the axial load $P_u = 0$, the interaction equation for biaxially loaded beam-columns with low axial load ratio (Eq. 11.14.10b) reduces to:

$$\frac{9}{8} m M_{ux} + \frac{9}{8} n M_{uy} \leq 1.0 \qquad (11.16.3)$$

Values of the parameters m and n are given in LRFDM Table 6-2, for a large number of W-shapes for a range of unbraced lengths (L_b) for steels with $F_y = 50$ ksi.

Analysis of a Biaxially Bent Beam

EXAMPLE 11.16.1

A W24×68 of A992 steel is used as a simply supported beam with regard to both principal axes. It has a span of 16 ft. Lateral braces are provided only at the supports. The beam weight is negligible, and the only load is 60 kips concentrated at midspan. The load is located in a plane perpendicular to the longitudinal axis of the member and passes through the shear center of the cross section. The load makes an angle of $\theta = 16.7°$ with the vertical plane of symmetry. Check the adequacy of the beam as per LRFD Specification.

Solution
a. Data

Span, $L = 16$ ft; Unbraced length, $L_b = 16$ ft
Factored load, $Q_u = 60$ kips; Inclination, $\theta = 16.7°$
Vertical component, $Q_{uy} = 60 \cos 16.7° = 57.5$ kips
Horizontal component, $Q_{ux} = 60 \sin 16.7° = 17.2$ kips
Maximum major-axis moment,

$$M_{x,\max} = \frac{Q_{uy}L}{4} = \frac{57.5(16)}{4} = 230 \text{ ft-kips}$$

Maximum minor-axis moment,

$$M_{y,\max} = \frac{Q_{ux}L}{4} = \frac{17.2(16)}{4} = 68.8 \text{ ft-kips}$$

From LRFDM Table 5-2, for a W24×68 of $F_y = 50$ ksi steel, $\phi_b M_{py} = 88.3$ ft-kips.
From beam selection plots for W-shapes (LRFDM Table 5-5), for a W24 ×68 of $F_y = 50$ ksi steel, $C_b = 1.0$ and $L_b = 16$ ft; $M_{dx}^o = 488$ ft-kips. A simply supported beam under a central concentrated load and laterally braced at the ends only has a C_b value of 1.32 from Table 5-1 of the LRFDM. As the unbraced length, L_b, is greater than L_p, the design major-axis bending strength of the member is:

$$M_{dx} = \min [C_b M_{dx}^o, \phi_b M_{px}] = \min [1.32(488), 664] = 644 \text{ ft-kips}$$

Design minor axis bending strength of the member, $M_{dy} = \phi_b M_{py} = 88.3$ ft-kips

$$\frac{M_{ux}}{M_{dx}} + \frac{M_{uy}}{M_{dy}} = \frac{230}{644} + \frac{68.8}{88.3} = 0.357 + 0.779 = 1.14 > 1.00$$

So, the W24×68 is inadequate.   (Ans.)

EXAMPLE 11.16.2   **Design of a Biaxially Bent Beam**

Select the lightest W10 section of A992 steel to be used in an inclined position such that the plane of loading makes an angle of 30° with the plane of
the web. The beam has lateral support only at the ends of the 20-ft simple
span. The uniform gravity load is 0.15 klf dead load and 0.4 klf snow load.
The dead load includes provision for self-weight of the beam.

Solution

a. Data
    Length of beam, $L = 20$ ft; Unbraced length, $L_b = 20$ ft
    Dead load, $D = 0.15$ klf; Live load, $L = 0.40$ klf
    Factored load, $q_u = 1.2(0.15) + 1.6(0.40) = 0.82$ klf
    Inclination of the plane of loading with the plane of the web, $\theta = 30°$
    Component, $q_{uy} = 0.82 \cos 30° = 0.71$ klf
    Component, $q_{ux} = 0.82 \sin 30° = 0.41$ klf

$$\text{Maximum major-axis moment, } M_{ux} = \frac{q_{uy} L^2}{8} = 0.71 \times \frac{20^2}{8} = 35.5$$

ft-kips

$$\text{Maximum minor-axis moment, } M_{uy} = \frac{q_{ux} L^2}{8} = 0.41\left(\frac{20^2}{8}\right) = 20.5 \text{ ft-kips}$$

b. Preliminary selection
    From Table 6-1, for W10-shapes, the median value of m is 13.0×10^{-3} ft-kips, which when substituted in Eq. 11.16.3 results in:

$$\frac{9}{8}[m(35.5) + (13.0 \times 10^{-3})(20.5)] = 1.0 \rightarrow m = 17.5 \times 10^{-3}$$

From LRFDM Table 6-2, for a W10×30 having an unbraced length $L_b = 20$ ft,

$$m = 13.7 \times 10^{-3} < 17.5 \times 10^{-3} \quad \text{O.K.}$$

$$n = 27.5 \times 10^{-3}$$

$$\text{LHS} = \frac{9}{8}(13.7 \times 10^{-3})(35.5) + \frac{9}{8}(27.5 \times 10^{-3})(20.5)$$

$$= 0.547 + 0.634 = 1.18 > 1.0 \quad \text{N.G.}$$

Consider the next heavier section: a W10×33 with $m = 8.61 \times 10^{-3}$ and
$n = 17.2 \times 10^{-3}$.

$$\text{LHS} = \frac{9}{8}(8.61 \times 10^{-3})(35.5) + \frac{9}{8}(17.2 \times 10^{-3})(20.5)$$

$$= 0.344 + 0.397 = 0.741 < 1.0 \quad \text{O.K.}$$

So, select a W10×33 and check.

c. Check W10×33

From Eq. 11.14.12, $M_{dx}^o = \dfrac{8}{9}\left(\dfrac{1}{m}\right) = \dfrac{8}{9}\left(\dfrac{1}{8.61 \times 10^{-3}}\right) = 103$ ft-kips

From Fig. 10.4.3, $C_b = 1.14$ for a simply supported beam with lateral supports at the ends only under uniformly distributed loads.
From LRFDM Table 6-2, for a W10×33,

$L_p = 6.85$ ft; $\phi_b M_{px} = 146$ ft-kips; $\phi_b M_{py} = 51.8$ ft-kips

resulting in:

$M_{dx} = \min [C_b M_{dx}^o; \phi_b M_{px}] = \min [1.14(103), 146] = 117$ ft-kips

$\dfrac{M_{ux}}{M_{dx}} + \dfrac{M_{uy}}{M_{dy}} = \dfrac{35.5}{117} + \dfrac{20.5}{51.8} = 0.303 + 0.396 = 0.699$ O.K.

So, select a W10×33 of A992 steel.   (Ans.)

11.16.2   Asymmetrical Bending

When bending takes place about one or both principal axes of a beam simultaneously with torsion, the beam is said to be in *asymmetrical bending.* Different types of asymmetrical bending problems are encountered in practice.

- The loads make an angle with the principal axes and are applied at the midwidth of the top flange of the beam.

  An example is that of the purlin resting on the inclined top chord of a roof truss. Here the loads are vertical and the principal axes of the beam are inclined (Fig. 11.16.3a). The vertical load Q_V may be resolved into components Q_y and Q_x parallel to the y and x axis respectively using Eq. 11.16.1. The component Q_y passes through the shear center and produces bending only. The component Q_x produces, in addition to bending, a torsional moment $M_T = (Q_x d)/2$, where d is the depth of the beam. The equivalent load system will, therefore, consist of forces Q_y, Q_x, and a torsional moment M_T, applied at the shear center.
- The loads are applied in two different directions at the top flange, one lying in the plane of the y axis and the other at right angles to it (Fig. 11.16.3b).

  An example is that of the crane runway girder in industrial buildings, the horizontal load being due to the lateral thrust from the

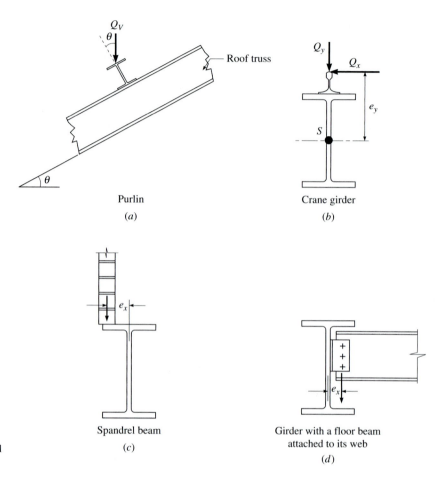

Figure 11.16.3: Beams subjected to asymmetrical bending.

moving crane. Again, the component Q_y passes through the shear center and produces bending only. The component Q_x produces, in addition to bending a torsional moment, $M_T = (Q_x\, e_y)$, where e_y is the distance from the shear center of the girder to the top of the rail. The equivalent load system will again consist of forces Q_y, Q_x, and a torsional moment M_T, applied at the shear center.

• The load consists of an eccentric vertical concentrated load.

  An example is a spandrel beam for which the dead load from the wall (or cladding) is applied eccentrically (Fig. 11.16.3c). Another example is that of a girder with floor beams attached to one side of its web (Fig. 11.16.3d). In both these cases, the applied loads Q_y do not pass through the shear centers. In these cases, the member is subjected to a torsional moment, $M_T = Q_y\, e_x$, in addition to major-axis bending.

The equivalent force system will consist of force, Q_y, and a torsional moment, M_T, applied at the shear center.

In all these cases the member sustains torsion in addition to flexure, and thus is under asymmetrical bending. A detailed treatment of this topic can be found in *Torsional Analysis of Structural Steel Members* [AISC, 1997].

Simplified procedures, wherein the applied torsion is replaced by bending of flange or flanges, are often used in practice. Thus, if torsion is caused by an applied eccentric horizontal force (Figs. 11.16.3a and b), then this horizontal force is assumed to cause bending only in the flange nearest to it. In the case of a crane girder or a purlin, the upper flange alone has to resist a horizontal force,

$$Q_f = \frac{M_T}{y_S} \qquad (11.16.4)$$

where y_S is the distance from the shear center, S, to the center of the upper flange ($= (d - t_f)/2$ for rolled I-shapes). Thus, the relatively thin web and the bottom flange are disregarded in resisting the applied horizontal force (Fig. 11.16.4a).

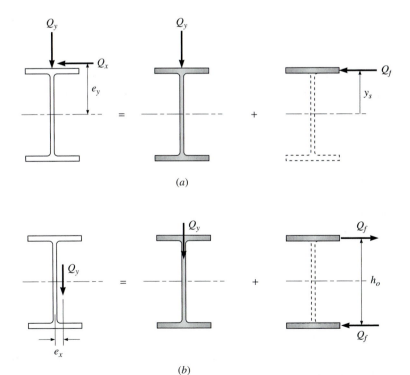

Figure 11.16.4: Design simplifications for asymmetrical bending.

In cases where the torsion is caused by an applied vertical eccentric load (Fig. 11.16.3c and d), the twisting moment is replaced by two equal and opposite flange forces,

$$Q_f = \frac{M_T}{h_o} \tag{11.16.5}$$

where h_o is the center-to-center distance between the flanges. Apart from the vertical force acting on the section, each flange is exposed to bending by the force, Q_f, with such bending occurring in the plane of that flange. Thus, the relatively thin web is disregarded in resisting the torsional moment (Fig. 11.16.4b). These simplified procedures are sufficiently accurate for design purposes, if the applied torsion is not excessive.

The interaction formula (Eq. 11.16.2) may now be rewritten as:

$$\frac{M_{ux}}{M_{dx}} + \frac{M_{uf}}{0.5M_{dy}} = 0 \;\rightarrow\; \frac{M_{ux}}{M_{dx}} + \frac{2M_{uf}}{M_{dy}} = 0 \tag{11.16.6}$$

where M_{ux} = factored bending moment about the x axis
M_{uf} = factored bending moment about the y axis of the flange
M_{dx} = design bending strength of the section about the x axis
M_{dy} = design bending strength of the section about the y axis

Purlins

As discussed in Section 3.4, purlins are flexural members supporting the roof decks of industrial buildings, and span from one bent to the other. When used on sloping roofs, purlins are subjected to asymmetrical bending. As the sections used for purlins (usually C-, Z-, or S-shapes) are weak about their web axes, sag rods may be necessary to reduce the span lengths for bending about those axes. Sag rods are placed at midspan or at the third-points, depending on the span of the purlins, the weight of the roof, and the pitch of the roof truss. The sag rod should be connected near the top flange (at minimum gage distance below the top, usually). The sag rod serves as an intermediate support for the purlin in its weaker direction. Thus, it supports the tangential component of the gravity load applied at the top flange of the purlin. Sag rods will also act as lateral bracing with respect to major axis bending of the purlin.

If the purlins are simply supported at the trusses and no sag rods are provided, the maximum bending moments for which the purlin is designed are:

$$M_{ux} = \frac{1}{8} q_{uy} L^2; \quad M_{uf} = \frac{1}{8} q_{ux} L^2 \tag{11.16.7}$$

If the purlins are simply supported at the trusses and a single line of sag rods is provided, the purlin acts as a beam continuous over two equal spans of $L/2$ for bending about the y axis. The maximum bending moment occurs at the support (location of sag rod). The maximum bending moments for which the purlin is designed are:

$$M_{ux} = \frac{1}{8} q_{uy} L^2; \quad M_{uf} = \frac{1}{32} q_{ux} L^2 \qquad (11.16.8)$$

If the purlins are simply supported at the trusses and two lines of sag rods are provided, the purlin acts as a beam continuous over three equal spans of $L/3$ for bending about the y axis. The maximum bending moment occurs at the support (location of sag rod). The maximum bending moments for which the purlin is designed are:

$$M_{ux} = \frac{1}{8} q_{uy} L^2; \quad M_{uf} = \frac{1}{90} q_{ux} L^2 \qquad (11.16.9)$$

Design of an Asymmetrically Bent Beam

EXAMPLE 11.16.3

A roof consists of trusses 20 ft on centers, the top chords of which make an angle of 26° 34′ with the horizontal. The purlins are 10 ft apart horizontally and carry a roof that weighs 16 psf. Assume that the snow load is 30 psf, and the wind pressure is 15 psf. Sag rods will be located at the center of each purlin. Use A36 steel and select a channel shape for the purlins.

Solution

Inclination of roof, $\theta = 26.57°$
Spacing of trusses, $L = 20$ ft
Horizontal spacing of purlins $= 10$ ft

Spacing of purlins along roof $= \dfrac{10}{\cos(26.57°)} = 11.2$ ft

Uniform service loads per foot of purlin:

$$\text{Dead load, } D = \frac{11.2(1.0)(16.0)}{1000} = 0.18 \text{ klf}$$

$$\text{Snow load, } S = \frac{10.0(1.0)(30.0)}{1000} = 0.30 \text{ klf}$$

$$\text{Wind load, } W = \frac{11.2(1.0)(15.0)}{1000} = 0.17 \text{ klf}$$

With only dead, snow, and wind loads acting on the purlin, load combination LC-3, namely, $[1.2D + 1.6S] + 0.8W$, controls the design. Of these, the dead and snow loads are vertical, while the wind pressure acts normal to the roof surface. The components of the factored loads acting

[continues on next page]

Example 11.16.3 continues ...

on the purlin are:

$$q_{uV} = 1.2D + 1.6S = 1.2(0.18) + 1.6(0.30) = 0.70 \text{ klf}$$

$$q_{uN} = 0.8W = 0.8(0.17) = 0.14 \text{ klf}$$

The components of the factored loads acting in y- and x-directions at the midwidth of the purlin are:

$$q_{uy} = q_{uV} \cos \theta + q_{uN} = 0.70 \cos (26.57°) + 0.14 = 0.77 \text{ klf}$$

$$q_{ux} = q_{uV} \sin \theta = 0.70 \sin (26.57°) = 0.31 \text{ klf}$$

Note that q_{uy} produces major-axis moment, while q_{ux} produces minor-axis moment. The purlin acts as a simple beam of span L (= 20 ft) for major-axis bending. It is laterally supported at midspan by the sag rod. So, L_b = 10 ft. With sag rods placed at the midpoint of each purlin, the purlins are two-span continuous beams with respect to weak axis bending, with total length, L = 20 ft.
The required bending strengths are:

$$M_{ux} = \frac{q_{uy} L^2}{8} = \frac{0.77(20^2)}{8} = 38.5 \text{ ft-kips}$$

$$M_{uf} = \frac{q_{ux} L^2}{32} = \frac{0.31(20^2)}{32} = 3.88 \text{ ft-kips}$$

To select a trial shape, use the beam selection plots for C-shapes (LRFDM Table 5-11), and choose a shape with a relatively large margin of strength with respect to major-axis bending. For an unbraced length, L_b = 10 ft, F_y = 36 ksi, and C_b = 1.0, an MC10×28.5 provides M_{dx}^o = 71.6 ft-kips.
Also, from LRFDM Table 5-10, for an MC10×28.5,

$$L_p = 4.83 \text{ ft}; \qquad \phi_b M_{px} = 81.0 \text{ ft-kips}$$

From Figure 10.4.3, C_b = 1.30 for a simply supported beam under a uniformly distributed load, with lateral supports at the ends and midspan only.
Design flexural strength for major-axis bending,

$$M_{dx} = \min [1.30(71.6), 81.0] = 81.0 \text{ ft-kips}$$

For C-shapes bent about their minor axis, the shape factor is greater than 1.5. So, the design flexural strength for minor-axis bending,

$$M_{dy} = \phi_b (1.5S_y) F_y = \frac{0.9(1.5)(3.99)(36)}{12} = 16.2 \text{ ft-kips}$$

Substituting in the interaction Eq. 11.16.6 results in:

$$\text{LHS} = \frac{38.5}{81.0} + \frac{2(3.88)}{16.2} = 0.475 + 0.479 = 0.954 < 1.0 \qquad \text{O.K.}$$

The MC10×28.5 shape is compact as there is no foot note in the factored uniform load tables for C-shapes (LRFDM Table 5-10).

Use a MC10×28.5   (Ans.)

References

11.1 AISC [1994]: *Manual of Steel Construction: Load and Resistance Factor Design,* vol. 1, 2nd ed., AISC, Chicago.

11.2 AISC [1997]: *Torsional Analysis of Structural Steel Members, Steel Design Series Guide 9,* by Seaburg, P. A. and Carter, C. J., AISC, Chicago.

11.3 Aminmansour, A. [2000]: "A New Approach for Design of Steel Beam-Columns," *Engineering Journal,* vol. 37, no. 2, pp. 41–72, AISC, Chicago.

11.4 ASCE [1971]: *Plastic Design in Steel—A Guide and Commentary,* Manual of Engineering Practice, ASCE, no. 41, 2nd ed.

11.5 Austin, W. J. [1961]: "Strength and Design of Metal Beam-Columns," *Journal of the Structural Division*, ASCE, vol. 87, no. ST4, pp. 1–32, April.

11.6 Campus, F., and Massonnet, C. [1956]: "Recherches sur le flambement de colonnes en acier A37 a profil en double te sollicitées obliquement" (Research on the Lateral Buckling of Obliquely Loaded A-37 Steel I-Columns), C.R. Research Report, IRSIA, Liege, vol. 1, pp. 119–338, April.

11.7 Chen, W. F. and Atsuta, T. [1977]: *Theory of Beam-Columns,* vol. 2, McGraw-Hill, NY.

11.8 Driscoll, G. C., et al. [1965]: "Plastic Design of Multi-Story Frames: Design Aids," Fritz Engineering Laboratory, Lehigh University, Department of Civil Engineering, Report No. 273.24, Summer.

11.9 Galambos, T. V. [1968]: *Structural Members and Frames,* Prentice-Hall, NY.

11.10 Galambos, T.V., ed. [1988]: *Guide to Stability Design Criteria for Metal Structures,* 4th ed., John Wiley & Sons.

11.11 Haaijer, G., and Thurlimann, B. [1958]: "On Inelastic Buckling in Steel," *Proceedings ASCE,* vol. 84, no. EM2, April.

11.12 Massonnet, Ch. [1976]: "Forty Years of Research on Beam-Columns in Steel," *Solid Mechanics Archives*, vol. 1, no. 1, pp. 27–157.

11.13 Uang, C. W., Wattar, S. W., and Leet, K. M. [1990]: "Proposed Revision of the Equivalent Axial Load Method for LRFD Steel and Composite Beam Column Design," *Engineering Journal,* AISC, vol. 27, no. 4, pp. 150–157.

11.14 Vinnakota, S. [1977]: "Finite Difference Method for Plastic Beam Columns," in *Theory of Beam-Columns,* vol. 2, by W. F. Chen and T. Atsuta, McGraw-Hill, NY, chapter 10.

11.15 Yura, J. A. [1988]: *Elements for Teaching Load and Resistance Factor Design,* AISC, Chicago, April.

▐ P R O B L E M S

P11.1. A W12×40 of A992 steel is used as a 13-ft column in a braced frame to carry an axial factored load of 160 kips. Will it be adequate to carry moments about the strong axis of 100 ft-kips at each end, bending the member in single curvature? For buckling about the minor axis consider the column to be pinned at both ends and provided with a brace at midheight.

P11.2. A W14×61 is used as a 15-ft-long column in a braced frame to carry an axial factored load of 200 kips. Determine the maximum moment that may be applied about the strong axis on the upper end when the lower end is hinged.

P11.3. Figure P11.3 shows a 15-ft-long W12×96 column with pinned ends. Two girders and a beam bring in 200 kips of axial load. A column from above delivers 90 kips at an eccentricity of 18 in. through a bracket connected to the web of the column. The structure is braced in both *xx* and *yy* planes. The loads are factored loads. Is the column adequate?

P11.4. An interior column of a building structure has floor girders framing into it at top and bottom with moment resisting connections. It must carry a factored axial load of 800 kips, including the girder and beam reactions, and the self-weight of the column. Live load imbalance in checkerboard loading causes a potential maximum moment at the top and bottom of 180 ft-kips, as shown in Figure P11.4. *K* for the weak axis is 1.0, and *K* for the strong axis is estimated as 0.9. The column is 12 ft 6 in. long. Is a W12×106 of A992 steel adequate to carry the load?

P11.5. A W14×22 tension member of A572 Grade 42 steel is subjected to a factored tensile load T_u = 60 kips and factored bending moments M_{ux} of 28 ft-kips and M_{uy}

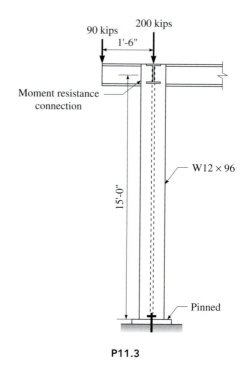

P11.3

of 6 ft-kips. Is the member satisfactory if L_b = 8.0 ft and C_b = 1.67?

P11.6. A W10×68 of A992 steel is used as an exterior column of a building structure (see Fig. 5.2.5c for a plan view). It must carry 310 kips from the girders connected to the web, plus a 100-kip reaction from the floor beam that frames into the column flange with an eccentricity of 8 in. The column is 15-ft long and is assumed to be pinned about both axes at both ends. The structure is braced in both *xx* and *yy* planes of the column. Is the member adequate?

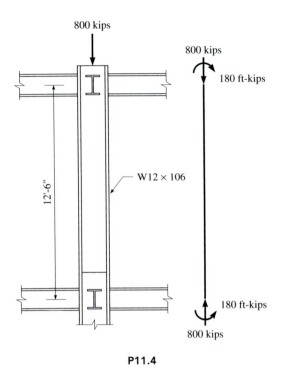

P11.4

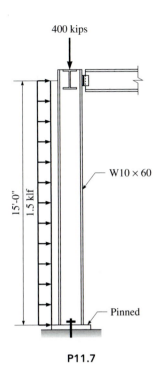

P11.7

P11.7. Figure P11.7 shows a 15-ft-long hinged base column. It must carry a factored axial load of 400 kips and a uniformly applied wind load of 1.5 klf that causes bending about major axis. The structure is braced in both xx and yy planes. Check if a W10×60 of A992 steel will be adequate.

P11.8. An 18-ft-long hinged base column carries a factored axial load of 160 kips as shown in Figure P11.8. The column has its weak axis braced at third-points by channel girts. The wind load on the wall is picked up by the girts and transferred to the column as point loads (8 kips each, under factored wind). The structure is braced in both xx and yy planes. Check if a W10×33 of A992 is adequate.

P11.9. Investigate the adequacy of a W8×24 beam-column used with its web vertical, as the top chord member of a truss. Under factored loads of a load combination, the member is subjected to an axial force of 110 kips and a uniformly distributed lateral load of 0.32 klf. The member

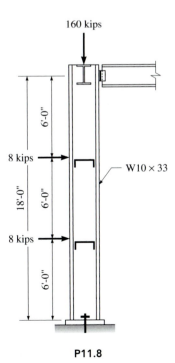

P11.8

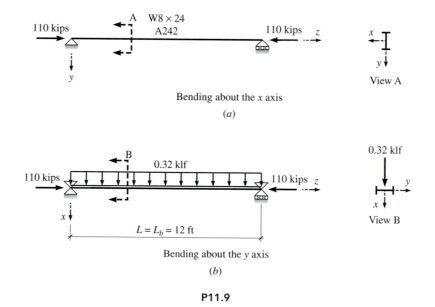

Bending about the x axis

(a)

$L = L_b = 12$ ft

Bending about the y axis

(b)

P11.9

is 12-ft long and could be assumed to be pinned about both axes at both Ends. There is no intermediate bracing present. Assume A242 steel. Neglect self-weight.

P11.10. A W12×50 of A992 steel is used as the bottom chord member of a welded roof truss. The 12-ft-long member, between panel points, is subjected to an axial tensile force of 200 kips under factored loads. Determine the maximum factored concentrated load that could be suspended at midspan. Assume that the load causes bending about the major axis and that lateral supports are provided at panel points only.

P11.11. A W14×82 is used as a column in an unbraced portal frame. It is 13-ft long and pinned at the base. Under factored loads of a load combination, it is subjected to an axial load of 191 kips, $M_{nt} = 156$ ft-kips and $M_{lt} = 104$ ft kips at the top. Assume $K_x = 1.7$ and $K_y = 1.0$. Check the adequacy of the member.

P11.12. An unbraced frame contains a 14-ft column which is subjected to an axial load of 428 kips, a no-translation moment $M_{nt} = 205$ ft-kips, and a translation-permitted moment $M_{lt} = 123$ ft-kips at the top-end, with half of these moments at the bottom-end causing reverse curvature. The forces given are under factored loads of a load combination, and the bending is about the major axis.

Check if a W12×96 will be able to carry this loading. Assume that the effective length factor in the plane is $K_x = 1.8$, and $K_y = 1.0$.

P11.13. Check the adequacy of a W14×48 as a 12-ft column in an unbraced frame. The member is subjected to a factored axial load of 200 kips, a no-translation moment $M_{nt} = 100$ ft-kips, and a translation-permitted moment $M_{lt} = 108$ ft-kips at the top of the member. One-half of these moments are applied at the other end of the member, bending in single-curvature. All moments are applied about the strong axis. Assume $K_x = 1.6$ and $K_y = 0.9$.

P11.14. Select a 9-ft long W10 to support a factored tensile load of 164 kips applied with an eccentricity of 6 in. with respect to the x axis. The member is to be bolted to gusset plates at the ends and braced laterally only at its ends. Assume C_b is equal to 1.0, $A_n = 0.85 A_g$ and U = 0.9.

P11.15. Select a W12-shape for a pin-ended column with a length of 16 ft to carry a factored load of 360 kips and end moments of 480 ft-kips producing symmetric single curvature bending about the major axis. The member is part of a braced frame in its xx and yy planes.

P11.16. Select a W14-shape for a pin-ended column with a length of 24 ft to carry a factored load of 640 kips and end moments of 240 ft-kips producing symmetric single curvature bending about the major axis. The member is part of a braced frame in its *xx* and *yy* planes.

P11.17. Repeat Problem P11.16, if the member is provided with a brace at midlength.

P11.18. Select a W14 section of A992 steel for a 14-ft-long beam-column in a framed structure braced in both directions. Factored axial load, P_u = 840 kips. The first-order, symmetric, single curvature end moments are M_{ntx} = 280 ft-kips and M_{nty} = 40 ft-kips. Assume $K_x = K_y$ = 1.0.

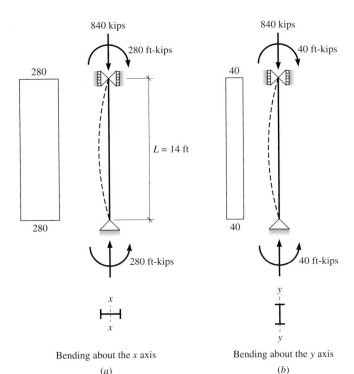

Bending about the *x* axis
(a)

Bending about the *y* axis
(b)

P11.18

P11.19. A pin-ended column in a braced frame must carry a factored axial load of 222 kips along with a factored uniformly distributed transverse load of 1.2 klf. The column is 14-ft long and is braced at midlength. The transverse load is applied to put bending about the strong axis. Select the lightest W10-shape of A992 steel.

P11.20. Solve Example 11.15.1, if the solution is limited to W12-shapes.

P11.21. Solve Example 11.5.1, if the member is 16-ft long.

P11.22. Solve Example 11.15.2, if the solution is limited to W14-shapes.

P11.23. Solve Problem P11.18, if the solution is limited to W12-shapes.

P11.24. Solve Example 11.15.4, if the solution is limited to W12-shapes.

P11.25. Solve Example 11.15.5, if the solution is limited to W12-shapes.

P11.26. Solve Example 11.15.6, if the solution is limited to W12-shapes.

P11.27. Solve Example 11.15.7, if the solution is limited to W12-shapes.

P11.28. Solve Example 11.15.8, if the solution is limited to W14-shapes.

P11.29. Solve Example 11.15.9, if the solution is limited to W12-shapes.

P11.30. Solve Example 11.15.10, if the solution is limited to W14-shapes.

P11.31. Solve Example 11.15.11, if the portal frame is fixed at the base. Limit the selection of columns to W14-shapes.

P11.32. A 24-ft-long W21×93 of A992 steel is used as a simply supported beam with regard to both principal axes. Lateral braces are provided only at the supports. The beam is subjected to a single concentrated load Q at the midspan. The load passes through the shear center of the cross section, and is inclined at an angle of 15° with the vertical axis (web axis). Neglect the self-weight of the beam and determine the maximum factored load Q_u as per LRFDS.

P11.33. Redesign the crane runway beam of Example 10.4.10, if it has to carry a lateral force of 3 kips in addition to the loads given there. The lateral force acts perpendicular to the beam, at each wheel, $4\frac{1}{4}$ in. above the top flange.

P11.34. A crane runway girder 24 ft in length is to be designed to carry the two end wheels of a 5-ton crane. Lateral supports are provided at the ends only. The wheels of the crane are 8 ft on centers, and each wheel transfers a maximum load of 12.5 kips to the top of a 60-lb rail. Assume 10 percent of the wheel load acting as a lateral load applied at $4\frac{1}{4}$ in. above the top of the compression flange. Select the lightest W14-shape of A992 steel.

P11.35. Select the lightest MC-shape of A36 steel to be used as purlins in an industrial building. The roof pitch is 5 on 12, the rafters are 18 ft on centers, and the purlins are spaced at 8-ft intervals along the roof. The roof covering, including purlins, weighs 16 psf of the roof surface, and the snow load is 24 psf of the horizontal surface. The purlins should be designed without sag rods.

P11.36. Solve Problem P11.35, if one line of sag rods is used.

P11.37. Solve Problem P11.35, if W-shapes of A992 steel are used.

P11.38. Select the lightest W-section of A992 steel to carry 0.4 klf dead load, in addition to the weight of the beam and a live load of 1.5 klf. The superimposed loads are applied eccentrically 4 in. from the web plane. The beam is simply supported and has a span of 28 ft. Assume that lateral bracing is provided at the end only.

Steel connection sculpture, Marquette University.

The steel connection sculpture—whose prototype was conceived and designed by Prof. Duane Ellifritt, Univ. of Florida, Gainesville in 1986—functions as a visual aid in teaching students about the many ways steel members may be connected to one another in typical steel construction practice. Since then, scaled down versions of the sculpture were installed at several US universities. It can be used to show students real connections and real steel members in full scale. It also helps them visualize the three dimensional nature of steel connections considered in Chapters 6, 12, and 13.

Photo courtesy Jyothi Vinnakota Robertson

CHAPTER 12

Joints and Connecting Elements

12.1 Introduction

The individual fabricated members of a steel structure are assembled together by connections to form a rigid structure capable of transferring the applied loads to the foundations. **Connections** consist of, in addition to the primary structural members, connectors such as bolts, welds, and occasionally pins, and quite often, connecting elements. Typical **connecting elements** are short pieces of angle or tee sections. Small pieces of plate material are also often used as connecting elements (gusset plates, splice plates, cover plates, and bracket plates). The primary purpose of connecting elements is to aid in the transfer of force from one structural member to another. However, not all joints necessarily use connecting elements. Typically, the connecting elements are much smaller in terms of the absolute dimensions than are the primary members.

The force transfer between members at a connection is through one or more planar surfaces called *joints*. A *joint* is a planar area where two or more structural member ends or surfaces are connected and across which force transfer takes place. Joints may be classified by the type of connector, for example, bolted, welded, or pinned, or by the method of force transfer, such as direct shear, eccentric shear, or direct tension.

12.2 Joint Loading Cases

Let us consider a bolt group composed of N bolts (Fig. 12.2.1). Let A_{bi} be the cross-sectional area and (x_i', y_i') the x and y coordinates of the ith bolt with respect to an arbitrary orthogonal coordinate system $x'y'$. The bolt group area is considered as an elastic cross section. The coordinates $\bar{x}'$, $\bar{y}'$ of the center of gravity G of the bolt group are determined by the method of moments used to locate the center of gravity of built-up sections. Thus:

$$\bar{x}' = \frac{\Sigma A_{bi} x_i'}{\Sigma A_{bi}}; \quad \bar{y}' = \frac{\Sigma A_{bi} y_i'}{\Sigma A_{bi}} \tag{12.2.1}$$

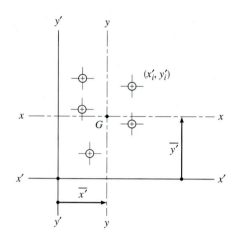

Figure 12.2.1: Center of gravity of a bolt group.

We will assume that all of the bolts in the group are of the same size and made of the same material, a situation that typically occurs in most steel building and bridge structures for reasons of economy. If all of the bolts are of the same size, the area terms cancel out in Eq. 12.2.1, giving the simpler relations:

$$\bar{x}' = \frac{\sum x_i}{N}; \quad \bar{y}' = \frac{\sum y_i}{N} \qquad (12.2.2)$$

We now consider a rectangular, right-handed coordinate system XYZ. Its origin coincides with the centroid G of the bolt group (Fig. 12.2.2). The X and Y axes lie in the plane of the joint while the Z axis is normal to it. In addition, the X and Y axes coincide with the principal axes of the bolt group.

Depending on the relative position of the connected members and the loads being transmitted, joints are subject to various actions. In a general three-dimensional space frame, joint loading may have six components—three forces and three moments. Note that the term *joint load* refers to the vectors formed from all of the force and moment components acting on the joint. The symbols used for such an oriented set are as follows (Fig. 12.2.2):

$$
\begin{array}{lll}
P_x, P_y & = \text{in-plane or shear forces} & \\
P_z & = \text{normal force} & (T \text{ tension, or } C \text{ compression}) \\
M_{yz}, M_{zx} & = \text{bending moment} & (M_x, M_y) \\
M_{xy} & = \text{twisting moment} & (M_z \text{ or } M_T)
\end{array}
$$

This systematic method of labeling the loading components makes it easy to classify any joint of a structure, however complex. Also, this form of specification is helpful if the connections are analyzed and designed by a general computer program. The most general loading on a typical bolt consists of a tensile force B_t and a shear force B_v having components B_{vx} and B_{vy} in the x and y direction, respectively, as shown in Fig. 12.2.2 and as previously discussed in Section 6.3.1.

Loading case $[P_x\ P_y]$, shown in Fig. 12.2.2, occurs when the resultant joint load is in the plane of the joint, and its line of action passes through the centroid of the bolt group. Particular cases of this loading case are the $[P_x]$ and $[P_y]$ types. These loading cases are referred to as *direct shear loadings,* as the bolts are subjected to shear forces only. Depending on the joint configuration, the bolts may be in either single shear or double shear. Lap joints with bolts in single shear and butt joints with bolts in double shear are two examples (Figs. 6.1.2*a* and *b*). Other examples of this case can be found in gusset plate connections to tension members (see Chapter 7) and in splices to tension members (Figs. 12.2.3). The behavior and design of bolted joints in direct shear will be considered in Section 12.3, while welded joints in direct shear will be considered in Section 12.8.

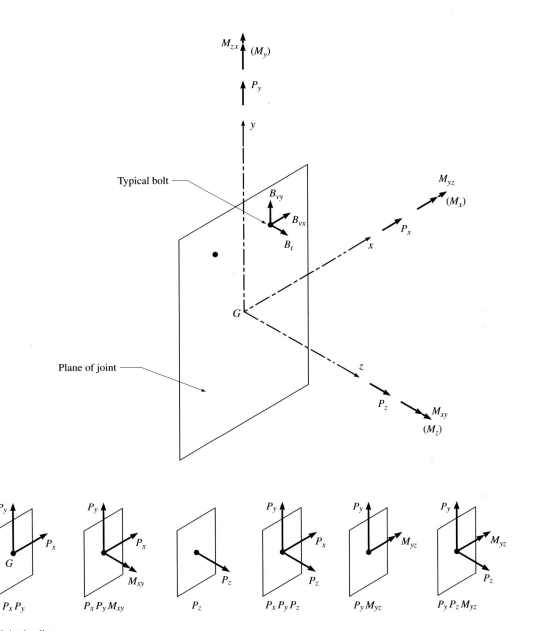

Figure 12.2.2: Joint loading cases.

Loading case $[P_x\, P_y\, M_{xy}]$, shown in Fig. 12.2.2, occurs when the resultant joint load is in the plane of the joint, but its line of action does not pass through the centroid of the bolt group. Such a condition is called ***eccentric shear loading.*** An example is shown in Fig. 12.2.4a. By adding a pair of equal, oppositely directed, and collinear forces of magnitude P at the centroid

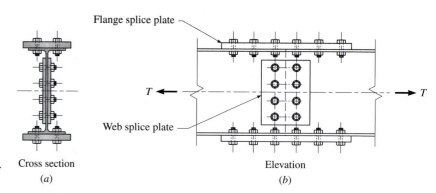

Figure 12.2.3: Examples of joints under direct shear loading.

Cross section
(a)

Elevation
(b)

of the bolt group (as shown in Fig. 12.2.4b), the applied eccentric load P is replaced by a central load P and the torsional moment $M_T = Pe$ as shown in Figs. 12.2.4c and d. The central load P could further be resolved into its components P_x and P_y along the axes x and y, respectively. It is seen that the individual bolts are again subjected to shear forces only. Splice connections of plate girder webs and bracket connections to column flanges (Fig. 6.1.2c and Fig. 6.1.3f) are typically subjected to this type of loading. Note that this type of bracket loading introduces bending moments about the minor axis of the column. Behavior and design of bolted joints in eccentric shear will be considered in Section 12.4, while welded joints in eccentric shear will be considered in Section 12.9.

Loading case [P_z], shown in Fig. 12.2.2, occurs when the line of action of the resultant joint load is perpendicular to the plane of the joint and passes through the centroid of the bolt group. Only the case where P_z is a tensile force is of interest in the design of connectors. The bolts are assumed to be in direct tension only. A hanger connection is a good example of this loading case (Fig. 6.1.2d). The behavior and design of bolted joints in direct tension are considered in Section 12.5.

Loading case [$P_x\,P_y\,P_z$], shown in Fig. 12.2.2, occurs when the line of action of the resultant joint load passes through the centroid of the bolt group but is inclined to the plane of the joint at an angle between 0° and 90°. Again the case where P_z is a tensile force is of greater interest since

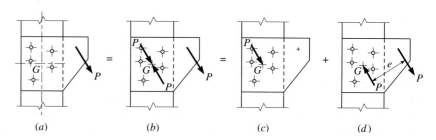

Figure 12.2.4: Joint under eccentric shear loading.

(a)   (b)   (c)   (d)

the bolts are subjected to combined shear and tension. Particular cases of this loading are $[P_x\,P_z]$ or $[P_y\,P_z]$ when the resultant load is in the horizontal or vertical plane, respectively. Diagonal braces in frames are generally attached to column flanges through a T-stub (or double angles and gusset plate). The joint connecting this T-stub (or the connection angles) to the column flange (Fig. 6.1.2*e*) is typically subjected to loading case $[P_y\,P_z]$. Note that this type of bracket loading introduces bending moments about the major axis of the column. Bolted joints in shear and tension are considered in Section 12.6, while welded joints in shear and tension are studied in Section 12.10.

Loading case $[P_y\,M_{yz}]$, shown in Fig. 12.2.2, occurs when the resultant joint load is in the yz plane and is parallel to the plane of the joint. In such bolted joints, all of the bolts are subjected to direct shear by the component P_y, while certain bolts are subjected, in addition, to tension by the applied moment. Bracket connections shown in Fig. 6.1.2*f* and Fig. 6.1.3*h* are two examples of such joints. Loading case $[P_x\,M_{zx}]$ is similar to this case. The behavior and design of bolted joints in shear and bending are considered in Section 12.7, while welded joints in shear and bending are considered in Section 12.11.

Loading case $[P_y\,P_z\,M_{yz}]$, shown in Fig. 12.2.2, occurs when the resultant joint load is in the symmetry plane yz of the bolt group but does not pass through the centroid of the bolt group. The bolts in the group are subject to shear by P_y and tension by P_z; some bolts are also subjected to additional nonuniform tension by M_{yz}. Thus, the bolts are again subject to combined shear and tension. Loading case $[P_x\,P_z\,M_{zx}]$ is similar to this case.

12.3   Bolted Joints in Direct Shear (Loading Case $P_x\,P_y$)

As mentioned earlier, connections of gusset plates to tension members, and of splice plates to the flanges and webs of spliced tension members are some examples of bolted joints in direct shear.

12.3.1   Joints in Tension Members

The simplest type of structural connection with joints in direct shear is the splice of a flat plate tension member using lap joints or butt joints, as shown in Fig. 6.4.1. The applied load passes through the centroid of the bolt group. A butt splice is preferred over a lap splice because the symmetry of the shear planes prevents bending of the plate material, and the bolts are in double shear. In a lap splice, the bolts act in single shear. The forces in the plates of thickness t_1 and t_2 act under small eccentricities $e_1 = t_1/2$ and $e_2 = t_2/2$, with respect to the shear plane. These eccentricities cause bending of the members as the loads attempt to align themselves axially.

12.3.2   Horizontal Shear in Built-Up Bolted Beams

When a beam with cover plates (Fig. 12.3.1a) is loaded, there is a tendency for longitudinal slip to occur between the cover plate and the beam proper, as shown in exaggerated fashion by Fig. 12.3.1b. This slip is prevented by the flange bolts (or welds). In resisting the slip at a given point along the flange, the bolts are subjected to horizontal shear at the faying surface between the cover plate and the beam flange. If p is the pitch of the bolts, the horizontal shear over a length p, or over the tributary shearing area shown cross-hatched in Fig. 12.3.1d, has to be resisted by the bolts in that area. Two bolts are shown in Fig. 12.3.1d, but there might be four in the case of a large wide-flange beam. In either case, the capacity of a bolt B_d is the single shear strength or the bearing strength, whichever is smaller. If R_d denotes the design strength of n bolts in the shaded area, then:

$$R_d = nB_d \geq q_{sv}\,p \qquad (12.3.1)$$

where q_{sv} = shear flow at the faying surface

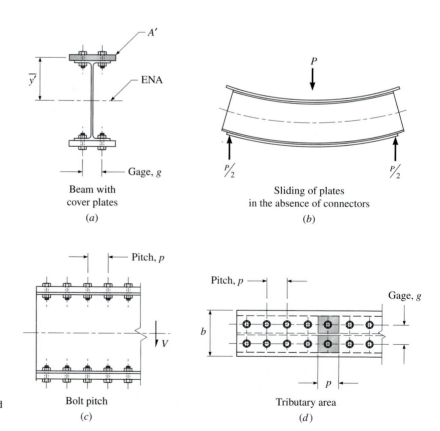

Figure 12.3.1: Horizontal shear in a built-up bolted beam.

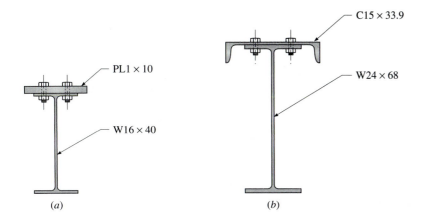

C15 × 33.9

PL1 × 10

W24 × 68

W16 × 40

(a)   (b)   **Figure 12.3.2:** Examples of built-up bolted beams.

Substituting the value for q_{sv} (= VQ'/I) from mechanics [Timoshenko and Young, 9.10] in Eq. 12.3.1 and rearranging, we obtain:

$$p \leq \frac{nB_d}{q_{sv}} = \frac{nB_d I}{VQ'} = \frac{nB_d I}{VA' \, \overline{y}'} \qquad (12.3.2)$$

where p = longitudinal pitch of the cover plate bolts
   I = moment of inertia of the cross section including cover plates
   V = total factored shear at that point along the beam where the pitch is desired
   A' = cross-sectional area of the cover plate or plates at one flange
   $\overline{y}'$ = distance from the elastic neutral axis (ENA) of the built-up section to the centroid of the area A'
   Q' = static moment of the area A' about the neutral axis of the built-up section
   B_d = design strength of a single bolt
   n = number of bolts in pitch length p at one flange

Figure 12.3.2 shows two examples of built-up crane girders having bolted joints in horizontal shear.

Analysis of Bolted Built-Up Beam   EXAMPLE 12.3.1

Calculate the required spacing of ¾-in.-diameter A325-N high-strength bolts for the beam composed of a W24×68 of A992 steel and a C15×33.9 of A36 steel, as shown in Fig. 12.3.2b, under a factored shear load of 186 kips.

[*continues on next page*]

Example 12.3.1 continues ...

Solution

For a W24×68 shape: $d_W = 23.7$ in., $t_{fW} = 0.585$ in.
For a C15×33.9 shape: $A_C = 9.95$ in.^{2}, $t_{wC} = 0.400$ in., $\bar{x}_C = 0.788$ in.
From LRFDM Table 1-17, for the combination section given,

$$I_x = 2710 \text{ in.}^4; \quad y_1 = 15.7 \text{ in.}$$

where I_x is the moment of inertia of the section and y_1 is the distance to the center of gravity of the built-up section from the bottom fiber of the W-shape.

The bolts connecting the web of the channel to the flange of the W-shape are subjected to the shear flow at the interface. The shear flow at this interface can be calculated using:

$$V = 186 \text{ kips}; \quad I_x = 2710 \text{ in.}^4; \quad A' = A_C = 9.95 \text{ in.}^2$$

$$\bar{y}' = d_W - y_1 + t_{wC} - \bar{x}_C = 23.7 - 15.7 + 0.400 - 0.788$$
$$= 7.61 \text{ in.}$$

So,

$$q_{sv} = \frac{VA'\bar{y}'}{I_x} = \frac{186(9.95)(7.61)}{2710} = 5.20 \text{ kli}$$

The bolts are in single shear. From LRFDM Table 7-10, the shear strength of a ¾-in.-diameter A325-N bolt, $B_{dv} = 15.9$ kips.

The bolts are in bearing on the flange of the W-shape ($t_f = 0.585$ in., $F_u = 65$ ksi) and on the web of the channel ($t_w = 0.440$ in., $F_u = 58$ ksi), which controls. Assume pitch, $p \geq p_{full}$. The design bearing strength of a ¾-in. bolt on 0.44-in.-thick A36 plate material can be written with the help of LRFDM Table 7-12 as:

$$B_{dbo} = 78.3(0.440) = 34.5 \text{ kips}$$

The design strength of a single bolt, $B_d = \min [15.9, 34.5] = 15.9$ kips Assuming that the bolts are placed two at a section, and without stagger at pitch p, we have:

$$pq_{sv} \leq 2B_d \rightarrow p \leq \frac{2(15.9)}{5.20} = 6.12 \text{ in.}$$

Use a pitch of 6 in. $> 3d = 2^1/_4$ in.   O.K.   (Ans.)

12.4   Bolted Joints in Eccentric Shear (Loading Case $P_x P_y M_{xy}$)

When the resultant of the applied loads on a bolted joint lies in the plane of the joint, but does not pass through the center of gravity of the bolt group, the joint is said to be a ***bolted joint in eccentric shear.*** Some common examples

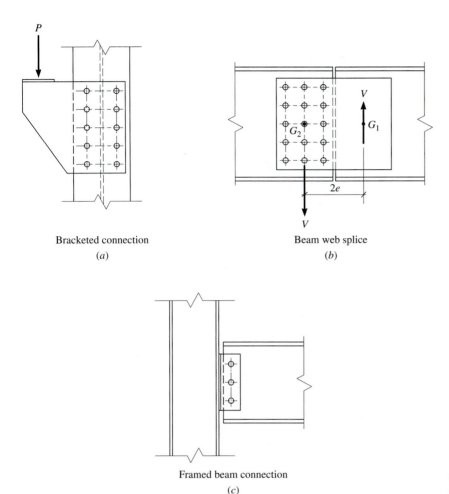

Bracketed connection

(*a*)

Beam web splice

(*b*)

Framed beam connection

(*c*)

Figure 12.4.1: Examples of bolted joints in eccentric shear.

of eccentrically loaded bolted joints are shown in Fig. 12.4.1. Column brackets (Fig. 12.4.1*a*) consist of a pair of plates straddling the column, with one plate bolted to each flange. They project beyond the column flanges to support the load and are stiffened outside the column flanges, if necessary, by some type of diaphragm between them. They subject the column to bending about its weaker axis. Beam web splices (Fig. 12.4.1*b*) and simple beam-to-column connections (Fig. 12.4.1*c*) are other examples of joints in eccentric shear. Beam web splices will be considered in Section 12.15, while framed beam connections will be studied in Chapter 13. The theories developed in this section for bracket plates are applicable to these members as well.

At present there are two different procedures for analyzing eccentrically loaded bolted joints. The ***elastic method*** makes use of simple concepts learned in elementary mechanics of materials. It does not take advantage of any ductility of the bolts or any corresponding redistribution of the load. The method is approximate and generally gives conservative results. This approach is presented in Section 12.4.1. A more rational method, known as the **ultimate strength method,** utilizes the load-deformation relationship of a single bolt and better predicts the strength of an eccentrically loaded bolt group. This approach is presented in Section 12.4.2.

12.4.1   Elastic Method

Let us consider a bracket plate bolted to a supporting element as shown in Fig. 12.4.2a. The bolt group is composed of N bolts of the same size, each having a cross-sectional area A_b (eight bolts are shown in Fig. 12.4.2). The bolt group area is considered as an elastic cross section and has its center of gravity at G. In the case of an asymmetrical group of bolts, G is located using the method of moments described in Section 12.2. In the example considered, because of the double symmetry of the bolt pattern, the centroid can readily be located by inspection as occurring at the intersection of the two lines of symmetry.

The external load P is acting on the bracket plate with an eccentricity e. The load P has a vertical component P_y and a horizontal component P_x. By adding a pair of equal, oppositely directed, and collinear forces of magnitude P at the centroid of the bolt group (as described in Section 12.2 and shown in Fig. 12.2.4b), the eccentrically applied load may be replaced by a centroidal load P and a torsional moment $M_T \equiv M_{xy} = Pe$ as shown in Fig. 12.4.2b. (If the load P passes through a point x_e, y_e, the torsional moment may also be calculated from the relation, $M_T = P_y x_e - P_x y_e$.) Note that counterclockwise moments are considered positive. Under the action of the applied eccentric load, once the static friction has been overcome, the bolts slip into bearing and are stressed in shear by the two direct components and by the torsional moment. We assume that the bracket plate is perfectly rigid and remains undeformed under load, while the bolts are assumed to be deformable and perfectly elastic.

Let B_{xD} and B_{yD} represent the shear force components in a single bolt due to the centroidal (direct) forces P_x and P_y, respectively, in the x and y directions (Fig. 12.4.2c). Assuming that the effect of the centroidal forces is shared equally among all bolts in the group, we have:

$$B_{xD} = \frac{P_x}{N}; \quad B_{yD} = \frac{P_y}{N} \qquad (12.4.1)$$

The torque M_T is assumed to cause the bracket plate to rotate about the center of gravity of the bolt group and produce additional shear in each bolt that acts normal to the radius vector extending from the center of gravity G to the bolt under consideration. Let B_{iT} be the shear force on the ith bolt, with coordinates (x_i, y_i) and radial distance r_i, due to the torque M_T (Fig. 12.4.2d). The shear forces B_{iT} are directed so that they produce a torque about the centroid in the same sense as the external torque. The variation of the bolt forces B_{iT}

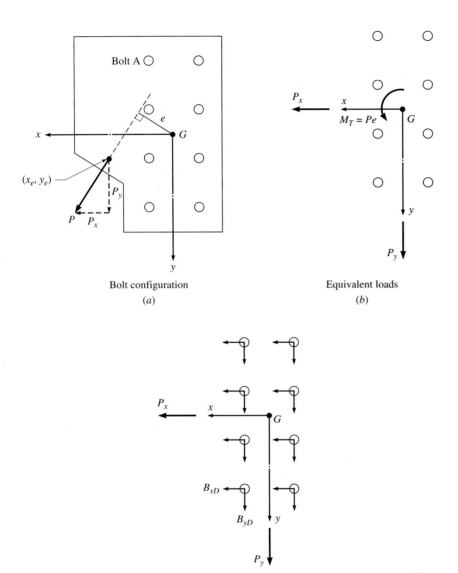

Bolt configuration

(a)

Equivalent loads

(b)

Direct components of the bolt forces

(c)

Figure 12.4.2: Elastic analysis of a bolted joint in eccentric shear.

and their orientation is shown in Fig. 12.4.2e. If we neglect friction between the plates, the applied torque can be equated to the sum of these individual contributions by the bolts. That is:

$$\sum B_{iT}\, r_i \;=\; M_T \qquad\qquad (12.4.2)$$

Now, let B_T^* be the shear force due to torque on a reference bolt situated at a coordinate point (x^*, y^*) with a radius vector r^* (Fig.12.4.2e). We assume that,

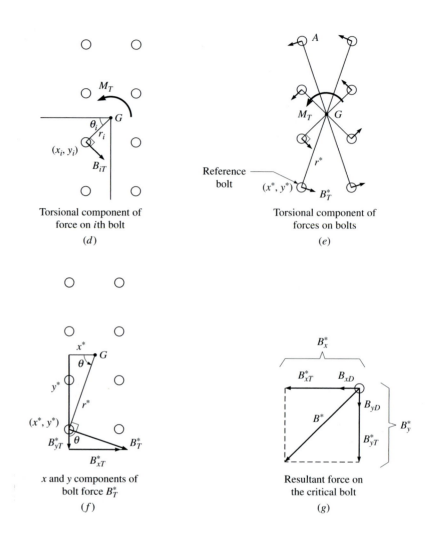

Torsional component of
force on ith bolt

(d)

Torsional component of
forces on bolts

(e)

x and y components of
bolt force B_T^*

(f)

Resultant force on
the critical bolt

(g)

Figure 12.4.2 (continued).

under torque M_T, the rotational displacement of any bolt is proportional to its distance from the center of gravity of the bolt group. Further, we assume that the bolts behave linearly and elastically. Hence, the effect of the torque is to cause shear stress on each bolt in direct proportion to its radial distance from the center of gravity, G. For configurations where all of the bolts are the same size, we can write:

$$\frac{e_i}{e^*} = \frac{r_i}{r^*} \rightarrow \frac{f_{iT}}{f_T^*} = \frac{r_i}{r^*} \rightarrow \frac{B_{iT}}{B_T^*} = \frac{r_i}{r^*} \rightarrow B_{iT} = \frac{r_i}{r^*} B_T^* \quad (12.4.3)$$

Here e_i and f_{iT} are the displacement and the stress of the ith bolt, while e^* and f_T^* are the corresponding terms for the reference bolt. This relation defines the shear force on any bolt due to M_T as a function of the corresponding shear force on the reference bolt. Substituting this value of B_{iT} into Eq. 12.4.2 results in:

$$M_T = \frac{B_T^*}{r^*} \Sigma r_i^2 \qquad (12.4.4)$$

The shear force on the reference bolt due to M_T is therefore:

$$B_T^* = \frac{M_T r^*}{\Sigma r_i^2} \qquad (12.4.5)$$

The corresponding shear stress on the reference bolt is:

$$f_T^* = \frac{B_T^*}{A_b} = \frac{M_T r^*}{\Sigma A_b r_i^2} = \frac{M_T r^*}{J} \qquad (12.4.6)$$

This relation is the same as the familiar mechanics of materials formula for torsion on a circular shaft, Tr/J in which the twisting moment T, corresponds to M_T; the radius from the center of the shaft to the point at which the shear stress is computed, r, corresponds to r^*; and the polar moment of inertia, J, corresponds to $\Sigma A_b r_i^2$.

The resultant load on the reference bolt is obtained as the vector sum of the components B_{xD}, B_{yD}, and B_T^*. This vector addition is easier to perform analytically if we resolve B_T^* also into its x and y components. By observing from Fig. 12.4.2f that the angle θ between the radius r^* and x^* equals the angle between B_T^* and B_{yT}^*, we obtain from similar triangles:

$$\frac{B_{xT}^*}{B_T^*} = \frac{-y^*}{r^*}; \quad \frac{B_{yT}^*}{B_T^*} = \frac{x^*}{r^*} \qquad (12.4.7)$$

Replacing B_T^* in these relations by its value from Eq. 12.4.5, results in:

$$B_{xT}^* = \frac{M_T(-y^*)}{\Sigma r_i^2}; \quad B_{yT}^* = \frac{M_T x^*}{\Sigma r_i^2} \qquad (12.4.8)$$

These are the horizontal and vertical components of the force acting on a selected bolt due to the torsional moment M_T only. Hence, the resultant of these horizontal and vertical force components acting on all bolts in the group must be zero.

The x and y components of the total shear force on the reference bolt, due to both the direct load and the torsional moment M_T are:

$$B_x^* = B_{xD} + B_{xT}^* = \frac{P_x}{N} + \frac{M_T(-y^*)}{\Sigma r_i^2} \qquad (12.4.9a)$$

$$B_y^* = B_{yD} + B_{yT}^* = \frac{P_y}{N} + \frac{M_T(x^*)}{\Sigma r_i^2} \qquad (12.4.9b)$$

In calculating $\sum r_i^2$, it is generally more convenient to use the x and y coordinates of the bolts:

$$\sum r_i^2 \;=\; \sum x_i^2 \;+\; \sum y_i^2 \tag{12.4.10}$$

Finally, the magnitude of resultant force on the reference bolt is obtained as the vector resultant of the x and y components. Thus:

$$B^* \;=\; \sqrt{(B_x^*)^2 \;+\; (B_y^*)^2} \tag{12.4.11}$$

According to the LRFDS, the joint design is adequate if:

$$B^* \leq B_d \tag{12.4.12}$$

where B^* is the resultant force on the critical (most highly loaded) bolt when it is considered as the reference bolt, and B_d is the design strength of the bolt used in the connection.

Most eccentric bolted connections contain one or more parallel lines of bolts at a uniform pitch. Such arrangements allow the critical bolt, for which B_{xD} and B_{xT}^* maximum are additive, as well as B_{yD} and B_{yT}^* maximum, are also additive, to be readily located. For example, for the group shown in Fig. 12.4.2a, the following reasoning indicates that the critical bolt is at A:

1. B_{xD} and B_{yD} both act in the positive directions.
2. The bolt at A is one of the four bolts located farthest from the center of gravity G where B_{iT} is a maximum (Fig. 12.4.2e).
3. Bolt A is the only bolt for which both B_{xT}^* and B_{yT}^* act in the same respective directions as B_{xD} and B_{yD} and thereby increase the magnitude of both B_{xD} and B_{yD} (Fig. 12.4.2g).

However, in irregularly placed bolted joints, the bolt most remote from the centroid may not be the most highly loaded. So, it may not be possible to locate the most critically loaded bolt by inspection prior to analysis. It may be necessary to check all possible candidates to identify the most highly loaded bolt.

For eccentric bolted joints, the number and arrangement of bolts is generally first assumed. For detailing and fabrication, the edge distance is usually made $1\tfrac{1}{4}$ in., $1\tfrac{1}{2}$ in., or 2 in. and the pitch 3 in. and multiples thereof, for bolts up to 1 in. in diameter. The workable gages for W-shapes, channels, tees (given in the tables in Part 1 of the LRFDM) and for angles (given in Table 6.2.1) should be used.

12.4.2   Ultimate Strength Method

For information on the ultimate strength method for bolted joints in eccentric shear and its application, refer to Section W12.1 on the website http://www.mhhe.com/Vinnakota.

12.4.3   LRFDM Design Tables

For a given pattern of bolts in a bolted joint under an eccentric shear load P, the ultimate load P_u is proportional to the nominal strength of a single bolt B_n. Hence, the design strength of a bolt group in eccentric shear can be written in the form:

$$P_d = CB_d \qquad\qquad (12.4.13)$$

where B_d = design strength of a single bolt = ϕB_n
        C = nondimensional coefficient

The value of B_d is determined from the limit states of bolt shear strength, the bearing strength at bolt holes, and the slip resistance, if the joint is slip-critical.

Parametric studies were made based upon the ultimate strength method, for several widely used bolt patterns and load eccentricity conditions. The parameters chosen were number of bolts in a vertical row, n; bolt spacing in a vertical row, s; number of vertical rows; and the gages of vertical rows. Each fastener combination was analyzed for loads inclined at 0°, 15°, 30°, 45°, 60°, and 75° with the vertical axis and for different intercepts e_x of the load line with the x axis. It was observed that the C coefficients do not vary greatly for bolt characteristics such as bolt material, bolt diameter, and type of joint. Consequently, it is customary to tabulate and apply one set of nondimensional coefficients to all cases of eccentrically loaded joints regardless of the bolt diameter. The results are provided in LRFDM Tables 7-17 to 7-24, Coefficients for Eccentrically Loaded Bolt Groups. The values are in fact slightly conservative when used with A490 bolts. Design strengths given by these tables lead to a factor of safety equivalent to that for bolts in joints less than 50 in. long and subjected to shear produced by a concentric load on either bearing-type or slip-critical joints. Linear interpolation within a given table between adjacent values of e_x is permitted; however, linear interpolation between C values for different load inclinations α may be highly unconservative. Therefore it is recommended to use the C values for the next lower angle. Although the procedure used to develop the tables is based on connections which were expected to slip under load (that is, bearing-type connections), both load tests and analytical studies by Kulak [1975] indicated that the procedure may be conservatively extended to slip-critical connections.

Multiplying the value of C for a given fastener pattern by the design strength of a single bolt, B_d, gives the design strength of the connection, P_d. Or, if the factored load P_u is given, dividing it by B_d gives the minimum coefficient C_{req}. A bolt group can then be selected for which the tabulated coefficient C is of that magnitude or greater.

EXAMPLE 12.4.1

Analysis of Bolted Joint in Eccentric Shear

Determine the factored load, P_u, that the nine-bolt eccentrically loaded bracket connection can transmit to the web of a C10×15.3 channel. The ⅜-in.-thick bracket plate and the channel are of A36 steel. The ¾-in.-dia. A325-X type bolts are in standard holes. Use (*a*) the elastic vector method, (*b*) the ultimate strength tables in the LRFDM.

Solution

Number of bolts, $N = 9$
Diameter of the bolt, $d = $ ¾ in. Assuming standard holes,

$$d_h = \frac{13}{16} \text{ in.}; \quad 2d + 0.5d_h = 1.91 \text{ in.}; \quad 2d + d_h = 2.31 \text{ in.}$$

The bolts are in single shear. From LRFDM Table 7-10 design shear strength of a single ¾-in.-dia. A325-X type bolt, $B_{dv} = 19.9$ kips.

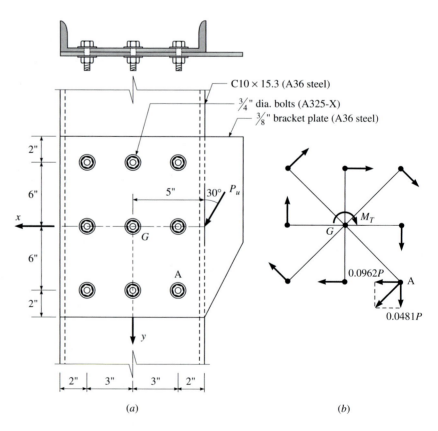

Figure X12.4.1

(*a*)

(*b*)

As the end distance, 2 in. $> 2d + 0.5d_h = 1.91$ in., and the spacing s, 3 in. $> 2d + d_h = 2.31$ in., the bearing capacity of all the bolts is determined by the limit state of the ovalization of the hole. As the web thickness, 0.240 in., of the C10 $\times$ 15.3 channel is less than the bracket plate thickness, 0.375 in., the bolt bearing strength is controlled by the web. The design bearing strength of a bolt can be obtained with the help of LRFDM Table 7-12 as,

$$B_{db} = 78.3(0.240) = 18.8 \text{ kips}$$

The design strength of a single bolt is:

$$B_d = \min[B_{dv}, B_{db}] = \min[19.9, 18.8] = 18.8 \text{ kips}$$

a. Elastic vector method
The centroid G of the bolt group is located from symmetry.

$$I_p = \sum x_i^2 + \sum y_i^2 = 6(3^2) + 6(6^2) = 270 \text{ in.}^2$$

$$P_x = P \sin 30° = 0.500P; \qquad P_y = P \cos 30° = 0.866P$$

Noting that counterclockwise moments are considered positive,

$$M_T = e_x P_y = -5.0(0.866P) = -4.33P$$

$$B_{xD} = \frac{P_x}{N} = \frac{0.500P}{9} = 0.0556P; \, B_{yD} = \frac{P_y}{N} = \frac{0.866P}{9} = 0.0962P$$

By inspection, bolt A in the right-hand lower corner is the most critical bolt. For this bolt,

$$x^* = -3.0 \text{ in. and } y^* = 6.0 \text{ in. With the help of Eq. 12.4.8:}$$

$$B_{xT}^* = -\frac{M_T y^*}{I_p} = -\frac{(-4.33P)(6.0)}{270} = 0.0962P$$

$$B_{yT}^* = \frac{M_T x^*}{I_p} = \frac{(-4.33P)(-3.0)}{270} = 0.0481P$$

Hence,

$$B_x^* = 0.0556P + 0.0962P = 0.152P$$

$$B_y^* = 0.0962P + 0.0481P = 0.144P$$

The resultant bolt force is therefore:

$$B^* = \sqrt{(B_x^*)^2 + (B_y^*)^2} = \sqrt{(0.152P)^2 + (0.144P)^2} = 0.209P$$

According to the LRFD Specification,

$$B_{\text{req}} = 0.209P_u \leq B_d = 18.8 \text{ kips}$$

$$P_u \leq \frac{18.8}{0.209} = 90.0 \text{ kips} \qquad \text{(Ans.)}$$

[*continues on next page*]

Example 12.4.1 continues ...

b. Ultimate strength tables

Select LRFDM Table 7-21 entitled "Coefficients C for Eccentrically Loaded Bolt Groups" which is valid for bolt groups with three vertical rows of bolts with a gage of 3 in., and subjected to a load P_u inclined at 30° with the vertical. From this table, read $C = 6.14$, for a bolt group with a vertical bolt spacing s of 6 in.; number of bolts in a vertical row, n, of 3; and an e_x value of 5 in. Substituting,

$$P_u = C\phi r_n = CB_d = 6.14(18.8) = 115 \text{ kips} \qquad \text{(Ans.)}$$

EXAMPLE 12.4.2

Analysis of Bolted Joint in Eccentric Shear

Determine the factored load P_u that can be carried by the ½-in.-thick A36 steel bracket plate shown in Fig. X12.4.2. The ¾-in.-dia. bolts are of A325-X type. Use (*a*) the elastic vector method, (*b*) the LRFDM tables.

Solution

Number of bolts, $N = 5$

a. Elastic vector method

The *xy* coordinate system is shown in Fig. X12.4.2*a* and counterclockwise moments are considered positive.

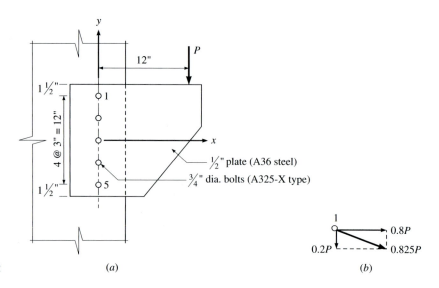

Figure X12.4.2   (*a*)   (*b*)

The center of gravity is located from inspection and coincides with the middle bolt.

$$I_p = \sum (x_i^2 + y_i^2) = 0 + [2(3^2) + 2(6^2)] = 90.0 \text{ in.}^2$$

Also,

$$P_x = 0; \quad P_y = -P; \quad M_T = -12P$$

$$B_{xD} = 0; \quad B_{yD} = -\frac{P}{5} = -0.20P$$

From observation, bolt 1 (or bolt 5) is the critical bolt with $x^* = 0$; $y^* = 6.0$ in.

$$B_{xT}^* = -\frac{M_T y^*}{I_p} = \frac{12P(6.0)}{90.0} = 0.80P; \quad B_{yT}^* = 0.0$$

The resultant force on the critical bolt, using Eqs. 12.4.9 and 12.4.11, is:

$$B^* = \sqrt{(0 + 0.80P)^2 + (0.20P + 0)^2} = 0.825P$$

The bolts are in single shear. From LRFDM Table 7-10, the shear strength of a ¾-in.-dia. A325-X bolt in single shear, B_{dv}, is 19.9 kips. Equating the resultant force on the reference bolt to the design strength of the bolt, we obtain:

$$0.825P_u \leq 19.9 \rightarrow P_u \leq 24.1 \text{ kips} \qquad \text{(Ans.)}$$

b. LRFDM tables

From LRFDM Table 7-17, corresponding to a pitch, $s = 3$ in., number of bolts in a vertical row, $n = 5$, eccentricity, $e_x = e = 12.0$ in., and angle $= 0°$, the coefficient C is read as 1.40. The design strength of the bolted joint is therefore:

$$P_u = \phi r_n C = 19.9(1.40) = 27.9 \text{ kips} \qquad \text{(Ans.)}$$

For an application of the ultimate strength method to determine P_u, see Web Section W12.1.

12.5   Bolted Joints in Direct Tension (Loading Case P_z)

An example of a bolt group under direct (concentric) tension is the T-stub hanger connection shown in Fig. 6.1.2d, with a single line of bolts parallel and on each side of the web of the tee. The applied load is concentric, and assuming perfect symmetry the bolts are equally loaded in tension. LRFDS Section J3.1 stipulates that A490 bolts in connections subjected to tension loads (such as hanger connections) shall be tightened to a bolt tension not less than that given in LRFDS Table J3.1, regardless of whether the connection is slip-critical or not.

If the flange of the tee stub is flexible, loads applied to the hanger will cause significant bending of the flange. Due to this bending, contact stresses, more or less concentrated at the flange tips, develop. The summation of these stresses is known as the ***prying force.*** As the resultant force in any bolt of the hanger connection equals the applied load plus any prying force that develops due to the flexibility of the connection, the useful capacity of the bolts is now decreased. ***Prying*** is a phenomenon whereby the deformation of a connection element under a tensile force increases the tensile force in the bolt over and above that due to the direct tensile force alone. Prying phenomenon occurs in bolted joints only and for tensile bolt forces only. In recognition of this, LRFDS Section J3.6 requires that prying action be included in the computation of tensile loads applied to bolts.

For information on the behavior, analysis, and design of bolted, T-section hanger connections including prying forces, refer to Section W12.2 on the website http://www.mhhe.com/Vinnakota.

12.6   Bolted Joints in Shear and Tension (Loading Case P_y, P_z)

Bolted joints may be loaded in such a way as to produce a direct shear as well as a tensile force on the bolts. A structural tee segment bolted to the face of a wide flange column for the purpose of receiving a diagonal brace (Fig. 6.1.2e) is an example. The line of action of the brace force passes through the center of gravity of the bolted joint. The vertical component (P_y) of the load puts the bolts in shear, and the horizontal component (P_z) induces tension in bolts. As the load passes through the center of gravity of the joint, each bolt can be assumed to receive an equal share of each component. The column flange and the tee flange are assumed to be thick enough that prying can be neglected.

12.6.1   Bearing-Type Joints

Let

N = number of bolts in the joint
P_{uy} = shear component of applied force under factored loads
P_{uz} = tension component of applied force under factored loads
B_{uv} = shear component of applied factored load in a bolt = P_{uy}/N
B_{dv} = design strength of a bolt in single shear (LRFDM Table 7-10)
B_{dt} = design strength of a bolt in direct tension (LRFDM Table 7-14)

B'_{dt} = design tensile strength of a bolt simultaneously subjected to shear force, B_v.

= $C_1 - C_2 B_{uv} \leq B_{dt}$

Values of C_1 and C_2 may be obtained with the help of LRFDS Table J3-5 (see Web Section W6.4 and Eq. W6.4.12).

Let N_1, N_2, and N_3 be the number of bolts required to satisfy the limit states of shear, tension, and combined shear and tension, respectively. The limit state of shear requires:

$$N_1 B_{dv} \geq P_{uy} \rightarrow N_1 \geq \frac{P_{uy}}{B_{dv}} \qquad (12.6.1)$$

The limit state of direct tension requires:

$$N_2 B_{dt} \geq P_{uz} \rightarrow N_2 \geq \frac{P_{uz}}{B_{dt}} \qquad (12.6.2)$$

The limit state of combined shear and tension requires:

$$N_3 B'_{dt} \geq P_{uz} \rightarrow N_3[C_1 - C_2 B_{uv}] \geq P_{uz} \rightarrow$$

$$N_3 \left[C_1 - C_2 \frac{P_{uy}}{N_3} \right] \geq P_{uz} \rightarrow N_3 C_1 - C_2 P_{uy} \geq P_{uz}$$

$$N_3 \geq \frac{P_{uz} + C_2 P_{uy}}{C_1} \qquad (12.6.3)$$

The number of bolts required to transmit the factored load is, therefore, given by:

$$N \geq \max [N_1, N_2, N_3] \qquad (12.6.4)$$

12.6.2 Slip-Critical Joints

If the joint considered is a slip-critical joint, the joint must provide adequate slip resistance under shear loads. This resistance could be checked using the factored loads (using LRFDS Sections J3.8a and J3.9a) or using the service loads (using LRFDS Sections J3.8b and J3.9b).

For information on bolted joints in shear and tension designed as slip-critical joints, including numerical examples, refer to Section W12.3 on the website http://www.mhhe.com/Vinnakota.

Analysis of Bolted Joint in Shear and Tension   E X A M P L E   1 2 . 6 . 1

Determine the adequacy of the brace connection to the flange of a W141×74 as shown in Fig. X12.6.1. The brace is subjected to a wind load of 20 kips

[continues on next page]

Example 12.6.1 continues ...

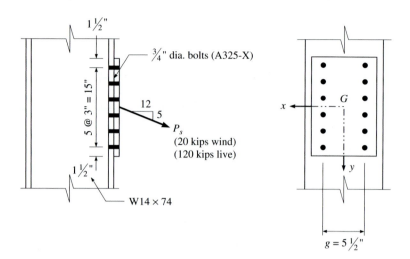

Figure X12.6.1

and a roof live load of 120 kips. Twelve ¾-in.-dia. A325 bolts are used with threads excluded from the shear plane. Consider the joint as a bearing-type joint.

Solution

The bolt group is subjected to both a direct shear load and an axial tensile load.

The factored load in the brace, $P_u = 1.6L_r + 0.8W = 1.6(120) + 0.8(20) = 208$ kips

The factored load components on the joint are:

$$P_{uy} = \frac{5}{13}(208) = 80.0 \text{ kips} \downarrow; \quad P_{uz} = \frac{12}{13}(208) = 192 \text{ kips} \rightarrow$$

The joint is to be analysed as a bearing-type joint.

Factored shear force on a bolt, $B_{vu} = \dfrac{P_{uy}}{N} = \dfrac{80.0}{12} = 6.67$ kips

Factored tensile force on a bolt, $B_{tu} = \dfrac{P_{uz}}{N} = \dfrac{192}{12} = 16.0$ kips

From LRFDM Table 7-10, design strength of a ¾-in.-dia. A325-X bolt in single shear:

$$B_{dv} = 19.9 \text{ kips} > B_{vu} = 6.67 \text{ kips} \quad \text{O.K.}$$

From LRFDM Table 7-14, design strength of a ¾-in.-dia. A325 bolt in direct tension:

$$B_{dt} = 29.8 \text{ kips} > B_{tu} = 16.0 \text{ kips} \quad \text{O.K.}$$

Design tensile strength of ¾-in.-dia. A325-X type bolt in the presence of shear load:

$$B'_{dt} = \min\,[0.75(117 - 2.0f_{vu})A_b,\ B_{dt}]$$

$$= \min\,\{[0.75(117)(0.442) - 0.75(2.0B_{vu})],\ B_{dt}\}$$

$$= \min\,\{[38.8 - 1.5(6.67)],\ 29.8\}$$

$$= 28.8\text{ kips} > B_{tu} = 16.0\text{ kips}\quad\text{O.K.}$$

Thus the bolts are adequate for use as a bearing-type joint.

Design of Bolted Joint in Shear and Tension

EXAMPLE 12.6.2

A WT12×51.5 is used as a bracket to transmit a 120 kip service load (25 percent dead load and 75 percent live load) to the flange of a W14×90, as shown in Fig. X12.6.2. Both the column and the bracket are of A992 steel. Determine the required number of ⅞-in.-dia. A325-N bolts assuming a bearing-type joint with threads excluded from the shear plane.

Solution

The joint is to be analyzed as a bearing-type connection.
Factored load, $P_u = 1.2D + 1.6L = 1.2(0.25)(120) + 1.6(0.75)(120)$
$$= 180\text{ kips}$$

Shear load, $P_{uy} = \dfrac{3}{5}(180) = 108$ kips

Tension load, $P_{uz} = \dfrac{4}{5}(180) = 144$ kips

Diameter of bolt, $d = \dfrac{7}{8}$ in. $\rightarrow A_b = 0.601$ in.; $d_h = \dfrac{15}{16}$ in.

Figure X12.6.2

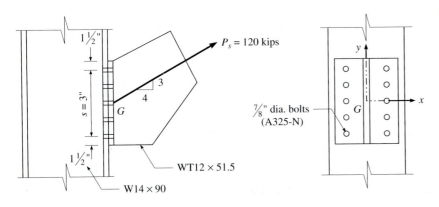

[continues on next page]

Example 12.6.2 continues ...

Design strength of $\frac{7}{8}$-in.-dia. A325-N bolt in single shear,

$$B_{dv} = 0.75(0.40F_{ub})\, A_b = 0.75(0.40)(120)(0.601) = 21.6 \text{ kips}$$

Thickness of flange of WT12×51.5 = 0.98 in.
Thickness of flange of W14×90 = 0.710 in.
It can be shown that the bearing strength of these connected elements of A992 steel is greater than the shear strength of the bolts.
Design strength of the bolt in direct tension,

$$B_{dt} = 0.75(0.75F_{ub})\, A_b = 0.75(0.75)(120)(0.601) = 40.6 \text{ kips}$$

Design strength of bolt in combined shear and tension,

$$B'_{dt} = 0.75\,[117 - 2.5\,f_v]\,A_b = 0.75(117)(0.601) - 0.75(2.5B_{vu})$$

$$= 52.8 - 1.88B_v \equiv C_1 - C_2 B_{vu}$$

Number of bolts required for the limit states of:

Shear (Eq. 12.6.1): $N_1 \geq \dfrac{P_{uy}}{B_{dv}} = \dfrac{108}{21.6} = 5.0$, say 5

Tension (Eq. 12.6.2): $N_2 \geq \dfrac{P_{uz}}{B_{dt}} = \dfrac{144}{40.6} = 3.55$, say 4

Combined shear and tension (Eq. 12.6.3):

$$N_3 \geq \frac{P_{uz} + C_2 P_{uy}}{C_1} = \frac{144 + 1.88(108)}{52.8} = 6.6, \text{ say } 8$$

Provide 8 bolts, 4 in each vertical row, using a gage, $g = 5\frac{1}{2}$ in., pitch $p = 3$ in., and an end distance of $1\frac{1}{2}$ in.   (Ans.)

12.7   Bolted Joints in Shear and Bending (Loading Case $P_y\, M_{yz}$)

Bolted joints in steel building and bridge structures are often subjected to shears and bending moments. Beam-to-column moment connections and bracket connections are some examples. In beam-to-column moment connections, the bending moment acting on a joint results from continuous structural action. The amount of continuity achieved, and hence the bending moment on the joint, depends upon the strength and stiffness of the connection. Their behavior and design will be considered in Chapter 13. Crane brackets are often attached to the flange face of the column, to take advantage of the stronger bending axis of the column (Figs. 6.1.2f and 11.11.1a). In practice, these brackets consist of a tee-section or a cut W-section with its flange bolted against that of the column.

Let us consider a tee-stub bracket whose flange is connected to the flange of a column by two vertical lines of bolts as shown in Figs. 12.7.1a and b.

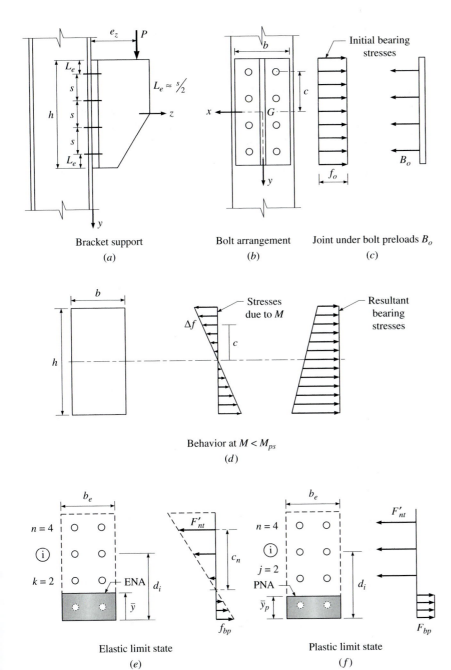

Bracket support
(a)

Bolt arrangement
(b)

Joint under bolt preloads B_o
(c)

Behavior at $M < M_{ps}$
(d)

Elastic limit state
(e)

Plastic limit state
(f)

Figure 12.7.1: Behavior of a bolted joint in shear and bending.

If a vertical load, P, is applied at a distance, e_z, from the face of the bracket support, the bolted joint (faying surface between the flange of the bracket and the flange of the column) is subjected to direct shear load P_y $(= P)$ acting in the plane of the joint and a bending moment M_{yz} $(= Pe_z)$ resulting from the eccentric, out-of-plane load P. All of the bolts in the joint are subjected to shear by the direct load P_y while the bolts in the upper part are also subjected to tension by the moment M_{yz}. LRFDS J3.1 stipulates that A490 high-strength bolts in such joints subject to tension loads shall be installed with a prescribed initial tension in them, regardless of whether the joint is considered as a bearing-type or as a slip-critical type. Also, since both shear and tension are present in the bolts, their design strengths are governed by the interaction formulas of the LRFDS Section J3.

When a moment is applied to such a bolted joint, a rotation of the bracket flange plate takes place such that the upper part of the plate will tend to pull away and thus experience tension, while the lower part of the plate presses against the column. The rotation is accommodated by deformation of the connection elements (bracket plate and column flange) and, once the prestress has been overcome, by elongation of the bolts. These joints have traditionally been analysed using either the elastic concrete analogy or the plastic (Whitney) contact stress block coupled with bolts acting at their ultimate strengths. The elastic analogy assumes a linear stress distribution over the depth of the joint with the elastic neutral axis (ENA) displaced towards the compression end of the joint. The elastic limit state corresponds to the load for which the tension in the extreme bolt reaches its nominal tensile strength, or the bearing pressure at the bottom edge reaches the nominal bearing pressure. The plastic analogy (or plastic limit state) sets the stresses of all connection components at yield values. Balancing the tensile and compressive forces places the plastic neutral axis (PNA) quite far down the joint, usually below the location of the ENA.

The direct shear load $P_y (= P)$ is assumed to be distributed equally among all the bolts of the group. Thus,

$$B_v = \frac{P}{N} \tag{12.7.1}$$

where N is the number of bolts in the joint, and B_v is the shear force in any bolt.

Let B_o be the pretension, if any, applied to each of the N bolts in the bolt group. This pretension will precompress the flange plates together, resulting in equilibrating contact or bearing stresses, f_o (Fig. 12.7.1c). Assuming that the initial bearing stress is uniform over the contact area bh, we have:

$$f_o = \frac{NB_o}{bh} = \frac{nn_r B_o}{bh} \tag{12.7.2}$$

where b is the width of the bracket flange plate; h is the height of the plate; n_r is the number of bolts in each horizontal row; n is the number of bolts in each vertical line; and N is the total number of bolts in the bolt group ($= nn_r$). If s is the vertical spacing (pitch) of bolts, the contact area tributary to each bolt is (bs/n_r), and the resultant compressive force acting on this area is:

$$C_o = \frac{bs}{n_r} f_o = B_o \qquad (12.7.3)$$

12.7.1   Limit State of Plate Separation at Top Edge

When a moment M is applied to this assembly, the distribution of bolt loads is a function of the stiffness of the flange plates, the bolt pretension, and the level of applied moment. If the flange plates are stiff, the deformations due to transverse bending of the plates, and hence the prying forces that develop in the bolts, may be neglected. We further assume that plane sections (flange faces) remain plane. So, for low levels of moment M, the bracket flange plate rotates about the center of gravity of the contact area, bh. This rotation will change the contact stresses and hence the forces in the bolts. In effect, the compression at the top will be relieved, and the compression at the bottom will increase (Fig. 12.7.1d). The change Δf in the bearing stress, at the extreme bolt located at a distance c from the x axis, can be calculated using the beam formula, as:

$$\Delta f = M \frac{c}{I} = \frac{6M}{bh^3} (h - s) \qquad (12.7.4a)$$

The reduction ΔC in the contact force C_o, on the area tributary to the extreme bolt is:

$$\Delta C = \Delta f \frac{bs}{n_r} = \frac{6M}{n_r h^3} (h - s)s \qquad (12.7.4b)$$

As in the case of a single pretensioned bolt under external tension load, studied in Web Section W6.6, the increase ΔB in bolt tension B_o will be much smaller than the change ΔC in the plate compression (see Web Eq. W6.6.10). This is true as long as there is no plate separation, that is, as long as the force, ΔC, induced in the end bolt by the moment is less than C_o. Denoting the corresponding applied moment as M_{ps}, we have from Fig. W6.6.2 and Eqs. W6.6.10 and 12.7.4b:

$$\frac{6M_{ps}}{n_r h^3} (h - s)s = \Delta C = C_o = B_o$$

or, rewriting:

$$M_{ps} = \frac{n_r h^3}{6(h - s)\, s} B_o \qquad (12.7.5)$$

Here,

B_o = bolt pretension, if any, for snug tightened joints

= T_b given in LRFDS Table J3.1, for bolts in pretensioned and slip-critical joints

M_{ps} = bending moment on the bracket at the start of plate separation

The design load on the bracket, P_{dps}, corresponding to the limit state of separation of plates at the top edge may be determined from the relation:

$$P_{dps} = \phi \frac{n_r h^3}{6(h - s)s} \frac{T_b}{e_z}$$ (12.7.6a)

In slip-critical joints, although the friction that provides shear resistance is reduced at the top of the bracket, it is increased at the bottom, and the total frictional (shear) resistance remains virtually unchanged. For this reason, the reduction in design strength B_{ds} required by LRFD Specification Section J3.9 is not applicable. Therefore, the design shear strength and design tension strength must conform to the provisions of LRFDS Sections J3.8 and J3.6, respectively. Let B_{dsf} be the design resistance of a bolt to shear at service loads using factored loads (LRFDM Table 7-15). The design load on the bracket, corresponding to the limit state of slip is given by:

$$P_{dsf} = N B_{dsf}$$ (12.7.6b)

The design load on the bracket connection, considered as a slip-critical connection, is given by:

$$P_{dsc} = \min [P_{dps}, P_{dsf}]$$ (12.7.6c)

12.7.2   Elastic Limit State (Yielding of the Extreme Bolt)

For values of M larger than M_{ps}, the precompression on the upper portion of the faying surfaces is released (the bolts are in phase II described in Web Section W6.6 for a single bolt). Thus, as the bracket tends to deform under the applied moment M, the top bolts are subjected to tension from the moment and to shear from the direct load P_y. The compression due to the moment is resisted by the bearing stresses at the faying surface at the bottom part of the bracket plate with the column flange. Holes are generally disregarded so that the compressed part may be considered as rectangular. The contact or bending section between the bracket and the face of the column therefore consists of tensile bolt areas above the neutral axis and a rectangular-shaped compression area below the neutral axis. The elastic neutral axis (ENA) is generally located well below the center of gravity of the bolt group. Due to flexural deformation of the flanges, the bearing stress at the lateral edges of the flange plate is less than the uniform bearing stress across the width of the flange plate indicated by the flexure formula. To account for the effect of such deformation, an effective width b_e may be used instead of the entire width b of the bracket flange. The effective flange width for bearing will depend upon the bending stiffness of the bracket flange. The LRFD Manual suggests that the effective flange width be limited to $8t_f$, where t_f is the thickness of the thinner flange. The tensile stress in the upper bolts and the compressive stress at the bottom will vary in direct proportion to their distances from the ENA; bolts below the ENA are assumed to resist shear only. The assumed triangular variation in the pressure zone is probably far from the truth, but the zone is small and this assumed distribution of pressure will not affect the results greatly.

To determine between which rows of bolts the ENA falls and to initiate the iteration, the distance from the bottom of the bracket to the ENA is often initially assumed to be one-sixth to one-seventh of the height h of the bracket flange plate. The location of the ENA is then verified from the fact that the static moment of the nominal areas of the bolts above the ENA should equal the static moment of the compressive area below the ENA, both moments being taken about an arbitrary reference axis. For example, taking moments about the assumed ENA, we obtain (Fig. 12.7.1e):

$$\frac{b_e \bar{y}^2}{2} = \sum_{i=k}^{n} n_r A_b (d_i - \bar{y}) \qquad (12.7.7)$$

where
$\bar{y}$ = depth of compression block from the bottom edge
A_b = nominal area of one bolt
b_e = effective width of the compression block
   = min $[b, 8t_f]$
k = the number of the first row of bolts immediately above the ENA ($k = 2$ in Fig. 12.7.1e)
d_i = distance from the bottom edge to the ith row of bolts

If the neutral axis as calculated from Eq. 12.7.7 proves to be between rows other than those assumed in the trial calculations, the equation must be rewritten with a new trial position. Once $\bar{y}$ is found, the moment of inertia of the effective area, about its neutral axis, is calculated from:

$$I = \frac{b_e \bar{y}^3}{3} + \sum_{i=k}^{n} (n_r A_b)(d_i - \bar{y})^2 \qquad (12.7.8)$$

The critical tension load will always be in the top row of bolts. If the distance from the elastic neutral axis to the top row of bolts is denoted by c_n, the tensile load in the critical bolt is:

$$B_t = \frac{Mc_n}{I} A_b \qquad (12.7.9a)$$

As the bolts are subjected to combined shear and tension, their design strength is governed by the interaction formulas studied in Web Section W6.4. Let B'_{nt}

be the nominal tensile strength of a bolt in a bearing-type joint, when the bolt is subjected to combined shear and tension loading. If M_{els1} is the moment corresponding to the limit state of yielding of the extreme bolt, then from Eq. 12.7.9a:

$$M_{els1} = \frac{B'_{nt}}{A_b} \frac{I}{c_n} \qquad (12.7.9b)$$

Also, the maximum bearing stress at the bottom edge of the bracket plate is:

$$f_{bp} = \frac{M\bar{y}}{I} \qquad (12.7.10a)$$

Let M_{els2} represent the moment, when f_{bp} equals the nominal bearing pressure for the limit state of bearing of the plate on the support, then:

$$M_{els2} = \frac{F_{bp}I}{\bar{y}} \qquad (12.7.10b)$$

where F_{bp} = nominal bearing pressure = F_y

Thus, the controlling elastic limit state is reached when the extreme bolt reaches its nominal tensile strength, B'_{nt}, or when the bearing stress reaches the nominal bearing pressure, F_{bp}, whichever occurs first. Hence,

$$M_{els} = \min [M_{els1}, M_{els2}] \qquad (12.7.11a)$$

The design strength of the bearing-type joint, corresponding to the limit state of yielding is:

$$P_{dels} = \phi \frac{M_{els}}{e_z} \qquad (12.7.11b)$$

12.7.3   Plastic Limit State

The moment corresponding to the plastic limit state of the bolted connection may be determined by assuming the limiting stress distribution shown in Fig. 12.7.1f. Here, the forces in the bolts are assumed to be at their nominal tensile strength B'_{nt} (which is the nominal strength in tension, B_{nt}, reduced because of the presence of shear). The bearing area is assumed to be compressed uniformly to the nominal bearing pressure, F_{bp}. The location of the plastic neutral axis (PNA) is determined from the force equilibrium:

$$b_e \bar{y}_p F_{bp} = n_r \sum_{i=j}^{n} B'_{nt} \qquad (12.7.12)$$

where $\bar{y}_p$ = distance from the bottom edge of the compression zone to the PNA.

j = the number of the first row of bolts immediately above the PNA ($j = 2$ in Fig. 12.7.1f)

F_{bp} = nominal bearing pressure ($= F_y$)

b_e = effective width of the compression block

As b_e, F_y, n_r, and B'_{nt} are known, the position of the plastic neutral axis, $\bar{y}_p$, is easily found after making a proper estimate of j. The moment corresponding to the plastic limit state of the connection is:

$$M_{pls} = \frac{1}{2} b_e \bar{y}_p^2 F_{bp} + \sum_{i=j}^{n} n_r B'_{nt} (d_i - \bar{y}_p) \qquad (12.7.13a)$$

The design strength of the joint corresponding to the plastic limit state is:

$$P_{dpls} = \phi \frac{M_{pls}}{e_z} \qquad (12.7.13b)$$

Remarks

1. The method is necessarily approximate in assuming that plane sections will remain plane as the end plate will necessarily deform during the transmission of the bending moment to the plates. In addition, the effect of prying action has been ignored.

2. Section 12.7.1 applies to slip-critical joints. Sections 12.7.2 and 3 apply to bearing-type joints. In joints considered in Sections 12.7.2 and 3, bolts above the neutral axis are subjected to the shear force B_{vu}, the tensile force B_{tu}, and the effect of prying action; bolts below the neutral axis are subjected to the shear force B_{vu} only.

3. To design a bolted joint to transmit an eccentric load, a trial-and-error procedure is generally used. The bolt pattern and the number of bolts are estimated. Then an analysis is made on the trial joint to determine the design load on the joint using the appropriate limit states described in Sections 12.7.2 and 3. If the design load is less than the factored load P_u on the bracket, the number of bolts is increased, or their arrangement is changed to provide greater resistance to the applied loads. If the computed load is far greater than the factored load, the number of bolts may be reduced. The connection is revised and reanalysed until a satisfactory number and arrangement of bolts is determined.

12.7.4   Plastic Neutral Axis at C.G. of the Bolt Group

Eccentrically loaded bolt groups of the type $[P_y M_{yz}]$ considered in this section (Fig. 12.7.1a) may be conservatively designed by using the method presented in Part 7 of the LRFDM. Here, the plastic neutral axis is assumed to be located at the centroid of the bolt group. The bolts above the neutral axis are in tension, and the bolts below the neutral axis are *said to be in compression*.

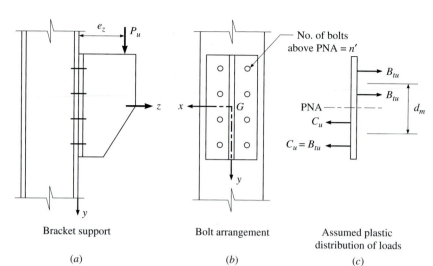

Figure 12.7.2: Bolted joint in shear and bending, PNA at C.G. of the bolt group.

Bracket support

(a)

Bolt arrangement

(b)

Assumed plastic distribution of loads

(c)

A plastic stress distribution as shown in Fig. 12.7.2 is assumed. The tensile force in each bolt above the neutral axis caused by the eccentricity of the factored load about the bolt group is given by:

$$B_{tu} = \frac{P_u e_z}{n' d_m} \qquad (12.7.14)$$

where P_u = factored applied load on the joint, kips

e_z = eccentricity of applied load about the bolt group, in.

n' = number of bolts above the neutral axis

d_m = moment arm between resultant tensile force and resultant compressive force, in.

The shear force in each bolt due to the factored applied load is given by:

$$B_{vu} = \frac{P_u}{N} \qquad (12.7.15)$$

Bolts above the neutral axis are subjected to the tensile force B_{tu}, the shear force B_{vu}, and the effect of prying action; bolts below the neutral axis are subjected to the shear force B_{vu} only. The LRFDS requires:

$$B_{tu} \leq B'_{dt}; \qquad B_{vu} \leq B_{dv} \qquad (12.7.16)$$

E X A M P L E   1 2 . 7 . 1

Analysis of Bolted Joint in Shear and Bending

A section cut from a W18×71 shape is attached to a column flange with 10 fully tightened $\frac{7}{8}$-in.-dia. A325-X bolts in two vertical lines with a pitch of

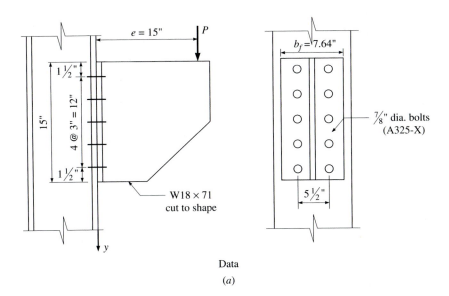

Data

(a)

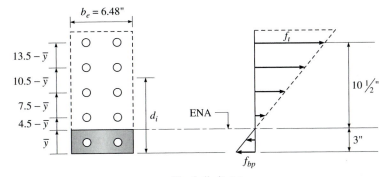

Elastic limit state

(b)

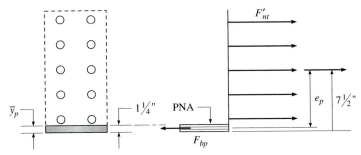

Plastic limit state

(c)

Figure X12.7.1

[*continues on next page*]

Example 12.7.1 continues ...

3 in. and an end distance of $1\frac{1}{2}$ in. The bracket is subjected to a vertical load P with an eccentricity of 15 in. Determine the capacity P: (a) assuming no separation at factored loads, (b) if separation is permitted, and (c) at the ultimate strength. Prying action may be neglected.

Solution

For a W18×71: $b_f = 7.64$ in.; $t_f = 0.810$ in.
Design strength of A325-X bolt of $\frac{7}{8}$-in.-dia. in:

$$\text{Shear, } B_{dv} = 0.75(0.50F_{ub}) \; m \; A_b = 0.75(0.50)(120.0)(1)(0.601)$$
$$= 27.1 \text{ kips}$$

$$\text{Tension, } B_{dt} = 0.75(0.75F_{ub}) \; A_b = 0.75(0.75)(120.0)(0.601)$$
$$= 40.6 \text{ kips}$$

$$\text{Combined shear and tension, } B'_{dt} = 0.75[117 - 2.0f_v] \; A_b$$
$$= 0.75(117)(0.601) - 0.75(2.0) \; B_{vu} = 52.7 - 1.5B_{vu}$$

Here, B_{vu} is the shear force in the bolt under factored loading. From LRFDS Table J3-1, minimum bolt pretension T_b for $\frac{7}{8}$-in.-dia. A325 bolts equals 39 kips.

a. Limit state of plate separation

From Eq. 12.7.6a, the design load on the bracket corresponding to the limit state of separation of the plates is:

$$P_{dps} = \phi \frac{n_r h^3}{6(h-s)s} \frac{T_b}{e_z} = \frac{0.75(2)(15.0)^3}{6(15.0 - 3.0)(3)} \frac{39.0}{15.0} = 60.9 \text{ kips}$$

From Eq. 12.7.6b, the design load on the bracket, corresponding to the limit state of slip is:

$$P_{dsf} = NB_{dsf} = 10(14.5) = 145 \text{ kips}$$

So, the design strength of the connection, P_{dsf}, is 60.9 kips, if separation of plates is to be prevented at factored loads.   (Ans.)

b. Elastic limit state
$b_e = \min [b_f, 8t_f] = \min [7.64, 6.48] = 6.48$ in.
Assuming that the elastic neutral axis is approximately one-sixth the total height of the bracket ($\approx \frac{1}{6}(15) = 2.5$) from the bottom, it would seem to fall between the two lower horizontal rows of bolts. By equating the sum of the static moments taken about the assumed position of the neutral axis of the bolt areas above to that of the compressed area of the bracket flange below the neutral axis, we obtain the expression derived

by applying Eq. 12.7.7:

$$\frac{1}{2}(6.48)\bar{y}^2 = 2(0.601)[(4.5 - \bar{y}) + (7.5 - \bar{y}) + (10.5 - \bar{y})$$

$$+ (13.5 - \bar{y})]$$

$$\rightarrow \bar{y}^2 + 1.485\,\bar{y} - 13.36 = 0$$

$$\rightarrow \bar{y} = 3.0 \text{ in.} \quad \text{O.K.,} \quad \text{as } 1.5 < \bar{y} = 3.0 < 4.5$$

From Eq. 12.7.8:

$$I = \frac{1}{3}(6.48)(3.0^3) + 2(0.601)(1.5^2 + 4.5^2 + 7.5^2 + 10.5^2)$$

$$= 286 \text{ in.}^4$$

From Eq. 12.7.9a, the tensile load in the critical bolt,

$$B_t = \frac{P(15.0)}{286}(10.5)(0.601) = 0.331P$$

According to the LRFDS,

$$B_v \leq B_{dv} \rightarrow 0.10P \leq 27.1 \rightarrow P \leq 271 \text{ kips}$$

$$B_t \leq B_{dt} \rightarrow 0.331P \leq 40.6 \rightarrow P \leq 122 \text{ kips}$$

$$B_t \leq B'_{dt} \rightarrow 0.331P \leq 52.7 - 1.5B_v$$

$$\rightarrow 0.331P \leq 52.7 - 1.5(0.10)P \rightarrow P = 110 \text{ kips}$$

From Eq. 12.7.10a, maximum bearing pressure,

$$f_{bp} = \frac{(110)(15.0)}{286}(3.0) = 17.3 \text{ ksi}$$

$$< \phi F_y = 0.75(50) = 37.5 \text{ ksi} \qquad \text{O.K.}$$

So, if separation is permitted, and elastic stress distribution is assumed, the design strength of the connection, P_{dels}, is 110 kips.   (Ans.)

c. Plastic limit state
   If the interaction of the shear and tension is taken into account,

$$B'_{dt} = 52.7 - 1.5B_v = 52.7 - 0.15P$$

Assume that the PNA is below the bottom row of bolts as shown in Fig. X12.7.1c. The PNA can be located by using Eq. 12.7.12:

$$6.48(\bar{y}_p)(0.75)(50) = 10\,[52.7 - 0.15P]$$

$$\bar{y}_p = 2.17 - \frac{P}{162}$$

[continues on next page]

Example 12.7.1 continues ...

The design strength can be obtained by computing the moment of the couple formed by the compressive and tensile forces. Taking the moment of the bolt tensile forces about the resultant compressive force C:

$$M_d = 10(52.7 - 0.15P)\left[7.5 - \frac{1}{2}\left(2.17 - \frac{P}{162}\right)\right]$$

$$= 3385 - 7.99P - 0.00463P^2$$

As per the LRFD,

$$3385 - 7.99P - 0.00463P^2 \geq 15P \rightarrow P \leq 143 \text{ kips}$$

Also,

$$\bar{y}_p = 2.17 - \frac{P}{162} = 2.17 - \frac{143}{162} = 1.29 \text{ in.}$$

Assumption about the location of the PNA is O.K. as, $\bar{y}_p = 1.29$ in. $<$ 1.5 in.
The design strength of the connection corresponding to the plastic limit state is given by:

$$P_{dpls} = \min[271, 143] = 143 \text{ kips} \qquad \text{(Ans.)}$$

EXAMPLE 12.7.2

Analysis of Bolted Joint in Shear and Bending

Determine the capacity P_u of the eccentrically loaded bracket considered in Example 12.7.1, using the simplified method suggested in Part 7 of the LRFDM (Section 12.7.4). Prying action may be neglected.

Solution

For the bolt group given, from Figs. X12.7.2a and b:

$$e_z = 15 \text{ in.}; \quad B'_{dt} = 52.7 - 0.15P_u$$

Number of bolts above the neutral axis, $n' = 4$
Moment arm between resultant compressive and tensile forces, $d_m = 9$ in.
Equating B_{tu} from Eq. 12.7.14 to B'_{dt} and rearranging, we obtain:

$$P_d = \frac{B'_{dt}\, n'\, d_m}{e_z} = \frac{(52.7 - 0.15P_u)(4)(9.0)}{15.0} \rightarrow P_d = 93.0 \text{ kips}$$

Compare this result (conservative) with the values of 110 kips and 143 kips obtained for P_{dels} and P_{dpls}, respectively, using the methods developed in Sections 12.7.2 and 3.

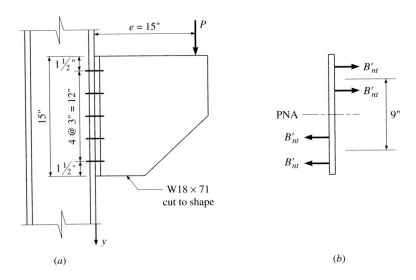

(a) $\qquad\qquad\qquad\qquad\qquad\qquad (b)$ $\qquad\qquad$ **Figure X12.7.2**

12.8   Welded Joints in Direct Shear (Loading Case P_x P_y)

12.8.1   Balanced Welds

For members designed as axially loaded tension members, LRFDS Section J1.8 requires that the C.G. axis of the end-connection welds coincides with the center of gravity axis of the member; otherwise provision has to be made in the member design for the resultant eccentricity (see Section 12.9). This requirement is not applicable, however, to end connections of statically loaded single angle, double angle, and similar members. In effect, tests conducted by Gibson and Wake [1942] have shown that welded angle connections subjected to static loads do not need to have the weld group balanced. The tests were made on members connected by balanced and unbalanced welding. For single-angle tension members, the ultimate strength of the balanced connection was about 3 percent greater than for the unbalanced connection. For double-angle tension members, the test results for the two types were similar.

Let us consider an angle member subjected to an axial tensile force T_u under factored loads, and welded to a gusset plate at its end. Let c be the size of the connected leg. Also, let W_d be the design strength per unit-length of the selected weld and L_{req} the total length of the weld required to transfer the factored load T_u. We have:

$$L_{\text{req}} = \frac{T_u}{W_d} \qquad\qquad (12.8.1)$$

Figure 12.8.1 shows four different arrangements of providing this required weld length. Figures 12.8.1a and b provide two symmetric welds having

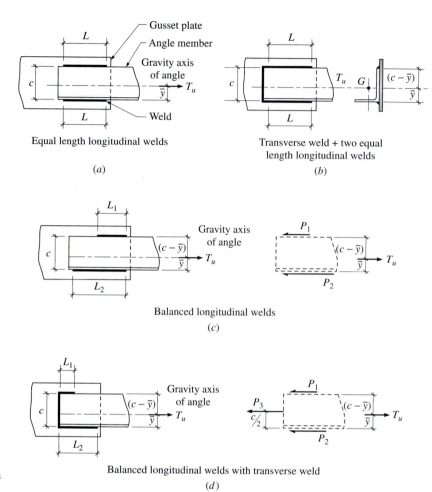

Figure 12.8.1: Welding of an angle member to a gusset plate.

lengths $L = \frac{1}{2} L_{\text{req}}$ and $L = \frac{1}{2} (L_{\text{req}} - c)$, respectively. Note that the tensile force does not pass through the C.G. of the weld. As mentioned already, for static loading, the resulting eccentricity could be neglected in design. Figures 12.8.1c and d provide two asymmetric welds with weld lengths adjusted such that the tensile force passes through the C.G. of the weld.

Let L_1, L_2 be the lengths of longitudinal welds at the toe and heel of the angle that result in a balanced end connection. Fig. 12.8.1c also shows the forces that keep the angle in equilibrium. Here, P_1 and P_2 are the forces exerted by the welds on the angle at the toe and heel, respectively. They are assumed to act along the edges of the angle. The force T_u acts along the centroidal axis of the angle ($\bar{y}$ from the heel and $c - \bar{y}$ from toe). We have:

$$P_1 = L_1 W_d; \qquad P_2 = L_2 W_d \qquad (12.8.2)$$

The balanced welds yield a resultant weld force that lies along the same line of action as the applied load and has the same magnitude. From Eqs. 12.8.1, 12.8.2, and from equilibrium of forces,

$$\sum F_x = 0 \rightarrow P_1 + P_2 - T_u = 0 \rightarrow L_1 + L_2 = L_{req} \quad (12.8.3)$$

From equilibrium of moments about the line of action of P_2, we obtain:

$$cP_1 - T_u\bar{y} = 0 \rightarrow cL_1W_d - L_{req}W_d\bar{y} = 0 \rightarrow L_1 = \frac{\bar{y}}{c}L_{req} \quad (12.8.4)$$

Values of c and $\bar{y}$ are taken from Part 1 of the LRFDM for the angle considered. From equilibrium of moments about the line of action of P_1 we have:

$$cP_2 - T_u(c - \bar{y}) = 0 \rightarrow L_2 = \frac{(c - \bar{y})}{c}L_{req} \quad (12.8.5)$$

As can be seen from Eq. 12.8.3, the total length of the weld is unchanged; only the placement of the weld is different for balanced welds (L_1 and L_2, instead of $\frac{1}{2}L_{req}$ each). The length of the connection (not to be confused with the length of the weld, which does not change) may be reduced if, in addition to the longitudinal welds, a fillet weld of length c is placed along the transverse end of the angle, as shown in Fig. 12.8.1d.

One situation where reducing the length of the connection may be advantageous occurs for the case of welded roof trusses and open web steel joists. In the fabrication of these trussed structures, tee-sections are often used for the chords, and angle sections are welded to the stem of the tee and serve as the web members. Unless a transverse weld is used, the stem of the tee may not be deep enough to provide for the requisite total length of weld. Welds L_1, L_2, and c are of the same size, and Eq. 12.8.1 is still valid. In addition, we have:

$$P_1 = L_1W_d; \quad P_2 = L_2W_d; \quad P_3 = cW_d \quad (12.8.6)$$

From Eqs. 12.8.1, 12.8.6, and from equilibrium of forces, we have:

$$\sum (F)_x = 0 \rightarrow P_1 + P_2 + P_3 - T_u = 0$$
$$\rightarrow L_1 + L_2 + c = L_{req} \quad (12.8.7a)$$

Noting that the force P_3 will act at the centroid of the transverse weld located at $\frac{1}{2}c$, equilibrium of the moments about the line of action of P_2 results in:

$$P_1c + P_3\frac{c}{2} - T_u\bar{y} = 0 \rightarrow L_1 = \frac{\bar{y}}{c}L_{req} - \frac{c}{2} \quad (12.8.8a)$$

Similarly, equilibrium of the moments about the line of action of P_1 gives:

$$P_2c + P_3\frac{c}{2} - T_u(c - \bar{y}) = 0 \rightarrow L_2 = \frac{(c - \bar{y})}{c}L_{req} - \frac{c}{2} \quad (12.8.9a)$$

LRFDS Appendix Section J2.4 permits a 50 percent increase in design strength for a transverse weld compared to a longitudinal weld. If this increased strength for a transverse weld is taken into account, the relations 12.8.7a, 12.8.8a, and 12.8.9a for the balanced weld derived earlier become:

$$L_1 + L_2 + 1.5c = L_{req} \tag{12.8.7b}$$

$$L_1 = \frac{\bar{y}}{c} L_{req} - \frac{1.5c}{2} \tag{12.8.8b}$$

$$L_2 = \frac{(c - \bar{y})}{c} L_{req} - \frac{1.5c}{2} \tag{12.8.9b}$$

If Eq. 12.8.8a or Eq. 12.8.8b results in a negative value for L_1, this indicates that the load T is not sufficiently large to justify the use of the transverse weld c, at least not for the full width.

As mentioned earlier, balanced welds, though desirable, are not required by the LRFDS for statically loaded single-angle and double-angle members designed as axially loaded tension members. Sometimes the weld is made a sufficient length to develop the full strength of the member being connected, even though the actual factored load in the member may be somewhat less than this amount.

E X A M P L E   1 2 . 8 . 1

Design of Balanced Welds

The long leg of a $7 \times 4 \times \frac{3}{8}$ in. angle is welded to a $\frac{3}{8}$-in.-thick gusset plate at its end. Assume A36 steel, and design the shortest balanced weld to transmit a factored tensile load of 120 kips (see Fig. 12.8.1d).

Solution

Minimum weld size, $w_{min} = \frac{3}{16}$ in.   (LRFDS Table J2.4)

Maximum weld size, $w_{max} = \frac{3}{8} - \frac{1}{16} = \frac{5}{16}$ in. (LRFDS Section J2.2b)
Since A36 steel is assumed, the appropriate electrode is E70XX. So, use $\frac{5}{16}$-in. fillet welds with E70 electrodes. Shear capacity per inch length of weld,

$$W_d = \phi(0.6F_{uw})(0.707\ w)(1) = 0.75(0.6)(70)(0.707)(\tfrac{5}{16})(1)$$
$$= 6.96 \text{ kli}$$

Required length of weld, $L_{req} = \dfrac{T_u}{W_d} = \dfrac{120}{6.96} = 17.2$ in.

From LRFDM Table 1-7 for a L7×4×$\frac{3}{8}$, $\bar{y}_L = 2.37$ in.

The lengths L_1 and L_2 of longitudinal welds for a balanced weld are given by Eqs. 12.8.8a and 12.8.9a as:

$$L_1 = \frac{\bar{y}}{c} L_{\text{req}} - \frac{c}{2} = \frac{2.37}{7.0}(17.2) - \frac{7.0}{2} = 2.34 \text{ in., say } 2\frac{1}{2} \text{ in.}$$

$$L_2 = \frac{c - \bar{y}}{c} L_{\text{req}} - \frac{c}{2} = \frac{(7.0 - 2.37)}{7.0}(17.2) - \frac{7.0}{2}$$
$$= 7.90 \text{ in., say } 8 \text{ in.}$$

Total length of weld provided, $L = L_1 + L_2 + c = 2.5 + 8.0 + 7.0 = 17.5$ in.

From LRFDM Table 1-7, for a L7×4×⅜: $A_L = 4.00$ in.^{2}, $\bar{x}_L = 0.861$ in. $= \bar{x}_{\text{con}}$

So, $U = 1 - \dfrac{\bar{x}_{\text{con}}}{L_{\text{con}}} = 1 - \dfrac{0.861}{8.0} = 0.892 < 0.9$

Design tensile strength corresponding to the limit state of yield on gross area,

$$T_{d1} = 0.9 A_g F_y = 0.9(4.00)(36) = 130 \text{ kips} > 120 \text{ kips} \quad \text{O.K.}$$

Design tensile strength corresponding to the limit state of fracture on effective net area,

$$T_{d2} = 0.75 A_n U F_u = 0.75(4.00)(0.892)(58) = 155 \text{ kips} > 120 \text{ kips} \quad \text{O.K.}$$

So, the angle with the balanced weld, consisting of a transverse weld and two longitudinal welds $L_1 = 2\frac{1}{2}$ in. and $L_2 = 8$ in. shown in Fig. X12.8.1, has adequate tensile strength.   (Ans.)

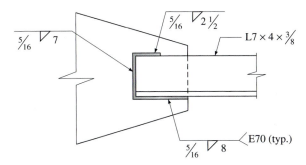

Figure X12.8.1 Solution.

12.8.2   Horizontal Shear in Built-Up Welded Beams

Figure 12.8.2 illustrates several examples of built-up welded beams. The welds connecting a cover plate to the flange of an I-beam, or the flange of a plate girder to the web, are designed in exactly the same way as the bolts of

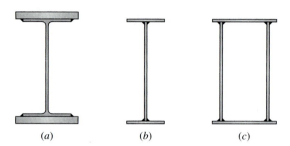

(a)   (b)   (c)

Figure 12.8.2: Examples of built-up welded beams.

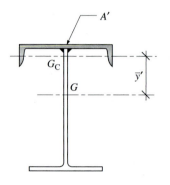

Figure 12.8.3: Parameters to calculate horizontal shear in built-up welded beams.

a bolted cover plated beam (Section 12.3.2). That is, the welds are proportioned to transfer the horizontal shear at the faying surface.

The built-up section shown in Fig. 12.8.3 is obtained by welding the web of a tee to the web of a channel shape to form a singly symmetric section. Let G_C represents the center of gravity of the channel and G the center of gravity of the built-up section. Let w be the size of the fillet weld along each side of the web of the tee. Also, let W_d be the design strength of each weld per unit length (values tabulated in Tables 6.19.1 and 6.19.2). If q_{sv} is the shear flow on the faying surface,

$$2W_d \geq q_{sv} = \frac{VQ'}{I} = \frac{VA'\bar{y}'}{I} \qquad (12.8.10)$$

where I = moment of inertia of the built-up section, in.[4]
   V = total factored shear force at that point along the beam where the weld size is desired, kips
   A' = cross-sectional area of the channel, cover plate, or flange plate at one flange, in.[2]
   $\bar{y}'$ = distance form the neutral axis of the built-up section to the centroid of the area A', in.
   Q' = static moment of the area A', in.[3]
   W_d = design strength of the weld selected, per unit length, kli
   q_{sv} = shear flow, kli

The smallest practical size of a fillet weld is usually considered to be $^3/_{16}$ in. Also, a minimum fillet weld size is stipulated in the LRFDS Section J2.2b as a function of the thickness of the thicker of the two parts joined (cover plate and beam flange). Therefore, whenever Eq. 12.8.10 indicates a weld of less than $^3/_{16}$ in. or w_{min}, it is better to use an intermittent weld of larger but convenient size (see Section 6.16.6). Equation 12.3.2, developed to determine the pitch of bolts, can be adopted to determine the spacing of intermittent welds:

$$p_{iw} \leq \frac{(2W_d L_{iw})I}{(A'\bar{y}')V} \qquad (12.8.11)$$

where p_{iw} = longitudinal pitch, center to center, of intermittent fillet
welds, in.
   L_{iw} = length of the intermittent fillet weld ($\geq 4w$; $\geq 1\frac{1}{2}$ in.)

Design of Welded Built-Up Beam

E X A M P L E   1 2 . 8 . 2

A W21×68 has $\frac{7}{8}$×16 in. cover plates attached to the flanges by fillet welds
as shown in Fig. 12.8.2a. Assume A572 Grade 50 steel and E70 electrodes.
Determine the required weld size at a point where the external shear is
220 kips under factored loads, if (a) continuous welds are required, (b) intermittent fillet welds are permitted.

Solution

W21×68 → d_W = 21.1 in.; t_{fW} = 0.685 in.; I_{xW} = 1480 in.^{4}

PL $\frac{7}{8}$×16 → A_{PL} = $\frac{7}{8}$(16) = 14.0 in.^{2}, $\bar{y}_{PL}$ = $\frac{21.1}{2}$ + $\frac{7}{16}$ = 11.0 in.

For the built-up section:

I_x = (1480 + 0) + 2 [0 + 14.0(11.0^2)] = 4870 in.^{4}

Q' = $A'\bar{y}'$ = 14.0(11.0) = 154 in.^{3}; V = 220 kips

The shear flow at the connection of cover plate to flange is:

q_{sv} = $\frac{VQ'}{I_x}$ = $\frac{220(154)}{4870}$ = 6.96 kli/2 welds = 3.48 kli/weld

a. If continuous welds are required

Required weld size, D = $\frac{3.48}{1.392}$ = 2.5, say 3, → w = $\frac{3}{16}$ in.

Thickness of thicker plate = $\frac{7}{8}$ in.
From Table J2.4 of the LRFDS, minimum size of fillet weld is $\frac{5}{16}$ in. for
a thickness of $\frac{3}{4}$ in. or more for the thicker part joined. So, use $\frac{5}{16}$-in. E70
continuous weld.   (Ans.)

b. If intermittent fillet welds are permitted
Assume $\frac{5}{16}$-in. welds (minimum size permitted by LRFDS Table J2.4).
Design strength of unit length weld, W_d = 1.392(5) = 6.96 kli

[*continues on next page*]

Example 12.8.2 continues ...

If 12-in. spacing is selected for intermittent welds, length of intermittent welds:

$$L_{iw} = \frac{p_{iw}\, q_{sv}}{W_d} = \frac{12(3.48)}{6.96} = 6.00 \text{ in.}$$

Use $^5/_{16}$-in. welds 6 in. long, 12 in. on center. (Ans.)

12.9 Welded Joints in Eccentric Shear (Loading Case $P_x\, P_y\, M_{xy}$)

As in the case of eccentrically loaded bolted joints, either an elastic method described in Section 12.9.1 or an ultimate strength method described in Section 12.9.2 may be used to analyze/design eccentrically loaded weld groups. The LRFD Specification does not prescribe which method of analysis is to be used for fillet weld configurations eccentrically loaded in shear. However, the LRFD Manual provides design tables for some common weld groups using the ultimate strength method.

12.9.1 Elastic Method

Figure 12.9.1a shows a welded joint subjected to an eccentric load in the plane of the joint. (A channel shaped weld ABCEF with its center of gravity located at point G is shown here). The eccentric load P is equivalent to a direct load P through the center of gravity of the weld and a torsional moment $M_T = Pe$ about the same point (Fig. 12.9.1b). The centroidal load P can be resolved into a vertical component P_y and a horizontal component P_x. Again, counterclockwise moments will be considered positive. In the elastic vector method described here, the direct loads and the torsional moment are considered separately, and the results are then superimposed.

The following assumptions are made in the analysis:

1. The plate elements connected by the welds are rigid and their deformations negligible. Thus, all deformations occur in the weld.
2. Only nominal stresses due to external loads are considered. Thus, effects of residual stresses and stress concentrations are neglected.
3. The weld behavior is considered linearly elastic.

The direct force is assumed uniformly distributed over the effective area of the weld. For the general case of inclined load shown in Fig. 12.9.1a, the components of stress due to direct shear are (Fig. 12.9.1c):

$$f_{xD} = \frac{P_x}{A} = \frac{P_x}{Lt_e}\,; \quad f_{yD} = \frac{P_y}{A} = \frac{P_y}{Lt_e} \qquad (12.9.1)$$

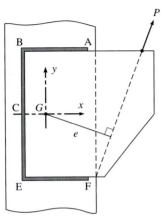

Welded joint and loading

(a)

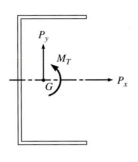

Load components

(b)

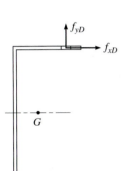

Stresses in element i due
to P_x and P_y

(c)

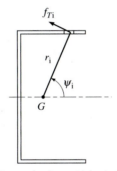

Stresses in element i due to M_T

(d)

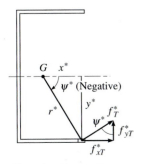

Stress in reference element
due to M_T

(e)

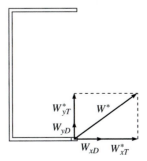

Forces in reference element

(f)

Figure 12.9.1: Elastic analysis of a welded joint in
eccentric shear.

where A = effective area of the weld

$\quad\quad L$ = overall length of the weld

$\quad\quad t_e$ = effective throat thickness of the weld

$\quad\quad P_x, P_y$ = components of the applied force P in the x and y directions, respectively

Under the action of the torsional moment M_T, the plate is assumed to rotate about the center of gravity of the weld configuration. Also, the magnitude of the shear stress in a weld element due to M_T is assumed to vary linearly with its radial distance from the centroid. Thus, if f_{Ti} is the shear stress in an element located at a point (x_i, y_i) and f_T^* is the shear stress in an element located at a reference point (x^*, y^*), we have:

$$\frac{f_{Ti}}{r_i} = \frac{f_T^*}{r^*} \tag{12.9.2}$$

where r_i and r^* are the radial distances of the points (x_i, y_i) and (x^*, y^*), respectively, from the centroid G. The calculated stress on the weld is assumed to be shear on the throat regardless of its actual direction. These shear stresses act perpendicular to the line connecting the centroid to the point where the stress is being calculated (Fig. 12.9.1d) and are directed so that they produce a torque about the centroid in the same sense as the external torque M_T. Hence, from equilibrium:

$$M_T = \int_A f_{Ti}\, dA\, r = \frac{f_T^*}{r^*}\int_A r^2\, dA = \frac{f_T^*}{r^*} I_p \tag{12.9.3}$$

where I_p is the polar moment of inertia of the weld area. Thus, the shear stress in the reference element due to the torsional moment is obtained as:

$$f_T^* = \frac{M_T r^*}{I_p} \tag{12.9.4}$$

Superposition of direct and torsional effects is simplified if the x and y components of the torsional shear stress are calculated. These components of f_T^* are obtained with the help of the two similar triangles shown in Fig. 12.9.1e, as:

$$f_{xT}^* = \frac{M_T(-y^*)}{I_p}; \quad f_{yT}^* = \frac{M_T x^*}{I_p} \tag{12.9.5}$$

The resultant stress in the reference element located at (x^*, y^*) is then given by:

$$f^* = \sqrt{\left(f_{xD} + f_{xT}^*\right)^2 + \left(f_{yD} + f_{yT}^*\right)^2} \tag{12.9.6}$$

It should be noted that all stresses acting on a fillet weld are considered to be shearing stresses regardless of their actual direction, and the critical section is always considered to be through the throat of the weld, although other sections may actually be subjected to higher stresses.

In practice, the weld size and hence the effective throat thickness, t_e, are kept constant for a given welded structural steel joint. In this case the weld characteristics could be rewritten as:

$$A = \int_A dA = t_e \int_L dL = t_e L \equiv t_e A_l \tag{12.9.7a}$$

$$I_p = \int_A r_i^2\, dA = t_e \int_L r_i^2\, dL \equiv t_e I_{pl} \tag{12.9.7b}$$

$$I_{pl} = \int_L r_i^2 dL = \int_L \left(x_i^2 + y_i^2\right) dL = I_{xl} + I_{yl}$$

$$= \sum_i^m \left[\frac{1}{12} L_i^3 + L_i\left(\bar{x}_i^2 + \bar{y}_i^2\right)\right] \tag{12.9.7c}$$

The notations A_l and I_{pl} stand for the area and polar moment of inertia of the weld, with the weld considered as a line (without width). Hence, the area A_l has the dimension of length (in.) while the moments of inertia I_{xl}, I_{yl} and the polar moment of inertia I_{pl} have the dimensions of in.^{3}. The last relation in Eq. 12.9.7c can be used to calculate I_{pl} of a weld configuration composed of m straight segments. L_i is the length of the ith segment, and $\bar{x}_i$ and $\bar{y}_i$ refer to distances from the center of gravity G of the weld group to the center of gravity G_i of the weld segment i (Fig. 12.9.2a). When the weld segment is parallel to one of the axes the last relation simplifies further to:

$$I_{pli} = \frac{1}{12} b_i^3 + b_i\left(\bar{x}_i^2 + \bar{y}_i^2\right) \text{ for a weld segment parallel to } x \text{ axis}$$

$$\tag{12.9.7d}$$

$$I_{pli} = \frac{1}{12} d_i^3 + d_i\left(\bar{x}_i^2 + \bar{y}_i^2\right) \text{ for a weld segment parallel to } y \text{ axis}$$

$$\tag{12.9.7e}$$

To simplify the calculation of these geometrical constants, each weld segment is assumed as a line coincident with the edge of the plate along which the fillet weld is placed rather than at the center of its effective throat. As the throat dimensions are relatively small compared to the weld lengths, the resulting error will be quite small.

If W is the force in the unit-length weld element, then:

$$W = f t_e \tag{12.9.8}$$

With the help of definitions introduced in Eqs. 12.9.7 and 12.9.8 and Fig. 12.9.1f, Equations 12.9.1, 12.9.5, and 12.9.6 could be rewritten as:

$$W_{xD} = \frac{P_x}{A_l}; \quad W_{yD} = \frac{P_y}{A_l} \tag{12.9.9}$$

$$W_{xT}^* = \frac{M_T(-y^*)}{I_{pl}}; \quad W_{yT}^* = \frac{M_T x^*}{I_{pl}} \tag{12.9.10}$$

$$W^* = \sqrt{\left(W_{xD} + W_{xT}^*\right)^2 + \left(W_{yD} + W_{yT}^*\right)^2} \tag{12.9.11}$$

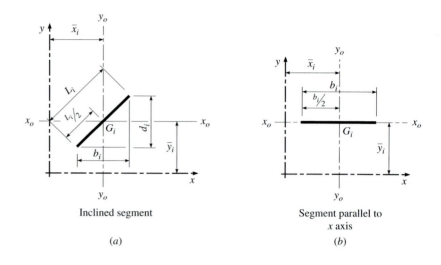

Inclined segment

(a)

Segment parallel to
x axis

(b)

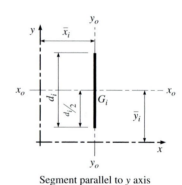

Segment parallel to y axis

(c)

Figure 12.9.2: Properties of weld segments treated
as line elements.

where A_l = area of the weld considered as a line, in.

I_{pl} = polar moment of inertia of the weld considered as a line, in.^{3}

P_x, P_y = x and y components of the applied load P, kips

M_T = torsional moment about the C.G. of the weld = Pe, in.-kips

W_{xD} = shear force component in x direction, in any unit-length element, due to the direct load P, kli

W_{yD} = shear force component in y direction, in any unit-length element, due to the direct load P, kli

W_{xT}^* = shear force component in x direction, in the unit-length element located at (x^*, y^*), due to the torsional moment M_T

W_{yT}^* = shear force component in y direction, in the unit-length element located at (x^*, y^*), due to the torsional moment M_T, kli

W^* = resultant force, in the unit-length element located at (x^*, y^*) due to the eccentric load P on the welded joint, kli

Observe that when the size of the weld is constant, it does not enter into computation of W^*, which depends only on the magnitude and eccentricity of the load and on the length and configuration of the weld. This procedure is particularly advantageous in design where the required size of weld is to be determined.

For a safe design, the maximum force anywhere in the weld, under factored load P_u, should not exceed the design strength of the weld. That is:

$$W_u^* \leq W_d \qquad (12.9.12)$$

Table 12.9.1 gives the properties of some typical weld configurations where the welds are treated as lines. It contains the cross-sectional area (A_l), the polar moment of inertia (I_{pl}) for use with twisting moments, and the section modulus (S_{xl}) for use with bending moments (to be discussed in Section 12.11.1).

The design procedure for a welded joint in eccentric shear, using the elastic method, consists of the following steps.

1. For the given loads, select a suitable weld line geometry.
2. Locate the centroid G of the weld configuration. Calculate the section properties A_l and I_{pl} of the weld treated as a line (see Table 12.9.1).
3. Determine the x and y components of the resultant force P, the eccentricity of the weld e, and the torsional moment M_T acting at the centroid G.
4. Determine the x and y components W_{xD}, W_{yD}, W_{xT}^*, W_{yT}^* of the forces W^* on a unit-length weld, at reference point (x^*, y^*), under the action of direct force P and the torsional moment M_T, using Eqs. 12.9.9 and 10.
5. Combine the individual force components vectorially and determine the resultant weld force W_u^* on unit-length weld at (x^*, y^*), using Eq. 12.9.11.
6. Determine the weld size w that provides a design shear strength W_d greater than the required strength W_u^*, using Eq. 12.9.12 and Table 6.19.1 or 2.

12.9.2   Ultimate Strength Method

For information on the ultimate strength method for welded joints in eccentric shear, refer to Section W12.4 on the website http://www.mhhe.com/ Vinnakota.

TABLE 12.9.1

Properties of Welds Treated as Lines

No.	Section	C.G.	A_l/d	I_{pl}/d^3	S_{xl}/d^2
1.			1	$\dfrac{1}{12}$	$\dfrac{1}{6}$
2.			2	$\dfrac{3\alpha^2 + 1}{6}$	$\dfrac{1}{3}$
3.			2α	$\dfrac{\alpha(3 + \alpha^2)}{6}$	α
4.			$2\alpha + 1$	$\dfrac{(\alpha + 1)^3}{6}$	$\alpha + \dfrac{1}{3}$
5.			$2\alpha + 1$	$\dfrac{\alpha^3 + 3\alpha + 1}{6}$	$\alpha + \dfrac{1}{3}$
6.		$\dfrac{\bar{y}}{d} = \dfrac{1}{\alpha + 2}$	$\alpha + 2$	$\dfrac{\alpha^3 + 8}{12} - \dfrac{1}{\alpha + 2}$	$\dfrac{2\alpha + 1}{3}$
7.		$\dfrac{\bar{y}}{d} = \dfrac{1}{\alpha + 2}$	$\alpha + 2$	$\dfrac{\alpha^3 + 6\alpha^2 + 8}{12} - \dfrac{1}{\alpha + 2}$	$\dfrac{2\alpha + 1}{3}$
8.		$\dfrac{\bar{x}}{d} = \dfrac{\alpha^2}{2\alpha + 1}$	$2\alpha + 1$	$\dfrac{8\alpha^3 + 6\alpha + 1}{12} - \dfrac{\alpha^4}{2\alpha + 1}$	$\alpha + \dfrac{1}{6}$
9.		$\dfrac{\bar{x}}{d} = \dfrac{\alpha^2}{2(\alpha + 1)}; \dfrac{\bar{y}}{d} = \dfrac{1}{2(\alpha + 1)}$	$\alpha + 1$	$\dfrac{(\alpha + 1)^4 - 6\alpha^2}{12(\alpha + 1)}$	$\dfrac{4\alpha + 1}{6}$

d = depth; b = width; $\alpha = b/d$
$\bar{x}$ = x coordinate of C.G.; $\bar{y}$ = y coordinate of C.G.
A_l = area of weld treated as a line
I_{pl} = polar moment of inertia of weld (treated as a line) about center of gravity
S_{xl} = section modulus of weld (treated as a line) about x axis

12.9.3   LRFDM Design Tables

The LRFD Manual contains tables of coefficients for use in calculating the ultimate strength of eccentrically loaded shear connections for the weld configurations and loading conditions shown in Fig. 12.9.3. The coefficients, given as LRFDM Tables 8-5 through 8-12, are based on computer solutions to the ultimate strength of these connections using the instantaneous center of rotation method (in accordance with LRFDS Appendix J2.4) described in Section 12.9.2. Values are tabulated for loads inclined at 0°, 15°, 30°, 45°, 60° and 75° with the vertical. The tables may be used for either analysis or design.

For any of the weld group geometries shown, the design strength of the eccentrically loaded weld group is:

$$P_d = \phi R_n = C C_1 D l \qquad (12.9.13)$$

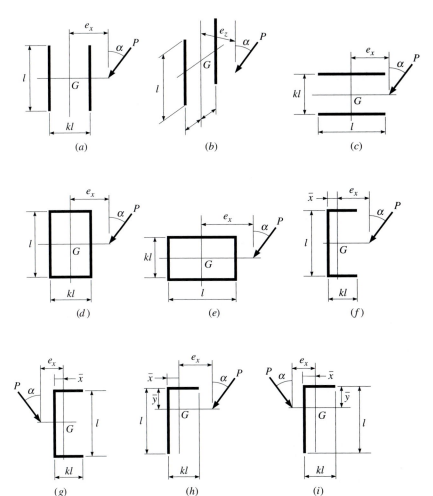

Figure 12.9.3: Eccentrically loaded weld groups presented in LRFDM tables.

where P_d = design strength of the weld group, kips

R_n = nominal strength of the weld group, kips

ϕ = resistance factor = 0.75

C = tabular value of coefficient (which includes the resistance factor ϕ of 0.75)

C_1 = electrode coefficient which adjusts tabular value, which is based on E70 XX electrodes, for other electrodes

D = number of sixteenths-of-an-inch in the weld size, w (where

$$ w = D\left(\frac{1}{16}\right) \text{in.)} $$

l = length of the reference segment of the weld group considered, in.

When the weld configuration is asymmetric (with respect to horizontal and/or vertical axes through the center of gravity of the weld configuration), coefficients to locate the center of gravity are tabulated in the last one or two lines at the bottom of the tables (designated as coefficients x and y). The coefficients C are tabulated as a function of two parameters, k and a ($= e_x/l$). The first line in each table (with $a = 0$) gives the design strength of a concentrically loaded weld group, as per LRFDS Appendix J2.4a. The tabulated values for C are based upon the strength through the throat of an equal-legged fillet weld having a leg size of one-sixteenth inch. Tabulated values are valid for weld metal with a strength level equal to or matching the base material. Values for electrode coefficient C_1 are given in LRFDM Table 8-4. When fillet welds are made on base material of strength less than matching, strength should be based on consideration of the shear strength of the base material on the diagrammatic fusion surface ($0.6F_{yBM} \times \frac{1}{16}D$).

Linear interpolation within a given table between adjacent a and k values is allowed. However, linear interpolation between C values for different load inclinations is likely to be significantly unconservative. So, it is recommended to use only the values tabulated for the next lower angle. Since the coefficients in these tables were derived by ultimate strength analysis, they should be used only for the weld patterns indicated and not in combination with any additional loading.

As mentioned earlier, elastic vector analysis on weld groups gives varying factors of safety depending on the length and configuration of the weld. Ultimate strength design gives a more uniform safety factor. However, to design (analyze) eccentrically loaded weld groups not conforming to one of the configurations tabulated in the LRFD Manual, the elastic vector method described in Section 12.9.1 may still be used.

E X A M P L E   1 2 . 9 . 1   **Design of Welded Joint in Eccentric Shear**

Determine the fillet weld size to transfer an eccentric vertical load of 24 kips, applied to a C-shaped welded joint shown in Fig. X12.9.1a. Assume SMAW

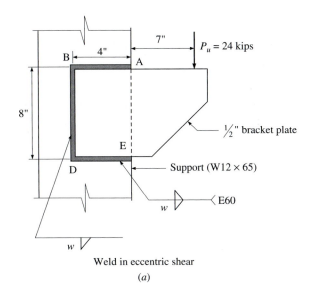

½" bracket plate

Support (W12 × 65)

E60

Weld in eccentric shear

(a)

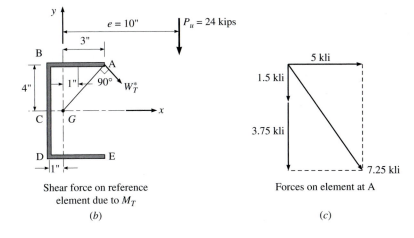

Shear force on reference
element due to M_T

(b)

Forces on element at A

(c)

Figure X12.9.1

fillet welds using E60 electrodes. Use (a) the elastic vector method, (b) the
LRFDM tables.

Solution

Factored load, $P_u = 24$ kips
For SMAW welds, $F_{EXX} = 60$ ksi

a. Elastic vector method
Properties of welds treated as lines:

$$b = 4 \text{ in.}; \quad d = 8 \text{ in.}; \quad A_l = 2(4) + 8 = 16 \text{ in.}$$

[continues on next page]

Example 12.9.1 continues ...

From Table 12.9.1:

$$\alpha = \frac{b}{d} = \frac{4.0}{8.0} = 0.5$$

$$\frac{\bar{x}}{d} = \frac{\alpha^2}{2\alpha + 1} \quad \rightarrow \quad \bar{x} = \frac{0.5^2(8)}{2(0.5) + 1} = 1.0 \text{ in.}$$

$$I_{pl} = \left[\frac{8\alpha^3 + 6\alpha + 1}{12} - \frac{\alpha^4}{2\alpha + 1} \right] d^3 = 192 \text{ in.}^3$$

$e = 4.0 + 7.0 - 1.0 = 10.0 \text{ in.} = e_x$

Factored load, $P_u = 24$ kips acting vertically downwards on the bracket.

So, $\rightarrow P_x = 0; \quad P_y = 24$ kips $\downarrow$

The components of shear on any unit-length weld element due to direct shear are:

$$W_{xD} = 0; \quad W_{yD} = \frac{P_y}{A_l} = \frac{24.0}{16.0} = 1.50 \text{ kli} \quad \downarrow$$

By observation, the maximum weld force will occur in unit-length weld elements at A and E.

Torsional moment about the centroid, $M_T = P_y \, e_x = 24.0(10.0) = 240$ in.-kips (clockwise)

The horizontal and vertical components of shear force due to torsional moment on a unit-length weld element at reference point A are, using Eqs. 12.9.10:

$$W_{xT}^* = \frac{M_T y^*}{I_{pl}} = \frac{240(4.0)}{192} = 5.00 \text{ kli} \quad \rightarrow$$

$$W_{yT}^* = \frac{M_T x^*}{I_{pl}} = \frac{240(3.0)}{192} = 3.75 \text{ kli} \quad \downarrow$$

Next combining the direct and torsional moment components,

$$W_x^* = 0.00 + 5.00 = 5.00 \text{ kli} \quad \rightarrow$$

$$W_y^* = 1.50 + 3.75 = 5.25 \text{ kli} \quad \downarrow$$

Hence,

$$W_A^* = \sqrt{5.00^2 + 5.25^2} = 7.25 \text{ kli}$$

From Table 6.19.1: Design Shear Strength of Unit-length SMAW Fillet Welds, for a $\frac{7}{16}$-in. SMAW fillet weld using E60 electrodes,

$$W_d = 8.35 \text{ kli} > W_{req} = W_A^* = 7.25 \text{ kli}$$

So, select a $\frac{7}{16}$-in. SMAW fillet weld, using E60 electrodes.   (Ans.)

b. Ultimate strength method (LRFDM tables)
   Referring to Fig. 12.9.3*f*:

$$l = d = 8 \text{ in.}$$

$$kl = b = 4 \text{ in.} \quad \rightarrow \quad k = 0.5$$

$$al = e_x = 10 \text{ in.} \quad \rightarrow \quad a = 1.25$$

From LRFDM Table 8-9: Coefficients C for Eccentrically Loaded Weld Groups, for channel-shaped welds and angle $= 0°$, corresponding to $k = 0.5$, $a = 1.25$, coefficient $C = 1.03$. From LRFDM Table 8-4, corresponding to E60 welds, coefficient $C_l = 0.857$. So, from Eq. 12.9.13:

$$D = \frac{P_u}{CC_l l} = \frac{24.0}{1.03(0.857)(8)} = 3.40, \quad \text{say } 4$$

So, provide a $\frac{1}{4}$-in. SMAW fillet weld using E60 electrodes.   (Ans.)

Remark
The weld size $\frac{1}{4}$ in. obtained using the tables may be compared to the $\frac{7}{16}$-in. weld required when using the elastic vector method. This example shows that major savings can be realized for common weld configurations found in the Manual, when using the ultimate strength method rather than the elastic method.

Design of Welded Joint in Eccentric Shear

EXAMPLE 12.9.2

A 4-in.-deep A36 steel bracket plate is attached to the flange of a W12×58 column by two horizontal fillet welds AB and DE as shown in Fig. X12.9.2*a*. The bracket carries a factored, vertical load of 16 kips, 4 in. from the vertical line BE. Determine the size of the fillet welds and the thickness of the plate required. Assume E70 electrodes.

Solution

The area of the weld considered as a line, $A_l = 6 + 4 = 10$ in.
The centroid of the weld lines, with respect to an origin at A, is located at:

$$\bar{x} = \frac{\Sigma A_i x_i}{\Sigma A_i} = \frac{6(3) + 4(2 + 2)}{10} = 3.4 \text{ in.}$$

$$\bar{y} = \frac{\Sigma A_i y_i}{\Sigma A_i} = \frac{0 + 4(4)}{10} = 1.6 \text{ in.}$$

Using these coordinates, the centroid of the weld configuration is as shown in Fig. X12.9.2*b*. Let L_i be the length of the weld segment i, and

[continues on next page]

Example 12.9.2 continues ...

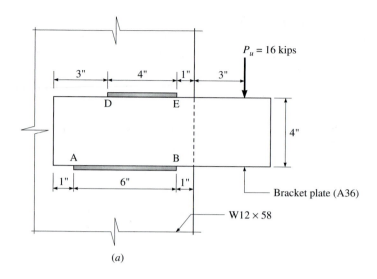

(a)

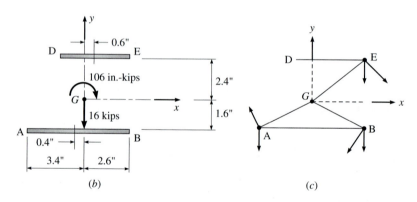

(b)   (c)

Figure X12.9.2

$\bar{x}_i$ and $\bar{y}_i$ be the coordinates of its center relative to the centroid of the weld configuration. So,

for segment AB: $L_1 = 6$; $\bar{x}_1 = -0.4$; $\bar{y}_1 = -1.6$

for segment DE: $L_2 = 4$; $\bar{x}_2 = +0.6$; $\bar{y}_2 = +2.4$

The polar moment of inertia of the weld relative to the centroid G (from Eq. 12.9.7c):

$$I_{pl} = \sum I_{pli} = \sum L_i \left[\frac{1}{12} L_i^2 + \bar{x}_i^2 + \bar{y}_i^2 \right]$$

$$= 6 \left[\frac{1}{12} (6)^2 + (-0.4)^2 + (-1.6)^2 \right] + 4 \left[\frac{1}{12} (4)^2 + 0.6^2 + 2.4^2 \right]$$

$$= 34.3 + 29.8 = 64.1 \text{ in.}^3$$

Eccentricity of the vertical load, $e_x = 2.6 + 4 = 6.6$ in.
Factored load, $P_u = 16$ kips; $P_x = 0$; $P_y = -16$ kips
Considering counterclockwise moments as positive, torsional moment about the centroid,

$$M_T = P_y e_x = -16.0(6.6) = -106 \text{ in.-kips}$$

The components of shear force on a unit-length weld due to direct shear loads P_x, P_y are:

$$W_{xD} = 0; \quad W_{yD} = \frac{P_y}{A_l} = \frac{-16.0}{10.0} = -1.60 \text{ kli}$$

Using Eq. 12.9.10, the horizontal and vertical components of shear force on a unit-length weld at reference point (x^*, y^*), due to torsional moment are:

$$W_{xT}^* = -\frac{M_T y^*}{I_{pl}} = \frac{106 y^*}{64.1} = 1.65 y^*$$

$$W_{yT}^* = \frac{M_T x^*}{I_{pl}} = \frac{-106 x^*}{64.1} = -1.65 x^*$$

In Table X12.9.2, these values are calculated for the reference points A, B, and E. Next, combining the shear forces due to the direct and torsional moment components, it is seen that the critical loading occurs in the weld element at E, where:

$$W_E^* = \sqrt{(W_x^*)^2 + (W_y^*)^2} = \sqrt{3.93^2 + 5.86^2} = 7.06 \text{ kli}$$

From Table 6.19.1: Design Shear Strength of Unit-Length SMAW Fillet Welds, for a $\frac{3}{8}$-in. SMAW fillet weld using E70 electrodes,

$$W_d = 8.35 \text{ kli} > W_{req} = W_E^* = 7.06 \text{ kli}$$

So, tentatively select $\frac{3}{8}$-in. fillet welds.

TABLE X12.9.2

Point	(x^*, y^*) in.	$W_{xT}^* = 1.65 y^*$ kli	$W_{yT}^* = -1.65 x^*$ kli	W_x^* kli	W_y^* kli
A	$(-3.4, -1.6)$	-2.62	5.57	-2.62	3.97
B	$(2.6, -1.6)$	-2.62	-4.26	-2.62	-5.86
E	$(2.6, 2.4)$	3.93	-4.26	3.93	-5.86

[continues on next page]

Example 12.9.2 continues ...

The factored moment in the bracket plate is a maximum at vertical line BE, and its value is:

$$M_{max} = 16(4) = 64.0 \text{ in.-kips}$$

If the plate thickness is t, the limit state of yielding of the plate results in:

$$M_{dBE} = \phi_b S_x F_y = 0.90 \left[\tfrac{1}{6}(t)(4^2) \right] 36.0$$

$$\geq M_{req} = 64.0 \rightarrow t \geq 0.74 \text{ in.}$$

Provide ¾-in.-thick plate.
Minimum size of weld with ¾-in.-thick plate, $w_{min} = \tfrac{1}{4}$ in.

Maximum size of fillet weld, $w_{max} = \tfrac{3}{4} - \tfrac{1}{16} = \tfrac{11}{16}$ in.

So, select a ¾-in.-thick A36 plate and provide ⅜-in. SMAW fillet welds using E70 electrodes, as shown in Fig. X12.9.2a.   (Ans.)

EXAMPLE 12.9.3

Design of Welded Joint in Eccentric Shear

Determine the size of fillet weld required for the bracket connection shown in Fig. X12.9.3a. The 16-kip load, inclined at 45° with the vertical, is a factored load. The ½-in.-thick bracket plate and the W10×39 column are of A36 steel. Assume E70 electrodes. Neglect the end return in strength calculations. Use (*a*) the elastic vector method, (*b*) the LRFDM tables based on ultimate strength method. Check the adequacy of the bracket plate.

Solution

a. Elastic vector method
The angle-shaped weld consists of two line elements AB and BC. First locate the centroid of the weld, taking AB and BC as reference lines. Next calculate the polar moment of inertia of the weld, with the help of Eqs. 12.9.7d and e. From Fig. X12.9.3b:

$$A_l = 3.0 + 6.0 = 9.0 \text{ in.}$$

$$\bar{x} = \frac{\sum A_i x_i}{\sum A_i} = \frac{3.0(1.5) + 0}{9.0} = 0.5 \text{ in.}$$

$$\bar{y} = \frac{\sum A_i y_i}{\sum A_i} = \frac{0 + 6.0(3.0)}{9.0} = 2.0 \text{ in.}$$

$$\bar{x}_1 = 1.0 \text{ in.;} \quad \bar{y}_1 = 2.0 \text{ in.;} \quad \bar{x}_2 = -0.5 \text{ in.;} \quad \bar{y}_2 = -1.0 \text{ in.}$$

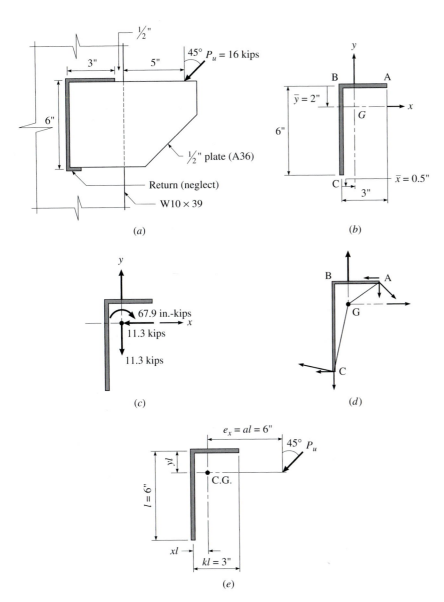

Figure X12.9.3

$$I_p = (I_{pl1} + I_{pl2})$$

$$= \left[\frac{1}{12}(3.0^3) + 3.0(1.0^2) + 3.0(2.0^2) \right]$$

$$+ \left[\frac{1}{12}(6.0^3) + 6.0(0.5^2) + 6.0(1.0^2) \right]$$

$$= 17.3 + 25.5 = 42.8 \text{ in.}^3$$

[continues on next page]

Example 12.9.3 continues ...

Transferring the applied load to the center of gravity, we obtain (Fig. X12.9.3c):

$$P_x = -16 \sin 45° = -11.3 \text{ kips}$$

$$P_y = -16 \cos 45° = -11.3 \text{ kips}$$

$$M_T = 11.3(2.0) - 11.3(3.0 + 0.5 + 5.0 - 0.5)$$

$$= -67.9 \text{ in.-kips}$$

The shear forces in a unit-length weld, due to the centroidal forces P_x and P_y are:

$$W_{xD} = \frac{P_x}{A_l} = \frac{-11.3}{9.0} = -1.26 \text{ kli}$$

$$W_{yD} = \frac{P_y}{A_l} = \frac{-11.3}{9.0} = -1.26 \text{ kli}$$

The shear force components in a unit-length weld, located at a reference point (x^*, y^*) due to the torsional moment are:

$$W_{xT}^* = -\frac{M_T y^*}{I_{pl}} = \frac{67.9 y^*}{42.8} = 1.59 y^*$$

$$W_{yT}^* = \frac{M_T x^*}{I_{pl}} = \frac{-67.9}{42.8} x^* = -1.59 x^*$$

The unit-length weld forces are calculated at points A ($x^* = 2.5$ in., $y^* = 2.0$ in.) and C ($x^* = -0.5$ in., $y^* = -4.0$ in.). See Figure X12.9.3d for force orientation. We have at point A:

$$W_{xT}^* = 1.59(2.0) = 3.18 \text{ kli}$$

$$W_{yT}^* = -1.59(2.5) = -3.98 \text{ kli}$$

$$W_x^* = -1.26 + 3.18 = 1.92 \text{ kli}$$

$$W_y^* = -1.26 - 3.98 = -5.24 \text{ kli}$$

Hence, $W_A^* = \sqrt{1.92^2 + 5.24^2} = 5.58$ kli
at point C:

$$W_{xT}^* = 1.59(-4.0) = -6.36 \text{ kli}$$

$$W_{yT}^* = -1.59(-0.5) = -0.80 \text{ kli}$$

$$W_x^* = -1.26 - 6.36 = -7.62 \text{ kli}$$

$$W_y^* = -1.26 + 0.80 = -0.46 \text{ kli}$$

Hence, $W_C^* = \sqrt{7.62^2 + 0.46^2} = 7.63$ kli

The most stressed point of the weld is therefore at point C of the weld configuration, with a required shear strength W^* of 7.63 kli. From Table 6.19.1, observe that a $\frac{3}{8}$-in. fillet weld using E70 electrodes, with a design shear strength of 8.35 kli provides adequate strength. The weld connects the $\frac{1}{2}$-in.-thick bracket plate to the 0.53-in.-thick flange plate. According to LRFDS Table J2.4, the minimum size of the fillet weld is $\frac{1}{4}$ in. and the maximum size of the fillet weld is $\frac{7}{16}$ in., from LRFDS Section J2.2*b*.

So, select a $\frac{3}{8}$-in. SMAW fillet weld, and provide a weld return at point C of 1 in. length ($> 2w = \frac{3}{4}$ in.).   (Ans.)

b. Ultimate strength tables

The weld configuration (L shape) and inclination of the load (45° with the vertical) correspond to LRFDM Table 8-11. For the given weld configuration,

$$l = 6 \text{ in.}; \quad kl = 3.0 \quad \rightarrow \quad k = 0.50$$

In Table 8-11 the value of x under $k = 0.5$ is 0.083 leading to $xl = 0.5$ in. Hence, $e_x = (3.0 + 0.5 + 5.0 - 2 \tan 45° - xl) = 6.0$ in. $= al$, resulting in $a = 1.0$.

For $a = 1.0$ and $k = 0.50$, $C = 1.07$ from LRFDM Table 8-11. For E70 electrodes, $C_1 = 1.0$, from LRFDM Table 8-4. So, from Eq. 12.9.13:

$$D = \frac{P_u}{CC_1 l} = \frac{16.0}{1.07(1.0)(6.0)} = 2.49, \text{ say } 3$$

Weld size, $w = D(\frac{1}{16}) = \frac{3}{16}$ in.

Therefore, provide the minimum size weld required, namely, $w = \frac{1}{4}$ in.

(Ans.)

This may be compared to the $\frac{3}{8}$-in. weld required when using elastic vector method. This example again shows that major savings can be realized for common weld configurations found in the Manual, when using the ultimate strength method rather than the elastic method.

c. Check adequacy of the bracket plate

Assume conservatively that the plate acts as a cantilever beam and that the critical section is a vertical section through the end of the weld return. This (rectangular) section is subjected to a shear force V, a normal force N, and a bending moment M, namely:

$$V = P_y = 11.3 \text{ kips}; \quad N = P_x = 11.3 \text{ kips}$$

$$M = 11.3(7.5) - 11.3(3.0) = 50.9 \text{ in.-kips}$$

[*continues on next page*]

Example 12.9.3 continues ...

For the ½-in.-thick, 6-in.-deep plate provided:

$$A = \frac{1}{2}(6.0) = 3.0 \text{ in.}^2; \quad S_x = \frac{1}{6}td^2 = \frac{1}{6}\left(\frac{1}{2}\right)(6.0^2) = 3.0 \text{ in.}^3$$

$$\text{Shear stress, } f_v = \frac{V}{A} = \frac{11.3}{3.0} = 3.77 \text{ ksi}$$

$$\text{Normal stress, } f_a = \frac{N}{A} = \frac{11.3}{3.0} = 3.77 \text{ ksi}$$

Noting that the plate is subjected simultaneously to axial and bending stresses, the design bending strength of the plate corresponding to the limit state of yielding, can be written as:

$$M_d = \phi S_x(F_y - f_a) = 0.90(3.0)(36.0 - 3.77)$$

$$= 87.1 \text{ in.-kips} > M_{\text{req}} = 50.9 \text{ in.-kips} \quad \text{O.K.}$$

12.10 Welded Joints in Shear and Tension (Loading Case P_y, P_z)

The welded joint at the base of gusset plate-to-support connection for an inclined bracing member (Fig. 12.10.1) is a common example of a welded joint under P_yP_z loading case. The factored load P_u in the brace can be resolved into a direct shear P_{uy} acting at the faying surface of the joint and a direct tension P_{uz} acting normal to the faying surface of the joint. Each unit length segment of weld is then assumed to resist an equal share of the direct

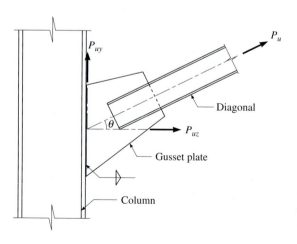

Figure 12.10.1:   Welded joint in shear and tension.

forces P_{uy} and P_{uz}, resulting in:

$$W_{uy} = \frac{P_{uy}}{2L}\,; \quad W_{uz} = \frac{P_{uz}}{2L} \tag{12.10.1}$$

where W_{uy} = shear force on the throat of a unit-length weld
  W_{uz} = tensile force on the throat of a unit-length weld
  L = length of each fillet weld connecting the gusset to the support

In contrast to bolts, where the interaction of shear and tension must be considered for welds, shear and tension can be combined vectorially into a resultant shear on the throat area. Thus:

$$W_u = \sqrt{W_{uy}^2 + W_{uz}^2} \tag{12.10.2}$$

For a safe design, the LRFDS requires:

$$W_d \ge W_{req} = W_u \tag{12.10.3}$$

where W_d is the design shear strength of a unit-length fillet weld provided.

12.11 Welded Joints in Shear and Bending (Loading Case $P_y\,M_{yz}$)

When the resultant applied load on a welded joint is parallel to, but outside of, the plane of the joint, the weld is subjected to shear and bending. For example, loads are frequently supported on brackets welded to the flange or web faces of columns as shown in Fig. 6.1.3*h*. Downward movement of the bracket under the applied load P is resisted by shear in the welds which connect the bracket to the column. The downward load P and the resultant upward shear in the weld constitute a couple, of moment Pe_z, which must be resisted by an equal couple consisting of tension in upper parts of the welded joint and compression of the lower part of the weld against the column. The LRFD Specification gives the design strength of welds, but does not specify a method of analysis for joints loaded in shear and bending. The method used is open to the discretion of the designer.

12.11.1 Elastic Method

First, let us consider a plate supporting an eccentric load P, fillet welded to the face of a column as shown in Fig. 12.11.1*a*. The effective throat is a rectangle of width $2t_e$ and height L, as shown in Fig. 12.11.1*b*. Even in this relatively simple case, the actual distributions of shear and bending stresses in the weld on the throat plane are not known exactly. In what follows, it is conservatively assumed that there is no contact between the connected parts so that all the load is carried through the weld only. Figure 12.11.1*c* shows the distribution of shear stress, parabolic with a maximum value $1\frac{1}{2}$ times the average value,

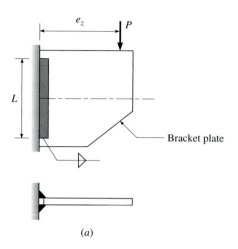

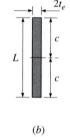

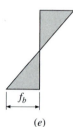
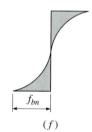

Figure 12.11.1:   Distribution of stresses in a weld subjected to shear and bending.

obtained using common beam theory. However, in the analysis of welds by the elastic method, common practice considers the shear stresses uniformly distributed as shown in Fig.12.11.1d. Also, Fig. 12.11.1e shows the commonly accepted (linear) variation of bending stresses obtained using elementary beam theory. Note that according to the beam theory, the maximum shear stress and the maximum bending stress occur at different points of the weld and need not be combined. However, it is highly improbable that elementary beam theory is correct when applied at a location with an abrupt change in cross section. It is more likely true that the bending stress in the extreme fiber of the weld is greater than that given by the flexural formula, with the stress variation being somewhat nonlinear as shown in Fig. 12.11.1f. This concentration of bending stress in the extreme fibers will be compensated for by the assumption of a uniform distribution of shear which gives a greater shearing stress on the extreme fiber than the actual. Therefore, the commonly used practice of vectorially combining maximum bending stresses using beam theory and average shearing stresses is generally safe.

Figure 12.11.2 shows a weld consisting of two vertical weld segments and two horizontal segments, symmetric about the vertical axis y-y. G is the

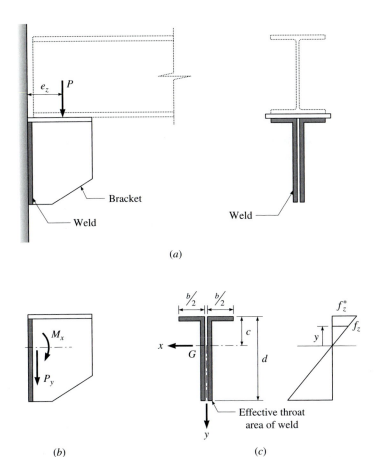

Figure 12.11.2:   Welds subjected to shear and bending.

center of gravity of the weld. An eccentric vertical load P is shown acting in the vertical plane of symmetry, outside the plane of the joint. The eccentric load P in Fig. 12.11.2a may be replaced by its equivalent, namely, a concentrated vertical load P_y through G and a moment $M_{yz} \equiv M_x = P_y e_z$ as shown in Fig. 12.11.2b.

The direct component P_y is assumed to be distributed uniformly over the effective weld throat area A. The shear stress applied to the weld due to the direct load P_y is therefore:

$$f_y = \frac{P_y}{A} = \frac{P_y}{L t_e} \qquad (12.11.1)$$

where L = total length of the weld = $2d + b$
t_e = effective throat thickness of the weld

Under the action of the bending moment M_x, normal stresses that vary linearly with distance from the neutral axis, xx, develop in the weld. If f_z is the normal stress in the weld at a distance y from the NA, and f_z^* is the normal stress at the farthest point in the weld located at a distance c from the NA, the following relations can be written with the help of Fig. 12.11.2c:

$$\frac{f_z}{y} = \frac{f_z^*}{c} \quad \rightarrow \quad f_z = \frac{f_z^*}{c} y \qquad (12.11.2)$$

For the problem under consideration, the f_z stresses are tensile for all weld portions above the neutral axis, and they are compressive for all weld portions below that axis.

Let dL be an elemental length of weld located a distance y from the x axis and dM the moment about the xx axis of the force on element dL; then:

$$dM = f_z dAy = \frac{f_z^*}{c} (y^2 dA)$$

The sum of the effect of all such elemental portions should, for equilibrium, equal the applied moment M_x. Therefore:

$$M_x = \frac{f_z^*}{c} \int_A y^2 dA = \frac{f_z^*}{c} I_x$$

or

$$f_z^* = \frac{M_x}{I_x} c \qquad (12.11.3)$$

where I_x is the moment of inertia of the effective weld throat area A about the centroidal axis.

$$I_x = \int_A y^2 dA$$

The usual practice is to take c as the distance from the centroidal axis to the *farthest point on the tension side*. The combined effect of the direct force and the moment on the reference element at the extreme fiber will be the vectorial sum of the stresses f_y and f_z^*, which act at right angles to one another. The resultant stress at the extreme fiber is therefore given by:

$$f^* = \sqrt{f_y^2 + f_z^{*2}} \qquad (12.11.4)$$

The weld size and hence the effective throat thickness t_e are almost always kept constant for a given welded joint. In this case the weld characteristics A, I_x, and S_x could be written in the alternate form:

$$A = \int_A dA = t_e \int_L dL = t_e L = t_e A_l \qquad (12.11.5a)$$

$$I_x = \int_A y^2 dA = t_e \int_L y^2 dL = t_e I_{xl} \qquad (12.11.5b)$$

$$S_x = \frac{I_x}{c} = t_e \frac{I_{xl}}{c} = t_e S_{xl} \qquad (12.11.5c)$$

The notations A_l, I_{xl}, and S_{xl} stand for the area, moment of inertia, and elastic section modulus of the weld considered as a line (without width). Hence, the area A_l has dimension of length (in.), the moment of inertia I_{xl} has dimensions of length³ (in.³), while S_{xl} has the dimensions of length squared (in.²). Table 12.9.1 gives A_l, and S_{xl} values for several weld configurations treated as line elements.

Let W represent the force per unit length of weld, at any point. Then,

$$W = f t_e$$

With the help of the definitions introduced in Eq. 12.11.5, Eqs. 12.11.1, 12.11.3, and 12.11.4 could be rewritten as:

$$W_y = \frac{P}{A_l}; \quad W_z^* = \frac{M_x}{I_{xl}} c = \frac{M_x}{S_{xl}} \tag{12.11.6}$$

$$W^* = \sqrt{(W_y)^2 + (W_z^*)^2} \tag{12.11.7}$$

For a safe design, the maximum force anywhere in the weld should not exceed the design strength of the weld as obtained by the size and type of weld. Or:

$$W_{req} = W^* \leq W_d \tag{12.11.8}$$

12.11.2   Ultimate Strength Method (Using the LRFDM Tables)

The ultimate strength of a weld group consisting of two vertical lines of welds subject to load not in plane of the weld group (see Fig. 12.9.3b) may be obtained from Table 8-5 of the LRFDM. Using the notations given in Table 8-5 of the LRFDM, the factored eccentric load is given by:

$$P_u = C C_1 D l$$

with

C_1 = coefficient for electrode used (from LRFDM Table 8-4)
    = 1 for E70 electrodes
l = length of each vertical weld, in.
D = number of sixteenths-of-an-inch in fillet weld size
al = load eccentricity, e

The coefficient C can be obtained from LRFDM Table 8-5 corresponding to $k =$ and a.

Welded framed beam connections and unstiffened and stiffened seated beam connections (considered in Chapter 13) use welds subjected to shear and bending, studied in this section.

EXAMPLE 12.11.1   **Welded Joint in Shear and Bending**

A bracket plate is fillet welded to the flange of a column (Fig. X12.11.1) to transfer a factored load of 108 kips applied with an eccentricity of 6 in. Calculate the required weld size. Use A36 steel and E70 electrodes. Assume the column flange and the bracket plate do not control. Use (a) the elastic method, (b) the ultimate strength method.

Solution

a. Elastic method

Factored load, P_u = 108 kips;   Eccentricity, e_z = 6 in.

Moment, M_{ux} = 108(6) = 648 in.-kips

There are two lines of vertical welds, 18 in. long each. → d = 18.0 in.

From symmetry the C.G. (and xx axis) are located at middepth.

Characteristics of the weld, considered as two line elements are:

$$A_l = L = 2d = 2(18.0) = 36.0 \text{ in.}$$

$$S_{xl} = 2(1/6)d^2 = 2(1/6)(18.0^2) = 108 \text{ in.}^2$$

The forces on a unit-length weld at the extreme fiber ($c = d/2 = 9$ in.) are:

$$W_y = \frac{P_u}{A_l} = \frac{108}{36} = 3.0 \text{ kli} \downarrow; \quad W_z^* = \frac{M_{ux}}{S_{xl}} = \frac{648}{108} = 6.0 \text{ kli} \rightarrow$$

$$W^* = \sqrt{(W_y)^2 + (W_y^*)^2} = \sqrt{3.0^2 + 6.0^2} = 6.71 \text{ kli} = W_{req}$$

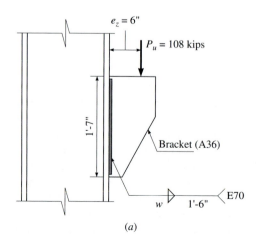

Figure X12.11.1   (a)   (b)

The design strength of an E70 electrode unit-length fillet weld with a leg size w is:

$$W_d = \phi\,(0.6F_{uw})(0.707w) = 0.75(0.6)(70)(0.707)w = 22.3w \text{ kli}$$

The fillet weld size required is such that:

$$W_d \geq W_{\text{req}} \;\rightarrow\; 22.3w \geq 6.71 \;\rightarrow\; w \geq 0.30 \text{ in.}$$

Provide $\frac{5}{16}$ in. welds using E70 electrodes.   (Ans.)

b. Ultimate strength method (using LRFDM Tables)
Using the notations given in Table 8-5 of the LRFDM, the factored eccentric load is:

$$P_u = C\,C_1 D l$$

with

$$
\begin{aligned}
C_1 &= \text{coefficient for electrode used} \\
&= 1 \text{ for E70 electrodes, from LRFDM Table 8-4} \\
l &= \text{length of each vertical weld} = 18.0 \text{ in.} \\
D &= \text{number of sixteenths-of-an-inch in fillet weld size} \\
al &= \text{load eccentricity} = 6.0 \text{ in.} \;\rightarrow\; a = 6/18 = 0.333
\end{aligned}
$$

Entering LRFDM Table 8-5 with $k = 0$ and $a = 0.333$, gives, by interpolation, $C = 2.22$. Hence,

$$D = \frac{P_u}{CC_1 l} = \frac{108}{2.22(1)(18)} = 2.71, \text{ say } 3$$

Provide $\frac{3}{16}$ in. welds using E70 electrodes.   (Ans.)

12.12   Connecting Elements

12.12.1   Block Shear Rupture

When beams are attached to intersecting girders such that the flanges of both are at the same elevation, the beams framing-in must be notched out at the top to prevent interference with the flange of the girder (Fig. 12.12.1a). Such a notch is called a *cope, block,* or *cut.* These cuts are rectangular in shape, even though some supporting beams may have sloping inner flange faces. Removal of flange material eliminates any containment that the flange has upon the web and thus may lead to rupture of the web material in a vertical direction. Tests by Birkemoe and Gilmor [7.1] have shown that, in coped high-strength bolted beam end connections, the bearing stress may be high enough to cause failure by web tear out through the bolt holes as shown in Fig. 12.12.1b. This

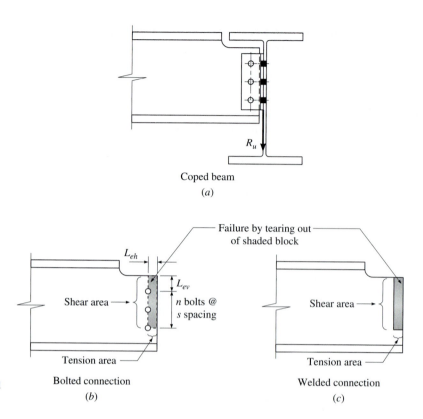

Figure 12.12.1: Block shear rupture in coped beams.

failure mode is referred to as ***block shear rupture,*** and is covered by LRFD Specification Section J4.3. The rectangular block in the beam web is bound by a horizontal section at the lowest bolt hole and a vertical section through all the bolt holes. The failure path is defined by the center lines of the bolt holes. Block shear combines tensile strength on one plane and shear strength on a perpendicular plane. Further, block shear failure involves a combination of fracture of the beam web through one plane, along with a simultaneous yielding or fracture on a perpendicular plane. This problem develops when high reactions are imposed on relatively thin material through a short connection length. Thus connections detailed for thin webbed beams using relatively few large diameter bolts—which thus results in a shorter connection length than if a greater number of smaller diameter bolts were used—should be checked for adequate block shear strength.

Based on results of tests conducted at Toronto [Birkemoe and Gilmor, 7.1], Texas [Rickles and Yura, 7.10], and Alberta and Arizona [Hardash and Bjorhovde, 7.7] universities, the LRFD Specification has adopted a model to predict the block shear strength wherein shear strength on one plane is added to the tension strength of the perpendicular plane. Therefore, two possible block shear strengths R_{bs1} and R_{bs2} can be calculated. R_{bs1} is the fracture strength on the net tensile section combined with the smaller of yielding or fracture strength on the shear plane. R_{bs2} is the fracture strength on the net shear area combined with the smaller of yielding or fracture strength on the tensile area. Gross areas are used for the limit state of yielding while net areas are used for the limit state of fracture, so that these formulas are consistent with the philosophy in LRFDS Chapter D for tension members. The controlling equation is the one which contains the larger fracture component. Thus:

$$R_{dbs} = R_{fnt} + \min[R_{ygv}, R_{fnv}] \quad \text{if} \quad R_{fnt} \geq R_{fnv} \qquad (12.12.1a)$$

$$R_{dbs} = R_{fnv} + \min[R_{ygt}, R_{fnt}] \quad \text{if} \quad R_{fnv} > R_{fnt} \qquad (12.12.1b)$$

with

$$R_{fnt} = \phi F_u A_{nt} \qquad (12.12.2a)$$

$$R_{ygv} = \phi(0.6F_y)A_{gv} \qquad (12.12.2b)$$

$$R_{fnv} = \phi(0.6F_u)A_{nv} \qquad (12.12.2c)$$

$$R_{ygt} = \phi F_y A_{gt} \qquad (12.12.2d)$$

where ϕ = resistance factor = 0.75
A_{gv} = gross area subjected to shear, in.^{2}
A_{gt} = gross area subjected to tension, in.^{2}
A_{nv} = net area subjected to shear, in.^{2}
A_{nt} = net area subjected to tension, in.^{2}
F_y = yield stress in tension, ksi
F_u = ultimate stress in tension, ksi
R_{dbs} = design block shear rupture strength, kips
R_{ygv} = shear yielding component, kips
R_{fnt} = tension rupture component, kips
R_{fnv} = shear rupture component, kips
R_{ygt} = tension yielding component, kips

Or, explicitly with the help of Fig. 12.12.1b:

$$R_{fnt} = 0.75F_u(L_{eh} - 0.5d_e)t_w \qquad (12.12.3a)$$

$$R_{ygv} = 0.75(0.6F_y)[(n - 1)s + L_{ev}]t_w \qquad (12.12.3b)$$

$$R_{fnv} = 0.75(0.6F_u)[(n - 1)s + L_{ev} - nd_e]t_w \qquad (12.12.3c)$$

$$R_{ygt} = 0.75F_y L_{eh}t_w \qquad (12.12.3d)$$

where d_e = diameter of hole (diameter of bolt + $\frac{1}{8}$ in. for standard punched holes), in.
L_{eh} = horizontal distance from center of hole to beam end, in.
L_{ev} = vertical distance from center of hole to top edge of web, in.
n = number of bolts

$$s \quad = \text{bolt spacing, in.}$$
$$t_w \quad = \text{thickness of beam web, in.}$$

For welded coped beams, block shear rupture is treated as for bolted connections (Fig. 12.12.1c); the only difference is that, as there are no bolt holes, $A_{nv} = A_{gv}$ and $A_{nt} = A_{gt}$. LRFDM Tables 9-3a and b and 9-4a and b list the four block shear components R per inch of plate thickness for three bolt sizes ($\frac{3}{4}$, $\frac{7}{8}$, and 1 in.), bolt spacing (s) of 3 in., different grades of steel ($F_y = 36$ and 50 ksi and $F_u = 58, 65, 70$ ksi), and a range of end and edge distances. Their use in determining block shear strength will be illustrated in Chapter 13 (see Example 13.12.1).

As mentioned earlier, the block shear model is not limited to coped ends of beams. Various types of tension connections (member ends, gusset plates, splice plates, cover plates), some of which are considered in Chapter 7, are also susceptible to block shear failure.

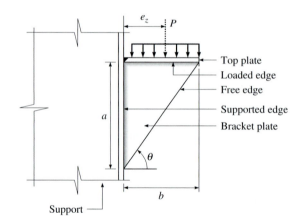

Triangular bracket plate

(a)

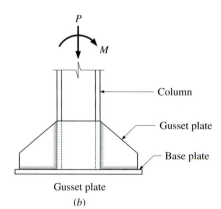

Gusset plate

(b)

Figure 12.13.1:   Bracket plates.

12.13 Bracket Plates

Triangular bracket plates are used in structures as support brackets to transfer load from an offset beam to a column (Fig. 12.13.1*a*), as gussets in heavy column bases (Fig. 12.13.1*b*), and as stiffeners. They have rigid built-in edges on the shorter sides, while the longer (diagonal) side is free. They act as short cantilever beams to transfer load from one edge *(loaded edge)* to the other edge *(supported edge)*. The function of the *top plate* is to distribute the applied load to the loaded edge of the bracket plate and to support (stabilize) the bracket plate. The bracket plate is fillet welded to the top plate and either fillet welded or groove welded along the supported edge to the supporting column. For heavy loads, plate or angle stiffeners may be required along the diagonal edge.

For information on the behavior and design of bracket plates, refer to Section W12.5 on the website http://www.mhhe.com/Vinnakota.

12.14 Gusset Plates

In welded trusses, connections between adjoining members at a joint can sometimes be made by directly butting or lapping, while in bolted trusses the connection must usually be made to a plate called *gusset plate.* A single plate at each joint is generally adequate for lightly loaded trusses and truss joists *(single-plane trusses),* while two parallel gusset plates are required at each joint of heavier long-span trusses *(double-plane trusses).* The practical problems of cutting the members precisely to the right length and of making them meet so that a weld can be deposited are so difficult that some sort of connecting piece is usually necessary, even in welded trusses.

The lateral dimensions of a gusset plate are essentially determined by the connector requirements of the members. That is, the number of bolts or length of welds required in each of the connecting members is calculated, and the plate is simply made large enough to accommodate them. Gusset plates must be of adequate thickness to resist shear, direct, and flexural loads acting on the weakest or critical section(s). The design strengths are calculated based on the assumption that the elementary formulas for beams apply (plane sections remain plane, etc.). Often, it is not possible to tell by inspection what section of the gusset plate will experience the critical loading, so several sections cut parallel to and normal to members are selected. If large compressive stresses occur along any free edge, some thought should be given to stiffening the plate by making it thicker, or by adding a stiffener along the edge. Tests on gusset plates [Whitmore, 1952] indicated that bending stresses are not distributed linearly, and that the neutral axis of a cross section does not coincide with the centroidal axis. The maximum bending stress was found to occur at an interior point rather than on the extreme fiber, and the shearing stresses were not distributed according to the parabolic law which corresponds to a linear distribution of bending stress.

Deflection of a bolted or welded truss under load tends to change the angle between its members. Thus, if the angle between adjacent members connected to a gusset plate decreases, compressive stresses may develop parallel to and at the free edge of that portion of the gusset plate. Therefore, the width/thickness ratio of the free edge of the gusset plate in that portion should be limited, to avoid premature plate local buckling. For example, the AASHTO Specification requires that an unsupported edge of a gusset plate be stiffened if it is longer than $(2.04t)\sqrt{E/F_y}$, with the yield stress of the gusset plate F_y expressed in ksi. This results in a value of $58t$ for A36 steel and $49t$ for Grade 50 steel.

Gusset plates are traditionally made rectangular because it is easier (and therefore a little less costly) for the shop to fabricate them that way. For bolted gusset plates, the plate thickness is selected such that the bolt bearing strength does not limit the design strength of the bolt.

When connection elements, such as gusset plates, are large in comparison to the bolted or welded joints within them, the effective width of the gusset plate may limit the gross and net areas of the gusset plate to less than the full width. The width of this section, called the *Whitmore section* [Whitmore, 1952; Bjorhovde and Chakrabarti, 1985] is determined at the end of the joint by spreading the force from the start of the joint, 30° to each side in the connection element along the line of force, as shown in Fig. 12.14.1. The Whitmore section is shown with dashed lines in Fig. 12.14.1. We have:

$$L_{WM} = a + 2b\sin 30° \qquad (12.14.1)$$

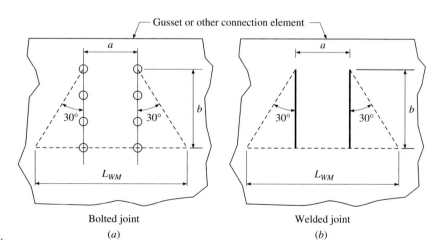

Bolted joint

(a)

Welded joint

(b)

Figure 12.14.1:   Length of the Whitmore section.

The Whitmore section may spread across the joint between connection elements but cannot spread beyond an unconnected edge. If the member force is a compressive force, gusset buckling has to be checked. For additional information on gusset plate design refer to Gross [1990], Thornton [1991], and Part 13 of the LRFDM.

12.15 Beam Splices

Splices in beams and girders may be classified as either shop splices or field splices. ***Shop splices*** are made during the fabrication of the member in the shop. The location of a shop splice in a member is often determined by the designer to provide a change in section to reflect the variation in strength required along the span, or sometimes by the fabricator to use the available lengths of material (plates and shapes). ***Field splices*** are necessary when the lengths of members are limited by the means of transportation from the shop to the construction site, by the capacity (maximum size or weight) of the available erection equipment in the field, or by the construction process utilized.

Shop splices are nearly always groove welded. A shop splice using full penetration welds for both flanges and web is shown in Web Section W6.12 (Fig. W6.12.1c). Since full penetration groove welds are as strong as the base metal, such splices require no further discussion. A commonly used field-bolted splice is shown in Fig. 12.15.1. Rectangular splice plates are lapped across the joint and bolted to the flanges and web of the beam in order to transfer the load. This type of splice is usually called a ***web-flange splice.*** In bolted splices, the web and the flanges are spliced at the same location. A single shear splice plate on each flange is often sufficient. For large shapes, heavy flange splice plates may be required on both sides of the flanges to reduce the number of bolts by providing a double shear condition and to reduce the splice plate thicknesses. Two web splice plates, one on either side of the web, are generally used. This not only creates a symmetric

Photo 12.15:   All-bolted web-flange beam splice. *Photo by S. Vinnakota.*

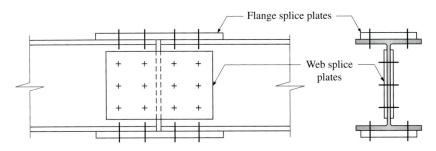

Figure 12.15.1: Bolted web-flange beam splice.

load transfer with respect to the plane of the web, but also the bolts are subjected to double shear conditions which reduces the required number of bolts. It also eliminates the inherent eccentricity.

LRFDS Section J7 stipulates that groove welded splices in beams and plate girders shall develop the full strength of the smaller spliced section. Other types of splices in cross sections of beams and plate girders shall develop the strength required by the forces at the point of splice. However, Section J7 does not provide guidance as to how the eccentric effect of the shear force should be accounted for in the design of the web splice, or how the moment at the section should be proportioned between the web splice and flange splices. Garretts and Madsen [1941] carried out experiments to investigate the behavior of riveted and bolted web-flange girder splices. Tests indicate that flange splices alone can be assumed to transfer the moment. At ultimate load, plastic hinges formed in the constant moment region, and failure occurred by local buckling of the compression flange.

More recently Kulak and Green [1990] presented a rational approach to predict the ultimate capacity of web-flange beam bolted splices. It satisfies the equations of static equilibrium and uses the true shear load versus the shear deformation response of the bolts. Hence it allows the use of ultimate strength tables for eccentrically loaded bolted connections available in the LRFD Manual. In this method, a simple beam that contains a bolted web-flange splice located at a section L_o, where both shear (V_o) and moment (M_o) are present, is shown in Fig. 12.15.2a. A free-body diagram taken by cutting the beam through one set of bolts is shown in Fig. 12.15.1b. The forces in these bolts are assumed to rotate about an instantaneous center, I, as shown in this figure. We can write the following three equations of equilibrium [Kulak and Green, 1990]:

$$\sum F_x = 0 \ \rightarrow \ \sum_{i=1}^{N} B_{ix} = 0 \tag{12.15.1}$$

$$\sum F_y = 0 \ \rightarrow \ \sum_{i=1}^{N} B_{iy} - V_o = 0 \tag{12.15.2}$$

$$\sum M_I = 0 \ \rightarrow \ \sum_{i=1}^{N} B_i r_i - [V_o (e_o + r_o) + (M_o - F_f d_s)] = 0 \tag{12.15.3}$$

Equation 1 is automatically satisfied because there are no external horizontal loads present. Equation 2 is satisfied when the sum of the vertical components of the bolt forces is equal to the shear V_o acting at the section. Equation 3 identifies how the moment transferred across the splice is shared between the bolts in the web splice and the bolts in the flange splice. In these equations,

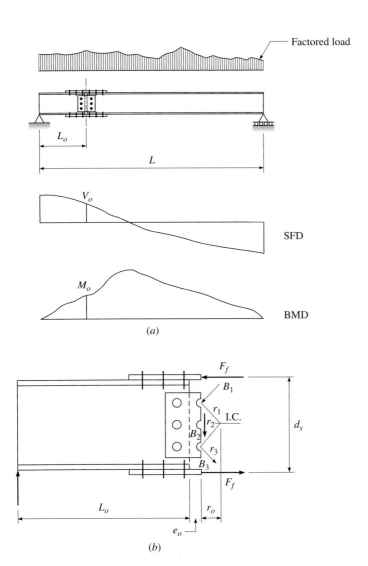

Figure 12.15.2: Analytical model for a web-flange beam splice.

B_i = resultant force in the ith bolt
B_{ix} = horizontal component of the bolt force B_i
B_{iy} = vertical component of the bolt force B_i
N = number of bolts on one side of the web splice
V_o = beam shear at the splice line
M_o = beam moment at the splice line
d_s = distance between the centroids of the top and bottom flange
splice plates
F_f = force in the top or bottom flange bolts on one side of splice line

e_o = distance from the centerline of the splice to the centroid of the bolt group on one side of the splice line (eccentricity of the shear force)

r_o = distance from the centroid of one bolt group to its instantaneous center of rotation.

For the special case of a beam in which the splice is located at a point of contraflexure, Eq. 12.15.3 reduces to:

$$\sum M_I = 0 \;\rightarrow\; \sum_{i=1}^{N} B_i r_i - [V_o(e_o + r_o)] = 0 \qquad (12.15.4)$$

The design of a bolt group on one side can proceed on the same basis as that for a bolt group acting under an eccentric load equal to the shear at splice and located at the centerline of the splice. LRFDM Tables 7-17 to 7-24 with $\theta = 0$, can be used for this purpose.

For the case of a beam in which the web and flanges are spliced, the designer will have to make an assumption regarding the portion of the moment at the location of the splice that the flange splices will be designed to resist. Traditionally, the flange splices have been designed to resist either 100 percent of the moment at the centerline of the splice (a conservative assumption) or the portion of the moment that the flanges in the beam were designed to resist. Either one of these approaches may be used as long as the equilibrium Eqs. 12.15.2 and 12.15.3 are satisfied.

Equation 12.15.3 can be rewritten as:

$$\sum M_I = 0 \;\rightarrow\; \sum_{i=1}^{N} B_i r_i - [V_o(e_o^* + r_o)] = 0 \qquad (12.15.5)$$

with

$$e^* = e_o + \frac{(M_o - F_f d_s)}{V_o} \qquad (12.15.6)$$

So the design of a bolt group on one side of the splice can proceed on the same basis as that of a bolt group acted on by a vertical load equal to the shear at the splice line but having an eccentricity e^* defined by the relation 12.15.6. Again LRFDM Tables in Part 7 can be used for design purposes.

The flange splice plates in the tension region should be treated as tension members and are subject to the design recommendations of LRFDS Section J5.2. The moment capacity of the beam is not affected by the reduction in cross-sectional area caused by the bolt holes in the tension flange, if LRFDS Eq. B10.1 is satisfied. The overall dimensions of the web splice plates depend on the selected bolt pattern. The thickness of the splice plate can be determined from the applied eccentric shear load and the applicable shear, bending, bearing and block shear strengths.

Moment end-plate connections (see Section 13.13) are sometimes used as beam splices. Such end-plate splices are most economical in relatively light constructional steel work because they require less material and fewer bolts than conventional web-flange splices. Satisfactory behavior up to the plastic limit load of the beam can be achieved if the bolts are adequately designed. As beam sizes are increased or large shear forces are transferred, the end-plate splice loses much of its economy and is replaced by the conventional beam splice studied earlier. Note that the bolts in the end-plate splice are generally subjected to combined axial tension and shear force, whereas the bolts in the web-flange splice are subjected to shear alone.

Design of a Bolted Web-Flange Beam Splice

EXAMPLE 12.15.1

Design a bolted web-flange splice for a W30×108 beam steel, to be located where the factored moment is 790 ft-kips and the factored shear is 180 kips. Use A572 Gr 50 splice plates and $7/8$-in.-dia. A325-X type bolts. Assume 90 percent of the moment is carried by the flange splice plates and the remaining 10 percent by the web-splice plates. All the shear is carried in the web-splice plates.

Solution

a. Data

For a W30×108:

d = 29.8 in.; t_w = 0.545 in.; T = 26.5 in.
b_f = 10.5 in.; t_f = 0.760 in.; A = 31.7 in.^{2}
Z_x = 346 in.^{3}

For a W30×108 of A992 steel, from LRFDM Table 5-3:

$$\phi_b M_{px} = 1300 \text{ ft-kips}; \quad \phi_v V_n = 439 \text{ kips}$$

Factored loads at the splice are:

$$\text{moment} = 790 \text{ ft-kips} = 61\% \text{ of } \phi_b M_{px}$$

$$\text{shear} = 180 \text{ kips} = 41\% \text{ of } \phi_v V_n$$

Even though LRFDS does not require any minimum proportion of the strength to be developed by a splice (LRFDS Section J1.1), it is advisable to design splices for a significant proportion (say, at least 50 percent) of the member strength).

Bolts: $7/8$-in.-dia. A325-X type

From LRFDM Table 7-10, design strength of a single bolt is 27.1 kips in single shear and 54.1 kips in double shear. Also, assuming $s > s_{\text{full}}$, from LRFDM Table 7-12 bearing strength on 1-in.-thick A572 Gr 50 ($F_u = 65$) plate is 102 kips.

[*continues on next page*]

Example 12.15.1 continues ...

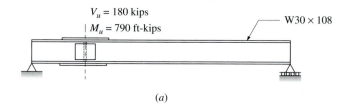

(a)

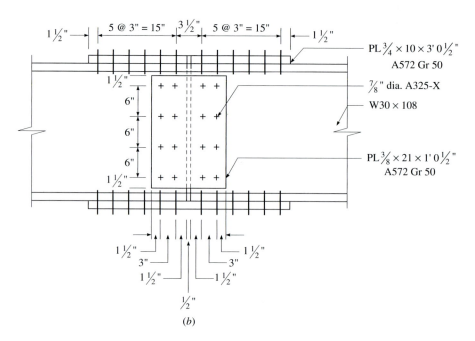

(b)

Figure X12.15.1

b. Flange splice

Moment to be resisted by flange splice = 90%(790) = 711 ft-kips

The flange splice plate in tension is designed as a tension member, and the same arrangement is used on the compression side. We estimate the thickness of the flange splice plate as $\frac{3}{4}$ in., about the same as the flange thickness of the beam.

Flange force = $T = C = \dfrac{M}{d + \text{est. } t} = \dfrac{711(12)}{(29.8 + 0.75)} = 279$ kips

Number of bolts required, in single shear = $\dfrac{279}{27.1} = 10.3$

Provide 12 bolts, 6 in each of two rows on each side of splice line. Assume splice plate width of 10 in.; slightly less than the beam flange width of 10.5 in.

Gross width, $w_g = 10.0$ in.

Net width, $w_n = w_g - 2d_e = 10.0 - 2(1.0) = 8.0$ in.
Corresponding to the limit state of yield on gross area,

$$0.9F_y w_g t \geq T_u \quad \rightarrow \quad t \geq \frac{279}{0.9(50)(10.0)} = 0.62 \text{ in.}$$

Corresponding to the limit state of fracture on effective net area,

$$0.75F_u U w_n t \geq T_u \quad \rightarrow \quad t \geq \frac{279}{0.75(65)(1.0)(8.0)} = 0.72 \text{ in.}$$

So, select a $\frac{3}{4} \times 10$ in. plate, resulting in,

$$A_g = 10.0(\tfrac{3}{4}) = 7.5 \text{ in.}^2; \quad A_n = 8.0(\tfrac{3}{4}) = 6.0 \text{ in.}^2$$

$$A_n = 0.8A_g < 0.85A_g \text{ (LRFDS Section J5.2b)} \qquad \text{O.K.}$$

Use a pitch of 3 in. $\left(> 3d + \dfrac{1}{16} = 2\dfrac{11}{16} \text{ in.} \right)$, and an end distance of

$1\frac{1}{2}$ in. $\left(< 2\dfrac{1}{2}d + \dfrac{1}{32} = 2.22 \text{ in.} \right)$.

Select a gage of $5\frac{1}{2}$ in. resulting in a side distance of 2.25 in.
$\left(> \dfrac{3}{2}d = 1\dfrac{15}{16} \text{ in.} \right)$.

The total length of the splice plate is $2[1.5 + 5(3.0) + 1.75] = 3'\,0\,\frac{1}{2}''$.

Bearing strength of an interior bolt on $\frac{3}{4}$-in. splice plate,

$$B_{dbi} = 0.75(102) = 76.5 \text{ kips}$$

As the design bearing strength of the 10 interior bolts exceeds the required strength of 279 kips, bearing will not control the design.
So, for the flange splice, use two A572 Gr 50 steel plates $\frac{3}{4} \times 10 \times 3'\,0\,\frac{1}{2}''$ long.   **(Ans.)**

c. Web splice
Shear to be resisted by web splice $= V_o = 180$ kips
Moment to be resisted by web splice $= 10\%$ of moment at splice line

$$= 0.10(790) = 79.0 \text{ ft-kips}$$

Assume there are going to be 2 vertical lines of bolts separated by 3 in. and with a vertical spacing, s, of 6 in. Equivalent eccentricity of the vertical load is (Eq. 12.15.6):

$$e^* = (1.5 + 1.75) + \frac{79.0(12.0)}{180} = 8.52 \text{ in.}$$

[continues on next page]

Example 12.15.1 continues ...

The bolts are in double shear and the design shear strength of a single bolt is 54.1 kips. So, the coefficient C required for the eccentrically loaded bolt group for ultimate strength design is:

$$C_{\text{req}} \geq \frac{P_u}{B_d} = \frac{180}{54.1} = 3.33$$

From LRFDM Table 7-19, corresponding to $s = 6$ in., $n = 4$, and $e_x = e^* = 8.5$ in., we obtain:

$$C = 4.67 > 3.33 \quad \text{O.K.}$$

So, use 8 bolts on each side of the splice line.

Depth of web splice plate $3(6.0) + 2(1.5) = 21.0 < T = 26.5$ O.K.

Width of splice plate $= 2(1.5 + 3.0 + 1.75) = 2(5.25) = 1'\,0\tfrac{1}{2}''$

Bending moment at the critical section (bolt line) $= 180(8.5 - 1.5) = 1260$ in.-kips

$$Z_{\text{req}} = \frac{M_u}{\phi F_y} = \frac{1260}{0.9(50)} = 28.0 \text{ in.}^3$$

$$Z = 2\left[\frac{1}{4}t(21^2)\right] \geq 28.0 \quad \rightarrow \quad t \geq 0.127 \text{ in.}$$

Provide two $\tfrac{3}{8}$-in.-thick plates.

Check shear:

Shear yield $= \phi\,(0.6F_y)A_g = 0.9(0.6)(50)(2)(\tfrac{3}{8})(21) = 425$ kips

Shear fracture $= \phi\,(0.6F_u)A_n = 0.75(0.6)(65)[(21 - 4(1.0)](2)(\tfrac{3}{8})$

$$= 373 \text{ kips} > 180 \quad \text{O.K.}$$

Bolt bearing strength of web $= 102(0.545) = 55.6 > 54.1$

So, for the web splice use two PL$\tfrac{3}{8}$×21×1'\,0\tfrac{1}{2}''$ long, A572 Gr 50 steel plates connected with 8 bolts on each side.   (Ans.)

12.16   Splicing of Compression Members

Splices in members are required whenever the available length of shapes and/or plates is smaller than the member length. Splices are also used whenever the member is too long to ship or to erect in one piece. There are two distinct types of compression splices: those that depend either fully or partly on contact bearing to transfer the compressive load and those that do not depend on contact bearing. In splices of the first category, the spliced ends of the adjacent segments of the member are prepared by milling, sawing, or other suitable means to provide virtually flat surfaces. In the finished structure, these two ends are placed firmly in contact with each other and are held together by splice material (flange plates, butt plates, and/or web plates) connected to the two segments. At least a part of the axial load is assumed to be transmitted from one member to the other by the bearing pressure between the milled surfaces. Splices in columns in tier buildings fall into this category.

In splices of the second category, the ends of the spliced members are cut by ordinary methods. The spliced ends of the adjacent segments of the member are kept slightly apart (i.e., with a gap) in the finished structure, and splice material (such as plates, angles, etc.) is connected to each element (web, flange, etc.) of the members. The splice material is designed to transmit, from one segment to the other, the load of the element to which it is connected. The connection components are proportioned so that their design strength equals or exceeds the required strength determined by structural analysis for factored loads acting on the structure or a specified proportion of the strength of the connected members, whichever is appropriate (LRFDS Section J1.1). Splices in compression chords of trusses generally fall into this category.

12.16.1   Column Splices

Columns in buildings are commonly fabricated in two-story lengths that are spliced from 2 to 4 ft above the floor at a height where the column splice plates will not interfere with the girder-to-column connection, and at a location where erection becomes easy. However, to accommodate the attachment of safety cables which may be required at floor edges or openings, column splices in tier buildings are preferably located about 4 ft above the finished floor. As the column loading changes at each floor level, the two-story column length uses a small amount of excess material that could be saved by changing the column section at every floor level, but the saving in erection costs and the material and fabrication cost of the splice saved more than offsets this small waste of material. Three-story columns are generally difficult to erect, but have been used in some high-rise buildings where special erection equipment is available.

For the W-shapes most frequently used as columns and of a given nominal depth, the distance $(h - 2t_f)$ between the inner faces of the flanges is constant. This distance is 8.86, 10.91, and 12.60 inches, for column sizes W10×33 to 112, W12×40 to 336, and W14×43 to 730, respectively (see LRFDM Fig. 14-13). Thus, for a nominal depth selected, as the weight per foot increases, it is the flange and web thicknesses which increase. Thus, full contact bearing is always obtained when lighter sections are centered over heavier sections of the same nominal depth. From LRFDS

All-bolted flange-plated column splice. Note the filler plate.

Photo by S. Vinnakota.

Section J8, the design bearing strength of the contact area of a milled surface is:

$$P_{dp} = \phi P_n = 0.75[1.8F_y A_{pb}] = (1.35F_y)A_{pb} \qquad (12.16.1)$$

where
P_{dp} = design bearing strength of the contact area, kips
A_{pb} = contact or bearing area, in.^{2}
P_n = nominal bearing strength of the contact area, kips
F_y = yield stress, ksi
ϕ = resistance factor = 0.75

When the connection is between W-shapes of the same nominal depth, the bearing strength given by Eq. 12.16.1 is much greater than the axial strength of the column above and will seldom become critical in the member design. Hence, LRFDS permits the assumption for tier-building columns that all of the compressive load is transmitted by the bearing of the milled ends against each other and requires that splice material be placed on all sides, so as to hold the abutting elements firmly in line with each other. Bolts or welds are required by LRFDS Section J1.4 sufficient to hold all parts securely in place. However, the specification leaves it to the judgement of the designer as to what it takes to hold all parts securely in place. If columns of different nominal depths must bear on each other or if the upper column is not centered over the lower column, some areas of the upper column will not be in contact with the lower column (i.e., $A_{pb} < A_{uc}$, where A_{ue} is the area of the upper column). Filler plates that are shop-bolted or shop-welded to the upper column, finished to bear on the lower section, and of sufficient area so that the total area in bearing satisfies Eq. 12.16.1 may be required. An alternative arrangement is to provide a butt plate shop-welded to the lower column and field-welded to the upper column. The different loads for which the column splices have to be designed are described below.

Direct Tension

LRFDS Section J1.4 stipulates that the column splices should be proportioned to resist any tension developed by the factored load combinations. Typically, this is load combination LC-6 of Chapter 4, which is $0.9D \pm (1.6W$ or $1.0E)$. Thus, for dead and wind loads acting, the required strength is $0.9D - 1.6W$. Here, D is the compressive force due to the dead load, and W is the tensile force due to wind load. If $9D \geq 1.6W$, the splice is not subjected to tension and a nominal splice may be selected from LRFDM Table 14-3. When $0.9D < 1.6W$, the splice will be subjected to a factored tensile force $T_u = (1.6W - 0.9D)$. The nominal splices from LRFDM Table 14-3 are acceptable if the design tensile strength of the splice plates $T_d(= \phi T_n)$ is greater than or equal to the required tensile strength T_u; otherwise a splice must be designed with sufficient area and connectors. Note that, in these calculations, it is not permissible to use forces due to live load to offset the tensile forces from wind or seismic loads.

Tension Due to Bending

If the columns are designed to transmit lateral loads in buildings (as part of a moment resisting frame, for example), large moments may be acting on the column, and the column splices must be designed to transmit the combination of axial load and bending moments. If the bending is sufficiently large to produce tension in some parts of the section, then the part subjected to tension (usually a flange) must be spliced so that it transmits all its tensile load. The compressive forces resulting from the moment are of little concern on those splices designed for bearing on finished surfaces. However, splices of this kind are made the same for both flanges, to preserve the symmetry and also because in many cases the moment will be in the opposite direction for other conditions of loading (wind from right and wind from left, for example). It is advantageous to locate splices in such members at sections of low bending moments (i.e., near the points of inflection). Where two members of different depths are spliced, it is recommended that they be located with their centers of gravity in the same line. If this is not possible, the moment caused by the eccentricity should be included in the splice design.

All-bolted flange-plated column splice. Note the extra line of bolts in the filler plate.
Photo by S. Vinnakota.

Bolts or welds transfer the tensile load from the element of the upper segment to the splice plate and from the splice plate to the corresponding element of the lower segment. Note that the bolts connecting the splice plate to the flanges are in single shear and the bolts connecting the web plates to the webs are in double shear.

Shear

Shear loads in columns of tier buildings are, generally, relatively small and can usually be resisted by friction on the contact bearing surfaces and/or by the flange plates. When the shear force in the direction of the web is large, a pair of splice plates to resist it can be either bolted or welded to the column webs across the splice line.

Erection Loads

Column splices should provide for safety and stability during erection when the column might be subjected to wind loads, construction loads, and/or accidental loading prior to the placing of the floor system. In the calculation of wind loads during erection, the tributary areas that may be provided by the beams, girders, or other surfaces that are attached to the structure should be included. Overturning moments become more severe at the upper levels, where the wind loads are larger and the smaller column sizes usually provide a smaller resisting lever arm. The overturning moments should be checked about both axes of the column.

LRFDM Table 14-3 gives typical column splice details for W-shape columns. These details are not splice standards, but rather typical column splices in accordance with the LRFD Specification provisions. These nominal splices have been developed to withstand accidental or construction loading prior to placing of the floor system. In practice, the designer selects from this table the

All-bolted flange-and web-plated column splice. Note the pin-hole in the web-plate for lifting hitch *Photo @ Hoffman Construction Company, courtesy of KPFF Consulting.*

column splice detail that appears reasonable for the size of columns to be connected and checks the adequacy under different load combinations described earlier. In most cases, the nominal splice designs given in LRFDM Table 14-3 will be adequate for the gravity loads and wind moments in a typical building.

LRFDS Section J2.11 stipulates that fully pretensioned high-strength bolts or welds shall be used for the following splice connections:

1. Column splices in all tier structures with the height H and the least horizontal dimension B of the structure satisfying one of the criteria below:

    a. $H \geq 200$ ft
    b. $100 < H < 200$ ft,   if $H/B > 2.5$
    c. $H < 100$ ft,   if $H/B > 4.0$

2. Roof truss splices and column splices in all structures carrying cranes over 5-ton capacity.

In all other cases, splice connections are permitted to be made with A307 bolts or snug-tight high-strength bolts. Except for conditions where full penetration welds are required for loads, every effort should be made to design the splice using bolted flange plates or partial penetration welds for economy.

References

12.1 Bjorhovde, R., and Chakrabarti, S. K. [1985]: "Tests of Full-Size Gusset Plate Connections," *Journal of Structural Engineering,* ASCE, vol. 111, no. 3, March pp. 667–684.

12.2 Garretts, J. M., and Madsen, I. E. [1941]: "An Investigation of Plate Girder Web Splices," *Transactions,* American Society of Civil Engineering, June.

12.3 Gibson, G. J., and Wake, B. T. [1942]: "An Investigation of Welded Connections for Angle Tension Members," *Welding Journal,* vol. 21, no. 1, January p. 44s.

12.4 Gross, J. L. [1990]: "Experimental Study of Gusseted Connections," *Engineering Journal,* AISC, vol. 27, no. 3, pp. 89–97.

12.5 Kulak, G. L. [1975]: "Eccentrically Loaded Slip-Resistant Connections," *Engineering Journal,* AISC, vol. 12, no. 2, pp. 52–55.

12.6 Kulak, G. L., and Green, D. L. [1990]: "Design of Connectors in Web Flange Beam or Girder Splices," *Engineering Journal,* AISC, vol. 27, no. 2, pp. 41–48.

12.7 Thornton, W. A. [1991]: "On the Analysis and Design of Bracing Connections," National Steel Construction Conference Proceedings, AISC, Chicago, pp. 26.1–26.33.

12.8 Whitmore, R. E. [1952]: "Experimental Investigation of Stresses in Gusset Plates," University of Tennessee Engineering Experiment Station Bulletin 16, May.

P R O B L E M S

Bolted Joints

P12.1. A $\frac{1}{2}\times6$-in. plate and a $\frac{3}{8}\times7$-in. plate are joined by four $\frac{7}{8}$-in.-dia. A325-N type bolts, as shown in Fig. P12.1. Determine the factored load T_u.

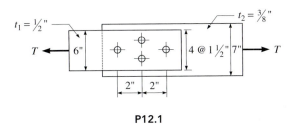

P12.1

P12.2. For the two $\frac{3}{8}\times9$-in. plates in the lap joint shown in Fig. P12.2, find the service load T, consisting of 25 percent dead load, and 75 percent live load, using the LRFD Specification. Assume A36 steel and $\frac{3}{4}$-in.-dia. A325 bolts with threads included in the shear plane. Consider the joint as (*a*) a bearing-type joint, (*b*) a slip-critical joint with type A surface preparation. Block shear need not be considered.

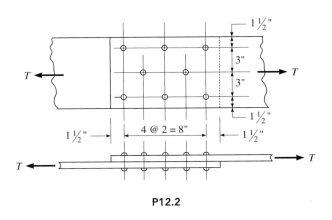

P12.2

P12.3. A tension member consists of two $3\times3\times\frac{3}{8}$ in. angles connected to opposite sides of a $\frac{3}{8}$-in. gusset plate using $\frac{3}{4}$-in.-dia. bolts. Design a bolted connection to develop the strength of the member. Use (*a*) A307 bolts,

(*b*) A325-N type bolts. Include the limit state of block shear rupture.

P12.4. Design a connection to develop the full tensile value of a member consisting of two C8×11.5 channels placed $\frac{1}{2}$ in. back-to-back and bolted with two rows of $\frac{1}{2}$-in. bolts to a $\frac{1}{2}$-in. gusset plate. Assume A572 Grade 42 steel and A325-N type bolts.

P12.5. A W18×50 has a C12×20.7 channel bolted to the top flange by A325-N, $\frac{3}{4}$-in.-dia. bolts (Fig. P12.5). Design the bolts at a location where the shear is 184 kips under factored loads. Assume A572 Grade 50 steel.

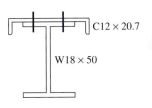

P12.5

P12.6. (*a*) A $\frac{3}{8}$-in.-thick bracket plate is connected to the flange of a column by three $\frac{3}{4}$-in.-dia. A325-X type bolts as shown in Fig. P12.6. A load P is applied at point A. Determine the design strength P_u of the joint, if the load is applied horizontally as shown. Use (*i*) the elastic method; (*ii*) the ultimate strength method.

(*b*) Repeat the calculations, if the load is applied vertically at point A.

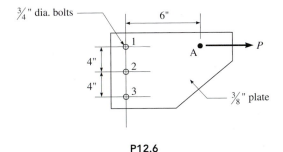

P12.6

P12.7. Determine the factored load P that can be carried by the $\frac{1}{2}$-in.-thick, A36 steel bracket plate shown in Fig. P12.7. The $\frac{3}{4}$-in.-dia. bolts are of A325-X type. Use (a) the elastic vector method, (b) the LRFDM tables, (c) the ultimate strength method.

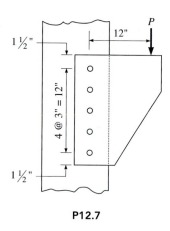

P12.7

P12.8. Determine the size of A490-N type bolts to transfer the factored load of 120 kips shown in Fig. P12.8. The $\frac{5}{8}$-in. plate is of A36 steel.

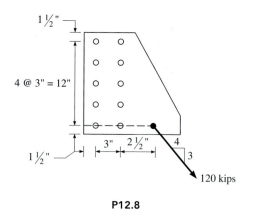

P12.8

P12.9. For the joint shown in Fig. P12.9, determine how many $\frac{3}{4}$-in.-dia. A490-X type bolts are required. The spacing is 3 in., end distance $1\frac{1}{2}$ in., gage distance 4 in. Check the adequacy of $\frac{1}{2}$-in.-thick bracket plate.

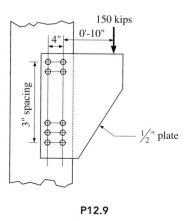

P12.9

P12.10. A bracket plate used in a mill building is shown in Fig. P12.10. The plate of A572 Grade 50 steel is $\frac{1}{2}$ in. thick. The loads shown are factored loads. The bolts are of A325-X type. Determine the size of bolts.

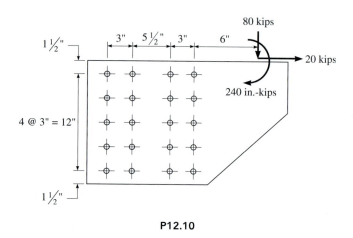

P12.10

P12.11. A bracket plate subjected to a factored load of 160 kips is connected to a column flange by 16 bolts of A325-X type as shown in Fig. P12.11. Use elastic vector method and select a suitable diameter of the bolt. The bracket plate is $\frac{1}{2}$ in. thick and of A572 Grade 50 steel.

P12.12. Determine the design strength of the eccentric bolted connection shown in Fig. P12.12. The $\frac{3}{4}$-in.-dia. bolts are of A490-N type. Use (a) the elastic vector method, (b) the ultimate strength method.

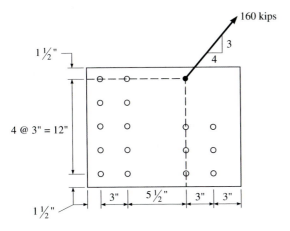

P12.11

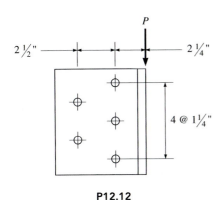

P12.12

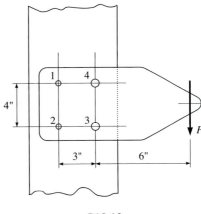

P12.13

P12.13. Determine the design strength of the bolted connection shown in Fig. P12.13. Bolts 1 and 2 are $\frac{1}{2}$ in. dia., and bolts 3 and 4 are $\frac{7}{8}$ in. dia. All bolts are of A325-N type. The plate is $\frac{3}{8}$ in. thick. Use (a) the elastic vector method, (b) the ultimate strength method.

P12.14. A WT16.5 $\times$ 65 $\times$ 1′ 0″ hanger is connected to the bottom flange of a W33×318 beam using six $\frac{7}{8}$-in.-dia. bolts, using a gage of $5\frac{1}{2}$ in., and a longitudinal spacing of 4 in. to transfer a factored tensile load of 142 kips (see Fig. W12.2.1). Check the adequacy of the hanger connection. Include prying action.

P12.15. A concentrically loaded connection of a diagonal brace to column flange is subjected to service load shear force of 60 kips and a service load tensile force of

100 kips. The loads are 10 percent dead load, 30 percent live load and 60 percent wind load. The bolts will be in single shear, and bearing strength will be controlled by a $\frac{1}{2}$-in. connected part. Determine the required number of $\frac{3}{4}$-in.-dia. A325 bolts. Assume threads are included in the shear plane, standard holes, and A992 steel.

    (a)   Bearing type connection.
    (b)   Slip-critical connection designed at service loads (class A surface).
    (c)   Slip-critical connection designed at factored loads (class A surface).

P12.16. Check the adequacy of the bracket connection shown in Fig. P12.16. The loads are service loads. The vertical load of 90 kips is 30 percent dead load and 70 percent live load, while the 20 kips horizontal load is 100 percent

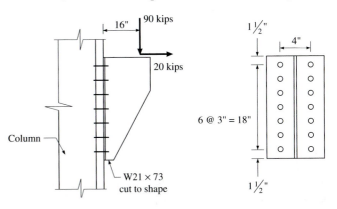

P12.16

live load. The ¾-in.-dia. bolts are of A325-X type. Assume an effective width of $8t_f$ for the bracket flange and A992 steel.

P12.17. For the connection of the bracket to the column shown in Fig. P12.17, determine the number of ⅞-in.-dia. A325 bolts required to transmit the eccentric load shown. The 90 kip load is 40 percent dead load and 60 percent live load. Assume 3-in. spacing and 1½-in. end distance. Threads are excluded from the shear plane. Use (a) a bearing-type connection, (b) a slip-critical connection.

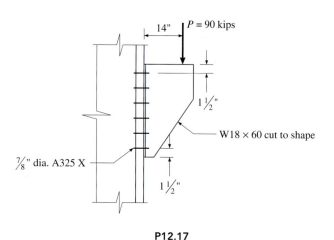

P12.17

Welded Joints

P12.18. The 6-in. leg of a 6×4×⅜-in. angle is connected to a gusset plate by maximum size fillet welds using E60 electrodes as shown in Fig. 12.8.1. The welds are required to develop the full tensile strength of the A36 steel member using balanced welds. Use (a) longitudinal welds only, (b) longitudinal welds and a transverse weld.

P12.19. The ⅝×8-in. plate of A36 steel is to be connected to a gusset plate using 5⁄16 in., E60 fillet welds. Due to clearance limitations, the two members can overlap only 6½ in. as shown in Fig. P12.19. Select suitable plug welds to develop the full strength of the bar in tension.

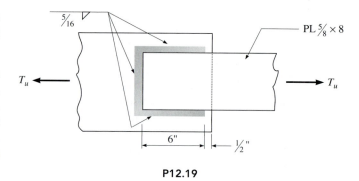

P12.19

P12.20. A ⅞×10-in. hanger plate of A36 steel is to be connected to another member using ⅜-in. E60 fillet welds. Due to clearance limitations, the two members can overlap only 7 in. as shown in Fig. P12.20. If the factored load on the connection is 250 kips, determine the minimum value of the dimension b.

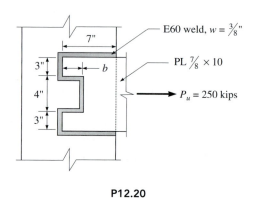

P12.20

P12.21. A MC13×35 channel of A572 Grade 50 steel is to be connected to another member using E70 fillet welds.

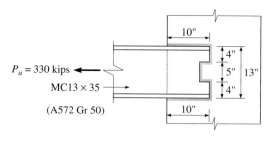

P12.21

Due to clearance limitations, the two members can overlap only 10 in. as shown in Fig. P12.21. If the factored load on the connection is 330 kips, determine the distance b so the connection is adequate.

P12.22. A MC13×35 channel of A572 Grade 50 steel is to be connected to the stem of a tee using E70 fillet welds. Due to clearance limitations, the two members can overlap only 10 in. as shown in Fig. P12.22. If the factored axial load on the connection is 330 kips, design suitable slot welds and fillet welds to carry the load.

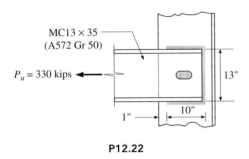

P12.22

P12.23. A built-up beam consists of two 2×18-in. flange plates welded to a $\frac{1}{2}$×44-in. web plate (see Fig. 12.8.2b). The beam is subjected to a maximum shear of 320 kips under factored loads. Assume E70 electrodes and determine the fillet weld sizes for (a) continuous welds, (b) intermittent welds.

P12.24. A welded built-up beam (see Fig. 12.8.2b) fabricated from two $1\frac{1}{2}$×16-in. flange plates and $\frac{5}{16}$×44-in. web plate is subjected to a maximum factored shear of 80 kips. Assume SMAW welds using E70 electrodes and determine the fillet weld size for (a) a continuous weld, (b) intermittent welds.

P12.25. A W27×129 has 1×14-in. cover plates attached to the flanges by E70 fillet welds (see Fig. 12.8.2a). Determine the size of welds for a location where a 240-kips shear occurs under factored loads. Use (a) continuous welds, (b) intermittent welds.

P12.26. A 6-in.-long, $\frac{3}{8}$-in. E70 fillet weld is used to carry an eccentric load P_u acting 8 in. from the weld as

shown in Fig. P12.26. Neglect end returns and determine the factored load P_u. Use (a) the elastic vector method, (b) the ultimate strength tables in the Manual.

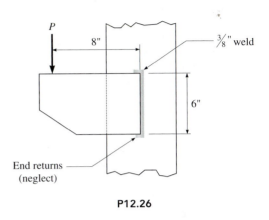

P12.26

P12.27. Using the LRFD specification, determine the effective length L of the fillet welds shown in Fig. P12.27. The 60-kip load is the service load (25 percent dead load, 75 percent live load). Assume E80 electrodes. Use (a) the ultimate strength tables, (b) the elastic vector method.

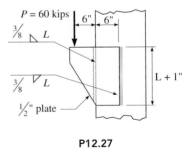

P12.27

P12.28. Determine the weld size required for the bracket of Fig. P12.28 when the service load P is 16 kips (75 percent live load and 25 percent dead load). Compare the results using (a) the elastic vector method, (b) the LRFDM tables.

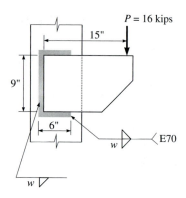

P12.28

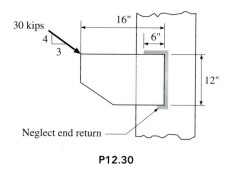

P12.30

P12.29. A channel-shaped weld is used to transmit a factored load of 60 kips from the bracket plate to the column flange shown in Fig. P12.29. Assume E70 electrodes and determine the minimum size of the weld. Use (*a*) the elastic vector method, (*b*) the LRFDM tables.

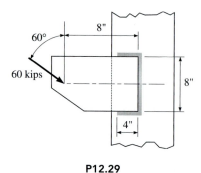

P12.29

P12.30. Determine the minimum size of the angle-shaped weld required to resist the factored load of 30 kips shown in Fig. P12.30. Assume E70 electrodes. Neglect the end return. Use (*a*) the elastic vector method, (*b*) the LRFDM tables.

P12.31. What size E70 fillet weld is required for the bracket plate shown in Fig. P12.31? The loads are factored loads. The horizontal load can act to the left or to the right. Determine the thickness of the bracket plate.

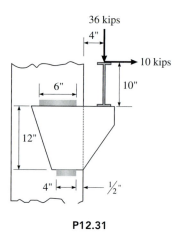

P12.31

P12.32. Determine the size of the fillet weld required for the bracket plate shown in Fig. P12.32, connecting the 10-in.-deep plate with 9-in.-long fillet welds using the SMAW process and E70 electrodes. Assume that the column and bracket do not control.

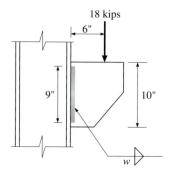

P12.32

P12.33. A tee bracket is welded to a column flange by two ⅜-in. fillet welds over its entire depth of 12 in. with 1-in.-long return welds, as shown in Fig. P12.33. The bracket it supports a vertical load P with an eccentricity of 6 in. Assume E70 electrodes and A572 Grade 50 steel.

Determine the factored load P_u using the elastic vector method.

  (a) Include returns in your calculations.

  (b) Neglect end returns.

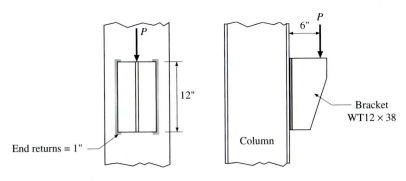

P12.33

Aerial view of the Athens' Summer 2004 Olympic Stadium.

The 18,000-ton, Santiago Calatrava designed structure features two elegantly bowed movable steel arches crossing over the sides of the stadium holding thousands of translucent glass plates over 72,000 seats. The arches—more than 300 m long and 80 m high—are made of tubular steel 3.6 m in diameter.
Yiorgos Karahalis/Reuters/Corbis

Connections

13.1 Introduction

Steel members studied in Chapters 7 through 11 are held in place by means of connections at the supports (usually at the ends), and loads are applied through other connections. Here, a **connection** may be defined as the aggregate of component parts used to join members, and may consist of **affected elements** (such as column flanges, beam webs, and beam flanges), **connecting elements** (such as gussets, plates, angles, and tees), and **connectors** (such as bolts, welds, and pins). Connections in steel structures may have to transmit member end forces consisting of axial forces (tension or compression), bending moments, shear forces, or torsion moments, that are applied individually or in combination. Most connections are statically indeterminate, and the distribution of internal forces and stresses depends upon the relative deformations of the component parts and the connectors themselves. Now, a connection is often made up of small components (plate elements), each capable of a significant influence on the overall deformation pattern. Bolt heads, washers, and welds also influence the total deformation. Stress concentrations that develop due to discontinuities such as bolt holes and the ends of welds further complicate the situation. The behavior of the connections is often nonlinear even when the connected members are in an elastic state of stress. Thus, the behavior of a connection is often so complex that it is practically impossible to analyze most connections by a rigorous and exact mathematical procedure. During the last three decades, attempts have been made to utilize computer programs such as NASTRAN and ANSYS to study the behavior of connections [Krishnamurthy, 1978]. However, it is still not possible to accurately predict the behavior of many of these connections because of their complexity.

Most of the analyses used in the design of connections are approximate and based on simplifying assumptions. The design procedures use simple formulas, which are often based on theories such as simple bending of beams and thin plate models modified appropriately to bring them into agreement with test results.

The ductile (plastic) behavior of structural steel is extremely important in connections because plastic yielding relieves the stress concentrations and prevents premature failure. The connecting elements are often of A36 steel (even when the members themselves are of higher strength steel), as long as they can transfer the required factored loads.

It should be noted here that the LRFD Specification only prescribes the design strengths of the individual connectors for particular types of forces on the connector such as shear and bearing forces on bolts, but does not prescribe the method of analysis to determine the connector forces under factored loads. The LRFD Manual gives design data and design tables for a variety of standard connections. However, the designer should understand the basis and limitations of the methods of analysis used in the development of these design aids.

Connections are one of the highest unit cost items in a steel structure. Hence, the cost of fabricated steel construction is influenced to a large degree by the choice of the connections. One of the design objectives is to have as many like pieces of material and framing conditions as possible in the overall job rather than optimizing each connection on its individual requirements. Also, the number of different shop operations performed (such as sawing, punching, and coping) on any single piece should be minimized.

13.1.1   Types of Steel Construction

All connections provide some restraint to the rotation of the member ends connected. Based on the degree of restraint provided, beam-to-column connections may be classified as rigid, simple shear, or semi-rigid. *Rigid connections* are connections with rigidity sufficient to maintain the original angle between intersecting members virtually unchanged under the design load. *Simple shear connections* are those connections that provide zero rotational restraint at the connection. *Semi-rigid connections* are connections that have a dependable and known moment capacity intermediate in degree between the rigidity of rigid connections and the flexibility of simple shear connections. They possess an insufficient rigidity to hold the original angles between the connected members. Rigid connections are designed to develop full resistance to shear and bending moment. Simple shear connections are designed to transfer shear only, assuming there is no bending moment present at the connection. Semi-rigid connections are designed to resist shear and moments whose values are intermediate between the values for simple and fully rigid connections.

Figure 13.1.1 compares the bending moment diagrams for a given beam, supporting a uniformly distributed load, for each of the three classes of end connections. (Column deformations are not shown for simplicity.) The disadvantage of a beam with simple shear connections (a simple supported beam) is that the entire moment requirement applies to one portion of the beam; the central section has the greatest moment equal to $M_C = M_S = QL/8$. Here, L is the span and Q is the total uniformly distributed load on the beam. Rigid end

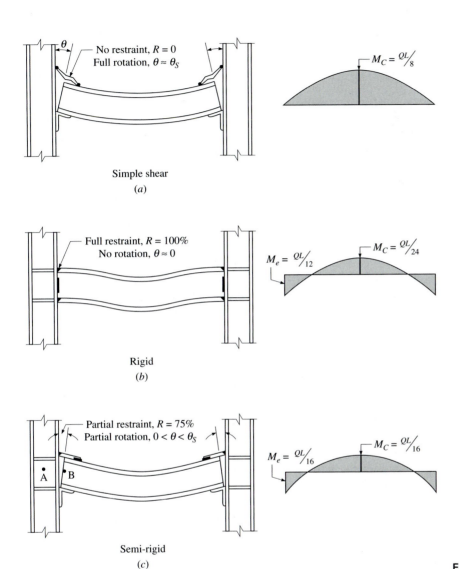

Figure **13.1.1:** Types of steel construction.

connections (fixed-end beam) reduce the simple beam moment at the central portion of the beam, with a corresponding increase in the moment at the ends. Thus, fully rigid connections for a uniformly loaded beam results in the end moment $M_e = M_F = QL/12$ and a central moment $M_C = QL/24$. If semi-rigid end connections having an end restraint R of 75 percent are used instead, both the end moment and the midspan moments have equal values of $M_e = M_C = QL/16$. Equalization of the end moment and midspan moment in the semi-rigid case will normally result in economies in beam design. Moment-resistant end connections also increase the stiffness of the structure.

Completely simple and completely rigid behavior are, of course, ideal conditions which can only be approached—never attained. In practice, it is necessary to accept something less than ideal, since real structural connections perform in the broad range between simple shear and fully rigid connections. Thus, connections with rotational restraint on the order of 90 percent or more of that necessary to prevent any angle change may be considered fully rigid. Similarly, a connection may be considered simple when the original angle between connected members may change by approximately 80 percent or more of the amount it theoretically would change if a frictionless hinged connection could be used. A connection may be treated as semi-rigid connection when rotational restraint provided by the connection is generally between 20 percent and 90 percent of that necessary to prevent relative angle change.

For beams subjected to vertical loads only, simple shear connections used for beam-to-column connections are sufficient for force transfer, and are the easiest to fabricate and assemble. Thus, in steel buildings, in which resistance to lateral loads is obtained through diagonal bracing or shear walls, many if not most of the connections will be simple shear connections.

In buildings of moderate height (10 to 15 stories) it may be more economical to resist lateral forces through the bending of beams and columns, which requires moment-resistant beam-to-column connections. Also, in plastically designed buildings of any height, either braced or unbraced, inelastic moment redistribution is essential to the realization of the full capacity of the frame. Connections, therefore, must be capable of developing the plastic resisting moments of the supported beams, and they must have substantial inelastic rotation capacity.

Remarks

1. In a frame provided with simple shear connections, the compression flanges of the beams are, for the most part, laterally restrained by the roof and the floor slabs they support. By introducing rigid or semi-rigid connections and transferring moments from beams to columns, one relieves moments from beams not subject to lateral-torsional buckling and increases moments in the columns which are usually not so restrained. Also, the compressed unrestrained bottom flanges of the beams adjacent to the columns require detailed examination for lateral-torsional buckling; in some cases this limit state may govern the beam section size. As a result, the overall structural weight will, in most cases, be reduced by only a small proportion, but in light of the greater cost of joints of adequate continuity, any savings in cost is likely to be insignificant.

2. In high-rise buildings located in hurricane areas, where the magnitude of elastic horizontal deflection (drift) tends to govern the design, full joint rigidity is desirable.

3. The design of a fully rigid frame or a semi-rigid frame is more complex than one that is braced and provided with simple shear connections. This

is exemplified in the problem of determining the distribution of bending moments throughout the frame under various load combinations, pattern floor loading, live load reduction, sway stability, and so on.

Structures using rigid connections have long been called *Type 1 Construction* in ASD Section A2.2 and are called *Type FR Construction* in LRFD Section A2. They are also called *Fully Restrained*, *Rigid Frame,* or *Continuous Frame Construction.* Construction utilizing Type FR connections may be designed in LRFD using either elastic or plastic analysis provided the appropriate specification provisions are satisfied. Also, the Type FR construction is unconditionally permitted by the AISC Specifications.

Structures using simple shear connections have long been called *Type 2 Construction* in ASD-A2.2. Also, structures using semi-rigid connections have long been called *Type 3 Construction* in ASD-A2.2. LRFD includes both Type 2 and Type 3 constructions of ASD-A2.2 under a more general classification called *Type PR Construction.* The designation PR (for partially restrained) for these connections is in recognition of the fact that some connection restraint is always present. PR connections have insufficient rigidity to maintain the original angles between intersecting members virtually unchanged. When the rotational restraint of the PR connections is used in the design of the connected members, or for the stability of the structure as a whole, AISC Specifications require that the capacity of the connections to provide the needed restraint must be established by analytical or empirical means or must be documented in the technical literature. Type PR construction may necessitate some inelastic but self-limiting deformation of a structural steel part. The term *self-limiting deformation* is included in the statement to prevent the use of simple and semi-rigid connections for cantilever beams, whose deflections are not limited by the rotation of the connection but may continue to progress, resulting in the failure of such connections.

Although ASD and LRFD Specifications offer a semi-rigid connection option to the designer, it has been rarely used because of the difficulty of obtaining a reliable analytical model to predict the rather complex response of the connection. The expense of laboratory verification of each semi-rigid connection type has put a brake on the use of this type of construction. Analysis of semi-rigid frames is also more cumbersome than rigid-frame or simple-frame analysis. Further, when incorporating connection restraint into the design, the designer should take into account the reduced connection stiffness on the stability of the structure and its effect on the magnitude of second-order effects.

Some general requirements for structural connections can now be stated. (1) They must be strong enough; (2) they should be ductile (provide adequate rotational capacity); (3) they should behave in a predicable fashion so as to provide the intended degree of restraint or lack of restraint; and (4) they should be as simple to fabricate and assemble as the other requirements permit.

Under the designation **flexible framing,** specifications permit the practice of ignoring beam end moments generated by a connection's resistance to gravity load in gravity load analyses, while counting on the same connection to resist wind moments resulting from calculations based on the assumption of fully rigid behavior in the lateral load analysis. This is a time-tested procedure, which has been found to be safe provided the actual end moment, which can be higher than the design moment, does not overstress the connectors. In reality, connections designed under this procedure are generally semi-rigid, with components that deform inelastically and thus prevent connector overloading. Evidently, it is difficult to calculate the true combined moment in such cases. Reliance must be placed on joint configurations of demonstrable ductility.

For "flexible framing" LRFDS Section A2 requires:

1. The connections and the connected members must be adequate to resist the factored gravity loads as simple beams.
2. The connections and the connected members must be adequate to resist the factored lateral loads.
3. The connections must have sufficient inelastic rotational capacity to avoid overload of connectors under combined factored gravity and lateral loading.

The type of connection, whether simple shear, semi-rigid, or rigid, is a primary determinant of a structure. Selection of the connection type should therefore go hand in hand with the planning of the framing system. In addition, the structural performance of the connection itself in tall buildings is greatly influenced by whether gravity or horizontal loads are dominant. This indicates that connection details themselves should vary qualitatively as well as quantitatively as the relative influence of gravity and lateral force varies throughout the height of the building.

13.2 Connection Behavior

13.2.1 Moment-Rotation Characteristics

The most important behavioral characteristics of a building connection are rotational stiffness and moment resistance as represented on a moment-rotation ($M_c - \theta_c$) diagram (Fig. 13.2.1). Generally, accurate $M_c - \theta_c$ diagrams can only be obtained experimentally. They are markedly nonlinear as a result of early yielding of component elements. The moment-rotation graph for a simply supported beam (pinned end) coincides with the axis of θ_c, since M_c remains zero for any rotation. Similarly, the graph for a perfectly fixed beam coincides with the axis of M_c, since there is no rotation at any moment. Completely simple and completely rigid behavior are, of course, ideal conditions which can only be approached, never attained. Figure 13.2.1 indicates

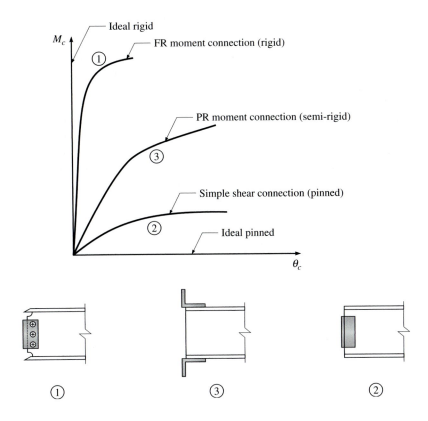

Figure 13.2.1: Connection moment-rotation curves.

some reasonable proportions of restraint that could be expected with each of three indicated connection types, where Example 1 is considered "rigid," Example 2 considered "simple shear;" and Example 3 considered "semi-rigid." It is impossible to define precise practical boundaries between types, but the response of any given connection should be sufficiently clear to permit the assignment of a meaningful and useful behavioral label. Thus, we observe that the directly welded connection (Example 1) is quite rigid; the single plate connection (Example 2) is fairly flexible; while the top and seat angle connection (Example 3) is semi-rigid.

Figure 13.1.1c shows the positions (points A and B) at which rotations are measured in full scale tests. Point B is positioned as close as is physically possible to the end section of the beam such that the flexural component of the beam rotation is insignificant. The rotations of points A and B are measured relative to space, and the true rotation of the connection is given by:

$$\theta_c = \theta_B \pm \theta_A \qquad (13.2.1)$$

The rotation of points A and B are generally measured by securing independent arms at both points.

The moment-rotation characteristics of a connection depend upon many physical parameters such as the type of connection, the size of angles, end plates, top and bottom plates, and the gage for bolt location. Stress concentrations around bolt holes, at the ends of welds, and at bends in connection components, coupled with typical residual or erection stresses, often result in local yielding under service loads. The rigidity of a connection is also influenced by the rigidity of its support. For beams framing into column flanges, a decrease in rigidity will occur if the column flanges are too thin, or if stiffeners are not used between the column flanges in line with the beam flanges. For a single beam framing into a column web, a decrease in rigidity may occur unless the beam flange is also welded directly to the column flanges or attached with suitable connecting plates. Research into the behavior of connections and their moment-rotation characteristics was first carried out by Wilson and Moore [1917]. This experimental investigation attempted to determine the rigidity of riveted joints in steel structures. Since then, several data bases for connection moment-rotation curves based on experimental results have been developed [Goverdhan, 1984; Kishi and Chen, 1986].

13.2.2   Beam-Line Method

Let us consider a beam ij of length, L, and elastic rigidity, EI, supported such that there is no relative translation of the ends of the beam (Fig. 13.2.2). The beam is subjected to end moments M_{ij} and M_{ji} and an arbitrary system of transverse loads, all lying in the vertical plane of symmetry of the cross section. Let the end slopes be θ_{ij} and θ_{ji}. The end moments and end rotations are considered positive clockwise. Using the slope-deflection equations [5.4], we can write:

$$M_{ij} = M_{ij}^F + \frac{2EI}{L}(2\theta_{ij} + \theta_{ji}) \qquad (13.2.2a)$$

$$M_{ji} = M_{ji}^F + \frac{2EI}{L}(\theta_{ij} + 2\theta_{ji}) \qquad (13.2.2b)$$

where, M_{ij}^F and M_{ji}^F are the fixed-end moments of the beam at i and j, respectively. If symmetrical loading is considered, then:

$$M_{ji} = -M_{ij}; \qquad \theta_{ji} = -\theta_{ij}; \qquad M_{ji}^F = -M_{ij}^F \qquad (13.2.3)$$

and Eq. 13.2.2a simplifies to:

$$M_{ij} = M_{ij}^F + \frac{2EI}{L}\theta_{ij} \qquad (13.2.4)$$

In particular, for a beam supporting the total uniformly distributed load Q $(= qL)$, we obtain:

$$M_{ij} = -\frac{QL}{12} + \frac{2EI}{L}\theta_{ij} \qquad (13.2.5)$$

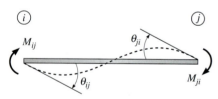

Final end moments and slopes

(a)

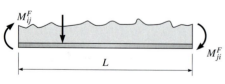

Fixed-end moments

(b)

Figure 13.2.2: End rotations of a beam under end moments and transverse load.

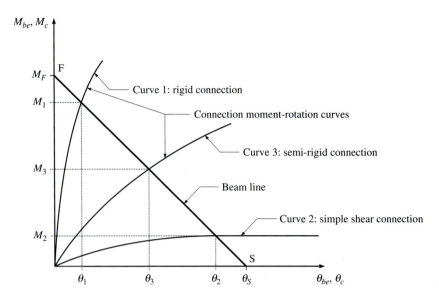

Figure 13.2.3: Beam-line method.

Equation 13.2.5 can be rewritten, with slight change in notation, as follows:

$$M_{be} = M_{be}^F - \frac{2EI}{L} \theta_{be} \tag{13.2.6}$$

where θ_{be} = beam end-rotation
M_{be} = beam end-moment corresponding to θ_{be}
M_{be}^F = fixed-end moment under uniformly distributed transverse load = $QL/12$

Thus, the relationship between the end moment, M_{be}, and end rotation, θ_{be}, of a beam is a straight line called the **beam line.** The beam line is represented graphically by the straight line FS in Fig. 13.2.3, between point F (rigidly fixed beam, no rotation, end moment M_F) and point S (simply supported beam, end rotation θ_s, end moment zero). From Eq. 13.2.6 we obtain for:

$$\theta_{be} = 0 \rightarrow M_{be} = M_{be}^F = \frac{QL}{12} = M_F \tag{13.2.7a}$$

and for

$$M_{be} = 0 \rightarrow \theta_{be} = \frac{M_{be}^F L}{2EI} = \frac{QL^2}{24EI} = \theta_S \tag{13.2.7b}$$

Thus, the ordinate of point F on the beam line is the end moment of a completely restrained beam, and the abscissa of the point S is the rotation at the end of the beam when the beam has zero restraint at the end. Also, from Eq. 13.2.6:

$$\frac{dM_{be}}{d\theta_{be}} = -\frac{2EI}{L} \tag{13.2.8}$$

Thus, the slope of the beam line is the rotational stiffness of the given beam (rigidity EI and length L), under symmetric loads. For increased factored load Q on the beam, the beam line moves out parallel to the first line, with correspondingly increased values of end moment (M_F) and end rotation (θ_s).

If a moment-rotation curve of a connection is plotted together with the beam line of a selected beam and loading, there will be one point of intersection where the curve crosses the beam line. This point defines the amount of end moment in the particular beam for that particular connection and loading, and the corresponding end rotation. The dependence of the beam behavior on the rigidity of the end connections can be studied by using this diagram. In the construction of this diagram, it is assumed that the behavior of the two end connections is the same, the beam is subjected to loads placed symmetrically on the beam, and the beam behaves elastically.

The behavior of the types of connections mentioned previously, namely rigid connection, simple shear connection, and semi-rigid connection, can be studied using the beam line diagram (Fig. 13.2.3):

- Curve 1 represents the moment-rotation characteristics of a rigid connection. The coordinates of the intersection point (θ_1, M_1) are such that:

$$\theta_1 \approx 0 \quad \text{and} \quad M_1 \geq 90\% \, M_F$$

  That is, such connections are detailed so as to allow virtually no rotation at the ends and to develop the full beam end moment.

- Curve 2 represents a simple shear connection. When provided with such a connection the beam end rotates through an angle θ_2, which is very nearly equal to the rotation θ_S of a completely unrestrained beam. Corresponding to this rotation a moment M_2 is generated at the ends which signifies that even with the so-called simple shear connections, some end moment does develop. Normally the bending moment developed with simple shear connections is from 5 to 20 percent of the fully fixed moment, M_F.

- Curve 3 represents a semi-rigid connection. The restraint offered by this type of connection can vary anywhere from a low of 20 percent to a high of 90 percent of the full fixity. That is M_3 equal to 20 to 90 percent of M_F.

All three of these connections have ample reserve carrying capacity, as shown by where their curves intersect the beam line. And while connections considered to be fully restrained seldom provide for zero rotation between members, the small amount of flexibility present is usually neglected and the connection is idealized to prevent relative rotation.

To sum up, the beam-line method graphically determines the equilibrium position of a beam, assuming elastic behavior. Two curves are plotted: the beam line, which is computed, and the moment vs. rotation ($M_c - \theta_c$) curve, which is experimentally determined.

In this text, we will study the behavior and design of only two types of connections: simple shear connections designed for gravity loads (with lateral frame stability provided by positive bracing system) in Sections 13.3 to 13.9, and rigid connections in Sections 13.10 to 13.14.

13.3   Simple Shear Connections

Simple shear connections are those connections that are capable of transferring only shear (and axial load, if any), but no moment. So, the ends of members with simple shear connections are assumed to be free to rotate under load. Of course, this is on the safe side as far as the beam is concerned, since the maximum positive moment (at the center) is reduced by the amount of the end moment developed. However, this conservative approach results in uncomputed stresses in the connection itself, since it is the standard procedure to design the framed beam connections for only the end shear. Simple shear connections are extensively used in steel construction to connect a beam to the web of a girder, or a beam to the flange or web of a column, when simple support of the beam has been assumed in the analysis. Thus, beam-to-column connections in braced frames are generally provided with simple shear connections. Simple shear connections should be able to rotate through the end rotation of a simply supported beam under factored gravity loads without fracturing.

    Simple shear connections may be broadly classified into *framed beam connections* and *seated beam connections,* each of which may be further subdivided as given below (the numbers within brackets represent maximum member end reaction for which standard connection details are available in Tables in Part 10 of the LRFDM):

Framed beam connections.
    Double-angle connections (574 kips).
    Shear end-plate connections (431 kips).
    Single-plate connections (330 kips).
    Single-angle connections (260 kips).
    Tee connections.
Seated beam connections.
    Unstiffened seated connections (156 kips).
    Stiffened seated connections (795 kips).

Bolted welded single angle shear connection of a roof beam to column flange.
Photo courtesy OPUS Corporation

It should be noted that vertical welding in the field is costly and usually should be avoided. So, in today's fabrication practice, field connections are usually bolted, while the shop connections are often welded. The choice of connection type is often a matter of personal preference (of the fabricator, subject to the approval of the EOR), but there are situations where the choice is dictated by framing conditions, fabricator's equipment, fabrication costs, and convenience in field erection. In selecting the type of simple shear connection to be used, the following points should be kept in mind:

Two-Sided Connections

Two-side connections, such as double-angle and shear end-plate connections, have the following advantages:

1. They are suitable for use when the end reaction is large.
2. They result in compact connections (usually, the entire connection is contained within the flanges of the supported beam).
3. Eccentricity perpendicular to the beam axis need not be considered for workable gages.

One-Sided Connections

One-sided connections, such as single-plate, single-angle, and tee connections, offer the following advantages:

1. They simplify fabrication and erection processes by allowing for shop attachment of connection elements to supports.
2. They require the handling of fewer connection elements and thus reduce shop labor.
3. They provide ample erection clearance.
4. They provide excellent safety during erection, since double connections can be eliminated.

Seated Beam Connections

Seated beam connections have the following advantages over framed beam connections:

1. They permit fabrication of plain punched beams.
2. Seats can be shop attached to the support, simplifying erection.
3. They result in better erection clearances when a beam connects to a column web. The end clearance allowed at each end when using framed beam connections in tier buildings is only $\frac{1}{16}$ in., whereas $\frac{1}{2}$-in. clearance can be used at seated connections. Since the overall length of an unframed punched beam is less than the back-to-back distance of connection angles for a framed beam, the beam can be easily lowered into the trough formed by the flanges of the supporting column.
4. They provide the erector a means of supporting the beam (thus freeing the crane) while aligning the field holes and inserting erection bolts. Hence the erection is fast, safe, and simple.
5. The bay length of the structure is easily maintained.
6. In the case of large size beams, seated connections reduce the number of field bolts, resulting in an overall economy.

Simple computer programs are now available for the design and checking of connections. Despite the increasing use of computer software in fabricating shops and design offices, design aids for connection design provided in the LRFD Manual continue to be used extensively by the beginning detailers, as well as engineers.

13.4 Double-Angle Connections

13.4.1 General

A **double-angle connection** is made with a pair of angles, called **framing angles** or **clip angles,** that are generally shop-connected to the web of the supported beam, the outstanding legs of the angles being field-connected to the web of another beam (generally called a girder) or to the flange or web of a column (Fig. 13.4.1). It is one of the most commonly used connections

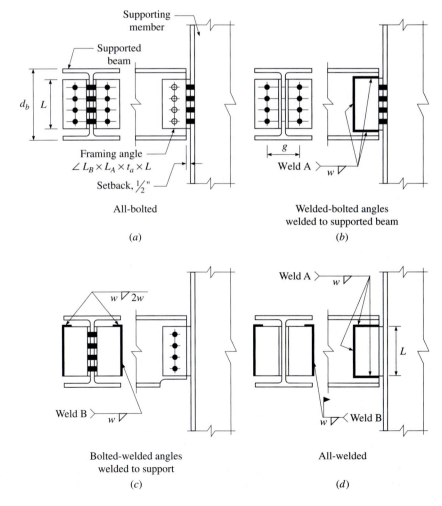

(a) All-bolted

(b) Welded-bolted angles welded to supported beam

(c) Bolted-welded angles welded to support

(d) All-welded

Figure 13.4.1: Double-angle connections.

to transfer end-reactions of beams in tier buildings. It is not unusual to use a combination of bolts and welds. Thus, the framing angles may be bolted to the supported beam web and bolted to the supporting member as shown in Fig. 13.4.1a; shop welded to the supported beam web and field bolted to the supporting member as shown in Fig. 13.4.1b; shop welded to the supporting member and field bolted to the supported beam as shown in Fig. 13.4.1c; or shop welded to the supported beam and field welded to the supporting member as shown in Fig. 13.4.1d. A single vertical line of bolts is generally used in each leg of the angles. The ordered length of the supported beam is such that its end will stop approximately $\frac{1}{2}$ in. short of the back of the framing angles (Fig. 13.4.1a). This is known as **set back,** which allows for tolerances in length, any inaccuracies in cutting of the beam length at the mill and in the shop, and thus eliminates possible recutting or trimming. The angles may be relocated, if necessary, without cutting off a piece of the beam.

When the beam and the girder to which it is attached are flush top (that is, their tops are at the same elevation), the end of the supported beam must be notched out at the top (Figs. 13.4.2a and b) to prevent interference with the flange of the girder. Such a notch is called a **cope.** The copes are generally rectangular in shape and a segment, centered at the intersection of the horizontal and vertical cuts, is shaped to a smooth radius to provide a fillet.

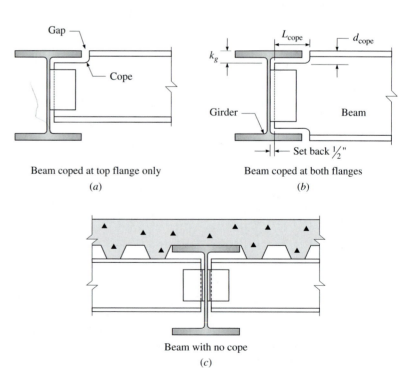

Beam coped at top flange only

(a)

Beam coped at both flanges

(b)

Beam with no cope

(c)

Figure 13.4.2: Coped and uncoped beam-to-girder connections.

When copes are needed, the minimum depth of cut d_{cope} should at least be equal to the k distance for the girder so as to clear its web fillet (Note: For this purpose, use the fractional dimension given for k in Table 1-1 of the LRFDM, not the decimal value.) The length of cut L_{cope} should provide $\frac{1}{2}$ to $\frac{3}{4}$ in. of clearance from the toe of the girder-flange. Thus:

$$d_{cope} \approx k_g; \quad L_{cope} = \frac{1}{2}b_{fg} - \frac{1}{2}t_{wg} - \frac{1}{2} + \{\frac{1}{2} \text{ to } \frac{3}{4} \text{ in.}\}$$

$$(13.4.1)$$

where b_{fg} is the flange width, t_{wg} the web thickness and k_g the k dimension of the girder. The dimensions d_{cope} and L_{cope} are usually rounded to the next $\frac{1}{4}$ in. Material removal is costly and should be avoided when possible. In some cases, it may be possible to do so by setting the elevation of the tops of the supported beams a sufficient distance below the tops of the girders to clear the girder fillet radius (Fig. 13.4.2c). Copes can reduce the design strength of the beam and may require web reinforcement.

Bolted Double-Angle Framing Connections

Let $L_B \times L_A \times t_a \times L$ be the dimensions of the framing angles. Here, L_A is the length of the angle-leg connected to the supported beam; L_B, the length of the angle-leg connected to the supporting member; t_a, thickness of the angle; and L, the length of the angle (Fig. 13.4.1a and Fig. 13.4.3). Further, let g_A be the gage used in leg A of the framing angle; g_B, the gage used in leg B of the framing angle; and g, the gage used in the supporting element (Fig. 13.4.3a). We have:

$$g \approx 2g_B + t_{wb} \qquad (13.4.2)$$

where t_{wb} is the thickness of the web of the supported beam (use the appropriate fractional dimension given in Table 1-1 of the LRFDM).

Double-angle connections are traditionally considered as resisting shear only, and the beams that they support are considered to be simply supported. In reality, these connections develop some end restraint and hence some end moment in the beam (generally not more than 20 percent of full fixed end moment), but it is neglected in design. When the beam end-rotation takes place, the upper part of the connection is in tension while the lower part is compressed against the supporting member. The end rotation, therefore, results in part through the bending of the outstanding legs in the upper part of the framing angles, as well as through the elongation of the upper bolts in leg B once the pretension (if any) in bolts has been overcome. To assure the flexibility assumed in the design, rather thin angles with wide outstanding legs are selected and wide gages for the bolts in the outstanding legs are used. Thus, the dimension, g, actually used should be as large as possible and consistent with the workable gage for the supporting element, if appropriate. Ordinarily, the framing angles need only be thick enough to develop, in bearing, the single shear value of the bolt.

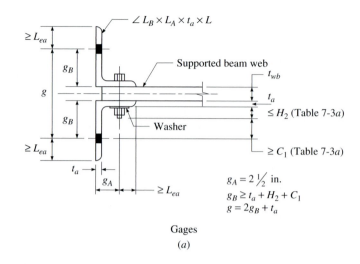

$$g_A = 2\tfrac{1}{2} \text{ in.}$$
$$g_B \geq t_a + H_2 + C_1$$
$$g = 2g_B + t_a$$

Gages

(a)

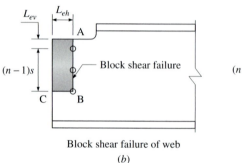

Block shear failure of web

(b)

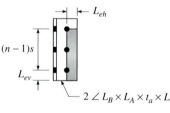

Block shear failure of angles

(c)

Figure 13.4.3: All-bolted double-angle shear connection.

All bolted, double-angle, double-sided, shear connection of a floor beam to girder. Note the extra line of bolts on the far side to facilitate erection. Also note that the floor beam is coped at the top.
Photo courtesy OPUS Corporation

The following remarks should be kept in mind in the design of double-angle framed beam connections:

1. When the beam is connected to the web of a column, it is necessary to have a column not less than 10 in. deep, so that the two connection angles can be fitted between the flanges of the column.
2. The maximum length of the framing angles must be compatible with the T-dimension for an uncoped beam, and the remaining web depth, exclusive of fillets, for a coped beam.
3. The minimum length of the framing angles should at least be equal to half the T-dimension of the supported beam, so as to provide adequate stability for the beam during erection.
4. Vertical fastener spacing is arbitrarily chosen as 3 in. It is common practice to place holes for connections along horizontal beam (web) gage lines spaced 3 in. vertically, with the upper most gage line set 3 in. below the top of the beam, when practicable.

5. The end distance on angles is set at $1\frac{1}{4}$ in. as permitted by LRFDS Table J3.4. When using oversize or slotted holes, the edge and end distances must be adjusted over that for standard holes as required.

6. The standard connection angle lengths (L values in the tables) vary from $5\frac{1}{2}$ to $35\frac{1}{2}$ in. using increments of 3 in.

7. To assure the flexibility assumed in the design, the angles used are rather thin; $\frac{1}{2}$ in. is the arbitrary maximum thickness tabulated in the LRFDM.

If a single row of bolts connecting the framing angles to the beam web is insufficient, angles with legs wide enough for two rows of bolts can be used. The outstanding legs of the framing angles can also be made wide enough to accommodate two rows each. For the usual gages used, and standard or short-slotted holes, eccentricity in bolted double-angle framed connections may be neglected, except in the case of double vertical row of bolts through the web of the supported beam. The tensile force in the upper bolts in the outstanding legs is also neglected. The bolts connecting the angles to the web of the beam are in double shear. The bolts connecting the angles to the supporting member are, usually, in single shear and in bearing.

In the design of all-bolted double-angle connections, the following limit states should be investigated:

Strength of the bolt group connecting the framing angles to the beam web.
    Double shear strength of the bolts.
    Bearing strength on the web of the beam.
    Bearing strength on the angles.
Strength of the bolt group connecting the framing angles to the supporting element.
    Single shear strength of the bolts.
    Bearing strength on the supporting element.
Block shear rupture of the beam web (Fig. 13.4.3b).
Shear yielding of the angles.
Shear rupture of the angles.
Block shear rupture of the angles (Fig. 13.4.3c).
Shear strength of the coped section.
Bending and buckling strength of the coped section.

The design strength of the connection, corresponding to the limit state of bolt shear is:

$$R_{dbv} = nB_{dv} \qquad (13.4.3)$$

Here,

    n = number of bolts in the leg A
    B_{dv} = design shear strength of a bolt in double shear

The design bearing strength of the connection corresponding to the limit state of bolts bearing on the supported beam web is given by:

$$R_{dbw} = [B_{dbe} + (n-1)B_{dbi}]_w \qquad (13.4.4a)$$

Here,

B_{dbi} = design bearing strength of an interior bolt, bearing on a plate material of thickness t_{wb} and ultimate tensile stress, F_{ub}

B_{dbe} = design bearing strength of the end (top) bolt

The design strength of the connection, corresponding to the limit state of bolts bearing on the framing angles is:

$$R_{dba} = 2[B_{dbe} + (n-1)B_{dbi}]_a \qquad (13.4.4b)$$

where B_{dbi} = design bearing strength of an interior bolt, bearing on a plate material of thickness t_a and ultimate tensile stress, F_{ua}

B_{dbe} = design bearing strength of the end (bottom) bolt

The design strength of the connection corresponding to the limit state of shear yielding of the framing angles (from LRFDS Eq. J5-3) is :

$$R_{davy} = 2[0.90(0.6F_{ya})Lt_a] \qquad (13.4.5)$$

Here,

F_{ya} = yield stress of the angle material (usually, 36 ksi).

The design strength corresponding to the limit state of shear rupture of the framing angles is obtained from LRFDS Eq. J4-1. The net shear area is calculated by passing a vertical section through the framing angle holes. We obtain:

$$R_{davr} = 2\{0.75(0.6F_{ua})[L - nd_e]t_a\} \qquad (13.4.6)$$

Here,

F_{ua} = ultimate tensile stress of the angle material (usually, 58 ksi)

d_e = effective diameter of the bolt hole

A coped web subject to high bearing stresses in a high-strength bolted beam end connection may fail in a block shear failure mode along a line through bolt holes (Fig. 13.4.3b). The problem may particularly arise when there are only a few large diameter high-strength bolts through a relatively thin beam web, and they do not extend uniformly over the entire web depth. The limit state of block shear in coped beams was studied by Birkemoe and Gilmor [7.1], and Ricles and Yura [7.10] among others. This block shear failure can occur by a combination of (Fig. 13.4.3b):

1. Shear fracture along a vertical plane from the cope through the bolt holes to the bottom hole in the supported beam web, plus a tensile (yield or fracture) failure along a horizontal plane from the bottom hole to the end of the beam.

Shop-welded, field-bolted, double-angle shear connection of a floor beam to a column flange. Note that the bottom flange of the beam is cut to facilitate erection.

Photo courtesy OPUS Corporation

2. Tension fracture along the horizontal plane plus shear (yield or fracture) failure along the vertical plane.

As mentioned in Section 12.12.1, the result with the larger fracture term must be used in design for block shear rupture strength of the web. LRFDM Tables 9-3 and 9-4 give coefficients to be used for evaluating the block shear strength for several common beam framing conditions of edge distance, L_{eh}; end distance, L_{ev}; different bolt diameters ($\frac{3}{4}$ in., $\frac{7}{8}$ in., and 1 in.) in standard size holes; and different number of bolts ($n = 2$ to 12) at 3 in. vertical spacing. The results are given for two different yield stresses ($F_y = 36$ ksi and 50 ksi), and for three different ultimate tensile stresses ($F_u = 58$ ksi, 65 ksi, and 70 ksi). Values are given for the tension rupture (Table 9-3a), shear yield (Table 9-3b), shear rupture (Table 9-4a), and tension yield (Table 9-4b) components. The case with the larger rupture term governs the design. To this strength, add the smaller strength (yield or rupture) of the perpendicular plane. For conditions that differ from those stipulated in the tables, the general equations given in Section 12.12.1 can be utilized.

The design strength of an unreinforced coped beam may also be controlled by the limit states of flexural yielding, local buckling, or lateral-torsional buckling, if applicable. Local buckling of coped beams has been reported by Gupta [1984], Cheng and Yura [1986], and lateral-torsional buckling of coped beams has been reported by Cheng et al. [1988].

Welded Double-Angle Framing Connections

The welded double-angle framing connection is made by using two fillet welds: weld A connects the framing angle to the supported beam web, and weld B connects the outstanding (usually, longer) leg of the framing angle to the supporting member (Figs. 13.4.4a and b). Welds A and B are in two perpendicular planes. Let $L_B \times L_A \times t_a \times L$ be the dimensions of the framing angles. As indicated in Fig. 13.4.4, the fillet welds connecting the framing angle to beam web should be continuous along the ends of the angle, at both the top and bottom for the distance b, where $b = L_A - \frac{1}{2}$ in. They should not be continued around the end of the web because of the danger of creating a notch in the beam web. Eccentricity must be taken into account in the design of welds A and B of the double-angle shear connections.

We assume that the angles are subjected only to a vertical shear, equal to the end reaction R_u, acting at the intersection of the beam web plane with the face of the supporting member. From Figs. 13.4.4a and b, it is seen that the force acting on one angle, $\frac{1}{2}R_u$, is eccentric with respect to both the welds A and B and causes both twisting and bending moments on the welds. To simplify the analysis, we assume that:

- Weld A is subjected to a direct force $\frac{1}{2}R_u$, plus a torsional moment of $\frac{1}{2}R_u e_A$, where $e_A = (L_A - \bar{x})$ (see Fig. 13.4.4c). The distance $\bar{x}$ from the vertical weld to the centroid of the channel-shaped weld can be obtained from Table 8-9 of the LRFDM.

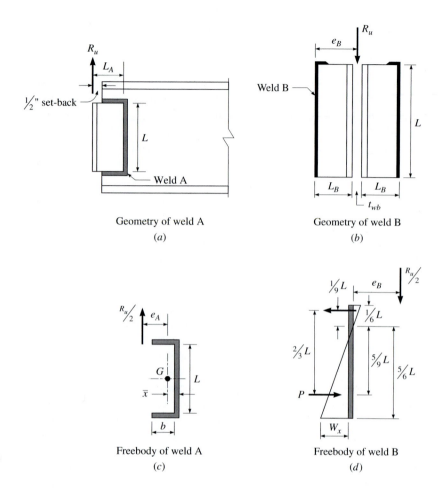

Figure 13.4.4: Analysis of welds A and B of a double-angle all-welded shear connection.

- Weld B is subjected to a direct force $\frac{1}{2}R_u$, plus a torsional moment of $\frac{1}{2}R_u e_B$, where $e_B = L_B$ (see Fig. 13.4.4d).

The procedure for the analysis of the weld A is basically the same as that outlined in Section 12.9.2. Thus, the design strengths given in LRFDM Tables 10-2 and 10-3 for welds A, are determined by the instantaneous center of rotation method for a channel shaped weld having: vertical leg length, L; horizontal leg length, b; applied force, $\frac{1}{2}R_u$; eccentricity, e_A; and inclination of the load with the vertical, $\theta = 0°$ (that is, using LRFDM Table 8-9).

For a cope depth of $1\frac{3}{4}$ in. and the connected angle edge distance of $1\frac{1}{4}$ in., for the field bolts in leg B, no weld can be placed across the top edge of the connection angles unless the ***punch down,*** the vertical distance from the top flange of the beam to the center of the top bolt, is increased to about $3\frac{1}{4}$ in. Some fabricators will prefer to maintain the punch down at 3 in.

In that case, only an L-shaped weld can be made to the supported beam. The design strength of this weld can be determined from Table 8-11 of the LRFDM.

The minimum thickness of the supported beam web for weld A is determined by matching the shear rupture strength of the beam web base material with the shear strength of the weld. Thus, for SMAW process using the E70 electrodes:

$$0.75(0.6F_{ub})t_{wb} \times 1 \geq 0.75(0.6F_{EXX}) \times (0.707w)(2) \times 1 \quad (13.4.7a)$$

$$t_{wb} \geq \frac{0.75(0.6F_{EXX}) \times (0.707w)(2)}{0.75(0.6F_{ub})} = \frac{6.19D}{F_{ub}} \equiv t_{wb\,min} \quad (13.4.7b)$$

where D is the weld dimension, w, expressed in sixteenths-of-an-inch. For example, for a $\frac{3}{16}$ in. ($D = 3$) weld A connecting a W-shape of A992 steel ($F_{ub} = 65$ ksi), $t_{wb\,min} = 0.286$ in. This is also the value tabulated in LRFDM Tables 10-2 and 10-3.

Next, let us consider the behavior of weld B. The length of the return at the top of the angle, typically $2w$, is omitted in computing the weld strength. The reaction R_u, acting along the line intersecting the beam web plane with the face of the support, tends to force the two angles to rotate in opposite directions. Consequently, the angles press against the beam web at the top and spread apart at the bottom. Resistance to the twisting moment $\frac{1}{2}R_u e_B$ on a framing angle is provided by the bearing pressure between the angle and the supported beam web at the upper end of the angle as is shown in Fig. 13.4.4d; and by the horizontal shear stresses that develop between the weld and the angle over the lower part of weld B. The usual practice is to assume that the contact between the angle and the web extends about one-sixth of the length of the angle, from the top. Also, the neutral axis is assumed to be at a distance $L/6$ from the top of the angle, and the bearing stresses are assumed to vary linearly from the neutral axis. The horizontal shear stresses in the weld, that develop to resist the moment on the angle, are also assumed to vary linearly over the remaining $\frac{5}{6}$ of the length L. The effective length of the fillet weld is the overall length of the fillet including end returns; however, the returns are often omitted to simplify calculations.

Moment equilibrium in the plane of the factored load R_u and weld B requires:

$$\left(\frac{1}{2}W_x\right)\left(\frac{5}{6}L\right)\left(\frac{2}{3}L\right) = \frac{R_u}{2}e_B \rightarrow W_x = \frac{9}{5}\left(\frac{R_u\,e_B}{L^2}\right)$$

where W_x is the horizontal shear force that develops on the critical unit-length weld which is always the element farthest from the neutral axis. The direct (vertical) shear component on a unit-length weld is,

$$W_y = \frac{R_u}{2L}$$

Since the two shearing components are at right angles, they can be combined vectorially to obtain the maximum resultant force on the critical unit-length weld as

$$W_u = \sqrt{\left(\frac{9}{5}\frac{R_u\,e_B}{L^2}\right)^2 + \left(\frac{R_u}{2L}\right)^2}$$

If D is the number of sixteenths-of-an-inch in the weld size, the design strength of the fillet weld—assuming SMAW process and E70 electrodes—is:

$$W_d = 1.392D$$

According to LRFDS,

$$W_u \leq W_d \rightarrow \frac{R_u}{2L}\sqrt{1 + \left(\frac{18}{5}\right)^2\left(\frac{e_B}{L}\right)^2} \leq 1.392D$$

$$R_u \leq \frac{2(1.392D)L}{\sqrt{1 + 12.96(e_B/L)^2}} \equiv R_{dw} \tag{13.4.8}$$

For example, for $L = 11\frac{1}{2}$ in., $D = 5$, $e_B = 4$ in., $R_{dw} = 99.9$ kips, from Eq. 13.4.8. This is also the value given in LRFDM Table 10-2 for weld B. The length of the angle (L) and the size of the weld (D) may be varied to arrive at an adequate and economical design. The length of an all-welded connection usually varies from one-half to two-thirds the T-distance of the supported beam. Equation 13.4.8 neglects eccentricity, e_A, on weld B, which tends to cause tension at the top of the weld line.

13.4.2   Design Tables

LRFDM Table 10-1 for All-Bolted Double-Angle Connections
All-bolted double-angle connections can be easily designed as simple shear connections to beams by using LRFDM Table 10-1. Separate tables are provided for three bolt diameters ($\frac{3}{4}$ in., $\frac{7}{8}$ in., and 1 in.), and for 2 through 12 horizontal rows of bolts in a single vertical row. Values are tabulated for supported and supporting member material with $F_y = 50$ ksi and $F_u = 65$ ksi, and angle material with $F_y = 36$ ksi and $F_u = 58$ ksi. Each page lists, in the upper left-hand corner, the applicable beam depths based on a maximum and a minimum angle length of T and $T/2$, respectively. The first part of the table gives the total shear that the bolts and angles can carry for different angle thicknesses. For calculation purposes angle edge distances L_{ev} and L_{eh} are assumed to be $1\frac{1}{4}$ in. All values, including slip-critical bolt design strengths, are for comparison with factored loads. The second part of the table gives the supported-beam web design strengths, per inch of web thickness, for three conditions (uncoped, coped at top flange only, and coped at both flanges), for standard, oversized, and short slotted holes. Values for

coped beams are tabulated for beam web edge distances L_{ev} from $1\frac{1}{4}$ to 3 in., and for beam end distances L_{eh} of $1\frac{1}{2}$ and $1\frac{3}{4}$ in. For calculation purposes, these beam end distances have been reduced to $1\frac{1}{4}$ and $1\frac{1}{2}$ in, respectively, to account for possible mill underrun in beam length. The third part of the tables give the design bearing strength of the supporting member to which the outstanding legs are attached (per inch of flange or web thickness).

Tabulated bolt and angle design strengths have included the limit states of bolt shear (double shear), bolt bearing on the angles, shear yielding of the angles, shear rupture of the angles, and block shear rupture of the angles. Tabulated supported-beam-web design strengths consider the limit state of bolt bearing on the beam web. For the beams coped at the top flange only, the limit state of block shear rupture is also considered. In addition, for beams coped at the top and bottom flanges, the tabulated values consider the limit states of shear yielding and shear rupture of the beam web. However, note that, for coped members, the limit states of flexural yielding and local buckling must be checked separately by the designer. Tabulated supporting member design strengths consider the limit state of bolt bearing on the supporting member. The effect of deep and/or long cope on beam performance has not been included in the tables and must be checked independently [LRFDM, Part 9; Cheng et al., 1988].

It should be noted that the length and thickness of the angle can be determined from Table 10-1 of the LRFDM, but the width of the angle is not given. The width depends upon the gages used, which in turn depend upon the diameter and the type of bolts and the clearance required for tightening field bolts in the outstanding legs (see Fig. 13.4.3a).

LRFDM Table 10-2 for Bolted/Welded Double-Angle Connections
LRFDM Table 10-2 permits substitution of welds for bolts in double-angle connections designed with Table 10-1. Welds A may be used in place of bolts through the supported beam web legs of the double angles. Welds B may be used in place of bolts through the supporting member legs of the double angles. Weld strengths are tabulated for angle lengths corresponding to 2 through 12 rows of bolts provided in the other leg. E70 electrodes are assumed. Design strengths for welds A are determined by the instantaneous center of rotation (ultimate strength) method using LRFDM Table 8-9 with $\theta = 0°$. Design strengths for welds B, on the other hand, are determined by the elastic method (Eq. 13.4.8) using an e_B value of 4 in.

The tabulated minimum thicknesses of the supported beam web for welds A and the supporting-member web or flange for welds B are obtained by matching the shear rupture strength of these elements (base metal strength) with the strength of the weld metal using Eq. 13.4.7. The minimum angle thickness, when LRFDM Table 10-2 is used, is the weld size plus $\frac{1}{16}$ in., but not less than the angle thickness determined from LRFDM Table 10-1 for the bolted leg. In general, a 2L4×3½ will accommodate usual gages, with the 4-in. leg attached to the supporting member. The width of the web legs (leg A) in Case I may be optionally reduced from $3\frac{1}{2}$ to 3 in. The width of out-standing legs (leg B) in Case II may be optionally reduced from 4 to 3 in. for values of angle lengths L from $5\frac{1}{2}$ through $17\frac{1}{2}$. Again, a small thickness of angle is preferred to minimize restraint to the beam rotation at the support.

LRFDM Table 10-3 for All-Welded Double-Angle Connections

LRFDM Table 10-3 is a design aid for all-welded double-angle connections. Although welds A and welds B can be combined from LRFDM Table 10-2, an economical all-welded connection, that provides greater flexibility in angle length selection and connection capacity, can be achieved through the use of the LRFDM Table 10-3. Weld design strengths for the limit state of weld shear are tabulated for different practical values of angle lengths L varying from 4 to 36 in. and three different weld sizes ($\frac{5}{16}$, $\frac{1}{4}$, and $\frac{3}{16}$ in.). Electrode strength is assumed to be 70 ksi. Again, design strengths for welds A are determined by the instantaneous center of rotation method (ultimate strength method) using LRFDM Table 8-9 with $\theta = 0°$. Design strengths for welds B are determined by the elastic method. A 2L4×3 must be used for angle lengths equal to or greater than 18 in.; a 2L3×3 must be used otherwise. Normally, the framing angles are welded to the beam web in the shop prior to shipping to the field and then field welded to the supporting girder or column. Erection bolts are often used for erecting the beams.

EXAMPLE 13.4.1   All-Bolted, Double-Angle Connection

Select an all-bolted, double-angle connection for a W16×45 beam to a W24×55 girder web connection. The beam carries a total factored uniformly distributed load of 180 kips over its length. The beam and the girder are of A992 steel. Use $\frac{3}{4}$-in.-dia. A325-N type bolts in standard holes and A36 steel connection angle. The top flanges of the beam and girder are at the same level.

Solution

$$\text{W16×45 beam} \rightarrow \quad d_b = 16.1 \text{ in.}; \quad t_{wb} = 0.345 \text{ in.}$$

$$\frac{t_{wb}}{2} = \frac{3}{16} \text{ in.}; \quad k_b = 1\frac{1}{4} \text{ in.}$$

$$\text{W24×55 girder} \rightarrow \quad b_{fg} = 7.01 \text{ in.}; \quad d_g = 23.6 \text{ in.}; \quad t_{wg} = 0.395 \text{ in.}$$

$$\frac{t_{wg}}{2} = \frac{3}{16} \text{ in.}; \quad k_g = 1\frac{7}{16} \text{ in.}$$

a. Cope the top flange of the beam $1\frac{3}{4}$ in. deep by $3\frac{1}{2}$ in. long.
Depth of cope, $d_{cope} = 1\frac{3}{4}$ in. $> k_g = 1\frac{7}{16}$ in.   O.K.
Cope length, $L_{cope} = 3\frac{1}{2}$ in.

$$\text{Gap} = 3.5 + \frac{1}{2} + \frac{3}{16} - \frac{7.01}{2} = 0.683 \text{ in.}$$

As the gap is between $\frac{1}{2}$ in. and $\frac{3}{4}$ in., the cope length is acceptable. Since the typical distance from the top of the flange to the center of the top bolt (punch down) is 3 in., the end distance L_{ev} is $1\frac{1}{4}$ in. This satisfies the minimum edge distance requirement in LRFDS Table J3.4, which is 1 in. at gas cut edges for $\frac{3}{4}$-in.-dia. bolts. The depth of the beam web, clear of the cope at the top flange and of the fillet at the bottom flange is $16.1 - 1.75 - 1.25 = 13.1$ in. So, the maximum framing angle length is $L_{max} \approx 13$ in.

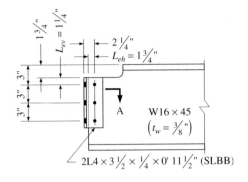

b. Select number of bolts and angle thickness
Total factored distributed load on beam = 180 kips

End reaction, $R_u = \dfrac{180}{2} = 90$ kips

From LRFDM Table 10-1, corresponding to $\frac{3}{4}$-in.-dia. bolts, A36 steel angles, W16 shapes, select four rows of A325-N bolts and $\frac{1}{4}$-in. angle thickness, for which:

$$R_d = \phi R_n = 104 \text{ kips} > 90 \text{ kips} \qquad \text{O.K.}$$

c. Check supported beam web
From Table 10-1 of the LRFDM, for four rows of $\frac{3}{4}$-in.-dia. bolts and a beam web material with $F_y = 50$ ksi and $F_u = 65$ ksi, $L_{eh} = 1\frac{3}{4}$ in. and $L_{ev} = 1\frac{1}{4}$ in., we obtain design bearing strength for a 1-in.-thick web as 262 kips. This corresponds to the limit state of ovalization of the bolt hole. So, for a W16×45 beam with a web thickness of 0.345 in., we obtain:

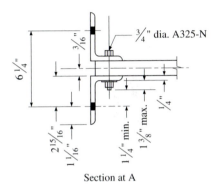

Section at A

Figure X13.4.1

$$R_{dbb} = \phi R_n = (262 \text{ kips/in.})(0.345 \text{ in.}) = 90.4 > 90 \text{ kips} \quad \text{O.K.}$$

d. Check supporting girder web
From Table 10-1 of the LRFDM for four rows of $\frac{3}{4}$-in.-dia. bolts and a girder material having $F_u = 65$ ksi and a web thickness of 0.395 in., the girder web design bearing strength is:

$$R_{dbg} = \phi R_n = 2(351 \text{ kips/in.})(0.395 \text{ in.}) = 277 \text{ kips} > 90 \text{ kips} \qquad \text{O.K.}$$

So, cope top flange of beam $1\frac{3}{4}$ in. deep by $3\frac{1}{2}$ in. long. Select 2L4× $3\frac{1}{2}\times\frac{1}{4}\times0'$ $11\frac{1}{2}''$ (SLBB) of A36 steel, and use four $\frac{3}{4}$-in.-dia. A325-N bolts in standard holes.   (Ans.)

Welded Bolted, Double-Angle Connection

EXAMPLE 13.4.2

Rework Example 13.4.1 using welds to connect the supported-beam web legs of the double-angle connection (welds A).

[continues on next page]

Example 13.4.2 continues ...

Solution

Thickness of the web = $\frac{3}{8}$ in.
Thickness of the angle = $\frac{1}{4}$ in.
Minimum size of fillet weld (LRFDS Table J2-4) is $\frac{3}{16}$ in.

Maximum size of fillet weld is: $t_p - \frac{1}{16} = \frac{1}{4} - \frac{1}{16} = \frac{3}{16}$ in.

So, select a weld size, $w = \frac{3}{16}$ in.

From Table 10-2 of the LRFDM, for four rows of bolts (an angle length of $11\frac{1}{2}$ in.), a $\frac{3}{16}$-in. weld provides $R_d = \phi R_n = 142$ kips. Also from this table, for beam web material with $F_y = 50$ ksi, the minimum web thickness is 0.286 in. Since $t_{wb} = 0.345$ in. > 0.286 in., no reduction in the tabulated value is required. Thus:

$$R_d = \phi R_n = 142 \text{ kips} > R_{\text{req}} = 90 \text{ kips.} \qquad \text{O.K.}$$

So, cope the top flange of the beam $1\frac{3}{4}$ in. deep by $3\frac{1}{2}$ in. long. Select 2L4× $3\frac{1}{2} \times \frac{1}{4} \times 0'\ 11\frac{1}{2}''$ (SLBB) of A36 steel and use $\frac{3}{16}$-in. fillet welds (welds A) to connect the short legs to the beam web. Use four $\frac{3}{4}$-in.-dia. A325-N bolts in standard holes in the outstanding leg.   (Ans.)

EXAMPLE 13.4.3

Bolted/Welded, Double-Angle Connection

Rework Example 13.4.1 using welds to connect the support legs of the double-angle connection (welds B) to the girder.

Solution

Minimum size of fillet weld = $\frac{3}{16}$ in.

From Table 10-2 of the LRFDM, corresponding to four rows of bolts in the short leg, an angle length of $11\frac{1}{2}$ in. with a $\frac{5}{16}$-in. weld (weld B) provides a design strength of 99.9 kips, which is greater than the required strength of 90 kips. For girder material with $F_y = 50$ ksi, the minimum girder web thickness is 0.238 in. Since $t_w = 0.395$ in. > 0.238 in., no reduction of the tabulated value is required.

$$R_d = 99.9 \text{ kips } > 90 \text{ kips} \qquad \text{O.K.}$$

The minimum angle thickness from Table 10-2 of the LRFDM is the weld size plus $\frac{1}{16}$ in., but not less than the thickness determined from Table 10-1 of the LRFDM.

$$t_{\text{min}} = \frac{5}{16} + \frac{1}{16} = \frac{3}{8} \text{ in. } > \frac{1}{4} \text{ in.} \qquad \text{O.K.}$$

Thus, the angle thickness must be increased to $\frac{3}{8}$ in. to accommodate the welded legs of the double angle connection.

So, cope the top flange of the beam $1\frac{3}{4}$ in. deep by $3\frac{1}{2}$ in. long. Select $2L4\times3\frac{1}{2}\times\frac{3}{8}\times0'\ 11\frac{1}{2}''$ with four A325-N $\frac{3}{4}$-in.-dia. bolts in standard holes connecting the beam web, and $\frac{5}{16}$-in. E70 weld (weld B) connecting the framing angle to the girder web.   (Ans.)

All-Welded, Double-Angle Connection

EXAMPLE 13.4.4

Rework Example 13.4.1 using an all-welded, double-angle connection for the W16×45 beam with the web of W24×55 girder.

Solution

From Example 13.4.1 maximum length of framing angle, L_{max} = 13 in. From Table 10-3 of the LRFDM, we observe that the requirements for weld B are more demanding. So, entering the table for welds B, we observe that for L = 10 in., weld size of $\frac{5}{16}$ in. provides a design strength $R_d = \phi R_n$ = 94.6 kips > 90 kips. Since t_w of the girder 0.395 in., is greater than the 0.238 in. required, no reduction of the tabulated value is required. For welds A corresponding to L = 10 in., and a minimum tabulated weld size of $\frac{3}{16}$ in.,

$$R_d = \phi R_n = 126 \text{ kips} > 90 \text{ kips} \qquad \text{O.K.}$$

Base the minimum angle thickness on the larger (welds B) size plus $\frac{1}{16}$ in.

$$t_{min} = \tfrac{5}{16} + \tfrac{1}{16} = \tfrac{3}{8} \text{ in.}$$

So, cope the top flange of the beam $1\frac{3}{4}$ in. deep by $3\frac{1}{2}$ in. long. Use 2L4 $\times3\times\frac{3}{8}\times0'\ 10''$ of A36 steel with $\frac{5}{16}$-in. welds B and $\frac{3}{16}$-in. welds A.
                        (Ans.)

13.5   Shear End-Plate Connections

In a **shear end-plate connection,** a rectangular plate called the **end-plate** is prepunched and shop welded to the end of the supported beam with the plane of the plate perpendicular to the longitudinal axis of the beam (Fig. 13.5.1). The length of the plate is limited to the T-dimension of the beam web (= d_b – $2k_b$) so that the welding is to the beam web only and not to the flanges. The plate-beam assembly is, generally, field bolted to the supporting member, as shown in Figure 13.5.1. Large gages and thin plates are used to achieve flexibility. Such shear end-plate connections have proven economical, if the beam reactions are in the light to medium range (35 to 430 kips under factored loads). Shear end-plate connections may be made to the flanges of supporting columns and to the webs of supporting girders. Because of bolting

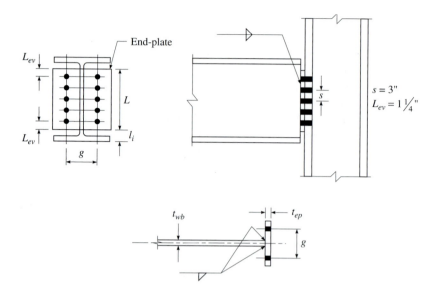

Figure 13.5.1: Shear end-plate connection.

and welding clearances, they may not be suitable for connections to the webs of W8 columns, and may be impossible for W6 columns. The design of shear end-plate connections has been studied by Kennedy [1969].

To ensure adequate flexibility and end rotation capacity, the end-plate should be in the thickness range of $\frac{1}{4}$ to $\frac{3}{8}$ in., inclusive. The transverse spacing of holes in the end-plate, gage g, should be $3\frac{1}{2}$ to $5\frac{1}{2}$ in., and the edge distances at the top, bottom, and sides should be $1\frac{1}{4}$ in. Lesser values of edge distance should be avoided.

The end plate should be attached on the supported beam web as close as possible to the compression flange of the beam in order to provide lateral stability for that flange. The end-plate is always shop welded to the supported beam web with a fillet weld on each side of the beam web. These welds should not be returned across the thickness of the beam web, at the top or bottom of the end-plate, because of the danger of creating a notch in the beam web. Hence, the effective length of weld is the depth of the plate minus twice the weld size. Since the weld is placed on both sides of the beam web, the design shear rupture strength of the base metal of the beam web should be checked, in addition to the shear capacity of the weld. The end-plate must have sufficient shear capacity on the net or gross section to transmit the load from the fasteners.

Fabrication of this type of connection requires close control in cutting the beam to length. Also, adequate consideration must be given to squaring the beam ends such that the plates at both ends are parallel and the effect of beam camber does not result in out-of-square end-plates. Otherwise, erection will be difficult. Shims may be required in the field to compensate for mill and shop tolerances.

The following limit states are to be investigated:

- Shear strength of bolts.
- Bearing strength of bolts on the end-plate and on the supporting element.
- Shear yielding of the end plate.
- Shear rupture of the end plate.
- Strength of welds.
- Shear rupture strength of the base metal (supported beam web).
- Block shear rupture strength of the end plate.
- In the case of beams with coped ends, it is necessary to investigate the shear and bending strengths of the reduced beam section at the copes.

In addition, the designer should ensure that the beam compression flange does not bear on the supporting beam or column. In practice, the end slope of a simply supported practical beam is not likely to exceed 0.03 radians, irrespective of the type of loading. Hence, with a N.A. close to the lower edge of the end-plate [Pask, 1992], if the expression

$$l_1 \leq 33t_{ep} \tag{13.5.1}$$

is satisfied, then contact pressure will not develop between the beam compression flange and the support. Here, t_{ep} is the thickness of the end-plate, and l_1 is the vertical distance from the bottom edge of the end-plate to the bottom edge of the supported beam. Thus, for a $\frac{1}{4}$-in.-thick end-plate l_1 should be less than or equal to 8.25 in.

Eccentricity is not considered in the design of the bolts or the welds. Thus, the design strength of bolts, which are in single shear, is given by:

$$R_{d1} = 2nB_{dv} \tag{13.5.2}$$

where n is the number of rows of bolts and B_{dv} is the design shear strength of a bolt in single shear.

The end-plate and the member to which the end-plate connects (that is, the girder web or column web or flange) must be checked for bearing. For the case of bolts bearing on the end-plate, the two end bolts (the two top bolts) with smaller clear distance have lower bearing strength. We have:

$$R_{d2} = 2[(n-1)B_{dbi} + B_{dbe}]_{ep} \tag{13.5.3}$$

$$R_{d3} = 2nB_{dbg} \tag{13.5.4}$$

where B_{dbi} and B_{dbe} are the design bearing strength of an interior and end bolt, respectively, bearing on the end-plate; while B_{dbg} is the design bearing strength of a bolt bearing on the supporting girder.

The minimum supported beam web thickness, which matches the shear rupture strength of the web material to the shear strength of the two lines of fillet welds, is

$$t_{wb\,min} = \frac{6.19D}{F_{ub}} \tag{13.5.5}$$

where F_{ub} is the ultimate tensile stress of the supported beam material, and D the weld size in sixteenths-of-an-inch.

LRFDM Table 10-4 for Bolted/Welded Shear End-Plate Connections

LRFDM Table 10-4 facilitates the design of standard shear end-plate connections. Separate tables are provided for $3/4$-, $7/8$-, and 1-in.-dia. high-strength bolts, for 2 to12 rows of bolts in two vertical lines. Design strengths are tabulated for supported and supporting member material with $F_y = 50$ ksi and $F_u = 65$ ksi, and end-plate material with $F_y = 36$ ksi and $F_u = 58$ ksi. The first part of each table gives the bolt and end-plate design strengths for several end-plate thicknesses. These results are given for A325-N, A325-X, A490-N, and A490-X type bearing-type joints. Also given are the values for A325-SC and A490-SC slip-critical joints with Class A and Class B type surface preparation and standard, oversize, or short-slotted holes transverse to the direction of load. End-plate edge distances L_{ev} and L_{eh} are assumed to be $1\frac{1}{4}$ in. Vertical spacing of bolts is 3 in. Tabulated bolt and end-plate design strengths consider the limit states of bolt shear, bolt bearing on the end-plate, shear yielding of the end-plate, shear rupture of the end-plate, and block shear rupture of the end-plate. The second part of each table gives the weld design strengths and consider the limit state of weld shear assuming an effective weld length equal to the end-plate length minus twice the weld size. Also given here is the minimum beam web thickness required to match the strength of the weld material as per Eq. 13.5.5. If the thickness of the web to which the plate is welded is less than that given in the table, the tabulated weld capacity must be reduced by the ratio of the thickness supplied to the thickness required. A third part of the Table 10-4 gives the supporting member design strength for the limit state of bolt bearing, per inch of flange or web thickness for steel with $F_u = 65$ ksi.

EXAMPLE 13.5.1

Shear End-Plate Connection

Design a shear end-plate connection for the W21×44 beam to the flange of a W12×58 column. The factored end reaction is 110 kips. The girder and the column are of A992 steel. Assume $3/4$-in.-dia. A325-N type bolts and E70 electrodes.

Solution

1. Data
   W21×44 beam
   $$d_b = 20.7 \text{ in.}; \qquad t_{wb} = 0.350; \qquad T_b = 18\tfrac{3}{8} \text{ in.}$$
   $$b_{fb} = 6.50 \text{ in.}; \qquad t_{fb} = 0.450$$

   From LRFDM Table 5-4, $\phi V_n = 196$ kips > 110 kips   O.K.

W12×58 column

$$d_c = 12.2 \text{ in.}; \qquad t_{wc} = 0.360 \text{ in.}$$
$$b_{fc} = 10.0 \text{ in.}; \qquad t_{fc} = 0.640 \text{ in.}$$

Bolts

$d = \frac{3}{4}$-in., A325-N type, standard holes, single shear

Shear strength, $B_{dv} = 15.9$ kips

Bearing strength, on shear end-plate ($F_u = 58$ ksi)

Interior bolt, $s = 3$ in. (LRFDM Table 7-12) = 78.3 kips/in.

End bolt, $L_e = 1\frac{1}{4}$ in. (LRFDM Table 7-13) = 44.0 kips/in.

Bearing strength, on column flange ($F_u = 65$ ksi) = 87.7 kips/in.

2. Selection of shear end-plate

From Table 10-4 of the LRFDM, for $\frac{3}{4}$-in.-dia. bolts of A325-N type select five rows of bolts and $\frac{1}{4}$-in. plate thickness providing a bolt and end-plate design strength of:

$$R_d = \phi R_n = 132 \text{ kips} \; > \; R_u = 110 \text{ kips} \qquad \text{O.K.}$$

Also from Table 10-4, a $\frac{3}{16}$-in. weld provides a design shear strength of 118 kips. For a beam material with F_y equal to 50 ksi, the minimum web thickness is 0.286 in. Since the actual $t_w = 0.35$ in. > 0.286 in., no correction is needed. And thus, the base metal shear strength does not control the weld strength.

The standard (workable) gage for a W12×58 is $5\frac{1}{2}$ in. So, providing for an edge distance of $1\frac{1}{4}$ in., the width of the end-plate, b_{ep}, is 8 in. Also, providing a vertical spacing, $s = 3$ in. and end distance, $L_{ev} = 1\frac{1}{4}$ in. for the bolts, the vertical dimension of the plate, L_{ep}, is $4(3) + 2(1\frac{1}{4}) = 14\frac{1}{2}$ in. We have:

$$L_{ep} = 14\frac{1}{2} \text{ in.} > \frac{1}{2}T_b = 9.19 \text{ in.} \qquad \text{O.K.}$$

$$L_{ep} = 14\frac{1}{2} \text{ in.} < T_b = 18.4 \text{ in.} \qquad \text{O.K.}$$

So, use PL$\frac{1}{4}$×8×14$\frac{1}{2}$ with ten $\frac{3}{4}$-in. A325-N bolts in standard holes and $\frac{3}{16}$ in. E70 welds.   (Ans.)

3. Limit states of strength

Bolt shear strength = $10(15.9) = 159$ kips

Bolt bearing strength on end-plate = $[8(78.3) + 2(44.0)](\frac{1}{4}) = 179$ kips

Bolt bearing strength on column = $10(87.7)(0.64) = 561$ kips

Shear yielding of end-plate = $0.9(0.6)(36)(14.5)(\frac{1}{4})(2) = 141$ kips

Shear rupture of plate = $0.75(0.6)(58)[14.5 - 5(\frac{7}{8})](\frac{1}{4})(2)$

$$= 132 \text{ kips} \; \leftarrow \text{ controls}$$

Shear strength of weld = $2[14.5 - 2(\frac{3}{16})](3)(1.392)$

$$= 118 \text{ kips} \; \leftarrow \text{ controls}$$

13.6 Single-Plate Framing Connections

A *single-plate framing connection* consists of a rectangular plate, with pre-punched bolt holes, shop welded to the supporting member on both sides of the plate-edge (Fig. 13.6.1). The web of the supported beam is field bolted to the connection plate. This connection is often called the *shear tab.* Single-plate connections may be made to the webs of supporting girders or to the flanges of supporting columns. They may not be suitable for connections to the webs of supporting columns because of bolting clearances. The plate is shop welded to the support, permitting side erection of the supported beam. It is a simple, inexpensive, and popular shear connection for light to moderate end shear. Even though the bolts are in single shear and, therefore, require twice as many bolts as in the traditional double-angle connections, single-plate connections use less connection material compared to double-angle connections resulting in overall economy. Also, they permit safer and faster field erection when beams frame opposite sides of a girder web. Snug-tight rather than pretensioned bolts are highly recommended for these connections. This is because snug-tight bolts will allow the connection to slip during construction thus avoiding the possibility of a sudden, pronounced, and often very loud slip subsequent to building occupancy.

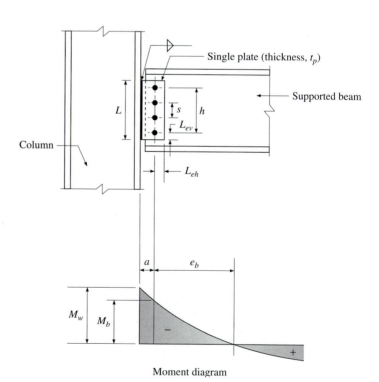

Figure 13.6.1: Single-plate framing connection.

Moment diagram

The design of single-plate framing connections has been studied by Hormby et al. [1984], Astaneh et al. [1989], and Richard et al. [1990]. At low load levels, the shear tab is quite stiff. This results in the development of some moment in the connection and the beam (Fig. 13.6.1), which is equivalent to an eccentric load on the bolts and welds. This eccentricity is based on the location of the point of inflection. As the load on the beam is increased and the force on the connection increases, the shear tab yields [Astaneh et al., 1989]. This yielding causes the inflection point to move toward the support and thus, at failure, the eccentricity is small. A design procedure based on this research has been adopted by the LRFDM for the design of single-plate framing connections. Tests have shown that the end moments developed depend on such geometric and material properties as the thickness of the plate; the number, size, and arrangement of the bolts; and the flexibility of the supporting member. The bolts in single-plate framing connections must, therefore, be designed to resist the shear R_u from the supported beam and an eccentric moment $R_u e_b$. The eccentricity on the bolts, e_b, which is the distance from the bolt line to the point of inflection of the beam, depends upon the support condition present (whether flexible or rigid) and whether standard or short-slotted holes are used in the plate [Richard et al., 1990]. A *flexible support* possesses relatively low rotational stiffness and permits beam-end-rotation of the attached simply supported beam primarily through this supporting member's rotation. Such an end condition may exist with one-sided beam-to-girder-web connections, or with deep beams connected to relatively light columns. In contrast, a *rigid support* possesses relatively high rotational stiffness which constrains beam-end-rotation of the attached simply supported beam to occur primarily within the end connection. Such an end condition may develop when a beam is attached to a column flange, or with two concurrent beam-to-girder-web connections. Values of e_b are given for flexible and rigid support conditions. Thus:

for a flexible support

$$e_b = \max\left[\,|(n-1) - a|, a\,\right] \quad \text{with standard holes} \qquad (13.6.1)$$

$$= \max\left[\,\left|\frac{2n}{3} - a\right|, a\,\right] \quad \text{with short-slotted holes} \qquad (13.6.2)$$

for a rigid support

$$e_b = |(n - 1) - a| \qquad\qquad \text{with standard holes} \qquad (13.6.3)$$

$$= \left|\frac{2n}{3} - a\right| \qquad\qquad \text{with short-slotted holes} \qquad (13.6.4)$$

where e_b = eccentricity on the bolts, in.
        a = distance between the bolt line and weld line (see Fig. 13.6.1), in.
        n = number of bolts

Whether or not a support is rigid or flexible is a matter of judgement by the engineer. When the support condition is intermediate between flexible and rigid, or cannot be readily classified as flexible or rigid, the larger value of e_b calculated may conservatively be used in design. Eqs. 13.6.1 to 13.6.4 are valid for single-plate connections satisfying the relation $2\frac{1}{2}$ in. $\leq a \leq 3\frac{1}{2}$ in. Note that the equations for eccentricity on the bolts are dimensionally inconsistent. These equations result from tests and are purely empirical.

The minimum plate length should at least be equal to half the T-dimension of the supported beam, so as to provide adequate stability for the supported beam during erection. If single-plate connections are used for laterally unsupported beams, for stability under service loading the minimum depth connection so determined should be increased by one row of bolts. The maximum length of the plate must be compatible with the T-dimension if the beam is uncoped, or with the remaining web depth exclusive of fillets, if the beam is coped.

To prevent local buckling of the plate in flexure, the minimum plate thickness should be such that:

$$t_{p\,min} = \max\left[\frac{L}{234}\sqrt{\frac{F_y}{k}},\ \tfrac{1}{4} \text{ in.}\right] \qquad (13.6.5)$$

were L is the length of the plate in inches and k is the plate buckling coefficient. The table on LRFDM page 9–9, partially reproduced here as Table 13.6.1, lists the values of k as a function of the plate aspect ratio, $2a/L$.

The minimum thickness is based on a simple conservative model which assumes that one-half the plate depth is subject to uniform compression from flexure.

To provide for rotational ductility in the single-plate connection, the maximum plate thickness should be such that:

$$t_{p\,max} = \max\left[\left(\frac{d}{2} + \frac{1}{16}\right), t_{p\,min}\right] \qquad (13.6.6)$$

where d is the bolt diameter in inches. This assures that bearing deformations will occur in the bolt holes prior to bolt shear.

TABLE 13.6.1

2a/L	k
0.25	16
0.30	13
0.40	10
0.50	6.0
0.75	2.5
1.00	1.3

The design strength of the connection corresponding to the limit state of shear yielding of the plate is:

$$R_{dvy} = 0.90(0.6F_{yp})Lt_p \qquad (13.6.7)$$

The design strength of the bolted joint, considering the eccentricity on the bolt line (e_b) that exists at the ultimate load of the connection, is:

$$R_{db} = CB_{dv} \qquad (13.6.8)$$

where B_{dv} is the design strength of a single bolt in single shear and C is the coefficient given in the LRFDM Table 7-17 for a single vertical row of bolts in eccentric shear. Similarly, the design strength of the welded joint, considering the eccentricity on the weld line ($e_w = e_b + a$) that exists at the ultimate load of the connection, is:

$$R_{dw} = CC_1 DL_w \qquad (13.6.9)$$

where C is the coefficient given in Table 8-5 of the LRFDM for eccentrically loaded weld group.

LRFDM Table 10-9 for Single-Plate Connections

LRFDM Tables 10-9 are design aids for single-plate connections welded to the supporting member and bolted to the supported beam. Design strengths are tabulated for plate material with $F_y = 36$ ksi and $F_u = 58$ ksi. Separate tables are provided for $\frac{3}{4}$-in., $\frac{7}{8}$-in., 1-in. and $1\frac{1}{8}$-in.-dia. bolts, provided in a single vertical row. Values are tabulated for two through twelve bolts of A325-N, A325-X, A490-N, and A490-X bolts for standard and for short-slotted holes, oriented transverse to direction of load. Values are tabulated for flexible and rigid supports. Vertical spacing of bolts is 3 in. For calculation purposes, plate edge distances L_{ev} and L_{eh} are assumed to be $1\frac{1}{2}$-in. Weld sizes tabulated are equal to $\frac{3}{4} t_p$. The tabulated values are valid for laterally supported beams in noncomposite and composite steel construction, for all types of loading, snug-tightened and pretensioned bolts, and for supported and supporting members of all grades of steel. The tabulated values are based on $a = 3$ in; however, they may conservatively be used for, a between $2\frac{1}{2}$ in. and 3 in. The tabulated bolt and plate design strengths consider the limit states of bolt shear, bolt bearing on the plate, shear yielding of the plate, shear rupture of the plate, block shear rupture of the plate, and weld shear.

Single-Plate Connection

EXAMPLE 13.6.1

Design a single-plate framing connection for a W18×35 floor beam to the web of a W21×44 girder. The factored end reaction is 52 kips. The beam and the girder are of A992 steel. There is a similar connection on the other side of the girder web. Use $\frac{3}{4}$-in.-dia. A325-N bolts in standard holes and E70 electrodes.

[*continues on next page*]

Example 13.6.1 continues ...

Solution

1. Data

    W18×35 beam

    $d_b = 17.7$ in.; $t_{wb} = 0.300$ in.; $k_b = 1\frac{1}{8}$ in.
    $b_{fb} = 6.00$ in.; $t_{fb} = 0.425$ in.; $T_b = 15\frac{1}{2}$ in.

    W21×44 girder

    $d_g = 20.7$ in.; $t_{wg} = 0.350$ in; $k_g = 1\frac{1}{8}$ in.
    $b_{fg} = 6.50$ in.; $t_{fg} = 0.450$ in.; $T_g = 18\frac{3}{8}$ in.

    Bolts

    $d = \frac{3}{4}$ in.; $1\frac{1}{2}(d) = 1\frac{1}{8}$ in.; $3d = 2\frac{1}{4}$ in.

    A325-N type in standard holes, in single shear
    Shear strength, $B_{dv} = 15.9$ kips (from LRFDM Table 7-10)
    Bearing strength,

    $B_{db} = 78.3$ kips/in. ($s = 3$ in., $F_u = 58$ ksi; from LRFDM Table 7-12)
    $\quad\ \ = 44.0$ kips/in. ($L_e = 1\frac{1}{4}$ in., $F_u = 58$ ksi; from LRFDM Table 7-13)
    $\quad\ \ = 87.7$ kips/in. ($s = 3$ in., $F_u = 65$ ksi; from LRFDM Table 7-12)
    $\quad\ \ = 49.4$ kips/in. ($L_e = 1\frac{1}{4}$ in., $F_u = 65$ ksi; from LRFDM Table 7-13)

2. Plate selection

    Assume that the support provided by the girder falls halfway between
    a flexible support and a rigid support. From LRFDM Table 10-9, for
    $\frac{3}{4}$-in.-dia. A325-N type bolts in standard holes, select four rows of bolts,
    $\frac{5}{16}$-in. plate and $\frac{1}{4}$-in. fillet welds. The design strength corresponding to
    a flexible support is 44.7 kips and to a rigid support, 63.6 kips. So, the
    design strength of the connection,

    $$R_d = \phi R_n = \tfrac{1}{2}\,(44.7 + 63.6) = 54.2 \text{ kips } > R_u = 52 \text{ kips}\quad \text{O.K.}$$

    From LRFDM Table 10-1, for a beam coped at top flange only, four rows of
    $\frac{3}{4}$-in.-dia. bolts, beam material with $F_y = 50$ *ksi* and $F_u = 65$ ksi, $L_{ev} =$
    $1\frac{1}{2}$ in. and $L_{eh} = 1\frac{1}{2}$ in. (assumed to be $1\frac{1}{4}$ in. for calculation purposes
    to account for possible under run in beam length), the design strength of
    the coped beam is:

    $$R_d = \phi R_n = 257(0.30) = 77.1 \text{ kips } > 52 \text{ kips}\qquad \text{O.K.}$$

    Assuming that the distance a is $2\frac{1}{2}$ in., the plate width is $2\frac{1}{2} + 1\frac{1}{2} =$
    4 in.
    So, select a PL$\frac{5}{16}$×4×12 of A36 steel for the shear tab and weld it
    to the web of the supporting girder by two $\frac{1}{4}$-in. fillet welds. Use four
    $\frac{3}{4}$-in.-dia. A325-N bolts in standard holes with a spacing of 3 in. and an
    end distance of $1\frac{1}{2}$ in. to connect the beam to the shear tab.   (Ans.)

3. Strength checks

Shear strength of bolts, $R_{dbv} = 4(15.9) = 63.6$ kips   ← controls

Bearing strength of bolts, $R_{dbb} = [3(78.3) + 1(44.0)]\,(^5\!/_{16})$
$= 87.2$ kips   O.K.

Shear yield of plate, $R_{dpvy} = 0.9(0.6)(36)(12)(^5\!/_{16}) = 72.9$ kips   O.K.

Shear rupture of plate, $R_{dpvr} = 0.75(0.6)(58)[12 - 4(^7\!/_8)](^5\!/_{16})$
$= 69.3$ kips   O.K.

Shear strength of weld, $R_{dw} = 2(1.392)(4)\,[12 - 2(^1\!/_4)] = 128$ kips   O.K.

Plate aspect ratio $= \dfrac{2a}{L} = \dfrac{2(2.5)}{12} = 0.42$

From Table 3.6.1, corresponding to this plate aspect ratio, $k = 9.32$

$$t_{p\,min} = \max\left[\frac{L}{234}\sqrt{\frac{F_y}{k}},\,{}^1\!/_4\right] = \max\left[\frac{12}{234}\sqrt{\frac{36}{9.32}},\,{}^1\!/_4\right]$$

$= \max[0.10,\,{}^1\!/_4] = {}^1\!/_4 < t_p = {}^5\!/_{16}$ in.   O.K.

$t_{p\,max} = \max\,[(d/2 + {}^1\!/_{16}),\,t_{p\,min}] = \max\{(^3\!/_4)(^1\!/_2) + {}^1\!/_{16},\,{}^5\!/_{16}\}$
$= {}^7\!/_{16} > t_p = {}^5\!/_{16}$ in.   O.K.

$e_b = \max\,[|(n-1) - a|,\,a]$
$= \max\{[(4-1) - 3],\,3\} = 3$ in. for a flexible support

$e_b = |(n-1) - a| = 0$ in. for a rigid support

Note that we used in these calculations a conservative value of 3 in. for a (value also used in the preparation of LRFDM Table 10-9) instead of the actual value ($2\frac{1}{2}$ in.) provided.

Entering LRFDM Table 7-17: Coefficients C for Eccentrically Loaded Bolt Groups (single vertical row of bolts) with $\theta = 0°$, $n = 4$, $s = 3$ in., and $e_x = e_b = 3$ in., the coefficient C is read as 2.81. The design strength of the eccentrically loaded bolt group is:

$$\phi R_n = C B_d = 2.81(15.9) = 44.7 \text{ kips}$$

As $e_b = 0$ for a rigid support, the design strength of the bolt group is simply $4(15.9) = 63.6$ kips. These two numbers, namely, 44.7 kips and 63.6 kips are the values read from LRFDM Table 10-9 as noted above.

Eccentricity on welds: $e_w = 3 + 3 = 6$ in.   for a flexible support
$= 3 + 0 = 3$ in.   for a rigid support

For $L_w = 11.5$ in.: $a_w = 6/11.5 = 0.522$ for a flexible support
$= 3/11.5 = 0.261$ for a rigid support

Entering LRFDM Table 8-5: Coefficients C for Eccentrically Loaded Weld Groups with $\theta = 0°$, $k = 0$, $e_x = e_w = 6$ in. ($a_w = 0.522$), the coefficient C is read as 1.68. The design strength of the eccentrically loaded weld group is:

$$R_{dw} = 1.68(1.0)(4)(11.5) = 77.3 \text{ kips}$$

[continues on next page]

Example 13.6.1 continues ...

Similarly, for $e_x = 3$ in., $C = 2.45$, resulting in:

$$R_{dw} = 2.45(1.0)(4)(11.5) = 112.7 \text{ kips}$$

So, $R_{dw} = \dfrac{1}{2}(77.3 + 112.7) = 95.0 > R_u$ O.K.

13.7   Single-Angle Framing Connections

A ***single-angle framing connection*** is made with an angle on one side of the web of the supported beam (Fig. 13.7.1). This angle is preferably shop bolted or welded to the supporting member and field bolted to the supported beam. Single-angle connections may be made to the webs of supporting girders and to the flanges of supporting columns. Shop attaching the angle to the supporting element offers the advantage of side erection of the beam.

Let $L_B \times L_A \times t_a \times L$ be the dimensions of the framing angle. Here, L_A is the length of the angle-leg connected to the supported beam; L_B, the length of the angle-leg connected to the supporting member; t_a, the thickness of the angle; and L, the length of the angle. The following remarks should be kept in mind in the design of single-angle framing connections:

1. The minimum length of the framing angle should at least be equal to half the T-dimension of the supported beam, so as to provide adequate stability for the beam during erection.
2. The maximum length of the framing angle must be compatible with the T-dimension for an uncoped beam, and the remaining web depth exclusive of fillets, for a coped beam.
3. Vertical fastener spacing is arbitrarily chosen as 3 in.

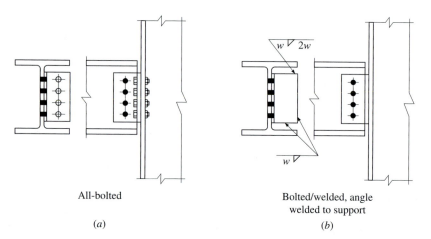

All-bolted	Bolted/welded, angle welded to support
(a)	*(b)*

Figure 13.7.1: Single-angle framing connections.

4. The end distance on angles is set at $1\frac{1}{4}$ in. as permitted by LRFDS Table J3.4. When using slotted holes, the edge and end distances must be adjusted over that for standard holes as required.

5. The standard connection angle lengths (L values in the tables) vary from $5\frac{1}{2}$ to $35\frac{1}{2}$ in. using increments of 3 in.

6. To assure the flexibility assumed in the design, the angles used are rather thin. However, a minimum angle thickness of $\frac{3}{8}$ in. for $\frac{3}{4}$-in.- and $\frac{7}{8}$-in.-dia. bolts, and $\frac{1}{2}$ in. for 1-in.-dia. bolts must be used.

7. It is common practice to place holes for connections on horizontal beam (web) gage lines spaced 3 in. vertically. When practicable, the upper most gage line is set 3 in. below the top of the beam.

8. A 4×3 angle is normally selected for a single-angle welded to the support with the 3-in. leg being the welded leg. The weld is placed along the toe and across the bottom of the angle, with only a return at the top for a distance not to exceed $4w$ (LRFDS Section J2.2b) to provide adequate connection flexibility.

For bolt Group B, the effect of eccentricity must be considered. For bolt Group A, eccentricity must be considered only for a double vertical row of bolts through the web of the supported beam or when the eccentricity exceeds 3 in. ($2\frac{3}{4}$-in. gage plus $\frac{1}{4}$-in. half web). However, eccentricity must always be considered in the design of welds for single-angle connections.

13.8   Unstiffened Seated Connections

13.8.1   General

Several types of seated connections are shown in Fig. 13.8.1. One advantage of seated connections is that the erector can immediately rest the beam on the seat. In an ***unstiffened seated connection,*** the beam end rests on the outstanding leg of an angle, called the ***seat angle,*** whose other leg is bolted or welded to the supporting member (Figs. 13.8.1b and c). The seat angle is usually shop attached to the supporting member. It is the principal type of connection used to connect a beam to a column web. Unstiffened seats are also used to connect beams to column flanges. They may also be used to connect beams to the webs of supporting girders when the girders are deep enough to provide space for the seat angle. The standard ***beam clearance,*** or ***set back,*** is $\frac{1}{2}$ in. To provide lateral and torsional support to the beam at the end, a small angle (called ***top*** or ***cap angle***) connecting the upper portion of the beam—either on the top flange or on the top part of the web of the supported beam—is required. The outstanding leg of the seat angle, acting approximately as a cantilever beam, is assumed to carry the entire end reaction of the supported beam and transfer it to the supporting member. The

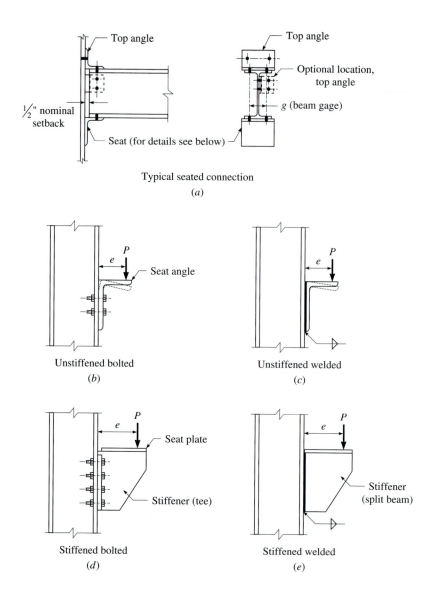

Figure 13.8.1: Seated beam connections.

unstiffened seat is suitable for supporting relatively small loads only. The largest seat angle used for a standard unstiffened seated connection is a L8× 4×1×0′ 8″; when bolted or welded to a column, its load capacity is approximately 150 kips.

Roeder and Dailey [1989] and Carter et al. [1997] discussed the behavior of seated beam connections. Garrett and Brockenbrough [1986] and Brockenbrough [1987] reviewed the design of seated beam connections and the background for the LRFDM load tables.

Bolted Unstiffened Seated Connections

Several details for all-bolted unstiffened seated connections are shown in Fig. 13.8.2. The following remarks should be kept in mind in the design of such unstiffened seated beam connections:

1. Nominally, the end of a seated beam is stopped approximately $\frac{1}{2}$ in. short of the face of the supporting member to which the seat angle is attached, to provide for possible mill overrun in the beam length (Fig.13.8.2a).

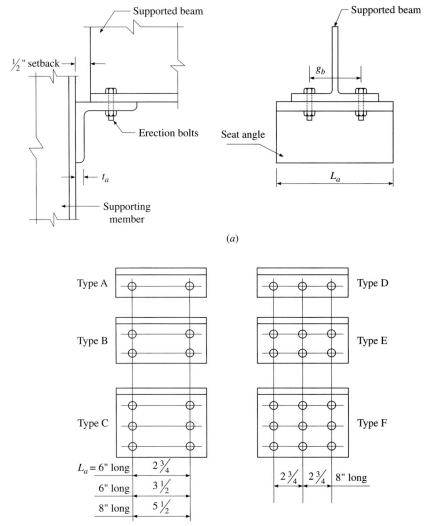

(a)

Bolt arrangements

(b)

Figure 13.8.2: All-bolted unstiffened seated connection details.

2. Although the supported beam is set back $\frac{1}{2}$ in. from the column face, the capacity of the outstanding leg is based on a set back of $\frac{3}{4}$ in. to provide for possible mill underrun in the beam length.

3. The outstanding leg size of an unstiffened seat angle should not be less than 3 in. It is usually 4 in. wide, thus providing $3\frac{1}{4}$ in. of bearing, measured longitudinally along the web of the supported beam. Greater leg widths would be of no advantage. Reactions requiring more than $3\frac{1}{4}$ in. of bearing or exceeding the capacity of unstiffened seats require the use of stiffened seats.

4. The thickness of the seat angle runs from $\frac{3}{8}$ in. to 1 in.

5. The length of the seat angle is generally taken as either 6 or 8 in. (Fig. 13.8.2a). These lengths are suitable, respectively, for the $3\frac{1}{2}$-in. and $5\frac{1}{2}$-in. gage distances that are generally used in detailing columns. An increase in the length of these seat angles would add some stiffness to the outstanding leg, but would not substantially increase the load carrying capacity of the connection, since this capacity is also limited by the web thickness of the supported beam.

6. The gage distances used in the vertical legs should provide adequate assembly clearances for the bolts. Three horizontal rows are the maximum number allowed, based on an 8-in. vertical leg size for the seat angle. The vertical leg size is determined by adding to the required minimum assembly clearances given in Tables 7-3a and 7-3b of the LRFDM, the vertical spacing of bolts, if applicable, plus the minimum edge distance for a rolled edge as specified in LRFDS Table J3.4.

7. The choice of a seated beam connection from Types A through F (Fig. 13.8.2b) is, to some extent, based on the supporting structural member. A deep connection, such as Type C, may not be practical for a beam-to-girder connection because of the limited depth of the girder. In the case of a seated connection to the web of W10 columns, a seat angle length of 6 in. is generally used. On the other hand, for webs of W12 and larger columns, the 8-in. seat angle length is commonly used.

8. The top angle is not assumed to transfer any loads. So, it should be flexible and capable of conforming to the beam's end rotation. In most connections, angles of the order of $\frac{1}{4}$ to $\frac{3}{8}$ in. thick will be adequate. A common top angle size selected is a L4×4×$\frac{1}{4}$, 6 in. long. This angle has two bolts in each of its legs, even in the case of large, heavy beams. Ordinary bolts with $\frac{3}{4}$-in. diameters, are generally specified to secure the top angle to the beam and to the supporting column or girder. Slotted holes provided in the vertical leg of the angle will accommodate both overrun and underrun in the beam depth. The top angle holds the top flange of the supported beam in position and therefore contributes significantly to the resistance of the web to web-crippling. The top angle also acts as a lateral support for the compression flange of the beam (brace point) without introducing any moment at the end. So, the top angle is an essential part of the connection.

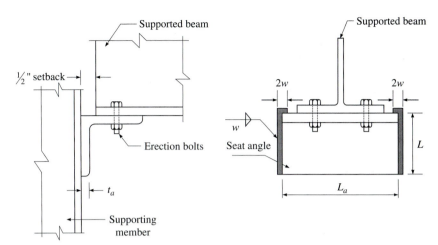

Figure 13.8.3: All-welded unstiffened seated connection details.

9. The top and bottom flanges of the beam are cut at the ends, if necessary, to provide at least $\frac{1}{2}$-in. erection clearance between the column flanges.

Welded Unstiffened Seated Connections

Several details of all-welded unstiffened seated connections are shown in Fig. 13.8.3. Here, two vertical welds, one on each end of the seat angle, are used to transfer the beam end reaction from the seat angle to the supporting member. These welds must be returned across the top of the seat angle for a distance of about $2w$ in order to eliminate craters at the top of the vertical welds. The length of the seat angle is made to extend at least $\frac{1}{2}$ in. on each side of the supported beam flange.

13.8.2   Limit States of Strength

Tests on unstiffened seated connections indicate that thinner (flexible) seats tend to distribute the reaction more towards the heel of the outstanding leg (Fig. 13.8.4a), while thicker (stiffer) seats tend to concentrate the reaction at the toe of the outstanding leg (Fig. 13.8.4b). For the design of an unstiffened seat angle, the reaction is assumed to be uniformly distributed over a length, measured from the end of the supported beam and sufficient to satisfy the web-yielding and web-crippling requirements of the beam (Fig. 13.8.4c).

    The different limit states of strength for an unstiffened seated beam connection are:

- Web local yielding of the supported-beam web.
- Web crippling of the supported-beam web.
- Bending strength of the seat angle leg.
- Shear strength of the seat angle leg.
- Strength of bolts or welds.

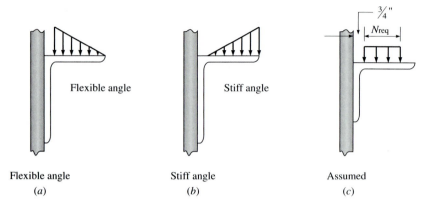

Figure 13.8.4: Distribution of bearing stresses on seat angle.

Flexible angle   Stiff angle   Assumed

(a)   (b)   (c)

Local Yielding of the Supported-Beam Web

From LRFDS Section K1-3, the design strength of the beam web corresponding to the limit state of web local yielding is:

$$R_{dWY} \equiv \phi R_n = \phi R_1 + N(\phi R_2) \qquad (13.8.1a)$$

where R_{dWY} = design strength of the beam web, corresponding to the limit state of web local yielding, kips

       N = length of bearing, in.

       ϕR_1 = $\phi(2.5 k_b t_{wb} F_{yb})$, kips $(13.8.1b)$

       ϕR_2 = $\phi(t_{wb} F_{yb})$, kips/in. $(13.8.1c)$

with

       ϕ = resistance factor = 1.0

       k_b = distance from outer face of the supported beam flange to the web toe of the fillet, in. (Note: For this purpose, use the decimal dimension given for k in Table 1-1 of the LRFDM, not the fractional value given there.)

       t_{wb} = web thickness of the supported beam, in.

       F_{yb} = yield stress of the beam web, ksi

For any rolled steel W-shape ($F_y = 50$ ksi), the design strength R_{dWY} may be determined from constants ϕR_1 and ϕR_2 given in Table 9-5 of the LRFDM. For a large beam with a small reaction, this equation may actually yield a negative value for N. To avoid this inconsistency, LRFDS requires that N be taken at not less than k_b. Thus, we obtain:

$$N_1 \geq \max\left[\frac{R_u - \phi R_1}{\phi R_2}, k_b\right] \qquad (13.8.2)$$

Web Crippling of the Supported-Beam Web

From LRFDS Section K1-4, the design strength of the beam web corresponding to the limit state of web crippling under the concentrated compressive force (end reaction) is:

$$R_{dWC} \equiv \phi_r R_n = \phi_r R_3 + N(\phi_r R_4) \quad \text{for} \frac{N}{d_b} \le 0.2 \quad (13.8.3a)$$

$$= \phi_r R_5 + N(\phi_r R_6) \quad \text{for} \frac{N}{d_b} > 0.2 \quad (13.8.3b)$$

where $\quad \phi_r R_3 = \phi_r \left(0.4 t_{wb}^2\right) \sqrt{\dfrac{EF_{yb}\, t_{fb}}{t_{wb}}} \quad (13.8.3c)$

$$\phi_r R_4 = \phi_r \left(0.4 t_{wb}^2\right)\left(\frac{3}{d_b}\right)\left(\frac{t_{wb}}{t_{fb}}\right)^{1.5}\sqrt{\frac{EF_{yb}\, t_{fb}}{t_{wb}}} \quad (13.8.3d)$$

$$\phi_r R_5 = \phi_r \left(0.4 t_{wb}^2\right)\left[1 - 0.2\left(\frac{t_{wb}}{t_{fb}}\right)^{1.5}\right]\sqrt{\frac{EF_{yb}\, t_{fb}}{t_{wb}}} \quad (13.8.3e)$$

$$\phi_r R_6 = \phi_r \left(0.4 t_{wb}^2\right)\left(\frac{4}{d_b}\right)\left(\frac{t_{wb}}{t_{fb}}\right)^{1.5}\sqrt{\frac{EF_{yb}\, t_{fb}}{t_{wb}}} \quad (13.8.3f)$$

with

R_{dWC} = web-crippling strength of the beam web, kips
ϕ_r = resistance factor = 0.75
d_b = overall depth of the supported beam, in.
t_{fb} = flange thickness of the supported beam, in.

For rolled steel W-shapes ($F_y = 50$ ksi), the design web-crippling strength may be determined from the coefficients $\phi_r R_3$, $\phi_r R_4$, $\phi_r R_5$, and $\phi_r R_6$ given in Table 9-5 of the LRFDM. Thus:

$$N_2 \ge \frac{R_u - \phi_r R_3}{\phi_r R_4} \quad \text{for} \quad \frac{N}{d_b} \le 0.2 \quad (13.8.4a)$$

$$N_2 \ge \frac{R_u - \phi_r R_5}{\phi_r R_6} \quad \text{for} \quad \frac{N}{d_b} > 0.2 \quad (13.8.4b)$$

Thus, we have:

$$N_{\text{req}} = \max[N_1, N_2] \quad (13.8.5)$$

where N_1 and N_2 are given by Eqs. 13.8.2 and 13.8.4, respectively, and N_{req} is the required bearing length (in.).

Web crippling is unlikely to occur with rolled wide flange beams under uniformly distributed loading unless the span is very short compared to the beam depth. However, it could be an important mode of failure for stiffened seated connections with supporting beams with thin webs. Table 9-5 of the LRFDM also gives the maximum reaction (= ϕR_n) for a $3\frac{1}{4}$-in. bearing, corresponding to a 4-in. seat angle length.

Flexural Strength of the Seat Angle

The unstiffened seat angle generally fails through bending of the seat angle. Experiments show that the top angle carried between 9 and 36 percent of the end reaction for connections which failed through web crippling or yielding of the seat angle [Roeder, 1987]. Also, the top angle stiffens the seat connection somewhat in the elastic range. This contribution of the top angle to transfer the end reaction is neglected in the design. The outstanding leg of the seat angle cannot bend as a simple cantilever because of the stiffening action of the beam bottom flange to which it is attached. Again, this contribution from the beam flange is neglected in the design. So, the seat angle should have the required flexural strength to transfer the entire beam reaction to the bolts or welds in the vertical leg.

The bottom side of the beam lengthens as the beam is loaded. As the supported beam bottom flange is always attached to the seat angle by a pair of bolts or welds, the beam-end-rotation creates a force that pins the heel of the seat angle against the supporting member and relieves the bending in the vertical leg. The critical section for bending is therefore located in the horizontal leg of the angle. It is taken at the base of the fillet on the outstanding leg of the seat angle, located at the k distance of the angle measured from the heel of the angle. To simplify design, this section is normally defined by assuming a fillet radius of $\frac{3}{8}$ in. for all angle sizes used as seat angles. That is, the critical section is assumed to be at a distance, $k_a = t_a + \frac{3}{8}$ in. from the heel of the angle (Fig. 13.8.5). The beam reaction acts at the centroid of the bearing stress distribution. A conservative approach to design is to assume the reaction acts at the center of the full contact width (usually $3\frac{1}{4}$ in.). A less conservative approach, and a common practice, is to assume the reaction acts at the center of the required bearing length N_{req} (given by Eq. 13.8.5), measured from the end of the beam. As mentioned earlier, although the supported beam is set back $\frac{1}{2}$ in. from the column face, the capacity of the outstanding leg is

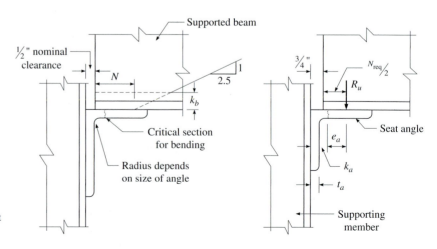

Figure 13.8.5: Limit state of bending of the seat angle.

based on a set back of $\frac{3}{4}$ in. to provide for possible mill underrun in the beam length. Let e_a be the eccentricity of the beam reaction R_u with respect to the critical section. We have (Fig. 13.8.5):

$$e_a = \left(\frac{N_{req}}{2} + \frac{3}{4}\right) - \left(t_a + \frac{3}{8}\right) = \left(\frac{N_{req}}{2} - t_a + \frac{3}{8}\right) \tag{13.8.6}$$

and the applied moment is:

$$M_u = R_u e_a \tag{13.8.7}$$

The outstanding leg of the seat angle acts approximately as a cantilever beam to support the beam end reaction. The design flexural strength of the angle's outstanding leg (a rectangular section of width L_a and thickness t_a) is:

$$M_{da} = \phi_b\left(\frac{1}{4} L_a t_a^2\right)F_{ya} \tag{13.8.8}$$

Thus, the limit state of flexure of the seat angle may be written as:

$$R_u e_a \le M_{da} \rightarrow R_u \le \frac{M_{da}}{e_a} \equiv R_{dfa} \tag{13.8.9a}$$

with

$$R_{dfa} = \frac{\phi_b}{4}\frac{L_a}{e_a} t_a^2 F_{ya} \tag{13.8.9b}$$

where
R_u = beam end reaction under factored loads, kips
R_{dfa} = design strength of the seat angle corresponding to the limit state of flexure, kips
t_a = seat angle leg thickness, in.
L_a = seat angle length, in.
e_a = eccentricity of the beam reaction with respect to the critical section (Eq. 13.8.6), in.
F_{ya} = yield stress of the seat angle, ksi
N_{req} = required bearing length (Eq. 13.8.5), in.

For example, for a 6 in. long L6×4×$\frac{3}{4}$ of A36 steel with a required bearing length of $2\frac{3}{4}$ in. (that is, for $n_{req} = 2\frac{3}{4}$ in., $L_a = 6$ in., $t_a = \frac{3}{4}$ in., $F_{ya} = 36$ ksi), $e_a = 1$ in., and R_{dfa} equals 27.3 kips. This is also the value tabulated in LRFDM Table 10-5. Observe that the procedure used is sensitive to small changes in the clearance at the end of the beam, the beam web thickness, and the values of k for the beam and the seat angle. For example, if the beam clearance is changed by only $\frac{1}{8}$ in. the design flexural strength is changed by 11 percent. The critical section for bending of the angle leg was assumed to be the same whether a bolted beam seat or a welded beam seat is used. Consequently, the first part of LRFDM Table 10-6 for welded beam seats is identical to the first part in LRFDM Table 10-5 for bolted beam seats.

Shear Strength of Seat Angle

The seat angle should have adequate shear strength to transfer the beam end reaction to the bolts or welds in the vertical leg. Thus,

$$R_{dva} \equiv \phi_v R_n \geq R_u \tag{13.8.10a}$$

where

$$R_{dva} = \phi_v (0.6 F_{ya}) L_a t_a \tag{13.8.10b}$$

with

R_{dva} = design shear strength of the seat angle leg, kips
ϕ_v = resistance factor (= 0.9)

For example, for a L6×4×¾×0′6″ of A36 steel with a required bearing length of $2\frac{3}{4}$ in. (that is, for L_a = 6 in., t_a = ¾ in., F_{ya} = 36 ksi), R_{dva} equals 87.5 kips. This value also is given in LRFDM Table 10-5.

Design Strength of Bolts

The bolts connecting the vertical leg of the seat angle to the support should have adequate strength to transmit the factored beam reaction R_u to the support. In the design of the bolts connecting the vertical leg to the support, the moment on the bolt group due to load eccentricity is neglected. The bolts are in single shear, bearing on the seat angle and the supporting element (girder web, column web, or column flange). The number of bolts required is determined from the relation:

$$n_{\text{req}} \geq \frac{R_u}{B_{dv}} \tag{13.8.11a}$$

where B_{dv} is the design shear strength of a bolt in single shear. A suitable bolt arrangement (among Types A to F) is selected. The design bearing strength of the plate material for the bolts bearing on the seat angle is checked next:

$$n_i B_{dbi} + n_e B_{dbe} \geq R_u \tag{13.8.11b}$$

where n_i is the number of interior bolts and n_e is the number of end bolts in the bolt arrangement type selected.

Design Strength of Welds

In the design of welds connecting the vertical legs of the seat angle to the support, bending due to load eccentricity is accounted for. The eccentricity is again based on the required bearing length, N_{req}, and not on the actual value provided, N (usually $3\frac{1}{4}$ in. with a seat angle size of 4 in.). The design strength of the vertical welds is computed on the assumption that the bottom part of the seat angle presses against the face of the column, and that the center of rotation of the weld is at its lower third point. The return of $2w$ at the top of the weld has been omitted in computing the weld strength. The size of the weld is based on the vector sum of the shearing and tensile bending stresses. Let W_y and W_z be the vertical and horizontal forces on the throat of a unit-length weld at the extreme top fibre of the weld. Shear and moment equilibrium of forces acting on the connection gives:

$$W_y = \frac{R_u}{2L}; \quad \frac{R_u}{2}e = \frac{W_z}{2} \times \frac{2L}{3} \times \frac{2L}{3} \quad \rightarrow \quad W_z = \frac{9}{4}\frac{R_u e}{L^2}$$

The resultant force on the unit-length weld is given by:

$$W_u = \sqrt{W_y^2 + W_z^2} = \frac{R_u}{2L}\sqrt{1 + 20.25(e/L)^2}$$

Let D be the dimension of the fillet weld in sixteenths-of-an-inch. Assuming the SMAW process and E70 electrodes, we have as per LRFD:

$$W_u \leq W_d = 1.392D \quad \rightarrow \quad R_u \leq R_{dw} \qquad (13.8.12a)$$

where

$$R_{dw} = \frac{2(1.392D)L}{\sqrt{1 + 20.25(e/L)^2}} \qquad (13.8.12b)$$

with

R_{dw} = design strength of (E70) welds connecting the seat angle to the support, kips

D = number of sixteenths-of-an-inch in the weld size

e = eccentricity of the beam end reaction with respect to the weld lines, in.

$\quad = \dfrac{L_B}{2} + \text{\textthreequarters}\ ^3\!/_8$ in.

L = vertical leg dimension of the seat angle, in.

L_B = outstanding leg length of the seat angle, in.

For example, for a L6×4 welded to the support with ¼-in., E70 welds (that is, for $D = 4$, $L = 6.00$ in., and $L_B = 4.00$ in.), eccentricity $e = 2.38$ in. and from Eq. 13.8.12b, $R_{dw} = 32.7$ kips, which is the value tabulated in Table 10-8 of the LRFDM. The size of the vertical leg of the welded beam seat is determined by the required length of the weld to resist direct shear and bending. The size of this leg (L) and the size of the weld (D) may be varied to arrive at an adequate and economical design. In Eq. 13.8.12b, the term in the nominator represents the design strength of the weld in shear alone, while the term in the denominator is a measure of the decrease in this strength due to load eccentricity.

13.8.3   Design Tables for Unstiffened Seated Connections

LRFDM Table 10-5 for All-Bolted Unstiffened Seated Connections

LRFDM Table 10-5 provides design data for all-bolted unstiffened seated connections. Seat design strengths are tabulated for angle material with $F_y = 36$ ksi and $F_u = 58$ ksi. These tables will be conservative when used for angle material with $F_y = 50$ ksi and $F_u = 65$ ksi. The first part of the table gives the outstanding angle leg design strengths for two different seat lengths (6 in. and 8 in.); five different thicknesses of the seat angle ($\frac{3}{8}$ in. to 1 in.); and several required bearing lengths for the supported beam ($\frac{1}{2}$ in. to $3\frac{1}{4}$ in.). Also given are the suggested outstanding leg sizes of the angle ($3\frac{1}{2}$ in. and 4 in.). For strength calculations, a set back of $\frac{3}{4}$ in. was used. The design strengths are obtained considering the limit states of flexural yielding and shear yielding of the outstanding leg of the seat angle. The required bearing length, to enter this table, is obtained from Eq. 13.8.5 considering the local yielding and crippling of the beam web under factored loads.

The second part of the LRFDM Table 10-5 gives the bolt design strengths for the seat connection Types A though F shown in Fig. 13.8.2, with $\frac{3}{4}$-in., $\frac{7}{8}$-in., and 1-in.-dia. A325 and A490 high strength bolts of N- and X-types. Vertical spacing of bolts and gages in seat angles may be arranged to suit conditions, provided they conform to the provisions of the LRFDS Section J3. Available angle sizes and thicknesses for different seat types are also identified in this table.

LRFDM Table 10-6 for All-Welded Unstiffened Seated Connections

LRFDM Table 10-6 is a design aid for all-welded unstiffened seated connections. The first part of this table is identical to the first part in LRFDM Table 10-5 for bolted beam seats.

The second part of the LRFDM Table 10-6 gives weld design strengths, calculated using the elastic vector method (Eq. 13.8.12b) and assuming a weld material with $F_{EXX} = 70$ ksi. They may also be used for other electrodes; simply multiply the tabulated weld strengths by the factor ($F_{EXX}/70$), where F_{EXX} is the strength of the weld used. Also given in the table are the minimum and maximum angle thickness available for the different seat angle sizes selected. No reduction of the tabulated weld capacities is required when unstiffened angle seats line up on opposite sides of the supporting web.

The largest-sized standard angle that is tabulated for the unstiffened beam seat is the L8×4×1 in. with a maximum capacity of 147 kips. Beams with reactions larger than 147 kip under factored loads therefore require the use of stiffened seats (see Section 13.9).

E X A M P L E   1 3 . 8 . 1   **Unstiffened Seated Beam Connection**

Design an unstiffened seated beam connection to connect a W18×40 floor beam to the web of a W12×58 column. The factored end reaction is 52 kips. The beam and the column are of A992 steel. There is a similar connection on the other side of the column web. Use $\frac{3}{4}$-in.-dia. A325-N type bolts.

Solution

1. Data

From LRFDM Table 1-1, for a W18×40 beam of A992 steel:

$$d_b = 17.9 \text{ in.;} \qquad t_{wb} = 0.315 \text{ in.;} \qquad g_b = 3\tfrac{1}{2} \text{ in.}$$
$$k_b = 0.927 \text{ in. } (1\tfrac{3}{16} \text{ in.);} \qquad t_{fb} = 0.525 \text{ in.;} \qquad b_{fb} = 6.02 \text{ in.}$$

$$\phi R_1 = \phi(2.5 k_b t_{wb} F_{yb}) = 1.0(2.5)(0.927)(0.315)(50) = 36.5 \text{ kips}$$

$$\phi R_2 = \phi F_{yb} t_{wb} = 1.0(50)(0.315) = 15.8 \text{ kips}$$

$0.2 d_b = 3.58 \text{ in.}$

From Table 9-5 of the LRFDM, corresponding to a W18×40 beam:

$$\phi V_n = 152 \text{ kips} \quad > \quad R_u = 52 \text{ kips} \qquad\qquad \text{O.K.}$$
$$\phi R_1 = 36.5 \text{ kips;} \qquad \phi R_2 = 15.8 \text{ kips/in.}$$
$$\phi_r R_3 = 46.3 \text{ kips;} \qquad \phi_r R_4 = 3.60 \text{ kips/in.}$$

From Eq. 13.8.2, the required bearing length N_1 for the limit state of web yielding is:

$$N_1 = \max\left[\frac{R_u - \phi R_1}{\phi R_2}, k_b\right] = \max\left[\frac{52 - 36.5}{15.8}, 0.927\right] = 0.981 \text{ in.}$$

Assuming $N/d_b < 0.2$, the required bearing length N_2 for the limit state of web crippling from Eq. 13.8.4a is:

$$N_2 = \frac{R_u - \phi_r R_3}{\phi_r R_4} = \frac{52 - 46.3}{3.60} = 1.58 \text{ in.}$$

$$N_{req} = \max [N_1, N_2] = \max [0.981, 1.58] = 1.58 \approx 1\tfrac{5}{8} \text{ in.}$$

$$N_{req} < 0.2 \, d_b = 3.58, \text{ as assumed.} \qquad\qquad \text{O.K.}$$

From Table 1-1 of the LRFDM, for a W12×58 column of A992 steel:

$$d_c = 12.2 \text{ in.;} \qquad t_{wc} = 0.360$$
$$T_c = 9\tfrac{1}{4} \text{ in.}$$

2. All-bolted unstiffened seated connection

Bolts: $d = \tfrac{3}{4}$-in. A325-N type in standard holes in single shear
From LRFDM Table 7-10, $B_{dv} = 15.9 \text{ kips}$
From LRFDM Table 7-12, $B_{dbo} = 78.3 \text{ kips/in. for } F_u = 58 \text{ ksi}$
$$= 87.7 \text{ kips/in. for } F_u = 65 \text{ ksi}$$
Usual gage for W18×40 beam, $g_b = 3\tfrac{1}{2}$ in.
So select a seat angle length, $L_a = 6$ in. $< T_c = 9\tfrac{1}{4}$ in. O.K.
Enter the first part of LRFDM Table 10-5 for All-Bolted Unstiffened Seated Connections, with angle length, $L_a = 6$ in.; required bearing length, $N_{req} = 1\tfrac{5}{8}$ in.; and a required strength, R_u, of 52.0 kips to select a

[continues on next page]

Example 13.8.1 continues ...

¾-in. angle thickness with a minimum angle leg of $3\frac{1}{2}$ in. From the table, the angle leg design strength is:

$$R_d = 62.5 \text{ kips} > R_u = 52.0 \text{ kips} \qquad \text{O.K.}$$

From the second part of LRFDM Table 10-5, corresponding to A325-N bolts of ¾-in. dia.; a connection angle length, $L_a = 6$ in. (thus limiting the selection to Types A, B); and a required shear strength, R_u of 52.0 kips. The bolt design strength provided by this connection is:

$$R_d = 63.6 \text{ kips} > R_u = 52.0 \text{ kips} \qquad \text{O.K.}$$

The Table also indicates that, for connection Type B, a $6 \times 4 \times \frac{3}{4}$ in. angle is available (4-in. outstanding leg).
The bearing strength of a ¾-in.-dia. A325 bolt on a ¾-in.-thick A36 steel angle leg is:

$$B_{db} = 78.3(\tfrac{3}{4}) = 58.7 \text{ kips} > B_{dv} = 15.9 \text{ kips} \qquad \text{O.K.}$$

The bearing strength of a ¾-in.-dia. A325 bolt on a 0.36-in.-thick A992 steel column web (noting that there is a similar connection on the other side of the column web) is:

$$B_{db} = 87.7\left(\frac{0.360}{2}\right) = 15.8 \text{ kips} < B_{dv} = 15.9 \text{ kips} \qquad \text{N.G.}$$

So, the reduced bolt design strength is:

$$R_{db} = 63.6\left(\frac{15.8}{15.9}\right) = 63.2 \text{ kips} > R_u = 52 \text{ kips} \qquad \text{O.K.}$$

So, select a Type B unstiffened seated beam connection using an L6×4×¾×0′ 6″ A36 steel angle with the 6-in. leg connected to the column web by four ¾-in.-dia. A325-N bolts. Use two ¾-in.-dia. A325 bolts to connect the beam to the seat angle. Also, use an L4×4×¼ with two ¾-in.-dia. A325-N bolts through the supported-beam leg of the angle and two more in the column-web leg of the angle. (Ans.)

3. All-welded unstiffened seated connection
Again select 6-in.-long seat angle.
Enter the first part of LRFDM Table 10-6 for All-Welded Unstiffened Seated Connections with $L_a = 6$ in., $F_{yb} = 50$ ksi, $N_{req} = 1\frac{5}{8}$ in. and a required strength, R_u of 52 kips, and select an angle leg thickness, t_a of ¾ in. giving a design strength of:

$$R_d = 62.5 \text{ kips} > R_u = 52 \text{ kips} \qquad \text{O.K.}$$

Also, the recommended minimum outstanding leg length is $3\frac{1}{2}$ in.
The welds connect ¾-in.-thick angle to 0.360-in.-thick column web.
Maximum size of weld $= \frac{3}{4} - \frac{1}{16} = \frac{11}{16}$ in.

Minimum size of weld (from LRFDM Table J.2-4) = $\frac{1}{4}$ in.
Select $\frac{5}{16}$-in. weld (single pass). From the second part of LRFDM Table 10-6, a $\frac{5}{16}$-weld to a $7 \times 4 \times \frac{3}{4}$ seat angle gives a design strength:

$$R_{dw} = 53.4 \text{ kips} > R_u = 52 \text{ kips} \qquad\qquad \text{O.K.}$$

The third part of Table 10-6 indicates that a $7 \times 4 \times \frac{3}{4}$ angle is available. So, select a L$7 \times 4 \times \frac{3}{4} \times 0'$ 6″ of A36 steel with the long leg vertical. Provide $\frac{5}{16}$-in. E70 fillet welds along the vertical legs of the angle with returns of $\frac{5}{8}$ in. at the top of the seat. (Ans.)

13.9   Stiffened Seated Connections

A **stiffened seated connection** consists of a **seat plate,** a **stiffening element,** and a **top angle.** Stiffened seats may be bolted or welded (Figs. 13.9.1 and 2). A bolted stiffened seated connection may consist of a seat plate stiffened by a pair of angles (Fig. 13.9.1) or a tee (Fig. 13.8.1*d*). A welded stiffened seated connection may consist of two plates (welded in the form of a tee as shown in Fig. 13.9.2), a structural tee, or a split beam (Fig. 13.8.1*e*). The stiffened seated connection is used to support beam reactions too large for a single standard thickness seat angle (unstiffened seated connection) to resist the resulting bending moments. In the case of stiffened seated connections, the stiffening element is so much more rigid than the outstanding leg of the seat that the stiffening element is assumed to carry all of the load. Stiffened seated connections are frequently used for beams framed to the web of columns. They are also used for beams framed to column flanges, if the seat stiffener

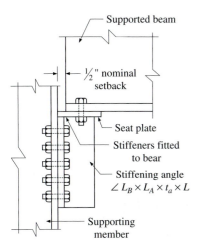

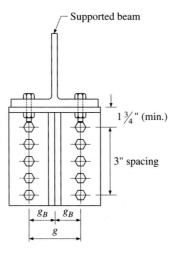

Figure 13.9.1: All-bolted stiffened seated connection.

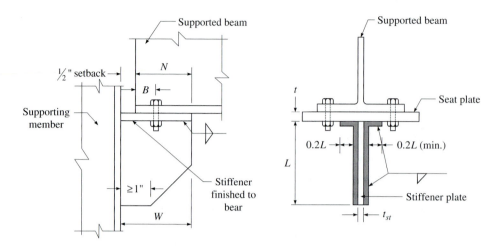

Figure 13.9.2: Bolted/welded stiffened seated connection.

will not interfere with fireproofing or architectural requirements. There are two distinct types of loading on stiffened seats. In the first type, the reaction is carried with the beam web directly in line with the plane of the stiffening element, as shown in Fig. 13.9.1. In the second type, the beam is oriented so that the plane of the beam web is at 90° to the plane of the stiffening element (a column bracket supporting a crane girder is an example, Fig. 11.11.1). Only, the first type will be considered in this section. Stiffened seated connections, like unstiffened seated connections, should only be used when the beam rotation about its longitudinal axis at the supports is prevented by a top angle. Stiffened seated connections, as discussed here, are treated as simple shear connections, under LRFD Type PR (ASD Type 2) as described in Section 13.1.1.

Once more, there is a difference in the treatment of eccentricity for bolted versus welded connections. Load eccentricity is neglected in the design of bolted stiffened seated connections but included in the calculations for welded stiffened seated connections.

Bolted Stiffened Seated Connection

Bolted stiffened seated connections often consist of a pair of angles, called *stiffener angles,* placed under a seat plate. The seat plate is usually made $\frac{3}{8}$ or $\frac{1}{2}$ in. thick. The *seat width* (dimension of the seat plate, parallel to the beam) is based on the required bearing length, N_{req}, to prevent web local yielding and web crippling, unless the web is provided with stiffeners. Since the stiffener angles relieve the seat of the bending, it is desirable to have their outstanding legs extend as near to the toe of the seat as possible. However, in calculating the contact area, the effective length of stiffener bearing is assumed $\frac{3}{4}$ in. less than length of outstanding leg. Thus, the design bearing strength of the stiffener angles may be written with the help of LRFDS Eq. J8-1 as:

$$R_{dbst} = 0.75(1.8F_{yst})(2)\left(b_{st} - \frac{3}{4}\right)(t_{st}) \qquad (13.9.1)$$

where b_{st} = stiffener angle outstanding leg size, in.
        t_{st} = stiffener angle thickness, in.
        F_{yst} = yield stress of the stiffener angle, ksi

For example, for a stiffened seated connection with a pair of $\frac{1}{2}$-in.-thick A36 angles with 4-in. outstanding legs, the design bearing strength using Eq. 13.9.1 is 158 kips. This is also the value given in Table 10-7 of the LRFDM. If a structural tee is used instead of angles, the 2 in the RHS of Eq. 13.9.1 would not be used. Equation 13.9.1 assumes no eccentricity of loading with respect to the center of the bearing contact area. The size of the connected leg is limited by the width between fillets of the column web. That is $2L_B \leq T_c$. The paired stiffener angles can be separated to accommodate to column gages. A filler or spacer plate must be inserted in the separation gap and stitch bolted in a manner similar to that of a double-angle column when the angles are not in contact (LRFDS Section E4.2).

Since the outstanding leg of the stiffener is in compression, it may fail by plate local buckling. The same limiting thickness used for outstanding elements of compression members is traditionally chosen. Thus, from LRFDM page 16.1–183:

$$\frac{b_{st}}{t_{st}} \leq 0.56 \sqrt{\frac{E}{F_{yst}}} \qquad (13.9.2)$$

This relation was developed for a long rectangular plate subjected to compressive stresses on two opposite edges, supported along a third edge, and free on the fourth edge. Note, however, that in the case of the stiffener outstanding leg under consideration, the beam reaction which is applied at the upper end of the outstanding leg is resisted along the supported (vertical) edge, while the other two adjacent edges are free.

As mentioned earlier, traditionally, the effect of load eccentricity is neglected in the design of the bolt group connecting the stiffener angles to the supporting member. The design strength of the bolt group is therefore given by:

$$R_{dbv} = nB_{dv} \qquad (13.9.3a)$$

$$R_{dbb} = nB_{dbo} \qquad (13.9.3b)$$

LRFDM Table 10-7 for All-Bolted Stiffened Seated Connections

The LRFDM Table 10-7 facilitates the design of bolted stiffened seats. Thus, the first part of Table 10-7 provides design bearing strengths for steels of $F_y = 36$ and 50 ksi in the stiffener angles, for stiffener outstanding lengths

(b_{st}) of $3\frac{1}{2}$, 4 and 5 in., and for five different thicknesses of stiffener out-standing legs. The second part of the Table 10-7 gives the design strengths of bolt groups based on single shear. Vertical spacing of bolts in stiffener angles may be arranged to suit conditions, provided they conform to LRFDS Sections J3.3 and J3.4 with respect to minimum spacing and minimum edge distance. Design strength of the connection is based on the lesser of the values from the two parts of Table 10-7.

In bolted stiffened seated connections, the question arises as to when the connection eccentricity may be neglected. That is, when does the connection cease to be an ordinary stiffened seat and become a bracket of the type shown in Fig. 12.13.1 subjected to the methods described in Section 12.13. Table 10-7 of the LRFD Manual points to an implied limit to this eccentricity. The maximum outstanding stiffener leg is 5 in. It would seem that for seat lengths (projections) greater than 5 in., the designer should investigate the influence of bending on the bolt group. Even for seat lengths somewhat less than 5 in., bending should be included in shallow connections, that is, those with a few horizontal rows of bolts.

Welded Stiffened Seated Connections

A widely used stiffened seat fabricated by welding two plates in the form of a tee is shown in Fig. 13.9.2. The *top plate,* also known as the *seat plate,* is usually made $\frac{3}{8}$ in. or $\frac{1}{2}$ in. thick. The length of the seat plate is slightly larger than the flange width of the supported beam. The thickness of the supporting vertical plate, known as *stiffener plate,* should not be less than the beam web thickness. The stiffened seated connection is shop welded to the supporting member in the flat or down hand position. The welds to the supporting element run the full length L, along each side of the stiffener plate and then run horizontally along the underside of the seat plate. The length of each horizontal weld will usually be one-fifth to one-half of the vertical weld; a value of $0.2L$ is used in the preparation of LRFDM Table 10-8. The top weld length is usually less than the bracket width. As the two vertical welds along the stiffener plate are close, the addition of these horizontal segments increases the torsional stiffness of the connection. Usually the seat plate is welded on the underside only. By placing the weld on the underside of the seat plate only, it does not interfere in anyway with the supported beam in case of beam overrun. However, some designers would prefer to place the fillet weld on the top of the bracket. In seated beam connections, where the seat and stiffener are two separate plates, the stiffener plate should be finished to bear against the seat plate. The welds connecting the seat plate to the top of the stiffener plate should have a strength equal to or greater than the horizontal welds to the support under the seat plate. As mentioned earlier, the seat may also be made from a tee or a split beam, thus eliminating the top plate and its welding to the vertical stiffening plate. The width of the seat plate (dimension W parallel to the beam) will also depend upon the set back (usually $\frac{1}{2}$ in.) between the beam end and the face of the supporting member. When loaded, the supported beam tends to pivot about the outer edge of the stiffened seat. This produces larger eccentricity than is found in an unstiffened seat. Thus, an unstiffened seated connection is preferred, when its capacity is large enough to carry the factored load.

The stiffened seated connection must be designed to provide adequate bearing length for the supported beam, prevent buckling of the stiffener plate, and resist safely the direct shear and bending moment on the weld attaching the seat to the supporting element. The length of the stiffener plate (L) is determined by the length of the vertical segment of the L-shaped weld.

The beam reaction R_u is assumed to act at a distance $\frac{1}{2}N_{req}$ measured from the free edge of the stiffener plate. Here N_{req} is the length of the bearing required from web local yielding and web crippling (Eq. 13.8.5). The resulting eccentricity is:

$$e = W - \frac{N_{req}}{2} \tag{13.9.4}$$

Under factored loads, the welded joint connecting the seat and stiffener plates to the supporting element is therefore subject to direct shear R_u and bending moment $R_u e$ (Loading case $P_y M_{yz}$, discussed in Section 12.11). The normal stresses in the weld are calculated on the assumption that the entire moment is resisted by the welds only. In addition, the stresses traditionally determined are the tensile stresses at the top of the weld group. These are computed using the traditional elastic vector method. While the compressive stresses at the lower, farther end of the vertical weld are theoretically larger, tests have shown that stiffened seats designed on this basis have ample margins of safety with respect to failure. The length of the vertical segment of the weld is assumed equal to the vertical length of the stiffener plate. For simplicity, the length of the top weld is assumed to be a certain percentage of the vertical weld length. Using $d = L$ and $b = 0.2L$ in the results for Case 6 of Table 12.9.1 gives, for the weld considered as a line element:

$$A_l = 2.4L; \quad S_{xl} = 0.6L^2 \tag{13.9.5}$$

The vertical and horizontal components and the resultant force on a unit-length weld, located at the top of the weld, are:

$$W_y = \frac{R_u}{2.4L}; \quad W_z^* = \frac{R_u e}{0.6L^2} \tag{13.9.6}$$

$$W_u^* = \sqrt{(W_y)^2 + (W_z^*)^2} = \frac{R_u}{2.4L}\sqrt{1 + 16\left(\frac{e}{L}\right)^2} \tag{13.9.7}$$

Here, W_u^* is the resultant force on the critical unit-length weld for which the weld is to be designed. If W_d is the design strength of the unit-length weld provided, we have:

$$W_u^* \leq W_d \tag{13.9.8}$$

Assuming E70 electrodes and a weld of size D, and making use of Eqs. 13.9.6, 7 and 8, the capacity of the stiffened seat may be expressed as:

$$R_{dw} \equiv \phi R_n \geq R_u \tag{13.9.9}$$

with

$$R_{dw} = \frac{(1.392D)(2.4L)}{\sqrt{1 + 16(e/L)^2}} \tag{13.9.10}$$

where

R_u = end reaction of the beam under factored loads, kips
R_{dw} = design strength of the weld, kips
e = moment arm (Eq. 13.9.4); taken as 0.8W in the preparation of LRFDM Table 10-8), in.

For example, for $L = 14$ in., $D = 5$ ($w = \frac{5}{16}$ in.), and $W = 8$ in., we obtain $e = 0.8(8) = 6.4$ in. and from Eq. 13.9.10 $R_{dw} = 112$ kips. This is also the value given in LRFDM Table 10-8 for this connection. Generally, a weld that can be placed in one pass would be preferred.

Bearing on the contact area of the stiffener plate must satisfy LRFDS Eq. J8.1. For the stiffener plate under eccentric load, the maximum bearing stress at the outer edge of the stiffener, under factored load and assuming elastic behavior, is given by:

$$f_p = \frac{P}{A} + \frac{M}{S} = \frac{R_u}{Wt_{st}} + \frac{6R_u(W - N)}{2t_{st} W^2} = \frac{R_u(4W - 3N)}{t_{st} W^2}$$

From LRFDS Section J8, the bearing stress f_p may not exceed the nominal bearing stress $(1.8F_{yst})$. Rearranging, we obtain:

$$t_{st} \geq \frac{R_u(4W - 3N)}{\phi(1.8F_{yst})W^2} \tag{13.9.11}$$

Here,

R_u = beam reaction under factored loads, kips
F_{yst} = yield stress of stiffener material, ksi
N = beam bearing length, in.
t_{st} = thickness of stiffener plate, in.
W = width of stiffener plate, in.
ϕ = resistance factor ($=0.75$)

The stiffener plate should have adequate bending resistance. That is:

$$M_{dst} \geq M_{ust} \tag{13.9.12a}$$

with

$$M_{ust} = R_u e; \quad M_{dst} = \phi_b S_{xst} F_{yst}; \quad S_{xst} = \frac{1}{6} t_{st} L^2 \tag{13.9.12b}$$

where M_{ust} = bending moment on the stiffener acting as a cantilever beam
    M_{dst} = bending resistance of the stiffener plate

Note that the moment M_{ust} used in the design of the stiffener plate and the stiffener-to-column weld is traditionally ignored when designing the column in which the connection is framed. The connection and the beam provide rotational restraint to the column, so that a column design based on an effective length factor of unity is conservative.

LRFDM Table 10-8 for Bolted/Welded Stiffened Seated Connections

The LRFD Manual provides design strength tables for a variety of welded stiffened beam seats for seat widths from 4 to 9 in., different weld sizes, and weld lengths, L. In addition, note the following:

1. Design loads in the table are based on the use of E70 electrodes. For other electrodes, multiply the tabular value by the factor $F_{EXX}/70$ where F_{EXX} is the strength of the weld material used, provided the weld and base metals meet the requirements of the LRFDS Section J2.
2. The length of the seat plate (dimension perpendicular to beam length) is slightly larger than the flange width of the supported beam. The thickness of the seat plate is either $\frac{3}{8}$ or $\frac{1}{2}$ in.
3. Since the stiffener plate relieves the seat of the bending, it is desirable to have its outstanding leg extend as near to the edge of the seat as possible. The effective width of the outstanding leg in bearing against the seat is taken $\frac{3}{4}$ in. less than its actual length for calculation purposes.
4. The minimum stiffener plate thickness should not be less than the beam web thickness, t_{wb}, multiplied by the ratio of the yield stress of the beam to the yield stress of the stiffener plate, or:

$$t_{st} \geq \frac{t_{wb} F_{yb}}{F_{yst}} \qquad (13.9.13)$$

5. According to the note in Table 10-8 of the LRFDM, the minimum stiffener plate thickness should be at least 2 times (1.5 times) the required E70 weld size when F_y of the bracket is 36 ksi (50 ksi). That is:

$$t_{st} \geq 2w \quad \text{for stiffeners with } F_y = 36 \text{ ksi} \qquad (13.9.14)$$

$$t_{st} \geq 1.5w \quad \text{for stiffeners with } F_y = 50 \text{ ksi}$$

This is based on the idea that the stiffener should be capable of resisting the full shear capacity of the two welds:

$$0.9(0.6)(36t_{st}) \geq 2(0.75)(0.6)(70)(0.707w) \rightarrow t_{st} \geq 2.29w \approx 2w$$

6. The width-to-thickness ratio of the outstanding leg of the stiffener should satisfy the local buckling criteria.

$$\frac{W}{t_{st}} \leq 0.56\sqrt{\frac{E}{F_{yst}}} \qquad (13.9.15)$$

7. Combinations of the material thicknesses and weld sizes selected from LRFDM Table 10-8 should meet the limitations on minimum weld size (LRFDS Table J2-4). If these limitations are not met, the weld size or material thickness should be increased as required.

8. For stiffener seats aligned on opposite sides of a column web of $F_y = 50$ ksi material, select a E70 weld size no greater than 0.67 of the column web thickness. This stipulation assures that when the weld is fully stressed, the connected (base) material is not over stressed.

9. The supported beam may be connected to the bracket seat by bolts or by field welds.

Additional Notes for Stiffened Seated Connections to Column Webs

The vertical stem of the stiffened seated connection when welded to a plate, such as a column web, produces bending in the web plate. If this web plate is relatively thin, it may be over stressed. So, for stiffened seated connections made on one side of a supporting column web, the following restrictions are imposed to ensure that weld failure will occur before column web failure:

1. This simplified approach is applicable to the following column sections:

   W14×43-808;   W12×40-336;   W10×33-112;
   W8×24-67;   W6×20-25;   W5×16-19

2. The supported beam must be bolted to the seat plate with A325 or A490 high-strength bolts to resist the prying forces developed by rotation of the connection at ultimate load. Welding the beam flange to the seat plate is not recommended because welds lack the required strength and ductility. These bolts should be located no more than the greater of $W/2$ or $2\frac{5}{8}$ in. from the column web face.

3. For seated connections where the seat width $W = 8$ in. or $W = 9$ in. and $3\frac{1}{2}$ in. $< B < W/2$, or where $W = 7$ in. and 3 in. $< B \leq W/2$ for a W14× 43 column, refer to Sputo and Ellifrit [1991]. These limitations are summarized at the bottom of LRFDM Table 10-8.

4. The seat plate should not be welded to the column flanges. To do so will introduce relatively large moments into the column section and will negate this design procedure.

5. Except as noted, the maximum weld size for E70 electrodes is limited to the column web thickness t_{wc} for connections to one side of the web. For connections in line on both sides of a column web, the maximum

weld size is $2t_{wc}/3$ for $F_{yc} = 50$ ksi. This approximately matches the shear yielding strength of the column web with the shear strength of the weld.

6. The top angle may be bolted or welded, but must have a minimum $\frac{1}{4}$-in. thickness. The rotation of the resulting stiffened seated beam connection and the column web makes this a very flexible connection, approximating a fully simple condition. The flexibility of the connection, coupled with the small eccentricity of the load makes it unnecessary to consider any eccentricity of load or applied moment in the design of the column [Ellifritt and Sputo, 1999].

Stiffened Seated Connection

EXAMPLE 13.9.1

Design a stiffened seated beam connection to connect a W21×62 spandrel beam to the web of a W12×58 corner column. The factored end reaction is 110 kips. The beam and the column are of A992steel. Use E70 electrodes.

Solution

1. Data
    From Table 1-1 of the LRFDM, for a W21×62 beam

    d_b = 21.0 in.; t_{wb} = 0.400 in. ($\frac{3}{8}$ in.)
    b_{fb} = 8.24 in.; t_{fb} = 0.615 in. ($\frac{5}{8}$ in.)
    k_b = 1.12 in. ($1\frac{5}{16}$ in.); g_b = $5\frac{1}{2}$ in.

    From Table 9-5 of the LRFDM, for a W21×62 beam of A992 steel

    ϕR_1 = 56.0 kips; ϕR_2 = 20.0 kips/in.
    ϕR_5 = 64.2 kips; ϕR_6 = 7.16 kips/in.
    $0.2d_b$ = 4.20 in.; $0.5d_b$ = 10.5 in.
    ϕV_n = 227 kips > R_u = 110 kips   O.K.

    W12×58 column of A992 steel

    d_c = 12.2 in.; t_{wc} = 0.360 in.
    T_c = $9\frac{1}{4}$ in.

2. Seat plate dimensions
    From Eq. 13.8.2, the required bearing length N_1 for the limit state of web yielding is:

    $$N_1 = \max\left[\frac{R_u - \phi R_1}{\phi R_2}, k_b\right] = \max\left[\frac{110 - 56.0}{20.0}, 1.12\right] = 2.70 \text{ in.}$$

[continues on next page]

Example 13.9.1 continues ...

Assuming $N/d_b > 0.2$, the required bearing length N_2 for the limit state of web crippling from Eq. 13.8.4b is:

$$N_2 = \frac{R_u - \phi R_5}{\phi R_6} = \frac{110 - 64.2}{7.16} = 6.40 \text{ in.}$$

So the stiffener width is controlled by web crippling. By adding $\frac{1}{2}$ in. for the beam set back and $\frac{1}{4}$ in. to account for possible underrun in beam length, the minimum stiffener width is:

$$W_{min} = 6.40 + 0.5 + 0.25 = 7.15 \text{ in.}$$

So use a seat width, W, of 8.0 in. giving an N value of 7.25 in. $> 0.2d_b = 4.20$ in.   O.K.

From LRFDM Table 10-8 for Bolted/Welded Stiffened Seated Connections, a stiffener with $L = 14$ in., width $W = 8$ in., and $\frac{5}{16}$-in. weld size provides:

$$R_d = \phi R_n = 112 \text{ kips} > R_u = 110 \text{ kips} \qquad \text{O.K.}$$

The length of seat plate-to-column web weld on each side of the stiffener plate is $0.2L = 2.80$ in., say, 3 in. Use 3 in. of $\frac{5}{16}$ in. horizontal weld on both sides of the stiffener plate for the seat-plate to column web welds; and 14 in. of $\frac{5}{16}$ in. vertical weld on both sides of the stiffener plate for the stiffener plate to column web welds.   (Ans.)

To accommodate two $\frac{3}{4}$-in.-dia. A325-N bolts on a $5\frac{1}{2}$-in. gage connecting the beam flange to the seat plate, a width of 8 in. is adequate. This is greater than the width required to accommodate the seat-plate-to-column-web welds.

So, use PL$\frac{3}{8}$×8×0′ 8″ for the seat plate.   (Ans.)

3. Stiffener plate

From Eq. 13.9.11, the thickness required to satisfy the limit state of bearing is:

$$t_{st} \geq \frac{110[4(8.00) - 3(6.40)]}{0.75(1.8)(36)(8.00)^2} = 0.453 \text{ in.}$$

From Eqs. 13.9.12a and b, the thickness required to satisfy the limit state of flexure may be written as:

$$0.9(\tfrac{1}{6})(14^2)(36t_{st}) \geq 110[8.00 - 0.5(6.40)] \rightarrow t_{st} \geq 0.499 \text{ in.}$$

For a stiffener with $F_y = 36$ ksi and beam with $F_y = 50$ ksi, the stiffener thickness is:

$$t_{st} \geq \frac{50}{36} t_{wb} = 1.4(0.400) = 0.560 \text{ in.}$$

To develop the stiffener-to-seat-plate welds, the stiffener thickness is:

$$t_{st} \geq 2w = 2(\tfrac{5}{16}) = \tfrac{5}{8} \text{ in.}$$

The later controls.
So, use $PL\tfrac{5}{8} \times 8 \times 1' \; 2''$ of A36 steel for the stiffener.   (Ans.)

4. Top angle
   Use $L4 \times 4 \times \tfrac{1}{4}$ with $\tfrac{3}{4}$-in.-dia. A325-N bolts through the supported beam leg of the angle. Use a $\tfrac{1}{8}$-in. fillet weld along the toe of the angle to the column web (minimum size from LRFDS Table J2.4).

13.10   Rigid-Frame Knees

In portal frames, the two rolled W-shapes used as beam and column (with their webs lying in the plane of the frame) are often rigidly joined together using what is known as a **knee joint** (or, more appropriately, a **knee connection**) or a **corner connection.** Typical knee connections shown in Fig. 13.10.1 are:

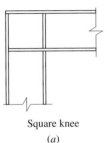

a. The square knee or straight corner connection,
b. The tapered haunched knee, and
c. The curved haunched knee.

Square knee
(*a*)

From the analysis of portal frames (see Section 5.3), it is evident that the corner connection must be capable of transferring all three possible components of force—namely, moment, shear, and thrust—from one member to the other. Also, the connection elements must be proportioned carefully so that their premature failure will not result in a reduced overall connection strength and/or connection rotational capacity. The corner connections are almost always welded, to reduce the number of connection elements. Note that corner connections are located at points of maximum moment and shear. So, if splices are needed (in case of large beam spans, for example), they should be provided away from the knee and at a location of lower moment and shear. Behavior and plastic design of corner connections are summarized in ASCE Manual 41 [1971] pp. 167–186, which forms the basis for much of what follows for LRFD of these connections.

Tapered haunched knee
(*b*)

13.10.1   Square Knees

Figure 13.10.2*a* shows the moment diagram for a typical portal frame loaded with uniformly distributed factored gravity loads on the beam. It is determined from a frame analysis, either elastic (as shown here) or plastic. A typical unreinforced square knee, wherein the beam is run through the connection, is shown in Fig. 13.10.2*b*. Connection element DA is made either as a continuation of the column outer flange or as an equivalent plate groove welded to the beam bottom flange. Full penetration groove welds are used at

Curved haunched knee
(*c*)

Figure 13.10.1: Knee or corner connections.

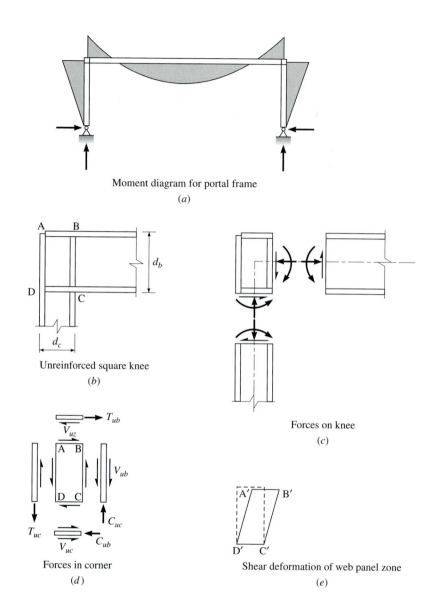

Moment diagram for portal frame

(a)

Unreinforced square knee

(b)

T_{ub}

V_{uz}

V_{ub}

C_{uc}

T_{uc}

V_{uc} C_{ub}

Forces in corner

(d)

Forces on knee

(c)

Shear deformation of web panel zone

(e)

Figure 13.10.2: Force transfer in a square knee.

the junction of the column flanges and the lower flange of the beam. Fillet welds can be used at the other locations to transfer the forces. Connection element CB can be made as a partial or full-depth stiffener. The moment, shear, and thrust acting on the corner connection are indicated in Fig. 13.10.2c.

To simplify the analysis of the connection it is generally assumed that the bending moment and thrust in a member are resisted by the flanges only, while the shear is resisted by the web only. The action of the applied forces on the elements of the corner and of the elements on each other are indicated by arrows in Fig. 13.10.2d. Thus, the tensile force in the outer flange of the beam (T_{ub}) is carried into the web in shear along line BA. Similarly, the tensile force in the outer flange of the column (T_{uc}) is carried through the end plate into the beam web as a shear along line DA. In each case, the tensile force in the flange is reduced from a maximum value at the edge of the corner (B or D) to zero at the external corner (A). The prolongation of the inner flange of the beam carries two external forces: the normal force due to bending and thrust in the beam (C_{ub}) and the shear of the column (V_{uc}). The resultant of these two forces is carried into the panel zone as a shear along line CD. A similar pair of force components, acting on the vertical stiffener, causes shear along line CB of the beam web.

As the horizontal beam continues into the connection, the segments AB and CD of the flange will be adequate in the connection if they were adequate in the beam. The end plate, AD, should have the same area as the flange of the column. The vertical stiffener, CB, must have sufficient strength to transfer the column compression flange force into the beam web. In all cases the welds must be adequate to transmit the required shear or normal force on the joint. Observe that the rectangular area ABCD, known as the **panel zone,** is subjected to shear forces along the four edges (Fig. 13.10.2d). These shear forces would tend to deform the panel as shown in Fig. 13.10.2e. The diagonal AC of the rectangle ABCD represents the compression diagonal of the panel zone.

Neglecting the bending resistance of the web, and approximating the distance between the flange centroids as $0.95d_b$, the beam flange force may be written as:

$$T_{ub} = \frac{M_u}{0.95d_b} \tag{13.10.1}$$

where d_b is the overall depth of the beam and M_u is the factored moment at the corner. The axial force in the beam is usually small, and of opposite sign to the axial force in the top flange of the beam. So, its contribution to shear equilibrium of the segment AD is usually neglected. If V_{uz} is the panel zone shear force acting on the web along AB, equilibrium in the horizontal direction of the forces acting on the portion AB of the flange gives:

$$V_{uz} = T_{ub} \tag{13.10.2}$$

The design shear strength of the web across AB is (LRFDS Eq. K1-9):

$$V_{dz} = \phi_v(0.6F_{yz})t_{wz}d_c \tag{13.10.3}$$

where ϕ_v = resistance factor in shear ($=0.9$)
        d_c = overall depth of the column, in.
        t_{wz} = thickness of the web in the panel zone, in.
        F_{yz} = yield stress of the web in the panel zone, ksi

Equation 13.10.3 assumes that interaction between column axial load P_u and column shear V_u does not control, that is, that $P_u \le 0.4P_y$ where P_y is the squash load of the column. According to the LRFDS:

$$V_{dz} \geq V_{uz} \tag{13.10.4}$$

With the help of Eqs. 13.10.1 to 3, this relation may be rewritten as:

$$t_{wz} \geq \frac{M_u}{(0.95)(0.9)(0.6)F_{yz} A_z} \equiv t_{wzreq} \tag{13.10.5}$$

where A_z = area of the pane zone (= $d_b d_c$), in.^{2}
t_{wz} = actual web thickness in the panel zone, in.
t_{wzreq} = required web thickness in the panel zone, in.

For a square knee obtained by extending the beam over the column, the web thickness provided by the W-section will generally be less than the required web thickness in the panel zone. So, the web in the panel zone needs to be reinforced. This reinforcement may take the form of a doubler plate or a pair of diagonal stiffeners. Diagonal stiffener is the preferred choice for portal frames. Equilibrium in the horizontal direction of the forces acting on the portion AB of the flange now gives:

$$V_{ur} + V_{dz} - T_{ub} = 0 \rightarrow V_{ur} = (T_{ub} - V_{dz}) \tag{13.10.6}$$

where T_{ub} = tension in the beam flange under factored loads (Eq. 13.10.1), kips
V_{dz} = design shear strength of the web in the panel zone (Eq. 13.10.3), kips
V_{ur} = horizontal shear to be provided by the web reinforcement (doubler plate or diagonal stiffeners) under factored load, kips

Doubler Plate

A **doubler plate** or **reinforcement plate** consists of a rectangular plate (dimensions $t_{dp} \times d_{dp} \times h_{dp}$, and yield stress F_{ydp}) welded to the web in the panel zone of a square connection, to increase the shear capacity of the panel zone. In design calculations, d_{dp} and h_{dp} are traditionally taken as d_c and d_b, respectively. The design shear strength of the doubler plate, corresponding to the limit state of shear yielding may be written as:

$$V_{ddp} = \phi_v (0.6 F_{ydp}) d_{dp} t_{dp} \tag{13.10.7}$$

LRFDS requires that:

$$V_{ddp} \geq V_{ur} \rightarrow t_{dp} \geq \frac{V_{ur}}{\phi_v (0.6 F_{ydp}) d_{dp}} \tag{13.10.8}$$

where ϕ_v = resistance factor (=0.9)
t_{dp} = thickness of the doubler plate, in.
d_{dp} = width of the doubler plate, in.

F_{ydp} = yield stress of the doubler plate material, ksi
V_{ur} = shear to be provided by the doubler plate, kips (Eq. 13.10.6)

To assure that shear buckling of the reinforcement plate does not precede shear yielding of the plate (assumed in Eq. 13.10.7), the plate width-to-thickness ratio should satisfy the relation (LRFDS Section A-F2):

$$\frac{d_{dp}}{t_{dp}} \le 1.10 \sqrt{\frac{k_v E}{F_{ydp}}} \qquad (13.10.9a)$$

where

$$k_v = 5 + \frac{5}{(h_{dp}/d_{dp})^2} \qquad (13.10.9b)$$

The fillet weld connecting the reinforcement plate should have adequate strength to transfer the factored load V_{ur} over length d_{dp}. If D is the number of sixteenths-of-an-inch in the weld size, we have (assuming E70 electrodes):

$$D \ge \frac{V_{ur}}{1.392\, d_{dp}} \qquad (13.10.10)$$

Further, the weld size provided must satisfy the minimum weld size requirement stipulated in Table J2.4 of the LRFDS. Finally, the thickness of the reinforcement plate provided should be adequate to provide the weld size selected (LRFDS Section J2.2b).

Diagonal Stiffeners
Diagonal stiffeners usually consist of a symmetric pair of rectangular plates (each of width, b_{st}; thickness, t_{st}; length, L_{st}; and yield stress, F_{yst}) welded to the web along the compression diagonal of the panel zone. These stiffeners act somewhat like the diagonals of a truss panel in preventing shear deformation. Let θ be the inclination of the compression diagonal with the horizontal, and let P_{ust} represent the compressive force in the stiffener under factored loads (Fig. 13.10.3). Equilibrium in the horizontal direction, of the forces acting on the portion AB of the flange now gives:

$$P_{ust} \cos\theta + V_{dz} - T_{ub} = 0 \rightarrow P_{ust} = \frac{V_{ur}}{\cos\theta} \qquad (13.10.11)$$

where V_{ur} = horizontal shear to be provided by the diagonal stiffener (Eq. 13.10.6)
   θ = inclination of the diagonal stiffener with the horizontal (tan $\theta \approx d_b/d_c$)

The stiffener may conservatively be designed as a pin ended column of length $L_{st} \approx d_c/\cos\theta$. The column having a rectangular cross section ($2b_{st} + t_{wz})(t_{st})$, is assumed to buckle about its major axis (out-of-plane of the web).

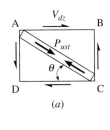

(a)

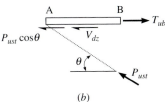

(b)

Figure 13.10.3: Diagonal stiffener.

The design compressive strength of the diagonal stiffener may be written as:

$$P_{dst} = \phi_c F_{crst} A_{st} \qquad (13.13.12)$$

where F_{crst} = nominal compressive stress of the diagonal stiffeners acting as a column (LRFDS Section E2), ksi

A_{st} = stiffener area, in.^{2}

$\quad = (2b_{st} + t_{wz}) t_{st}$

ϕ_c = resistance factor for compression elements (=0.85)

As per the LRFDS:

$$P_{dst} \geq P_{ust} \quad \rightarrow \quad A_{st} \geq \frac{V_{ur}}{0.85 F_{crst} \cos\theta} \qquad (13.10.13)$$

As the stiffener is a compression element, its width-to-thickness ratio should
be limited to avoid the possibility of premature local buckling (LRFDS
Section K3.9):

$$\frac{b_{st}}{t_{st}} \leq 0.56 \sqrt{\frac{E}{F_{yst}}} \qquad (13.10.14)$$

Diagonal stiffener AC must be developed by using full penetration groove
welds across the ends A and C. Also, minimum size fillet welds connecting the stiffener to the web must be used along the length of the diagonal
AC. These will prevent buckling of each individual stiffener plate ($b_{st} \times t_{st}$)
to act like a column, buckling about its minor axis ($r = 0.289t_{st}$). When a
diagonal stiffener is required, we would automatically provide a pair of
web stiffeners from C to B. The stiffener CB receives its load from the column interior flange by bearing on the beam interior flange, and transfers
its load to the beam web by the four welds along CB. The stiffener CB
must carry the difference between the compressive force from the column
interior flange and the resistance provided by the beam web. The later is
given by:

$$R_{dWY} = \phi R_1 + N(\phi R_2)$$

where R_{dWY} = design strength of the beam web, corresponding to the
limit state of web local yielding, kips

N = length of bearing (= t_{fc}, thickness of the column flange), in.

ϕR_1 = $\phi(2.5 k_b t_{wb} F_{yb})$

ϕR_2 = $\phi(t_{wb} F_{yb})$

with

ϕ = resistance factor = 1.0

k_b = distance from outer face of the supported beam flange to the
web toe of the fillet, in. (Note: For this purpose, use the decimal dimension given for k in Table 1-1 of the LRFDM, not the
fractional value given there.)

t_{wb} = beam web thickness, in.

F_{yb} = yield stress of the beam web, ksi

For any rolled steel W-shape (F_y = 50 ksi), the design strength R_{dWY} may be determined from constants ϕR_1 and ϕR_2 given in Table 9-5 of the LRFDM.

Straight Corner Connection

EXAMPLE 13.10.1

Design the straight corner connection joining a W30×90 beam to a W14×145 column of the portal frame designed in Example 11.15.11. Assume E70 electrodes for welds. The following loads acting on the corner are obtained from Example 11.15.11.

Factored 2nd order moment, M_u = 877 ft-kips

Factored axial load in column   = 148 kips

Factored axial load in beam   = 12 kips

Solution

1. Data

   W30×90 beam of A992 steel

   d_b = 29.5 in.; t_{wb} = 0.470 in.; T_b = 26½ in.

   b_{fb} = 10.4 in.; t_{fb} = 0.610 in.

   $\left(\dfrac{h}{t_w}\right)_b$ = 57.5 → h_b = 27.0 in.

   $\phi_b M_p$ = 1060 ft-kips; $\phi_v V_n$ = 374 kips

   W14×145 column of A992 steel

   d_c = 14.8 in.; t_{wc} = 0.680 in.; T_c = 10 in.

   b_{fc} = 15.5 in.; t_{fc} = 1.09 in.

   $\left(\dfrac{h}{t_w}\right)_c$ = 16.8 → h_c = 11.4 in.

   $\phi_b M_p$ = 975 ft-kips; $\phi_v V_n$ = 272 kips

2. Check the need for web reinforcement

   Tension in the beam flange under factored loads,

   $$T_{ub} = \frac{M_u}{0.95 d_b} = \frac{877(12)}{0.95(29.5)} = 376 \text{ kips}$$

   Design shear strength of the web in the panel-zone,

   $$V_{dz} = \phi_b(0.6F_y)t_{wb}d_c = 0.9(0.6)(50)(47)(14.8) = 188 \text{ kips}$$

   As $V_{dz} < T_{ub}$, the panel-zone needs to be reinforced by doubler plates or diagonal stiffeners. Shear to be resisted by the reinforcement,

   $$V_{ur} = T_{ub} - V_{dz} = 376 - 188 = 188 \text{ kips} \qquad \text{(Ans.)}$$

[continues on next page]

Example 13.10.1 continues ...

3. **Doubler plates**

Use two doubler plates, one on each side of web in the panel zone. Assume A36 steel and a doubler plate size of approximately $h_c \times h_b$ (say, 11 in. × 29 in.). These dimensions correspond to the planar web area within the knee. The required thickness of the doubler plates,

$$t_{dp} \geq \frac{V_{ur}}{2(0.9)(0.6F_{ydp})d_{dp}} = \frac{188}{2(0.9)(0.6)(36)(11)} = 0.440 \text{ in.}$$

Use a plate thickness, $t_{dp} = \frac{1}{2}$ in. > 0.440 in.   O.K.
Assume E70 electrodes. From LRFDS Table J2-4, minimum weld size $w_{min} = \frac{3}{16}$ in., to connect the $\frac{1}{2}$-in. doubler plate to the 0.47-in. web. Maximum weld size, $w_{max} = \frac{1}{2} - \frac{1}{16} = \frac{7}{16}$ in.
Weld size, number of sixteenths-of-an-inch, required to transfer the shear:

$$D \geq \frac{188}{2(1.392)(11)} = 0.614, \quad \text{say } 6.0$$

Use a weld size, $w = \frac{3}{8}$ in. $> \frac{3}{16}$ in.   O.K.
Check doubler plate for shear buckling:

$$\frac{h_{dp}}{d_{dp}} = \frac{27}{11} = 2.5; \quad k_v = 5 + \frac{5}{(a/h)^2} = 5.8$$

$$\frac{d_{dp}}{t_{tp}} = \frac{11}{1/2} = 22.0 < 1.10\sqrt{\frac{5.8(29,000)}{36}} = 75.2 \qquad \text{O.K.}$$

Weld size along the longer side of the doubler plate,

$$D = 6\left(\frac{11}{27}\right) = 2.44. \text{ So, use } w = w_{min} = \frac{3}{16} \text{ in.}$$

Select two PL$\frac{1}{2}$×11×27 of A36 steel, and use $\frac{3}{8}$-in. fillet welds on the short sides and $\frac{3}{16}$-in. fillet welds on the long sides.   (Ans.)

4. **Diagonal stiffeners**

$$\tan \theta = \frac{d_b}{d_c} = \frac{29.5}{14.8} \rightarrow \theta = 63.4° \rightarrow \cos \theta = 0.448$$

Compressive force in the diagonal stiffener, $P_{ust} = \frac{V_{ur}}{\cos \theta} = \frac{188}{0.448} = 420$ kips
Assume A572 Gr 50 steel stiffeners.

Diagonal stiffener area, $A_{st} \geq \frac{P_{ust}}{0.85F_{yst}} = \frac{420}{0.85(50)} = 9.89 \text{ in.}^2$

Select two plates 1 in. × 5 in. of A572 Gr 50 steel.

$$A_{st} = (2b_{st} + t_{wb})t_{st} = [2(5) + 0.470]1.0$$
$$= 10.5 \text{ in.}^2 > 8.8 \text{ in.}^2 \qquad \text{O.K.}$$

Length of stiffener, $L_{st} \approx d_c/(\cos \theta) = 33.0$ in.

The radius of gyration of the stiffener is:

$$r_{st} = \frac{1}{\sqrt{12}} (2b_{st} + t_{wb}) = \frac{10.5}{\sqrt{12}} = 3.03 \text{ in.}$$

The effective slender ratio of the stiffener acting as a column, assuming pinned ends,

$$\frac{KL_{st}}{r_{st}} = \frac{1.0(33.0)}{3.03} \approx 10.9$$

From LRFDS Table 3-50 corresponding to $KL/r = 10.9$, $\phi_c F_{cr} = 42.1$ ksi. This value is only slightly less than the value of 42.5 ksi used earlier for arriving at the required area for the stiffener. This will generally be true for stiffener design. Design compressive strength of the diagonal stiffener,

$$P_{dst} = 42.1(10.5) = 442 \text{ kips} > 420 \text{ kips} \qquad \text{O.K.}$$

Check stiffener plate for local buckling:

$$\lambda = \frac{b_{st}}{t_{st}} = 5.0 < 0.56 \sqrt{\frac{E}{F_{yst}}} = 0.56 \sqrt{\frac{29,000}{50}} = 13.5 \quad \text{O.K.}$$

From LRFDS Table J2.4, minimum weld size (with $t_2 = 1$ in.) is $\frac{5}{16}$ in. ($D = 5$).
Length of weld required,

$$L_{w \text{ req}} \geq \frac{P_{ust}}{2(2)(1.392D)} = \frac{420}{2(2)(1.392)(5)} = 15.1 \text{ in.}$$

So, select 2PL1×5 of A572 Gr 50 steel as diagonal stiffeners. Use CJP groove welds across the ends, and weld the stiffners to the web in the panel zone, using $\frac{5}{16}$-in. E70 fillet welds.

13.11   Fully Restrained Moment Connections

A FR beam-to-column moment connection must have adequate strength and stiffness to transfer the factored bending moments, axial force, and shear force at the beam ends to the column without any appreciable change in the angle between each beam and column. In an I-beam bent about its major axis, the shear is carried primarily by the web (see Section 9.3); it can therefore be transferred directly to the column, through a connection of the beam web to the column. The bending moment in an I-shaped beam bent about its major axis is essentially resisted by the beam flanges (see Section 9.2). The moment can therefore be resolved into an effective tension-compression couple acting as axial forces acting at the beam flanges. The flange force P_{uf} may be calculated as:

$$P_{uf} = \frac{M_u}{d_m} \tag{13.11.1}$$

where M_u = beam end moment under factored load, in.-kips
        P_{uf} = factored beam flange force, tensile or compressive, kips
        d_m = moment arm between flange forces, in.

For a negative moment at the beam end, as would be the case with gravity loading, the forces P_{uf} are directed as shown in Fig. 13.11.1a, with the top flange of the beam delivering a tensile force to the column, and the bottom flange delivering a compressive force.

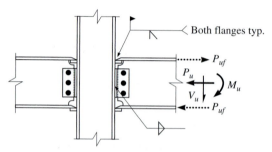

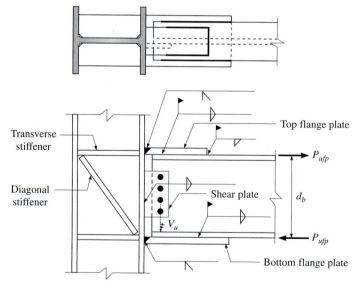

Directly welded moment connection
(a)

Flange-plated welded moment connection
(b)

Figure 13.11.1: Fully restrained welded moment connections.

The variety of arrangements for fully restrained beam-to-column connections is so great as to prevent any complete listing; however, those shown
in Figs 13.11.1 and 13.11.2 are the most common in current design practice.
The connections may be welded (Fig. 13.11.1) or bolted using high-strength
bolts (Fig. 13.11.2). Columns in frames may have beam attachments to
both flanges, as in Fig. 13.11.1a, or only to one flange, as in Figs. 13.11.1b
and 13.11.2.

The most direct connection of a beam to a column flange is shown in
Fig. 13.11.1a. Here, the beam flanges are welded to the column flange using
CJP groove welds. The plate connecting the beam web to the column flange is
shop welded to the column and field bolted to the beam. With this arrangement,

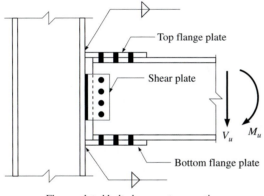

Flange-plated bolted moment connection

(a)

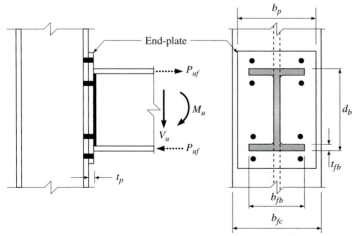

Extended end-plate moment connection

(b)

Figure 13.11.2: Fully restrained bolted moment
connections.

the beam is erected and held in position so that the beam flanges can be field welded to the column. The plate and its connections are designed to resist only shear. Research at Lehigh University and University of California at Berkeley has shown that the full plastic moment capacity of the supported beam can be developed with sufficient inelastic rotation and deformation capacity, through such a connection. Although this is the most direct moment connection that can be made, it requires a very close control on tolerances. In effect, the rolling mill supplies rolled W-shapes to a tolerance of $\pm\frac{3}{8}$ to $\frac{1}{2}$ in. or more, depending on the depth and length, on the ordered length of a beam (see LRFDM Table 1-54 for tolerances). For example, a 30-ft-long W16 beam has tolerance of $\pm\frac{3}{8}$ in. Also, the beam end may be out of square by not more than $\frac{1}{64}$ in. per inch of depth of the shape. Thus the W16 beam considered might show a gap which is $\frac{16}{64} = \frac{1}{4}$ in. greater at one flange than the other. In addition, the column to which the beam connects will itself be subjected to tolerances. For example, the depth of a W-shape column may vary $\pm\frac{1}{8}$ in. from its specified value. These tolerances must be foreseen in placing the field welds indicated in Fig. 13.11.1a.

Figure 13.11.1b shows a common type of fully restrained welded moment connection, wherein the tensile force in the top flange of the supported beam is transferred by fillet welds to a plate called the *top flange plate,* and by full penetration single bevel groove welds from the top flange plate to the column. A *bottom flange plate* is provided to transfer the bottom flange force to the column flange. The web connection plate is shop welded to the column flange and field bolted (Fig. 13.11.1b) to the beam web. Field welding should be arranged for welding in the flat or horizontal position and preference should be given to fillet welds over groove welds whenever possible. So, the width of the top flange plate is usually made narrower than the width of the beam flange, so that the weld can be placed on top of the flange. The bottom plate, however, is made wider than the beam flange to permit placing the weld on top of the plate, thereby avoiding overhead welding. Sometimes, the top flange plate is field welded in place. In that case a backing strip is placed between the beam top flange and the top flange plate to obtain a good quality, complete joint penetration groove weld to the column. In fabrication, the beams are cut approximately 1 in. shorter than the distance between their two supports for clearance purposes. A common practice in structures with welded moment connections is to directly butt weld the beam flanges with the column flange at one end of the beam (Fig. 13.11.1a) and connect the beam to the column at its other end with a connection of the type shown in Fig. 13.11.1b. This method allows for field adjustment for mill tolerances in the length of the beam and depth of the column, as discussed earlier.

Figure 13.11.2a shows a bolted, flange-plated fully-restrained beam-to-column moment connection. Here, the moment is assumed to be resisted by the flange plates, shop welded to the column flange, and field bolted to the beam flanges. The beam shear is assumed to be transferred to the column by a vertical plate, shop welded to the column and field bolted to the beam web.

Fully restrained, all welded, beam-to-column moment connection. Note the column web stiffeners.
Photo by S. Vinnakota

Oversized holes in the flange connection plates assist in the field assembly of this type of connection by compensating for rolling, fabrication, and erection tolerances. Behavior and design of flange-plated moment connections will be considered in Section 13.12.

The ***extended end-plate moment connection*** (Figure 13.11.2*b*) consists of a plate of length greater than the beam depth, shop welded to the end of the beam, and then field bolted to the column flange. High-strength bolts are arranged near the top flange to transfer the flexural tensile force in the beam flange, and are placed in other locations, as necessary, to help resist the beam end shear. The welds connecting the end plate to the beam are usually fillet welds, though they may be CJP groove welds. Behavior and design of extended end-plate FR moment connections will be considered in Section 13.13.

The student designer should realize that the choice of the lightest members carrying the main structural load can lead to local stress and local buckling problems with connections. For example, the use of light, thin flange columns to carry moment connections may require stiffening of the column web, as shown in Fig. 13.11.1*b*. This may well be a column which would otherwise have no weld fittings, and the cost of extra handling and workmanship to weld transverse and diagonal stiffeners can be substantial. Also, these stiffeners may interfere with weak axis framing. Often it is less costly to select a member with a thicker flange and/or web, or to use a higher yield strength material than it is to add the transverse stiffening [Thornton, 1992]. Column stiffening at beam-to-column moment connections will be discussed in Section 13.14.

13.12   Flange-Plated Moment Connections

Flange-plated moment connections consist of two plates—the ***top flange plate*** and the ***bottom flange plate***—which connect the flanges of the supported beam to the supporting column, A third plate, called a ***shear plate,*** connects the web of the supported beam to the column. The web plate is pre-punched, shop welded to the supporting column, and field bolted to the supported beam. The flange plates are shop-welded to the supporting column and may be either field welded (Fig. 13.11.1*b*) or field bolted (Fig. 13.11.2*a*) to the flanges of the supported beam. To provide for mill variations in beam depth ($\frac{1}{8}$ to $\frac{1}{4}$ in.) and to facilitate erection, the distance between flange plates is made larger than the nominal depth of the supported beam, usually by about $\frac{3}{8}$ in. If it is found that clearance is not necessary, this gap is filled at the top flange during erection with ***shims,*** which are thin strips of steel used for adjusting the fit at joints. A backup strip tack-welded to the column flange is sometimes provided for convenience in groove welding.

The flange-plated moment connections in FR construction must be designed to transfer the factored moments and shear and axial forces from the beam to the column. The beam end moment M_u is assumed to be resisted totally by the flange plates. Consequently, the end moment may be resolved into an effective tension-compression couple acting as axial forces in the flange plates. The axial force in the flange plate is given by:

$$P_{ufp} = \frac{M_u}{d_{fp}} \qquad (13.12.1)$$

where P_{ufp} = factored axial force in the flange plate, tensile or compressive, kips
    M_u = factored beam end moment, in.-kips
    d_{fp} = moment arm between the center lines of the flange plates (conservatively taken as the depth of the beam, d_b, in preliminary calculations), in.

Axial force P_u, if present, is normally assumed to be distributed equally to the two flanges of the beam and are additive algebraically to the flange forces obtained from Eq. 13.12.1. Bolts connecting the flange plates to the beam are subjected to single shear by the tensile force in the top flange plate and by the compressive force in the bottom flange plate. The top flange plate is designed as a short tension member. If welds are used to connect the flange plates to the beam flanges, the top plate is not usually welded full length to provide certain ductility in the connection (Fig. 13.11.1b). The unwelded length of the flange plate is about one-half to one times the plate width for FR connection and longer for semi-rigid connection. The axial force in the flange plates is transmitted from the plate to the column flange by fillet or groove welds. It is common practice to assume that the beam shear V_u is transferred entirely through the beam web shear connection. Since, by definition, the angle between the beam and column in an FR moment connection remains unchanged under loading, eccentricity may be neglected entirely in the design of the shear connection.

Some of the limit states to be considered in the design of flange-plated moment connections are:

Shear Plate
    Shear strength of bolts (single shear).
    Bearing strength of bolts on shear plate.
    Bearing strength of bolts on beam web.
    Shear yielding of the plate.
    Shear rupture of the plate.
    Block shear rupture of the plate.
    Strength of welds connecting the plate to the column.

Top Flange Plate
    Shear strength of the bolts.
    Bearing strength of the bolts on the beam flange.
    Bearing strength of the bolts on the flange plate.
    Tension yielding of the flange plate.
    Tension fracture of the flange plate.

Block shear rupture of the beam flange.
Block shear rupture of the flange plate.
Strength of welds connecting the top plate to the column.

Bottom Flange Plate
Column strength of the bottom flange plate.

When bolts are used to connect the flange plates, part of the beam area will be lost because of the bolt holes, and the moment capacity may be reduced. LRFDS Section B10 permits this reduction to be neglected when the fracture strength of the beam tension flange is greater than the yield strength of the tension flange (LRFDS Eq. B10.1). That is:

$$0.75 F_u A_{fn} \geq 0.9 F_y A_{fg} \qquad (13.12.2)$$

where A_{fg} = gross area of beam flange, in.^{2}
       A_{fn} = net area of beam flange, in.^{2}

If the above relation is not satisfied, the member flexural properties must be based on an effective tension flange area A_{fe} of:

$$A_{fe} = \frac{5}{6} \frac{F_u}{F_y} A_{fn} \qquad (13.12.3)$$

and the maximum flexural strength must be based on the elastic section modulus (LRFDS Section B10).

Flange-Plated Moment Connection

EXAMPLE 13.12.1

A W21×57 beam is connected to the flange of W14×90 exterior column, using a top flange plate, a bottom flange plate, and a shear plate. The shear plate is welded to the column flange and bolted to the beam. A frame analysis shows that the connection must transfer a factored load moment of 350 ft-kips and a factored load shear of 52 kips. Assume A992 steel for the structural members and A36 steel for the connecting material. Use E70 electrodes and $^3/_4$-in.-dia. A490-N bolts. Design the moment connection.

Solution

1. Data
    W21×57 beam of A992 steel
    A = 16.7 in.^{2}; Z_x = 129 in.^{3}
    b_{fb} = 6.56 in.; d_b = 21.1 in.
    t_{fb} = 0.650 in.; t_{wb} = 0.405 in.
    T_b = 18$^3/_8$ in.; I_x = 1170 in.^{4} ; g_b = 3$^1/_2$ in.
    $\phi_b M_p$ = 484 ft-kips   (LRFDM Table 5-3)
    $\phi_v V_n$ = 231 kips   (LRFDM Table 5-3)

[continues on next page]

Example 13.12.1 continues ...

W14×90 column of A992 steel

$A = 26.5$ in.²;	$k_c = 1.31$ in. (2 in.)
$b_{fc} = 14.5$ in.;	$d_c = 14.0$ in.
$t_{fc} = 0.71$ in.;	$t_{wc} = 0.44$ in.

Bolts: ¾-in.-dia. A490-N type

$d = \frac{3}{4}$ in.;	$3d = 2\frac{1}{4}$ in.;	$1\frac{1}{2}d = 1\frac{1}{8}$ in.
$d_h = \frac{13}{16}$ in.		

Assume two rows of ¾-in.-dia. A490-N bolts in standard holes to connect the flange plates and one vertical row in shear plate.
Design strength of a bolt in single shear (LRFDM Table 7-10), $B_{dv} = 19.9$ kips
Design bearing strength of ¾-in.-dia. bolt in STD holes (for $L_c > 2d$)

> A36 connected material = 78.3 kips/in.
> A992 connected material = 87.7 kips/in.

2. Check beam flexural strength at holes

$$A_{fg} = b_f t_f = 6.56(0.650) = 4.26 \text{ in.}^2$$

$$A_{fn} = A_{fg} - 2d_e t_f = 4.26 - 2(\tfrac{7}{8})(0.650) = 3.12 \text{ in.}^2$$

$$0.9F_y A_{fg} = 0.9(50)(4.26) = 192 \text{ kips}$$
$$0.75F_u A_{fn} = 0.75(65)(3.12) = 152 \text{ kips}$$

As the fracture strength of the tension flange, $0.75F_u A_{fn}$ (=152 kips) is less than its yield strength, $0.9F_y A_{fg}$ (=192 kips). Consequently the effective tension on flange area (from Eq. 13.12.3) is:

$$A_{fe} = \frac{5}{6}\frac{F_u}{F_y} A_{fn} = \frac{5}{6}\left(\frac{65}{50}\right)3.12 = 3.38 \text{ in.}^2$$

The effective elastic moment of inertia is:

$$I_{xe} = I_x - 2[A_{fg} - A_{fe}] \, \bar{y}_f^2$$

$$= 1170 - 2(4.26 - 3.38)\left(\frac{21.1}{2} - \frac{0.65}{2}\right)^2 = 986 \text{ in.}^4$$

The effective elastic section modulus, $S_{xe} = 986/10.55 = 93.5 \text{ in.}^3$

$$\text{Design flexural strength} = \frac{0.9(93.5)(50)}{12} = 351 \text{ ft-kips}$$

$$> M_{\text{req}} = 350 \text{ ft-kips}$$

Hence, the design flexural strength of the beam at holes is O.K.

3. Design of the shear plate
Number of $\frac{3}{4}$-in.-dia. A490-N bolts required to transfer the beam end shear:

$$n \geq \frac{V_u}{B_{dv}} = \frac{52.0}{19.9} = 2.6 \quad \text{use 3 bolts}$$

Assume pitch of bolts, $s = 3$ in.; end distance, $L_{ev} = 1\frac{1}{2}$ in. ($> 1\frac{1}{4}$ in. minimum required for sheared edges, as per LRFDS Table J3.4), and try PL$\frac{5}{16}\times3\frac{1}{2}\times9$. As the beam web thickness $t_{wb} = 0.405 > t_{pl} = \frac{5}{16}$, and as the beam is of a higher strength material than the shear plate, the bearing strength will be governed by the shear plate.
The clear distance for an interior bolt is:

$$L_{ci} = s - d_h = 2.19 > 2d = 1.5 \text{ in.}$$

So,

$$B_{dbi} = B_{dbo} = 78.3\left(\frac{5}{16}\right) = 24.5 \text{ kips}$$

The clear distance for an end bolt is:

$$L_{ce} = L_{ev} - 0.5d_h = 1.09 < 2d = 1.5 \text{ in.}$$

So, design bearing strength of an end bolt (bottom bolt) is:

$$= 0.75(1.2L_{ce}\,F_{up}\,t) = 0.75(1.2)(1.09)(58)(\tfrac{5}{16}) = 17.8 \text{ kips}$$

Design bearing strength of the shear plate

$$= 2(24.5) + 1(17.8) = 66.8 \text{ kips} > 52.0 \text{ kips} \qquad \text{O.K.}$$

Hence, bearing strength does not control the bolt design.
The shear yielding strength of the plate is:

$$R_{dvy} = 0.9(0.6F_y)A_g = 0.9(0.6)(36)(9)(\tfrac{5}{16}) = 54.7 > 52 \text{ kips} \quad \text{O.K.}$$

The shear rupture strength of the plate is:

$$R_{dvu} = 0.75(0.6F_u)A_n = 0.76(0.6)(58)[9 - 3(\tfrac{7}{8})]\,(\tfrac{5}{16})$$

$$= 52.0 \text{ kips, which equals the required strength of 52.0 kips} \quad \text{O.K.}$$

For the block defining block shear failure of the plate, we have:

$s = 3$ in.; $\qquad L_{eh} = 1\frac{1}{2}$ in.; $\qquad L_{ev} = 1\frac{1}{2}$ in.
$n = 3$; $\qquad\qquad d = \frac{3}{4}$; $\qquad\qquad F_y = 36$ ksi; $\qquad F_u = 58$ ksi

From LRFDM Table 9-3a: $\qquad \phi F_u A_{nt} = 46.2$ kips/in.
From LRFDM Table 9-4a: $\qquad \phi(0.6F_u)\,A_{nv} = 139$ kips/in.
From LRFDM Table 9-4b: $\qquad \phi F_y A_{gt} = 40.5$ kips/in.
From LRFDM Table 9-3b: $\qquad \phi(0.6F_y)\,A_{gv} = 122$ kips/in.

As $\phi(0.6F_u)A_{nv}$ is greater than $\phi F_u A_{nt}$ shear fracture controls the block shear strength.

[*continues on next page*]

Example 13.12.1 continues ...

Block shear strength,

$$R_{dbs} = [\phi(0.6F_u)A_{nv} + \min \{\phi F_y A_{gt}, \phi F_u A_{nt}\}] \, t_{pl}$$
$$= [139.0 + \min \{40.5, \; 46.2\}] \, (^5\!/_{16}) = 56.1 \text{ kips} \qquad \text{O.K.}$$

Design of weld connecting the shear plate to column:
From LRFDS Table J2.4, corresponding to a plate thickness $t_2 = t_{fc} = 0.71$ in. for the thickness of the thicker plate connected, the minimum fillet weld size is $^1\!/_4$ in. Use two $^1\!/_4$-in. fillet welds, 8 in. long (< 9 in.). The shear strength of two $^1\!/_4$-in. ($D = 4$) fillet welds using E70 electrodes is:

$$R_{dw} = 2(1.392)(4)(8) = 89.1 \text{ kips} > 52.0 \text{ kips} \qquad \text{O.K.}$$

4. Design of the top flange plate
Required tensile strength,

$$P_{ufp} = \frac{12M_u}{d_{fp}} \approx \frac{12(350)}{21.1} = 199, \text{ say, } 200 \text{ kips}$$

Note that we conservatively used the depth of the beam, d_b, for the lever arm since the thickness of the flange plates is not known at this stage.
The number of $^3\!/_4$-in.-dia. A490-N bolts in single shear required is:

$$n = \frac{P_{ufp}}{B_{dv}} = \frac{200}{19.9} = 10.0$$

Provide $n = 10$, or two rows of 5 bolts each, on a $3^1\!/_2$-in. gage. Let us consider an $8 \times ^7\!/_8$-in. plate of A36 steel. The width considered is slightly larger than beam flange width ($=6.56$ in.) and less than the width of the column flange ($=14.5$ in.). Also, the thickness of the plate is slightly greater than the beam flange thickness ($=0.650$ in.). Thus, the bolt bearing does not control the design.
Gross area of flange plate $= 8.0(0.875) = 7.00$ in.^{2}
Net area of flange plate $= [8.0 - 2(0.875)](0.875) = 5.47$ in.^{2}
Effective net area of flange plate (LRFDS Section J5.2b) $= \min [5.47, 0.85(7.0)] = 5.47$ in.^{2}
Tension yielding of flange plate,

$$T_{d1} = 0.9 \, F_y A_g = 0.9(36)(7.0) = 227 \text{ kips} > 200 \text{ kips} \qquad \text{O.K.}$$

Tension rupture of flange plate,

$$T_{d2} = 0.75F_u A_n = 0.75(58)(5.47) = 238 > 200 \text{ kips} \qquad \text{O.K.}$$

Use a spacing, $s = 3$ in. $> 3d$, end distance, $L_e = 2$ in. $> 1^1\!/_2 \, d$, gage of $3^1\!/_2$ in., and a side distance of $2^1\!/_4$ in. The first row of bolts is provided at a distance of 2 in. from column face, which allows for $^1\!/_2$-in. setback and

$1\frac{1}{2}$-in. end distance for bolts in the beam flange. Check the block shear rupture strength associated with the tear out of the block between the two rows of holes in the flange plate. Tables 9-3 and 9-4 may be adopted for this purpose by considering the $3\frac{1}{2}$ in. width to be comprised of two $1\frac{3}{4}$-in.-wide blocks. For each of these blocks, we have:

$$s = 3 \text{ in.; } L_{eh} = 1\frac{3}{4} \text{ in.; } \qquad L_{ev} = 2 \text{ in.}$$
$$n = 5; \qquad d = \frac{3}{4} \text{ in.; } \qquad F_y = 36 \text{ ksi; } \qquad F_u = 58 \text{ ksi}$$

From LRFDM Table 9-3a: $\phi F_u A_{nt} = 57.1$ kips/in.
From LRFDM Table 9-4a: $\phi(0.6F_u)A_{nv} = 263$ kips/in.
From LRFDM Table 9-4b: $\phi F_y A_{gt} = 47.3$ kips/in.
From LRFDM Table 9-3b: $\phi(0.6F_y)A_{gv} = 227$ kips/in.

As $\phi(0.6F_u)A_{nv}$ is greater than $\phi F_u A_{nt}$ shear fracture controls the block shear strength.
Block shear strength,

$$T_{dbs} = [\phi(0.6F_u)A_{nv} + \min\{\phi F_y A_{gt}, \phi F_u A_{nt}\}] \, t_{pl}$$
$$= 2[263 + \min\{47.3, 57.1\}] \, (\tfrac{7}{8}) = 543 \text{ kips}$$

$$T_{dbs} = 543 \text{ kips} > 200 \text{ kips} \qquad\qquad\qquad\qquad \text{O.K.}$$

Noting that the fillet welds connecting flange plate to column are transverse welds, from LRFDS Eq. A-J2-1:

$$D = \frac{P_{ufp}}{2(1.392)(1.5)L_w} = \frac{200}{2(1.392)(1.5)(8)} = 5.99 \rightarrow 6$$

Use two $\frac{3}{8}$-in. fillet welds, 8 in. long.

5. **Design of the bottom flange plate**
   Check design compressive strength of the flange plate assuming effective length factor, $K = 0.65$, length, $L = 2$ in.

$$r = 0.289 t_{fp} = 0.289(\tfrac{7}{8}) = 0.253 \text{ in.}$$

$$\frac{KL}{r} = \frac{0.65(2.0)}{0.253} = 5.1$$

From LRFDS Table 3-36, for $KL/r = 5.1$, $F_{dc} = \phi F_{cr} = 30.6$ ksi
So, the design compressive strength of the bottom flange plate is:

$$P_{dc} = F_{dc} A = 30.6(8)(0.875) = 214 \text{ kips} > 200 \text{ kips} \qquad \text{O.K.}$$

The compression flange plate will be identical to the tension flange plate, with ten bolts in two rows of five bolts on a $3\frac{1}{2}$-in. gage and two $\frac{3}{8}$-in. E70 fillet welds 8 in. long to the supporting column flange.

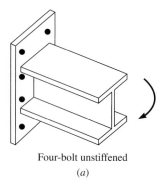

Four-bolt unstiffened

(*a*)

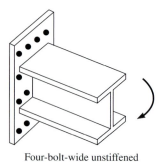

Four-bolt-wide unstiffened

(*b*)

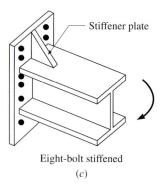

Stiffener plate

Eight-bolt stiffened

(*c*)

Figure 13.13.1: Types of extended end-plate FR moment connections.

13.13   Extended End-Plate FR Moment Connections

13.13.1   General

In ***extended end-plate moment connections,*** a single rectangular plate of a length greater than the beam depth is pre-punched and shop welded to the end of the supported beam, with the plane of the plate perpendicular to the longitudinal axis of the beam (Figs. 13.11.2*b* and 13.13.1). The ***end-plate*** is always shop welded to the web and both flanges of the beam on each side. The assembly is field bolted to the supporting member with high-strength bolts. Generally, the plate extends beyond the tension flange of the beam, and there are two rows of two bolts each, one row on each side of the tension flange. An essential requirement of the design model is that the upper two rows of bolts are equally spaced about the beam tension flange. The plate projection and arrangement of bolts on the compression side of the beam are not critical to the analysis or physical behavior of the connection. This is known as the ***four-bolt unstiffened extended end-plate moment connection*** (Fig. 13.13.1*a*). Its capacity is generally limited by bolt strength such that its economical use is restricted to less than one-half of the available I-shapes. The strength of this connection can be increased by increasing the number of bolts to eight. This could be achieved either by increasing the number of bolts per row to four (Fig. 13.13.1*b*), or the number of rows to four using a stiffened extended end-plate (Fig. 13.13.1*c*). Note however that A490 bolts should not be used in the eight-bolt stiffened configuration.

Extended end-plate moment connections have been investigated by Douty and McGuire [1963]; Nair, Birkemoe, and Munse [1969]; Kennedy, Vinnakota, and Sherbourne [1981]; Murray and Kukreti [1988]; and Murray [1990]. These studies indicate that extended end-plate moment connections with high-strength bolts develop the full plastic moment of the connected members, producing a plastic hinge with adequate rotation capacity within the member. The use of end-plated moment connection results in savings in material and ease of fabrication. It also results in the handling of fewer pieces in the field than other types of bolted moment connections and in the decrease in field erection time and labor costs. Moment end-plate connections are frequently used in gable frame metal buildings. Metal building manufacturers prefer this type of connection for use in beam-to-beam or in beam-to-column connections. Bolted end-plate connections are preferred for their excellent rotational restraint characteristics and because they require no field welding. However, the beam has to be cut and fabricated to the exact length as there is very little scope for longitudinal field adjustment during erection. So, the beam is frequently fabricated short to take care of the column overrun tolerances; shims are furnished to fill any gaps which might result.

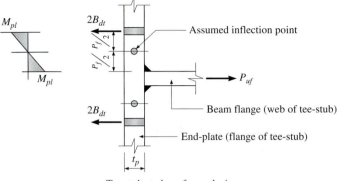

Tee-stub analogy for analysis

(a)

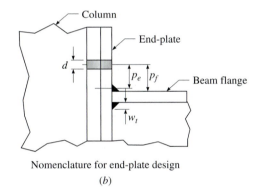

Nomenclature for end-plate design

(b)

Figure 13.13.2: Extended end-plate FR moment connection details.

Sherbourne [1961] and Douty and McGuire [1963] assumed that the end-plate deforms about the beam tension flange and thus develops a ***tee-stub analogy,*** in which the design of the end-plate is controlled by the stresses developed around the tension flange and neighboring bolts. This analogy assumes the flange of the supported beam to be equivalent to the web of the tee-stub, a part of the end-plate to be equivalent to the flange of the tee-stub, and presumes symmetry of bolt geometry about the beam flange (Fig. 13.13.2a). So only the part of the end-plate, symmetrically located about the tension flange of the supported beam, is used in determining the performance and capacity of the connection. Disque [1960] at the 1962 AISC National Engineering Conference discussed end-plate connections from a practical application viewpoint pointing out many of the economies of end-plates and the shop practices which are necessary to achieve the most satisfactory and desired results. Packer and Morris [1977] identified three failure mechanisms in a tee-stub: bolt fracture with no tee-stub yielding, yielding of the tee-stub at the flange followed by bolt fracture, and yielding of the tee-stub at both the flange and bolt line. Designs based on the prying force formulas were found to result in excessively thick plates and large bolt sizes. Krishnamurthy [1978] used two- and three-dimensional finite element analysis to predict end-plate connection behavior. Using a multiple regression analysis, the prediction equations developed compare favorably with test results. Through the use of modification factors, a design moment is found from the moment calculated using flange force and effective bolt distance. This complex relation was determined by a regression analysis of numerous finite element studies. No direct calculation of the prying force is given or even acknowledged.

The design procedure described in LRFDM (and used below) is based on Krishnamurthy [1978], Hendrick and Murray [1984], and Curtis and Murray [1989]. As mentioned earlier in Krishnamurthy's design procedure, prying forces are considered negligible, and the tensile flange force is distributed equally among the four tension bolts. Possible local yielding of the tension flange and tensile area of the web is neglected.

An extended end-plate connection is capable of resisting both the moment and shear at a joint. The beam end moment is transferred from the flanges through the end-plate and into the column flange. This results in tension forces acting on the column at bolt locations in the tension region, and compressive stresses acting over some bearing zone in the compression region. The bolts that are positioned close to the compression flange have been neglected in the analysis. It is considered that these bolts have little to offer towards the moment resistance of the connection and will only be useful to transmit the beam-end shear force. They have therefore been neglected in the moment transfer. The criteria for determining whether the column requires reinforcement depends on the ability of the tension and compression zones of the column flange to develop the necessary design beam moment within acceptable stress and deformation limits [see Vinnakota et al., 1987., Carter, 1999]. For thin end-plates the end-plate moment connection acts as a partially restrained (PR) connection. By increasing the plate thickness, which will result in only a small increase in cost, the connection could be moved into the rigid (i.e., FR) category.

13.13.2   Four-Bolt Unstiffened Extended End-Plate Moment Connection

These connections may be used only in statically loaded applications. Wind, snow, and temperature loads are considered static loads (for their application to buildings in seismic zones see Carter [1999]). When the applied moment is less than the design flexural strength of the beam, the bolts and end-plate may be designed for the applied moment only. The contribution of the beam web to the moment capacity of the beam section is ignored. Hence, the moment at the end of the beam is replaced by a pair of statically equivalent forces P_{uf} in the flanges of the beam given by:

$$P_{uf} = \frac{12M_u}{(d_b - t_{fb})} \qquad (13.13.1)$$

where M_u = beam end moment under factored loads, ft-kips
d_b = depth of beam, in.
t_{fb} = beam flange thickness, in.

Assessing the adequacy of the unstiffened extended end-plate moment connection involves checking the following limit states:

- Strength of bolts in the tension region of the connection.
- Strength of bolts in resisting shear forces in the beam web.
- Strength of the end-plate in bending.
- Strength of the beam-tension-flange to end-plate welds.
- Strength of the beam-web to end-plate welds.

Finally, the necessity for column stiffening should be considered [Carter, 1999].

Design of Bolts Near Tension Flange

The tensile force in the top flange is assumed to be transmitted to the column flange by the top four bolts. Any forces from prying action are neglected in design. So, the limit state of bolt tension results in:

$$4B_{dt} \geq P_{uf} \qquad (13.13.2)$$

where B_{dt} is the design tensile strength of a single bolt. A suitable bolt size is selected using LRFDM Table 7-14 to satisfy Eq. 13.13.2. Pretensioned A325 or A490 high-strength bolts, in diameters not greater than $1\frac{1}{2}$ in., must be used. If it is necessary to use snug-tightened bolts, see Murray et al. [1992]. The four bolt solution is for the tension flange; for those loading cases producing reversal of moment, additional bolts are required at the other flange. In addition, transfer of the beam's reaction through bolt shear must be accounted for in the design of the entire connection. The compression force in the lower flange is assumed to be transmitted by direct bearing on the column.

Design of End-Plate

Several assumptions have been made in the design procedures described in the LRFDM (used in this section) for four-bolt unstiffened extended end-plate connections:

1. The end-plate material should preferably be ASTM A36.
2. The end-plate must be designed to resist the bending moment and shear that develop in it, while transferring the flange force P_{uf} from the tension flange to the adjacent bolts. The thickness needed for end-plate is found

by assuming that there is a point of inflection half-way between the toe of the beam-flange-to-end-plate weld and the bolt (Fig. 13.13.2).

3. The end-plate width is 1.15 times the beam flange width, rounded up to the nearest $\frac{1}{2}$ in. The end-plate width, which is considered effective in resisting the applied moment, is not greater than the beam flange width b_f plus 1 in.

4. The recommended minimum distance from the face of the beam flange to the nearest bolt centerline is the bolt diameter, d, plus $\frac{1}{2}$ in. Although the smallest possible distance will generally result in the most economical connection, many fabricators prefer to use a standard dimension, usually 2 in., which is adequate for all bolt diameters.

5. The high-strength bolts in an end-plate moment connection are located a distance, p_f, above and approximately the same distance below the beam tension flange. The force applied to each of these four bolts by the tensile flange force P_{uf} can then be considered equal.

6. The gage of the tension bolts (horizontal distance between vertical bolt lines) should not exceed the beam tension flange width.

7. The end-plate is extended outside the beam compression flange by an amount equal to the end-plate thickness t_p.

The effective critical moment in the end-plate is given by:

$$M_{eu} = \frac{1}{4}\,\alpha_m P_{uf}\, p_e \tag{13.13.3}$$

with

$$\alpha_m = C_a C_b \left(\frac{A_f}{A_w}\right)^{\frac{1}{3}}\left(\frac{p_e}{d}\right)^{\frac{1}{4}} \tag{13.13.4a}$$

$$C_b = \left(\frac{b_f}{b_{pe}}\right)^{\frac{1}{2}} \tag{13.13.4b}$$

$$b_{pe} = \min\,[b_p,\ b_f + 1 \text{ in.}] \tag{13.13.4c}$$

$$p_e = p_f - \tfrac{1}{4}\,d - w_t \tag{13.13.4d}$$

where M_{eu} = effective critical moment in end-plate, in.-kips
P_{uf} = factored beam flange force, kips
p_e = effective pitch, in.
p_f = distance from the centerline of the bolt to nearer surface of the beam tension flange, in.
d = nominal bolt diameter, in.
w_t = fillet weld throat size or size of reinforcement for groove weld, in.
  = $0.707w$ for SMAW welds
C_a = coefficient depending on the yield stresses of the beam and end-plate material, and type of bolt

b_f = beam flange width, in.
b_p = end-plate width, in.
b_{pe} = effective end-plate width, in.
A_f = area of beam tension flange, in.^{2}
A_w = area of beam web, clear of flanges, in.^{2}

For beams with F_y = 50 ksi, values of C_a are tabulated for two end-plate material grades (F_y = 36 and 50 ksi), for A325 and A490 bolts in LRFDM Table 12.1. Thus, for example, for A325 bolts and end-plate material with F_y = 36 ksi, the coefficient C_a is 1.45 with beam material having yield stress of 50 ksi. Also, values of A_f/A_w for the W-shapes listed in LRFDM Table 1-1 are given in LRFDM Table 12-2. These values vary between 0.40 and 2.10.

A number of designers use the rule that the bolt pitch, p_f, be equal to the bolt diameter, d, plus $\frac{1}{2}$ in. This provides the minimum bolt pitch dimension consistent with reasonable entering and tightening clearances. However, many fabricators prefer to use a standard dimension such as 2 in. This is done to avoid errors in the drafting room and shop, to save cost of multiple drilling templates, and to allow greater driving clearances in the field. However, the student designer and fabricator should realize that the calculated end-plate thickness is dependent to a very great degree on the bolt pitch, p_f. A small increase in bolt pitch can result in a surprising increase in the thickness requirement for the end plate. Observe, from Fig. 13.13.2a, that the plate bending spans are only of the order of two to four times the bolt diameter, the weld size, and the plate thickness, indicating that each and every one of these minor details would play a major role in the connection behavior.

The required end-plate thickness t_p, corresponding to the limit state of the formation of a plastic hinge mechanism, is:

$$t_p \geq \sqrt{\frac{4M_{eu}}{\phi_b F_{yp} b_{pe}}} \tag{13.13.5}$$

where M_{eu} = effective critical moment in end-plate (Eq.13.3.4) in.-kips
b_{pe} = effective end-plate width, in.
t_p = thickness of the end-plate, in.
F_{yp} = yield stress of the end-plate material, ksi
ϕ_b = resistance factor = 0.90

The end-plate selected must be checked for the limit state of shear, using the relation:

$$V_d = \min [V_{dg}, V_{dn}] \geq V_u \tag{13.13.6}$$

with

$$V_u = \frac{1}{2} P_{uf} \tag{13.13.7}$$

$$V_{dg} = 0.75(0.6F_{yp}) b_p t_p \tag{13.13.8}$$

$$V_{dn} = 0.75(0.6F_{up})(b_p - 2d_e) t_p \tag{13.13.9}$$

where
V_d = design shear strength of the end-plate, kips
V_{dg} = design shear strength on the gross area of the end-plate, kips
V_{dn} = design shear strength on the net area of the end-plate, kips
V_u = shear in the end-plate, kips
d_e = effective diameter of the bolt hole, in.

Design of Welds

The following remarks should be noted in the design of welds connecting the beam to the end-plate:

1. The welds connecting the end-plate to the beam tension flange are either fillet welds completely surrounding the beam flange or CJP groove welds. The size of fillet welds at the tension flange should be large enough to develop the force P_{uf} resulting from factored moment. The maximum size of these fillet welds, for economical reasons, is arbitrarily limited to $\frac{1}{2}$ in. If greater capacity is required, a CJP groove weld, with reinforcement as per AWS, is recommended.

2. Welds at compression flange may be minimum size fillet welds as per LRFDS Table J2.4. However, some fabricators prefer to provide compression flange weld size the same as the tension flange weld, to ensure equal strength.

3. The welds to the beam web are usually fillet welds designed to match the web thickness. They should be adequate to resist the beam reaction. The beam-web-to-end-plate welds in the vicinity of tension bolts (a distance $p_f + 2d$ from the inner face of the tension flange) should be designed to develop 60 percent of the minimum specified yield strength of the beam web. This is recommended even if the full design flexural strength of the beam is not required for frame strength. For fillet welds using E70 electrodes, this requires:

$$2(1.392D) \geq 0.6F_{yb}t_{wb} \tag{13.13.10}$$

4. Only the web-to-end-plate weld over a distance L_v is considered effective in resisting the beam end shear, where:

$$L_v = \min(L_1, L_2) \tag{13.13.11a}$$

where

$$L_1 = \frac{d_b}{2} - t_{fb}; \quad L_2 = d_b - 2t_{fb} - (p_f + 2d) \tag{13.13.11b}$$

Here,
L_1 = distance between the mid-depth of the beam to the inside face of the beam compression flange, in.
L_2 = distance between the inner row of tension bolts plus $2d$ and the inside face of the beam compression flange, in.
d = diameter of the bolt, in.
d_b = depth of the beam, in.

End-Plate Moment Connection

A W21×57 beam is connected to the flange of a W14×90 column, using a four-bolt unstiffened extended end-plate FR moment connection. A frame analysis shows that the connection must transfer a factored load moment of 350 ft-kips and a factored load shear of 52 kips. Assume A992 steel for the structural members and A36 steel for the connecting material. Use E70 electrodes and A490-SC bolts (Class A surfaces). Design the connection.

Solution

1. Data

W21×57 beam of A992 steel

$$
\begin{aligned}
A &= 16.7 \text{ in.}^2; & Z_x &= 129 \text{ in.}^3 \\
b_{fb} &= 6.56 \text{ in.}; & d_b &= 21.1 \text{ in.} \\
t_{fb} &= 0.650 \text{ in.}; & t_{wb} &= 0.405 \text{ in.} \\
\phi_b M_{px} &= 484 \text{ ft-kips (LRFDM Table 5-3)} \\
\phi_v V_n &= 231 \text{ kips (LRFDM Table 5-3)}
\end{aligned}
$$

W14×90 column of A992 steel

$$
\begin{aligned}
A &= 26.5 \text{ in.}^2; & g_c &= 5\tfrac{1}{2} \text{ in.} \\
b_{fc} &= 14.5 \text{ in.}; & d_c &= 14.0 \text{ in.} \\
t_{fc} &= 0.710 \text{ in.}; & t_{wc} &= 0.440 \text{ in.}
\end{aligned}
$$

Also

$$
R_u = 52 \text{ kips}; \qquad M_u = 350 \text{ ft-kips}
$$

2. Beam shear and flexural strengths

$$
\begin{aligned}
\text{As} \quad R_u &= 52 \ < \ \phi_v V_n = 231 \text{ kips, and} & \text{O.K} \\
M_u &= 350 \ < \ \phi_b M_{px} = 484 \text{ ft-kips} & \text{O.K}
\end{aligned}
$$

the beam design shear and flexural strengths are O.K.

3. Bolt selection

The tensile flange force, $P_{uf} = \dfrac{12 M_u}{d_b - t_{fb}} = \dfrac{12(350)}{(21.1 - 0.650)} = 205$ kips

Assuming that this force is resisted by four fully pretensioned high-strength bolts at tension flange, the required tensile strength of each bolt is:

$$
B_{ut} = \frac{P_{uf}}{n} = \frac{205}{4} = 51.4 \text{ kips}
$$

It is seen from LRFDM Table 7-14 that a $\tfrac{7}{8}$-in.-dia. A490 bolt has a design tensile strength

$$
B_{dt} = 51.0 \text{ kips} \approx B_{ut} = 51.4 \text{ kips} \qquad\qquad \text{O.K.}
$$

[continues on next page]

Example 13.13.1 continues ...

From LRFDM Table 7-15, the design resistance to shear at service loads using factored loads, for a $\frac{7}{8}$-in.-dia. A490-SC bolt, is 18.3 kips. Thus, the number of bolts required to resist factored shear is:

$$n = \frac{R_u}{B_{dsf}} = \frac{52.0}{18.3} = 2.84, \text{ say, 3 bolts}$$

Provide six $\frac{7}{8}$-in.-dia. A490-SC bolts in standard holes, namely, four bolts at tension flange and two bolts at compression flange.   (Ans.)

$$d = \frac{7}{8} \text{ in.;} \quad 1\frac{1}{2}\,d = 1\frac{5}{16} \text{ in.;} \quad 3d = 2\frac{5}{8} \text{ in.}$$

4. **Design of the end-plate**
    Try an end-plate with:
    $L_e = 1\frac{1}{2}$ in.; $g = g_c = 5\frac{1}{2}$ in.; $p_f = d + \frac{1}{2} = 1\frac{3}{8}$ in.
    $b_p = g + 2L_e = 5.5 + 2(1.5) = 8.50$ in.
    $b_{pe} = \min[b_p,\ b_{fb} + 1 \text{ in.}] = \min[8.50,\ 6.56 + 1] = 7.56$ in.
    So, the effective width is less than the full width of the end-plate.

    The effective critical moment in the end-plate is: $M_{eu} = \alpha_m \dfrac{P_{uf}p_e}{4}$

    with

    $$\alpha_m = C_a C_b \left(\frac{A_f}{A}\right)^{\frac{1}{3}} \left(\frac{p_e}{d}\right)^{\frac{1}{4}}$$

    From LRFDM Table 12-1, $C_a = 1.48$, for A490 bolts, A36 end-plate material and Grade 50 beam material:

    $$C_b = \sqrt{\frac{b_f}{b_{pe}}} = \sqrt{\frac{6.56}{7.56}} = 0.932$$

    $$p_e = p_f - \frac{d}{4} - w_t \text{ (assuming } \frac{5}{16} \text{ fillet weld)}$$

    $$= 1\frac{3}{8} - (\tfrac{1}{4})(\tfrac{7}{8}) - 0.707(\tfrac{5}{16}) = 0.935 \text{ in.}$$

    $$\frac{A_f}{A_w} = 0.532 \quad \text{(from LRFDM Table 12-2 for a W21×57)}$$

    $$\alpha_m = \left[1.48(0.932)(0.532)^{\frac{1}{3}}\right]\left(\frac{0.935}{0.875}\right)^{\frac{1}{4}} = 1.14$$

    $$M_{eu} = 1.14(205)\left(\frac{0.935}{4}\right) = 54.6 \text{ in.-kips}$$

    End-plate thickness, $t_p \geq \sqrt{\dfrac{4M_{eu}}{\phi F_{yp}b_{pe}}} = \sqrt{\dfrac{4(54.6)}{0.9(36)(7.56)}} = 0.944$ in.

    Try a 1 in. × $8\frac{1}{2}$ in. end-plate of A36 steel.

Shear yielding of the end-plate, from Eq. 13.13.8:

$$V_{dg} = 0.9(0.6)(36)(8.5)(1.0) = 165 \text{ kips} > 103 \text{ kips} \qquad \text{O.K.}$$

The shear fracture of the end-plate, from Eq. 13.13.9:

$$V_{dn} = 0.75(0.6)(58)(8.5 - 2 \times 1.0)(1.0)$$
$$= 170 \text{ kips} > 103 \text{ kips} \qquad \text{O.K.}$$

So, provide a 1 in. $\times$ $8\frac{1}{2}$ in. end-plate of A36 steel.   (Ans.)

5. Weld design
   From LRFDS Table J2.4, the minimum size of weld to connect a 0.65-in.-thick beam flange to a 1-in.-thick end-plate is $\frac{5}{16}$ in.
   Length of fillet weld connecting beam tension flange to end-plate,

   $$L_w = 2(b_{fb} + t_{fb}) - t_{wb} = 2(6.56 + 0.65) - 0.405 = 14.0 \text{ in.}$$

   Required fillet weld size, noting that it is a transverse weld,

   $$D \geq \frac{P_{uf}}{1.392(1.5L_w)} = \frac{205}{1.392(1.5)(14.0)} = 7.01 \rightarrow 7 \text{ sixteenths}$$

   Use $\frac{7}{16}$-in. fillet welds at beam tension flange. Welds at compression flange may be $\frac{5}{16}$-in. fillet welds (minimum size from LRFDS Table J2.4).   (Ans.)
   The weld size required, to develop web flexural strength near the tension bolts using E70 electrodes, is:

   $$D \geq \frac{(60\%)(F_{yb}\,t_{wb})}{2(1.392)} = \frac{0.6(50)(0.405)}{2(1.392)} = 4.36 \rightarrow 5 \text{ sixteenths}$$

   Use $\frac{5}{16}$-in. fillet welds on both sides of the beam web.
   The beam end reaction R_u is resisted by the web welds over an effective length, L_v, calculated as follows:

   $$L_1 = \frac{d_b}{2} - t_{fb} = \frac{21.1}{2} - 0.650 = 9.90 \text{ in.}$$

   $$L_2 = 21.1 - 2(0.650) - 1.375 - 2(0.875) = 16.7 \text{ in.}$$

   $$L_v = \min(L_1, L_2) = \min(9.90, 16.7) \approx 10 \text{ in.}$$

   The design shear strength of the E70 electrode weld is:

   $$R_{dw} = 1.392D(2L_v) = 1.392(5)(2)(10.0) = 139 > R_u = 52 \text{ kips O.K.}$$

   Use $\frac{5}{16}$-in. fillet weld on both sides of the beam web below the tension bolt region.   (Ans.)
   Note:
   If the fabricator decides to use a standard dimension of 2 in. for p_f, instead of the minimum value of $1\frac{3}{8}$ in. used in the calculations above, by

[*continues on next page*]

Example 13.13.1 continues ...

repeating the calculations we observe that the effective critical moment in the end-plate increases to 108 in.-kips and that the plate thickness must be increased to $1\frac{1}{4}$ in.

From the description of beam-to-column connections given in Sections 13.11, 12, and 13, it is evident that the beam flanges or flange connection plates apply concentrated forces to the column and cause stresses in the flanges and web of the column. It is the transfer of these essentially concentrated forces to the column flange, for which provision is made in LRFDS Chapter K, that forms the subject of this section. Moment connections are called **double-concentrated forces** because one tensile flange force and one compressive flange force form a couple to act on the same side of the column (Fig. 13.14.1*a*). When opposing connected beams coincide, a **pair of double-concentrated forces** results, as shown in Fig. 13.14.1*b*.

The web of a column within the boundary of the column flanges and the tensile and compressive concentrated forces imposed by moment connections is called the **panel zone** (Fig. 13.14.1*c*). If the moments in two beams, connected to opposite flanges of an interior column, differ significantly in magnitude (or when a one-sided beam-to-column connection is encountered), large shear forces may develop in the column web within the panel zone. The panel zone undergoes both shear and moment deformations that are generally responsible for a major portion (up to 50 percent) of the total joint rotation of tier buildings under lateral loads even when stresses remain in the elastic domain.

The basic requirements that must be satisfied by any connection are related to its strength, stability, and deformations. On the basis of these three requirements the provisions for restrained members subjected to concentrated flange forces, given in Chapter K of the LRFD Specification, can be categorized as:

1. Strength requirement: Section K1.3. Web local yielding
2. Stability requirements: Section K1.4. Web crippling
                               Section K1.6. Web compression buckling
3. Deformation requirement: Section K1.2. Flange local bending

In addition, the column web safety is checked under shear forces that develop in the panel zone (Section K1.7).

For information on web local yielding, web crippling, web compression buckling, flange local bending, panel zone shear capacity, and column stiffening at moment connections, refer to Section W13.1 on the website http:// www.mhhe.com/Vinnakota.

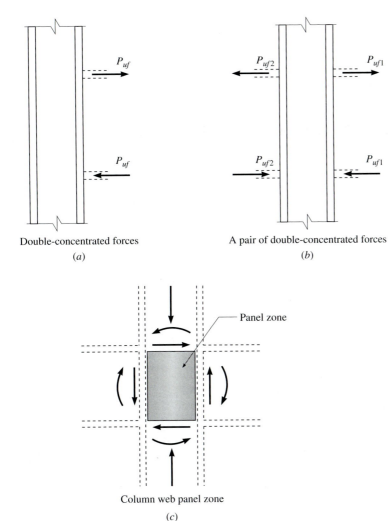

Double-concentrated forces

(a)

A pair of double-concentrated forces

(b)

Panel zone

Column web panel zone

(c)

Figure 13.14.1: Beam-to-column moment connection terminology.

References

13.1 ASCE [1971]: *Plastic Design in Steel: A Guide and Commentary,* Joint Committee of Welding Research Council and the American Society of Civil Engineers, ASCE Manual No. 41, NY.

13.2 Astaneh, A., Call, S. M., and McMullin, K. M. [1989]: "Design of Single-Plate Shear Connections," *Engineering Journal,* AISC, Chicago, vol. 26, no. 1, pp. 21–32.

13.3 Brockenbrough, R. L. [1987]: "Design Loads for Seated-Beam Connections," Proceedings National Engineering Conference, AISC, Chicago, IL, April, pp. 12–1 to 12–16.

13.4   Carter, C. J. [1999]: *Stiffening of Wide-Flange Columns at Moment Connections: Wind and Seismic Applications,* Steel Design Guide Series, no. 13, AISC, Chicago, IL.

13.5   Carter, C. J., Thornton, W. A., and Murray, T. M. [1997]: "The Behavior and Load Carrying Capacity of Unstiffened Seated Beam Connections," *Engineering Journal,* AISC, Chicago, vol. 34, no. 4, pp. 151–156.

13.6   Cheng, J. J. R., and Yura, J. A. [1986]: "Local Web Buckling of Coped Beams," *Journal of Structural Engineering,* ASCE, vol. 112, no. 10, October, pp. 2314–2331.

13.7   Cheng, J. J. R., Yura, J. A., and Johnson, C. P. [1988]: "Lateral Buckling of Coped Steel Beams," *Journal of Structural Engineering,* ASCE, vol. 114, no. 1, January, pp. 1–15.

13.8   Curtis, L. E., and Murray, T. M. [1989]: "Column Flange Strength at Moment End-Plate Connections," *Engineering Journal,* AISC, vol. 26, no. 2, pp. 41–50.

13.9   Disque, R. O. [1960]: "End-Plate Connections," *Proceedings National Engineering Conference,* AISC, pp. 30–37.

13.10 Douty, R. J., and McGuire, W. [1963]: "Research on Bolted Moment Connections—A Progress Report," *Proceedings of National Engineering Conference,* AISC, pp. 48–55.

13.11 Ellifritt, D. S., and Sputo, T. E. [1999]: "Design Criteria for Stiffened Seated Connections to Column Webs," *Engineering Journal,* AISC, Chicago, vol. 36, no. 4, pp. 160–167.

13.12 Garret, J. H., Jr., and Brockenbrough, R. L. [1986]: "Design Loads for Seated-Beam Connections in LRFD," *Engineering Journal,* AISC, Chicago, vol. 23, no. 2, pp. 84–88.

13.13 Goverdhan, A. V. [1984]: "A Collection of Experimental Moment Rotation Curves and Evaluation of Prediction Equations for Semi-Rigid Connections," Ph.D. Thesis, Vanderbilt University, Nashville, TN.

13.14 Gupta, A. K. [1984]: "Buckling of Coped Steel Beams," *Journal of Structural Engineering,* ASCE, vol. 110, no. 9, September, pp. 1977–1987.

13.15 Hendrick, R. A., and Murray, T. M. [1984]: "Column Web Compression Strength at End-Plate Connections," *Engineering Journal,* AISC, vol. 21, no. 3, pp. 161–169.

13.16 Hormby, D. E., Richard, R. M., and Kriegh, J. D. [1984]: "Single-Plate Framing Connections with Grade 50 Steel and Composite Construction," *Engineering Journal,* AISC, vol. 21, no. 3, pp. 125–138.

13.17 Kennedy, D. J. L. [1969]: "Moment-Rotation Characteristics of Shear Connections," *Engineering Journal,* AISC, Chicago, vol. 6, no. 4, pp. 105–115.

13.18 Kennedy, N. A., Vinnakota, S., and Sherbourne, A. N. [1981]: "The Split-Tee Analogy in Bolted Splices and Beam-Column Connections," in *Joints in Structural Steelwork,* Proceedings of the International

Conference on Joints in Steelwork, Middlesbrough, Cleveland, UK, Pentach Press, pp. 138–157.

13.19 Kishi, N., and Chen, W. F. [1986]: "Data Base of Steel Beam-to-Column Connections," CE-STR-86-26, Purdue University, School of Engineering, West Lafayette, IN (in two volumes).

13.20 Krishnamurthy, N. [1978]: "A Fresh Look at Bolted End-Plate Behavior and Design," *Engineering Journal,* AISC, vol. 15, no. 2, pp. 39–49.

13.21 Murray, T. M., Ed. [1990]: Extended End-Plate Moment Connections, Publication No. D804, AISC, Chicago, IL.

13.22 Murray, T. M., and Kukreti, A. [1988]: "Design of Eight-Bolt Stiffened Moment End-Plates," *Engineering Journal,* AISC, vol. 25, no. 2, pp. 45–52.

13.23 Murray, T. M., Kline, D. P., and Rojani, K. B. [1992]: "Use of Snug-Tightened Bolts in End-Plate Connections," *Connections in Steel Structures II,* R. Bjorhovde, A. Colson, G. Haaijer, and J. W. B. Stark, Editors, AISC, Chicago, IL.

13.24 Nair, R. S., Birkmoe, P. C. and Munse, W. H. [1969]: "Behavior of Bolts in Tee-Connections Subject to Prying Action," Structural Research Series, University of Illinois, Urbana, Illinois, 353.

13.25 Packer, J. A., and Morris, L. J. [1977]: "A Limit State Design Method for the Tension Region of Bolted Beam-to-Column Connections," *The Structural Engineer,* London, vol. 55, pp. 446–458.

13.26 Pask, J. W. [1992]: "Simple Beam-to-Column Connections," in *Connections in Steel Structures II: Behavior, Strength and Design,* AISC, Chicago.

13.27 Richard, R. M., Gillett, P. E., Kriegh, J. D., and Lewis, B. A. [1990]: "The Analysis and Design of Single Plate Framing Connections," *Engineering Journal,* AISC, vol. 17, no. 2, pp. 38–51.

13.28 Roeder, C. W. [1987]: "Results of Experiments on Seated-Beam Connections," Proceedings National Engineering Conference, AISC, Chicago, IL, pp. 43–1 to 43–12.

13.29 Roeder, C. W., and Dailey, R. W. [1989]: "The Results of Experiments on Seated Beam Connections," *Engineering Journal,* vol. 26, no. 3, AISC, Chicago, pp. 90–95.

13.30 Sherbourne, A. N. [1961]: "Bolted Beam-to-Column Connections," *The Structural Engineer,* London.

13.31 Sputo, T., and Ellifritt, D. [1991]: "Proposed Design Criteria for Stiffened Seated Connections to Column Webs," National Steel Construction Conference Proceedings, AISC, Chicago, pp. 8.1–8.26.

13.32 Thornton, W. A. [1992]: "Designing for Cost Efficient Fabrication," *Modern Steel Construction,* AISC, Chicago, IL, vol. 25, no. 2, February, pp. 12–20.

13.33 Vinnakota, S., Mallare, M. P., Jr., and Vinnakota, M. R. [1987]: "Load and Resistance Factor Design of Beam-to-Column Flange Connections for Restrained Members: Design Aids," *Journal of Structural Engineering Practice,* vol. 4, no. 182, pp. 19–66.

13.34 Wilson, W. M. and Moore, H. F. [1917]: "Tests to Determine the Rigidity of Riveted Joints of Steel Structures," Engineering Experiment Station, Bulletin 104, University of Illinois, Urbana.

PROBLEMS

Simple Shear Connections

P13.1. A W12×45 beam and a W18×60 beam frame on opposite sides of the web of a W21×68 girder. The top flanges of the beams and the girder are at the same elevation. The W12 beam spans 14 ft and supports dead and live loads of 2.4 and 4.2 klf, respectively. The W18 spans 16 ft and supports dead and live loads of 2.4 and 6.4 klf, respectively. Design a simple shear connection using $\frac{7}{8}$-in.-dia. A325-N bolts and/or E70 electrodes.

P13.2. A W16×45 floor beam is to be connected to the flange of a W14×74 column. The beam reaction under factored loads is 69 kips. Use A325-N high-strength bolts and E70 electrodes. Assume A36 steel for connection material and design a:

    a. Single-plate shear connection

    b. Shear end-plate connection.

P13.3. Repeat Problem 13.2, if the floor beam is to be connected instead to the web of a W21×62 spandrel girder. The tops of the beam and the girder are at the same elevation.

P13.4. Repeat Problem 13.2, if the floor beam is to be connected instead to the web of a W24×68 interior girder. The tops of the beam and the girder are at the same elevation. Assume that there is a similar floor beam to be connected on the girder web.

P13.5. Design unstiffened seat angles to support two W14×48 floor beams, one on each side of the web of a W12×58 column. Under factored loads each beam reaction is 62 kips. Use A36 steel for connection material.

    a. Bolted connection using $\frac{3}{4}$-in.-dia. A325-X bolts.

    b. Welded connection using E70 electrodes

P13.6. A W24×76 beam spanning 40 ft and supporting a factored uniform load of 12.4 klf (including its own weight) is connected at each end to the flange of a W14×68 column using double-angle framing connections. Use A36 steel connection material, E70 electrodes, and $\frac{7}{8}$-in.-dia. A490-N bolts and design the following:

    a. All-bolted connection

    b. Bolted/welded connection

    c. Welded/bolted connection

    d. All-welded connection

P13.7. Repeat Problem P13.6 if the connection is to the web of the column using:

    a. Bolted stiffened seated connection

    b. Welded stiffened seated connection

Moment Connections

P13.8. Design an extended end-plate FR moment connection for a W24×76 beam attached to the flange of a W14×90 column, for a moment of 480 ft-kips and a shear of 100 kips under factored loads. Use A490-X bolts in a bearing type connection.

P13.9. It is necessary to increase the area now available in an office mezzanine floor which is supported by two rows of columns, the columns 30 ft apart in the row. The additional area is obtained by cantilevering the floor 12 ft beyond one row of columns. The loads on the new floor are 50 psf DL and 80 psf LL. Design a fully restrained moment connection that will attach a W16 beam to the flange of each W14×109 column.

P13.10. A W18×60 beam, spanning between two W12× 120 columns, is part of the steel frame for a tall building. The columns are 24 ft on center. The beam ends are subjected to the following shears and moments from gravity and horizontal components of factored loads: gravity-load moment, 210 ft-kips; wind moment, ±180 ft-kips; gravity-load shear, 60 kips; and wind shear, ±16 kips. Design a four-bolted FR moment connection at the end.

Appendix to Chapter 5

··

A5.1 Introduction to Second-Order Moments
(*P*δ Effect and *P*Δ Effect)

A5.1.1 *P*δ Effect (Moment Magnification Factor B_1)

Let us consider three identical prismatic members AB of length L and bending rigidity EI which are simply supported at their ends as shown in Fig. A5.1.1. The member shown in Fig. A5.1.1*a* is subjected to equal and opposite end moments M^o, while the member in Fig. A5.1.1*b* is subjected, in addition to the end moments M^o, to axial compressive forces P acting at ends A and B. Finally, the member shown in Fig. A5.1.1*c* is subjected to axial tensile loads T acting at ends A and B, as well as the end moments M^o. In these figures, the left support is taken as the origin of the coordinate system. Point D represents the section located at distance z from the origin, and point C is the section at midspan.

By considering the equilibrium of the moments of all the forces acting on member AB, taken about either support, we find that the reactions R_B and R_A are zero for all three members under study. By considering moment equilibrium about point D, of all the forces acting on segment AD of length z, we obtain for all three members

$$\Sigma(M)_D = 0 \qquad M^o - M_z = 0 \rightarrow M_z = M^o \qquad \text{(A5.1.1)}$$

The bending moment diagram, valid for all three members, is shown in Fig. A5.1.1*d*. Note that we have considered the equilibrium of segment AD in its undeformed (or undeflected) state. Such calculations are known as **first-order calculations.** The moments calculated under this assumption are known as **first-order moments,** and the corresponding deflections are known as **first-order deflections.**

We next study the equilibrium of the three members shown in Figs. A5.1.1*a*, *b*, and *c*, in their deformed state as shown in detail in Figs. A5.1.2, A5.1.3, and A5.1.4, respectively.

The deflected shape of the beam under pure moment given in Fig. A5.1.2*a* is shown in Fig. A5.1.2*b*. Due to symmetry, the maximum deflection δ^o occurs at the midspan section C. The deflection at section D, located at a

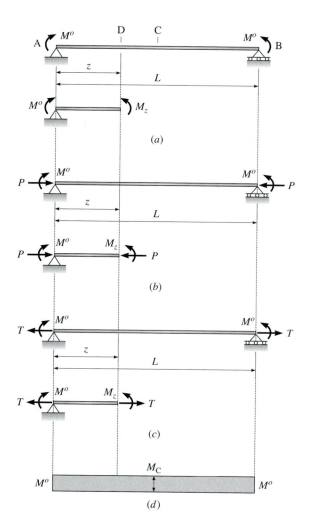

Figure A5.1.1: First-order moments in members.

distance z from A, is denoted as u^o. Equilibrium of moments about point D, of all the forces acting on segment AD in its deflected position (Fig. A5.1.2c), results in:

$$\Sigma(M)_D = 0 \qquad M^o - M_z^* = 0 \rightarrow M_z^* = M^o \qquad \text{(A5.1.2)}$$

The superscript * is added to indicate that the equilibrium is written for the deformed state of the member. The resulting bending moment diagram is shown in Fig. A5.1.2d.

The deflected shape of the beam-column of Fig. A5.1.3a is given in Fig. A5.1.3b. Here, δ_c^* is the maximum deflection which occurs at the midspan section due to symmetry, and u_c^* is the deflection at D located a distance z from the support A. The subscript c indicates that these values are calculated by taking into consideration the axial compressive loads P. The forces acting on

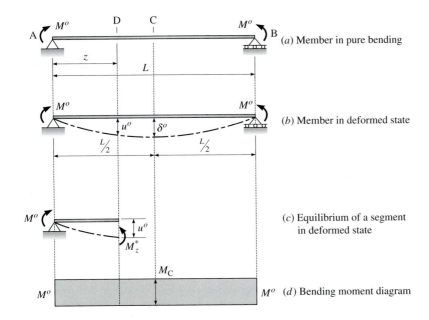

(a) Member in pure bending

(b) Member in deformed state

(c) Equilibrium of a segment in deformed state

(d) Bending moment diagram

Figure A5.1.2: Member under pure bending.

segment AD in its deformed state are indicated in Fig. A5.1.3c. Equilibrium of moments about point D, caused by the forces acting on segment AD in its deformed state, gives:

$$\Sigma(M)_D = 0 \qquad M^o - M_z^* + Pu_c^* = 0 \quad \rightarrow \quad M_z^* = M^o + Pu_c^*$$
$$(A5.1.3)$$

The additional bending moment Pu_c^* caused by the interaction of the axial load with the deflection is called the **secondary moment.** From Fig. A5.1.3d it can be seen that this extra bending moment will vary from end to end of the member and will have its maximum value ($= P\delta_c^*$) at the center. The total bending moment at any point is M_z^*. This **second-order bending moment** is found by adding the primary moment and secondary moment at that point. Its variation over the length of the beam-column is shown in Fig. A5.1.3e.

The deflected shape of the member under combined bending and tension shown in Fig. A5.1.4a is given in Fig. A5.1.4b. Here, δ_t^* is the maximum deflection which occurs at the midspan section due to symmetry, and u_t^* is the deflection at point D located at a distance z from the support A. The forces acting on segment AD in its deflected position are indicated in Fig. A5.1.4c. Equilibrium of the moments about point D, due to the forces acting on segment AD in its deformed state, results in:

$$\Sigma(M)_D = 0 \quad \rightarrow \quad M^o - M_z^* - Tu_t^* = 0$$

$$M_z^* = M^o - Tu_t^* \qquad (A5.1.4)$$

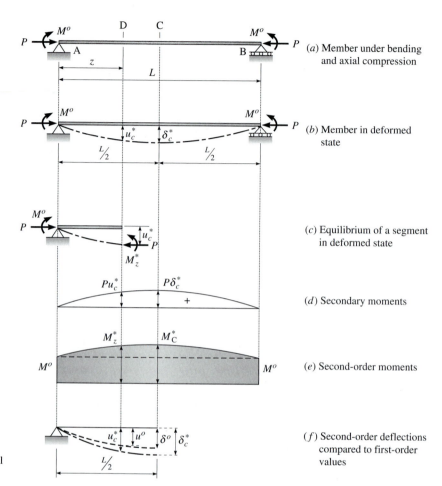

Figure A5.1.3: Member under bending and axial compression.

The secondary moment Tu_t^*, caused by the interaction of the axial tensile load with the deflection, varies from end to end of the member (Fig. A5.1.4d) and has its maximum value $T\delta_t^*$ at the center of the member. The total bending moment (second-order bending moment) M_z^* is obtained by subtracting the secondary bending moment from the primary bending moment. Its variation is shown in Fig. A5.1.4e. From Eqs. A5.1.1, A5.1.2, A5.1.3, and A5.1.4 observe that the effect of a compressive axial load is to magnify the first-order moments, while the effect of a tension load is to decrease the first-order moments.

Next, we will determine the central deflections δ^o, δ_c^* and δ_t^* for the beam, beam-column, and member under combined bending and tension considered earlier, using the second Moment-Area Theorem. (For details, see any standard textbook on structural analysis, [Leet and Uang, 2002].)

Let us consider the elastic curve of a simply supported member AB under symmetric loading shown in Fig. A5.1.5a. The supports at A and B are not

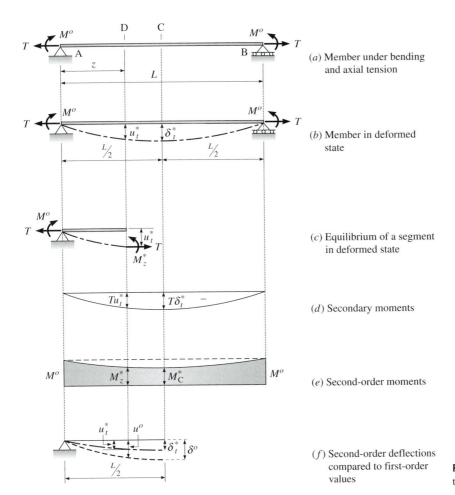

(a) Member under bending and axial tension

(b) Member in deformed state

(c) Equilibrium of a segment in deformed state

(d) Secondary moments

(e) Second-order moments

(f) Second-order deflections compared to first-order values

Figure A5.1.4: Member under bending and axial tension.

shown for clarity. Moment-Area Theorem 2 states that the deviation of any point A relative to the tangent extended from another point C, in a direction perpendicular to the original position of the member, is equal to the moment of the area under the M/EI diagram between points A and C and computed about point A.

The bending moment diagrams for the three members under consideration are reproduced in Figs. A5.1.5b, c, and d. These are identical to the bending moment diagrams shown in A5.1.2d, A5.1.3e, and A5.1.4e, respectively. As EI of each member is constant, the BMDs also represent the shape of the M/EI diagrams. Due to the symmetry of the structure and loading, the deflected shape is symmetric for all three members. Consequently the tangent at C is horizontal. From the construction shown in Fig. A5.1.5a we observe that the intercept on the vertical at A of the tangent from C, namely $t_{A/C}$, equals the deflection δ at the center. If we denote by A_F the area of the M/EI diagram due to first-order bending moment (that is, rectangular area ACC_1A_1), and by A_S,

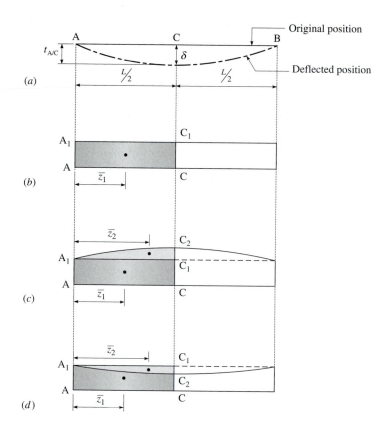

Figure A5.1.5: Central deflections of a beam, beam-column, and member under bending and tension.

the area of the M/EI diagram due to any secondary bending moment (that is, area $A_1C_1C_2$), we deduce from Moment-Area Theorem 2 that:

$$\delta^o = A_F \bar{z}_1 \tag{A5.1.5}$$

$$\delta_c^* = A_F \bar{z}_1 + A_{Sc} \bar{z}_2 = \delta^o + A_{Sc} \bar{z}_2 \tag{A5.1.6}$$

$$\delta_t^* = A_F \bar{z}_1 - A_{St} \bar{z}_2 = \delta^o - A_{St} \bar{z}_2 \tag{A5.1.7}$$

where $\bar{z}_1$ represents the distance to the center of gravity of area A_F from point A, and $\bar{z}_2$ the distance to the center of gravity of area A_S from A. The second term on the right-hand side of Eqs. A5.1.6 and A5.1.7 represents the contribution of the secondary moments to the deflection. We have, from Fig. A5.1.5b:

$$\delta^o = \left[\frac{M^o}{EI} \left(\frac{L}{2} \right) \right] \left[\frac{1}{2} \left(\frac{L}{2} \right) \right] = \frac{M^o L^2}{8EI} \tag{A5.1.8}$$

For small deflections, the deflected shape of the member can be assumed to be a parabola. We therefore have:

$$A_{Sc} \bar{z}_2 = \left[\frac{2}{3} \left(\frac{P\delta_c^*}{EI} \right) \left(\frac{L}{2} \right) \right] \left[\frac{5}{8} \left(\frac{L}{2} \right) \right] = \left(\frac{5PL^2}{48EI} \right) \delta_c^* \tag{A5.1.9}$$

With the help of Eqs. A5.1.8 and A5.1.9, Eq. A5.1.6 can be written as:

$$\delta_c^* = \delta^o + \left(\frac{PL^2}{9.6EI}\right)\delta_c^* \qquad (\text{A5.1.10})$$

where the second term represents the contribution of the secondary moments to the deflection. Or, rearranging:

$$\delta_c^* = \frac{\delta^o}{\left[1 - \dfrac{P}{(9.6\ EI/L^2)}\right]} \approx \frac{\delta^o}{\left[1 - \dfrac{P}{P_E}\right]} \qquad (\text{A5.1.11})$$

Here, L is the length of the member; EI, the flexural rigidity of the member; P, the axial compressive load in the member; δ_c^*, the deflection of the member at midspan due to the combined effects of bending and axial compressive load; δ^o, the deflection of the same member at the same point computed by conventional elementary methods, neglecting the effect of the axial load; and P_E is the buckling load of the member (See Eq. 8.4.14).

Similarly, Eq. A5.1.7 could be approximated by the relation:

$$\delta_t^* = \frac{\delta^o}{\left[1 + \dfrac{T}{9.6EI/L^2}\right]} \approx \frac{\delta^o}{\left[1 + \dfrac{T}{P_E}\right]} \qquad (\text{A5.1.12})$$

From Eqs. A5.1.6, A5.1.9, and A5.1.10, we observe that the deflection u_c^* of a beam-column is dependent on the distribution of moment M_z^*, which in turn depends on the distribution of deflections u_c^*. Consequently, the deflections and internal moments are no longer linear functions of the applied moment M^o since they also depend on the axial load P in a nonlinear manner (see Eq. A5.1.3). On the other hand, for a statically determinate system (like the simply supported members considered here) and for a given (constant) axial load P, the response is a linear function of the applied bending moment, M^o. In other words, when M^o is constant, the response (internal moments and deflections) of the system is a nonlinear function of P; when P is constant, the response is a linear function of M^o.

Equation A5.1.3 for beam-columns is usually written in the generic form:

$$M_C^* = M_C + P\delta \qquad (\text{A5.1.13}a)$$

or

$$M_{\text{II}} = B_1 M_{\text{I}} \qquad (\text{A5.1.13}b)$$

where P = axial compressive force in the member
M_C = first-order moment at point C ($\equiv M_{\text{I}}$)
M_C^* = second-order moment at point C ($\equiv M_{\text{II}}$)
$P\delta$ = secondary moment ($P\delta$ moment) developed due to the interaction of the axial load with the chord deflection
B_1 = moment magnification factor (≥ 1)

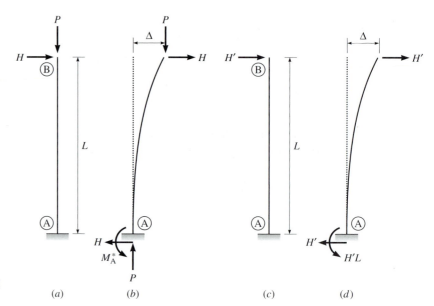

Figure A5.1.6: Cantilever column with a lateral load.

(a) (b) (c) (d)

The use of B_1 factor in the design of beam-columns is considered in Chapter 11.

A5.1.2 $P\Delta$ Effect (Moment Magnification Factor B_2)

Let us consider the vertical cantilever column AB of length L subjected to a lateral force H and vertical axial load P, shown in Fig. A5.1.6a. The deflected shape of the column is shown in Fig. A5.1.6b. By considering moment equilibrium about the base of the column A of the forces acting on the column in its deflected position, we obtain:

$$M_A^* = HL + P\Delta = (HL)\left[1 + \frac{P\Delta}{HL}\right] = M_A\left[1 + \frac{P\Delta}{HL}\right]$$

$$(A5.1.14)$$

Here M_A^* is the second-order moment at A, and M_A is the first-order moment at A (the moment with P set to zero). The term $P\Delta$ is known as the ***PΔ effect*** and represents the secondary moment due to the relative translation of the two ends of the column in the presence of the axial load.

An equivalent lateral load, one that will result in the same moment in the member at the base as the second-order moment of the beam-column, may be obtained from Fig. A5.1.6c as $H + (P\Delta/L)$. Here it is assumed, with minimal error, that the displacements at the top of the member (beam-column) in Fig. A5.1.6b and the free end of the beam in Fig. A5.1.6d are the same.

That is:

$$\Delta = \left(H + \frac{P\Delta}{L}\right)\frac{L^3}{3EI} = \left(1 + \frac{P\Delta}{HL}\right)\left(\frac{HL^3}{3EI}\right) \quad \text{(A5.1.15a)}$$

$$= \Delta_o\left(1 + \frac{P\Delta}{HL}\right) \quad \text{(A5.1.15b)}$$

where Δ_o is the first-order lateral deflection of the cantilever beam tip under the lateral load H. Solving the equation above for Δ results in:

$$\Delta = \frac{\Delta_o}{1 - \dfrac{P\Delta_o}{HL}} \quad \text{(A5.1.16)}$$

Substituting this expression for the second-order moment in (Eq. A5.1.14) and simplifying gives:

$$M_A^* = \frac{1}{\left(1 - \dfrac{P\Delta_o}{HL}\right)}M_A = B_2\,M_A \quad \text{(A5.1.17)}$$

Here, B_2 is known as the **sway moment amplification factor,** M_A is the first-order moment, and M_A^* is the second-order moment at the base of the cantilever column.

The use of the B_2 factor in the design of beam-columns is considered in Chapter 11.

INDEX

$$P_d = \phi_c \left[0.658^{\lambda_c^2} \right] F_y A_g \qquad \text{for } \lambda_c \leq 1.5 \tag{8.7.6}$$

$$P_d = \phi_c \frac{0.877}{\lambda_c^2} F_y A_g \qquad \text{for } \lambda_c > 1.5 \tag{8.7.7}$$

$$\lambda_c = \frac{KL/r}{\pi \sqrt{E/F_y}} = \sqrt{\frac{F_y}{F_e}} = \sqrt{\frac{P_y}{P_e}} \tag{8.7.8}$$

$$(K_x L_x)_y = \frac{K_x L_x}{(r_x/r_y)} \tag{8.7.12}$$

$$G_i = \frac{\Sigma(E_t I/L)_c}{\Sigma(EI/L)_g} = \frac{E_t}{E} G_e = \tau G_e \tag{8.8.1}$$

$$f_{cr} = \frac{\pi^2 E k_c}{12(1 - \mu^2)(b/t)^2} \tag{8.9.2}$$

Flange half: $\qquad \rightarrow \lambda_{rf} = 0.56 \sqrt{E/F_y}$ $\tag{8.9.6a}$

Web of an I-shape: $\qquad \rightarrow \lambda_{rw} = 1.49 \sqrt{E/F_y}$ $\tag{8.9.7a}$

$$M_{rx} = S_x(F_y - f_{rc}) \tag{9.2.16}$$

$$M_{yx} = S_x F_y \tag{9.2.17}$$

$$M_{px} = Z_x F_y \tag{9.2.18}$$

$$A_t = A_c \tag{9.2.19}$$

$$M_p = T_p e_p = (A_t \bar{y}_t + A_c |\bar{y}_c|) F_y = Z F_y \tag{9.2.20}$$

$$\lambda_f \leq \lambda_{pf} \text{ and } \lambda_w \leq \lambda_{pw} \qquad \text{for compact shapes} \tag{9.5.1}$$

$$\lambda_{pf} = 0.38 \sqrt{\frac{E}{F_y}}, \qquad \lambda_{pw} = 3.76 \sqrt{\frac{E}{F_y}} \tag{9.5.3}$$

$$M_{dx} \equiv \phi_b M_n = \phi_b M_{px} = \phi_b Z_x F_y \qquad \text{for compact I-shapes with } L_b \leq L_p \tag{9.7.2}$$

$$L_p = 1.76 r_y \sqrt{\frac{E}{F_y}} \tag{9.7.4}$$

$$V_d = \phi_v V_n = \phi_v(0.6F_y) dt_w \qquad \text{for} \quad \lambda_w \leq \lambda_{pv} \tag{9.7.9}$$

$$\lambda_{pv} = 2.45 \sqrt{\frac{E}{F_y}} \tag{9.7.10}$$

$$Z_{x \text{ req}} \geq \frac{12 M_u}{\phi_b F_y} \tag{9.7.18}$$

$$L_r = \frac{r_y X_1}{(F_y - F_r)} \sqrt{1 + \sqrt{1 + X_2(F_y - F_r)^2}} \tag{10.3.2}$$

$$X_1 = \frac{\pi}{S_x} \sqrt{\frac{EAGJ}{2}}, \qquad X_2 = \frac{4C_w}{I_y}\left(\frac{S_x}{GJ}\right)^2 \tag{10.3.3}$$

$$M_d \equiv \phi_b M_n \geq M_u \tag{10.4.1}$$

$$M_d = \phi_b M_{px} = \phi_b Z_x F_y \quad \text{for} \quad L_b \leq L_p \tag{10.4.5}$$

$$M_d = M_{dI} = \min[C_b M_{dI}^o, \phi_b M_{px}] \quad \text{for} \quad L_p < L_b \leq L_r \tag{10.4.6}$$

$$M_{dI}^o = \phi_b M_{px} - (\phi_b M_{px} - \phi_b M_{rx})\frac{(L_b - L_p)}{(L_r - L_p)} \tag{10.4.7}$$

$$= \phi_b M_{px} - BF\,(L_b - L_p) \tag{10.4.8}$$

$$BF = \frac{(\phi_b M_{px} - \phi_b M_{rx})}{(L_r - L_p)} \tag{10.4.9}$$

$$M_{rx} = S_x(F_y - F_r) \tag{10.4.10}$$

$$M_d = M_{dE} = \min[C_b M_{dE}^o, \quad \phi_b M_{px}] \quad \text{for} \quad L_b > L_r \tag{10.4.11}$$

$$M_{dE}^o = \phi_b M_{cr}^o \tag{10.4.12}$$

$$M_{cr}^o = \frac{\pi}{L_b} \sqrt{EI_y GJ + \frac{\pi^2}{L_b^2} EC_w EI_y} \tag{10.4.13a}$$

$$= \frac{S_x X_1 \sqrt{2}}{L_b/r_y} \sqrt{1 + \frac{X_1^2 X_2}{2(L_b/r_y)^2}} \tag{10.4.13b}$$

$$C_b = \frac{12.5 M_{max}}{2.5 M_{max} + 3M_A + 4M_B + 3M_C} \tag{10.4.14}$$

$$\min[C_b M_d^o, \phi_b M_{px}] \geq M_u \tag{10.4.18}$$

$$M_d^o \geq \frac{M_u}{C_b} \equiv M_{ueq}^o; \qquad \phi_b M_{px} \geq M_u \tag{10.4.19a, b}$$

$$M_u^* = B_1 M_{nt} + B_2 M_{lt} \tag{11.9.3}$$

$$B_1 = \max\left[\frac{C_m}{1 - \frac{P_u}{P_{e1}}}, 1.0\right] \tag{11.9.4}$$

$$P_{e1} = \frac{\pi^2 EI}{(KL)_{nt}^2} = \frac{\pi^2 EA}{(KL/r)_{nt}^2} = F_e A \tag{11.9.5}$$

$$C_m = 0.6 - 0.4 r_M \tag{11.9.6}$$

$$r_M = \pm \frac{|M_1|}{|M_2|} \quad \text{with} \quad -1.0 \leq r_M \leq 1.0 \tag{11.9.7}$$

$$B_2 = \frac{1}{1 - \frac{\Delta_{oh}}{L} \frac{\sum_{i=1}^{n} P_{ui}}{\sum_{j=1}^{m} H_j}} \tag{11.9.8}$$

$$B_2 = \frac{1}{1 - \frac{\sum_{i=1}^{n} P_{ui}}{\sum_{i=1}^{n} P_{e2}}} \tag{11.9.9}$$

$$P_{e2} = \frac{\pi^2 EI}{(KL)_{lt}^2} = \frac{\pi^2 EA}{(KL/r)_{lt}^2} = F_e A \tag{11.9.10}$$

$$\frac{P_u}{P_d} + \frac{8}{9}\left[\frac{M_{ux}^*}{M_{dx}} + \frac{M_{uy}^*}{M_{dy}}\right] \leq 1.0 \quad \text{for} \quad \frac{P_u}{P_d} \geq 0.2 \tag{11.9.11a}$$

$$\frac{1}{2}\frac{P_u}{P_d} + \frac{M_{ux}^*}{M_{dx}} + \frac{M_{uy}^*}{M_{dy}} \leq 1.0 \quad \text{for} \quad \frac{P_u}{P_d} < 0.2 \tag{11.9.11b}$$

$$\frac{T_u}{T_d} + \frac{8}{9}\left[\frac{M_{ux}}{M_{dx}} + \frac{M_{uy}}{M_{dy}}\right] \leq 1 \quad \text{for} \quad \frac{T_u}{T_d} \geq 0.2 \tag{11.12.1a}$$

$$\frac{1}{2}\frac{T_u}{T_d} + \left[\frac{M_{ux}}{M_{dx}} + \frac{M_{uy}}{M_{dy}}\right] \leq 1 \quad \text{for} \quad \frac{T_u}{T_d} < 0.2 \tag{11.12.1b}$$

$$P_{ueq} \leq P_d \tag{11.14.3}$$

$$P_{ueq} = P_u + m M_{ux}^* + mu M_{uy}^* \tag{11.14.5a}$$

$$P_{ueq} = \frac{1}{2} P_u + \frac{9}{8} m M_{ux}^* + \frac{9}{8} mu M_{uy}^* \tag{11.14.5b}$$

$$b P_u + m M_{ux}^* + n M_{uy}^* \leq 1.0 \quad \text{for} \quad b P_u \geq 0.2 \tag{11.14.10a}$$

$$\frac{1}{2} b P_u + \frac{9}{8} m M_{ux}^* + \frac{9}{8} n M_{uy}^* \leq 1.0 \quad \text{for} \quad b P_u < 0.2 \tag{11.14.10b}$$

$$\frac{M_{ux}}{\phi_b M_{nx}} + \frac{M_{uy}}{\phi_b M_{ny}} = 0 \rightarrow \frac{M_{ux}}{M_{dx}} + \frac{M_{uy}}{M_{dy}} = 0 \tag{11.16.2}$$

$$\frac{M_{ux}}{M_{dx}} + \frac{M_{uf}}{{}^1\!/_2 M_{dy}} = 0 \rightarrow \frac{M_{ux}}{M_{dx}} + \frac{2 M_{uf}}{M_{dy}} = 0 \tag{11.16.6}$$